Introduction to Finite Element Analysis and Design

Introduction to Finite Element Analysis and Design

Second Edition

Nam H. Kim, Bhavani V. Sankar, and Ashok V. Kumar
University of Florida

Registered Offices
John Wiley & Sons, Inc., 111 River Street, Hoboken, NJ 07030, USA
John Wiley & Sons Ltd, The Atrium, Southern Gate, Chichester, West Sussex, PO19 8SQ, UK

Editorial Office
The Atrium, Southern Gate, Chichester, West Sussex, PO19 8SQ, UK

For details of our global editorial offices, customer services, and more information about Wiley products visit us at www.wiley.com.

Wiley also publishes its books in a variety of electronic formats and by print-on-demand. Some content that appears in standard print versions of this book may not be available in other formats.

Library of Congress Cataloging-in-Publication Data

Names: Kim, Nam H., author. | Sankar, Bhavani V., author. | Kumar, Ashok V., author.
Title: Introduction to finite element analysis and design / by Nam H. Kim, Bhavani V. Sankar, Ashok V. Kumar.
Description: Second edition. | Hoboken, NJ : John Wiley & Sons, 2018. | Includes bibliographical references and index. |
Identifiers: LCCN 2018002473 (print) | LCCN 2018007147 (ebook) | ISBN 9781119078746 (pdf) | ISBN 9781119078739 (epub) | ISBN 9781119078722 (cloth)
Subjects: LCSH: Finite element method. | Engineering mathematics.
Classification: LCC TA347.F5 (ebook) | LCC TA347.F5 K56 2018 (print) | DDC 620.001/51825–dc23
LC record available at https://lccn.loc.gov/2018002473

Cover Design: Wiley
Cover Image: Courtesy of Nam H. Kim

Set in 10/12pt Times by SPi Global, Pondicherry, India

Printed and bound by CPI Group (UK) Ltd, Croydon, CR0 4YY

10 9 8 7 6 5 4 3 2 1

To our wives, Jeehyun, Mira, and Gouri

Contents

Preface

Finite Element Method (FEM) is a numerical method for solving differential equations that describe many engineering problems. One of the reasons for FEM's popularity is that the method results in computer programs versatile in nature that can solve many practical problems with a small amount of training. Obviously, there is a danger in using computer programs without proper understanding of the theory behind them, and that is one of the reasons to have a thorough understanding of the theory behind FEM.

Many universities teach FEM to students at the junior/senior level. One of the biggest challenges to the instructor is finding a textbook appropriate to the level of students. In the past, FEM was taught only to graduate students who would carry out research in that field. Accordingly, many textbooks focus on theoretical development and numerical implementation of the method. However, the goal of an undergraduate FEM course is to introduce the basic concepts so that the students can use the method efficiently and interpret the results properly. Furthermore, the theoretical aspects of FEM must be presented without too much mathematical niceties. Practical applications through several design projects can help students to understand the method clearly.

This book is suitable for junior/senior level undergraduate students and beginning graduate students in engineering mechanics, mechanical, civil, aerospace, biomedical and industrial engineering as well as researchers and design engineers in the above fields.

The textbook is organized into ten chapters. The Appendix at the end summarizes most mathematical preliminaries that are repeatedly used in the text. The Appendix is by no means a comprehensive mathematical treatment of the subject. Rather, it provides a common notation and the minimum amount of mathematical knowledge that will be required in using the book effectively. This includes basics of matrix algebra, minimization of quadratic functions, and techniques for solving linear equations that are commonly used in commercial finite element programs.

The book begins with the introduction of finite element concepts via the direct stiffness method using spring elements. The concepts of nodes, elements, internal forces, equilibrium, assembly, and applying boundary conditions are presented in detail. The spring element is then extended to the uniaxial bar element without introducing interpolation. The concept of local (elemental) and global coordinates and their transformations and element connectivity tables are introduced via two– and three–dimensional truss elements. Four design projects are provided at the end of the chapter, so that students can apply the method to real life problems. The direct method in Chapter 1 provides a clear physical insight into FEM and is preferred in the beginning stages of learning the principles. However, it is limited in its application in that it can be used to solve one–dimensional problems only.

The direct stiffness method becomes impractical for more realistic problems especially multidimensional problems. In Chapter 2, we introduce more general approaches, such as, the Weighted Residual Methods and, in particular, the Galerkin Method. Similarity to energy methods in solid and structural mechanics problems is discussed. We include a simple 1–D variational formulation in Chapter 2 using boundary value problems. The concept of polynomial approximation and domain discretization is introduced. The formal procedure of finite element analysis is also presented in this chapter. Chapter 2 is written in such way that it can be left out in elementary level courses.

The 1–D formulation is further extended to beams and plane frames in Chapter 3. At this point, the direct method is not useful because the stiffness matrix generated from the direct method cannot provide a clear physical interpretation. Accordingly, we use the principle of minimum potential energy to derive the matrix equation at the element level. The 1–D beam element is extended to 2–D frame element by using coordinate transformation. A 2–D bicycle frame design project is included at the end of this chapter. Buckling of beams and plane frames is included in the revised second edition. First, the concepts of linear buckling of beam is introduced using the Rayleigh-Ritz method. Then the corresponding energy terms are derived in the finite element context.

The finite element formulation is extended to the steady–state heat transfer problem in Chapter 4. Both direct and Galerkin's methods along with convective boundary conditions are included. Two-dimensional heat transfer problems are discussed in the second edition. Practical issues in modeling 2D heat transfer problems are also discussed.

Before proceeding to solid elements in Chapter 6, a review of solid mechanics is provided in Chapter 5. The concepts of stress and strain are presented followed by constitutive relations and equilibrium equations. We limit our interest to linear, isotropic materials in order to make the concepts simple and clear. However, advanced concepts such as transformation of stress and strain, and the eigen value problem for calculating the principal values, are also included. Since, in practice, FEM is used mostly for designing a structure or a mechanical system, failure/yield criteria are also introduced in this chapter.

In Chapter 6, we introduce 2–D solid elements. The governing variational equation is developed using the principle of minimum potential energy. The finite element concepts are explained in detail using only triangular and rectangular elements. Numerical performance of each element is discussed through examples. A new addition to the second edition is the axisymmetric element as it is essentially a plane problem.

The concept of isoparametric mapping is introduce in a separate chapter (Chapter 7) as most practical problems require irregular elements such as linear or higher order quadrilateral elements. Three-dimensional solid elements are introduced in this chapter. Numerical integration and FE modeling practices for isoparametric elements are also included.

Dynamic problems is another addition to the second edition. The concept of free vibration, calculation of natural frequencies and mode shapes, various time integration methods and mode superposition method, are all explained using 1-D structural elements such as uniaxial bars and beams.

In Chapter 9, we discuss traditional finite element analysis procedures, including preliminary analysis, pre-processing, solving matrix equations, and post-processing. Emphasis is on selection of element types, approximating the part geometry, different types of meshing, convergence, and taking advantage of symmetry. A design project involving 2–D analysis is provided at the end of the chapter.

As one of the important goals of FEM is to use the tool for engineering design, the last chapter (Chapter 10) is dedicated to the topic of structural design using FEM. The basic concept of design parameterization and the standard design problem formulation are presented. This chapter is self contained and can be skipped depending on the schedule and content of the course.

Each chapter contains a comprehensive set of homework problems, some of which require commercial FEA programs. A total of nine design projects are provided in the book.

We are thankful to several instructors across the country who used the first edition and provided feedback. We are grateful for their valuable suggestions especially regarding example and exercise problems.

September 2017
Nam H. Kim, Bhavani V. Sankar and Ashok V. Kumar

About the Companion Website

This book is accompanied by a companion website:

www.wiley.com/go/kim/finite_element_analysis_design

The website includes:

- Programs
- Exercise problems

Chapter 1

Direct Method – Springs, Bars, and Truss Elements

An ability to predict the behavior of machines and engineering systems in general is of great importance at every stage of engineering processes, including design, manufacture, and operation. Such predictive methodologies are possible because engineers and scientists have made tremendous progress in understanding the physical behavior of materials and structures and have developed mathematical models, albeit approximate, in order to describe their physical behavior. Most often the mathematical models result in algebraic, differential, or integral equations or combinations thereof. Seldom can these equations be solved in closed form, and hence numerical methods are used to obtain solutions. The finite difference method is a classical method that provides approximate solutions to differential equations with reasonable accuracy. There are other methods of solving mathematical equations that are covered in traditional numerical methods courses[1].

The finite element method (FEM) is one of the numerical methods for solving differential equations. The FEM, originated in the area of structural mechanics, has been extended to other areas of solid mechanics and later to other fields such as heat transfer, fluid dynamics, and electromagnetism. In fact, FEM has been recognized as a powerful tool for solving partial differential equations and integro-differential equations, and it has become the numerical method of choice in many engineering and applied science areas. One of the reasons for FEM's popularity is that the method results in computer programs versatile in nature that can solve many practical problems with the least amount of training. Obviously, there is a danger in using computer programs without proper understanding of the theory behind them, and that is one of the reasons to have a thorough understanding of the theory behind the FEM.

The basic principle of FEM is to divide or *discretize* the system into a number of smaller elements called finite elements (FEs), to identify the degrees of freedom (DOFs) that describe its behavior, and then to write down the equations that describe the behavior of each element and its interaction with neighboring elements. The element-level equations are assembled to obtain global equations, often a linear system of equations, which are solved for the unknown DOFs. The phrase *finite element* refers to the fact that the elements are of a finite size as opposed to the infinitesimal or differential element considered in deriving the governing equations of the system. Another interpretation is that the FE equations deal with a finite number of DOFs as opposed to the infinite number of DOFs of a continuous system.

---

[1] Atkinson, K. E. 1978. *An Introduction to Numerical Analysis*. Wiley, New York.

Introduction to Finite Element Analysis and Design, Second Edition. Nam H. Kim, Bhavani V. Sankar, and Ashok V. Kumar.
© 2018 John Wiley & Sons Ltd. Published 2018 by John Wiley & Sons Ltd.
Companion website: www.wiley.com/go/kim/finite_element_analysis_design

In general, solutions of practical engineering problems are quite complex, and they cannot be represented using simple mathematical expressions. An important concept of the FEM is that the solution is approximated using simple polynomials, often linear or quadratic, within each element. Since elements are connected throughout the system, the solution of the system is approximated using piecewise polynomials. Such approximation may contain errors when the size of an element is large. As the size of element reduces, however, the approximated solution will converge to the exact solution.

There are three methods that can be used to derive the FE equations of a problem: (a) direct method, (b) variational method, and (c) weighted residual method. The direct method provides a clear physical insight into the FEM and is preferred in the beginning stages of learning the principles. However, it is limited in its application in that it can be used to solve one-dimensional problems only. The variational method is akin to the methods of calculus of variations and is a powerful tool for deriving the FE equations. However, it requires the existence of a functional, whose minimization results in the solution of the differential equations. The Galerkin method is one of the popular weighted residual methods and is applicable to most problems. If a variational function exists for the problem, then the variational and Galerkin methods yield identical solutions.

In this chapter, we will illustrate the direct method of FE analysis using one-dimensional elements such as linear spring, uniaxial bar, and truss elements. The emphasis is on construction and solution of the finite element equations and interpretation of the results, rather than the rigorous development of the general principles of the FEM.

1.1 ILLUSTRATION OF THE DIRECT METHOD

Consider a system of rigid bodies connected by springs as shown in figure 1.1. The bodies move only in the horizontal direction. Furthermore, we consider only the static problem and hence the mass effects (inertia) will be ignored. External forces, F_2, F_3, and F_4, are applied on the rigid bodies as shown. The objectives are to determine the displacement of each body, forces in the springs, and support reactions.

We will introduce the principles involved in the FEM through this example. Notice that there is no need to discretize the system as it already consists of discrete elements, namely, the springs. The elements are connected at the nodes. In this case, the rigid bodies are the nodes. Of course, the two walls are also the nodes as they connect to the elements. Numbers inside the little circles mark the nodes. The system of connected elements is called the mesh and is best described using a connectivity table that defines which nodes an element is connected to as shown in table 1.1. Such a connectivity table is included in input files for finite element analysis software to describe the mesh.

Consider the free-body diagram of a typical element (e) as shown in figure 1.2. It has two nodes, nodes i and j. They will also be referred to as the first and second node or local node 1 (LN1) and local node 2 (LN2), respectively, as shown in the connectivity table. Assume a coordinate system going from left to right. The convention for first and second nodes is that $x_i < x_j$. The forces acting at the nodes are denoted by $f_i^{(e)}$ and $f_j^{(e)}$. In this notation, the subscripts denote the node numbers and the superscript the

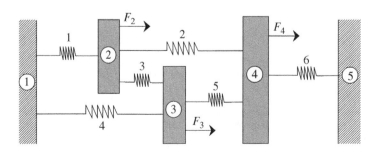

Figure 1.1 Rigid bodies connected by springs

Table 1.1 Connectivity table for figure 1.1

Element	LN1 (i)	LN2 (j)
1	1	2
2	2	4
3	2	3
4	1	3
5	3	4
6	4	5

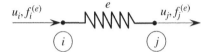

Figure 1.2 Spring element (e) connected by node i and node j

element number. This notation is adopted because multiple elements can be connected at a node, and each element may have different forces at the node. We will refer to them as *internal forces*. In figure 1.2, the forces are shown in the positive direction. The unknown displacements of nodes i and j are u_i and u_j, respectively. Note that there is no superscript for u, as the displacement is unique to the node denoted by the subscript. We would like to develop a relationship between the nodal displacements u_i and u_j and the internal forces $f_i^{(e)}$ and $f_j^{(e)}$.

The elongation of the spring is denoted by $\Delta^{(e)} = u_j - u_i$. Then the force of the spring is given by

$$P^{(e)} = k^{(e)}\Delta^{(e)} = k^{(e)}\left(u_j - u_i\right), \tag{1.1}$$

where $k^{(e)}$ is the spring rate or *stiffness* of element (e). In this text, the force in the spring, $P^{(e)}$, is referred to as *element force*. If $u_j > u_i$, then the spring is elongated, and the force in the spring is positive (tension). Otherwise, the spring is in compression. The spring element force is related to the internal force by

$$f_j^{(e)} = P^{(e)}. \tag{1.2}$$

Note that the sign of $f_i^{(e)}$ and $f_j^{(e)}$ is determined based on the direction that the force is applied, while the sign of $P^{(e)}$ is determined based on whether the element is in tension or compression. For equilibrium, the sum of the forces acting on element (e) must be equal to zero, i.e.,

$$f_i^{(e)} + f_j^{(e)} = 0 \ \text{ or } \ f_i^{(e)} = -f_j^{(e)}. \tag{1.3}$$

Therefore, the two forces are equal, and they are applied in opposite directions. When $f_j^{(e)}$ is positive, the element is in tension, and thus, $P^{(e)}$ is positive.

From eqs. (1.1)–(1.3), we can obtain a relation between the internal forces and the displacements as

$$f_i^{(e)} = k^{(e)}\left(u_i - u_j\right)$$
$$f_j^{(e)} = k^{(e)}\left(-u_i + u_j\right). \tag{1.4}$$

Equation (1.4) can be written in matrix forms as:

$$k^{(e)}\begin{bmatrix} 1 & -1 \\ -1 & 1 \end{bmatrix}\begin{Bmatrix} u_i \\ u_j \end{Bmatrix} = \begin{Bmatrix} f_i^{(e)} \\ f_j^{(e)} \end{Bmatrix}. \tag{1.5}$$

We also write eq. (1.5) in a shorthand notation as:

$$\left[\mathbf{k}^{(e)}\right]\left\{\begin{array}{c}u_i \\ u_j\end{array}\right\} = \left\{\begin{array}{c}f_i^{(e)} \\ f_j^{(e)}\end{array}\right\},$$

or,

$$\boxed{\left[\mathbf{k}^{(e)}\right]\left\{\mathbf{q}^{(e)}\right\} = \left\{\mathbf{f}^{(e)}\right\}}, \tag{1.6}$$

where $[\mathbf{k}^{(e)}]$ is the element stiffness matrix, $\{\mathbf{q}^{(e)}\}$ is the vector of DOFs associated with element (e), and $\{\mathbf{f}^{(e)}\}$ is the vector of internal forces. Sometimes we will omit the superscript (e) with the understanding that we are dealing with a generic element. Equation (1.6) is called the *element equilibrium equation*.

The element stiffness matrix $[\mathbf{k}^{(e)}]$ has the following properties:

1. It is square as it relates to the same number of forces as the displacements;
2. It is symmetric (a consequence of the Betti–Rayleigh Reciprocal theorem in solid and structural mechanics[2]);
3. It is singular, *i.e.*, its determinant is equal to zero, so it cannot be inverted; and
4. It is positive semidefinite.

Properties 3 and 4 are related to each other, and they have physical significance. Consider eq. (1.6). If the nodal displacements u_i and u_j of a spring element in a system are given, then it should be possible to predict the force $P^{(e)}$ in the spring from its change in length $(u_j - u_i)$, and hence the forces $\{\mathbf{f}^{(e)}\}$ acting at its nodes can be predicted. In fact, the internal forces can be computed by performing the matrix multiplication $[\mathbf{k}^{(e)}]\{\mathbf{q}^{(e)}\}$. On the other hand, if the two spring forces are given (they must have equal magnitudes but opposite directions), the nodal displacements cannot be determined uniquely, as a rigid body displacement (equal u_i and u_j) can be added without affecting the spring force. If $[\mathbf{k}^{(e)}]$ were to have an inverse, then it would have been possible to solve for $\{\mathbf{q}^{(e)}\} = \left[\mathbf{k}^{(e)}\right]^{-1}\left\{\mathbf{f}^{(e)}\right\}$ uniquely in violation of the physics. Property 4 has also a physical interpretation, which will be discussed in conjunction with energy methods.

In the next step, we develop a relationship between the internal forces $f_i^{(e)}$ and the known external forces F_i. For example, consider the free-body diagram of node 3 (or the rigid body in this case) in figure 1.1. The forces acting on the node are the external force F_3 and the internal forces from the springs connected to node 3 as shown in figure 1.3.

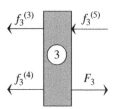

Figure 1.3 Free-body diagram of node 3 in the example shown in figure 1.1. The external force, F_3, and the forces, $f_3^{(e)}$, exerted by the springs attached to the node are shown. Note the forces $f_3^{(e)}$ act in the negative direction.

---

[2] Y. C. Fung. 1965. *Foundations of Solid Mechanics*. Prentice-Hall, Englewood Cliffs, NJ.

For equilibrium of the node, the sum of the forces acting on the node should be equal to zero:

$$F_i - \sum_{e=1}^{i_e} f_i^{(e)} = 0,$$

or,

$$F_i = \sum_{e=1}^{i_e} f_i^{(e)}, \quad i = 1, \ldots ND, \tag{1.7}$$

where i_e is the number of elements connected to node i, and ND is the total number of nodes in the model. Equation (1.7) is the equilibrium between externally applied forces at a node and internal forces from connected elements. If there is no externally applied force at a node, then the sum of internal forces at the node must be zero. Such equations can be written for each node including the boundary nodes, such as nodes 1 and 5 in figure 1.1. The internal forces $f_i^{(e)}$ in eq. (1.7) can be replaced by the unknown DOFs $\{\mathbf{q}\}$ by using eq. (1.6). For example, the force equilibrium for the springs in figure 1.1 can be written as

$$\begin{cases} F_1 = f_1^{(1)} + f_4^{(1)} = k^{(1)}(u_1 - u_2) + k^{(4)}(u_1 - u_3) \\ F_2 = f_2^{(1)} + f_2^{(3)} + f_2^{(2)} = k^{(1)}(u_2 - u_1) + k^{(3)}(u_2 - u_3) + k^{(2)}(u_2 - u_4) \\ F_3 = f_3^{(3)} + f_3^{(4)} + f_3^{(5)} = k^{(3)}(u_3 - u_2) + k^{(4)}(u_3 - u_1) + k^{(5)}(u_3 - u_4) \\ F_4 = f_4^{(2)} + f_4^{(5)} + f_4^{(6)} = k^{(2)}(u_4 - u_2) + k^{(5)}(u_4 - u_3) + k^{(6)}(u_4 - u_5) \\ F_5 = f_5^{(6)} = k^{(6)}(u_5 - u_4). \end{cases} \tag{1.8}$$

This will result in ND number of linear equations for the ND number of DOFs:

$$[\mathbf{K}_s] \begin{Bmatrix} u_1 \\ u_2 \\ \vdots \\ u_{ND} \end{Bmatrix} = \begin{Bmatrix} F_1 \\ F_2 \\ \vdots \\ F_{ND} \end{Bmatrix}. \tag{1.9}$$

Or, in shorthand notation $[\mathbf{K}_s]\{\mathbf{Q}_s\} = \{\mathbf{F}_s\}$ where $[\mathbf{K}_s]$ is the structural stiffness matrix, $\{\mathbf{Q}_s\}$ is the vector of displacements of all nodes in the model, and $\{\mathbf{F}_s\}$ is the vector of external forces, including the unknown reactions. The expanded form of eq. (1.9) is given in eq. (1.10) below:

$$\begin{bmatrix} k^{(1)} + k^{(4)} & -k^{(1)} & -k^{(4)} & 0 & 0 \\ -k^{(1)} & k^{(1)} + k^{(2)} + k^{(3)} & -k^{(3)} & -k^{(2)} & 0 \\ -k^{(4)} & -k^{(3)} & k^{(3)} + k^{(4)} + k^{(5)} & -k^{(5)} & 0 \\ 0 & -k^{(2)} & -k^{(5)} & k^{(2} + k^{(5)} + k^{(6)} & -k^{(6)} \\ 0 & 0 & 0 & -k^{(6)} & k^{(6)} \end{bmatrix} \begin{Bmatrix} u_1 \\ u_2 \\ u_3 \\ u_4 \\ u_5 \end{Bmatrix} = \begin{Bmatrix} F_1 \\ F_2 \\ F_3 \\ F_4 \\ F_5 \end{Bmatrix},$$

or,

$$[\mathbf{K}_s]\{\mathbf{Q}_s\} = \{\mathbf{F}_s\}. \tag{1.10}$$

The properties of the structural stiffness matrix $[\mathbf{K}_s]$ are similar to that of the element stiffness matrix: square, symmetric, singular, and positive semi-definite. In addition, when nodes are numbered properly, $[\mathbf{K}_s]$ will be a banded matrix. It should be noted that when the boundary displacements in $\{\mathbf{Q}_s\}$ are known (usually equal to zero[3]), the corresponding forces in $\{\mathbf{F}_s\}$ are unknown reactions. In the present illustration, $u_1 = u_5 = 0$, and corresponding forces (reactions) F_1 and F_5 are unknown. It should also be noted that when displacements in $\{\mathbf{Q}_s\}$ are unknown, the corresponding forces in $\{\mathbf{F}_s\}$ should be known (either a given value or zero when no force is applied).

We will impose the boundary conditions as follows. First, we ignore the equations for which the RHS forces are unknown and strike out the corresponding rows in $[\mathbf{K}_s]$. This is called *striking the rows*. Then we eliminate the columns in $[\mathbf{K}_s]$ that are multiplied by the zero values of displacements of the boundary nodes. This is called *striking the columns*. It may be noted that if the nth row is eliminated (struck), then the nth column will also be eliminated (struck). This process results in a system of equations given by $[\mathbf{K}]\{\mathbf{Q}\} = \{\mathbf{F}\}$, where $[\mathbf{K}]$ is the global stiffness matrix, $\{\mathbf{Q}\}$ is the vector of unknown DOFs, and $\{\mathbf{F}\}$ is the vector of known forces. The global stiffness matrix will be square, symmetric, and **positive definite** and hence nonsingular. Usually $[\mathbf{K}]$ will also be banded. In large systems, that is, in models with large numbers of DOFs, $[\mathbf{K}]$ will be a sparse matrix with a small proportion of nonzero numbers in a diagonal band.

After striking the rows and columns corresponding to zero DOFs (u_1 and u_5) in eq. (1.10), we obtain the global equations as follows:

$$\begin{bmatrix} k^{(1)} + k^{(2)} + k^{(3)} & -k^{(3)} & -k^{(2)} \\ -k^{(3)} & k^{(3)} + k^{(4)} + k^{(5)} & -k^{(5)} \\ -k^{(2)} & -k^{(5)} & k^{(2)} + k^{(5)} + k^{(6)} \end{bmatrix} \begin{Bmatrix} u_2 \\ u_3 \\ u_4 \end{Bmatrix} = \begin{Bmatrix} F_2 \\ F_3 \\ F_4 \end{Bmatrix},$$

or,

$$[\mathbf{K}]\{\mathbf{Q}\} = \{\mathbf{F}\}. \tag{1.11}$$

In principle, the solution can be obtained as $\{\mathbf{Q}\} = [\mathbf{K}]^{-1}\{\mathbf{F}\}$. Once the unknown DOFs are determined, the spring forces can be obtained using eq. (1.1). The support reactions can be obtained from either the nodal equilibrium equations (1.7) or the structural equations (1.10).

EXAMPLE 1.1 *Rigid body–spring system*

Find the displacements of the rigid bodies shown in figure 1.1. Assume that the only nonzero force is $F_3 = 1000$ N. Determine the element forces (tensile/compressive) in the springs. What are the reactions at the walls? Assume the bodies can undergo only translation in the horizontal direction. The spring constants (N/mm) are $k^{(1)} = 500$, $k^{(2)} = 400$, $k^{(3)} = 600$, $k^{(4)} = 200$, $k^{(5)} = 400$, and $k^{(6)} = 300$.

SOLUTION The element equilibrium equations are as follows:

$$\begin{Bmatrix} f_1^{(1)} \\ f_2^{(1)} \end{Bmatrix} = 500 \begin{bmatrix} 1 & -1 \\ -1 & 1 \end{bmatrix} \begin{Bmatrix} u_1 \\ u_2 \end{Bmatrix}; \quad \begin{Bmatrix} f_2^{(2)} \\ f_4^{(2)} \end{Bmatrix} = 400 \begin{bmatrix} 1 & -1 \\ -1 & 1 \end{bmatrix} \begin{Bmatrix} u_2 \\ u_4 \end{Bmatrix}$$

$$\begin{Bmatrix} f_2^{(3)} \\ f_3^{(3)} \end{Bmatrix} = 600 \begin{bmatrix} 1 & -1 \\ -1 & 1 \end{bmatrix} \begin{Bmatrix} u_2 \\ u_3 \end{Bmatrix}; \quad \begin{Bmatrix} f_1^{(4)} \\ f_3^{(4)} \end{Bmatrix} = 200 \begin{bmatrix} 1 & -1 \\ -1 & 1 \end{bmatrix} \begin{Bmatrix} u_1 \\ u_3 \end{Bmatrix} \tag{1.12}$$

$$\begin{Bmatrix} f_3^{(5)} \\ f_4^{(5)} \end{Bmatrix} = 400 \begin{bmatrix} 1 & -1 \\ -1 & 1 \end{bmatrix} \begin{Bmatrix} u_3 \\ u_4 \end{Bmatrix}; \quad \begin{Bmatrix} f_4^{(6)} \\ f_5^{(6)} \end{Bmatrix} = 300 \begin{bmatrix} 1 & -1 \\ -1 & 1 \end{bmatrix} \begin{Bmatrix} u_4 \\ u_5 \end{Bmatrix}.$$

The nodal equilibrium equations are:

---

[3] Nonzero or prescribed DOFs will be dealt with in chapter 4.

$$f_1^{(1)} + f_1^{(4)} = F_1 = R_1$$

$$f_2^{(1)} + f_2^{(2)} + f_2^{(3)} = F_2 = 0$$

$$f_3^{(3)} + f_3^{(4)} + f_3^{(5)} = F_3 = 1000 \tag{1.13}$$

$$f_4^{(2)} + f_4^{(5)} + f_4^{(6)} = F_4 = 0$$

$$f_5^{(6)} = F_5 = R_5,$$

where R_1 and R_5 are unknown reaction forces at nodes 1 and 5, respectively. In the above equation, F_2 and F_4 are equal to zero because no external forces act on those nodes. Combining eqs. (1.12) and (1.13) we obtain the equation $[\mathbf{K}_s]\{\mathbf{Q}_s\} = \{\mathbf{F}_s\}$,

$$100 \begin{bmatrix} 7 & -5 & -2 & 0 & 0 \\ -5 & 15 & -6 & -4 & 0 \\ -2 & -6 & 12 & -4 & 0 \\ 0 & -4 & -4 & 11 & -3 \\ 0 & 0 & 0 & -3 & 3 \end{bmatrix} \begin{Bmatrix} u_1 \\ u_2 \\ u_3 \\ u_4 \\ u_5 \end{Bmatrix} = \begin{Bmatrix} R_1 \\ 0 \\ 1000 \\ 0 \\ R_5 \end{Bmatrix}. \tag{1.14}$$

After implementing the boundary conditions at nodes 1 and 5 (striking the rows and columns corresponding to zero displacements), we obtain the following global equations $[\mathbf{K}]\{\mathbf{Q}\} = \{\mathbf{F}\}$:

$$100 \begin{bmatrix} 15 & -6 & -4 \\ -6 & 12 & -4 \\ -4 & -4 & 11 \end{bmatrix} \begin{Bmatrix} u_2 \\ u_3 \\ u_4 \end{Bmatrix} = \begin{Bmatrix} 0 \\ 1000 \\ 0 \end{Bmatrix}.$$

By inverting the global stiffness matrix, the unknown displacements can be obtained as: $u_2 = 0.854$ mm, $u_3 = 1.55$ mm, and $u_4 = 0.875$ mm.

The forces in the springs are computed using $P^{(e)} = k^{(e)}(u_j - u_i)$:

$$P^{(1)} = 427 \text{ N}; \quad P^{(2)} = \quad 8.3 \text{ N}; \quad P^{(3)} = \quad 419 \text{ N}$$

$$P^{(4)} = 310 \text{ N}; \quad P^{(5)} = -271 \text{ N}; \quad P^{(6)} = -263 \text{ N}.$$

Wall reactions, R_1 and R_5, can be computed either from eq. (1.14) after substituting for the displacements, or from eqs. (1.12) and (1.13) as $R_1 = -737$ N; $R_5 = -263$ N. Both reactions are negative meaning that they act on the structure (the system) from right to left. ■

1.2   UNIAXIAL BAR ELEMENT

The FE analysis procedure for the spring–force system in the previous section can easily be extended to uniaxial bars. Plane and space trusses consist of uniaxial bars, and hence a detailed study of uniaxial bar finite element will provide the basis for analysis of trusses. Typical problems that can be solved using uniaxial bar elements are shown in figure 1.4. A uniaxial bar is a slender two-force member where the length is much larger than the cross-sectional dimensions. The bar can have varying cross-sectional area, $A(x)$, and consists of different materials, that is, varying Young's modulus, $E(x)$. Both concentrated forces F and distributed force $p(x)$ can be applied. The distributed forces can be applied over a portion of the bar. The forces F and $p(x)$ are considered positive if they act in the positive direction of the x-axis. Both ends of the bar can be fixed making it a statically indeterminate problem. Solving this problem by solving the differential equation of equilibrium could be difficult, if not impossible. However, this problem can be readily solved using FE analysis.

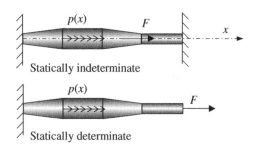

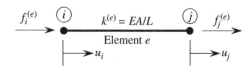

Figure 1.4 Typical one dimensional bar problems

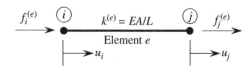

Figure 1.5 Uniaxial bar finite element

1.2.1 FE Formulation for Uniaxial Bar

The FE analysis procedures for the uniaxial bar are as follows:

1. Discretize the bar into a number of elements. The criteria for determining the size of the elements will become obvious after learning the properties of the element. It is assumed that each element has a constant axial rigidity, EA, throughout its length, although it may vary from element to element.

2. The elements are connected at nodes. Thus, more than one element can share a node. There will be nodes at points where the bar is supported.

3. External forces are applied only at the nodes, and they must be point forces (concentrated forces). If distributed forces are applied to the bar, they have to be approximated as point forces acting at nodes. At the bar boundary, if the displacement is specified, then the reaction is unknown. The reaction will be the external force acting on the boundary node. If a specified external force acts on the boundary, then the corresponding displacement is unknown. There will be no case when both displacement and force are unknown at a node.

4. The deformation of the bar is determined by the axial displacements of the nodes. That is, the nodal displacements are the DOFs in the FEM. Thus, the DOFs are $u_1, u_2, u_3, \ldots, u_N$, where N is the total number of nodes.

The objective of the FE analysis is to determine: (i) unknown DOF (u_i); (ii) axial force resultant ($P^{(e)}$) in each element; and (iii) support reactions. Once the axial force resultant, $P^{(e)}$, is available, the element stress can easily be calculated by $\sigma = P^{(e)}/A^{(e)}$ where $A^{(e)}$ is the cross-section of the element.

We will use the *direct stiffness method* to derive the element stiffness matrix. Consider the free-body diagram of a typical element (e), as illustrated in figure 1.5. Forces and displacements are defined as positive when they are in the positive x direction. The element has two nodes, namely, i and j. Node i will be the first node and node j will be called the second node. The convention is that the line i–j will be in the positive direction of the x-axis. The displacements of the nodes are u_i and u_j. The element has a stiffness of $k^{(e)} = (EA/L)^{(e)}$ where EA is the axial rigidity, and L is the length of the element. It will be shown later that the stiffness $k^{(e)}$ plays exactly the same role as in the stiffness of a spring element in the previous section.

The forces acting at the two ends of the free body are $f_i^{(e)}$ and $f_j^{(e)}$. The superscript denotes the element number, and the subscripts denote the node numbers. The (lowercase) force f denotes the internal force as opposed to the (uppercase) external force F_i acting on the nodes. Since we do not know the direction of f, we will assume that all forces act in the positive coordinate direction. It should be noted that the nodal displacements do not need a superscript, as they are unique to the nodes. However, the internal force acting at a node may be different for different elements connected to the same node.

First, we will determine a relation between the f's and u's of the element (e). For equilibrium of the free-body diagram, we have

$$f_i^{(e)} + f_j^{(e)} = 0, \qquad (1.15)$$

which means that the two forces acting on the two nodes of the element are equal and in opposite directions. Referring to figure 1.5, it is clear that when $f_j^{(e)} > 0$, the element is in tension, and when $f_j^{(e)} < 0$, the element is in compression.

From elementary mechanics of materials, the force is proportional to the elongation of the element. The elongation of the bar element is denoted by $\Delta^{(e)} = u_j - u_i$. Then, similar to the spring element, where $f = kx$, the force equilibrium of the one-dimensional bar element can be written, as

$$f_j^{(e)} = \left(\frac{AE}{L}\right)^{(e)} (u_j - u_i)$$

$$f_i^{(e)} = -f_j^{(e)} = \left(\frac{AE}{L}\right)^{(e)} (u_i - u_j),$$

where A, E, and L, respectively, are the area of the cross section, Young's modulus, and the length of the element. Using matrix notation, the above equations can be written as

$$\begin{Bmatrix} f_i^{(e)} \\ f_j^{(e)} \end{Bmatrix} = \left(\frac{AE}{L}\right)^{(e)} \begin{bmatrix} 1 & -1 \\ -1 & 1 \end{bmatrix} \begin{Bmatrix} u_i \\ u_j \end{Bmatrix}. \qquad (1.16)$$

Equation (1.16) is called the *element equilibrium equation*, which relates the nodal forces of element (e) to the corresponding nodal displacements. Note that eq. (1.16) is similar to eq. (1.5) of the spring element if $k^{(e)} = (EA/L)^{(e)}$. Equation (1.16) for each element can be written in a compact form as

$$\left\{ \mathbf{f}^{(e)} \right\} = \left[\mathbf{k}^{(e)} \right] \left\{ \mathbf{q}^{(e)} \right\}, \quad e = 1, 2, \dots, N_e, \qquad (1.17)$$

where $[\mathbf{k}^{(e)}]$ is the element stiffness matrix of element (e), $\{\mathbf{q}^{(e)}\}$ is the vector of nodal displacements of the element, and N_e is the total number of elements in the model.

Note that the element stiffness matrix in eq. (1.16) is singular. The fact that the element stiffness matrix does not have an inverse has a physical significance. If the nodal displacements of an element are specified, then the element forces can be uniquely determined by performing the matrix multiplication in eq. (1.16). On the other hand, if the forces acting on the element are given, the nodal displacements cannot be uniquely determined because one can always translate the element by adding a rigid body displacement without affecting the forces acting on it. Thus, it is always necessary to remove the rigid body motion by fixing some displacements at nodes.

1.2.2 Nodal Equilibrium

Consider the free-body diagram of a typical node i. It is connected to, say, elements (e) and $(e + 1)$. Then, the forces acting on the nodes are the external force F_i and reactions to the element forces as shown in

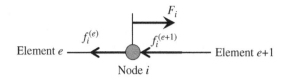

Figure 1.6 Force equilibrium at node i

figure 1.6. The internal forces are applied in the negative x direction because they are the reaction to the forces acting on the element. The sum of the forces acting on node i must be equal to zero:

$$F_i - f_i^{(e)} - f_i^{(e+1)} = 0,$$

or

$$f_i^{(e)} + f_i^{(e+1)} = F_i. \tag{1.18}$$

In general, the external force acting on a node is equal to sum of all the internal forces acting on different elements connected to the node, and eq. (1.18) can be generalized as

$$F_i = \sum_{e=1}^{i_e} f_i^{(e)}, \tag{1.19}$$

where i_e is the number of elements connected to node i, and the sum is carried out over all the elements connected to node i.

1.2.3 Assembly

The next step is to eliminate the internal forces from eq. (1.18) using eq. (1.17) in order to obtain a relation between the unknown displacements $\{Q_s\}$ and known forces $\{F_s\}$. This step results in a process called an *assembly* of the element stiffness matrices. We substitute for f's from eq. (1.17) into eq. (1.19) in order to find a relation between the nodal displacements and external forces. The force equilibrium in eq. (1.19) can be written for each DOF at each node yielding a relation between the external forces and displacements as

$$[\mathbf{K}_s]\{\mathbf{Q}_s\} = \{\mathbf{F}_s\}. \tag{1.20}$$

Equation (1.20) is called the *structural matrix equation*. In the above equation, $[\mathbf{K}_s]$ is the structural stiffness matrix, which characterizes the load-deflection behavior of the entire structure; $\{\mathbf{Q}_s\}$ is the vector of all nodal displacements, known and unknown; and $\{\mathbf{F}_s\}$ is the vector of external forces acting at the nodes including the unknown reactions.

There is a systematic procedure by which the element stiffness matrices $[\mathbf{k}^{(e)}]$ can be assembled to obtain $[\mathbf{K}_s]$. We will assign a row address and column address for each entry in $[\mathbf{k}^{(e)}]$ and $[\mathbf{K}_s]$. The column address of a column is the DOF that the column multiplies with in the equilibrium equation. For example, the column addresses of the first and second column in $[\mathbf{k}^{(e)}]$ are u_i and u_j, respectively. The column addresses of columns 1, 2, 3,... in $[\mathbf{K}_s]$ are $u_1, u_2, u_3,...$ respectively. The row addresses and column addresses are always symmetric. That is, the row address of the ith row is same as the column address of the ith column. Having determined the row and column addresses of $[\mathbf{k}^{(e)}]$ and $[\mathbf{K}_s]$, assembly of the element stiffness matrices can be done in a mechanical way. Each of the four entries (boxes) of an element stiffness matrix is transferred to the box in $[\mathbf{K}_s]$ with corresponding row and column addresses.

It is important to discuss the properties of the structural stiffness matrix $[\mathbf{K}_s]$. After assembly, the matrix $[\mathbf{K}_s]$ has the following properties:

1. It is square;
2. It is symmetric;

3. It is positive semi-definite;

4. Its determinant is equal to zero, and thus it does not have an inverse (it is singular);

5. The diagonal entries of the matrix are greater than or equal to zero.

For a given $\{\mathbf{Q}_s\}$, $\{\mathbf{F}_s\}$ can be determined uniquely; however, for a given $\{\mathbf{F}_s\}$, $\{\mathbf{Q}_s\}$ cannot be determined uniquely because an arbitrary rigid-body displacement can be added to $\{\mathbf{Q}_s\}$ without affecting $\{\mathbf{F}_s\}$.

1.2.4   Boundary Conditions

Before we solve eq. (1.20) we need to impose the displacement boundary conditions, that is, use the known nodal displacements in eq. (1.20). Mathematically, it means to make the global stiffness matrix positive definite so that the unknown displacements can be uniquely determined. Let us assume that the total size of $[\mathbf{K}_s]$ is $m \times m$. From the m equations, we will discard the equations for which we do not know the right-hand side (unknown reaction forces). This is called "striking-the-rows." The structural stiffness matrix becomes rectangular, as the number of equations is less than m. Now we delete the columns that will multiply into prescribed zero displacements in $\{\mathbf{Q}_s\}$. Usually, if the ith row is deleted, then the ith column will also be deleted. Thus, we will be deleting as many columns as we did for rows. This procedure is called "striking-the-columns." Now the stiffness matrix becomes square with size $n \times n$, where n is the number of unknown displacements. The resulting equations can be written as

$$[\mathbf{K}]\{\mathbf{Q}\} = \{\mathbf{F}\},\tag{1.21}$$

where $[\mathbf{K}]$ is the global stiffness matrix, $\{\mathbf{Q}\}$ are the unknown displacements, and $\{\mathbf{F}\}$ are the known external forces applied to nodes. Equation (1.21) is called the *global matrix equations*. In the structural matrix equations in eq. (1.20), the vector $\{\mathbf{Q}_s\}$ includes both known and unknown displacements. However, after applying boundary conditions, that is, striking the rows and striking the columns, the vector $\{\mathbf{Q}\}$ only includes unknown nodal displacements. For the same reason, the vector $\{\mathbf{F}\}$ only includes known external forces, not support reactions. The global stiffness matrix is always a positive definite matrix, which has an inverse. It is square symmetric and its diagonal elements are positive, that is, $K_{ii} > 0$, $i = 1, \ldots, n$. Thus, the displacements $\{\mathbf{Q}\}$ can be solved uniquely for a given set of nodal forces $\{\mathbf{F}\}$.

1.2.5   Calculation of Element Forces and Reaction Forces

Now that all the DOFs are known, the element force in element (e) can be determined using eq. (1.16). The axial force resultant $P^{(e)}$ in element (e) is given by

$$P^{(e)} = \left(\frac{AE}{L}\right)^{(e)} \Delta^{(e)} = \left(\frac{AE}{L}\right)^{(e)} (u_j - u_i).\tag{1.22}$$

The sign convention of axial force resultant is similar to that of stress. It is positive when the bar is in tension and negative when it is in compression. Another method of determining the axial-force resultant distribution along an element length is as follows. Consider the element equation (1.16). At the first node or node i, the axial force is given by $P_i = -f_i$. That is, if f_i acts in the positive direction, that end is under compression. If f_i is in the negative direction, the element is under tension. On the other hand, the opposite is true at the second node, node j. In that case, $P_j = +f_j$. Then, we can modify eq. (1.16) as

$$\left\{\begin{array}{c} -P_i^{(e)} \\ +P_j^{(e)} \end{array}\right\} = \left(\frac{AE}{L}\right)^{(e)} \begin{bmatrix} 1 & -1 \\ -1 & 1 \end{bmatrix} \left\{\begin{array}{c} u_i \\ u_j \end{array}\right\}.\tag{1.23}$$

It happens that $P_i^{(e)} = P_j^{(e)}$, and hence we use a single variable $P^{(e)}$ to denote the axial force in an element as shown in (1.22).

It is important to realize that according to the convention used in structural mechanics, the reactions are forces acting on the structure exerted by the supports. There are two methods of determining the support reactions. The straightforward method is to use eq. (1.20) to determine the unknown $\{\mathbf{F}_s\}$. However, in some FE programs, the structural stiffness matrix $[\mathbf{K}_s]$ is never assembled. The striking of rows and columns is performed at element level, and the global stiffness matrix $[\mathbf{K}]$ is assembled directly. In such situations, eq. (1.19) is used to compute the reactions. For example, the reaction at the ith node is obtained by computing the internal forces in the elements connected to node i and summing all the internal forces.

EXAMPLE 1.2   *Clamped-clamped uniaxial bar*

Use FEM to determine the axial force P in each portion, AB and BC, of the uniaxial bar shown in figure 1.7. What are the support reactions at A and C? Young's modulus is $E = 100$ GPa; the areas of the cross sections of the two portions AB and BC are, respectively, 1×10^{-4} m^2 and 2×10^{-4} m^2 and $F = 10,000$ N. The force F is applied at the cross section at B.

SOLUTION: Since the applied force is a concentrated or point force, it is sufficient to use two elements, AB and BC. The nodes A, B, and C, respectively, will be nodes 1, 2, and 3.

Using eq. (1.16), the element stiffness matrices for two elements are first calculated by

$$\left[\mathbf{k}^{(1)}\right] = \frac{10^{11} \times 10^{-4}}{0.25}\begin{bmatrix} 1 & -1 \\ -1 & 1 \end{bmatrix} = 10^7 \begin{array}{cc} u_1 & u_2 \\ \begin{bmatrix} 4 & -4 \\ -4 & 4 \end{bmatrix} & \begin{array}{c} u_1 \\ u_2 \end{array} \end{array},$$

$$\left[\mathbf{k}^{(2)}\right] = \frac{10^{11} \times 2 \times 10^{-4}}{0.4}\begin{bmatrix} 1 & -1 \\ -1 & 1 \end{bmatrix} = 10^7 \begin{array}{cc} u_2 & u_3 \\ \begin{bmatrix} 5 & -5 \\ -5 & 5 \end{bmatrix} & \begin{array}{c} u_2 \\ u_3 \end{array} \end{array}.$$

Note that the row addresses are written against each row in the element stiffness matrices, and column addresses are shown above each column. Using eqs. (1.19) and (1.20), the two elements are assembled to produce the structural equilibrium equations:

$$10^7 \begin{array}{ccc} u_1 & u_2 & u_3 \end{array} \\ \begin{bmatrix} 4 & -4 & 0 \\ -4 & 4+5 & -5 \\ 0 & -5 & 5 \end{bmatrix} \begin{Bmatrix} u_1 \\ u_2 \\ u_3 \end{Bmatrix} = \begin{Bmatrix} F_1 \\ 10,000 \\ F_3 \end{Bmatrix}. \tag{1.24}$$

Note that nodes 1 and 3 are fixed and have unknown reaction forces. After deleting the rows and columns corresponding to the fixed DOFs (u_1 and u_3), we obtain $[\mathbf{K}]\{\mathbf{Q}\} = \{\mathbf{F}\}$:

$$10^7 [9]\{u_2\} = \{10,000\} \quad \Rightarrow \quad u_2 = 1.111 \times 10^{-4} \text{m}.$$

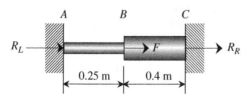

Figure 1.7 Two clamped uniaxial bars

Note that the final equation turns out to be a scalar equation because there is only one free DOF. By collecting all DOFs, the vector of nodal displacements can be obtained as: $\{\mathbf{Q}_s\}^T = \{u_1, u_2, u_3\} = \{0, \ 1.111 \times 10^{-4}, \ 0\}$. After solving for the unknown nodal displacements, the axial forces of the elements can be computed using $P = (AE/L)(u_j - u_i)$, as

$$P^{(1)} = 4 \times 10^7 (u_2 - u_1) = 4{,}444\text{N}$$
$$P^{(2)} = 5 \times 10^7 (u_3 - u_2) = -5{,}556\text{N}.$$

Note that the first element is under tension, while the second is under compressive force.

The reaction forces can be calculated from the first and third rows in eq. (1.24) using the calculated nodal DOFs, as

$$R_L = F_1 = -4 \times 10^7 u_2 = -4{,}444\text{N},$$
$$R_R = F_2 = -5 \times 10^7 u_2 = -5{,}556\text{N}.$$

Alternatively, from the equilibrium between internal and external forces [eq. (1.19)], the two reaction forces can be calculated using the internal forces, as

$$R_L = -P^{(1)} = -4{,}444N,$$
$$R_R = +P^{(2)} = -5{,}556N.$$

Note that both reaction forces are in the negative x direction, and the sum of reactions is the same as the external force at node 2 with the opposite sign. ◼

EXAMPLE 1.3   *Three uniaxial bar elements*

Consider an assembly of three two-force members as shown in figure 1.8. Motion is restricted to one dimension along the x-axis. Determine the displacement of the rigid member, element forces, and reaction forces from the wall. Assume $k^{(1)} = 50$ N/cm, $k^{(2)} = 30$ N/cm, $k^{(3)} = 70$ N/cm, and $F_1 = 40$ N.

SOLUTION   The assembly consists of three elements and four nodes. Figure 1.9 illustrates the free-body diagram of the system with node and element numbers.

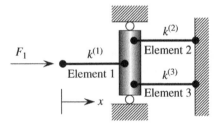

Figure 1.8 One-dimensional structure with three uniaxial bar elements

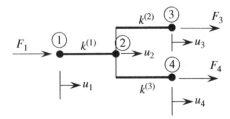

Figure 1.9 Finite element model

Write down the stiffness matrix of each element along with the row addresses. From now on, we will not show the column addresses over the stiffness matrices.

Element 1: $\left[\mathbf{k}^{(1)}\right] = k^{(1)} \begin{bmatrix} 1 & -1 \\ -1 & 1 \end{bmatrix} \begin{matrix} u_1 \\ u_2 \end{matrix}$.

Element 2: $\left[\mathbf{k}^{(2)}\right] = k^{(2)} \begin{bmatrix} 1 & -1 \\ -1 & 1 \end{bmatrix} \begin{matrix} u_2 \\ u_3 \end{matrix}$.

Element 3: $\left[\mathbf{k}^{(3)}\right] = k^{(3)} \begin{bmatrix} 1 & -1 \\ -1 & 1 \end{bmatrix} \begin{matrix} u_2 \\ u_4 \end{matrix}$.

After assembling the element stiffness matrices, we obtain the following structural stiffness matrix:

$$[\mathbf{K}_s] = \underbrace{\begin{bmatrix} k^{(1)} & -k^{(1)} & 0 & 0 \\ -k^{(1)} & \left(k^{(1)} + k^{(2)} + k^{(3)}\right) & -k^{(2)} & -k^{(3)} \\ 0 & -k^{(2)} & k^{(2)} & 0 \\ 0 & -k^{(3)} & 0 & k^{(3)} \end{bmatrix}}_{\text{Structural Stiffness Matrix}} \begin{matrix} u_1 \\ u_2 \\ u_3 \\ u_4 \end{matrix}.$$

The equation $[\mathbf{K}_s]\{\mathbf{Q}_s\} = \{\mathbf{F}_s\}$ takes the form:

$$\begin{bmatrix} k^{(1)} & -k^{(1)} & 0 & 0 \\ -k^{(1)} & \left(k^{(1)} + k^{(2)} + k^{(3)}\right) & -k^{(2)} & -k^{(3)} \\ 0 & -k^{(2)} & k^{(2)} & 0 \\ 0 & -k^{(3)} & 0 & k^{(3)} \end{bmatrix} \begin{Bmatrix} u_1 \\ u_2 \\ u_3 \\ u_4 \end{Bmatrix} = \begin{Bmatrix} F_1 \\ F_2 \\ F_3 \\ F_4 \end{Bmatrix}. \tag{1.25}$$

The next step is to substitute boundary conditions and solve for unknown displacements. At all nodes, either the externally applied load or the displacement is specified. Substituting for the stiffnesses $k^{(1)}$, $k^{(2)}$, and $k^{(3)}$, $F_1 = 40$ N and $F_2 = 0$, and $u_3 = u_4 = 0$ in eq. (1.25), we obtain

$$\begin{Bmatrix} F_1 = 40 \\ F_2 = 0 \\ F_3 = R_3 \\ F_4 = R_4 \end{Bmatrix} = \begin{bmatrix} 50 & -50 & 0 & 0 \\ -50 & (50+30+70) & -30 & -70 \\ 0 & -30 & 30 & 0 \\ 0 & -70 & 0 & 70 \end{bmatrix} \begin{Bmatrix} u_1 \\ u_2 \\ u_3 = 0 \\ u_4 = 0 \end{Bmatrix}. \tag{1.26}$$

Next, we delete the rows and columns corresponding to zero displacements. In this example, the third and fourth rows and columns correspond to zero displacements. Deleting these rows and columns, we obtain the global equations in the form $[\mathbf{K}]\{\mathbf{Q}\} = \{\mathbf{F}\}$, where $[\mathbf{K}]$ is the global stiffness matrix:

$$\begin{bmatrix} 50 & -50 \\ -50 & 150 \end{bmatrix} \begin{Bmatrix} u_1 \\ u_2 \end{Bmatrix} = \begin{Bmatrix} 40 \\ 0 \end{Bmatrix}.$$

The unknown displacements u_1 and u_2 can be obtained by solving the above equation as

$$u_1 = 1.2 \, \text{cm and } u_2 = 0.4 \, \text{cm}.$$

By collecting all DOFs, the vector of nodal displacements can be obtained as: $\{\mathbf{Q}_s\}^{\mathrm{T}} = \{u_1, u_2, u_3, u_4\} = \{1.2, 0.4, 0, 0\}$.

Next, we substitute u_1 and u_2 into rows 3 and 4 in eq. (1.26) to calculate the reaction forces F_3 and F_4:

$$F_3 = 0u_1 - 30u_2 + 30u_3 + 0u_4 = -12 \text{ N},$$
$$F_4 = 0u_1 - 70u_2 + 0u_3 + 70u_4 = -28 \text{ N}.$$

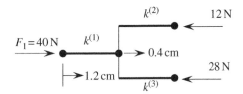

Figure 1.10 Free-body diagram of the structure

Based on the results obtained, we can now redraw the free-body diagram of the system, as shown in figure 1.10. Both reaction forces are in the negative x direction, and the sum of reactions is equal to the applied force in the opposite direction. ▄▄

1.2.6   FE Program Organization

As the finite element analysis follows a standard procedure as described in the preceding section, it is possible to make a general-purpose FE program. Commercial FE programs typically consist of three parts: preprocessor, FE solver, and postprocessor. A preprocessor allows the user to define the structure, divide it into a number of elements, identify the nodes and their coordinates, define connectivity between various elements, and define material properties and the loads. Developments in computer graphics and CAD technology have resulted in sophisticated preprocessors that let the users create models and define various properties interactively on the computer terminal itself. A postprocessor takes the FE analysis results and presents them in a user-friendly graphical form. Again, developments in software and graphics have resulted in very sophisticated animations to help the analysts better understand the results of an FE model. This book is mostly concerned with the principles involved in the development and operation of the core FE program, which computes the stiffness matrix and assembles and solves the final set of equations. More on this is discussed in chapter 9. In addition, a brief introduction is provided to perform finite element analysis using commercial programs in the companion website of the book, where various finite element analysis programs are introduced, including Abaqus, ANSYS, Autodesk Nastran, and MATLAB Toolbox.

1.3   PLANE TRUSS ELEMENTS

This section presents the formulation of stiffness matrix and general procedures for solving the two-dimensional or plane truss using the direct stiffness method. A truss consisting of two elements is used to illustrate the solution procedures.

Consider the plane truss consisting of two bar elements or members as shown in figure 1.11. Two bars are connected with each other and with the ground using a pin joint; that is, their motion is constrained but free to rotate. A horizontal force $F = 50$ N is applied at the top node. Although the elements

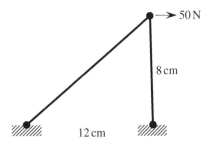

Figure 1.11 A plane truss consisting of two members

of the truss are uniaxial bars, the methods described in the previous section cannot be readily applied to this problem for two reasons: the two elements are not in the same direction but are inclined at different angles, and the external forces at a node can be applied in both x and y directions.

However, the element stiffness matrix of uniaxial bar elements will be applied to individual elements of the truss if we consider a local coordinate system. For a plane truss element, the following two coordinate systems can be defined:

1. The global coordinate system, $x-y$ for the entire structure.

2. A local coordinate system, $\bar{x}-\bar{y}$ for a particular element such that the $\bar{x}$-axis is along the length of the element.

Referring to figure 1.12, the force-displacement relation of a truss element can be written in the local coordinate system as

$$\left\{ \begin{array}{c} f_{1\bar{x}} \\ f_{2\bar{x}} \end{array} \right\} = \frac{EA}{L} \begin{bmatrix} 1 & -1 \\ -1 & 1 \end{bmatrix} \left\{ \begin{array}{c} \bar{u}_1 \\ \bar{u}_2 \end{array} \right\}, \tag{1.27}$$

where E, A, and L, respectively, are the Young's modulus, the area of the cross section, and the length of the element, and EA/L corresponds to the spring constant k in eq. (1.16).

Note that the forces and displacements are represented in the local coordinate system. In order to make the above equation more general, let us consider the transverse displacement $\bar{v}_1$ and $\bar{v}_2$ in the $\bar{y}$ direction. Corresponding transverse forces at each node can be defined as $f_{1\bar{y}}$ and $f_{2\bar{y}}$. However, in the truss element, these forces do not exist, and hence they are equated to zero. This is because the truss is a two-force member, where the member can only support a force in the axial direction. Then, the above stiffness matrix (system equations in matrix form) can be expanded to incorporate the forces and displacements in the $\bar{y}$ direction as shown below.

$$\left\{ \begin{array}{c} f_{1\bar{x}} \\ f_{1\bar{y}} \\ f_{2\bar{x}} \\ f_{2\bar{y}} \end{array} \right\} = \frac{EA}{L} \begin{bmatrix} 1 & 0 & -1 & 0 \\ 0 & 0 & 0 & 0 \\ -1 & 0 & 1 & 0 \\ 0 & 0 & 0 & 0 \end{bmatrix} \left\{ \begin{array}{c} \bar{u}_1 \\ \bar{v}_1 \\ \bar{u}_2 \\ \bar{v}_2 \end{array} \right\}. \tag{1.28}$$

The expanded local stiffness matrix in the above equation:

1. is a square matrix;

2. is symmetric; and

3. has diagonal elements that are greater than or equal to zero.

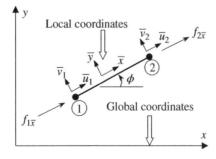

Figure 1.12 Local and global coordinate systems

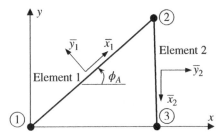

Figure 1.13 Local coordinate systems of the two-bar truss

The above stiffness matrix is valid only for the particular element 1 in the above example. It cannot be applied to other elements because the local coordinates $\bar{x}$–$\bar{y}$ are different for different elements. The local coordinates for element 2 are shown in figure 1.13.

In order to develop a system of equations that connect all elements in the truss, we need to transform the force-displacement relations, for instance, as shown in eq. (1.28), to the global coordinates, which is common for all elements of the truss. This requires the use of vector coordinate transformation.

1.3.1   Coordinate Transformation

As forces and displacements are vectors, we can use the vector transformation to find the relation between the displacements in local and global coordinates at a node. The global coordinate system, x–y, is fixed in space and common for the entire structure. On the other hand, the local coordinate system is parallel to the element. The local $\bar{x}$-axis is defined from node i to node j, while the local $\bar{y}$-axis is rotated by 90 degrees in the counterclockwise direction in the plane. Then, the angle ϕ for the element local coordinate is defined from the positive $\bar{x}$-axis to the positive x-axis.

Using the angle of the element, the relation between the displacements in local and global coordinates at node 1 can be written as

$$\begin{Bmatrix} \bar{u}_1 \\ \bar{v}_1 \end{Bmatrix} = \begin{bmatrix} \cos\phi & \sin\phi \\ -\sin\phi & \cos\phi \end{bmatrix} \begin{Bmatrix} u_1 \\ v_1 \end{Bmatrix}.$$

A similar relation for node 2 will be

$$\begin{Bmatrix} \bar{u}_2 \\ \bar{v}_2 \end{Bmatrix} = \begin{bmatrix} \cos\phi & \sin\phi \\ -\sin\phi & \cos\phi \end{bmatrix} \begin{Bmatrix} u_2 \\ v_2 \end{Bmatrix}.$$

Actually, we can combine the above relations for the two nodes to obtain the following relationship:

$$\underbrace{\begin{Bmatrix} \bar{u}_1 \\ \bar{v}_1 \\ \bar{u}_2 \\ \bar{v}_2 \end{Bmatrix}}_{\text{local}} = \begin{bmatrix} \cos\phi & \sin\phi & 0 & 0 \\ -\sin\phi & \cos\phi & 0 & 0 \\ 0 & 0 & \cos\phi & \sin\phi \\ 0 & 0 & -\sin\phi & \cos\phi \end{bmatrix} \underbrace{\begin{Bmatrix} u_1 \\ v_1 \\ u_2 \\ v_2 \end{Bmatrix}}_{\text{global}}.$$

The above relation between local and global displacements can be written using a shorthand notation as

$$\{\bar{\mathbf{q}}\} = [\mathbf{T}]\{\mathbf{q}\}, \tag{1.29}$$

where $\{\bar{\mathbf{q}}\}$ and $\{\mathbf{q}\}$ are the element DOFs in the local and global coordinates, respectively, and $[\mathbf{T}]$ is the *transformation matrix*. In some literature $[\mathbf{T}]$ is called the rotation matrix. Since forces are also vectors, the forces $\{\bar{\mathbf{f}}\}$ in element coordinates are related to $\{\mathbf{f}\}$ in global coordinates as

$$\underbrace{\begin{Bmatrix} f_{1\bar{x}} \\ f_{1\bar{y}} \\ f_{2\bar{x}} \\ f_{2\bar{y}} \end{Bmatrix}}_{\text{local}} = \begin{bmatrix} \cos\phi & \sin\phi & 0 & 0 \\ -\sin\phi & \cos\phi & 0 & 0 \\ 0 & 0 & \cos\phi & \sin\phi \\ 0 & 0 & -\sin\phi & \cos\phi \end{bmatrix} \underbrace{\begin{Bmatrix} f_{1x} \\ f_{1y} \\ f_{2x} \\ f_{2y} \end{Bmatrix}}_{\text{global}},$$

or in shorthand notation

$$\{\bar{\mathbf{f}}\} = [\mathbf{T}]\{\mathbf{f}\}. \tag{1.30}$$

In the following section, we will express the local element equation (1.28) in the global coordinate using transformation relations in eqs. (1.29) and (1.30). Once all element equations are expressed in the global coordinate, they can be assembled using procedures similar to that of uniaxial bar elements.

1.3.2 Element Stiffness Matrix in the Global Coordinates

A key concept in finite element method is to discretize the entire system into many elements and to assemble them to make connections between elements. In order to make the assembly process valid, it is necessary that all DOFs in elements are represented in the same coordinate system. In this section, the element stiffness matrix in the local coordinates will be transformed into the global coordinates using the transformation relationship in the previous section.

For a single truss element, using the above coordinate transformation equation, we can proceed to transform the element stiffness matrix from local to global coordinates. Consider the truss element arbitrarily positioned in two-dimensional space as shown in figure 1.14. The element is defined in such a way that node 1 is the first node and node 2 is the second node. Therefore, the local $\bar{x}$-axis is defined from node 1 to node 2. The angle ϕ is defined from the positive x-axis to the positive $\bar{x}$-axis. If the element connectivity is defined from node 2 to node 1, then the angle should be defined as $\phi + 180°$. The stiffness of the element is given as $k = EA/L$.

The force-displacement equations can be expressed in the local coordinates as:

$$\begin{Bmatrix} f_{1\bar{x}} \\ f_{1\bar{y}} \\ f_{2\bar{x}} \\ f_{2\bar{y}} \end{Bmatrix} = \underbrace{\frac{EA}{L} \begin{bmatrix} 1 & 0 & -1 & 0 \\ 0 & 0 & 0 & 0 \\ -1 & 0 & 1 & 0 \\ 0 & 0 & 0 & 0 \end{bmatrix}}_{\text{element stiffness matrix}} \begin{Bmatrix} \bar{u}_1 \\ \bar{v}_1 \\ \bar{u}_2 \\ \bar{v}_2 \end{Bmatrix}. \tag{1.31}$$

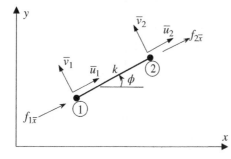

Figure 1.14 Definition of two-dimensional truss element

In the shorthand notation, eq. (1.31) takes the following form:

$$\{\bar{\mathbf{f}}\} = [\bar{\mathbf{k}}]\{\bar{\mathbf{q}}\}. \tag{1.32}$$

Substitution of eqs. (1.29) and (1.30) into eq. (1.31) yields

$$[\mathbf{T}]\{\mathbf{f}\} = [\bar{\mathbf{k}}][\mathbf{T}]\{\mathbf{q}\}.$$

Multiplying both sides of the equation by $[\mathbf{T}]^{-1}$,

$$\underbrace{\{\mathbf{f}\}}_{\text{global}} = [\mathbf{T}]^{-1}[\bar{\mathbf{k}}][\mathbf{T}]\underbrace{\{\mathbf{q}\}}_{\text{global}},$$

or

$$\{\mathbf{f}\} = [\mathbf{k}]\{\mathbf{q}\}. \tag{1.33}$$

The element stiffness matrix [**k**] in the global coordinates can now be expressed in terms of [**k̄**] as

$$[\mathbf{k}] = [\mathbf{T}]^{-1}[\bar{\mathbf{k}}][\mathbf{T}]. \tag{1.34}$$

It can be shown that the inverse of the transformation matrix [**T**] is equal to its transpose, and hence [**k**] can be written as

$$[\mathbf{k}] = [\mathbf{T}]^{\mathrm{T}}[\bar{\mathbf{k}}][\mathbf{T}]. \tag{1.35}$$

Performing the matrix multiplication in eq. (1.35), we obtain an explicit expression for [**k**] as

$$[\mathbf{k}] = \frac{EA}{L} \begin{bmatrix} \cos^2\phi & \cos\phi\sin\phi & -\cos^2\phi & -\cos\phi\sin\phi \\ \cos\phi\sin\phi & \sin^2\phi & -\cos\phi\sin\phi & -\sin^2\phi \\ -\cos^2\phi & -\cos\theta\sin\phi & \cos^2\phi & \cos\phi\sin\phi \\ -\cos\phi\sin\phi & -\sin^2\phi & \cos\phi\sin\phi & \sin^2\phi \end{bmatrix}. \tag{1.36}$$

From eq. (1.36), it is clear that the element stiffness matrix of a plane truss element depends on the length L, axial rigidity EA, and the angle of orientation. As mentioned earlier, the element stiffness matrix is symmetric. Its determinant is equal to zero, and hence it does not have an inverse. Furthermore, the element stiffness matrix is positive semi-definite, and its diagonal elements are either equal to zero or greater than zero.

EXAMPLE 1.4 *Two-bar truss*

The two-bar truss shown in figure 1.15 has circular cross sections with a diameter of 0.25 cm and Young's modulus $E = 30 \times 10^6 \, \text{N/cm}^2$. An external force $F = 50 \, \text{N}$ is applied in the horizontal direction at node 2. Calculate the displacement of each node and stress in each element.

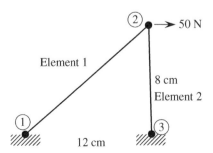

Figure 1.15 Two-bar truss structure

SOLUTION   **Element 1:**

In the local coordinate system shown in figure 1.16, the force-displacement equations for element 1 is given in eq. (1.28), which can be transformed to the global coordinates similar to the one in eq. (1.33), to yield

$$\left\{\mathbf{f}^{(1)}\right\} = \left[\mathbf{k}^{(1)}\right]\left\{\mathbf{q}^{(1)}\right\}.$$

Since the orientation angle of the element is $\phi_1 = 33.7°$, the element equations in the global coordinates can be obtained using the stiffness matrix in eq. (1.36), as

$$\begin{Bmatrix} f_{1x}^{(1)} \\ f_{1y}^{(1)} \\ f_{2x}^{(1)} \\ f_{2y}^{(1)} \end{Bmatrix} = 102,150 \begin{bmatrix} 0.692 & 0.462 & -0.692 & -0.462 \\ 0.462 & 0.308 & -0.462 & -0.308 \\ -0.692 & -0.462 & 0.692 & 0.462 \\ -0.462 & -0.308 & 0.462 & 0.308 \end{bmatrix} \begin{Bmatrix} u_1 \\ v_1 \\ u_2 \\ v_2 \end{Bmatrix}.$$

Element 2:

For element 2, the same procedure can be applied with the orientation angle of the element being $\phi_2 = -90°$ (see figure 1.17). In the global coordinates, the element equations become

$$\begin{Bmatrix} f_{2x}^{(2)} \\ f_{2y}^{(2)} \\ f_{3x}^{(2)} \\ f_{3y}^{(2)} \end{Bmatrix} = 184,125 \begin{bmatrix} 0 & 0 & 0 & 0 \\ 0 & 1 & 0 & -1 \\ 0 & 0 & 0 & 0 \\ 0 & -1 & 0 & 1 \end{bmatrix} \begin{Bmatrix} u_2 \\ v_2 \\ u_3 \\ v_3 \end{Bmatrix}.$$

Note that the orientation is measured in the counterclockwise direction from the positive x-axis of the global coordinates.

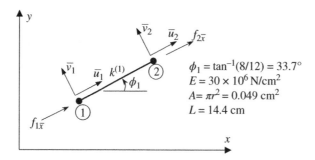

$\phi_1 = \tan^{-1}(8/12) = 33.7°$
$E = 30 \times 10^6 \, \text{N/cm}^2$
$A = \pi r^2 = 0.049 \, \text{cm}^2$
$L = 14.4 \, \text{cm}$

Figure 1.16 Local coordinates of element 1

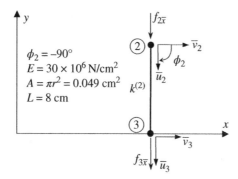

$\phi_2 = -90°$
$E = 30 \times 10^6 \, \text{N/cm}^2$
$A = \pi r^2 = 0.049 \, \text{cm}^2$
$L = 8 \, \text{cm}$

Figure 1.17 Local coordinates of element 2

Now we are ready to assemble the global stiffness matrix of the structure. Summing the two sets of force-displacement equations in the global coordinates:

Element 1

$$
\begin{Bmatrix} F_{1x} \\ F_{1y} \\ F_{2x} \\ F_{2y} \\ F_{3x} \\ F_{3y} \end{Bmatrix} = \begin{bmatrix} 70687 & 47193 & -70687 & -47193 & 0 & 0 \\ 47193 & 31462 & -47193 & -31462 & 0 & 0 \\ -70687 & -47193 & 70687 & 47193 & 0 & \\ -47193 & -31462 & 47193 & 215587 & 0 & -184125 \\ 0 & 0 & 0 & 0 & 0 & 0 \\ 0 & 0 & 0 & -184125 & 0 & 184125 \end{bmatrix} \begin{Bmatrix} u_1 \\ v_1 \\ u_2 \\ v_2 \\ u_3 \\ v_3 \end{Bmatrix}.
$$

Element 2

Note that two element stiffness matrices overlap at DOFs corresponding to u_2 and v_2 because the two elements are connected at node 2. Next, apply the following known boundary conditions:

(a) Nodes 1 and 3 are fixed; therefore, the displacement components of these two nodes are zero (u_1, v_1 and u_3, v_3).

(b) The only applied external forces are at node 2: $F_{2x} = 50$ N, and $F_{2y} = 0$ N.

$$
\begin{Bmatrix} F_{1x} \\ F_{1y} \\ 50 \\ 0 \\ F_{3x} \\ F_{3y} \end{Bmatrix} = \begin{bmatrix} 70687 & 47193 & -70687 & -47193 & 0 & 0 \\ 47193 & 31462 & -47193 & -31462 & 0 & 0 \\ -70687 & -47193 & 70687 & 47193 & 0 & 0 \\ -47193 & -31462 & 47193 & 215587 & 0 & -184125 \\ 0 & 0 & 0 & 0 & 0 & 0 \\ 0 & 0 & 0 & -184125 & 0 & 184125 \end{bmatrix} \begin{Bmatrix} 0 \\ 0 \\ u_2 \\ v_2 \\ 0 \\ 0 \end{Bmatrix}.
$$

We first delete the columns corresponding to zero displacements. In this example, the third and fourth columns correspond to nonzero displacements. We keep these two columns and strike out all other columns, to obtain

$$
\begin{Bmatrix} F_{1x} \\ F_{1y} \\ 50 \\ 0 \\ F_{3x} \\ F_{3y} \end{Bmatrix} = \begin{bmatrix} -70687 & -47193 \\ -47193 & -31462 \\ 70687 & 47193 \\ 47193 & 215587 \\ 0 & 0 \\ 0 & -184125 \end{bmatrix} \begin{Bmatrix} u_2 \\ v_2 \end{Bmatrix}. \tag{1.37}
$$

F_{1x}, F_{1y}, F_{3x}, and F_{3y} are unknown reaction forces, and therefore we will delete the rows that correspond to these unknown reaction forces. We delete these rows because we want to keep only unknown displacement and known forces. We will use the deleted rows later to calculate unknown reaction forces after all displacements are calculated. Then, finally we have the following 2×2 matrix equation for the nodal displacements u_2 and v_2:

$$
\begin{Bmatrix} 50 \\ 0 \end{Bmatrix} = \begin{bmatrix} 70687 & 47193 \\ 47193 & 215587 \end{bmatrix} \begin{Bmatrix} u_2 \\ v_2 \end{Bmatrix}.
$$

-60.2 N $\xleftarrow{\hspace{1.5cm}}$ ①————————② $\xrightarrow{\hspace{1.5cm}}$ 60.2 N   **Figure 1.18** Element force for element 1 in local coordinates

Note that the global matrix equations only include unknown displacements and known forces. Since the global stiffness matrix in the above equation is positive definite, it is possible to invert it to solve for the unknown nodal displacements:

$$u_2 = 8.28 \times 10^{-4} \text{ cm,}$$
$$v_2 = -1.81 \times 10^{-4} \text{ cm.}$$

Substituting the known u_2 and v_2 values into the matrix equation (1.37) and solve for the reaction forces:

$$\begin{Bmatrix} F_{1x} \\ F_{1y} \\ F_{3x} \\ F_{3y} \end{Bmatrix} = \begin{bmatrix} -70687 & -47193 \\ -47193 & -31462 \\ 0 & 0 \\ 0 & -184125 \end{bmatrix} \begin{Bmatrix} 8.28 \times 10^{-4} \\ -1.81 \times 10^{-4} \end{Bmatrix} = \begin{Bmatrix} -50 \\ -33.39 \\ 0 \\ 33.39 \end{Bmatrix} \text{N.}$$

Since the truss element is a two-force member, it is clear that the reaction force at node 3 is in the vertical direction, and the reaction force at node 1 is parallel to the direction of the element.

Once all displacements and forces are calculated in the global coordinates, it is necessary to go back to the element level in order to calculate element forces. However, it is important to note that the basic element behavior is expressed in the element local coordinates. Therefore, there are two steps involved in calculating element forces. First, among the vector of global displacements, $\{Q_s\}$, it is necessary to extract displacements that belong to the element, $\{q^{(e)}\}$. Then, it is necessary to transform the element displacements in the global coordinates into the local coordinates, $\{\bar{q}^{(e)}\}$. For example, the nodal displacements of element 1 in the local coordinate system can be obtained from eq. (1.29), as

$$\begin{Bmatrix} \bar{u}_1 \\ \bar{v}_1 \\ \bar{u}_2 \\ \bar{v}_2 \end{Bmatrix} = \begin{bmatrix} .832 & .555 & 0 & 0 \\ -.555 & .832 & 0 & 0 \\ 0 & 0 & .832 & .555 \\ 0 & 0 & -.555 & .832 \end{bmatrix} \begin{Bmatrix} 0 \\ 0 \\ u_2 \\ v_2 \end{Bmatrix} = \begin{Bmatrix} 0 \\ 0 \\ 5.89 \times 10^{-4} \\ -6.11 \times 10^{-4} \end{Bmatrix}.$$

Then, the local force-displacement equations (1.32) can be used to calculate the element forces, as

$$\begin{Bmatrix} f_{1\bar{x}} \\ f_{1\bar{y}} \\ f_{2\bar{x}} \\ f_{2\bar{y}} \end{Bmatrix} = \frac{EA}{L} \begin{bmatrix} 1 & 0 & -1 & 0 \\ 0 & 0 & 0 & 0 \\ -1 & 0 & 1 & 0 \\ 0 & 0 & 0 & 0 \end{bmatrix} \begin{Bmatrix} 0 \\ 0 \\ 5.89 \times 10^{-4} \\ -6.11 \times 10^{-4} \end{Bmatrix} = \begin{Bmatrix} -60.2 \\ 0 \\ 60.2 \\ 0 \end{Bmatrix} \text{N.} \qquad (1.38)$$

Equation (1.38) represents the forces acting on the element in the local coordinate system. As expected, there is no force component in the $\bar{y}$ direction (local y direction). In the $\bar{x}$ direction (local x direction), the two nodes have the same magnitude of internal forces but in the opposite direction. As can be seen in figure 1.18, two equal and opposite forces act on the truss element, which results in tensile stresses in the element. In general, the sign of the force $f_{j\bar{x}}$ at node j (second node) will be the same as the sign of the element force, P. In the present example, the element force of element 1 is positive and has the magnitude of 60.2 N. Therefore, the normal stress in element 1 is tensile, and it can be calculated as $(60.2 / 0.049) = 1228 \text{ N/cm}^2$. ■

Another method to calculate the element force in a truss element is described below. This follows the method used in deriving eq. (1.22). For a truss element at an arbitrary orientation, eq. (1.22) is modified as

$$P^{(e)} = \left(\frac{AE}{L}\right)^{(e)} \Delta^{(e)} = \left(\frac{AE}{L}\right)^{(e)} (\bar{u}_j - \bar{u}_i). \qquad (1.39)$$

Using the transformation relations in eq. (1.29), one can express the displacements in global coordinates. Let $l = \cos\phi$ and $m = \sin\phi$. Then, the expression for P takes the following form:

$$P^{(e)} = \left(\frac{AE}{L}\right)^{(e)} \left[\left(lu_j + mv_j\right) - \left(lu + mv_i\right)\right]$$
$$= \left(\frac{AE}{L}\right)^{(e)} \left[l\left(u_j - u_i\right) + m\left(v_j - v_i\right)\right]. \tag{1.40}$$

In the above example, the following basic principles of FE analysis were used: (1) derive the force-displacement relations of each truss member, (2) assemble the equations to obtain the global equations, and (3) solve for unknown displacements. However, in practical problems with a large number of elements, one need not write the equations of equilibrium for each element. We would like to develop a systematic procedure that is suitable for a large number of elements. In this method, each element is assigned a first node and second node. These node numbers are denoted by i and j, respectively. The choice of the first and second nodes is arbitrary; however, it has to be consistent throughout the solution of the problem. The orientation of the element is defined by the angle the direction i–j makes with the positive x-axis, and it is denoted by ϕ. The direction cosines of the element are: $l = \cos\phi, m = \sin\phi$. We assign the row and column addresses to the element stiffness matrix as shown below. The element stiffness matrix in eq. (1.36) can be written in terms of l and m with the row and column addresses as

$$[\mathbf{k}] = \left(\frac{EA}{L}\right)^{(e)}
\begin{array}{cccc}
 u_i & v_i & u_j & v_j
\end{array}
\begin{bmatrix}
l^2 & lm & -l^2 & -lm \\
lm & m^2 & -lm & -m^2 \\
-l^2 & -lm & l^2 & lm \\
-lm & -m^2 & lm & m^2
\end{bmatrix}
\begin{array}{c}
u_i \\ v_i \\ u_j \\ v_j
\end{array}. \tag{1.41}$$

As illustrated in the following example, the row and column addresses are useful in assembling the element stiffness matrices into the global stiffness matrix. Note that the row addresses are the transpose of the column addresses.

EXAMPLE 1.5   *Plane truss with three elements*

The plane truss shown in figure 1.19 consists of three members connected to each other and to the walls by pin joints. The members make equal angles with each other, and element 2 is vertical. The members are identical to each other with the following properties: Young's modulus $E = 206 \times 10^9$ Pa, cross-sectional area $A = 1 \times 10^{-4}$ m^2, and length

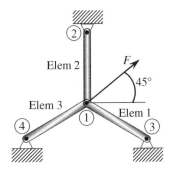

Figure 1.19 Plane structure with three truss elements

$L = 1$ m. An inclined force $F = 20{,}000$ N is applied at node 1. Solve for the displacements at Node 1 and stresses in the three elements.

SOLUTION Based on the above figure, a connectivity table that includes the element properties, node connectivity, and direction cosines can be calculated as shown in table 1.2.

Then, using eq. (1.36), the element stiffness matrices in the global coordinates system can be obtained as

$$\left[\mathbf{k}^{(1)}\right] = 206 \times 10^5 \begin{bmatrix} 0.750 & -0.433 & -0.750 & 0.433 \\ -0.433 & 0.250 & 0.433 & -0.250 \\ -0.750 & 0.433 & 0.750 & -0.433 \\ 0.433 & -0.250 & -0.433 & 0.250 \end{bmatrix} \begin{matrix} u_1 \\ v_1 \\ u_3 \\ v_3 \end{matrix},$$

$$\left[\mathbf{k}^{(2)}\right] = 206 \times 10^5 \begin{bmatrix} 0 & 0 & 0 & 0 \\ 0 & 1 & 0 & -1 \\ 0 & 0 & 0 & 0 \\ 0 & -1 & 0 & 1 \end{bmatrix} \begin{matrix} u_1 \\ v_1 \\ u_2 \\ v_2 \end{matrix},$$

$$\left[\mathbf{k}^{(3)}\right] = 206 \times 10^5 \begin{bmatrix} 0.750 & 0.433 & -0.750 & -0.433 \\ 0.433 & 0.250 & -0.433 & -0.250 \\ -0.750 & -0.433 & 0.750 & 0.433 \\ -0.433 & -0.250 & 0.433 & 0.250 \end{bmatrix} \begin{matrix} u_1 \\ v_1 \\ u_4 \\ v_4 \end{matrix}.$$

Note that row addresses are indicated on the RHS of the stiffness matrices, and they are useful in assembling the structural stiffness matrix. One can easily identify the column addresses, although they are not written above the stiffness matrices. The structural stiffness matrix and the finite element equations are obtained by assembling the three element stiffness matrices:

$$206 \times 10^5 \begin{bmatrix} 1.5 & 0 & 0 & 0 & -0.750 & 0.433 & -0.750 & -0.433 \\ & 1.5 & 0 & -1 & 0.433 & -0.250 & -0.433 & -0.250 \\ & & 0 & 0 & 0 & 0 & 0 & 0 \\ & & & 1 & 0 & 0 & 0 & 0 \\ & & & & 0.750 & -0.433 & 0 & 0 \\ & \text{Symmetric} & & & & 0.250 & 0 & 0 \\ & & & & & & 0.750 & 0.433 \\ & & & & & & & 0.250 \end{bmatrix} \begin{Bmatrix} u_1 \\ v_1 \\ u_2 \\ v_2 \\ u_3 \\ v_3 \\ u_4 \\ v_4 \end{Bmatrix} = \begin{Bmatrix} F_{x1} \\ F_{y1} \\ F_{x2} \\ F_{y2} \\ F_{x3} \\ F_{y3} \\ F_{x4} \\ F_{y4} \end{Bmatrix},$$

or

$$[\mathbf{K}_s]\{\mathbf{Q}_s\} = \{\mathbf{F}_s\}.$$

Table 1.2 Connectivity table with element properties for example 1.5

Element	AE/L	LN1 (i)	LN2 (j)	ϕ	$l = \cos\phi$	$m = \sin\phi$
1	206×10^5	1	3	$-\pi/6$	0.866	-0.5
2	206×10^5	1	2	$\pi/2$	0	1
3	206×10^5	1	4	$-5\pi/6$	-0.866	-0.5

Now, the known external forces and displacements are applied to the above matrix equation. First, the inclined force at node 1 is decomposed into x and y directions, as

$$F_{1x} = 20000 \cdot \cos(\pi/4) = 14,142$$
$$F_{1y} = 20000 \cdot \sin(\pi/4) = 14,142.$$

In addition, since nodes 2, 3, and 4 are fixed, their displacements are equal to zero:

$$u_2 = v_2 = u_3 = v_3 = u_4 = v_4 = 0.$$

We delete the rows and columns in matrix $[\mathbf{K}_s]$ corresponding to those DOFs that have zero displacements. Then, the global FE equations are written in the form $[\mathbf{K}]\{\mathbf{Q}\} = \{\mathbf{F}\}$:

$$206 \times 10^5 \begin{bmatrix} 1.5 & 0 \\ 0 & 1.5 \end{bmatrix} \begin{Bmatrix} u_1 \\ v_1 \end{Bmatrix} = \begin{Bmatrix} 14142 \\ 14142 \end{Bmatrix}.$$

The global stiffness matrix is now positive-definite, and thus invertible. After solving for unknown displacements, we have

$$u_1 = 0.458 \, \text{mm},$$
$$v_1 = 0.458 \, \text{mm}.$$

By including all other zero displacements, the vector of nodal displacements can be written as $\{\mathbf{Q}_s\}^{\text{T}} = \{0.458, 0.458, 0, 0, 0, 0, 0, 0\}$.

Force in each element can be obtained using eq. (1.40) and element properties given in table 1.2. For example, the force in element 1 is

$$P^{(1)} = 206 \times 10^5 (0.866(u - u_1) - 0.5(v - v_1)) = -3,450 \, \text{N}.$$

The same calculation can be repeated for other elements to obtain:

$$P^{(2)} = -9,440 \, \text{N},$$
$$P^{(3)} = 12,900 \, \text{N}.$$

The negative values of $P^{(1)}$ and $P^{(2)}$ indicate compressive forces in those elements. The stresses of the elements can be obtained by dividing the force by the area of cross section:

$$\sigma^{(1)} = -34.5 \, \text{MPa},$$
$$\sigma^{(2)} = -94.4 \, \text{MPa},$$
$$\sigma^{(3)} = 129 \, \text{MPa}.$$

Once element properties, node connectivity, and direction cosines of all elements are listed as in table 1.2, it is easy to make a computer program that can build the global stiffness matrix and solve for unknown displacements. ■

1.3.3 Method of Superposition

In deriving the truss equations, we have used two major assumptions. The displacements are small such that the elongation of an element or a member ΔL is much less than the original length L, that is, $\Delta L << L$, or the strain $\varepsilon = \Delta L / L << 1$. Another assumption is that the stress-strain relation is linear, that is, the Young's modulus is a constant. The above assumptions lead to a simple but useful principle called the superposition principle.

Consider a truss subjected to forces $\{\mathbf{F}\}$ at the nodes or joints. We assume that the truss is supported such that the rigid body displacements are completely constrained. The resulting displacements are $\{\mathbf{Q}\}$. The relation between the applied forces and displacements is given by

$$[\mathbf{K}]\{\mathbf{Q}\} = \{\mathbf{F}\}, \tag{1.42}$$

where $\{\mathbf{K}\}$ is the global stiffness matrix. Now if all the forces are multiplied by a nonzero factor α such that the applied forces are $\{\alpha\mathbf{F}\}$, then the corresponding displacements can be derived as $\{\alpha\mathbf{Q}\}$, that is,

$$[\mathbf{K}]\{\alpha\mathbf{Q}\} = \{\alpha\mathbf{F}\}. \tag{1.43}$$

That is, the relationship between displacements and applied forces is linear. If we double the forces, then the displacements will also be doubled. The above result is obvious as the coefficients of stiffness matrix $\{\mathbf{K}\}$ are constants and depend only the truss geometry and Young's modulus of the material. Note that the above factor α can be negative.

In addition to the linearity between displacements and applied forces, the relationship between displacement and strain is linear, as well as the relationship between strain and stress. Therefore, if the force is doubled, then the displacements, strains, and stresses will also be doubled. This can be a very useful tool for design. For example, let us assume that the maximum stress under the current load is 500 MPa, and we want to limit the maximum stress to 300 MPa. Then it is straightforward that we need to reduce the applied force by 40%.

The second principle that follows the above can be described as follows. Let $\{\mathbf{Q}^{(1)}\}$ and $\{\mathbf{Q}^{(2)}\}$ be the displacements for two different sets of loads $\{\mathbf{F}^{(1)}\}$ and $\{\mathbf{F}^{(2)}\}$, respectively. If the loads are applied together with different factors, that is, if the load is given by $\left\{\alpha\mathbf{F}^{(1)} + \beta\mathbf{F}^{(2)}\right\}$, then the resulting displacements will be the sum of the two sets of displacements, $\left\{\alpha\mathbf{Q}^{(1)} + \beta\mathbf{Q}^{(2)}\right\}$. Again the above result is obvious, as

$$[\mathbf{K}]\left\{\alpha\mathbf{Q}^{(1)} + \beta\mathbf{Q}^{(2)}\right\} = \left\{\alpha\mathbf{F}^{(1)} + \beta\mathbf{F}^{(2)}\right\}. \tag{1.44}$$

In the view of eq. (1.44), it is enough to apply a unit force and calculating resulting stresses, which is called stress influence coefficient. When the actual magnitude of the force is given, the actual stress can be calculated by multiplying the stress influence coefficients by the magnitude of the force.

EXAMPLE 1.6   *Superposition of multiple loads*

Consider the plane truss in example 1.5. Let $\{\mathbf{F}^{(1)}\}$ be a unit force in the x direction at node 1 and $\{\mathbf{F}^{(2)}\}$ be a unit force in the y direction. Show that the displacements using the superposition principle, $\{\mathbf{Q}\} = \alpha\left\{\mathbf{Q}^{(1)}\right\} + \beta\left\{\mathbf{Q}^{(2)}\right\}$, is the same as the actual displacements in Example 1.5, where $\alpha = \beta = 10{,}000\sqrt{2}$.

SOLUTION   The element stiffness matrices and the assembled global stiffness matrix can be found in Example 1.5. After applying the boundary conditions, the global matrix equations for $\{\mathbf{F}^{(1)}\}$ can be written as

$$206 \times 10^5 \begin{bmatrix} 1.5 & 0 \\ 0 & 1.5 \end{bmatrix} \begin{Bmatrix} u_1^{(1)} \\ v_1^{(1)} \end{Bmatrix} = \begin{Bmatrix} 1 \\ 0 \end{Bmatrix}.$$

After solving for unknown displacements, we have

$$\left\{\mathbf{Q}^{(1)}\right\} = \begin{Bmatrix} u_1^{(1)} \\ v_1^{(1)} \end{Bmatrix} = \begin{Bmatrix} 0.3236 \\ 0.0 \end{Bmatrix} \times 10^{-7}.$$

Also, when the vertical force $\{\mathbf{F}^{(2)}\}$ is applied,

$$206 \times 10^5 \begin{bmatrix} 1.5 & 0 \\ 0 & 1.5 \end{bmatrix} \begin{Bmatrix} u_1^{(2)} \\ v_1^{(2)} \end{Bmatrix} = \begin{Bmatrix} 0 \\ 1 \end{Bmatrix}.$$

After solving for unknown displacements, we have

$$\left\{\mathbf{Q}^{(2)}\right\} = \begin{Bmatrix} u_1^{(2)} \\ v_1^{(2)} \end{Bmatrix} = \begin{Bmatrix} 0.0 \\ 0.3236 \end{Bmatrix} \times 10^{-7}.$$

Therefore, the displacements caused by the combined force with $\alpha = \beta = 10{,}000\sqrt{2}$ can be obtained as

$$\{\mathbf{Q}\} = \alpha \left\{\mathbf{Q}^{(1)}\right\} + \beta \left\{\mathbf{Q}^{(2)}\right\} = \begin{Bmatrix} 0.458 \\ 0.458 \end{Bmatrix} \text{mm},$$

which is the same result than that of example 1.5. It is a good practice to show that the same superposition can be applied to the stress. ▰

1.4 THREE-DIMENSIONAL TRUSS ELEMENTS (SPACE TRUSS)

This section presents the formulation of the stiffness matrix and general procedures for solving the three-dimensional or space truss using the direct stiffness method. The procedure is similar to that of plane truss except that an additional DOF is added at each node.

1.4.1 Three-Dimensional Coordinate Transformation

The coordinate transformation used for the two-dimensional truss element can be generalized to the three-dimensional truss elements for space trusses.

A space-truss element has three DOFs, u, v, and w, at each node. Thus, the space–truss element is a 2-node 6-DOF element. Corresponding to the three displacements at each node, there are three forces, f_x, f_y, and f_z. Similar to the two-dimensional plane truss, space truss elements are connected with other elements or the ground using a ball-and-socket joint; that is, two elements connected at a node can rotate freely with respect to each other. The displacements and forces can also be expressed in a local or elemental coordinate system $\bar{x}$–$\bar{y}$–$\bar{z}$ as shown in figure 1.20. The forces and displacements in the local coordinate system are related by the same equation as for two dimensions and is therefore similar to eq. (1.27).

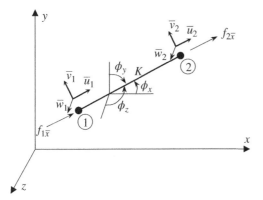

Figure 1.20 Three-dimensional coordinates transformation

$$\begin{Bmatrix} f_{i\bar{x}} \\ f_{j\bar{x}} \end{Bmatrix} = \frac{AE}{L} \begin{bmatrix} 1 & -1 \\ -1 & 1 \end{bmatrix} \begin{Bmatrix} \bar{u}_i \\ \bar{u}_j \end{Bmatrix},$$

or

$$\{\bar{\mathbf{f}}\} = [\bar{\mathbf{k}}]\{\bar{\mathbf{q}}\}. \tag{1.45}$$

In order to assemble truss elements, the element equation (1.45) must be transformed into the global coordinate system. For the plane truss element in eq. (1.31), we expanded the force-displacement relation to a 4×4 matrix equation by including two transverse displacements. In the same way, it is possible to expand the force-displacement relation to a 6×6 matrix equation for space truss, and the transformation matrix would also be a 6×6 matrix, which is a big matrix to handle. Instead, the same result can be obtained if we keep 2×2 matrix as in eq. (1.45) and the following 2×6 transformation matrix:

$$\begin{Bmatrix} \bar{u}_i \\ \bar{u}_j \end{Bmatrix} = \begin{bmatrix} l & m & n & 0 & 0 & 0 \\ 0 & 0 & 0 & l & m & n \end{bmatrix} \begin{Bmatrix} u_i \\ v_i \\ w_i \\ u_j \\ v_j \\ w_j \end{Bmatrix},$$

or

$$\{\bar{\mathbf{q}}\} = [\mathbf{T}]\{\mathbf{q}\}, \tag{1.46}$$

where l, m, and n are the direction cosines of the element connecting nodes i and j, which can be calculated from the element length L and the nodal coordinates as shown below:

$$l = \frac{x_j - x_i}{L}, \quad m = \frac{y_j - y_i}{L}, \quad n = \frac{z_j - z_i}{L},$$
$$L = \sqrt{(x_j - x_i)^2 + (y_j - y_i)^2 + (z_j - z_i)^2}. \tag{1.47}$$

The direction cosines in eq. (1.47) are basically the same as the direction cosines in the plane truss element, where we used the definition of $l = \cos\phi$ and $m = \sin\phi$. But, the actual definitions of direction cosines were $l = \cos\phi_x$ and $m = \cos\phi_y$ where ϕ_x is the angle between the positive local $\bar{x}$-axis to the positive global x-axis, and ϕ_y is the angle between the positive local $\bar{x}$-axis to the positive global y-axis. To compare with the plane truss case in figure 1.14, we can use the property that $\phi_x = \phi$ and $\phi_y = 90° - \phi$.

Similar to transformation for displacements, the nodal forces in the local coordinate system can be transformed as

$$\begin{Bmatrix} f_{ix} \\ f_{iy} \\ f_{iz} \\ f_{jx} \\ f_{jy} \\ f_{jz} \end{Bmatrix} = \begin{bmatrix} l & 0 \\ m & 0 \\ n & 0 \\ 0 & l \\ 0 & m \\ 0 & n \end{bmatrix} \begin{Bmatrix} f_{i\bar{x}} \\ f_{j\bar{x}} \end{Bmatrix},$$

or

$$\{\mathbf{f}\} = [\mathbf{T}]^{\mathrm{T}}\{\bar{\mathbf{f}}\}. \tag{1.48}$$

1.4.2 Element Stiffness Matrix in the Global Coordinates

Substituting for $\{\bar{q}\}$ from eq. (1.46) into eq. (1.45) and post-multiplying both sides of the equation by $[\mathbf{T}]^T$, we obtain

$$[\mathbf{T}]^T\{\bar{\mathbf{f}}\} = [\mathbf{T}]^T[\bar{\mathbf{k}}][\mathbf{T}]\{\mathbf{q}\},$$

or

$$\{\mathbf{f}\} = [\mathbf{k}]\{\mathbf{q}\}, \qquad (1.49)$$

where $[\mathbf{k}]$ is the element stiffness matrix that relates the nodal forces and displacements expressed in the global coordinates. From eq. (1.49) it is clear that $[\mathbf{k}]$ is obtained as the product $[\mathbf{T}]^T[\bar{\mathbf{k}}][\mathbf{T}]$. An explicit form of $[\mathbf{k}]$ is given below.

$$[\mathbf{k}] = \frac{EA}{L}\begin{bmatrix} l^2 & lm & ln & -l^2 & -lm & -ln \\ & m^2 & mn & -lm & -m^2 & -mn \\ & & n^2 & -ln & -mn & -n^2 \\ & & & l^2 & lm & ln \\ \text{sym} & & & & m^2 & mn \\ & & & & & n^2 \end{bmatrix}\begin{matrix} u_i \\ v_i \\ w_i \\ u_j \\ v_j \\ w_j \end{matrix}, \qquad (1.50)$$

where the row DOFs are shown next to the matrix. Assembling $[\mathbf{k}]$ into the structural stiffness matrix $[\mathbf{K_s}]$ and then deleting the rows and columns to obtain the global stiffness matrix $[\mathbf{K}]$ follow procedures similar to those of plane truss.

EXAMPLE 1.7   *Space truss*

Use the FEM to determine the displacements and forces in the space truss, shown in figure 1.21. The coordinates of the nodes in meter units are given in table 1.3. Assume Young's modulus $E = 70\,\text{GPa}$ and area of cross section $A = 1\,\text{cm}^2$. The magnitude of the downward force (negative z direction) at node 4 is equal to 10,000 N.

SOLUTION   The first step is to determine the direction cosines of the elements. Their length and direction cosines are calculated using the formulas in eq. (1.47). Table 1.4 shows the connectivity and the direction cosines of all the elements.

Using eq. (1.50), element stiffness matrices can be constructed. Since all nodes are fixed except for node 4, the rows and columns corresponding to zero DOF can be deleted at this stage, and the element stiffness matrix contains only non-fixed DOFs. In the case of element 1, for example, the 6 × 6 element stiffness matrix in eq. (1.50) involves

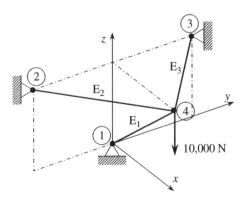

Figure 1.21 Three-bar space truss structure

Table 1.3 Nodal coordinates of space truss structure in example 1.6

Node	x	y	z
1	0	0	0
2	0	−1	1
3	0	1	1
4	1	0	1

Table 1.4 Element connectivity and direction cosines for truss structure in figure 1.21

Element	LN1 (i)	LN2 (j)	L (m)	l	m	n
1	1	4	$\sqrt{2}$	$\sqrt{2}/2$	0	$\sqrt{2}/2$
2	2	4	$\sqrt{2}$	$\sqrt{2}/2$	$\sqrt{2}/2$	0
3	3	4	$\sqrt{2}$	$\sqrt{2}/2$	$-\sqrt{2}/2$	0

two nodes, $i = 1$ and $j = 4$. Since node 1 is fixed, the three rows and columns that correspond to Node 1 can be deleted at the element level. Then, the 3×3 reduced element stiffness matrix can be obtained. By repeating the procedure for all three elements, we can obtain the following reduced element stiffness matrices:

$$\left[\mathbf{k}^{(1)}\right] = 35\sqrt{2} \times 10^5 \begin{bmatrix} 0.5 & 0 & 0.5 \\ 0 & 0 & 0 \\ 0.5 & 0 & 0.5 \end{bmatrix} \begin{matrix} u_4 \\ v_4 \\ w_4 \end{matrix},$$

$$\left[\mathbf{k}^{(2)}\right] = 35\sqrt{2} \times 10^5 \begin{bmatrix} 0.5 & 0.5 & 0 \\ 0.5 & 0.5 & 0 \\ 0 & 0 & 0 \end{bmatrix} \begin{matrix} u_4 \\ v_4 \\ w_4 \end{matrix},$$

$$\left[\mathbf{k}^{(3)}\right] = 35\sqrt{2} \times 10^5 \begin{bmatrix} 0.5 & -0.5 & 0 \\ -0.5 & 0.5 & 0 \\ 0 & 0 & 0 \end{bmatrix} \begin{matrix} u_4 \\ v_4 \\ w_4 \end{matrix}.$$

After assembly, we obtain the global equations in the form $[\mathbf{K}]\{\mathbf{Q}\} = \{\mathbf{F}\}$:

$$35\sqrt{2} \times 10^5 \begin{bmatrix} 1.5 & 0 & 0.5 \\ 0 & 1.0 & 0 \\ 0.5 & 0 & 0.5 \end{bmatrix} \begin{Bmatrix} u_4 \\ v_4 \\ w_4 \end{Bmatrix} = \begin{Bmatrix} 0 \\ 0 \\ -10,000 \end{Bmatrix}.$$

The above global stiffness matrix is positive definite as the displacement boundary conditions have already been implemented. By solving the global matrix equation, the unknown nodal displacements are obtained as

$$u_4 = 2.020 \times 10^{-3}\,\text{m},$$

$$v_4 = 0,$$

$$w_4 = -6.061 \times 10^{-3}\,\text{m}.$$

In order to calculate the element forces, the displacements of nodes of each element have to be transformed to the local coordinates using the relation $\{\bar{\mathbf{q}}\} = [\mathbf{T}]\{\mathbf{q}\}$ in eq. (1.46). Then, the element forces are calculated using

$\{\bar{\mathbf{f}}\} = [\bar{\mathbf{k}}]\{\bar{\mathbf{q}}\}$. The element force P can be obtained from the element force $f_{j\bar{x}}$ (the force in the second node) for the corresponding element. For element 1, the nodal displacements in the local coordinate system can be obtained as

$$\left\{ \begin{array}{c} \bar{u}_1 \\ \bar{u}_4 \end{array} \right\}^{(1)} = \frac{\sqrt{2}}{2} \begin{bmatrix} 1 & 0 & 1 & 0 & 0 & 0 \\ 0 & 0 & 0 & 1 & 0 & 1 \end{bmatrix} \left\{ \begin{array}{c} u_1 = 0 \\ v_1 = 0 \\ w_1 = 0 \\ u_4 \\ v_4 \\ w_4 \end{array} \right\} = \left\{ \begin{array}{c} 0 \\ -2.857 \end{array} \right\} \times 10^{-3} \mathrm{m}.$$

From the force-displacement relation, the element force can be obtained as

$$\left\{ \begin{array}{c} f_{1\bar{x}} \\ f_{4\bar{x}} \end{array} \right\}^{(1)} = \left(\frac{AE}{L} \right)^{(1)} \begin{bmatrix} 1 & -1 \\ -1 & 1 \end{bmatrix} \left\{ \begin{array}{c} \bar{u}_1 \\ \bar{u}_4 \end{array} \right\} = \left\{ \begin{array}{c} 14{,}141 \\ -14{,}141 \end{array} \right\}.$$

Thus, the element force is

$$P^{(1)} = \left(f_{4\bar{x}} \right)^{(1)} = -14{,}141 \, \mathrm{N}.$$

The above calculations are repeated for elements 2 and 3:

$$\left\{ \begin{array}{c} \bar{u}_2 \\ \bar{u}_4 \end{array} \right\}^{(2)} = \frac{\sqrt{2}}{2} \begin{bmatrix} 1 & 1 & 0 & 0 & 0 & 0 \\ 0 & 0 & 0 & 1 & 1 & 0 \end{bmatrix} \left\{ \begin{array}{c} u_2 \\ v_2 \\ w_2 \\ u_4 \\ v_4 \\ w_4 \end{array} \right\} = \left\{ \begin{array}{c} 0 \\ 1.428 \end{array} \right\} \times 10^{-3} \mathrm{m}$$

$$\left\{ \begin{array}{c} f_{2\bar{x}} \\ f_{4\bar{x}} \end{array} \right\}^{(2)} = \left(\frac{AE}{L} \right)^{(2)} \begin{bmatrix} 1 & -1 \\ -1 & 1 \end{bmatrix} \left\{ \begin{array}{c} \bar{u}_2 \\ \bar{u}_4 \end{array} \right\} = \left\{ \begin{array}{c} -7{,}070 \\ +7{,}070 \end{array} \right\}$$
$$P^{(2)} = \left(f_{4\bar{x}} \right)^{(2)} = +7{,}070 \mathrm{N}$$

$$\left\{ \begin{array}{c} \bar{u}_3 \\ \bar{u}_4 \end{array} \right\}^{(3)} = \frac{\sqrt{2}}{2} \begin{bmatrix} 1 & -1 & 0 & 0 & 0 & 0 \\ 0 & 0 & 0 & 1 & -1 & 0 \end{bmatrix} \left\{ \begin{array}{c} u_3 \\ v_3 \\ w_3 \\ u_4 \\ v_4 \\ w_4 \end{array} \right\} = \left\{ \begin{array}{c} 0 \\ 1.428 \end{array} \right\} 10^{-3} \mathrm{m}$$

$$\left\{ \begin{array}{c} f_{3\bar{x}} \\ f_{4\bar{x}} \end{array} \right\}^{(3)} = \left(\frac{AE}{L} \right)^{(3)} \begin{bmatrix} 1 & -1 \\ -1 & 1 \end{bmatrix} \left\{ \begin{array}{c} \bar{u}_2 \\ \bar{u}_4 \end{array} \right\} = \left\{ \begin{array}{c} -7{,}070 \\ +7{,}070 \end{array} \right\}$$

$$P^{(3)} = \left(f_{4\bar{x}} \right)^{(3)} = +7{,}070 \mathrm{N}.$$

Alternately, we can calculate the axial forces in an element using an equation similar to eq. (1.40). For a three-dimensional element, this equation takes the form:

$$P^{(e)} = \left(\frac{AE}{L} \right)^{(e)} \left(l(u_j - u_i) + m(v_j - v_i) + n(w_j - w_i) \right). \tag{1.51}$$

Note that element 1 is in compression, while elements 2 and 3 are in tension. ▪

1.5 THERMAL STRESSES

Thermal stresses in structural elements appear when they are subjected to a temperature change from the reference temperature. At the reference temperature, as shown in figure 1.22(a), if there are no external loads acting on the structure, then there will be no stresses; stresses and strains vanish simultaneously. When the temperature of one or more elements in a structure is changed as shown in figure 1.22(b), then the members tend to expand. However, if the expansion is partially constrained by the surrounding members, then stresses will develop due to the constraint. The constraining members will also experience a force as a reaction to this constraint. This reaction, in turn, produces thermal stresses in the members. The same idea can be extended to thermal stresses in a solid if we imagine the solid to contain many small elements, and each restraining others from expanding or contracting due to temperature change.

The linear relation between stresses and strains for linear elastic solids is valid only when the temperature remains constant and in the absence of residual stresses. In the presence of a temperature differential, that is, when the temperature is different from the reference temperature, we need to use the thermo-elastic stress-strain relations. Such a relation in one dimension is:

$$\sigma = E(\varepsilon - \alpha \Delta T), \tag{1.52}$$

where σ is the uniaxial stress, ε is the total strain, E is Young's modulus, α is the coefficient of thermal expansion (CTE) and ΔT is the difference between the operating temperature and the reference temperature. From eq. (1.52) it is clear that the *reference temperature* is defined as the temperature at which both stress and strain vanish simultaneously when there is no external load. Equation (1.52) states that the stress is caused by and proportional to the mechanical strain, which is the difference between the total strain $\varepsilon = \Delta L/L$ and the *thermal strain* $\alpha \Delta T$. The strain-stress relation now takes the following form:

$$\varepsilon = \frac{\sigma}{E} + \alpha \Delta T. \tag{1.53}$$

The total strain is the sum of the mechanical strain caused by the stresses and the thermal strain caused by the temperature rise. Similarly, thermo-elastic stress-strain relations can be developed in two and three dimensions[4]. It is noted that only the mechanical strain can produce stress, but not the thermal strain. However, the finite element method can only solve for the total strain, not the mechanical strain. Therefore, in order to calculate stress correctly, it is necessary to subtract the thermal strain from the total strain.

The above stress-strain relation can be converted into the force-displacement relation by multiplying eq. (1.52) by the area of cross section of the uniaxial bar, A:

$$P = AE \left(\frac{\Delta L}{L} - \alpha \Delta T \right) = AE \frac{\Delta L}{L} - AE\alpha \Delta T, \tag{1.54}$$

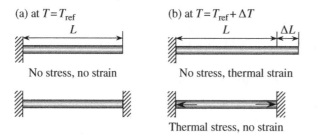

(a) at $T = T_{\text{ref}}$

L

No stress, no strain

Thermal stress, no strain

(b) at $T = T_{\text{ref}} + \Delta T$

L ΔL

No stress, thermal strain

Figure 1.22 Effects of temperature change on the structure

---

[4] A.P. Boresi and R.J. Schmidt. *Advanced Mechanics of Materials*, Sixth Edition. John Wiley & Sons, Inc., New York, NY.

where the first term in the parentheses is the total strain or simply strain and the second, the thermal strain.

Before we introduce the formal method of solving a thermal stress problem using FE, it will be instructive to discuss the method of superposition for solving thermal stress problems.

1.5.1 Method of Superposition

Consider the truss shown in figure 1.23(a). Assume that the temperature of element 2 is raised by ΔT, and elements 1 and 3 remain at the reference temperature. There are no external forces acting at node 4. The objective is to compute the nodal displacements and forces that will be developed in each member. First, if all three elements are disconnected, then only element 2 will expand due to temperature change. Imagine that we apply a pair of equal and opposite forces on the two nodes of element 2, such that the forces restrain the thermal expansion of the element (see figure 1.23(b)). This force can be determined by setting $\Delta L = 0$ in eq. (1.54) and it is equal to $-AE\alpha\Delta T$. That is, a compressive force is required to prevent element 2 from expanding. Hence the force resultant on element 2, $P^{(2)}$, is compressive with magnitude equal to $AE\alpha\Delta T$. If several members are subjected to temperature changes, then a corresponding pair of forces is applied to each element.

The solution to this problem, which will be called problem I, is obvious: the nodal displacements are all equal to zero because no element is allowed to expand or contract, and the force in element 2 is equal to $-AE\alpha\Delta T$. There are no forces in elements 1 and 3. However, this is not the problem we want to solve. The pair of forces applied to element 2 was not there in the original problem. Hence, we have to remove these extraneous forces.

To remove the force shown in figure 1.23(b), we superpose the results from a problem II, where there is no thermal effect, but the forces applied in problem I are all reversed. This is depicted in figure 1.23(c). Sometimes the forces acting in problem II are called fictitious thermal forces, as they do not actually exist. Problem II is a standard truss problem, and hence the FEM we have already discussed can be used to determine the nodal displacements and the element forces. The solution of the problem in figure 1.23(a) can be obtained by adding the solutions from both problems I and II.

EXAMPLE 1.8 *Thermal stresses in a plane truss*

Solve the nodal displacements and element forces of the plane truss problem in figure 1.23. Use the following numerical values: $AE = 10^7\,\text{N}$, $L = 1\,\text{m}$, $\alpha = 10^{-5}/°\text{C}$, $\Delta T = 100\,°\text{C}$.

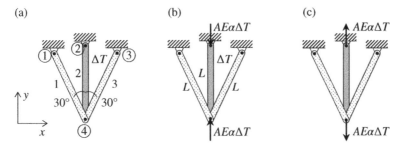

Figure 1.23 A three-element truss: (a) The middle element is subjected to a temperature rise. This is the given problem. (b) A pair of compressive forces is applied to element 2 to prevent it from expanding. This is called problem I. (c) The forces in problem I are reversed. No thermal stresses are involved in this problem. This is called problem II.

Table 1.5 Element connectivity and direction cosines for truss structure in figure 1.23

Element	LN1 (i)	LN2 (j)	AE/L (N/m)	$AE\alpha\Delta T$ (N)	ϕ (degrees)	$l = \cos\phi$	$m = \sin\phi$
1	1	4	10^7	0	−60	1/2	$-\sqrt{3}/2$
2	2	4	10^7	10,000	−90	0	−1
3	3	4	10^7	0	240	−1/2	$-\sqrt{3}/2$

SOLUTION   The solution to problem I is as follows:

$$u_4 = v_4 = 0,$$
$$P^{(1)} = 0, \ P^{(2)} = -AE\alpha\Delta T = -10{,}000\text{N}, \ P^{(3)} = 0.$$

Problem II is depicted in figure 1.23(c). The element properties and direction cosines are listed in table 1.5. The element stiffness matrices in the global coordinates are written below. For convenience, the rows and columns corresponding to zero DOFs are deleted, and only those corresponding to active DOFs are shown.

$$\left[\mathbf{k}^{(1)}\right] = \frac{10^7}{4} \begin{bmatrix} 1 & -\sqrt{3} \\ -\sqrt{3} & 3 \end{bmatrix} \begin{matrix} u_4 \\ v_4 \end{matrix},$$

$$\left[\mathbf{k}^{(2)}\right] = 10^7 \begin{bmatrix} 0 & 0 \\ 0 & 1 \end{bmatrix} \begin{matrix} u_4 \\ v_4 \end{matrix},$$

$$\left[\mathbf{k}^{(3)}\right] = \frac{10^7}{4} \begin{bmatrix} 1 & \sqrt{3} \\ \sqrt{3} & 3 \end{bmatrix} \begin{matrix} u_4 \\ v_4 \end{matrix}.$$

Assembling the element stiffness matrices, we obtain the global stiffness matrix [**K**]. The only external force for this problem is $F_{4y} = -10{,}000$ N.

$$\frac{10^7}{4} \begin{bmatrix} 2 & 0 \\ 0 & 10 \end{bmatrix} \begin{Bmatrix} u_4 \\ v_4 \end{Bmatrix} = \begin{Bmatrix} 0 \\ -10{,}000 \end{Bmatrix}.$$

Since the global stiffness matrix is positive definite, the solution for displacements can be obtained as:

$$u_4 = 0,$$
$$v_4 = -0.4 \times 10^{-3}\,\text{m}.$$

Table 1.6 Solution of thermal stresses in a truss using the superposition method

Variable	Problem I	Problem II	Final Solution
u_4	0	0	0
v_4	0	-0.4×10^{-3} m	-0.4×10^{-3} m
$P^{(1)}$	$-AE\alpha\Delta T^{(1)} = 0$	3,464 N	3,464 N
$P^{(2)}$	$-AE\alpha\Delta T^{(2)} = -10{,}000$ N	4,000 N	−6,000 N
$P^{(3)}$	$-AE\alpha\Delta T^{(3)} = 0$	3,464 N	3,464 N

The force resultants in the elements for problem II can be obtained from eq. (1.40). Substituting the element properties and displacements, we obtain:

$$P^{(1)} = 3{,}464\,\text{N},$$

$$P^{(2)} = 4{,}000\,\text{N},$$

$$P^{(3)} = 3{,}464\,\text{N}.$$

Then, the solution (displacements and forces) to the given problem is the sum of solutions to problems I and II as shown in table 1.6. Note that elements 1 and 3 are in tension, while element 2 is in compression.

If there were external forces acting at node 4 in the given problem, then they can be added to the fictitious forces in problem II. ▪

1.5.2   Thermal Stresses Using FEA

In using the FEM for thermal stress problems, we combine the two problems in the previous subsection as one problem and solve for displacements and forces simultaneously. This procedure is similar to the superposition method. Consider the element equilibrium equation in eq. (1.27) for the uniaxial bar element. It states that the forces acting on an element are the product of element stiffness matrix and the vector of nodal displacements, that is, $\{\bar{\mathbf{f}}\} = [\bar{\mathbf{k}}]\{\bar{\mathbf{q}}\}$. This is similar to the linear elastic stress-strain relation at the element level, where stresses are a linear combination of strains at a point, $\{\sigma\} = [\mathbf{E}]\{\varepsilon\}$. However, we notice that in the presence of a temperature differential, the stresses are not a linear combination of strains, [see eq. (1.52)]. A similar adjustment has to be made at the element level equation also. In the presence of thermal stresses, eq. (1.32) can be modified as

$$\left\{ \bar{\mathbf{f}}^{(e)} \right\} = \left[\bar{\mathbf{k}}^{(e)} \right] \left\{ \bar{\mathbf{q}}^{(e)} \right\} - \left\{ \bar{\mathbf{f}}_T^{(e)} \right\}, \tag{1.55}$$

where the element thermal force vector $\left\{ \bar{\mathbf{f}}_T^{(e)} \right\}$ in the local coordinate system is given by

$$\left\{ \bar{\mathbf{f}}_T^{(e)} \right\} = AE\alpha\Delta T \begin{Bmatrix} -1 \\ 0 \\ +1 \\ 0 \end{Bmatrix} \begin{matrix} \bar{u}_i \\ \bar{v}_i \\ \bar{u}_j \\ \bar{v}_j \end{matrix}. \tag{1.56}$$

Note that the row addresses or the DOFs corresponding to each force are indicated next to the force vector. Multiplying both sides of eq. (1.55) by the transpose of the transformation matrix, $[\mathbf{T}]^{\mathrm{T}}$, and also using $\{\bar{\mathbf{q}}\} = [\mathbf{T}]\{\mathbf{q}\}$, we obtain

$$\{\mathbf{f}\} = [\mathbf{k}]\{\mathbf{q}\} - \{\mathbf{f}_T\}, \tag{1.57}$$

where $[\mathbf{k}]$ is the element stiffness matrix defined in eq. (1.36) and $\{\mathbf{f}_T\}$ is the thermal force vector in the global coordinates given by

$$\{\mathbf{f}_T\} = AE\alpha\Delta T \begin{Bmatrix} -l \\ -m \\ +l \\ +m \end{Bmatrix} \begin{matrix} u_i \\ v_i \\ u_j \\ v_j \end{matrix}. \tag{1.58}$$

The vector $\{\mathbf{f}_T\}$ has four rows, and its row addresses are in the same order as those of $[\mathbf{k}]$.

If eq. (1.58) is substituted in the nodal equilibrium equations, we will obtain the global equations at the structural level as

$$[\mathbf{K}_s]\{\mathbf{Q}_s\} = \{\mathbf{F}_s\} + \{\mathbf{F}_{Ts}\}, \tag{1.59}$$

where $\{\mathbf{F}_{Ts}\}$ is the thermal load vector, which is obtained by assembling $\{\mathbf{f}_T\}$ of various elements. It is clear from eq. (1.59) that the increase in temperature is equivalent to adding an additional force to the member. After striking out the rows and columns corresponding to zero DOFs, we obtain the global equations as

$$[\mathbf{K}]\{\mathbf{Q}\} = \{\mathbf{F}\} + \{\mathbf{F}_T\}. \tag{1.60}$$

The assembly of $\{\mathbf{F}_T\}$ is similar to that of the stiffness matrix. Equation (1.60) is solved to obtain the unknown displacements $\{\mathbf{Q}\}$. In order to find the forces in elements, one must use eq. (1.54).

EXAMPLE 1.9   *Thermal stresses in a plane truss*

Solve the thermal stress problem in example 1.8 using the finite element method.

SOLUTION   The element stiffness matrices are already given in example 1.8. The thermal force vectors are written below. For convenience, the rows and columns corresponding to zero DOFs are deleted and only those corresponding to active DOFs are shown.

$$\left\{\mathbf{f}_T^{(1)}\right\} = AE\alpha\Delta T^{(1)} \begin{Bmatrix} 1/2 \\ -\sqrt{3}/2 \end{Bmatrix} = \begin{Bmatrix} 0 \\ 0 \end{Bmatrix} \begin{matrix} u_4 \\ v_4 \end{matrix},$$

$$\left\{\mathbf{f}_T^{(2)}\right\} = AE\alpha\Delta T^{(2)} \begin{Bmatrix} 0 \\ -1 \end{Bmatrix} = \begin{Bmatrix} 0 \\ -10{,}000 \end{Bmatrix} \begin{matrix} u_4 \\ v_4 \end{matrix}, \tag{1.61}$$

$$\left\{\mathbf{f}_T^{(3)}\right\} = AE\alpha\Delta T^{(3)} \begin{Bmatrix} 1/2 \\ -\sqrt{3}/2 \end{Bmatrix} = \begin{Bmatrix} 0 \\ 0 \end{Bmatrix} \begin{matrix} u_4 \\ v_4 \end{matrix}$$

Note that there is no thermal force vector for elements 1 and 3 because they are at the reference temperature. The row addresses are shown next to the thermal force vector in eq. (1.61).

Assembling the element stiffness matrices, we obtain the global stiffness matrix $[\mathbf{K}]$, and assembling the element thermal force vectors $\{\mathbf{f}_T\}$, we obtain the global thermal force vector $\{\mathbf{F}_T\}$ as

$$\{\mathbf{F}_T\} = \begin{Bmatrix} 0 \\ -10{,}000 \end{Bmatrix}. \tag{1.62}$$

The solution for displacements is obtained using the global equations

$$[\mathbf{K}]\{\mathbf{Q}\} = \{\mathbf{F}\} + \{\mathbf{F}_T\}. \tag{1.63}$$

Since there are no external forces in the present problem, $\{\mathbf{F}\} = \{\mathbf{0}\}$. Hence the global equations are:

$$\frac{10^7}{4} \begin{bmatrix} 2 & 0 \\ 0 & 10 \end{bmatrix} \begin{Bmatrix} u_4 \\ v_4 \end{Bmatrix} = \begin{Bmatrix} 0 \\ -10{,}000 \end{Bmatrix}. \tag{1.64}$$

The solution to the above equations is obtained as:

$$\boxed{\begin{aligned} u_4 &= 0, \\ v_4 &= -0.4 \times 10^{-3}\,\text{m}. \end{aligned}}$$

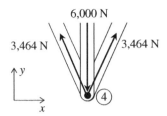

Figure 1.24 Force equilibrium at node 4

The force resultants in the elements are obtained from eq. (1.54):

$$P = AE\left(\frac{\Delta L}{L} - \alpha \Delta T\right)$$

$$= \frac{AE}{L}\left[l\left(u_j -\right) + m\left(v_j - v_i\right)\right] - AE\alpha\Delta T. \tag{1.65}$$

Substituting the element properties and displacements, we obtain:

$$\begin{aligned} P^{(1)} &= 3{,}464\,\text{N},\\ P^{(2)} &= -6{,}000\,\text{N},\\ P^{(3)} &= 3{,}464\,\text{N}. \end{aligned}$$

The FE solution for displacements and forces above can be compared with those obtained from the superposition method presented in table 1.6.

One can check the force equilibrium at node 4. The three forces acting on node 4 are shown in figure 1.24. Summing the forces in the x and y directions,

$$\sum F_x = -P^{(1)} \sin 30 + P^{(2)} \sin 30 = 0,$$

$$\sum F_y = P^{(1)} \cos 30 - P^{(2)} + P^{(3)} \cos 30$$

$$= 3464 \times \frac{\sqrt{3}}{2} - 6000 + 3464 \times \frac{\sqrt{3}}{2} \tag{1.66}$$

$$= 0.$$

Thermal stress analysis of space trusses follows the same procedures. The thermal force vector $\{\mathbf{f}_T\}$ is a 6×1 matrix and is given by

$$\{\mathbf{f}_T\}^{\mathrm{T}} = AE\alpha\Delta T\{ \overset{u_i}{-l} \quad \overset{v_i}{-m} \quad \overset{w_i}{-n} \quad \overset{u_j}{+l} \quad \overset{v_j}{+m} \quad \overset{w_j}{+n} \}. \tag{1.67}$$

In the above equation, $\{\mathbf{f}_T\}^{\mathrm{T}}$ is given as a row matrix with addresses shown above the elements of the matrix. Another difference between two- and three-dimensional thermal stress problems is in the calculation of force in an element. An equation similar to (1.65) for the three-dimensional truss element can be derived as

$$P = AE\left(\frac{\Delta L}{L} - \alpha\Delta T\right)$$

$$= \frac{AE}{L}\left[l\left(u_j - u_i\right) + m\left(v_j - v_i\right) + n\left(w_j - w_i\right)\right] - AE\alpha\Delta T, \tag{1.68}$$

where l, m, and n are the direction cosines of the element. ▪

EXAMPLE 1.10   *Thermal stresses in a space truss*

Use the FEM to determine the displacements and forces in the space truss, shown in figure 1.25. The coordinates of the nodes in meter units are given in table 1.7. The temperature of element 1 is raised by $100\,°\text{C}$ above the reference

Table 1.7   Nodal coordinates of space truss structure in example 1.10

Node	x	y	z
1	0	0	0
2	0	−1	1
3	0	1	1
4	1	0	1
5	0	0	1

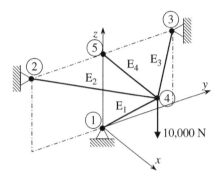

Figure 1.25   Three-bar space truss structure

temperature. Assume Young's modulus $E = 70$ Gpa and area of cross section $A = 1$ cm^2. Assume CTE $\alpha = 20 \times 10^{-6}/$°C. The magnitude of the downward force (negative z direction) at node 4 is equal to 10,000 N.

SOLUTION   The first step is to determine the direction cosines of the elements. The length and direction cosines of each element are calculated using the formulas in eq. (1.47).

Element	LN1 (i)	LN2 (j)	L (m)	l	m	n
1	1	4	$\sqrt{2}$	$\sqrt{2}/2$	0	$\sqrt{2}/2$
2	2	4	$\sqrt{2}$	$\sqrt{2}/2$	$\sqrt{2}/2$	0
3	3	4	$\sqrt{2}$	$\sqrt{2}/2$	$-\sqrt{2}/2$	0
4	5	4	1	1	0	0

Stiffness matrices of elements 1 through 3 are the same as in example 1.6. The stiffness matrix of element 4 is as follows:

$$\left[\mathbf{k}^{(4)}\right] = 70 \times 10^5 \begin{bmatrix} 1 & 0 & 0 \\ 0 & 0 & 0 \\ 0 & 0 & 0 \end{bmatrix} \begin{matrix} u_4 \\ v_4 \\ w_4 \end{matrix}.$$

We need to calculate the thermal force vector for element 1 only, as its temperature is different from the reference temperature. Using the formula in eq. (1.67) we obtain

$$\begin{matrix} u_1 & v_1 & w_1 & u_4 & v_4 & w_4 \end{matrix}$$
$$\left\{\mathbf{f}_T^{(4)}\right\}^{\mathrm{T}} = 7000\sqrt{2}\{-1 \quad 0 \quad -1 \quad 1 \quad 0 \quad 1\}. \tag{1.69}$$

After assembly, we obtain the global equations in the form $[\mathbf{K}]\{\mathbf{Q}\} = \{\mathbf{F}\} + \{\mathbf{F_T}\}$:

$$35\sqrt{2} \times 10^5 \begin{bmatrix} 1.5+\sqrt{2} & 0 & 0.5 \\ 0 & 1.0 & 0 \\ 0.5 & 0 & 0.5 \end{bmatrix} \begin{Bmatrix} u_4 \\ v_4 \\ w_4 \end{Bmatrix} = \begin{Bmatrix} 0 \\ 0 \\ -10,000 \end{Bmatrix} + \begin{Bmatrix} 9900 \\ 0 \\ 9900 \end{Bmatrix}.$$

Solving the above equation, the unknown nodal displacements are obtained as

$$u_4 = 0.8368 \times 10^{-3} \, \text{m},$$

$$v_4 = 0,$$

$$w_4 = -0.8772 \times 10^{-3} \, \text{m}.$$

The forces in the elements can be calculated using eq. (1.68), and they are as follows:

$$P^{(1)} = -14,141 \, \text{N},$$

$$P^{(2)} = +2,929 \, \text{N},$$

$$P^{(3)} = +2,929 \, \text{N},$$

$$P^{(4)} = +5,858 \, \text{N}.$$

One can verify that the force equilibrium is satisfied at node 4. ▄

1.6   FINITE ELEMENT MODELING PRACTICE FOR TRUSS

In this section, several analysis problems are used to discuss modeling issues as well as verifying the accuracy of analysis results with that of literature. The examples are presented in such a way that any finite element analysis program can be used to solve the problems. However, the analysis results can slightly be different because of implementation details of different finite element analysis programs.

1.6.1   Reaction Force of a Statically Indeterminate Bar[5]

A vertical prismatic bar shown in figure 1.26 is fixed at both ends. When two downward forces, $F_1 = 1000$ lb and $F_2 = 500$ lb, are applied as shown in the figure, calculate the reaction forces R_1 and R_2 at both ends. Use elastic modulus $E = 30 \times 10^6$ psi and a constant cross-sectional area $A = 0.1 \, \text{in}^2$.

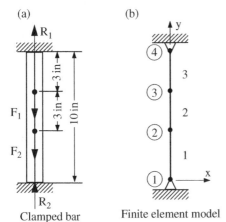

(a)   (b)

Clamped bar   Finite element model   **Figure 1.26** Statically indeterminate vertical bar

---

[5] Timoshenko, S. 1955. *Strength of Material, Part 1, Elementary Theory and Problems*, 3rd Edition. D. Van Nostrand Co., Inc., New York, NY. (Page 26, problem 10.)

Since the total applied load is 1500 lb, it is obvious that the sum of reaction forces will be 1500 lb. However, the individual reaction forces cannot be calculated from force equilibrium, as the structure is statically indeterminate. It is necessary to take into account compatibility in deformation to solve statically indeterminate systems. One way of solving the problem is by assuming that the bar is fixed only from the top and a force R_2 is applied at the bottom. Then, the unknown force R_2 can be calculated from the compatibility condition that the deflection at the bottom is zero. The deflection at the bottom due to F_1, F_2, and R_2 can be written as

$$\delta = \frac{(F_1 + F_2 - R_2) \times 3}{EA} + \frac{(F_2 - R_2) \times 3}{EA} - \frac{R_2 \times 4}{EA} = 0.$$

Note that each section of the bar is under different loads, which can easily be obtained from a free-body diagram. The reaction R_2 can be obtained by solving the above equation, $R_2 = 600$ lb. From the static equilibrium, the remaining reaction force $R_1 = 900$ lb can also be obtained.

One of the important advantages in finite element analysis is that there is no need to acknowledge the difference between statically determinate and indeterminate systems because the finite element analysis procedure automatically takes into account deformation. Figure 1.26(b) shows an example for finite element model for the statically indeterminate bar. Nodes 2 and 3 are located in order to apply the two nodes. It is possible to make more elements, but for this particular problem, the three elements will yield an accurate solution.

Even if all elements are vertically located, it is possible to model the bar using either 1D bar or 2D plane truss elements because deformation is limited along the axial direction. When 1D bar elements are used, the vertical direction is considered as the x coordinate, and forces are applied along the coordinate. In such a case, each node has a single DOF, u_I, and the total matrix size of the problem becomes 4×4. The assembled matrix equation becomes

$$10^6 \begin{bmatrix} 0.75 & -0.75 & 0 & 0 \\ -0.75 & 1.75 & -1 & 0 \\ 0 & -1 & 2 & -1 \\ 0 & 0 & -1 & 1 \end{bmatrix} \begin{Bmatrix} u_1 \\ u_2 \\ u_3 \\ u_4 \end{Bmatrix} = \begin{Bmatrix} R_2 \\ -F_2 \\ -F_1 \\ R_1 \end{Bmatrix}. \tag{1.70}$$

In the above equation, the applied forces and reaction forces in figure 1.26 are used with positive being the positive y coordinate direction. Since nodes 1 and 4 are fixed, the first and fourth columns and rows are deleted in the process of applying the boundary conditions. Therefore, only a 2×2 matrix needs to be solved.

$$10^6 \begin{bmatrix} 1.75 & -1 \\ -1 & 2 \end{bmatrix} \begin{Bmatrix} u_2 \\ u_3 \end{Bmatrix} = \begin{Bmatrix} -500 \\ -1000 \end{Bmatrix}.$$

The above matrix equation can be solved for nodal displacement $u_2 = -8 \times 10^{-4}$ in. and $u_3 = -9 \times 10^{-4}$ in. The calculated nodal displacement can be substituted into eq. (1.70) to calculated reaction forces. From the first and fourth rows of eq. (1.70), we have

$$R_2 = 10^6 \times (0.75 \times u_1 - 0.75 \times u_2) = 600 \text{ lb},$$
$$R_1 = 10^6 \times (-1 \times u_3 + 1 \times u_4) = 900 \text{ lb}.$$

Note that the above reaction forces are identical to the previous analytical calculation.

Many finite element analysis programs do not have 1D bar elements. Instead, they use 2D plane truss or 3D space truss elements. For example, ANSYS has LINK180, which supports uniaxial tension-compression with three DOFs at each node: translations in the nodal x, y, and z directions, UX, UY, and YZ. On the other hand, Abaqus support T2D2 element for the two-dimensional plane truss.

When the above problem is solved using 3D space truss element, all elements are located in the y-coordinate direction and all forces are applied in the same direction. For boundary conditions, it is necessary to fix all three DOFs at nodes 1 and 4. On the other hand, it is necessary to fix UX and UZ for nodes 2 and 3 so that the motion is limited to the y direction.

1.6.2   Thermally Loaded Support Structure[6]

Two copper wires and a steel wire are connected by a rigid body as shown in figure 1.27. The three wires are all of an equal length of 20 in. and in the same cross-sectional area of $A = 0.1$ in.^{2}. Initially, the structure was at a temperature of 70 °F. When a load $Q = 4000$ lb is applied and the temperature is increased to 80 °F simultaneously, find the stresses in the copper and steel wires. Note that the load Q is applied at the center of the rigid body. For the copper wire, use the elastic modulus $E_c = 1.6 \times 10^7$ psi and the thermal expansion coefficient $\alpha_c = 9.2 \times 10^{-6}$ in/in°F. For the steel wire, use the elastic modulus $E_s = 3.0 \times 10^7$ psi and the thermal expansion coefficient $\alpha_s = 7.0 \times 10^{-6}$ in/in°F.

Since the two copper wires have identical properties and the load is applied at the center of the rigid body, there will be no rotation, and all three wires will have the same amount of displacement. Therefore, the problem can be simplified into three 1D bars. The problem is statically indeterminate because not only we cannot calculate the internal load distribution between copper and steel from force equilibrium, but also the different thermal expansion coefficients cause different elongations.

1. ***Equilibrium under temperature change:***   In order to solve the problem analytically, we can use the method of superposition. That is, the stress caused by temperature change and the stress caused by the load can be calculated separately and then added to yield the final state of stresses. First, we ignore the rigid body for a moment and the three wires are free. When the temperature is increased by 10 °F, the elongation of copper and steel can be calculated by

$$\delta_c^1 = \alpha_c \Delta T L = 9.2 \times 10^{-6} \times 10 \times 20 = 0.00184 \text{ in,}$$

$$\delta_s^1 = \alpha_s \Delta T L = 7.0 \times 10^{-6} \times 10 \times 20 = 0.00140 \text{ in.}$$

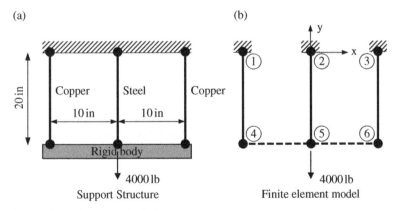

Figure 1.27 Thermally loaded three bars

[6] Timoshenko, S. 1955. *Strength of Material, Part 1, Elementary Theory and Problems*, 3rd Edition. D. Van Nostrand Co., Inc., New York, NY. (Page 30, problem 9.)

The superscript 1 stands for the stage when the temperature changes but the load Q is not applied. At this stage, there is no stress in the wires as they are not constrained. Connecting them with a rigid body means that the steel needs to be elongated further while the coppers need to be compressed. Let us assume that δ^1 is the final displacement after the rigid body is located. Then the internal forces can be written in terms of δ^1 as

$$
\begin{aligned}
F_c^1 &= \frac{E_c A}{L}\left(\delta^1 - \delta_c^1\right), \\
F_s^1 &= \frac{E_s A}{L}\left(\delta^1 - \delta_s^1\right).
\end{aligned}
\tag{1.71}
$$

The unknown displacement δ^1 can be calculated from the condition that there is no externally applied force. That is, the sum of internal forces must vanish.

$$
2F_c^1 + F_s^1 = 2\frac{E_c A}{L}\left(\delta^1 - \delta_c^1\right) + \frac{E_s A}{L}\left(\delta^1 - \delta_s^1\right) = 0.
$$

In the above equation, F_c is multiplied by two because there are two copper wires. The above equation can be solved for $\delta^1 = 0.00163$ in. The internal forces and stresses can be calculated by substituting δ^1 into eq. (1.71) as

$$
F_c^1 = \frac{E_c A}{L}\left(\delta^1 - \delta_c^1\right) = -17.03\text{lb} \qquad \sigma_c^1 = \frac{F_c^1}{A} = -170.3\text{psi},
$$

$$
F_s^1 = \frac{E_s A}{L}\left(\delta^1 - \delta_s^1\right) = 34.06\text{lb} \qquad \sigma_s^1 = \frac{F_s^1}{A} = 340.6\text{psi}.
$$

2. *Equilibrium under applied load Q:*   When $Q = 4000$ lb is applied at the center of the rigid body, all three wires will displace in the same amount. When the vertical displacement is δ^2, the sum of internal forces must be in equilibrium with the applied load as

$$
\left(2 \times \frac{E_c A}{L} + \frac{E_s A}{L}\right) \times \delta^2 = 4000,
$$

from which the displacement $\delta^1 = 0.0123$ in. can be obtained. Then, the internal forces and stresses can be calculated by

$$
F_c^2 = \frac{E_c A}{L}\delta^2 = 1032.3\text{lb} \qquad \sigma_c^2 = \frac{F_c^2}{A} = 10323\text{psi},
$$

$$
F_s^2 = \frac{E_s A}{L}\delta^2 = 1935.5\text{lb} \qquad \sigma_s^2 = \frac{F_s^2}{A} = 19355\text{psi}.
$$

3. *Combined internal forces and stresses:*   An important characteristic of a linear system is that the stresses from different loads can be superposed together to yield the stress at the combined loads. The final results become

$$
\delta = \delta^1 + \delta^2 = 0.01453 \text{ in},
$$

$$
F_c = F_c^1 + F_c^2 = 1015.2 \text{ lb} \quad \sigma_c = \sigma_c^1 + \sigma_c^2 = 10152 \text{ psi},
$$

$$
F_s = F_s^1 + F_s^2 = 1969.5 \text{ lb} \quad \sigma_s = \sigma_s^1 + \sigma_s^2 = 19695 \text{ psi}.
$$

4. *Finite element modeling and analysis:*   Figure 1.27(b) shows a finite element model, where three bar elements are used to represent the three wires. Similar to statically indeterminate systems, it is unnecessary to separate the temperature change from the applied load. The temperature change can be considered as a thermal load as shown in eq. (1.55). For simplicity, 1D bar

elements are used to model the thermally loaded support structure. In order to do that, the three element matrix equations with the thermal load can be written as

$$\text{Element 1: } \frac{E_c A}{L} \begin{bmatrix} 1 & -1 \\ -1 & 1 \end{bmatrix} \begin{Bmatrix} u_1 \\ u_4 \end{Bmatrix} = \begin{Bmatrix} f_1^{(1)} \\ f_4^{(1)} \end{Bmatrix} + A E_c \alpha_c \Delta T \begin{Bmatrix} -1 \\ 1 \end{Bmatrix}.$$

$$\text{Element 2: } \frac{E_s A}{L} \begin{bmatrix} 1 & -1 \\ -1 & 1 \end{bmatrix} \begin{Bmatrix} u_2 \\ u_5 \end{Bmatrix} = \begin{Bmatrix} f_2^{(2)} \\ f_3^{(2)} \end{Bmatrix} + A E_s \alpha_s \Delta T \begin{Bmatrix} -1 \\ 1 \end{Bmatrix}.$$

$$\text{Element 3: } \frac{E_c A}{L} \begin{bmatrix} 1 & -1 \\ -1 & 1 \end{bmatrix} \begin{Bmatrix} u_3 \\ u_6 \end{Bmatrix} = \begin{Bmatrix} f_3^{(3)} \\ f_6^{(3)} \end{Bmatrix} + A E_c \alpha_c \Delta T \begin{Bmatrix} -1 \\ 1 \end{Bmatrix}.$$

The assembly of the above three element matrix equations yields a 6×6 structural matrix equation. Since nodes 1, 2, and 3 are fixed, the first three rows and columns are deleted in the process of applying boundary conditions. Therefore, after applying boundary conditions, the following global matrix equation can be obtained:

$$\begin{bmatrix} 80000 & 0 & 0 \\ 0 & 150000 & 0 \\ 0 & 0 & 80000 \end{bmatrix} \begin{Bmatrix} u_4 \\ u_5 \\ u_6 \end{Bmatrix} = \begin{Bmatrix} F_4 \\ F_5 \\ F_6 \end{Bmatrix} + \begin{Bmatrix} 147.2 \\ 210 \\ 147.2 \end{Bmatrix}.$$

The condition of a rigid body can be written as $u_4 = u_5 = u_6$. There are many different ways of considering the effect of a rigid body; a simple method can be adding three equations together, which yields the following scalar equation:

$$(80000 + 150000 + 80000) \times u_5 = (F_4 + F_5 + F_6) + (147.2 + 210 + 147.2).$$

Considering the fact that $F_4 + F_5 + F_6 = Q = 4000$ lb., the above equation can be solved for $u_5 = 0.01453$ in. $= u_4 = u_6$. Now the calculated displacement can be substituted into the element matrix equation to calculate the element forces as

$$P^{(1)} = f_j^{(1)} = \frac{E_c A}{L} (u_4 - u_1) - A E_c \alpha_c \Delta T = 1015.2 \text{ lb,}$$

$$P^{(2)} = f_j^{(2)} = \frac{E_s A}{L} (u_5 - u_1) - A E_s \alpha_s \Delta T = 1969.5 \text{ lb,}$$

$$P^{(3)} = f_j^{(3)} = \frac{E_c A}{L} (u_6 - u_1) - A E_c \alpha_c \Delta T = 1015.2 \text{ lb.}$$

Note that the element forces are identical to the internal forces in the analytical calculation. Therefore, the stresses will also be identical.

1.6.3   Deflection of a Two-Bar Truss[7]

A structure consisting of two equal steel bars, each of length $L = 15$ ft. and cross-sectional area $A = 0.5$ in.^{2}, with hinged ends, is subjected to the action of a load $F = 5000$ lb. Determine the stress, σ, in the bars and the vertical deflection, δ, at node 2. Neglect the weight of the bars as a small quantity in comparison with the load F. For material property, use $E = 30 \times 10^6$ psi.

---

[7] Timoshenko, S. 1955. *Strength of Material, Part 1, Elementary Theory and Problems*, 3rd Edition. D. Van Nostrand Co., Inc., New York, NY. (Page 30, problem 9.)

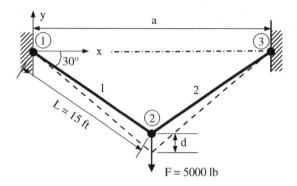

Figure 1.28 Two-bar truss

Since the truss is a two-force member, it can only support the force in the axial direction. In addition, due to symmetry, it is obvious that the force in member 1 will be the same as that of member 2. The sum of vertical components of the two member forces is in equilibrium with the externally applied load. That is,

$$2f\sin 30 = 5000 \quad \Rightarrow \quad f = 5000 \text{ lb}.$$

Therefore, the stress in both members are

$$\sigma^{(1)} = \sigma^{(2)} = \frac{f}{A} = 10{,}000 \text{ psi}.$$

The internal forces can cause elongation of the truss, which can be calculated by

$$\delta' = \frac{fL}{EA} = \frac{5000 \times 15 \times 12}{3 \times 10^7 \times 0.5} = 0.06 \text{ in}.$$

That is, the new length of the truss becomes 180.06 in. Using the geometry in figure 1.28, the new angle after deformation can be calculated by

$$\theta = \cos^{-1}\left(\frac{180 \times \cos(30)}{180.06}\right) = 30.033°.$$

Therefore, the vertical displacement, δ, in figure 1.28 can be calculated by

$$\delta = 180.06 \times \sin(30.033) - 180 \times \sin(30) = 0.12 \text{ in}.$$

For finite element modeling, two equal-length, 2D plane truss finite element can be used as shown in figure 1.28. In the case of plane truss elements, the following table can be used to define element connectivity and the element matrix equation:

Element	LN1 (i)	LN2 (j)	AE/L (N/m)	ϕ (degrees)	$l = \cos\phi$	$m = \sin\phi$
1	1	2	83333	−30	$\sqrt{3}/2$	−1/2
2	3	2	83333	−150	$-\sqrt{3}/2$	−1/2

In order to simplify the expression, the boundary conditions can be applied at the element level. That is, only free degrees of freedom are used in the expression. In this example, since nodes 1 and 3 are fixed, their degrees of freedom are removed at the element level. The two element matrix equations can be defined as

Element 1: $83333 \begin{bmatrix} 3/4 & -\sqrt{3}/4 \\ -\sqrt{3}/4 & 1/4 \end{bmatrix} \begin{Bmatrix} u_2 \\ v_2 \end{Bmatrix} = \begin{Bmatrix} f_{x2}^{(1)} \\ f_{y2}^{(1)} \end{Bmatrix}.$

Element 2: $83333 \begin{bmatrix} 3/4 & \sqrt{3}/4 \\ \sqrt{3}/4 & 1/4 \end{bmatrix} \begin{Bmatrix} u_2 \\ v_2 \end{Bmatrix} = \begin{Bmatrix} f_{x2}^{(2)} \\ f_{y2}^{(2)} \end{Bmatrix}.$

The assembled system of equation becomes

$$83333 \begin{bmatrix} 3/2 & 0 \\ 0 & 1/2 \end{bmatrix} \begin{Bmatrix} u_2 \\ v_2 \end{Bmatrix} = \begin{Bmatrix} 0 \\ -5000 \end{Bmatrix}.$$

The above equation solves for $u_2 = 0$ and $v_2 = -0.12$ in. Note that the vertical deflection $\delta = -v_2$. In order to calculate element stress, the formula in eq. (1.40) is used for the element force as

$$P^{(1)} = \left(\frac{AE}{L}\right)^{(1)} \left(l(u_j - u_i) + m(v_j - v_i)\right) = 83333 \left(-\frac{1}{2}(-0.12 - 0)\right) = 5000 \text{ lb},$$

and then, the element stress can be obtained by dividing the element force by the cross-sectional area:

$$\sigma^{(1)} = \frac{P^{(1)}}{A} = 10,000 \text{ psi}.$$

Since the truss is symmetric, element 2 will yield identical results.

1.7 PROJECTS

Project 1.1 Analysis and Design of a Space Truss

A space frame structure as shown in figure 1.29 consists of 25 truss members. Initially, all members have the same circular cross-sections with diameter $d = 2.0$ in. At nodes 1 and 2, a constant force $F = 60,000$ lb. is applied in the y direction. Four nodes (7, 8, 9, and 10) are fixed on the ground. The frame structure is made of a steel material whose properties are Young's modulus $E = 3 \times 10^7$ psi, Poisson's ratio $\nu = 0.3$, yield stress $\sigma_Y = 37,000$ psi, and density $\rho = 0.284$ lb./in^3. The safety factor $N = 1.5$ is used. Due to the manufacturing constraints, the diameter of the truss members should vary between 0.1 in. and 2.5 in.

1. Solve the initial truss structure using truss finite elements. Provide a plot that shows labels for elements and nodes along with boundary conditions. Provide deformed geometry of the structure and a table of stress in each element.

2. Minimize the structural weight by changing the cross-sectional diameter of each truss element, while all members should be safe under the given yield stress and the safety factor. You can use the symmetric geometry of the structure. Identify zero-force members. For zero-force members, use the lower bound of the cross-sectional diameter. Provide deformed geometry at the optimum design along with a table of stress at each element. Provide structural weights of initial and optimum designs.

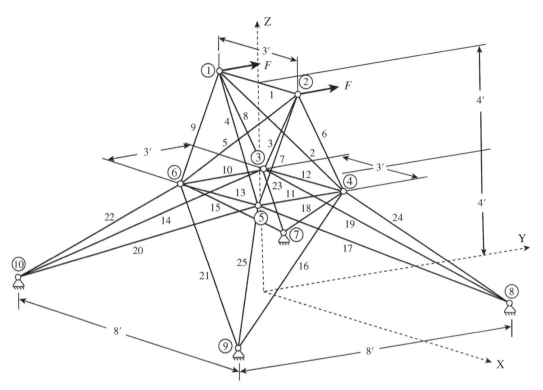

Figure 1.29 25–member space truss

Project 1.2   Analysis and Design of a Plane Truss 1

The truss shown in figure 1.30 has two elements. The members are made of the aluminum hollow square cross section. The outer dimension of the square is 12 mm, and the inner dimension is 9 mm. (The wall thickness is 1.5 mm on all four sides.) Assume Young's modulus $E = 70$ GPa, and yield strength $\sigma_Y = 70$ MPa. The magnitude of the force at node 1 (F) is equal to 1,000 N.

1. Use FEM to determine the displacements at node 1 and axial forces in elements 1 and 2. Use von Mises yield theory to determine if the elements will yield or not. Use Euler buckling load ($P_{cr} = \pi^2 EI/L^2$) to determine if the elements under compressive loads will buckle. In the above expression, P_{cr} is the axial compressive force, E is Young's modulus, I is the moment of inertia of the cross section given by $I = (a_o^4 - a_i^4)/12$, where a_o and a_i, respectively, are the outer and inner dimensions of the hollow square cross section, and L is the length of the element.

2. Redesign the truss so that both the stress and buckling constraints are satisfied with a safety factor of N not less than 2 for stresses, and N not less than 1.2 for buckling. Your design goal should be to reduce the weight of the truss as much as possible. The truss should be contained within the virtual rectangle shown by the dashed lines. Node 1 must be present to take the downward load $F = 1,000$ N. The nodes at the left wall have to be fixed completely. Nodes not attached to the wall have to be completely free to move in the x and y directions. Use the same cross section for all elements. Calculate the mass of the truss you have designed. Assume the density of aluminum as 2,800 kg/m^3.Draw the truss you have designed and provide the nodal coordinates and element connectivity in the form of a table. Results should also include the nodal displacements, forces in each member, and the safety factors for stresses and buckling for each element in the form of tables.

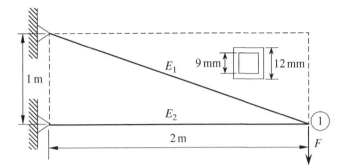

Figure 1.30 Plane truss and design domain for Project 1.2

Project 1.3 Analysis and Design of a Plane Truss 2

Consider a plane truss in figure 1.31. The horizontal and vertical members have length l, while inclined members have length $\sqrt{2}l$. Assume Young's modulus $E = 100\,\text{GPa}$, cross-sectional area $A = 1.0\,\text{cm}^2$, and $l = 0.3\,\text{m}$.

1. Use an FE program to determine the deflections and element forces for the following three load cases. Present you results in the form of a table.
   Load Case A) $F_{x13} = F_{x14} = 10{,}000\,\text{N}$
   Load Case B) $F_{y13} = F_{y14} = 10{,}000\,\text{N}$
   Load Case C) $F_{x13} = 10{,}000\,\text{N}$ and $F_{x14} = -10{,}000\,\text{N}$

2. Assuming that the truss behaves like a cantilever beam, one can determine the equivalent cross-sectional properties of the beam from the results for cases A through C above. The three beam properties are axial rigidity $(EA)_{eq}$ (this is *different* from the AE of the truss member), flexural rigidity $(EI)_{eq}$, and shear rigidity $(GA)_{eq}$. Let the beam length be equal to L ($L = 6 \times 0.3 = 1.8\,\text{m}$). The axial deflection of a beam due to an axial force F is given by:

$$u_{tip} = \frac{FL}{(EA)_{eq}}. \tag{1.72}$$

The transverse deflection due to a transverse force F at the tip is:

$$v_{tip} = \frac{FL^3}{3(EI)_{eq}} + \frac{FL}{(GA)_{eq}}. \tag{1.73}$$

In eq. (1.73) the first term on the RHS represents the deflection due to flexure and the second term, due to shear deformation. In the elementary beam theory (Euler-Bernoulli beam theory) we neglect the shear deformation, as it is usually much smaller than the flexural deflection. The transverse deflection due to an end couple C is given by:

$$v_{tip} = \frac{CL^2}{2(EI)_{eq}}. \tag{1.74}$$

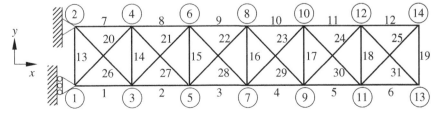

Figure 1.31 Plane truss and design domain for Project 1.3

Substitute the average tip deflections obtained in part 1 in eqs. (1.72)–(1.74) to compute the equivalent section properties: $(EA)_{eq}$, $(EI)_{eq}$, and $(GA)_{eq}$.

You may use the average of deflections at nodes 13 and 14 to determine the equivalent beam deflections.

3. Verify the beam model by adding two more bays to the truss ($L = 8 \times 0.3 = 2.4$ m). Compute the tip deflections of the extended truss for the three load cases A–C using the FE program. Compare the FE results with deflections obtained from the equivalent beam model (eqs. (1.72)–(1.74)).

Project 1.4 Fully Stressed Design of a Ten-Bar Truss

The fully stressed design is often used for truss structures. The idea is that we should remove material from members that are not fully stressed unless prevented by minimum cross-sectional area constraint. Practically, at every design cycle, the new cross-sectional area can be found using the following relation:

$$A_{new}^{(e)} = \frac{\sigma_{old}^{(e)}}{\sigma_{allowable}^{(e)}} A_{old}^{(e)}.$$

A ten-bar truss structure shown in figure 1.32 is under two loads, P_1 and P_2. The design goal is to minimize the weight, W, by varying the cross-sectional areas, A_i, of the truss members. The stress of the member should be less than the allowable stress with the safety factor. For manufacturing reasons, the cross-sectional areas should be greater than the minimum value. Input data are summarized in the table. Find optimum design using a fully stressed design.

Parameters	Values
Dimension, b	360 inches
Safety factor, S_F	1.5
Load, P_1	66.67 kips
Load, P_2	66.67 kips
Density, ρ	0.1 lb/in^3
Modulus of elasticity, E	10^4 ksi
Allowable stress, $\sigma_{allowable}$	25 ksi*
Initial area A_i	1.0 in^2
Minimum cross-sectional area	0.1 in^2

*for Element 9, allowable stress is 75 ksi

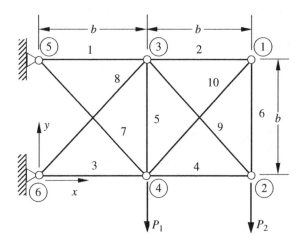

Figure 1.32 Ten-bar truss structure for project 1.4

1.8   EXERCISES

1. Answer the following descriptive questions

    (a) Write five properties of the element stiffness matrix.

    (b) Write five properties of the structural stiffness matrix.

    (c) Write five properties of the global stiffness matrix.

    (d) Will subdividing a truss element into many smaller elements improve the accuracy of the solution? Explain.

    (e) Explain when the element force $P^{(e)}$ is positive or negative.

    (f) For a given spring element, explain why we cannot calculate nodal displacement u_i and u_j when nodal forces $f_i^{(e)}$ and $f_j^{(e)}$ are given.

    (g) Explain what "striking the rows" and "striking the columns" are.

    (h) Once the gobal DOFs {**Q**} is solved, explain two methods of calculating nodal reaction forces.

    (i) Explain why we need to define the local coordinate in 2D truss element.

    (j) Explain how to determine the angle ϕ of 2D truss element.

    (k) When the connectivity of an element is changed from i → j to j → i, will the global stiffness matrix be the same or different? Explain why.

    (l) For the bar with fixed two ends in figure 1.22, when temperature is increased by ΔT, which of the following strains are not zero? Total strain, mechanical strain, or thermal strain?

    (m) Describe the finite element model you would use for a thin slender bar pinned at both ends with a transverse (perpendicular to the bar) concentrated load applied at the middle. (1) Draw a figure to show the elements, loads, and boundary conditions. (2) What type of elements will you use?

2. Calculate the displacement at node 2 and reaction forces at nodes 1 and 3 of the springs shown in the figure using two spring elements. A force $F = 1000\,\text{N}$ is applied at node 2. Use $k^{(1)} = 2000\,\text{N/mm}$ and $k^{(2)} = 3000\,\text{N/mm}$.

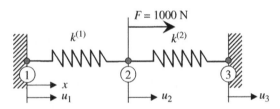

3. Repeat problem 2 by changing node numbers; that is, node 3 is now node 1, and node 1 is now node 3. Check if the results are the same as those of problem 2.

4. Three rigid bodies, 2, 3, and 4, are connected by four springs as shown in the figure. A horizontal force of 1,000 N is applied on body 4 as shown in the figure. Find the displacements of the three bodies and the forces (tensile/compressive) in the springs. What is the reaction at the wall? Assume the bodies can undergo only translation in the horizontal direction. The spring constants (N/mm) are: $k^{(1)} = 400$, $k^{(2)} = 500$, $k^{(3)} = 400$, $k^{(4)} = 200$.

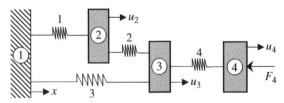

5. Three rigid bodies, 2, 3, and 4, are connected by six springs as shown in the figure. The rigid walls are represented by 1 and 5. A horizontal force $F_3 = 2000\,\text{N}$ is applied on body 3 in the direction shown in the figure. Find the displacements of the three bodies and the forces (tensile/compressive) in the springs. What are the reactions at the walls? Assume the bodies can undergo only translation in the horizontal direction. The spring constants (N/mm) are: $k^{(1)} = 200$, $k^{(2)} = 400$, $k^{(3)} = 600$, $k^{(4)} = 300$, $k^{(5)} = 500$, $k^{(6)} = 300$.

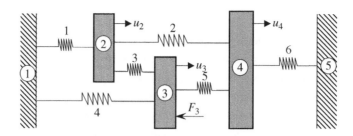

6. Consider the spring-rigid body system described in problem 5. What force F_2 should be applied on body 2 in order to keep it from moving? How will this affect the support reactions?

   Hint: Impose the boundary condition $u_2 = 0$ in the finite element model and solve for displacements u_3 and u_4. Then, the force F_2 will be the reaction at node 2.

7. Four rigid bodies, 1, 2, 3, and 4, are connected by four springs as shown in the figure. A horizontal force of 1,000 N is applied on body 1 as shown in the figure. Using FE analysis, (a) find the displacements of bodies 1 and 3, (2) find the element force (tensile/compressive) of spring 1, and (3) find the reaction force at the right wall (body 2). Assume the bodies can undergo only translation in the horizontal direction. The spring constants (N/mm) are: $k^{(1)} = 500$, $k^{(2)} = 300$, $k^{(3)} = 400$, and $k^{(4)} = 300$. Do not change node and element numbers.

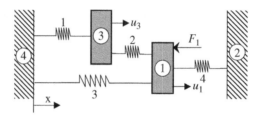

8. Determine the nodal displacements, element forces, and reaction forces using the direct stiffness method for the two-bar truss shown in the figure.

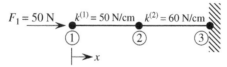

9. In the structure shown, rigid blocks are connected by linear springs. Imagine that only horizontal displacements are allowed. Write the global equilibrium equations $[\mathbf{K}]\{\mathbf{Q}\} = \{\mathbf{F}\}$ after applying displacement boundary conditions in terms of spring stiffness $k^{(i)}$, displacement DOFs u_i, and applied loads F_i.

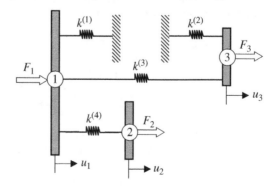

10. The spring-mass system shown in the figure is in equilibrium under the applied loads. The element connectivity table shows node numbers associated with each element and the stiffness of each element.

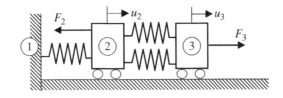

Element	Local Node 1	Local Node 2	Stiffness, k_i
1	1	2	2000 N/m
2	2	3	1000 N/m
3	2	3	500 N/m

(a) Write the structural matrix equations $[\mathbf{K}_s]\{\mathbf{Q}_s\} = \{\mathbf{F}_s\}$ by assembling the structural stiffness matrix $[\mathbf{K}_s]$ and the force vector $\{\mathbf{F}_s\}$. Use $F_3 = 5$ N and $F_2 = 2$ N.

(b) Apply the boundary conditions and solve for the displacement of the two masses.

11. A structure is composed of two one-dimensional bar elements. When a 10 N force is applied to node 2, calculate the displacement vector $\{\mathbf{Q}\}^T = \{u_1, u_2, u_3\}$ using the finite element method.

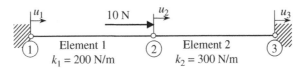

12. Two rigid masses, 1 and 2, are connected by three springs as shown in the figure. When gravity is applied with $g = 9.85$ m/sec^2, using FE analysis, (a) find the displacements of masses 1 and 2, (2) find the element forces of all the springs, and (3) find the reaction force at the wall (body 3). Assume the bodies can move only in the vertical direction and cannot rotate. The spring constants (N/mm) are $k^{(1)} = 400$, $k^{(2)} = 500$, and $k^{(3)} = 500$. The masses (kg) are $m_1 = 20$ and $m_2 = 40$.

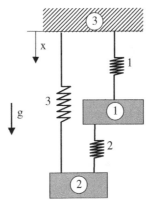

13. Use the finite element method to determine the axial force P in each portion, AB and BC, of the uniaxial bar. What are the support reactions? Assume: $E = 100$ GPa, the area of cross sections of the two portions AB and BC are, respectively, 10^{-4} m^2 and 2×10^{-4} m^2, and $F = 10,000$ N. The force F is applied at the cross section at B.

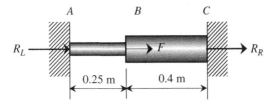

14. Consider a tapered bar of circular cross section. The length of the bar is 1 m, and the radius varies as $r(x) = 0.050 - 0.040x$, where r and x are in meters. Assume Young's modulus = 100 MPa. Both ends of the bar are fixed, and $F = 10,000$ N is applied at the center. Determine the displacements, axial force distribution, and the wall reactions using four elements of equal length.

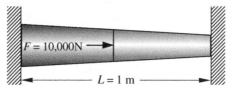

Hint: To approximate the area of cross section of a bar element, use the geometric mean of the end areas of the element, i.e., $A^{(e)} = \sqrt{A_i A_j} = \pi r_i r_j$.

15. The stepped bar shown in the figure is subjected to a force at the center. Use the finite element method to determine the displacement at the center and reactions R_L and R_R.

Assume: $E = 100$ GPa, the area of cross sections of the three portions shown are, respectively, 10^{-4} m^2, 2×10^{-4} m^2, and 10^{-4} m^2, and $F = 10,000$ N.

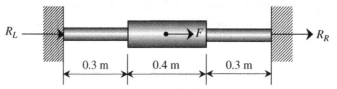

16. Using the direct stiffness matrix method, find the nodal displacements and the forces in each element and the reactions.

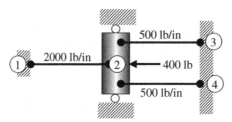

17. A stepped bar is clamped at one end and subjected to concentrated forces as shown.
Note: the node numbers are not in usual order!

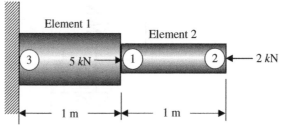

Assume: $E = 100$ GPa, small area of cross section = 1 cm^2, and large area of cross section = 2 cm^2.

(a) Write the element stiffness matrices of elements 1 and 2 showing the row addresses.

(b) Assemble the above element stiffness matrices to obtain the following structural level equations in the form $[\mathbf{K}_s]\{\mathbf{Q}_s\} = \{\mathbf{F}_s\}$.

(c) Delete the rows and columns corresponding to zero DOFs to obtain the global equations in the form of $[\mathbf{K}]\{\mathbf{Q}\} = \{\mathbf{F}\}$.

(d) Determine the displacements and element forces.

18. A stepped bar is clamped at both ends. A force of $F = 10{,}000$ N is applied as shown in the figure. Areas of cross sections of the two portions of the bar are 1×10^{-4} m^2 and 2×10^{-4} m^2, respectively. Young's modulus $= 10$ GPa. Use three bar elements to solve the problem.

(a) Sketch the displacement field $u(x)$ as a function of x on an x-y graph.

(b) Calculate the stress at a distance of 2.5 m from the left support.

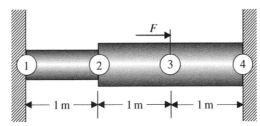

19. Repeat problem 18 for the stepped bar shown in the figure.

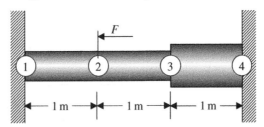

20. The finite element equation for the uniaxial bar can be used for other types of engineering problems if a proper analogy is applied. For example, consider the piping network shown in the figure. Each section of the network can be modeled using an FE. If the flow is laminar and steady, we can write the equations for a single pipe element as:

$$q_i = K\left(P_i - P_j\right)$$
$$q_j = K\left(P_j - P_i\right)$$

where q_i and q_j are fluid flow at nodes i and j, respectively; P_i and P_j are fluid pressure at nodes i and j, respectively; and K is

$$K = \frac{\pi D^4}{128 \mu L},$$

where D is the diameter of the pipe, μ is the viscosity, and L is the length of the pipe. The fluid flow is considered positive away from the node. The viscosity of the fluid is 9×10^{-4} Pa·s.

(a) Write the element matrix equation for the flow in the pipe element.

(b) The net flow rates into nodes 1 and 2 are 10 and 15 m^3/s, respectively. The pressures at the nodes 6, 7, and 8 are all zero. The net flow rate into the nodes 3, 4, and 5 are all zero. What is the outflow rate for elements 4, 6, and 7?

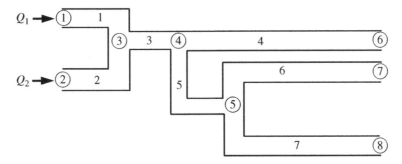

Elem	1	2	3	4	5	6	7
D(mm)	40	40	50	25	40	25	25
L(m)	1	1	1	4	2	3	3

21. The truss structure shown in the figure supports a force F. The finite element method is used to analyze this structure using two truss elements as shown in the figure. The cross-sectional area for both elements is $A = 2\,\text{in}^2$, and the lengths are $L^{(2)} = 10\,\text{ft}$. Young's Modulus of the material $E = 30 \times 10^6$ psi.

(a) Compute the transformation matrix [**T**] for element 2 that enables you to transform between global and local coordinates (as shown in equation below).

$$\left\{ \begin{array}{c} \bar{u}_1 \\ \bar{u}_3 \end{array} \right\} = [\mathbf{T}] \left\{ \begin{array}{c} u_1 \\ v_1 \\ u_3 \\ v_3 \end{array} \right\}.$$

(b) It is determined after solving the final equations that the displacement components of the node 1 are: $u_1 = 0.5 \times 10^{-2}$ in. and $v_1 = -1.5 \times 10^{-2}$ in. What are the strain, stress, and force in element 2?

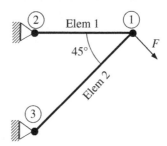

22. The properties of the two elements of a plane truss are given in the table below. Note that an external force of 10,000 N is acting on the truss at node 2.

Elem.	$i \rightarrow j$	ϕ	l	m	L (m)	A (cm^2)	E (GPa)	α (/°C)	ΔT (°C)
1	$1 \rightarrow 2$	90	0	1	1	1	100	20×10^{-6}	−100
2	$2 \rightarrow 3$	0	1	0	1	1	100	20×10^{-6}	0

(a) Write the thermal force vector for each element. Indicate row addresses clearly.

(b) Assemble the thermal force vectors to form the global thermal force $\{F_T\}$, which is a 2×1 matrix.

(c) The problem was solved using FEA to obtain the displacements as $u_2 = -1\,\text{mm}$, $v_2 = -2\,\text{mm}$. Determine the element force P in each element.

(d) Show that equilibrium is satisfied at node 2.

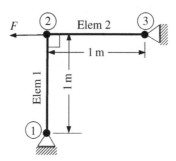

23. For a two-dimensional truss structure as shown in the figure, determine displacements of the nodes and normal stresses developed in the members using the direct stiffness method. Use $E = 30 \times 10^6$ N/cm^2, and the diameter of the circular cross-section is 0.25 cm.

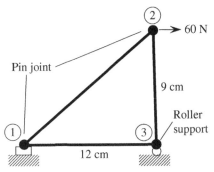

24. The 2D truss shown in the figure is assembled to build the global matrix equation. Before applying boundary conditions, the dimension of the global stiffness matrix is 8×8. Write (row, column) indices of locations corresponding to nonzero DOFs of the stiffness matrix, such as (1,1), (1,2), and so forth. The order of global DOFs is $\{\mathbf{Q}\} = \{u_1, v_1, u_2, v_2, u_3, v_3, u_4, v_4\}^{\mathsf{T}}$. Since the stiffness matrix is symmetric, only write its upper-right portion.

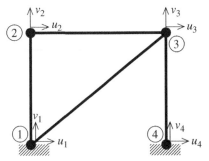

25. For a two-dimensional truss structure as shown in the figure, determine displacements of the nodes and normal stresses developed in the members using a commercial finite element analysis program. Use $E = 30 \times 10^6$ N/cm^2, and the diameter of the circular cross-section is 0.25 cm.

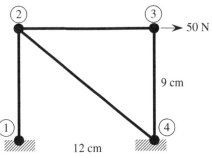

26. The truss shown in the figure supports force F at node 2. The finite element method is used to analyze this structure using two truss elements as shown.

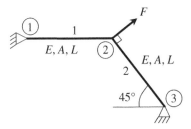

     (a) Compute the transformation matrix for elements 1 and 2.

     (b) Compute the element stiffness matrices for both elements in the global coordinate system.

     (c) Assemble the element stiffness matrices and force vectors to the structural matrix equation $[\mathbf{K}_s]\{\mathbf{Q}_s\} = \{\mathbf{F}_s\}$ before applying boundary conditions.

     (d) Solve the FE equation after applying the boundary conditions. Write nodal displacements in the global coordinates.

     (e) Compute stress in element 1. Is it tensile or compressive?

27. The truss structure shown in the figure supports the force F. The finite element method is used to analyze this structure using two truss elements as shown. The area of cross-section (for all elements) $A = 2\ \text{in}^2$, and Young's modulus $E = 30 \times 10^6$ psi. Both elements are of equal length $L = 10$ ft.

     (a) Compute the transformation matrix for elements 1 and 2 between the global coordinate system and the local coordinate system for each element.

     (b) Compute the stiffness matrix for the elements 1 and 2.

     (c) Assemble the structural matrix equation $[\mathbf{K}_s]\{\mathbf{Q}_s\} = \{\mathbf{F}_s\}$ (without applying the boundary conditions).

     (d) It is determined after solving the final equations that the displacement components of node 1 are: $u_{1x} = 1.5 \times 10^{-2}$ in., $u_{1y} = -0.5 \times 10^{-2}$ in. Compute the applied load F.

     (e) Compute stress and strain in element 1.

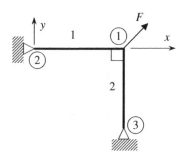

28. In the finite element model of a plane truss in problem 21, the lengths of elements 1 and 2 are 1 m and $\sqrt{2}$ m, respectively. The axial rigidity of the elements is $EA = 10^7$ N.

     (a) Fill in the connectivity information of your choice

Element	1st Node i	2nd Node j	φ, deg	l	m
1					
2					

     (b) Write down the element stiffness matrix $[k^{(2)}]$ of element 2 with DOFs.

$$\left[k^{(2)}\right] = \begin{bmatrix} & & & \end{bmatrix} \begin{matrix} \square \\ \square \\ \square \\ \square \end{matrix} .$$

     (c) Show where $[k^{(2)}]$ will be placed in the structural stiffness matrix $[K_s]$.

$$
[K_s] = \begin{bmatrix} & & & & \\ & & & & \\ & & & & \\ & & & & \\ & & & & \\ & & & & \end{bmatrix} \begin{matrix} u_1 \\ v_1 \\ u_2 \\ v_2 \\ u_2 \\ v_3 \end{matrix} .
$$

(d) The displacements are calculated as: $u_1 = +5$ mm, $v_1 = -5\sqrt{2}$ mm. Use your connectivity information to calculate the element force P in element 2.

29. Use the finite element method to solve the plane truss shown below. Assume $AE = 10^6$ N, $L = 1$ m. Determine the nodal displacements, element forces in each element, and the support reactions.

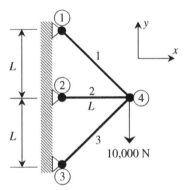

30. The plane truss shown in the figure has two elements and three nodes. Calculate the 4×4 element stiffness matrices. Show the row addresses clearly. Derive the final equations (after applying boundary conditions) for the truss in the form of $[\mathbf{K}]\{\mathbf{Q}\} = \{\mathbf{F}\}$. What are nodal displacements and the element forces? Assume: $E = 10^{11}$ Pa, $A = 10^{-4}$ m^2, $L = 1$ m, $F = 14{,}142$ N.

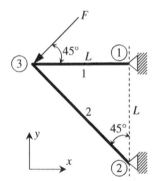

31. Two bars are connected as shown in the figure. Assume all joints are frictionless pin joints. At node 2, a vertical spring is connected as shown. Both bars are of length L and have the same properties: Young's modulus $= E$ and area of cross section $= A$. The spring stiffness $= k$.

(a) Set up the stiffness matrices for the two truss elements and the spring element.

(b) Assemble the stiffness matrices to form the global stiffness matrix.

(c) Compute the deflections at node 2, if a force $\mathbf{F} = \{F_x, F_y\}^T$ is applied at node 2.

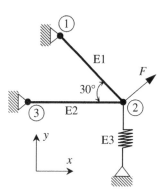

32. The truss structure shown in the figure supports the force F. The finite element method is used to analyze this structure using three truss elements as shown. The area of cross-section (for all elements) $A = 2$ in^2. Young's Modulus $E = 30 \times 10^6$ psi. The lengths of the elements are: $L_1 = L_3 = 10$ ft., $L_2 = 14.14$ ft.

    (a) Determine the stiffness matrix of element 2.

    (b) It is determined after solving the final equations that the displacement components of node 1 are: $u_1 = -0.5 \times 10^{-2}$ in. and $v_1 = -1.5 \times 10^{-2}$ in. Using the four equations for this element, compute the forces acting on element 2 at nodes 1 and 3. Are these two forces collinear?

    (c) What is the change in length of element 2?

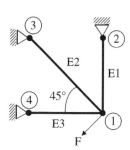

33. It is desired to use the finite element method to solve the two plane truss problems shown in the figure below. Assume $AE = 10^6$ N, $L = 1$ m. Before solving the global equations $[\mathbf{K}]\{\mathbf{Q}\} = \{\mathbf{F}\}$, find the determinant of $[\mathbf{K}]$. Does $[\mathbf{K}]$ have an inverse? Explain your answer.

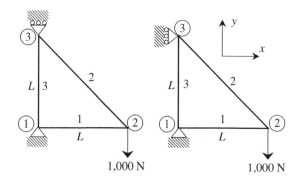

34. Determine the member force and axial stress in each member of the truss shown in the figure using a commercial finite element analysis program. Assume that Young's modulus is 10^4 psi and all cross-sections are circular with a diameter of 2 in. Compare the results with the exact solutions that are obtained from the free-body diagram.

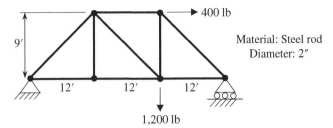

35. Determine the normal stress in each member of the truss structure. All joints are ball-joint, and the material is steel whose Young's modulus is $E = 210$ GPa. Nodes 1, 3, and 4 are pinned on the ground, while node 2 is free to move.

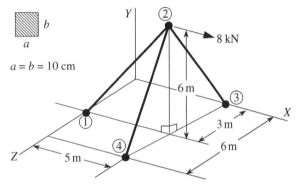

36. The space truss shown has four members. Determine the displacement components of node 5 and the force in each member. The node numbers are numbers in the circle in the figure. The dimensions of the imaginary box that encloses the truss are $1\,\text{m} \times 1\,\text{m} \times 2\,\text{m}$. Assume $AE = 10^6$ N. The coordinates of the nodes are given in the table below:

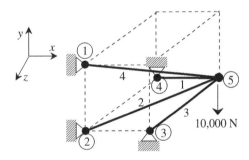

Node	x	y	z
1	0	1	2
2	0	0	2
3	1	0	2
4	0	0	0
5	1	0	0

37. The uniaxial bar shown below can be modeled as a one-dimensional bar. The bar has the following properties: $L = 1$ m, $A = 10^{-4}$ m^2, $E = 100$ GPa, and $\alpha = 2 \times 10^{-6}/°$C. From the stress-free initial state, a force of 5,000 N is applied at node 2, and the temperature is lowered by 50 °C below the reference temperature.

(a) Calculate the global matrix equation after applying boundary conditions.

(b) Solve for the displacement u_2.

(c) What is the element force P in the bar?

A, E, L 5,000 N

38. In the structure shown below, the temperature of element 2 is 50 °C above the reference temperature. An external force of 20,000 N is applied in the x direction (horizontal direction) at node 2. Assume $E = 10^{11}$ Pa, $A = 10^{-4}$ m^2, and $\alpha = 2 \times 10^{-6}$/°C.

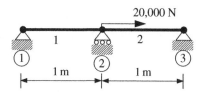

20,000 N

1 m   1 m

(a) Write down the stiffness matrices and thermal force vectors for each element.

(b) Write down the global matrix equations.

(c) Solve the global equations to determine the displacement at node 2.

(d) Determine the forces in each element. State whether it is tension or compression.

(e) Show that force equilibrium is satisfied at node 2.

39. The element properties of a plane truss are given in the table below. Derive the global equations in the form: $[\mathbf{K}]_{(2\times2)} \{\mathbf{Q}_{(2\times1)}\} = \{\mathbf{F}_{(2\times1)}\} + \{\mathbf{F}_{T(2\times1)}\}$. Do not solve the equations.

Elem.	$i \rightarrow j$	ϕ	l	m	L	A	E	α	ΔT
1					$\sqrt{2}$	1	10	2	10
2					1	1	10	2	−20

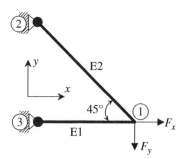

40. The three-bar truss problem in figure 1.23 is under a vertical load of $F_y = -5,000$ N at node 4 in addition to the temperature change. Solve the problem by (a) applying F_y and ΔT at the same time, and (b) superposing two solutions: one from applying F_y and the other from changing temperature ΔT.

41. Use the finite element method to determine the nodal displacements in the plane truss shown in figure (a). The temperature of element 2 is 200 °C above the reference temperature, that is, $\Delta T^{(2)} = 200$ °C. Compute the force in each element. Show that the force equilibrium is satisfied at node 3. Assume $L = 1$ m, $AE = 10^7$ N, $\alpha = 5 \times 10^{-6}$/°C.

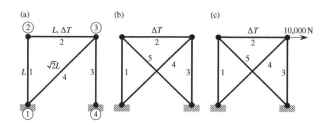

42. Repeat problem 41 for the new configuration with element 5 added, as shown in figure (b)

43. Repeat problem 42 with an external force added to node 3, as shown in figure (c).

44. The properties of the members of the truss in the left side of the figure are given in the table. Calculate the nodal displacement and element forces. Show that force equilibrium is satisfied as node 3.

Elem	L (m)	A (cm^2)	E (GPa)	α (/°C)	ΔT (°C)
1	1	1	100	20×10^{-6}	0
2	1	1	100	20×10^{-6}	0
3	1	1	100	20×10^{-6}	0
4	1	1	100	20×10^{-6}	0
5	$\sqrt{2}$	1	100	20×10^{-6}	-100

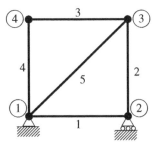 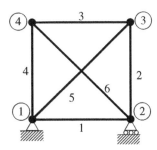

45. Repeat problem 44 for the truss on the right side of the figure. The properties of element 6 are same as those of element 5, but $\Delta T = 0$ °C.

46. The truss shown in the figure supports the force $F = 1,000$ N. Both elements have the same axial rigidity of $AE = 10^7$ N, thermal expansion coefficient of $\alpha = 2 \times 10^{-6}$/°C, and length $L = 1$ m. While the temperature at element 1 remains constant, that of element 2 is dropped by 50 °C.

(a) Write the 4×4 element stiffness matrices $[\mathbf{k}]$ and the 4×1 thermal force vectors $\{\mathbf{f}_T\}$ for elements 1 and 2. Show the row addresses clearly.

(b) Assemble two elements and apply the boundary conditions to obtain the global matrix equation in the form of $[\mathbf{K}]\{\mathbf{Q}\} = \{\mathbf{F}\} + \{\mathbf{F}_T\}$.

(c) Solve for the nodal displacements.

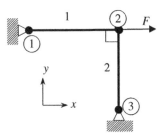

47. The finite element method was used to solve the truss problem shown below. The solution for displacements was obtained as $u_2 = 1.5$ mm, $v_2 = -1.5$ mm, $u_3 = 3$ mm, and $v_3 = -1.5$ mm. Note: the displacements are given in mm.

(a) Determine the axial forces P in elements 2 and 4.

(b) The forces in elements 3 and 5 are found to be as follows: $P^{(3)} = -2,000$ N, $P^{(5)} = 7,070$ N. Determine the support reactions R_{y4} at node 4 using the nodal equilibrium equations.

The element properties are listed in the following table.

Elem	$i \rightarrow J$	AE [N]	L [m]	ΔT [°C]	α [1/°C]	ϕ
1	1, 2	10^7	1	−100	10^{-6}	90°
2	2, 3	10^7	1	0	10^{-6}	0°
3	3, 4	10^7	1	+100	10^{-6}	−90°
4	1, 3	10^7	1.414	+200	10^{-6}	45°
5	2, 4	10^7	1.414	+200	10^{-6}	−45°

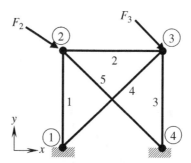

48. Use the finite element method to solve the plane truss shown in problem 29. Assume $AE = 2 \times 10^6$ N, $L = 1$ m, $\alpha = 10 \times 10^{-6}/°C$. The temperature of element 1 is 100 °C below the reference temperature, while elements 2 and 3 are at the reference temperature. Determine the nodal displacements, forces in each element, and the support reactions. Show that the nodal equilibrium is satisfied at node 4.

Chapter 2

Weighted Residual Methods for One-Dimensional Problems

In the previous chapter, we developed the finite element equations for the truss element using the direct stiffness method. However, this method becomes impractical when element formulations are complicated or when multidimensional problems are considered. Thus, we need to develop a more systematic approach to constructing the finite element equations for general engineering problems. In fact, the finite element method can be applied to any engineering problem that is governed by a differential equation. There are two other methods of deriving the finite element equations: weighted residual method and energy method. In the first part of this chapter, we will consider ordinary differential equations that occur commonly in engineering problems, and we will derive the corresponding finite element equations through the weighted residual method, in particular using the Galerkin method. Energy methods are alternative methods that are very powerful and amenable for approximating solutions when solving structures that are more realistic. In the second part of this chapter, we will use the principle of minimum potential energy to derive finite element equations of discrete systems and uniaxial bars.

2.1 EXACT VS. APPROXIMATE SOLUTION

2.1.1 Exact Solution

Many engineering problems such as the deformation of a beam and heat conduction in a solid can be described using a differential equation. The differential equation along with boundary conditions is called the *boundary value problem*. A simple, one-dimensional example of a boundary-value problem is:

$$\frac{\mathrm{d}^2 u}{\mathrm{d}x^2} + p(x) = 0, \quad 0 \le x \le 1$$

$$\left.\begin{array}{l} u(0) = 0 \\[2mm] \dfrac{\mathrm{d}u}{\mathrm{d}x}(1) = 1 \end{array}\right\} \quad \text{Boundary conditions.} \tag{2.1}$$

Introduction to Finite Element Analysis and Design, Second Edition. Nam H. Kim, Bhavani V. Sankar, and Ashok V. Kumar.
© 2018 John Wiley & Sons Ltd. Published 2018 by John Wiley & Sons Ltd.
Companion website: www.wiley.com/go/kim/finite_element_analysis_design

The above differential equation describes the displacements in a uniaxial bar subjected to a distributed force $p(x)$ along its axis. The first boundary condition prescribes the value of the solution at a given point and is called the *essential boundary condition*. The term-displacement boundary condition, or kinematic boundary condition, is also used in the context of solid mechanics. On the other hand, the second boundary condition prescribes the value of the derivative, du/dx, at $x = 1$, and is called the *natural boundary condition*. In solid mechanics, the term *force boundary condition* or *stress boundary condition* is also used. The solution $u(x)$ of eq. (2.1) satisfies the differential equation at every point in the domain, and it needs to be, at least, a twice-differentiable function because the differential equation includes the second-order derivative of $u(x)$. In this text, the solution $u(x)$ of eq. (2.1) will be referred to as the *exact solution*.

When the geometry is complicated, it is not trivial to solve for $u(x)$ analytically. Since the solution that satisfies the differential equation and boundary conditions can have a complicated expression, an infinite series solution is often employed.

2.1.2   Approximate Solution

In using the weighted residual method, we seek an approximate solution $\tilde{u}(x) \approx u(x)$ for eq. (2.1). *We choose $\tilde{u}(x)$ such that it satisfies the essential boundary conditions of the problem but not necessarily the natural boundary conditions.* In the present example, $\tilde{u}(x)$ must satisfy the condition $\tilde{u}(0) = 0$. Since $\tilde{u}(x)$ is an approximation of the exact solution $u(x)$, it will not satisfy the differential equation in eq. (2.1), and hence, the differential equation with the approximate solution will not be identically equal to zero everywhere in the domain. We will call the resulting function as the *residual* and denote it by $R(x)$:

$$\frac{d^2 \tilde{u}}{dx^2} + p(x) = R(x).$$

Our goal now is to minimize the error or the residual as much as possible. Instead of trying to make $R(x)$ vanish everywhere, we will make $R(x)$ equal to zero in some average sense. Consider the following integral:

$$\int_0^1 R(x) W(x) dx = 0, \tag{2.2}$$

where $W(x)$ is an arbitrary function called the *weight function*. The above equation requires the integral of the weighted residual to be equal to zero. This statement is weaker than the original differential equation because it can be satisfied even if there is an error in the solution since only the weighted average over the domain needs to be zero. Therefore, it is often referred to as the "weak form" while the original differential equation in eq. (2.1) is the "strong form". If eq. (2.2) is satisfied for any arbitrary weight function, then it can be said that the residual function is zero everywhere, and the approximate solution is identical to the exact solution. If eq. (2.2) is satisfied for several different linearly independent weight functions, then $R(x)$ will approach zero, and the approximate solution $\tilde{u}(x)$ will approach the exact solution $u(x)$. The choice of the weight functions leads to different weighted residual methods such as the least squares error method, the collocation method, Petrov-Galerkin method, and Galerkin method[1]. In this book, we will discuss only the Galerkin method, which is popular in deriving the finite element equations for many engineering problems.

---

[1] Cook, R. D. *Concepts and Applications of Finite Element Analysis*, 4th Ed. John Wiley & Sons, New York.

EXAMPLE 2.1   *Weighted residual method*

Calculate the approximate solution of eq. (2.1) using the weighted residual method. Use (a) $W(x) = 1$ and (b) $W(x) = x$. Compare these solutions with the exact solution. Assume the approximate solution to be $\tilde{u}(x) = c_0 + c_1 x + c_2 x^2$ and $p(x) = x$.

SOLUTION   First, the exact solution can be calculated by integrating the differential eq. (2.1) twice:

$$u_{exact}(x) = -\frac{1}{6}x^3 + a_1 x + a_2.$$

The integral constants, a_1 and a_2, can be calculated by applying the two boundary conditions, $u(0) = 0$ and $du(1)/dx = 1$; that is, $a_1 = 3/2$ and $a_2 = 0$. Therefore, the exact solution becomes

$$u_{exact}(x) = -\frac{1}{6}x^3 + \frac{3}{2}x.$$

(a) When $W(x) = 1$: using the approximate solution, we have $du/dx = c_1 + 2c_2 x$ and $d^2 u/dx^2 = 2c_2$. By applying the two boundary conditions[2], the unknown coefficients can be determined as

$$u(0) = 0 \;\Rightarrow\; c_0 = 0,$$

$$\frac{du(1)}{dx} = 1 \;\Rightarrow\; c_1 + 2c_2 = 1.$$

Using the weight function $W(x) = 1$, the weighted residual can be defined as

$$\int_0^1 \left(\frac{d^2 u}{dx^2} + x\right) \cdot W \, dx = \int_0^1 (2c_2 + x) \cdot 1 \, dx = 0.$$

By solving the above equation, the unknown coefficients can be determined as $c_2 = -1/4$ and $c_1 = 3/2$. Therefore, the approximate solution becomes

$$\tilde{u}(x) = -\frac{1}{4}x^2 + \frac{3}{2}x.$$

Note that the approximate solution satisfies the two boundary conditions.

(b) When $W(x) = x$: by applying the two boundary conditions, the unknown coefficients can be determined just as in (a): $c_0 = 0$ and $c_1 + 2c_2 = 1$. Using the weight function $W(x) = x$, the weighted residual can be defined as

$$\int_0^1 \left(\frac{d^2 u}{dx^2} + x\right) \cdot W \, dx = \int_0^1 (2c_2 + x) \cdot x \, dx = 0.$$

By solving the above equation, the unknown coefficients can be determined as $c_2 = -1/3$ and $c_1 = 5/3$. Therefore, the approximate solution becomes

$$\tilde{u}(x) = -\frac{1}{3}x^2 + \frac{5}{3}x.$$

Note that the approximate solution also satisfies the two boundary conditions. ■

Note that the approximate solution is different from the exact solution, because the approximate solution is quadratic in x, while the exact solution is a cubic polynomial. However, figure 2.1 shows that the approximate solutions are close to the exact solution. At $x = 0$ they are identical because the function value is prescribed at this point, while at $x = 1$ the slopes of all solutions are same. Figure 2.1 also shows that the weight function plays a role to determine the approximate solutions.

---

[2] The approximate solution does not satisfy the natural boundary condition in the weighted residual method, but we use this condition for the moment until we introduce integration by parts.

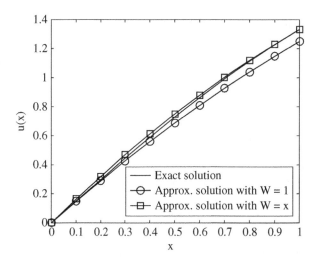

Figure 2.1 Comparison of exact solution and approximate solutions for example 2.1

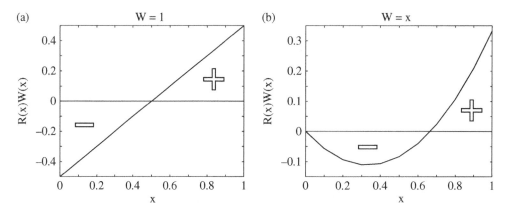

Figure 2.2 Weighted residual for differential equation in example 2.1

In this particular case, the case with $W(x) = x$ yield a better approximation than the case with the constant weight function.

It would be informative to see how the weighted residual method determines those coefficients in the example. Because of the two boundary conditions, there will only be one unknown coefficient, c_2, in the problem. Then, the residual, $R(x) = 2c_2 + x$, cannot vanish everywhere in the domain. Therefore, the weighted residual method determines the coefficient so that the integral of the weighted residual term is equal to zero. In the case when $W(x) = 1$, the weighted residual term is plotted in figure 2.2(a). Since the weighted residual is a linear function, the coefficient is determined such that the positive area is equal to the negative area, and the coefficient is determined as $c_2 = -1/4$ such that the sign changes at $x = 0.5$. When $W(x) = x$, the weighted residual becomes a quadratic function, $R(x)W(x) = 2c_2x + x^2$. In this case, the coefficient is determined as $c_2 = -1/3$ such that the sign changes at $x = 2/3$ as shown in figure 2.2(b). Note that the positive area is also the same as the negative area.

EXAMPLE 2.2 *Weighted residual method with exact form*

Solve the same problem in example 2.2 using $\tilde{u}(x) = c_0 + c_1 x + c_2 x^3$.

SOLUTION

(a) When $W(x) = 1$: using the approximate solution, we have $du/dx = c_1 + 3c_2x^2$ and $d^2u/dx^2 = 6c_2x$. By applying the two boundary conditions, the unknown coefficients can be determined as

$$u(0) = 0 \quad \Rightarrow \quad c_0 = 0,$$

$$\frac{du(1)}{dx} = 1 \quad \Rightarrow \quad c_1 + 3c_2 = 1.$$

Using the weight function $W(x) = 1$, the weighted residual can be defined as

$$\int_0^1 \left(\frac{d^2u}{dx^2} + x \right) W(x) \, dx = \int_0^1 (6c_2x + x) 1 \, dx = 0.$$

By solving the above equation, the unknown coefficients can be determined as $c_2 = -1/6$ and $c_1 = 3/2$. Therefore, the approximate solution becomes

$$\tilde{u}(x) = -\frac{1}{6}x^3 + \frac{3}{2}x.$$

Note that the approximate solution is identical to the exact solution.

(b) When $W(x) = x$: by applying the two boundary conditions, the unknown coefficients can be determined just as in (a): $c_0 = 0$ and $c_1 + 3c_2 = 1$. Using the weight function $W(x) = x$, the weighted residual can be defined as

$$\int_0^1 \left(\frac{d^2u}{dx^2} + x \right) \cdot W \, dx = \int_0^1 (6c_2x + x) \cdot x \, dx = 0.$$

By solving the above equation, the unknown coefficients can be determined as $c_2 = -1/6$ and $c_1 = 3/2$. Therefore, the approximate solution becomes identical to the exact solution. ■

This example shows that when the functional form of the approximate solution is the same as the exact solution, the weighted residual method can find the correct coefficients and yield the exact solution. When the approximate solution is a cubic polynomial, it is possible to make the residual to vanish. In the example, the weighted residual becomes $(6c_2x + x) \cdot W(x)$. Therefore, by choosing $c_2 = -1/6$, the residual becomes zero no matter what weight function is used. In general, the weighted residual method can find the exact solution if the approximate solution has the capability of representing the exact solution. Otherwise, the method will yield an approximate solution.

As we have seen in example 2.1, the approximate solution depends on the selection of weight functions. Therefore, it is important to select appropriate weight functions. In fact, different methods have been proposed by selecting different weight functions. In the following section, we will discuss one of the most popular weighted residual methods.

2.2 GALERKIN METHOD

The approximate solution $\tilde{u}(x)$ can be expressed as a sum of a number of functions called *trial functions*:

$$\tilde{u}(x) = \sum_{i=1}^{N} c_i \phi_i(x), \tag{2.3}$$

where N is the number of terms used, $\phi_i(x)$ are known *trial functions*, and c_i are coefficients to be determined using the weighted residual method. This form of approximation is common in engineering; for example, Taylor series expansion of a function or Fourier series. Rigorous mathematical studies show that these series representations converge to the exact solution as N increases when the exact solution is continuous.

Since the approximate solution is a linear combination of the trial functions, the accuracy of approximation depends on them. *The trial functions and coefficients are chosen such that $\tilde{u}(x)$ must satisfy the essential boundary conditions of the problem.* In the present example in eq. (2.1), $\tilde{u}(x)$ must satisfy $\tilde{u}(0) = 0$. However, the trial functions do not need to satisfy the natural boundary conditions. Natural boundary conditions will be a part of the solution process.

The *Galerkin method* differs from other weighted residual methods in that the N weight functions, $\phi_i(x)$, are the same as N trial functions. Thus, we obtain the following N number of weighted residual equations:

$$\int_0^1 R(x)\phi_i(x)dx = 0, \quad i = 1,\ldots,N. \tag{2.4}$$

Substituting for $R(x)$ from eq. (2.2) we obtain

$$\int_0^1 \left(\frac{d^2\tilde{u}}{dx^2} + p(x) \right) \phi_i(x)dx = 0, \quad i = 1,\ldots,N. \tag{2.5}$$

Since the function $p(x)$ is known, we will take the term containing it to the right-hand side (RHS) to obtain

$$\int_0^1 \frac{d^2\tilde{u}}{dx^2}\phi_i(x)dx = -\int_0^1 p(x)\phi_i(x)dx, \quad i = 1,\ldots,N. \tag{2.6}$$

In the view of approximation in eq. (2.3), the above N equations can be used to solve for the N unknown coefficients c_i. However, we will use integration by parts as shown below to reduce the order of differentiation of $\tilde{u}$:

$$\frac{d\tilde{u}}{dx}\phi_i \Big|_0^1 - \int_0^1 \frac{d\tilde{u}}{dx}\frac{d\phi_i}{dx}dx = -\int_0^1 p(x)\phi_i(x)dx, \quad i = 1,\ldots,N. \tag{2.7}$$

Note that the boundary terms on the left-hand side (LHS) of the above equation has the term $d\tilde{u}/dx$. For this, we will not use the approximation but use the actual boundary condition given in eq. (2.1). After rearrangement, eq. (2.7) takes the following form:

$$\int_0^1 \frac{d\phi_i}{dx}\frac{d\tilde{u}}{dx}dx = \int_0^1 p(x)\phi_i(x)dx \\ + \frac{du}{dx}(1)\phi_i(1) - \frac{du}{dx}(0)\phi_i(0), \quad i = 1,\ldots,N. \tag{2.8}$$

Note that the orders of differentiation for both the approximate solution and the trial functions are identical. To rewrite the above equation explicitly in terms of unknown coefficients c_i, the approximation in eq. (2.3) is substituted to obtain

$$\int_0^1 \frac{d\phi_i}{dx}\sum_{j=1}^N c_j \frac{d\phi_j}{dx}dx = \int_0^1 p(x)\phi_i(x)dx \\ + \frac{du}{dx}(1)\phi_i(1) - \frac{du}{dx}(0)\phi_i(0), \quad i = 1,\ldots,N. \tag{2.9}$$

Note that the approximation is applied only to the LHS, not to the boundary terms on the RHS. The above N equations can be written in a compact form as

$$\sum_{j=1}^N K_{ij}c_j = F_i, \quad i = 1,\ldots,N, \tag{2.10}$$

or in matrix form as

$$\underset{(N \times N)}{[\mathbf{K}]} \underset{(N \times 1)}{\{\mathbf{c}\}} = \underset{(N \times 1)}{\{\mathbf{F}\}}, \tag{2.11}$$

where the matrices $[\mathbf{K}]$ and $\{\mathbf{F}\}$ are defined as

$$K_{ij} = \int_0^1 \frac{d\phi_i}{dx} \frac{d\phi_j}{dx} dx, \tag{2.12}$$

and

$$F_i = \int_0^1 p(x)\phi_i(x)dx + \frac{du}{dx}(1)\phi_i(1) - \frac{du}{dx}(0)\phi_i(0). \tag{2.13}$$

In eq. (2.11) the numbers in parentheses below the matrices refer to the size of the respective matrices. Note that $[\mathbf{K}]$ is symmetric, as $K_{ij} = K_{ji}$. The equations (2.10) are solved to determine the N unknown coefficients c_i in the approximate solution. We will demonstrate the method for the problem defined by eq. (2.1) in the following example.

EXAMPLE 2.3 *Galerkin solution of a second-order differential equation, case 1*

Solve the differential equation of eq. (2.1) for $p(x) = 1$. Use two trial functions, $\phi_1(x) = x$ and $\phi_2(x) = x^2$.

SOLUTION Since $N = 2$, the approximate solution takes the following form:

$$\tilde{u}(x) = \sum_{i=1}^2 c_i \phi_i(x) = c_1 x + c_2 x^2. \tag{2.14}$$

We will discuss the method of choosing the trial functions later. Note that the approximation above satisfies the essential boundary condition $\tilde{u}(0) = 0$. On the other hand, the natural boundary condition, $\tilde{u}'(1) = c_1 + 2c_2$, may or may not be satisfied, depending on the two coefficients. The derivatives of the trial functions are $\phi_1'(x) = 1$ and $\phi_2'(x) = 2x$. Substituting for ϕ_i and its derivatives in eq. (2.12), the components of matrix $[\mathbf{K}]$ can be obtained as

$$K_{11} = \int_0^1 \left(\phi_1'\right)^2 dx = 1$$

$$K_{12} = K_{21} = \int_0^1 \left(\phi_1'\phi_2'\right)dx = 1$$

$$K_{22} = \int_0^1 \left(\phi_2'\right)^2 dx = \frac{4}{3}.$$

In a similar way, the components of vector $\{\mathbf{F}\}$ can be calculated by substituting for ϕ_i and the boundary condition $du(1)/dx = 1$ in eq. (2.13) as

$$F_1 = \int_0^1 \phi_1(x)dx + \phi_1(1) - \frac{du}{dx}(0)\phi_1(0) = \frac{3}{2}$$

$$F_2 = \int_0^1 \phi_2(x)dx + \phi_2(1) - \frac{du}{dx}(0)\phi_2(0) = \frac{4}{3}.$$

Note that the value of $du(0)/dx$ is not required because the trial functions are zero at $x = 0$. Thus, the following matrix $[\mathbf{K}]$ and vector $\{\mathbf{F}\}$, respectively, can be obtained

$$[\mathbf{K}] = \frac{1}{3}\begin{bmatrix} 3 & 3 \\ 3 & 4 \end{bmatrix},$$

and

$$\{\mathbf{F}\} = \frac{1}{6}\begin{Bmatrix} 9 \\ 8 \end{Bmatrix}.$$

Solving for $\{\mathbf{c}\} = [\mathbf{K}]^{-1}\{\mathbf{F}\}$, we obtain $c_1 = 2$ and $c_2 = -\frac{1}{2}$. Thus, the approximate solution becomes

$$\tilde{u}(x) = 2x - \frac{x^2}{2}. \tag{2.15}$$

The exact solution for this case can be obtained by integrating eq. (2.1) twice and using the boundary conditions to evaluate the arbitrary constants of integration. One can note that this is also the exact solution to the problem. This happened because the approximate solution contains the two terms x and x^2, which are also in the exact solution. ▄▄

EXAMPLE 2.4   *Galerkin solution of a second-order differential equation, case 2*

Solve the differential equation of eq. (2.1) for $p(x) = x$. Use the same trial functions as in example 2.3.

SOLUTION   We will use the same form of approximation as given in eq. (2.14). This will change only vector $\{\mathbf{F}\}$, and the matrix $[\mathbf{K}]$ will remain the same. Vector $\{\mathbf{F}\}$ is calculated as

$$\{\mathbf{F}\} = \frac{1}{12}\begin{Bmatrix} 16 \\ 15 \end{Bmatrix}.$$

Solving for $\{\mathbf{c}\}$ we obtain $c_1 = \frac{19}{12}$ and $c_2 = -\frac{1}{4}$. Then the approximate solution becomes

$$\tilde{u}(x) = \frac{19}{12}x - \frac{x^2}{4}.$$

The exact solution for this case can be obtained by integrating eq. (2.1) twice and using the boundary conditions to evaluate the arbitrary constants of integration. The exact solution is

$$u(x) = \frac{3}{2}x - \frac{x^3}{6}.$$

The exact and approximate solutions and their derivatives are compared in figure 2.3. From figure 2.3, one can note that the results for $u(x)$ and $\tilde{u}(x)$ agree quite well in the entire domain of the problem. Nevertheless, there is some difference in the results for the derivatives of $u(x)$ and $\tilde{u}(x)$, the maximum error being about 8%. In general, the approximate solution is exact at the essential boundary because the trial functions of the approximate solution satisfy the essential boundary condition. On the other hand, the approximate solution may not satisfy the

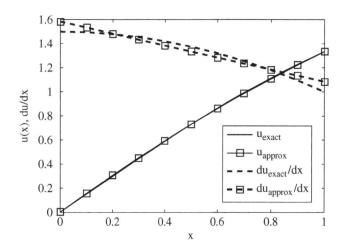

Figure 2.3 Comparison of exact solution and approximate solution and their derivatives for example 2.4

natural boundary condition because it is a part of the weighted residual. In this particular example, $\{\tilde{u}'(1) = 13/12\} \neq \{u'(1) = 1\}$. ■

EXAMPLE 2.5   *Galerkin solution of a second-order differential equation, case 3*

Solve the differential equation given below for $p(x) = x^2$.

$$\frac{d^2u}{dx^2} + p(x) = 0, \quad 0 \le x \le 1$$

$$\left.\begin{array}{l} u(0) = 0 \\ u(1) = 0 \end{array}\right\} \quad \text{Boundary conditions.}$$

SOLUTION   Note that both boundary conditions are essential boundary conditions, and so the approximate solution must satisfy both boundary conditions. Let the approximate solution be

$$\tilde{u}(x) = \sum_{i=1}^{2} c_i \phi_i(x) = c_1 x(x-1) + c_2 x^2(x-1). \tag{2.16}$$

From eq. (2.3) one can recognize that the trial functions are

$$\phi_1(x) = x(x-1), \quad \phi_2(x) = x^2(x-1).$$

These trial solutions are chosen in order to satisfy the two essential boundary conditions simultaneously; $\tilde{u}(0) = \tilde{u}(1) = 0$. The derivatives of the trial functions are

$$\phi_1'(x) = 2x - 1, \quad \phi_2'(x) = 3x^2 - 2x.$$

Substituting for ϕ_i' in eq. (2.12) we derive the components of matrix $[\mathbf{K}]$ as

$$K_{11} = \int_0^1 (\phi_1')^2 dx = \frac{1}{3}$$

$$K_{12} = K_{21} = \int_0^1 (\phi_1' \phi_2') dx = \frac{1}{6}$$

$$K_{22} = \int_0^1 (\phi_2')^2 dx = \frac{2}{15}.$$

Substituting $p(x) = x^2$ in eq. (2.13), we derive vector $\{\mathbf{F}\}$ as

$$F_1 = \int_0^1 x^2 \phi_1(x) dx + \frac{du}{dx}(1)\phi_1(1) - \frac{du}{dx}(0)\phi_1(0) = -\frac{1}{20}$$

$$F_2 = \int_0^1 x^2 \phi_2(x) dx + \frac{du}{dx}(1)\phi_2(1) - \frac{du}{dx}(0)\phi_2(0) = -\frac{1}{30}.$$

Note that we do not know the boundary values of du/dx. Still, we are able to evaluate F_1 and F_2 because the values of ϕ_1 and ϕ_2 at the boundaries are equal to zero.

Solving $[\mathbf{K}]\{\mathbf{c}\} = \{\mathbf{F}\}$, we obtain $c_1 = -\frac{1}{15}$ and $c_2 = -\frac{1}{6}$. Substituting for c_i in eq. (2.16) we obtain the expression for the approximate solution as

$$\tilde{u}(x) = \frac{x}{15} + \frac{x^2}{10} - \frac{x^3}{6}.$$

The exact solutions are as follows:

$$u(x) = \frac{1}{12}x(1 - x^3).$$

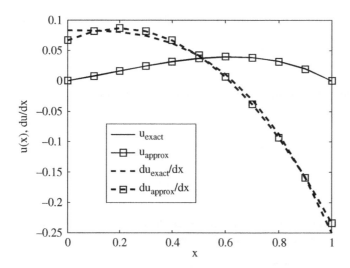

Figure 2.4 Comparison of $u(x)$ and its derivative obtained by the Galerkin method for example 2.5

It may be seen that the exact solution is a quartic polynomial, whereas the approximate solution is cubic in x. The two solutions and their respective derivatives are compared in figure 2.4.

The comparison between $u(x)$ and $\tilde{u}(x)$ is excellent. However, the derivatives deviate from one another. This is typical of approximate methods such as Galerkin method and in general, the finite element method. If we assume a three-term solution that includes an x^4 term in the approximation, then the approximate solution will be the same as the exact solution. ■

2.3 HIGHER-ORDER DIFFERENTIAL EQUATIONS

In the second-order differential equation in eq. (2.1), we perform integration by parts once so that the orders of differentiation for both the approximate solution and the trial functions are identical. We will now illustrate the Galerkin method for a fourth-order ordinary differential equation. Consider the following equation:

$$\frac{d^4 w}{dx^4} - p(x) = 0, \quad 0 \le x \le L, \tag{2.17}$$

which is the governing equation for deflection $w(x)$ of a beam of length L subjected to transverse loading $p(x)$. The essential boundary conditions for this problem are specifying w or the derivative dw/dx at $x = 0$ and/or $x = L$. The natural boundary conditions are known values of d^2w/dx^2 and d^3w/dx^3. Since it is a fourth-order equation, we need four boundary conditions, two at each end of the domain. For the purpose of illustration, assume the boundary conditions as follows:

$$
\left.
\begin{aligned}
w(0) &= 0 \\[4pt]
\frac{dw}{dx}(0) &= 0
\end{aligned}
\right\} \quad \text{Essential boundary conditions}
$$

$$
\left.
\begin{aligned}
\frac{d^2 w}{dx^2}(L) &= M \\[4pt]
\frac{d^3 w}{dx^3}(L) &= -V
\end{aligned}
\right\} \quad \text{Natural boundary conditions.}
\tag{2.18}
$$

In eq. (2.18) the first two are essential boundary conditions, and the last two are natural boundary conditions. Physically, the beam is clamped at $x=0$, and moment M and shear force V are applied at $x=L$.

As before we assume the approximate solution as $w(x) = \tilde{w}(x)$. Substituting in the governing equation, we obtain the residual as

$$\frac{d^4\tilde{w}}{dx^4} - p(x) = R(x). \tag{2.19}$$

We assume the approximate solution in a series form as

$$\tilde{w}(x) = \sum_{i=1}^{N} c_i \phi_i(x), \tag{2.20}$$

such that $\tilde{w}(x)$ satisfies the essential boundary conditions. The integral of the weighted residual takes the form

$$\int_0^L \left(\frac{d^4\tilde{w}}{dx^4} - p(x) \right) \phi_i(x) dx = 0, \ i = 1, \ldots, N, \tag{2.21}$$

where ϕ_i are used as the weight functions. We will take the term containing $p(x)$ to the RHS, and use integration by parts to reduce the order of differentiation of $\tilde{w}$:

$$\frac{d^3\tilde{w}}{dx^3} \phi_i \Big|_0^L - \frac{d^2\tilde{w}}{dx^2} \frac{d\phi_i}{dx} \Big|_0^L + \int_0^L \frac{d^2\tilde{w}}{dx^2} \frac{d^2\phi_i}{dx^2} dx$$
$$= \int_0^L p(x) \phi_i(x) dx, \ i = 1, \ldots, N. \tag{2.22}$$

It should be noted that we have used integration by parts twice to reduce the order of differentiation from four to two. Equation (2.22) can be rewritten as

$$\int_0^L \frac{d^2\tilde{w}}{dx^2} \frac{d^2\phi_i}{dx^2} dx = \int_0^L p(x)\phi_i(x) dx$$
$$- \frac{d^3\tilde{w}}{dx^3} \phi_i \Big|_0^L + \frac{d^2\tilde{w}}{dx^2} \frac{d\phi_i}{dx} \Big|_0^L, \ i = 1, \ldots, N. \tag{2.23}$$

Substituting for $\tilde{w}(x)$ from eq. (2.20) into eq. (2.23), we obtain

$$\int_0^L \sum_{j=1}^{N} c_j \frac{d^2\phi_j}{dx^2} \frac{d^2\phi_i}{dx^2} dx = \int_0^L p(x)\phi_i(x) dx$$
$$- \frac{d^3\tilde{w}}{dx^3} \phi_i \Big|_0^L + \frac{d^2\tilde{w}}{dx^2} \frac{d\phi_i}{dx} \Big|_0^L, \ i = 1, \ldots, N. \tag{2.24}$$

Note that the assumed solution is substituted only on the LHS of the above equations. For the boundary terms involving $\tilde{w}$, we will use the actual boundary conditions. Equations (2.24) can be written in a compact form as

$$\underset{N \times N}{[\mathbf{K}]} \underset{N \times 1}{\{\mathbf{c}\}} = \underset{N \times 1}{\{\mathbf{F}\}}, \tag{2.25}$$

where $\{\mathbf{c}\}$ is the column vector of unknown coefficients c_j, and the elements of matrices $[\mathbf{K}]$ and $\{\mathbf{F}\}$, respectively, are defined as

$$K_{ij} = \int_0^L \frac{d^2\phi_i}{dx^2} \frac{d^2\phi_j}{dx^2} dx, \tag{2.26}$$

and

$$F_i = \int_0^L p(x)\phi_i(x)\mathrm{d}x - \frac{\mathrm{d}^3 w}{\mathrm{d}x^3}\phi_i\bigg|_0^L + \frac{\mathrm{d}^2 w}{\mathrm{d}x^2}\frac{\mathrm{d}\phi_i}{\mathrm{d}x}\bigg|_0^L. \tag{2.27}$$

EXAMPLE 2.6 *Galerkin solution of a fourth-order differential equation*

Solve the fourth-order differential equation in eq. (2.17) with the boundary conditions in eq. (2.18). Assume $L = 1$, $p(x) = 1$, $V = 1$, and $M = 2$.

SOLUTION   To proceed further with the illustration, we assume that $N = 2$ and assume the trial functions ϕ_i as follows:

$$\phi_1 = x^2, \quad \phi_2 = x^3.$$

Note that the functions ϕ_i are such that $\tilde{w}(x)$ satisfies the essential boundary conditions given in eq. (2.18). The second derivatives of the trial functions become $\phi_1'' = 2$ and $\phi_2'' = 6x$. Substituting these two derivatives in eq. (2.26), we obtain the components of the matrix $[\mathbf{K}]$ as

$$K_{11} = \int_0^1 \left(\phi_1''\right)^2 \mathrm{d}x = 4,$$

$$K_{12} = K_{21} = \int_0^1 \left(\phi_1''\phi_2''\right)\mathrm{d}x = 6,$$

$$K_{22} = \int_0^1 \left(\phi_2''\right)^2 \mathrm{d}x = 12.$$

Thus, the matrix $[\mathbf{K}]$ becomes

$$[\mathbf{K}] = \begin{bmatrix} 4 & 6 \\ 6 & 12 \end{bmatrix}.$$

Similarly, substituting for ϕ_i and the natural boundary conditions in eq. (2.27), we obtain the components of vector $\{\mathbf{F}\}$ as

$$F_1 = \int_0^1 x^2 \mathrm{d}x + V\phi_1(1) + \frac{\mathrm{d}^3 w(0)}{\mathrm{d}x^3}\phi_1(0) + M\phi_1'(1) - \frac{\mathrm{d}^2 w(0)}{\mathrm{d}x^2}\phi_1'(0) = \frac{16}{3},$$

$$F_2 = \int_0^1 x^3 \mathrm{d}x + V\phi_2(1) + \frac{\mathrm{d}^3 w(0)}{\mathrm{d}x^3}\phi_2(0) + M\phi_2'(1) - \frac{\mathrm{d}^2 w(0)}{\mathrm{d}x^2}\phi_2'(0) = \frac{29}{4}.$$

Thus, vector $\{\mathbf{F}\}$ becomes

$$\left\{ \begin{matrix} F_1 \\ F_2 \end{matrix} \right\} = \left\{ \begin{matrix} \dfrac{16}{3} \\ \dfrac{29}{4} \end{matrix} \right\}.$$

Solving for c_j using eq. (2.25) and substituting in eq. (2.20), we obtain the approximate solution as

$$\tilde{w}(x) = \frac{41}{24}x^2 - \frac{1}{4}x^3. \tag{2.28}$$

The exact solution is obtained by integrating eq. (2.17) four times and applying the boundary conditions to evaluate the arbitrary constants of integration:

$$w(x) = \frac{1}{24}x^4 - \frac{1}{3}x^3 + \frac{7}{4}x^2. \tag{2.29}$$

Table 2.1 Comparison of approximate and exact solutions

	$w(x)$			$w'(x)$			$w''(x)$			$w'''(x)$		
x	exact	approx.	% error	exact	approx.	% error	exact	approx.	% error	exact	approx.	% error
0.0	0.000	0.000	0.00	0.000	0.000	0.00	3.500	3.42	−2.38	−2.0	−1.5	−25.0
0.1	0.017	0.017	−1.97	0.340	0.334	−1.76	3.305	3.27	−1.16	−1.9	−1.5	−21.1
0.2	0.067	0.066	−1.58	0.661	0.653	−1.21	3.120	3.12	−0.11	−1.8	−1.5	−16.7
0.3	0.149	0.147	−1.23	0.965	0.958	−0.73	2.945	2.97	0.74	−1.7	−1.5	−11.8
0.4	0.260	0.257	−0.92	1.251	1.247	−0.32	2.780	2.82	1.32	−1.6	−1.5	−6.3
0.5	0.398	0.396	−0.65	1.521	1.521	0.00	2.625	2.67	1.59	−1.5	−1.5	0.0
0.6	0.563	0.561	−0.43	1.776	1.780	0.23	2.480	2.52	1.48	−1.4	−1.5	7.1
0.7	0.753	0.751	−0.24	2.017	2.024	0.35	2.345	2.37	0.92	−1.3	−1.5	15.4
0.8	0.966	0.965	−0.11	2.245	2.253	0.36	2.220	2.22	−0.15	−1.2	−1.5	25.0
0.9	1.202	1.202	−0.03	2.462	2.468	0.24	2.105	2.07	−1.82	−1.1	−1.5	36.4
1.0	1.458	1.458	0.00	2.667	2.667	0.00	2.000	1.92	−4.17	−1.0	−1.5	50.0

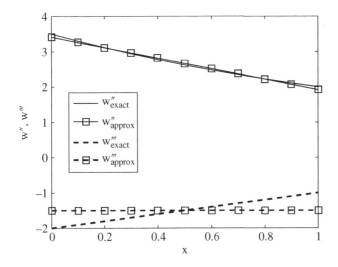

Figure 2.5 Comparison of w'' and w''' for the beam problem in example 2.6

The approximate and exact solutions are compared in table 2.1 by calculating the functions and its derivatives. One can note that the solution obtained using the Galerkin method compares very well to the exact solution up to the second derivative. The error in the third derivative is significant, as high as 50%. In fact, the third derivative is a constant for the approximate solution whereas for the exact solution it is a linear function in x. The second and third derivatives are compared in figure 2.5. ▪

2.4 FINITE ELEMENT APPROXIMATION

There are different ways of explaining the finite element method, but it can be explained as *a technique for obtaining approximate solutions of boundary-value problems by partitioning the domain into a set of simple shapes, called finite elements*. In particular, we use simple polynomials, such as a linear polynomial, within an element to approximate the solution. Since elements are connected to each other, these

(a)

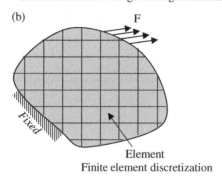

$$\begin{cases} \dfrac{\partial \sigma_{xx}}{\partial x} + \dfrac{\partial \tau_{xy}}{\partial y} + b_x = 0 \\[2mm] \dfrac{\partial \tau_{xy}}{\partial x} + \dfrac{\partial \sigma_{yy}}{\partial y} + b_y = 0 \end{cases}$$

Structural domain with governing differential equations

(b)

Element
Finite element discretization

Figure 2.6 Boundary-value problem in solid mechanics

polynomials are connected to each other at the element boundary and yield piecewise polynomials to approximate the solutions.

A boundary-value problem is a differential equation together with a set of additional constraints, called the boundary conditions. Many engineering problems are expressed in the form of differential equations, such as solid mechanics, electrostatics, wave propagation, and so forth. Figure 2.6(a) shows an illustration of a boundary-value problem in solid mechanics where the deformation behavior of the solid is expressed in terms of partial differential equations. In the boundary-value problem, conditions have to be given in all boundaries. In the figure, some boundary has prescribed displacements (essential boundary condition), while another boundary has prescribed forces (natural boundary condition).

In general, the governing differential equations, such as those shown in figure 2.6(a), can be solved analytically only for a simple geometry, such as a rectangular domain. Therefore, it would be difficult, if not impossible, to solve the governing differential equations for arbitrary geometry with complicated boundary conditions. In the finite element method, the domain is divided into many simple-shaped elements as shown in figure 2.6(b). Within a single element, it might be easier to solve the governing differential equations, but still, it is not trivial to solve the boundary-value problem because we do not know the boundary conditions for each element. Instead of solving the governing differential equations directly, the finite element method approximates the solution using simple polynomials and applies the weighted residual method in the previous section to find unknown coefficients. In this section, we will explain the above-mentioned concept using one-dimensional boundary-value problems, such as the one given in eq. (2.1). The extension to higher-dimensional problems will be presented in later chapters.

2.4.1 Domain Discretization

In the previous sections, we approximated the solution to the differential equation by a series of functions in the entire domain of the problem. In such a case, it is difficult to obtain the trial functions that satisfy the essential boundary conditions. An important idea of the finite element method is to divide the entire

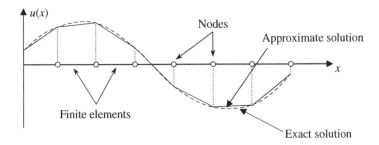

Figure 2.7 Piecewise linear approximation of the solution for a one-dimensional problem

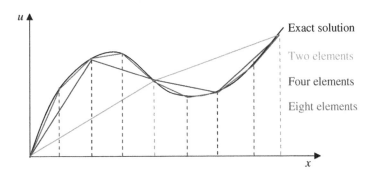

Figure 2.8 Convergence of one-dimensional finite element solution

domain into a set of simple sub-domains or *finite elements*. The finite elements are connected with adjacent elements by sharing their nodes. Then, within each finite element, the solution is approximated in a simple polynomial form. For example, let us assume that the domain is one-dimensional, and the exact solution $u(x)$ is given as a dashed curve in figure 2.7. When the entire domain is divided into sub-domains (finite elements), it is possible to approximate the solution using piecewise continuous linear polynomials as shown in figure 2.7. Within each element, the approximate solution is linear. Two adjacent elements have the same solution value at the shared node.

An important property of the finite element method is that the approximate solution converges to the exact solution as the number of elements increases. As can be seen in figure 2.8, when a larger number of elements are used, the approximate piecewise linear solution will converge to the exact solution. In addition, the approximation can be more accurate if higher-order polynomials are used in each element. We will discuss the convergence of finite element solutions detail in chapter 9.

Even if we only show the one-dimensional problem in this section, various types of finite elements can be used depending on the domain that needs to be discretized and the order of polynomials used to approximate the solution. Table 2.2 illustrates several types of finite elements that are often used in one-, two-, and three-dimensional problems.

After dividing the domain into finite elements, the integrations to compute $[\mathbf{K}]$ and $\{\mathbf{F}\}$, for instance, as in eqs. (2.26) and (2.27), are performed over each element. For example, let us assume that the one-dimensional domain $(0, 1)$ is divided into 10 equal-sized finite elements. Then, the integral in eq. (2.26) can be written as a summation of integrals over each element:

$$\int_0^1 \square \, dx = \int_0^{0.1} \square \, dx + \int_{0.1}^{0.2} \square \, dx + \cdots + \int_{0.9}^1 \square \, dx, \tag{2.30}$$

where $\square$ is the integrand. Because the integral over the domain is divided into integrals over each element, it is important not to overlap elements and not to leave any empty space in the domain.

Table 2.2   Different types of finite elements

Element	Name
	1D linear element
	2D triangular element
	2D rectangular element
	3D tetrahedron element
	3D hexahedron element

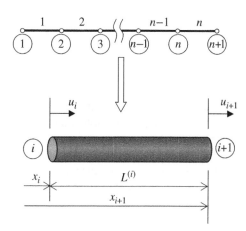

Figure 2.9 Domain discretization of one-dimensional problem

2.4.2   Trial Solution

After the domain is divided into a set of simple-shaped elements, the solution within an element is approximated in the form of simple polynomials. Let us consider a one-dimensional problem in eq. (2.1) in which the domain is discretized by n number of elements, as shown in figure 2.9. For this specific example, each element is composed of two end nodes. The trial solution is constructed in the element using the solution values at these nodes.

For example, the i-th element connects two nodes at $x = x_i$ and $x = x_{i+1}$. If we want to interpolate the solution using two nodal values, then the linear polynomial is the appropriate choice because it has two unknowns. Thus, the solution is approximated by

$$\tilde{u}(x) = a_0 + a_1 x, \quad x_i \leq x \leq x_{i+1}. \tag{2.31}$$

Note that the trial solution in the above equation is only defined in the i-th element. Although we can determine two coefficients a_0 and a_1, they do not have a physical meaning. Instead, the unknown coefficients a_0 and a_1 in eq. (2.31) will be expressed in terms of the nodal solutions $\tilde{u}(x_i)$ and $\tilde{u}(x_{i+1})$. By substituting these two nodal values, we have

$$\begin{cases} \tilde{u}(x_i) = u_i = a_0 + a_1 x_i \\ \tilde{u}(x_{i+1}) = u_{i+1} = a_0 + a_1 x_{i+1}, \end{cases} \tag{2.32}$$

where u_i and u_{i+1} are the solution values at the two end nodes. Then, by solving eq. (2.32), two unknown coefficients, a_0 and a_1, can be represented by nodal solution, u_i and u_{i+1}. After substituting the two coefficients into eq. (2.31), the approximate solution can be expressed in terms of the nodal solutions as

$$\tilde{u}(x) = \underbrace{\frac{x_{i+1} - x}{L^{(i)}}}_{N_i(x)} u_i + \underbrace{\frac{x - x_i}{L^{(i)}}}_{N_{i+1}(x)} u_{i+1}, \tag{2.33}$$

where $L^{(i)} = x_{i+1} - x_i$ is the length of the i-th element. Now, the approximate solution for $u(x)$ in eq. (2.31) can be rewritten as

$$\tilde{u}(x) = N_i(x)u_i + N_{i+1}(x)u_{i+1}, \quad x_i \le x \le x_{i+1}, \tag{2.34}$$

where the functions $N_i(x)$ and $N_{i+1}(x)$ are called *interpolation functions* for obvious reasons. The expression in eq. (2.34) shows that the solution $\tilde{u}(x)$ is interpolated using its nodal values u_i and u_{i+1}. $N_i(x) = 1$ at $x = x_i$ and $N_i(x) = 0$ at $x = x_{i+1}$, while $N_{i+1}(x) = 1$ at $x = x_{i+1}$ and $N_{i+1}(x) = 0$ at $x = x_i$. The interpolation functions $N_i(x)$ and $N_{i+1}(x)$ are also called *shape functions*, a term used in solid mechanics, as the functions describe the deformed shape of a solid or structure.

Note that the approximate solution in eq. (2.34) is similar to that of the weighted residual method in eq. (2.3). In this case, the interpolation function corresponds to the trial function. The difference is that the approximation in eq. (2.34) is written in terms of solution values at nodes, whereas the coefficients c_i in the approximation in eq. (2.3) do not have any physical meaning. The detailed relation between interpolation and trial functions will be discussed in the following section.

In order to explain the accuracy of approximation, the interpolated solution and its gradients for two continuous elements are illustrated in figure 2.10. Note that in this particular interpolation, the solution is approximated by a piecewise linear function, and so its gradient is constant within an element. Accordingly, the gradients are not continuous at the element interface. In structural problems, the solution $u(x)$ often represents the displacement of the structure, and its gradient is stress or strain. Thus, the approximation yields a continuous displacement but discontinuous stress and strain between elements. Many commercial finite element programs provide the stress values at nodes and display a smooth change of stresses across elements. However, users must be careful because these nodal stress values are the average of values for different elements connected to a node.

2.4.3 Galerkin Method

Before we apply the Galerkin method in the finite element scheme, we have to identify the relation between the interpolation functions $N_i(x)$ and the trial functions $\phi_i(x)$. One major difference is that

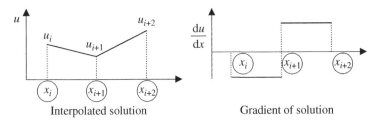

Figure 2.10 Interpolated solution and its gradient

the interpolation functions in eq. (2.34) are only defined within an element, while the trial functions are defined in the entire domain. Remember we expressed the approximate solution in terms of trial functions $\phi_i(x)$ with unknown coefficients c_i. The finite element approximation in eq. (2.34) can also be written as

$$\tilde{u}(x) = \sum_{i=1}^{N_D} u_i \phi_i(x),\qquad(2.35)$$

where u_i are the nodal values of the functions, and N_D is the total number of nodes in the finite element model. In the case of one-dimensional problems with linear interpolation between nodes, $N_D = N_E + 1$, where N_E is the number of elements. Comparing eqs. (2.34) and (2.35), the expression for $\phi_i(x)$ can be written as

$$\phi_i(x) = \begin{cases} 0, & 0 \leq x \leq x_{i-1} \\ N_i^{(i-1)}(x) = \dfrac{x - x_{i-1}}{L^{(i-1)}}, & x_{i-1} < x \leq x_i \\ N_i^{(i)}(x) = \dfrac{x_{i+1} - x}{L^{(i)}}, & x_i < x \leq x_{i+1} \\ 0, & x_{i+1} < x \leq x_{N_D}. \end{cases} \qquad (2.36)$$

In the above equations, the superscripts $(i-1)$ and (i) denote the element numbers. Thus the major difference between the weighted residual method described in the previous section and the finite element method is that the function $\phi_i(x)$ does not exist in the entire domain, but it exists only in elements connected to node i.

The derivative of $\phi_i(x)$ is also derived as

$$\frac{\mathrm{d}\phi_i(x)}{\mathrm{d}x} = \begin{cases} 0, & 0 \leq x \leq x_{i-1} \\ \dfrac{1}{L^{(i-1)}}, & x_{i-1} < x \leq x_i \\ -\dfrac{1}{L^{(i)}}, & x_i < x \leq x_{i+1} \\ 0, & x_{i+1} < x \leq x_{N_D}. \end{cases} \qquad (2.37)$$

Function $\phi_i(x)$ and its derivative are plotted in figure 2.11. Note that the trial functions are continuous even if they are nonzero only for two neighboring elements. On the other hand, the derivatives of trial functions are discontinuous at the node. Since the approximate solution is a linear combination of the trial functions, the approximate solution will also be continuous, while the derivative of the solution will be discontinuous at the node.

Galerkin's method in the finite element scheme is in fact the weighted residual method where the weighting function is interpolated within each element using the shape function. A weak form for any boundary-value problem can be obtained by integrating the weighted residual over the entire domain and setting it equal to zero. The weighting function is interpolated as

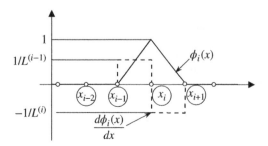

Figure 2.11 Function $\phi_i(x)$ and its derivative

$$W(x) = \sum_{i=1}^{N_D} w_i \phi_i(x).$$

Using this definition of the weighting function, the weak form can be derived using the weighted residual method as

$$\int_0^1 R(x) \left(\sum_{i=1}^{N_D} w_i \phi_i(x) \right) dx = 0.$$

This equation must be valid for arbitrary values of w_i which is only possible if

$$\int_0^1 R(x) \phi_i(x) dx = 0, \ i = 1, \dots, N.$$

Now we have N equations that must be satisfied for each $\phi_i(x)$. In Galerkin's method, the weak form is further expanded using integration by parts as we have seen earlier.

EXAMPLE 2.7   *Finite element solution of a differential equation*

Solve the differential equation given in eq. (2.1) for $p(x) = 1$ using the Galerkin finite element method. Assume that the domain $0 \le x \le 1$ is divided into two equal-length elements.

SOLUTION   Since the domain is divided into two elements of equal length, there exist three nodes located at $x = 0, 0.5$, and 1.0. The unknown coefficients—analogous to c_i in the previous examples—are the values of u at the nodes: u_1, u_2, and u_3. Then the approximate solution is written as

$$\tilde{u}(x) = u_1 \phi_1(x) + u_2 \phi_2(x) + u_3 \phi_3(x), \tag{2.38}$$

and the trial functions ϕ_i are defined as follows using eq. (2.36):

$$\phi_1(x) = \begin{cases} 1 - 2x, & 0 \le x \le 0.5 \\ 0, & 0.5 < x \le 1 \end{cases}$$

$$\phi_2(x) = \begin{cases} 2x, & 0 \le x \le 0.5 \\ 2 - 2x, & 0.5 < x \le 1 \end{cases}$$

$$\phi_3(x) = \begin{cases} 0, & 0 \le x \le 0.5 \\ -1 + 2x, & 0.5 < x \le 1. \end{cases}$$

The three functions are plotted in figure 2.12. It may be noted that trial functions ϕ_i are such that the variation of $\tilde{u}(x)$ is linear between nodes.

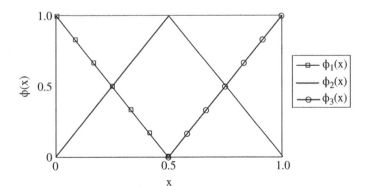

Figure 2.12 Trial function $\phi_i(x)$ for two equal-length finite elements

Because the trial functions are piecewise linear polynomials, the derivatives of ϕ_i are as follows:

$$\frac{d\phi_1(x)}{dx} = \begin{cases} -2, & 0 \leq x \leq 0.5 \\ 0, & 0.5 < x \leq 1 \end{cases}$$

$$\frac{d\phi_2(x)}{dx} = \begin{cases} 2, & 0 \leq x \leq 0.5 \\ -2, & 0.5 < x \leq 1 \end{cases}$$

$$\frac{d\phi_3(x)}{dx} = \begin{cases} 0, & 0 \leq x \leq 0.5 \\ 2, & 0.5 < x \leq 1. \end{cases}$$

We will evaluate $[\mathbf{K}]$ and $\{\mathbf{F}\}$ using eq. (2.12) for the case $p(x) = 1$. Note that $N = 3$ for this case. We will show the procedures for K_{12} and K_{22}. Evaluation of other terms of $[\mathbf{K}]$ is left as an exercise for the reader. Since two elements are involved, all integrations are divided into two parts.

$$\begin{aligned} K_{12} &= \int_0^1 \frac{d\phi_1}{dx}\frac{d\phi_2}{dx}dx \\ &= \int_0^{0.5} (-2)(2)dx + \int_{0.5}^1 (0)(-2)dx \\ &= -2, \end{aligned}$$

$$\begin{aligned} K_{22} &= \int_0^1 \frac{d\phi_2}{dx}\frac{d\phi_2}{dx}dx \\ &= \int_0^{0.5} 4dx + \int_{0.5}^1 4dx \\ &= 4. \end{aligned}$$

The first component of $\{\mathbf{F}\}$ can be calculated from eq. (2.13),

$$\begin{aligned} F_1 &= \int_0^{0.5} 1 \times (1 - 2x)dx + \int_{0.5}^1 1 \times (0)dx + \frac{du}{dx}(1)\phi_1(1) - \frac{du}{dx}(0)\phi_1(0) \\ &= 0.25 - \frac{du}{dx}(0). \end{aligned}$$

It should be noted that F_1 could not be computed as we do not know the boundary condition $du(0)/dx$. We will use F_1 as an unknown term. Computation of F_2 and F_3 is straightforward as shown below:

$$F_2 = \int_0^{0.5} 2xdx + \int_{0.5}^1 (2 - 2x)dx + \frac{du}{dx}(1)\phi_2(1) - \frac{du}{dx}(0)\phi_2(0) = 0.5,$$

$$F_3 = \int_0^{0.5} 0dx + \int_{0.5}^1 (-1 + 2x)dx + \frac{du}{dx}(1)\phi_3(1) - \frac{du}{dx}(0)\phi_3(0) = 1.25.$$

Note that we have used the boundary condition $du(1)/dx = 1$ in computing F_3.

After calculating all necessary components, the final matrix equation for the finite element analysis becomes

$$\begin{bmatrix} 2 & -2 & 0 \\ -2 & 4 & -2 \\ 0 & -2 & 2 \end{bmatrix} \begin{Bmatrix} u_1 \\ u_2 \\ u_3 \end{Bmatrix} = \begin{Bmatrix} F_1 \\ 0.5 \\ 1.25 \end{Bmatrix}, \tag{2.39}$$

where the nodal solutions u_1, u_2, and u_3 are required. We will discard the first of the three equations, as we do not know the RHS of that equation. This is called striking the row. Refer to the explanation above eq. (1.11) in chapter 1.

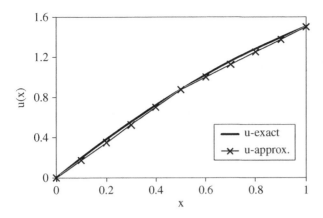

Figure 2.13 Exact solution $u(x)$ and finite element solution $\tilde{u}(x)$

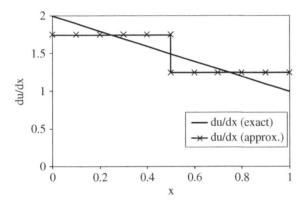

Figure 2.14 Derivatives of the exact and finite element solutions

Furthermore, since the essential boundary condition $u_1 = 0$ is given, we can eliminate the first column (striking the column). Then the last two equations take the form:

$$\begin{bmatrix} 4 & -2 \\ -2 & 2 \end{bmatrix} \begin{Bmatrix} u_2 \\ u_3 \end{Bmatrix} = \begin{Bmatrix} 0.5 \\ 1.25 \end{Bmatrix}. \tag{2.40}$$

Note that the matrix in eq. (2.39) is not positive definite, but the one in eq. (2.40) is. Solving eq. (2.40), we obtain the nodal solutions as: $u_2 = 0.875$ and $u_3 = 1.5$. Note that the nodal solution at node 1 is given as the essential boundary condition; that is, $u_1 = 0.0$. Once the nodal solutions are available, the approximate solution can be obtained from eq. (2.38) as

$$\tilde{u}(x) = \begin{cases} 1.75x, & 0 \le x \le 0.5 \\ 0.25 + 1.25x, & 0.5 \le x \le 1. \end{cases} \tag{2.41}$$

Due to the linear trial functions, the approximate finite element solution $\tilde{u}(x)$ is composed of piecewise linear polynomials. It is coincidental that the exact solution and the FE solution agree at the nodes ($x = 0.5$ and $x = 1$). The approximate solution from eq. (2.41) and the exact solution from eq. (2.15) are compared in figure 2.13. The maximum error is about 8%.

The derivative du/dx is plotted in figure 2.14, and one can note a large discrepancy in the derivatives of the two solutions. Since the approximate solution is a piecewise linear polynomial, its derivative is constant in each element. Accordingly, the derivative is not continuous at the element boundary. ◼

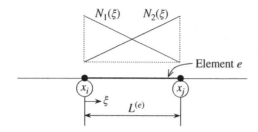

Figure 2.15 One-dimensional finite element with interpolation functions

2.4.4 Formal Procedure

Although example 2.7 illustrated the implementation of the Galerkin method in the finite element scheme, it is not still general enough to implement in computer code. Hence, we will present a slightly modified form of the Galerkin finite element formulation.

In this method, we apply the Galerkin method to one element at a time. Let us consider a general element, say element e in figure 2.15. It has two nodes, say, i and j. The choice of nodes i and j is such that $x_j > x_i$. We also introduce a local coordinate ξ such that $\xi = 0$ at node i, and $\xi = 1$ at node j. The mapping relation between x and ξ for element e is given by

$$x = x_i(1-\xi) + x_j\xi \quad \text{or} \quad \xi = \frac{x-x_i}{x_j-x_i} = \frac{x-x_i}{L^{(e)}}, \tag{2.42}$$

where $L^{(e)}$ denotes the length of the element. As will be seen later, the use of the local coordinate ξ is a matter of convenience as it helps in expressing the interpolation functions and their integrations in an elegant manner. The Jacobian relation between x and ξ can be obtained as

$$\frac{dx}{d\xi} = x_j - x_i = L^{(e)}, \quad \frac{d\xi}{dx} = \frac{1}{L^{(e)}}. \tag{2.43}$$

The approximate solution within element e is given by

$$\tilde{u}(x) = u_i N_1(x) + u_j N_2(x), \tag{2.44}$$

where the interpolations functions N_1 and N_2 can be conveniently expressed as a function of the variable ξ

$$\begin{aligned} N_1(\xi) &= (1-\xi) \\ N_2(\xi) &= \xi. \end{aligned} \tag{2.45}$$

In terms of the variable x, the interpolation functions take the form

$$\begin{aligned} N_1(x) &= \left(1 - \frac{x-x_i}{L^{(e)}}\right) \\ N_2(x) &= \frac{x-x_i}{L^{(e)}}. \end{aligned} \tag{2.46}$$

We will use the interpolation functions in eq. (2.45). Using the chain rule of differentiation, the derivatives of N_i are:

$$\begin{aligned} \frac{dN_1}{dx} &= \frac{dN_1}{d\xi}\frac{d\xi}{dx} = -\frac{1}{L^{(e)}} \\ \frac{dN_2}{dx} &= \frac{dN_2}{d\xi}\frac{d\xi}{dx} = +\frac{1}{L^{(e)}}. \end{aligned} \tag{2.47}$$

An important advantage of introducing the local coordinate ξ is that all elements will have the same shape functions as in eq. (2.45). Therefore, there is no need to build different shape functions for different elements. The only difference is the Jacobian relationship given in eq. (2.43), which is, in this case, the length of the element.

One can easily verify that the interpolation functions satisfy the following relation:

$$
\begin{aligned}
N_1(x_i) &= 1, \quad N_1(x_j) = 0 \\
N_2(x_i) &= 0, \quad N_2(x_j) = 1.
\end{aligned}
\tag{2.48}
$$

In addition, $N_1(x)$ linearly decreases from x_i to x_j, while $N_2(x)$ linearly increases. One can also verify that the above functions yield

$$
\begin{aligned}
\tilde{u}(x_i) &= u_i \\
\tilde{u}(x_j) &= u_j,
\end{aligned}
\tag{2.49}
$$

where u_i and u_j are a nodal solution at nodes i and j, respectively. Equation (2.49) is an important property of interpolation.

From eq. (2.44) the derivative of $\tilde{u}(x)$ is obtained as

$$
\frac{d\tilde{u}}{dx} = u_i \frac{dN_1}{dx} + u_j \frac{dN_2}{dx}.
\tag{2.50}
$$

The above equation can be written in matrix form, using the chain rule of differentiation in eq. (2.47), as

$$
\frac{d\tilde{u}}{dx} = \left\{ \frac{dN_1}{dx} \; \frac{dN_2}{dx} \right\} \left\{ \begin{array}{c} u_1 \\ u_2 \end{array} \right\} = \frac{1}{L^{(e)}} \left\{ \frac{dN_1}{d\xi} \; \frac{dN_2}{d\xi} \right\} \left\{ \begin{array}{c} u_1 \\ u_2 \end{array} \right\}.
\tag{2.51}
$$

We apply the Galerkin method described in section 2.1 at the element level. Then, N_1 and N_2 will be the two trial functions. A set of two equations similar to eq. (2.9) can be written for element e:

$$
\int_{x_i}^{x_j} \frac{dN_i}{dx} \frac{d\tilde{u}}{dx} dx = \int_{x_i}^{x_j} p(x) N_i(x) dx
\tag{2.52}
$$
$$
+ \frac{du}{dx}(x_j) N_i(x_j) - \frac{du}{dx}(x_i) N_i(x_i), \quad i = 1,2.
$$

Now, the variable x can be changed to ξ by substituting for $d\tilde{u}/dx$ from eq. (2.51) and dN_i/dx from eq. (2.47). The integral domain can also be changed by using the relation $dx = L^{(e)} d\xi$. After changing the variable, we obtain

$$
\frac{1}{L^{(e)}} \int_0^1 \frac{dN_i}{d\xi} \left\{ \frac{dN_1}{d\xi} \; \frac{dN_2}{d\xi} \right\} d\xi \cdot \left\{ \begin{array}{c} u_1 \\ u_2 \end{array} \right\} = L^{(e)} \int_0^1 p(x) N_i(\xi) d\xi
\tag{2.53}
$$
$$
+ \frac{du}{dx}(x_j) N_i(1) - \frac{du}{dx}(x_i) N_i(0), \quad i = 1,2.
$$

The second advantage of using the local coordinate is that the integral in eq. (2.53) can be performed in the local coordinates. Since the local coordinate is the same for all elements, it is convenient to standardize the integration process for different elements.

Note that the variable x still remains in $p(x)$. When a specific form of $p(x)$ is given, it will be converted into the function of ξ using the relation in eq. (2.42). We do not need to convert $du(x_j)/dx$ and $du(x_i)/dx$ because the boundary conditions do not use the approximation scheme. The two equations in eq. (2.53) can be written in a matrix form as

$$\left[\mathbf{k}^{(e)}\right]\left\{\mathbf{q}^{(e)}\right\} = \left\{\mathbf{f}^{(e)}\right\} + \left\{\begin{array}{c} -\dfrac{du}{dx}(x_i) \\[2mm] +\dfrac{du}{dx}(x_j) \end{array}\right\}, \tag{2.54}$$

where

$$\left[\mathbf{k}^{(e)}\right]_{2\times 2} = \frac{1}{L^{(e)}}\int_0^1 \begin{bmatrix} \left(\dfrac{dN_1}{d\xi}\right)^2 & \dfrac{dN_1}{d\xi}\dfrac{dN_2}{d\xi} \\[4mm] \dfrac{dN_2}{d\xi}\dfrac{dN_1}{d\xi} & \left(\dfrac{dN_2}{d\xi}\right)^2 \end{bmatrix} d\xi$$

$$= \frac{1}{L^{(e)}}\begin{bmatrix} 1 & -1 \\ -1 & 1 \end{bmatrix}, \tag{2.55}$$

$$\left\{\mathbf{f}^{(e)}\right\} = L^{(e)}\int_0^1 p(x)\left\{\begin{array}{c} N_1(\xi) \\ N_2(\xi) \end{array}\right\}d\xi, \tag{2.56}$$

and

$$\left\{\mathbf{q}^{(e)}\right\} = \left\{\begin{array}{c} u_i \\ u_j \end{array}\right\}. \tag{2.57}$$

In arriving at eq. (2.54) from eq. (2.53), we have used the boundary values of the interpolation functions given in eq. (2.48). Equation (2.54) is the element-level equivalent of the global equation derived in eq. (2.11). One can derive equation similar to eq. (2.54) for each element $e = 1, 2, \ldots, N_E$, where N_E is the number of elements.

The RHS of these equations contain terms that are derivatives at the nodes $du(x_i)/dx$ and $du(x_j)/dx$, which are not generally known. However, the second equation for element e can be added to the first equation of element $e + 1$ to eliminate the derivative term. To illustrate this point, consider the equations for elements 1 and 2. Two element matrix equations are

$$\begin{bmatrix} k_{11} & k_{12} \\ k_{21} & k_{22} \end{bmatrix}^{(1)}\left\{\begin{array}{c} u_1 \\ u_2 \end{array}\right\} = \left\{\begin{array}{c} f_1 \\ f_2 \end{array}\right\}^{(1)} + \left\{\begin{array}{c} -\dfrac{du}{dx}(x_1) \\[2mm] +\dfrac{du}{dx}(x_2) \end{array}\right\}, \tag{2.58}$$

and

$$\begin{bmatrix} k_{11} & k_{12} \\ k_{21} & k_{22} \end{bmatrix}^{(2)}\left\{\begin{array}{c} u_2 \\ u_3 \end{array}\right\} = \left\{\begin{array}{c} f_2 \\ f_3 \end{array}\right\}^{(2)} + \left\{\begin{array}{c} -\dfrac{du}{dx}(x_2) \\[2mm] +\dfrac{du}{dx}(x_3) \end{array}\right\}. \tag{2.59}$$

We want to combine these two matrix equations into one, which is called the *assembly process*. The assembled matrix equation will have three unknowns: u_1, u_2, and u_3. Equation (2.58) will be added to the first two rows, while eq. (2.59) will be added to the last two rows. When the second equation in eq. (2.58) and the first equation in eq. (2.59) are added together, the boundary term, $du(x_2)/dx$, is canceled. Thus, the assembled matrix equation becomes

$$\begin{bmatrix} k_{11}^{(1)} & k_{12}^{(1)} & 0 \\ k_{21}^{(1)} & k_{22}^{(1)}+k_{11}^{(2)} & k_{12}^{(2)} \\ 0 & k_{21}^{(2)} & k_{22}^{(2)} \end{bmatrix}\left\{\begin{array}{c} u_1 \\ u_2 \\ u_3 \end{array}\right\} = \left\{\begin{array}{c} f_1^{(1)} \\ f_2^{(1)}+f_2^{(2)} \\ f_3^{(2)} \end{array}\right\} + \left\{\begin{array}{c} -\dfrac{du}{dx}(x_1) \\[2mm] 0 \\[2mm] \dfrac{du}{dx}(x_3) \end{array}\right\}. \tag{2.60}$$

This process can be continued for successive elements, and the $2 \times N_E$ equations for the N_E elements will reduce to $N_E + 1$ number of equations. In fact, $N_E + 1 = N_D$, which is equal to the number of nodes. The N_D equations will take the form

$$
\begin{bmatrix}
k_{11}^{(1)} & k_{12}^{(1)} & 0 & \cdots & 0 \\
k_{21}^{(1)} & k_{22}^{(1)} + k_{11}^{(2)} & k_{12}^{(2)} & \cdots & 0 \\
0 & k_{221}^{(2)} & k_{22}^{(2)} + k_{11}^{(2)} & \cdots & 0 \\
\vdots & \vdots & \vdots & \ddots & \vdots \\
0 & 0 & 0 & k_{21}^{(N_E)} & k_{22}^{(N_E)}
\end{bmatrix}_{(N_D \times N_D)}
\begin{Bmatrix}
u_1 \\ u_2 \\ u_3 \\ \vdots \\ u_N
\end{Bmatrix}_{(N_D \times 1)}
=
\begin{Bmatrix}
f_1^{(1)} \\ f_2^{(1)} + f_2^{(2)} \\ f_3^{(2)} + f_3^{(3)} \\ \vdots \\ f_N^{(N_E)}
\end{Bmatrix}_{(N_D \times 1)}
+
\begin{Bmatrix}
-\dfrac{du}{dx}(x_1) \\ 0 \\ 0 \\ \vdots \\ +\dfrac{du}{dx}(x_N)
\end{Bmatrix}_{(N_D \times 1)}.
\tag{2.61}
$$

In compact form, the above equation is written as

$$
[\mathbf{K}_s]\{\mathbf{Q}_s\} = \{\mathbf{F}_s\}.
\tag{2.62}
$$

In general, the global matrix $[\mathbf{K}_s]$ will be singular, and hence the equations cannot be solved directly. However, the matrix will be nonsingular after implementing the boundary conditions. It may be noted that there are N_D unknowns in the N_D equations. At the boundaries ($x = 0$ and $x = 1$) either u (the essential boundary condition) or du/dx (the natural boundary condition) will be specified. We will illustrate the method in the following example.

EXAMPLE 2.8 *Three-element solution of a differential equation*

Using three elements of equal length, solve the differential equation given below for $p(x) = x$.

$$
\frac{d^2 u}{dx^2} + p(x) = 0, \quad 0 \le x \le 1
$$

$$
\left.\begin{aligned}
u(0) &= 0 \\
u(1) &= 0
\end{aligned}\right\} \quad \text{Boundary conditions.}
$$

SOLUTION Since the three elements are of equal length, each element has the length of $L^{(e)} = \frac{1}{3}$. Substituting in eq. (2.55) the element stiffness matrices for the three elements can be derived as

$$
\left[\mathbf{k}^{(e)}\right]_{2 \times 2} = \frac{1}{L^{(e)}}\begin{bmatrix} 1 & -1 \\ -1 & 1 \end{bmatrix} = \begin{bmatrix} 3 & -3 \\ -3 & 3 \end{bmatrix}, \quad (e = 1,2,3).
$$

Note that the element stiffness matrices for the three elements are identical. Now the variable in $p(x) = x$ can be changed to ξ using eq. (2.42), as

$$
p(\xi) = x_i(1 - \xi) + x_j \xi.
$$

Substituting this expression in eq. (2.56), the element force vectors for the three elements can be derived as

$$
\begin{aligned}
\{\mathbf{f}^{(e)}\} &= L^{(e)} \int_0^1 p(x) \begin{Bmatrix} N_1(\xi) \\ N_2(\xi) \end{Bmatrix} d\xi \\
&= L^{(e)} \int_0^1 \left[x_i(1 - \xi) + x_j \xi \right] \begin{Bmatrix} 1 - \xi \\ \xi \end{Bmatrix} d\xi \\
&= L^{(e)} \begin{Bmatrix} \dfrac{x_i}{3} + \dfrac{x_j}{6} \\ \dfrac{x_i}{6} + \dfrac{x_j}{3} \end{Bmatrix}, \quad (e = 1,2,3).
\end{aligned}
$$

Substituting for the element lengths and nodal coordinates

$$\left\{ \begin{matrix} f_1^{(1)} \\ f_2^{(1)} \end{matrix} \right\} = \frac{1}{54} \left\{ \begin{matrix} 1 \\ 2 \end{matrix} \right\}, \quad \left\{ \begin{matrix} f_2^{(2)} \\ f_3^{(2)} \end{matrix} \right\} = \frac{1}{54} \left\{ \begin{matrix} 4 \\ 5 \end{matrix} \right\}, \quad \left\{ \begin{matrix} f_3^{(3)} \\ f_4^{(3)} \end{matrix} \right\} = \frac{1}{54} \left\{ \begin{matrix} 7 \\ 8 \end{matrix} \right\}.$$

Now, the global matrix $[\mathbf{K}_s]$ and vector $\{\mathbf{F}_s\}$ can be assembled using eq. (2.61) as

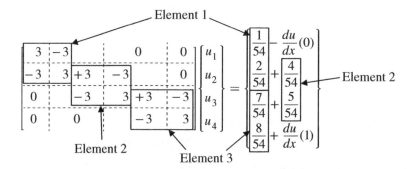

We discard the first and last rows, as we do not know the RHS of these equations (striking the rows). Furthermore, we note that $u_1 = u_4 = 0$. Thus, these two variables are removed, and the first and last columns of matrix $[\mathbf{K}_s]$ are deleted (striking the columns). Then, the four global equations reduce to two equations

$$\begin{bmatrix} 6 & -3 \\ -3 & 6 \end{bmatrix} \left\{ \begin{matrix} u_2 \\ u_3 \end{matrix} \right\} = \frac{1}{9} \left\{ \begin{matrix} 1 \\ 2 \end{matrix} \right\}.$$

Solving the above matrix equation, we obtain $u_2 = 4/81$ and $u_3 = 5/81$. Then using the interpolation functions in eq. (2.44), the approximate solution at each element can be expressed as

$$\tilde{u}(x) = \begin{cases} \dfrac{4}{27}x, & 0 \le x \le \dfrac{1}{3} \\[2mm] \dfrac{4}{81} + \dfrac{1}{27}\left(x - \dfrac{1}{3}\right), & \dfrac{1}{3} \le x \le \dfrac{2}{3} \\[2mm] \dfrac{5}{81} - \dfrac{5}{27}\left(x - \dfrac{2}{3}\right), & \dfrac{2}{3} \le x \le 1. \end{cases} \tag{2.63}$$

The exact solution can be obtained by integrating the governing differential equation twice and applying the two essential boundary conditions to solve for the constants:

$$u(x) = \frac{1}{6}x\left(1 - x^2\right). \tag{2.64}$$

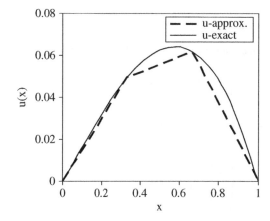

Figure 2.16 Comparison of exact and approximate solution for example 2.8

The exact and approximate solutions are plotted in figure 2.16. The value of the approximate solution at nodes 2 and 3 coincide with that of the exact solution, and it is actually a coincidence. Otherwise, one can note that the three-element solution is a poor approximation of the exact solution, and more elements are needed to obtain a more accurate solution. This is because the finite element solution is a linear function between nodes whereas the exact solution is a cubic polynomial in x. ▄▄▄

Sections 2.4.3 and 2.4.4 describe two different methods of implementing the Galerkin method in the finite element scheme. Although they appear slightly different, they result in the same matrix equations. The first method described in section 2.4.3 mimics the analytical method described in the preceding section 2.1. In the global finite element method (section 2.1) the global matrix $[\mathbf{K}]$ is directly calculated using eq. (2.12). However, in the local finite element method, we first compute the element matrix $[\mathbf{k}]$ for each element, and they are assembled to form the global matrix. Furthermore, the methods described in section 2.4.4 are amenable to easy coding in computer programs. The element matrix $[\mathbf{k}]$ is calculated for each element using the element properties such as the length of the element [eq. (2.55)]. This procedure could be automated easily as the interpolation functions expressed in the local coordinate ξ have the same form for all elements. The right-hand side matrix $\{\mathbf{f}\}$ can also be calculated easily for each element. Then, the assembling of the element matrices $[\mathbf{k}]$ and $\{\mathbf{f}\}$ to obtain the global matrices $[\mathbf{K}]$ and $\{\mathbf{F}\}$, respectively, can be performed in a more mechanical way as described in chapter 1. Such considerations are important in using the finite element method for large problems involving tens of thousands of elements.

2.5 ENERGY METHODS

Energy methods are alternative methods that are very powerful and amenable to approximate solutions when solving structures that are more realistic. Finite element equations can also be derived using energy methods. Some of the energy methods are Castigliano's theorems, the principle of minimum potential energy, the principle of minimum complementary potential energy, unit load method, and Rayleigh-Ritz method. The principle of virtual work is the fundamental principle from which all of the aforementioned methods are derived.

2.5.1 Principle of Virtual Work for a Particle

The principle of virtual work for a particle states that for a particle in equilibrium, the virtual work is identically equal to zero. *Virtual work is the work done by the (real) external forces through the virtual displacements. Virtual displacement is any small arbitrary (imaginary, not real) displacement that is consistent with the kinematic constraints of the particle.* Virtual displacement is not experienced but only assumed to exist so that various possible equilibrium positions can be compared to determine the correct one.

Consider a particle of mass m that is in equilibrium with four springs at the current position as shown in figure 2.17. For a particle in equilibrium, the sum of the forces acting on it in each coordinate direction must be equal to zero:

$$\sum F_x = 0, \quad \sum F_y = 0, \quad \sum F_z = 0. \tag{2.65}$$

Then, a small arbitrary perturbation, $\delta \mathbf{r} = \{\delta u, \delta v, \delta w\}^{\mathrm{T}}$, can be assumed. In this text, the symbol δ is used to denote a virtual function. For example, δu is the virtual displacement in the x direction, and δW is the virtual work. The key point is that the virtual displacement is not prescribed but arbitrary.

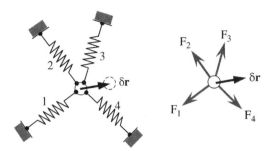

Figure 2.17 A particle in equilibrium with four springs

Therefore, $\delta\mathbf{r}$ is small but can be any arbitrary direction. Since $\delta\mathbf{r}$ is so small, the member forces of springs are assumed unchanged due to the virtual displacement. Then it is obvious that

$$\delta u \sum F_x = 0, \ \delta v \sum F_y = 0, \ \delta w \sum F_z = 0, \tag{2.66}$$

or in other words the *virtual work* is equal to zero:

$$\delta W = \delta u \sum F_x + \delta v \sum F_y + \delta w \sum F_z = 0. \tag{2.67}$$

In the case of the particle in figure 2.17, the work done by virtual displacement is

$$\delta W = \mathbf{F}_1 \cdot \delta\mathbf{r} + \mathbf{F}_2 \cdot \delta\mathbf{r} + \mathbf{F}_3 \cdot \delta\mathbf{r} + \mathbf{F}_4 \cdot \delta\mathbf{r} = (\mathbf{F}_1 + \mathbf{F}_2 + \mathbf{F}_3 + \mathbf{F}_4) \cdot \delta\mathbf{r} = 0. \tag{2.68}$$

Conversely, if the virtual work of a particle is equal to zero for arbitrary virtual displacements, then the particle is in equilibrium under the applied forces. In fact, the principle is another statement of equilibrium equations given in eq. (2.65). This seemingly trivial principle leads to some important results when applied to deformable bodies.

As a mechanical example, consider a static equilibrium of a mass-spring system under gravity as shown in figure 2.18(a). At equilibrium, the spring force, $F(u)$, is equal to the applied load, mg; that is, $F(u) - mg = 0$. In order to generate the spring force, the spring is elongated by u. In the case of a linear spring, the spring force is proportional to spring constant, $F(u) = k \cdot u$. This force is called the *internal force*, in contrast to the external force, mg. At equilibrium, let the position of the mass be perturbed by δu as shown in figure 2.18(b). Since both internal and external forces remain constant during a small, arbitrary perturbation by virtual displacement, we have the following relationship:

$$F(u) \cdot \delta u = mg \cdot \delta u, \tag{2.69}$$

where the LHS is the work done by internal force, which is referred to as internal virtual work, while the RHS is the external virtual work. Equation (2.69) means that the external work done by the external forces, mg, during the application of a small virtual displacement is equal to the internal work done

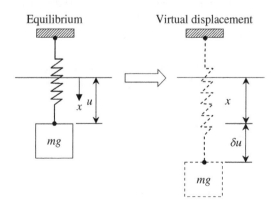

Figure 2.18 Equilibrium of mass-spring system

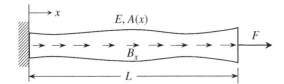

Figure 2.19 Uniaxial bar under body force B_x and concentrated force F

by the spring force during the application of that small virtual displacement. Therefore, in the principle of virtual work, the equilibrium is stated in terms of work, or equivalently, the energy.

2.5.2 Principle of Virtual Work for Deformable Bodies

As an example, we will consider the one-dimensional case, uniaxial bar as shown in figure 2.19). The bar is fixed at $x = 0$, and under the body force B_x per unit volume in the x direction as well as a concentrated force at F at $x = L$. In general, the cross-sectional area $A(x)$ can vary along the length of the bar L.

In the case of one-dimensional bar, the stress equilibrium equation can be written as

$$\frac{d\sigma_x}{dx} + B_x = 0, \tag{2.70}$$

where B_x is the body force expressed in force per unit volume. Although eq. (2.70) looks like a first-order differential equation, it is actually a second-order differential equation of displacement $u(x)$ because $\sigma_x = E\varepsilon_x$, and the strain is defined as the derivative of displacement; $\varepsilon_x = du/dx$. In the uniaxial bar, the stress is constant over the cross section.

We will now apply the weighted residual method to derive a weak form using a weighting function $\delta u(x)$. In solid mechanics, this weighting function is referred to as the virtual displacement field $\delta u(x)$ that is defined along the length of the bar, and it can be interpreted as a small variation from equilibrium. Since eq. (2.70) is true at every point in every cross section of the bar, the following must be true for any $\delta u(x)$:

$$\int_0^L \iint_A \left(\frac{d\sigma_x}{dx} + B_x \right) \delta u(x) dA dx = 0. \tag{2.71}$$

The quantity in the parentheses in the integrand is constant over the cross section. Therefore, it can be integrated over the cross-sectional area. By using the definition of axial force resultant P, that is, $P = A\sigma_x$, eq. (2.71) can be written as

$$\int_0^L \left(\frac{dP}{dx} + b_x \right) \delta u(x) dx = 0, \tag{2.72}$$

where the body force b_x is force per unit length of the bar.

This weak form is in fact the principle of virtual work. In the viewpoint of the principle of virtual work, $dP/dx + b(x) = 0$ is basically equilibrium between internal and external forces. Therefore, eq. (2.72) states that the work done by virtual displacement is zero when the bar is in equilibrium.

We will convert eq. (2.72) into a more convenient form for computation. Using integration by parts in eq. (2.72) we obtain

$$P\delta u \Big|_0^L - \int_0^L P \frac{d(\delta u)}{dx} dx + \int_0^L b_x \delta u(x) dx = 0. \tag{2.73}$$

Since the bar is fixed at $x = 0$, $u(0) = 0$. That means, $\delta u(0)$ is also equal to zero because *the virtual displacement should be consistent with the displacement constraints of the body*. This is similar to the requirement that the trial functions must satisfy the essential boundary conditions in the weighted residual method. Since F is the force applied at $x = L$, $P(L) = F$. We also define virtual strains $\delta\varepsilon(x)$ that are created by the virtual displacement field $\delta u(x)$:

$$\delta\varepsilon(x) = \frac{d(\delta u)}{dx}. \tag{2.74}$$

This is the same definition of strain if the virtual displacement is considered as the displacement. Then using eq. (2.74), eq. (2.73) can be written as:

$$F\delta u(L) + \int_0^L b_x \delta u(x) dx = \int_0^L P\delta\varepsilon(x) dx. \tag{2.75}$$

Note that the LHS is the work done by external forces, F and b_x, while the RHS is the work done by the internal force resultant, $P(x)$. We define the external virtual work δW_e and internal virtual work δW_i as follows:

$$\delta W_e = F\delta u(L) + \int_0^L b_x \delta u(x) dx, \tag{2.76}$$

$$\delta W_i = -\int_0^L \int_A \sigma_x \delta\varepsilon(x) dA dx = -\int_0^L P\delta\varepsilon(x) dx. \tag{2.77}$$

The negative sign is added in internal virtual work because it is positive when work is done to the system. Then, eq. (2.75) can be written as

$$\delta W_e + \delta W_i = 0. \tag{2.78}$$

Equation (2.78) states that for a bar in equilibrium, the sum of external and internal virtual work is zero for every virtual displacement field, and it constitutes the principle of virtual work for a one-dimensional deformable body.

This principle can be derived for three-dimensional bodies also, except the definition of δW_e and δW_i take different forms. The external virtual work will be the sum of work done by all (real) external forces through the corresponding virtual displacements. Considering both distributed forces (surface tractions t_x, t_y, and t_z) and concentrated forces (F_x, F_y, and F_z) we obtain

$$\delta W_e = \int_S \left(t_x \delta u + t_y \delta v + t_z \delta w\right) dS + \sum_i \left(F_{xi}\delta u_i + F_{yi}\delta v_i + F_{zi}\delta w_i\right). \tag{2.79}$$

In eq. (2.79) the integration is performed over the surface of the three-dimensional solid. The internal virtual work is defined by

$$\delta W_i = -\int_V \left(\sigma_x \delta\varepsilon_x + \sigma_y \delta\varepsilon_y + ... + \tau_{xy}\delta\gamma_{xy}\right) dV, \tag{2.80}$$

where the integration is performed over the volume of the body, V.

2.5.3   Variation of a Function

Virtual displacements in the previous section can be considered as a variation of real displacements. To explain it further, consider an x-directional displacement $u(x)$, and those neighboring displacements that are described by arbitrary virtual displacement $\delta u(x)$ and small parameter $\tau > 0$, as

$$u_\tau(x) = u(x) + \tau \delta u(x). \tag{2.81}$$

Thus, for a given virtual displacement, these neighboring displacements are controlled by one parameter τ. $u_\tau(x)$ is the perturbed displacement. Similar to the first variation in calculus, the variation of displacement can be obtained by

$$\left. \frac{du_\tau(x)}{d\tau} \right|_{\tau=0} = \delta u(x). \tag{2.82}$$

It is obvious that the virtual displacement $\delta u(x)$ is the *displacement variation*.

An important requirement of the virtual displacement is that it must satisfy kinematic constraints, that is, essential boundary conditions. It means that when the value of displacement is prescribed at a point, the perturbed displacement must have the same value. In order to satisfy this requirement, the displacement variation must be equal to zero at that point. For example, let the displacement boundary condition be given as $u(0) = 1$. Then, the perturbed displacement also satisfies the boundary condition, $u_\tau(1) = u(1) + \tau \delta u(1) = 1$. Therefore, the displacement variation must satisfy $\delta u(0) = 0$.

The variation of a complex function can be obtained using the chain rule of differentiation. For example, let a function $f(u)$ depend on displacement. The variation of $f(u)$ can be obtained using

$$\delta f = \left. \frac{df(u_\tau)}{d\tau} \right|_{\tau=0} = \frac{df}{du} \delta u. \tag{2.83}$$

An important property of the variation is that it is independent of differentiation with respect to space coordinates. For example, consider the variation of strain:

$$\delta \varepsilon_x = \delta \left(\frac{du}{dx} \right) = \frac{d(\delta u)}{dx}.$$

The concept of variation plays an important role in understanding structural equilibrium, which will be derived in the following section.

2.5.4 Principle of Minimum Potential Energy

Consider the strain energy density in a one–dimensional body in figure 2.19, which is given by:

$$U_0 = \frac{1}{2}\sigma_x \varepsilon_x = \frac{1}{2} E \varepsilon_x^2. \tag{2.84}$$

Suppose that the displacement field is perturbed slightly by $\delta u(x)$, then it will cause a slight change in the strain field also. Then the variation in the strain energy density is given by:

$$\delta U_0 = \frac{dU_0}{d\varepsilon_x} \delta \varepsilon_x = E \varepsilon_x \delta \varepsilon_x = \sigma_x \delta \varepsilon_x. \tag{2.85}$$

The change in strain energy of the bar is expressed as

$$\delta U = \int_0^L \int_A \delta U_0 dA dx = \int_0^L \int_A \sigma_x \delta \varepsilon_x dA dx = \int_0^L P \delta \varepsilon_x dx. \tag{2.86}$$

Comparing eq. (2.77) and eq. (2.86), we obtain

$$\delta U = -\delta W_i. \tag{2.87}$$

Next, we define the potential energy of external forces. Consider a force F at $x = L$, and the corresponding displacement $u(L)$. If there were an additional displacement of $\delta u(L)$, then the force would have done additional work of $F\delta u(L)$. Then we can claim that the force F has lost some potential to do work or its potential has been reduced by $F\delta u(L)$. If we denote the potential of the force by V, then the change in potential due to the variation $\delta u(L)$ is given by

$$\delta V = -F\delta u(L). \tag{2.88}$$

Since the force F does not vary through this change in displacement, we can write δV as

$$\delta V = -\delta(Fu(L)). \tag{2.89}$$

Thus, the potential energy of external forces is the negative of the external forces multiplied by displacements at that point. Since the body is distributed in $0 \le x \le L$, the potential energy must be integrated over the domain. Then the potential energy takes the following form:

$$V = -Fu(L) - \int_0^L b_x u(x)\mathrm{d}x. \tag{2.90}$$

The negative sign in front of the work term in the above equation is sometimes confusing to students. It can be easily understood if one considers the gravitational potential energy on the surface of the earth. From elementary physics, we know that the potential energy of a body of mass m at a height h from the datum is given by $V = mgh$. Since gravity acts downward and height is positive upward, the potential energy should be written as $V = -Fh$, where F is the force of gravity or the weight of the body, and the magnitude of F is equal to mg. If we replace the gravitational force by an external force, then one obtains eq. (2.90).

Comparing eqs. (2.76) and (2.90), one can note that the change in the potential of external forces is equal to the negative of the external virtual work:

$$\delta V = -\delta W_e. \tag{2.91}$$

Substituting for the external and internal virtual work terms from eqs. (2.87) and (2.91) into eq. (2.78), the principle of virtual work takes the form

$$\delta U + \delta V = 0 \quad \text{or} \quad \delta(U + V) = 0. \tag{2.92}$$

We define the total potential energy Π as the sum of the strain energy and the potential of external forces, that is, $\Pi = U + V$. Then, eq. (2.92) takes the form

$$\delta\Pi = 0. \tag{2.93}$$

Equation (2.93) is the *Principle of Minimum Potential Energy*, and it can be stated as follows:

> Of all displacement configurations of a solid consistent with its displacement (kinematic) constraints, the actual one that satisfies the equilibrium equations is given by the minimum value of total potential energy.

The variation of total potential energy is expressed in terms of differentiation with respect to displacements. Using the chain rule of differentiation,

$$\delta\Pi = \frac{\mathrm{d}\Pi}{\mathrm{d}u}\delta u = 0. \tag{2.94}$$

Since the above equation must be satisfied for all displacement variations that are consistent with kinematic constraints, the equilibrium can be found by putting the derivative of total potential energy equal to zero, that is, $\mathrm{d}\Pi/\mathrm{d}u = 0$. In practice, the displacement is often approximated using discrete parameters, such as coefficients c_i in eq. (2.3) or nodal solution u_i in eq.(2.34). In such a case, the differentiation in the above equation can be performed with respect to the unknown coefficients. This process will be explained in the following section.

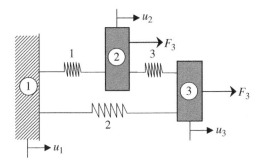

Figure 2.20 Example of a discrete system with finite number of degrees of freedom

2.5.5 Application of Principle of Minimum Potential Energy to Discrete Systems

Deformation of discrete systems can be defined by a finite number of variables. Consider the system of springs shown in figure 2.20. In the context of the finite element method, the springs are called elements, and the masses that are connected to the elements are called the nodes. External forces are applied only at the nodes. At least one node has to be fixed to prevent rigid-body displacement and thus obtain a unique solution for displacements. The strain energy of the system and the potential energy of the forces are expressed in terms of the displacements. Then the total potential energy is minimized with respect to the displacements, which yields as many linear equations as the number of unknown displacements. The procedures are illustrated in the following example.

EXAMPLE 2.9 *Energy method for a discrete system*

Consider a system of rigid bodies connected by springs as shown in figure 2.20. The bodies are assumed to move only in the horizontal direction. Further, we consider only the static problem, and hence the actual mass effects will not be considered. External forces are applied on the rigid bodies 2 and 3 as shown. The objectives are to determine the displacement of each body, forces in the springs, and support reaction. Use the following values: $k^{(1)} = 100\,\text{N/mm}$, $k^{(2)} = 200\,\text{N/mm}$, $k^{(3)} = 150\,\text{N/mm}$, $F_2 = 1{,}000\,\text{N}$, and $F_3 = 500\,\text{N}$.

SOLUTION The discrete degrees of freedom (DOFs) in the above problem are the displacements u_1, u_2, and u_3. However, the displacement at node 1 is known, and hence the number of DOFs in this problem is equal to two. It should be noted that the force F_1 corresponding to the known displacement u_1 is the reactions exerted by the support, and it is unknown. As mentioned in chapter 1, either the displacement or the force is unknown at a node, not both. The total potential energy of the system consists of two parts: strain energy of the system U, and potential energy V of the external forces F_2 and F_3.

The strain energy stored in any of the springs, say spring 1, can be expressed in terms of the displacements of its nodes, u_1 and u_2. The strain energy $U^{(1)}$ of the spring is then

$$U^{(1)} = \frac{1}{2}k^{(1)}(u_2 - u_1)^2.$$

The above expression can be written in a matrix form:

$$U^{(1)} = \frac{1}{2}\underset{(1\times2)}{\lfloor u_1\ u_2 \rfloor}\underset{(2\times2)}{\begin{bmatrix} k^{(1)} & -k^{(1)} \\ -k^{(1)} & k^{(1)} \end{bmatrix}}\underset{(2\times1)}{\begin{Bmatrix} u_1 \\ u_2 \end{Bmatrix}}.$$

Note that the matrix in the above equation is the same as the element stiffness matrix of a spring in chapter 1. In the above equation, the numbers in parentheses indicate the size of the matrix. Similarly, the strain energies in springs 2 and 3 can be written as

$$U^{(2)} = \frac{1}{2} \lfloor u_1 \ \ u_3 \rfloor \begin{bmatrix} k^{(2)} & -k^{(2)} \\ -k^{(2)} & k^{(2)} \end{bmatrix} \begin{Bmatrix} u_1 \\ u_3 \end{Bmatrix},$$

$$U^{(3)} = \frac{1}{2} \lfloor u_2 \ \ u_3 \rfloor \begin{bmatrix} k^{(3)} & -k^{(3)} \\ -k^{(3)} & k^{(3)} \end{bmatrix} \begin{Bmatrix} u_2 \\ u_3 \end{Bmatrix}.$$

Since the strain energy is a scalar, we can add energy in all springs to obtain the strain energy of the system:

$$U = \sum_{e=1}^{3} U^{(e)}.$$

It can be shown that the sum of the strain energy terms can be written again in a matrix form that includes all the DOFs in the system:

$$U = \frac{1}{2} \lfloor u_1 \ \ u_2 \ \ u_3 \rfloor \begin{bmatrix} k^{(1)} + k^{(2)} & -k^{(1)} & -k^{(2)} \\ -k^{(1)} & k^{(1)} + k^{(3)} & -k^{(3)} \\ -k^{(2)} & -k^{(3)} & k^{(2)} + k^{(3)} \end{bmatrix} \begin{Bmatrix} u_1 \\ u_2 \\ u_3 \end{Bmatrix}, \tag{2.95}$$

or symbolically

$$U = \frac{1}{2} \{\mathbf{Q}\}^{\mathrm{T}} [\mathbf{K}] \{\mathbf{Q}\}, \tag{2.96}$$

where the column vector $\{\mathbf{Q}\}$ represents the three DOFs, and $[\mathbf{K}]$ is the square matrix in eq. (2.95). Note that the above process is identical to the assembly in chapter 1. The potential energy of the external forces is given by

$$V = -\left(F_1 u_1 + F_2 u_2 + F_3 u_3\right)$$

$$= -\lfloor u_1 \ \ u_2 \ \ u_3 \rfloor \begin{Bmatrix} F_1 \\ F_2 \\ F_3 \end{Bmatrix} \tag{2.97}$$

$$= -\{\mathbf{Q}\}^{\mathrm{T}} \{\mathbf{F}\},$$

where $\{\mathbf{F}\}$ is the column vector of external forces. The total potential energy is the sum of the strain energy and the potential of external forces:

$$\Pi = U + V = \frac{1}{2} \{\mathbf{Q}\}^{T} [\mathbf{K}] \{\mathbf{Q}\} - \{\mathbf{Q}\}^{T} \{\mathbf{F}\}. \tag{2.98}$$

In the above equation, the total potential energy is written in terms of the vector of nodal displacements $\{\mathbf{Q}\} = \{u_1, \ u_2, \ u_3\}^{\mathrm{T}}$. Thus, the differentiation in eq. (2.94) can be applied to $\{\mathbf{Q}\}$. According to the principle of minimum potential energy, of all possible u_i, the displacements that will satisfy the equilibrium equations minimize the total potential energy. The total potential energy is minimized with respect to the DOFs:

$$\frac{\partial \Pi}{\partial u_1} = 0, \quad \frac{\partial \Pi}{\partial u_2} = 0, \quad \frac{\partial \Pi}{\partial u_3} = 0$$

$$\text{or,} \quad \frac{\partial \Pi}{\partial \{\mathbf{Q}\}} = 0. \tag{2.99}$$

The above minimization procedure results in the following equations:

$$[\mathbf{K}] \begin{Bmatrix} u_1 \\ u_2 \\ u_3 \end{Bmatrix} = \begin{Bmatrix} F_1 \\ F_2 \\ F_3 \end{Bmatrix}. \tag{2.100}$$

In the above system of equations, the three unknowns are u_2, u_3, and F_1. Substituting numerical values in eq. (2.100) we obtain

$$\begin{bmatrix} 300 & -100 & -200 \\ -100 & 250 & -150 \\ -200 & -150 & 350 \end{bmatrix} \begin{Bmatrix} 0 \\ u_2 \\ u_3 \end{Bmatrix} = \begin{Bmatrix} F_1 \\ 1{,}000 \\ 500 \end{Bmatrix}. \tag{2.101}$$

The solution to the above system of equations is given by $u_2 = 6.538$mm, $u_3 = 4.231$mm, and $F_1 = -1{,}500$N. The forces in the springs are calculated using

$$P^{(e)} = k^{(e)} \left(u_j - u_i \right). \tag{2.102}$$

Using the results for displacements, the forces in the springs are obtained as:

$$\begin{aligned} P^{(1)} &= k^{(1)} \left(u_2 - u_1 \right) = 654\text{N} \\ P^{(2)} &= k^{(2)} \left(u_3 - u_1 \right) = 846\text{N} \\ P^{(3)} &= k^{(3)} \left(u_3 - u_2 \right) = -346\text{N}. \end{aligned} \tag{2.103}$$

Note that springs 1 and 2 are in tension, while spring 3 is in compression.

2.5.6 Rayleigh-Ritz Method

The principle of minimum potential energy is easily applied to discrete systems (e.g., a number of springs connected together), wherein the unknown DOFs are finite. In that case, the method yields the exact solution. For continuous systems, for instance, beam, uniaxial bar, where the DOFs are infinite, an approximate method has to be used. The Rayleigh-Ritz method is one such method for continuous systems.

In the Rayleigh-Ritz method, a continuous system is approximated as a discrete system with finite number of DOFs. This is accomplished by approximating the displacements by a function containing a finite number of coefficients to be determined. The total potential energy is then evaluated in terms of the unknown coefficients. Then the principle of minimum potential energy is applied to determine the best set of coefficients by minimizing the total potential energy with respect to the coefficients. The solution thus obtained may not be exact. It is the best solution from among the family of solutions that can be obtained from the assumed displacement functions. In the following, we demonstrate the method to a uniaxial bar problem.

The steps involved in solving the uniaxial bar problem using the Rayleigh Ritz method are as follows.

1. Assume displacement of the bar in the form $u(x) = c_1 f_1(x) + \cdots + c_n f_n(x)$, where the c_i are coefficients to be determined and $f_i(x)$ are basis functions whose expressions are known. The above $u(x)$ *must* satisfy the displacement boundary conditions, for example, $u(0) = 0$, or $u(L) = 0$, where L is the length of the bar. That means, if some $f_i(x)$ does not satisfy the displacement boundary conditions, the corresponding c_i must be zero.

2. Determine the strain energy U in the bar using the formula in eq. (2.107) in terms of c_i.

3. Find the potential of external forces V using formulas of types given in eq. (2.108).

4. The total potential energy is obtained as $\Pi(c_1, c_2, \ldots c_n) = U + V$

5. Apply the principle of minimum potential energy to determine the coefficients $c_1, c_2, \ldots, c_n$.

6. After solving for the constants in the assumed displacement field, find the axial force resultant using $P(x) = AE du/dx$.

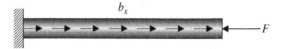

Figure 2.21 Uniaxial bar subject to distributed and concentrated forces

EXAMPLE 2.10 *Uniaxial bar using the Rayleigh–Ritz method*

The uniaxial bar shown in figure 2.21 is of uniform cross sectional area A and length L. It is clamped at the left end and subjected to a concentrated force F at the right end as shown. In addition, a uniformly distributed load $b_x(x)$ acts along the length of the bar. Use the Rayleigh-Ritz method to determine the displacements $u(x)$, the axial force resultant $P(x)$, and the support reaction. The Young's modulus of the material of the bar is E. Provide numerical results for the case: $L = 1$ m, $A = 100$ mm^2, $E = 100$ GPa, $F = 10$ kN, and $b_x = 10$ kN/m.

SOLUTION   In applying the Rayleigh-Ritz method, we approximate the displacement field using a simple but physically reasonable function. The assumed displacement must satisfy the displacement boundary conditions; in the present case, $u(0) = 0$. Actually, one needs some experience to make reasonable approximations. Let us assume a quadratic polynomial in x as the approximate displacement field:

$$u(x) = c_0 + c_1 x + c_2 x^2, \tag{2.104}$$

where c_i are coefficients to be determined. However, we have to make sure that the assumed displacement field satisfies the displacement boundary conditions *a priori*. In the present case, $u(0) = 0$, and this is accomplished by setting $c_0 = 0$. Then the assumed displacements take the form

$$u(x) = c_1 x + c_2 x^2. \tag{2.105}$$

Actually, we have reduced the continuous system to a discrete system with two DOFs. The coefficients c_1 and c_2 are the two DOFs as they determine the deformed configurations of the bar. The next step is to express the strain energy in the bar in terms of the unknown coefficients. The strain energy of the bar is obtained by integrating the strain energy per unit length U_L over the length of the bar:

$$U = \int_0^L U_L(x)dx = \int_0^L \frac{1}{2} A E \varepsilon_x^2 dx = \int_0^L \frac{1}{2} A E \left(\frac{du}{dx}\right)^2 dx. \tag{2.106}$$

In eq. (2.106) ε_x is the strain in the bar as a function of x. The strain in the uniaxial bar is given by $\varepsilon_x = du/dx$. Using the assumed displacements in eq. (2.105), we express the strains in terms of c_i as $\varepsilon_x = c_1 + 2c_2 x$. Substituting in eq. (2.106), the strain energy in the bar is obtained as

$$
\begin{aligned}
U(c_1, c_2) &= \frac{1}{2} A E \int_0^L (c_1 + 2c_2 x)^2 dx \\
&= \frac{1}{2} A E \left(L c_1^2 + 2L^2 c_1 c_2 + \frac{4}{3} L^3 c_2^2 \right).
\end{aligned}
\tag{2.107}
$$

The potential energy of the forces acting on the bar can be derived as follows:

$$
\begin{aligned}
V(c_1, c_2) &= -\int_0^L b_x(x) u(x)dx - (-F)u(L) \\
&= -\int_0^L b_x \left(c_1 x + c_2 x^2\right)dx + F\left(c_1 L + c_2 L^2\right) \\
&= c_1 \left(FL - b_x \frac{L^2}{2} \right) + c_2 \left(FL^2 - b_x \frac{L^3}{3} \right).
\end{aligned}
\tag{2.108}
$$

The total potential energy is then $\Pi(c_1, c_2) = U + V$. According to the principle of minimum potential energy, the best set of coefficients c_i is obtained by minimizing Π. We take the partial derivatives of Π with respect to c_i and equate them to zero:

$$\frac{\partial \Pi}{\partial c_1} = AELc_1 + AEL^2 c_2 + FL - b_x \frac{L^2}{2} = 0,$$

$$\frac{\partial \Pi}{\partial c_2} = AEL^2 c_1 + \frac{4}{3} AEL^3 c_2 + FL^2 - b_x \frac{L^3}{3} = 0. \tag{2.109}$$

The c_i are obtained by solving the two equations in eq. (2.109). Using the numerical values for the various bar properties and the forces, we obtain

$$10^7 c_1 + 10^7 c_2 = -5{,}000,$$

$$10^7 c_1 + \frac{4 \times 10^7}{3} c_2 = -6{,}667. \tag{2.110}$$

Solving the above equations, we obtain $c_1 = 0$ and $c_2 = -0.5 \times 10^{-3}$. Substituting for c_i in eq. (2.105), the displacement field is obtained as $u(x) = -0.5 \times 10^{-3} x^2$. The axial force in the bar is obtained using $P(x) = AEdu/dx = -10{,}000x$. The support reaction is given by $R = -P(0) = 0$. It should be noted that the above solution obtained using the Rayleigh-Ritz method is indeed the exact solution to the problem. The reason is that the exact solution is a quadratic polynomial in x, and we have assumed the same form of a solution in the Rayleigh-Ritz method. ■

2.6 EXERCISES

1. Answer the following descriptive questions.

   (a) Explain the difference between the essential and natural boundary conditions.

   (b) The beam-bending problem is governed by a fourth-order differential equation. Explain what are the essential and natural boundary conditions for the beam-bending problem.

   (c) What is the basic requirement of the trial functions in the Galerkin method?

   (d) What is the advantage of the Galerkin method compared to other weighted residual methods?

   (e) A tip load is applied to a cantilevered beam. When $w(x) = c_1 x^2 + c_2 x^3$ is used as a trial function, will the Galerkin method yield the accurate solution or an approximate one?

   (f) A uniformly distributed load is applied to a cantilevered beam. When $w(x) = c_1 x^2 + c_2 x^3$ is used as a trial function, will the Galerkin method yield the accurate solution or an approximate one?

   (g) For a simply supported beam, can $w(x) = c_0 + c_1 x + c_2 x^2 + c_3 x^3$ be a trial function or not? Explain why.

   (h) List at least three advantages of the finite element method compared to the weighted residual method.

   (i) Explain why the displacement is continuous across element boundary, but stress is not.

   (j) If a higher-order element, such as a quadratic element, is used, will stress be continuous across element boundary?

   (k) A cantilevered beam problem is solved using the Rayleigh-Ritz method with assumed deflection function $w(x) = c_0 + c_1 x + c_2 x^2 + c_3 x^3$. What are the values of c_0 and c_1? Explain why.

   (l) For the Rayleigh-Ritz method, you can assume the form of solution, such as a combination of polynomials. What is the condition that this form needs to satisfy?

2. Use the Galerkin method to solve the following boundary-value problem using a: (a) one-term approximation and (b) two-term approximation. Compare your results with the exact solution by plotting them on the same graph.

$$\frac{d^2u}{dx^2} + x^2 = 0, \quad 0 \le x \le 1$$

$$\left.\begin{array}{l} u(0) = 1 \\ u(1) = 0 \end{array}\right\} \quad \text{Boundary conditions.}$$

Hint: Use the following one- and two-term approximations

One-term approximation :

$$\tilde{u}(x) = (1-x) + c_1\phi_1(x)$$
$$= (1-x) + c_1 x(1-x),$$

Two-term approximation :

$$\tilde{u}(x) = (1-x) + c_1\phi_1(x) + c_2\phi_2(x)$$
$$= (1-x) + c_1 x(1-x) + c_2 x^2(1-x).$$

The exact solution is $u(x) = 1 - x(x^3 + 11)/12$.

The approximate solution is split into two parts. The first term satisfies the given essential boundary conditions exactly, i.e., $u(0) = 1$ and $u(1) = 0$. The rest of the solution containing the unknown coefficients vanishes at the boundaries.

3. Solve the differential equation in problem 2 using (a) two and (b) three finite elements. Use the finite element approximation described in section 2.4. Plot the exact solution and two- and three-element solutions on the same graph. Similarly, plot the derivative du/dx. *Note:* The boundary conditions are not homogeneous. The boundary condition $u(0) = 1$ has to be used in solving the final equations.

4. The following differential equation is going to be solved using a finite-element equation:

$$\frac{d^2u}{dx^2} = 0, \quad 1 \le x \le 2 \quad u(1) = 0, \quad \frac{du}{dx}(2) = 1.$$

   Answer the following questions:

   (a) When one element with two nodes is used, write the expression of approximate solution $\tilde{u}(x)$ in terms of two nodal values, u_1 and u_2.

   (b) Write the two equations of weighted residuals using the Galerkin method. Use the approximate solution obtained from part (a)

   (c) Solve the equations in part (b) for nodal values, u_1 and u_2, and approximate solution $\tilde{u}(x)$.

5. Using the Galerkin method, solve the following differential equation with the approximate solution in the form of $\tilde{u}(x) = c_1 x + c_2 x^2$. Compare the approximate solution with the exact one by plotting them on a graph. Also, compare the derivatives du/dx and d$\tilde{u}$/dx.

$$\frac{d^2u}{dx^2} + x^2 = 0, \quad 0 \le x \le 1$$

$$\left.\begin{array}{l} u(0) = 0 \\ \frac{du}{dx}(1) = 1 \end{array}\right\} \quad \text{Boundary conditions.}$$

6. A one-dimensional heat conduction problem can be expressed by the following differential equation:

$$k\frac{d^2T}{dx^2} + Q = 0, \quad 0 \le x \le L,$$

where k is the thermal conductivity, $T(x)$ is the temperature, and Q is heat generated per unit length. Q, the heat generated per unit length, is assumed constant. Two essential boundary conditions are given at both ends:

$T(0) = T(L) = 0$. Calculate the approximate temperature $T(x)$ using the Galerkin method. Compare the approximate solution with the exact one. *Hint:* Start with an assumed solution in the following form: $\tilde{T}(x) = c_0 + c_1x + c_2x^2$, and then make it satisfy the two essential boundary conditions.

7. Solve the one-dimensional heat conduction problem 6 using the Rayleigh-Ritz method. For the heat conduction problem, the total potential can be defined as

$$\Pi = \int_0^L \left[\frac{1}{2}K\left(\frac{dT}{dx}\right)^2 - QT \right] dx.$$

Use the approximate solution $\tilde{T}(x) = T_1\phi_1(x) + T_2\phi_2(x) + T_3\phi_3(x)$, where the trial functions are given in eq. (2.37) with $N_D = 3$ and $x_1 = 0$, $x_2 = L/2$, and $x_3 = L$. Compare the approximate temperature with the exact one by plotting them on a graph.

8. Consider the following differential equation:

$$\frac{d^2u}{dx^2} + u + x = 0, \quad 0 \le x \le 1$$

$$u(0) = 0$$

$$\left.\frac{du}{dx}\right|_{x=1} = 1.$$

Assume a solution of the form: $\tilde{u}(x) = c_1x + c_2x^2$. Calculate the unknown coefficients using the Galerkin method. Compare $u(x)$ and $du(x)/dx$ with the exact solution: $u(x) = 3.7\sin x - x$ by plotting the solution.

9. Solve the differential equation in problem 8 for the following boundary conditions using the Galerkin method:

$$u(0) = 1, \quad u(1) = 2.$$

Assume the approximate solution as:

$$\tilde{u}(x) = \phi_0(x) + c_1\phi_1(x),$$

where $\phi_0(x)$ is a function that satisfies the essential boundary conditions, and $\phi_1(x)$ is the weight function that satisfies the homogeneous part of the essential boundary conditions, that is, $\phi_1(0) = \phi_1(1) = 0$. Hence, assume the functions as follows:

$$\phi_0(x) = 1 + x, \quad \phi_1(x) = x(1-x).$$

Compare the approximate solution with the exact solution by plotting their graphs. The exact solution can be derived as:

$$u(x) = 2.9231\sin x + \cos x - x.$$

10. When the solution of the following differential equation is approximated by $\tilde{u}(x) = a_0 + a_1x^2$, calculate the unknown coefficient(s) using the Galerkin method. Note that the approximate solution must satisfy the essential boundary conditions

$$\frac{d^2u}{dx^2} + u = 1, \quad u(0) = 0, \quad du(1)/dx = 0, \quad 0 \le x \le 1.$$

11. Consider the following boundary-value problem:

$$2\frac{d^2u}{dx^2} + 3u = 0, \quad 0 < x < 2$$

$$u(0) = 1 \quad \text{and} \quad \frac{du}{dx}(2) = 1.$$

Using two equal-length finite elements, calculate unknown $u(x)$ and its derivative. Compare the finite element solution with the exact solution.

12. Consider the following boundary-value problem:

$$\frac{d}{dx}\left(x\frac{du}{dx}\right) = \frac{2}{x^2} \qquad 1 \leq x \leq 2$$

$$u(1) = 2$$

$$\frac{du}{dx}(2) = -\frac{1}{4}.$$

(a) When two equal-length finite elements are used to approximate the problem, write interpolation functions and their derivatives.

(b) Calculate the approximate solution using the Galerkin method.

13. Using the Galerkin method, calculate the approximate solution of the following differential equation in the form of $u(x) = c_1 x + c_2 \frac{1}{2}x^2$.

$$\frac{d}{dx}\left(x\frac{du}{dx}\right) = x \qquad 0 \leq x \leq 1$$

$$u(0) = 0$$

$$\frac{du}{dx}(1) = 1.$$

14. The boundary-value problem for a clamped-clamped beam can be written as

$$\frac{d^4 w}{dx^4} - p(x) = 0, \ 0 \leq x \leq 1$$

$$w(0) = w(1) = \frac{dw}{dx}(0) = \frac{dw}{dx}(1) = 0 \bigg\} \text{boundary conditions.}$$

When a uniformly distributed load is applied, that is, $p(x) = p_0$, calculate the approximate beam deflection $\tilde{w}(x)$ using the Galerkin method. Hint: Assume the approximate deflection as $\tilde{w}(x) = c\phi(x) = cx^2(1-x)^2$.

15. The boundary-value problem for a cantilevered beam can be written as

$$\frac{d^4 w}{dx^4} - p(x) = 0, \ 0 \leq x \leq 1$$

$$w(0) = \frac{dw}{dx}(0) = 0, \ \frac{d^2 w}{dx^2}(1) = 1, \ \frac{d^3 w}{dx^3}(1) = -1.$$

Assume $p(x) = x$. Assuming the approximate deflection in the form of $\tilde{w}(x) = c_1\phi_1(x) + c_2\phi_2(x) = c_1 x^2 + c_2 x^3$. Solve for the boundary-value problem using the Galerkin method. Compare the approximate solution to the exact solution by plotting the solutions on a graph.

16. Repeat problem 15 by assuming $\tilde{w}(x) = \sum_{i=1}^{3} c_i\phi_i(x) = c_1 x^2 + c_2 x^3 + c_3 x^4$.

17. Consider a finite element with three nodes, as shown in the figure. When the solution is approximated using $u(x) = N_1(x)u_1 + N_2(x)u_2 + N_3(x)u_3$, calculate the interpolation functions $N_1(x)$, $N_2(x)$, and $N_3(x)$. *Hint:* Start with an assumed solution in the following form: $u(x) = c_0 + c_1 x + c_2 x^2$.

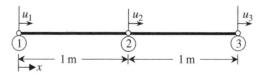

18. A vertical rod of elastic material is fixed at both ends with constant cross-sectional area A, Young's modulus E, and height of L under the distributed load f per unit length. The vertical deflection $u(x)$ of the rod is governed by the following differential equation:

$$AE\frac{d^2u}{dx^2}+f=0.$$

Using three elements of equal length, solve for $u(x)$ and compare it with the exact solution. Use the following numerical values: $A = 10^{-4}\,\text{m}^2$, $E = 10\,\text{GPa}$, $L = 0.3$ m, $f = 10^6$ N/m.

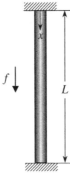

19. A bar in the figure is under the uniformly distributed load q due to gravity. For a linear elastic material with Young's modulus E and uniform cross-sectional area A, the governing differential equation can be written as

$$AE\frac{d^2u}{dx^2}+q=0,$$

where $u(x)$ is the downward displacement. The bar is fixed at the top and free at the bottom. Using the Galerkin method and two equal-length finite elements, answer the following questions.

(a) Starting from the above differential equation, derive an integral equation using the Galerkin method.

(b) Write the expression of boundary conditions at $x = 0$ and $x = L$. Identify whether they are essential or natural boundary conditions.

(c) Derive the assembled finite element matrix equation, and solve it after applying boundary conditions.

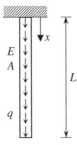

20. Consider a tapered bar of circular cross section. The length of the bar is 1 m, and the radius varies as $r(x) = 0.050 - 0.040x$, where r and x are in meters. Assume Young's modulus $= 100$ MPa. Both ends of the bar are fixed, and a uniformly distributed load of 10,000 N/m is applied along the entire length of the bar. Determine the displacements, axial force distribution, and the wall reactions using:

(a) three elements of equal length; and

(b) four elements of equal length.

Compare your results with the exact solution by plotting $u(x)$ and $P(x)$ curves for each case (a) and (b) and the exact solution. How do the finite element results for the left and right wall reactions, R_L and R_R, compare with the exact solution?

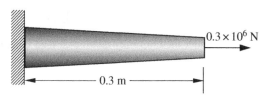

Hint: To approximate the area of the cross section of a bar element, use the geometric mean of the end areas of the element, that is, $A^{(e)} = \sqrt{A_i A_j} = \pi r_i r_j$. The exact solution is obtained by solving the following differential equation with the boundary conditions $u(0) = 0$ and $u(1) = 0$:

$$\frac{d}{dx}\left(A(x)E\frac{du}{dx}\right) = -p(x) = -10,000.$$

The axial force distribution is found from $P(x) = A(x)E\, du/dx$. The wall reactions are: $R_L = -P(0)$ and $R_R = P(1)$.

21. A tapered bar with circular cross section is fixed at $x = 0$, and an axial force of 0.3×10^6 N is applied at the other end. The length of the bar (L) is 0.3 m, and the radius varies as $r(x) = 0.03 - 0.07x$, where r and x are in meters. Use three equal-length finite elements to determine the displacements, axial force resultants, and support reactions. Compare your FE solutions with the exact solution by plotting u vs. x, and P (element force) vs. x. Use $E = 10^{10}$ Pa.

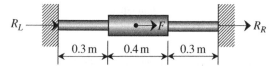

22. The stepped bar shown in the figure is subjected to a force at the center. Use the finite element method to determine the displacement field $u(x)$, axial force distribution $P(x)$, and reactions R_L and R_R.

Assume: $E = 100$ GPa, the areas of cross sections of the three portions shown are, respectively, 10^{-4} m^2, 2×10^{-4} m^2, and 10^{-4} m^2, and $F = 10,000$ N.

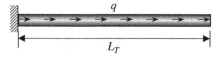

23. A bar shown in the figure is modeled using three equal-length bar elements. The total length of the bar is $L_T = 1.5$ m, and the radius of the circular cross section is $r = 0.1$ m. When Young's modulus $E = 207$ GPa, and distributed load $q = 1,000$ N/m, calculate displacement and stress using a commercial finite element analysis program. Compare your finite element solution with the exact solution. Provide XY graphs of displacement and stress with respect to bar length. Explain why the finite element solutions are different from the exact solutions.

24. Consider the tapered bar in problem 17. Use the Rayleigh-Ritz method to solve the same problem. Assume the displacement in the form of $u(x) = a_0 + a_1 x + a_2 x^2$. Compare the solutions for $u(x)$ and $P(x)$ with the exact solution given below by plotting them.

$$u(x) = \frac{Fx}{\pi E r(0) r(x)}, \quad P(x) = EA(x)\frac{du}{dx} = F.$$

25. Consider the tapered bar in problem 21. Use the Rayleigh-Ritz method to solve the same problem. Assume the displacement in the form of $u(x) = (x-1)(c_1 x + c_2 x^2)$.

26. Consider the uniform bar in the figure. Axial load q is linearly distributed along the length of the bar according to $q = cx$, where c is a constant that has units of force divided by the square of length. Calculate the axial displacement $u(x)$ and axial stress σ_x using the Rayleigh-Ritz method. Assume the following form of displacement: $u(x) = a_0 + a_1 x$.

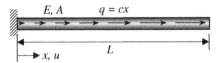

27. Determine shape functions of a bar element shown in the figure by assuming the following form of displacement: $u(x) = a_1 x + a_2 x^2$; that is, obtain $N_1(x)$ and $N_2(x)$ such that $u(x) = N_1(x)u_1 + N_2(x)u_2$. Calculate axial strain $\varepsilon_{xx} = du/dx$ when $u_1 = u_2 = 1$ (rigid body motion). Explain why strain is not zero under the rigid-body motion.

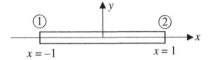

28. Consider a finite element with three nodes, as shown in the figure. When the solution is approximated using $u(x) = N_1(x)u_1 + N_2(x)u_2 + N_3(x)u_3$,

 (a) calculate the interpolation functions $N_1(x)$, $N_2(x)$, and $N_3(x)$ if it is intended to obtain the displacement field in the following form: $u(x) = c_0 + c_1 \sqrt{x} + c_2 x$; and

 (b) when the nodal displacements are given as: $u_1 = 0, u_2 = 0.5$, and $u_3 = 1$, sketch the function $u(x)$.

Chapter 3

Finite Element Analysis of Beams and Frames

In chapter 1, the finite element equations of a truss were obtained using the direct stiffness method. Similar direct methods for beams are possible but quite complicated, and such methods are impossible for plates and two-dimensional and three-dimensional solids. In chapter 2, we introduced the Galerkin method and the principle of minimum potential energy for different engineering problems. In this chapter, we will formally derive the finite element equations for beams using the energy method. The same finite element equation can be obtained using the principle of virtual work.[1]

In chapter 2, we learned that in the finite element method, the displacements in an element are interpolated using an expression of the form $u(x) = \{\mathbf{N}(x)\}^T\{\mathbf{q}\}$, in which $\{\mathbf{N}(x)\}$ is the column vector of shape functions, and $\{\mathbf{q}\}$ is the vector of nodal displacements or in general nodal degrees of freedom (DOFs). In the case of beam finite element, the nodal DOFs include the vertical (or transverse) deflection as well as the rotation (or slope). Using this interpolation scheme, the stiffness matrix and applied load vector are derived and solved for the nodal DOFs.

After a review of the elementary beam theory in section 3.1, we will first present the Rayleigh-Ritz method in section 3.2. The formal development of the interpolation functions for the beam finite elements is presented in section 3.3. The principle of minimum potential energy for beam elements is presented in the same section. In section 3.4, we present the two-dimensional frame finite element, which combines the action of a uniaxial bar and a beam. Section 3.5 presents the finite element formulation of beam buckling problems, which is further extended to buckling of frame finite element in section 3.6. Some modeling practices for beams are presented in section 3.7.

3.1 REVIEW OF ELEMENTARY BEAM THEORY

Unlike the uniaxial bar, a beam can carry a transverse load, and the slope of the beam can change along its span. In fact, the shape of a bar and that of a beam are similar, but their usage is different due to the type of loading. If a slender member carries a force in its axial direction, it can be modeled as a bar, while if the same slender member carries a force in the transverse direction, it should be modeled as a beam.

Let us consider a beam with its longitudinal axis parallel to the *x*-axis. We will consider beam cross sections that are symmetric about the plane of loading (*xy*-plane), and all applied loads will reside in this plane. The origin of the local coordinate system is assumed to be at the centroid of the cross section with the *x*-axis aligned along the length and is called the *neutral axis*. In elementary beam theory, which is

---

[1] Hughes, T. J. R. 1987. *The Finite Element Method*. Prentice-Hall, Englewood Cliffs, NJ.

Introduction to Finite Element Analysis and Design, Second Edition. Nam H. Kim, Bhavani V. Sankar, and Ashok V. Kumar.
© 2018 John Wiley & Sons Ltd. Published 2018 by John Wiley & Sons Ltd.
Companion website: www.wiley.com/go/kim/finite_element_analysis_design

also called *Euler-Bernoulli beam* theory, we assume that the transverse deflection is independent of y and is a function of x only. That is, the deflection of the beam is represented by $v(x)$, which is also called the *deflection curve*. The displacement in the x direction is represented by $u(x, y)$ because its value will change at different locations of y as the beam bends. Euler-Bernoulli beam theory is based on the assumption that plane sections normal to the beam axis remain plane and normal to the axis after deformation. Then, the displacement field $u(x, y)$ can be written as

$$u(x,y) = u_0(x) - y\frac{dv}{dx}, \tag{3.1}$$

where u_0 is the x-directional displacement of the beam along the neutral axis, and $\theta = dv/dx$ is the slope of the beam (see figure 3.1). The first term on the right-hand side is due to axial deformation (bar), and the second term is due to bending (beam). As the beam rotates, $u(x, y)$ is linearly proportional to y, and the proportionality constant is the slope of rotation.

From eq. (3.1) the normal strain in the beam is derived as

$$\varepsilon_{xx} = \frac{\partial u}{\partial x} = \frac{du_0}{dx} - y\frac{d^2v}{dx^2}. \tag{3.2}$$

The term du_0/dx or ε_0 represents the strain along the beam axis (x-axis) at the location of $y = 0$, that is, at the neutral axis. One may note that the strain varies linearly in y at a given cross section of the beam. The term $-d^2v/dx^2$ is an approximation for the *curvature* of the deflection curve. The normal strain ε_{yy} vanishes everywhere as we have assumed v is independent of y. It is interesting to note that the transverse deflection v causes the axial strain ε_{xx} because the second-order derivative is related to the change in the slope of the cross section.

We assume a state of plane stress normal to the z-axis as the depth of the beam is relatively small. Then, the stress-strain relationship is the same as the uniaxial problem. The normal stress σ_{xx} in the beam cross section is given by

$$\sigma_{xx} = E\varepsilon_{xx} = E\varepsilon_0 - Ey\frac{d^2v}{dx^2}. \tag{3.3}$$

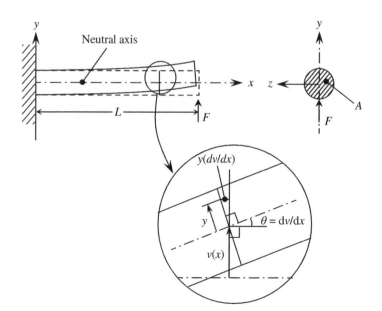

Figure 3.1 Deflection of a plane Euler-Bernoulli beam

Note that the normal stress σ_{xx} varies linearly in y at a given cross section of the beam and is offset by the stress from the uniaxial bar. Therefore, the maximum or minimum stress always occurs at the top or bottom end of the cross section.

The axial force resultant P and the bending moment M at a cross section can be obtained by integrating the axial stress, as

$$P = \int_A \sigma_{xx} \mathrm{d}A, \ M = -\int_A y\sigma_{xx} \mathrm{d}A, \tag{3.4}$$

where integration is performed over the cross-sectional area, A. Substituting for σ_{xx} from eq. (3.3), the axial force and bending moment in the above equations take the following forms:

$$P = E\varepsilon_0 \int_A \mathrm{d}A - E\frac{\mathrm{d}^2v}{\mathrm{d}x^2} \int_A y\mathrm{d}A,$$

$$M = -E\varepsilon_0 \int_A y\mathrm{d}A + E\frac{\mathrm{d}^2v}{\mathrm{d}x^2} \int_A y^2\mathrm{d}A. \tag{3.5}$$

In the above equation, ε_0 and the curvature terms are outside the integral because they are a function of the x-coordinate only. Since the choice of the beam axis (x-axis) is such that it passes through the centroid of the cross section, the first moment of the area, $\int y\mathrm{d}A$, vanishes, and we can recognize the second moment of inertia of the cross section, $I = \int y^2\,\mathrm{d}A$, in the expression for bending moment M in the above equation. Now the expressions for P and M take the following forms

$$P = EA\varepsilon_0,$$

$$M = EI\frac{\mathrm{d}^2v}{\mathrm{d}x^2}. \tag{3.6}$$

where A and I are, respectively, the area and moment of inertia of the cross section. It should be noted that the moment of inertia I is about the z–axis passing through the centroid of the cross section, which is usually denoted by I_{zz} or I_z. The terms EA and EI are called, respectively, the *axial rigidity* and *flexural rigidity* of the beam cross section. The first part of eq. (3.6) corresponds to the uniaxial bar in chapter 1. In this section, we will assume the beam does not have any net axial force, that is, $P = 0$ [2]. Then from eq. (3.6), $\varepsilon_0 = 0$. Thus the only equation we need is $M = EI\left(\mathrm{d}^2v/\mathrm{d}x^2\right)$. The second part of eq. (3.6) will be called the *beam constitutive relation*, or the *moment–curvature relation*. The relationship between the bending moment and the curvature is linear with the proportionality constant of flexural rigidity.

Since P and M are derived from the stress σ_{xx}, their sign conventions are similar to those of stresses rather than those of forces and couples. Positive P, V_y, and M are illustrated in figure 3.2. Note that V_y is

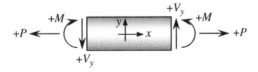

Figure 3.2 Positive directions for axial force, shear force, and bending moment of a plane beam

---

[2] When an axial force is present in addition to the bending moment, we call it a frame element in this text. Most finite element analysis programs refer to it as a beam element. We separate the frame element from the beam element for the purpose of progressive explanation. The frame finite element will be discussed in section 3.4.

the transverse shear force acting on the beam cross section. It is emphasized that these are not externally applied forces or moment; they are internally generated due to the deformation.

A beam can be subjected to concentrated forces and couples, F_i and C_i, and distributed transverse force $p(x)$ as shown in figure 3.3. These are externally applied loads. Note that F and p are considered positive when they act in the positive y direction, whereas a counterclockwise couple is considered positive. The common units of the distributed force p are N/m and lb/in.

Another set of equations that will complement the moment-curvature relation of eq. (3.6) are the beam equilibrium equations. Consider the free-body diagram of the infinitesimal beam shown in figure 3.3. The shear force acting at a cross section is denoted by V_y. Force equilibrium in the y direction requires

$$\sum f_y = 0 \Rightarrow p(x)\mathrm{d}x + \left(V_y + \frac{\mathrm{d}V_y}{\mathrm{d}x}\mathrm{d}x \right) - V_y = 0,$$

or

$$\boxed{\frac{\mathrm{d}V_y}{\mathrm{d}x} = -p(x)}. \tag{3.7}$$

Similarly, taking moments about z-axis passing through the right face of the element yields

$$-M + \left(M + \frac{\mathrm{d}M}{\mathrm{d}x}\mathrm{d}x \right) - (p\mathrm{d}x)\frac{\mathrm{d}x}{2} + V_y\mathrm{d}x = 0,$$

or

$$\boxed{V_y = -\frac{\mathrm{d}M}{\mathrm{d}x}}. \tag{3.8}$$

In deriving eqs. (3.7) and (3.8), the term that includes $(\mathrm{d}x)^2$ is ignored because it is a higher-order term. The equations in the boxes in eqs. (3.7) and (3.8) are the equilibrium equations of the beam. Combining these two equations with the beam constitutive relation $M = EI\left(\mathrm{d}^2v/\mathrm{d}x^2\right)$, we obtain the governing differential equation of the beam

$$\boxed{EI\frac{\mathrm{d}^4v}{\mathrm{d}x^4} = p(x)}. \tag{3.9}$$

The above equation is a fourth-order differential equation, and it requires four boundary conditions to determine integral constants.

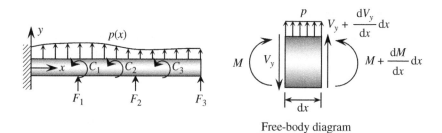

Free-body diagram

Figure 3.3 Equilibrium of infinitesimal beam section under various loadings

The stresses in the beam in a cross section can be determined from the bending moment and shear force resultants. The expression for stress σ_{xx} in the absence of axial force P can be obtained from eq. (3.3) as

$$\sigma_{xx} = -Ey\frac{\mathrm{d}^2 v}{\mathrm{d}x^2}. \tag{3.10}$$

Substituting for the curvature from eq. (3.6) we obtain

$$\boxed{\sigma_{xx}(x,y) = -\frac{M(x)y}{I}}. \tag{3.11}$$

Because of the Euler-Bernoulli beam theory assumptions that $v(x)$ is independent of y and the assumed form of $u(x, y)$ in eq. (3.1), we obtain the shear strain γ_{xy} as

$$\gamma_{xy} = \frac{\partial u}{\partial y} + \frac{\partial v}{\partial x} = -\frac{\partial v}{\partial x} + \frac{\partial v}{\partial x} = 0. \tag{3.12}$$

That is, Euler-Bernoulli beam theory predicts zero shear strain. However, we know that a beam cross section is subjected to shear stresses, which results in the transverse shear force resultant V_y as derived in eq. (3.8). According to the theory of elasticity, there are nonzero shear strains in a beam, but they are small compared to the normal strains. The average shear stress in a cross section is given by V_y/A, although the maximum shear stress depends on the cross-sectional geometry. For example, in a rectangular cross section, the shear stress has a parabolic variation through the thickness with maximum value at the center equal to 1.5 times the average shear stress.

We will use the principle of minimum potential energy, the Rayleigh-Ritz method to be specific, to derive the finite element equations. The potential energy is the net amount of energy that is stored in the structure during its deformation. Referring to chapter 2, the potential energy is defined as

$$\Pi = U + V, \tag{3.13}$$

where U is the strain energy, and V is the potential energy of external forces. In the Rayleigh-Ritz method, the deflection of the beam is expressed in terms of unknown coefficients. The objective is then to represent the potential energy in terms of these coefficients, and then to differentiate it with respect to the coefficients in order to minimize the potential energy.

The energy method requires an expression for strain energy in the beam, which is derived as follows. The strain energy density at a point in the beam is given by

$$U_0 = \frac{1}{2}\sigma_{xx}\varepsilon_{xx} = \frac{1}{2}E(\varepsilon_{xx})^2 = \frac{1}{2}E\left(-y\frac{\mathrm{d}^2 v}{\mathrm{d}x^2}\right)^2 = \frac{1}{2}Ey^2\left(\frac{\mathrm{d}^2 v}{\mathrm{d}x^2}\right)^2, \tag{3.14}$$

where we have substituted for strain from eq. (3.2) with $\varepsilon_0 = 0$. In the context of beams, we define another strain energy density term, which is called the strain energy per unit length of the beam, U_L. It is derived by integrating U_0 over the entire cross section at a given x:

$$U_L(x) = \int_A U_0(x, y, z)\,\mathrm{d}A. \tag{3.15}$$

Substituting for U_0 from eq. (3.14) in eq. (3.15) yields an expression for strain energy per unit length of the beam

$$U_L(x) = \int_A \frac{1}{2}Ey^2\left(\frac{\mathrm{d}^2 v}{\mathrm{d}x^2}\right)^2\,\mathrm{d}A = \frac{1}{2}E\left(\frac{\mathrm{d}^2 v}{\mathrm{d}x^2}\right)^2\int_A y^2\,\mathrm{d}A,$$

or

$$U_L(x) = \frac{1}{2}EI\left(\frac{d^2v}{dx^2}\right)^2.$$

(3.16)

The units for U_L are J/m or Nm/m or lbf·in/in. By substituting the moment–curvature relation of eq. (3.6) we obtain an expression for U_L in terms of M as $U_L = M^2/2EI$.

The strain energy U in the beam can be derived as

$$U = \int_0^L U_L(x)\,dx = \frac{1}{2}\int_0^L EI\left(\frac{d^2v}{dx^2}\right)^2 dx.$$

(3.17)

Figure 3.3 shows positive directions of concentrated forces, couples, and distributed loads. Using these notations, the potential energy of external forces and moments can be represented by

$$V = -\int_0^L p(x)v(x)\,dx - \sum_{i=1}^{N_F} F_i v(x_i) - \sum_{i=1}^{N_C} C_i \theta(x_i),$$

(3.18)

where N_F and N_C are, respectively, the number of concentrated forces and couples applied to the beam, and $\theta(x)$ is the rotation or slope of the beam.

Thus, the potential energy in eq. (3.13) can be represented using the transverse deflection and slope (derivative of the deflection), as

$$\Pi = U + V$$

$$= \frac{1}{2}\int_0^L EI\left(\frac{d^2v}{dx^2}\right)^2 dx - \int_0^L p(x)v(x)\,dx - \sum_{i=1}^{N_F} F_i v(x_i) - \sum_{i=1}^{N_C} C_i \theta(x_i).$$

(3.19)

The principle of minimum potential energy in chapter 2 says that the beam is in equilibrium when the potential energy has its minimum value. We will present the Rayleigh–Ritz method, followed by the finite element method.

3.2   RAYLEIGH-RITZ METHOD

As discussed in chapter 2, the Rayleigh–Ritz method can be used for continuous systems. In the Rayleigh–Ritz method, a continuous system is approximated as a discrete system with finite number of DOFs. This is accomplished by approximating the displacements by a function containing a finite number of coefficients to be determined. The total potential energy is then evaluated in terms of the unknown coefficients. Then the principle of minimum potential energy is applied to determine the best set of coefficients by minimizing the total potential energy with respect to the coefficients. The solution thus obtained may not be exact. It is the best solution from among the family of solutions that can be obtained from the assumed displacement functions. In the following, we demonstrate the method to beam problems. The steps involved in solving the beam problem using the Rayleigh–Ritz method are as follows.

1. Assume a deflection shape for the beam in the following form: $v(x) = c_1 f_1(x) + c_2 f_2(x) \ldots\ldots + c_n f_n(x)$, where c_i are coefficients to be determined, and $f_i(x)$ are known bases. The deflection curve $v(x)$ *must* satisfy the displacement boundary conditions, *for instance*, deflection $v = 0$ or slope $dv/dx = 0$.

2. Determine the strain energy U in the beam using the formula in eq. (3.17) in terms of c_i.

3. Find the potential energy of external forces V using formulas of types given in eq. (3.18).

4. The total potential energy is obtained as $\Pi(c_1, c_2, \ldots c_n) = U + V$.

5. Apply the principle of minimum potential energy to determine the coefficients $c_1, c_2, \ldots c_n$.

EXAMPLE 3.1 *Rayleigh–Ritz method for a simply supported beam*

Consider a simply supported beam of length L subjected to a uniformly distributed transverse load $p(x) = p_0$. Use the Rayleigh–Ritz method to determine the transverse deflection $v(x)$ of the beam. Assume the flexural rigidity of beam cross section is EI.

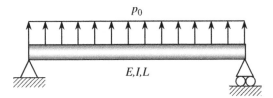

p_0

E,I,L

Figure 3.4 Simply supported beam under uniformly distributed load

SOLUTION As in the case of a uniaxial bar, we assume the deflection of the beam using some convenient functions that satisfy the displacement boundary conditions. In the case of simply supported beam, the deflection must be zero at both ends of the beam. We will deviate from typical polynomial forms and assume a sinusoidal function as follows:

$$v(x) = C \sin \frac{\pi x}{L}, \tag{3.20}$$

where C is the coefficient to be determined by minimizing the total potential energy. It is important to note that the function satisfies the displacement boundary condition independent of the choice of coefficient C, that is, $\sin 0 = \sin \pi = 0$. If a function does not satisfy the displacement boundary condition, its coefficient must be zero.

Here we have approximated the beam as a one-DOF system. The strain energy of the beam is given in eq. (3.17). Substituting for the deflection from eq. (3.20) into eq. (3.17), we obtain

$$U = \frac{C^2 EI \pi^4}{4L^3}. \tag{3.21}$$

The potential energy of the external forces is derived as

$$V = -\int_0^L p(x)v(x)\mathrm{d}x = -\int_0^L p_0 C \sin\frac{\pi x}{L}\mathrm{d}x = -\frac{2p_0 L}{\pi}C. \tag{3.22}$$

Then the total potential energy is derived as

$$\Pi = U + V = \frac{EI\pi^4}{4L^3}C^2 - \frac{2p_0 L}{\pi}C. \tag{3.23}$$

The principle of minimum potential energy requires $\mathrm{d}\Pi/\mathrm{d}C = 0$. That is,

$$\frac{\mathrm{d}\Pi}{\mathrm{d}C} = \frac{EI\pi^4}{2L^3}C - \frac{2p_0 L}{\pi} = 0 \ \Rightarrow \ C = \frac{4p_0 L^4}{EI\pi^5}. \tag{3.24}$$

The maximum deflection at the center is equal to the value of the coefficient C, which can be written as $C = P_0 L^4/76.5EI$. The exact deflection is given by $C = P_0 L^4/76.8EI$. The bending moment and the shear force are derived from the approximate deflection using the following relations:

$$\begin{aligned} M(x) &= EI\frac{\mathrm{d}^2 v}{\mathrm{d}x^2} = -EIC\frac{\pi^2}{L^2}\sin\frac{\pi x}{L} = -\frac{4p_0 L^2}{\pi^3}\sin\frac{\pi x}{L}, \\ V_y(x) &= -EI\frac{\mathrm{d}^3 v}{\mathrm{d}x^3} = EIC\frac{\pi^3}{L^3}\cos\frac{\pi x}{L} = \frac{4p_0 L}{\pi^2}\cos\frac{\pi x}{L}. \end{aligned} \tag{3.25}$$

The exact solution for this problem can be found in any mechanics of materials book[3]. The exact expressions for deflection, bending moment, and shear force are as follows:

$$v(x) = \frac{1}{EI}\left(\frac{p_0 L^3}{24}x - \frac{p_0 L}{12}x^3 + \frac{p_0}{24}x^4\right),$$

$$M(x) = -\frac{p_0 L}{2}x + \frac{p_0}{2}x^2, \tag{3.26}$$

$$V_y(x) = \frac{p_0 L}{2} - p_0 x.$$

Let us define non-dimensional deflection, bending moment, and shear force as

$$\bar{x} = \frac{x}{L}, \quad \bar{v} = \frac{384EI}{5p_0 L^4}v, \quad \bar{M} = \frac{1}{p_0 L^2}M, \quad \bar{V} = \frac{1}{p_0 L}V_y. \tag{3.27}$$

The approximate solution is compared with the exact solution by plotting the non-dimensional values of deflection and force and moment resultants over the length of the beam in figure 3.5(a)–(c). It turns out that the approximate deflection is close to the exact one. However, the error increases for the bending moment as well as the shear force.

(a)

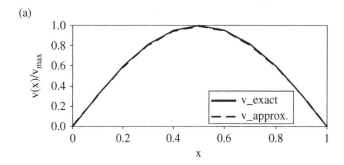

(b)

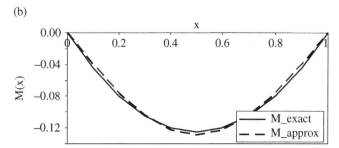

(c)

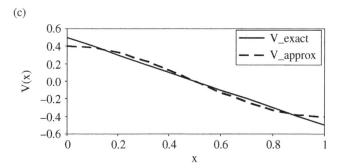

Figure 3.5 Comparison of finite element results with exact ones for a simply supported beam; (a) deflection, (b) bending moment, and (c) shear force

[3] Boresi, A. P., and Schmidt, R. J. 2003. *Advanced Mechanics of Materials*, 6th Ed. John Wiley & Sons, Hoboken, NJ.

Note that the exact shear force is a linear, but the approximate shear force has a cosine term because the deflection is approximated by a sine function. ▄▄

EXAMPLE 3.2 *Rayleigh–Ritz method for a cantilevered beam*

Consider a cantilevered beam subjected to a distributed load $p(x) = -p_0$, a transverse force F, and a couple C at the right end. Use the Rayleigh–Ritz method to obtain the deflection, shear force, and bending moment distribution along the length of the beam. Assume the following numerical values: $E = 100$ GPa, $I = 10^{-7}$ m^4, $L = 1$ m, $p_0 = 300$ N/m, $F = 500$ N, and $C = 100$ N·m.

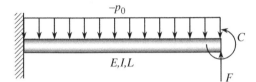

Figure 3.6 Simply supported beam under a uniformly distributed load

SOLUTION We will assume a polynomial in x to approximate the deflection $v(x)$. We will use a cubic polynomial, $v(x) = a + bx + c_1 x^2 + c_2 x^3$, to describe the beam deflection. The essential boundary conditions of the beam are as follows: at $x = 0$, $v(0) = 0$, $dv(0)/dx = 0$. These conditions must be satisfied by the assumed deflections. This requires: $a = b = 0$. Hence, the assumed deflection takes the form

$$v(x) = c_1 x^2 + c_2 x^3. \tag{3.28}$$

Using the expression for strain energy in a beam given in eq. (3.17), we obtain

$$U = \frac{EI}{2} \int_0^L (2c_1 + 6c_2 x)^2 \, dx. \tag{3.29}$$

It is not necessary to perform the integration in eq. (3.29) at this stage. Actually, we require the derivatives of U with respect to c_i, and they can be obtained by performing the differentiation under the integral as shown below:

$$\frac{\partial U}{\partial c_1} = 2EI \int_0^L (2c_1 + 6c_2 x) \, dx = EI \left(4Lc_1 + 6L^2 c_2 \right),$$

$$\frac{\partial U}{\partial c_2} = 6EI \int_0^L (2c_1 + 6c_2 x) x \, dx = EI \left(6L^2 c_1 + 12L^3 c_2 \right). \tag{3.30}$$

The potential energy of the external forces can be derived as follows:

$$
\begin{aligned}
V(c_1, c_2) &= -\int_0^L (-p_0) v(x) \, dx - Fv(L) - C \frac{dv}{dx}(L) \\
&= -\int_0^L (-p_0) \left(c_1 x^2 + c_2 x^3 \right) dx - F \left(c_1 L^2 + c_2 L^3 \right) \\
&\quad - C \left(2c_1 L + 3c_2 L^2 \right) \\
&= c_1 \left(\frac{p_0 L^3}{3} - FL^2 - 2CL \right) + c_2 \left(\frac{p_0 L^4}{4} - FL^3 - 3CL^2 \right).
\end{aligned}
\tag{3.31}
$$

The principle of minimum potential energy requires Π be minimum with respect to the coefficient c_i. Thus, the potential energy is differentiated with respect to two unknown coefficients to obtain:

$$\frac{\partial \Pi}{\partial c_1} = \frac{\partial U}{\partial c_1} + \frac{\partial V}{\partial c_1} = 0,$$
$$\frac{\partial \Pi}{\partial c_2} = \frac{\partial U}{\partial c_2} + \frac{\partial V}{\partial c_2} = 0. \tag{3.32}$$

Substituting from eqs. (3.30) and (3.31) into eq. (3.32) we obtain two equations for c_i:

$$EI(4Lc_1 + 6L^2 c_2) = -\frac{p_0 L^3}{3} + FL^2 + 2CL,$$
$$EI(6L^2 c_1 + 12L^3 c_2) = -\frac{p_0 L^4}{4} + FL^3 + 3CL^2. \tag{3.33}$$

Substituting the numerical values for the beam properties and loads, we obtain

$$10^4(4c_1 + 6c_2) = 600,$$
$$10^4(6c_1 + 12c_2) = 725. \tag{3.34}$$

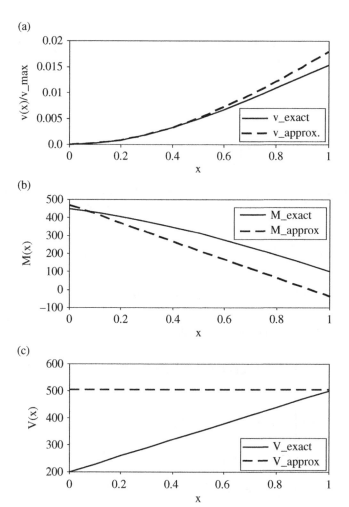

Figure 3.7 Comparison of finite element results with exact ones for a cantilevered beam; (a) deflection, (b) bending moment, and (c) shear force

The solution is: $c_1 = 23.75 \times 10^{-3}$, $c_2 = -5.833 \times 10^{-3}$. Substituting for c_i in eq. (3.28), we obtain the beam deflection $v(x)$ as

$$v(x) = 10^{-3}\left(23.75x^2 - 5.833x^3\right). \tag{3.35}$$

The exact solution for deflection is given by

$$v(x) = \frac{1}{24EI}\left(5400x^2 - 800x^3 - 300x^4\right).$$

The deflection, bending moment, and shear force resultants are compared with the exact solution in figure 3.7. It turns out that the exact solution is a quartic polynomial, while the approximate solution is cubic. This explains the error in the deflection curve in figure 3.7(a). If a fourth-order polynomial is used in eq. (3.28), the approximate solution will be the same as the exact solution. The errors usually increase for the bending moment and the shear force, as they are derivatives of the deflection curve. ▪

The Rayleigh-Ritz method is really a powerful tool to obtain an approximate solution. However, this method has two challenges in order to solve complex problems: (a) it is difficult to find functions that satisfy the displacement boundary conditions, especially when the geometry is complicated; and (b) it is difficult to find functions that can accurately represent the complicated deformation of the entire structure. In the following sections, we develop the finite element version of the Rayleigh-Ritz method that can resolve these two challenges.

3.3   FINITE ELEMENT FORMULATION FOR BEAMS

3.3.1   Finite Element Interpolation

The finite element method differs from the Rayleigh-Ritz method in that the approximation is performed within an element, rather than the entire structure. By using the property of interpolation functions, it is trivial to satisfy the essential boundary conditions in this approach. When many elements are used to discretize the structure, simple polynomial-type functions may be enough to approximate the solution within the element with an acceptable accuracy. In this section, we will present the interpolation method similar to that in chapter 2, but specialized for the beam element. This interpolation will be used in approximating the potential energy in the following section.

Consider a beam shown in figure 3.3. It is subjected to a distributed force and several concentrated or point forces and couples. Our goal is to determine the deflection curve $v(x)$ of the beam, the bending moment distribution $M(x)$, and shear force resultant $V_y(x)$ along the length of the beam. The first step in finite element analysis is to divide the beam into a number of elements. An element is connected to the adjacent element at nodes. Concentrated forces and couples can be applied only at nodes; that means, there should be nodes at points where concentrated forces and couples are applied. In this text, we will consider a beam element that consists of two end nodes. In general, however, it is possible to have more than two nodes in an element. The positive directions of applied forces and couples are shown in figure 3.8. The distributed load $p(x)$ can change along the x-axis. However, we will only consider the cases of either constant or linear distribution.

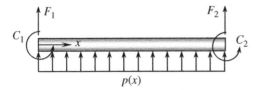

Figure 3.8 Positive directions for forces and couples in a beam element

The DOFs in beam elements are the transverse deflection v and the rotation θ of the cross section that is also equal to the slope dv/dx. The transverse deflection is positive in the positive y direction, whereas a counterclockwise rotation of the cross section is considered positive. Consider a typical element shown in figure 3.9. Our goal is to interpolate the deflection at any point on the element in terms of nodal DOFs $v_1, \theta_1, v_2,$ and θ_2. We first define a vector of nodal DOFs, as

$$\{\mathbf{q}\} = \{v_1 \quad \theta_1 \quad v_2 \quad \theta_2\}^{\mathrm{T}}. \tag{3.36}$$

It is convenient to define a parameter s that varies from 0 to 1 within the element so that a unified derivation is possible for an element of any length. The parameter s can be defined as (see figure 3.9)

$$s = \frac{x - x_1}{L}, \quad ds = \frac{1}{L}dx, \quad dx = Lds, \quad \frac{ds}{dx} = \frac{1}{L}, \tag{3.37}$$

where x_1 is the x-coordinate of node 1 (first node of the element). Thus, the deflection curve can be written in terms of the parameter s, that is, $v(s)$ [4]. The relation in eq. (3.37) can be considered as a mapping between the physical coordinate x and parametric coordinate s. In that regard, the last relation, $ds/dx = 1/L$, is called the Jacobian of the mapping.

Our goal is to interpolate the deflection $v(s)$ in terms of the nodal DOFs. Since the beam element has four nodal values, it is appropriate to use a cubic function to approximate the deflection with four unknown coefficients:

$$v(s) = a_0 + a_1 s + a_2 s^2 + a_3 s^3, \tag{3.38}$$

where $a_0, a_1, a_2,$ and a_3 are the constants to be determined in terms of the nodal DOFs. In addition to the deflection, the slope or rotation θ is also included as the nodal values. Thus, it is necessary to differentiate $v(s)$ with respect to x. However, $v(s)$ in eq. (3.38) is expressed in terms of the parameter s. The expression for the slope can be obtained using the chain rule of differentiation as

$$\frac{dv}{dx} = \frac{dv}{ds}\frac{ds}{dx} = \frac{1}{L}\left(a_1 + 2a_2 s + 3a_3 s^2\right). \tag{3.39}$$

Note that the relation $ds/dx = 1/L$ is from eq. (3.37).

The expression of the strain energy in eq. (3.16) contains the second derivative of the transverse displacement, whose expression can be obtained by differentiating the above equation as

$$\frac{d^2 v}{dx^2} = \frac{d}{ds}\left(\frac{dv}{dx}\right)\frac{ds}{dx} = \frac{1}{L^2}(2a_2 + 6a_3 s). \tag{3.40}$$

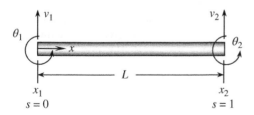

Figure 3.9 Nodal displacements and rotations for the beam element

---

[4] It is also possible to define the parameter as $-1 \le s \le 1$, which is commonly used in commercial finite element programs. It is a good practice to derive the shape functions of a beam element using this parameter.

Now, the four coefficients ($a_0, a_1, a_2,$ and a_3) need to be expressed in terms of nodal values. By definition, the vertical displacement at the left end of the element ($s=0$) is v_1, and the slope is θ_1. By evaluating eqs. (3.38) and (3.39) at $s=0$, we can calculate a_0 and a_1, as

$$v(0) = v_1 = a_0,$$

$$\frac{dv}{dx}(0) = \theta_1 = \frac{a_1}{L}, \quad a_1 = L\theta_1.$$

In the same way, we can evaluate (3.38) and (3.39) at $s=1$ to obtain the following simultaneous system equations, as

$$v(1) = v_2 = v_1 + L\theta_1 + a_2 + a_3,$$

$$\frac{dv}{dx}(1) = \theta_2 = \frac{1}{L}(L\theta_1 + 2a_2 + 3a_3).$$

By solving the above two equations for a_2 and a_3, we have

$$a_2 = -3v_1 - 2L\theta_1 + 3v_2 - L\theta_2,$$

$$a_3 = 2v_1 + L\theta_1 - 2v_2 + L\theta_2.$$

Thus, all unknown coefficients are expressed in terms of nodal values. By substituting these coefficients into eq. (3.38), we have

$$\begin{aligned} v(s) = v_1 + L\theta_1 s + (-3v_1 - 2L\theta_1 + 3v_2 - L\theta_2)s^2 \\ + (2v_1 + L\theta_1 - 2v_2 + L\theta_2)s^3. \end{aligned} \tag{3.41}$$

Since our goal is to express $v(s)$ in terms of nodal values, eq. (3.41) can be rearranged to obtain

$$\begin{aligned} v(s) = \left(1 - 3s^2 + 2s^3\right)v_1 \\ + L\left(s - 2s^2 + s^3\right)\theta_1 \\ + \left(3s^2 - 2s^3\right)v_2 \\ + L\left(-s^2 + s^3\right)\theta_2. \end{aligned} \tag{3.42}$$

An important concept in finite element approximation is the definition of the *shape functions*, which are the coefficients of the nodal values. The deflection $v(s)$ in (3.42) can be written in the form $v(s) = N_1(s)v_1 + N_2(s)\theta_1 + N_3(s)v_2 + N_4(s)\theta_2$. The coefficients of the nodal DOFs in eq. (3.42) are called the *shape functions* of the beam element:

$$\boxed{\begin{aligned} N_1(s) &= 1 - 3s^2 + 2s^3, \\ N_2(s) &= L(s - 2s^2 + s^3), \\ N_3(s) &= 3s^2 - 2s^3, \\ N_4(s) &= L(-s^2 + s^3). \end{aligned}} \tag{3.43}$$

Equation (3.41) can then be written in matrix form as

$$v(s) = \{N_1(s) \ \ N_2(s) \ \ N_3(s) \ \ N_4(s)\}\begin{Bmatrix} v_1 \\ \theta_1 \\ v_2 \\ \theta_2 \end{Bmatrix},$$

or

$$v(s) = \{\mathbf{N}\}^{\mathrm{T}}\{\mathbf{q}\},$$

(3.44)

where $\{\mathbf{N}\}$ is the column vector of shape functions of the beam element. Equation (3.44) approximates the deflection of the beam using the values of deflections and slope at the two nodes. For example, evaluation of eq. (3.44) at $s = 1/2$ will provide the beam deflection at the center of the element.

Figure 3.10 shows the plot of the shape functions. These shape functions are also called the *Hermite polynomials*. Note that when $s = 0$, all shape functions are zero except for N_1, which is equal to unity. Thus, eq. (3.41) yields $v(0) = v_1$, which is the desired result. When $s = 1$, the only nonzero shape function is N_3. Thus, the approximation in eq. (3.41) yields $v(1) = v_2$.

Note that the interpolation of the beam deflection in eq. (3.44) is valid within an element. If the beam consists of more than one element, the interpolation must be performed in each element. Two adjacent elements will have continuous deflection and slope, as they share the nodal values.

The second derivative of the deflection in eq. (3.40) can be derived as

$$\frac{d^2v}{dx^2} = \frac{1}{L^2}\frac{d^2v}{ds^2} = \frac{1}{L^2}\{-6+12s,\ L(-4+6s),\ 6-12s,\ L(-2+6s)\}\begin{Bmatrix} v_1 \\ \theta_1 \\ v_2 \\ \theta_2 \end{Bmatrix},$$

or

$$\frac{d^2v}{dx^2} = \frac{1}{L^2}\underset{1\times4}{\{\mathbf{B}\}^{\mathrm{T}}}\underset{4\times1}{\{\mathbf{q}\}},$$

(3.45)

where the column vector $\{\mathbf{B}\}$ is, in general, called the *strain–displacement vector*. In the context of beams, it relates the curvature of the beam to the nodal displacements. Note that the vector $\{\mathbf{B}\}$ is a linear function of the parameter s. Thus, the curvature varies in a linear fashion within an element. The reader can verify that the curvature term in (3.45) can also be written as

$$\frac{d^2v}{dx^2} = \frac{1}{L^2}\underset{1\times4}{\{\mathbf{q}\}^{\mathrm{T}}}\underset{4\times1}{\{\mathbf{B}\}}.$$

(3.46)

Equation (3.45) provides some information about the quality of interpolation. If the loads acting on a beam results in a constant or linear variation of curvature or bending moment, then the interpolation in eq. (3.45) will represent it accurately. However, when a problem requires a higher-order variation of

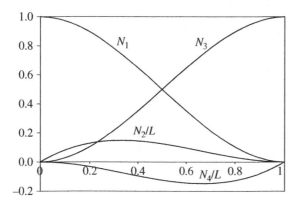

Figure 3.10 Shape functions of the beam element

curvature (bending moment) with respect to x, then the interpolation is only an approximation. In such a case, several beam elements will be required to approximate the higher-order variation of the bending moment with a reasonable accuracy.

Using eq. (3.45), the bending moment and shear force in the beam element can also be calculated in terms of nodal DOFs, as

$$M(s) = EI \frac{d^2 v}{dx^2} = \frac{EI}{L^2} \{\mathbf{B}\}^T \{\mathbf{q}\}, \qquad (3.47)$$

$$V_y = -\frac{dM}{dx} = -EI \frac{d^3 v}{dx^3} = \frac{EI}{L^3} \{-12 \quad -6L \quad 12 \quad -6L\} \{\mathbf{q}\}. \qquad (3.48)$$

It is interesting to note that the bending moment is a linear function of s, and thus, x, while the shear force is constant throughout the element. This is because the assumed deflection is a cubic function as in eq. (3.38). Since the stress is proportional to the bending moment as shown in eq. (3.11), the maximum stress occurs at the location of the maximum bending moment. Since the bending moment varies linearly, the maximum stress always occurs at either nodes of the beam element. However, caution is required because the bending moments between two adjacent elements are discontinuous at the node.

Equation (3.45) can be used to approximate the strain energy in the following section.

EXAMPLE 3.3 *Interpolation in a beam element*

Consider a cantilevered beam as shown in figure 3.11. The beam is approximated using a one-beam finite element. The nodal values of the beam element are given as $\{\mathbf{q}\} = \{0, \ 0, \ -0.1, \ -0.2\}^T$. Calculate the deflection and slope at the midpoint of the beam. Assume $L = 1$ m.

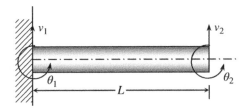

Figure 3.11 Cantilevered beam element with nodal displacements

SOLUTION The value of the parameter s in the middle of the element is $1/2$. Thus, the shape functions in eq. (3.43) are evaluated with $s = 1/2$:

$$N_1 \left(\frac{1}{2}\right) = \frac{1}{2}, \quad N_2 \left(\frac{1}{2}\right) = \frac{L}{8}, \quad N_3 \left(\frac{1}{2}\right) = \frac{1}{2}, \quad N_4 \left(\frac{1}{2}\right) = -\frac{L}{8}.$$

Thus, from eq. (3.44), the vertical deflection at the midpoint of the element is

$$v \left(\frac{1}{2}\right) = N_1 \left(\frac{1}{2}\right) v_1 + N_2 \left(\frac{1}{2}\right) \theta_1 + N_3 \left(\frac{1}{2}\right) v_2 + N_4 \left(\frac{1}{2}\right) \theta_2$$

$$= \frac{1}{2} \times 0 + \frac{L}{8} \times 0 + \frac{1}{2} \times v_2 - \frac{L}{8} \times \theta_2$$

$$= \frac{v_2}{2} - \frac{L\theta_2}{8}$$

$$= -0.025.$$

Next, the slope of the element is defined as $\theta = dv/dx$. Using (3.39), the slope can be expressed as

$$\frac{dv}{dx} = \frac{1}{L}\frac{dv}{ds} = \frac{1}{L}\left(v_1\frac{dN_1}{ds} + \theta_1\frac{dN_2}{ds} + v_2\frac{dN_3}{ds} + \theta_2\frac{dN_4}{ds}\right)$$

$$= v_1\frac{1}{L}\left(-6s + 6s^2\right) + \theta_1\left(1 - 4s + 3s^2\right)$$

$$+ v_2\frac{1}{L}\left(6s - 6s^2\right) + \theta_2\left(-2s + 3s^2\right).$$

The slope in the middle of the beam can be obtained by substituting $s = 1/2$ in the above equation. Then

$$\theta\left(\tfrac{1}{2}\right) = -\frac{3}{2L}v_1 - \frac{1}{4}\theta_1 + \frac{3}{2L}v_2 - \frac{1}{4}\theta_2 = -0.1.$$

3.3.2   Finite Element Equation for the Beam Element

As the reader may recall, one of the steps in finite element analysis is to express the strain energy of the solid in terms of nodal DOFs. In this section, we will derive the finite element equation using the principle of minimum potential energy. Let us consider a beam, shown in figure 3.12, with the total length of L_T. The beam is divided into NEL number of beam elements with equal length L. The elements do not have to be of the same length, but the assumption makes the explanation simple. We further assume that the cross-sectional area remains constant within an element. The beam is under concentrated forces and couples at the nodes and distributed load $p(x)$.

The strain energy in a beam can be formally written in terms of strain energy per unit length as

$$U = \int_0^{L_T} U_L(x)\,dx = \sum_{e=1}^{NEL}\int_{x_1^{(e)}}^{x_2^{(e)}} U_L(x)\,dx = \sum_{e=1}^{NEL} U^{(e)}, \qquad (3.49)$$

where $U^{(e)}$ is the strain energy of element e. The integration is performed over each beam element and summed over NEL number of elements. Note that $x_1^{(e)}$ and $x_2^{(e)}$, respectively, are the x-coordinates of the first and second nodes of element e. Substituting for U_L from eq. (3.16) we obtain

$$U^{(e)} = EI\int_{x_1^{(e)}}^{x_2^{(e)}} \frac{1}{2}\left(\frac{d^2v}{dx^2}\right)^2 dx = \frac{EI}{L^3}\int_0^1 \frac{1}{2}\left(\frac{d^2v}{ds^2}\right)^2 ds. \qquad (3.50)$$

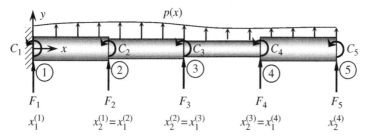

Figure 3.12 Finite element models using four beam elements

In the above equation, we have used the relation $dx = L\,ds$; see eq. (3.37). The expression of the strain energy in eq. (3.50) contains the second-order derivative of deflection $v(x)$. Using the interpolation in eqs. (3.45) and (3.46), we can write the second-order derivative term as

$$\left(\frac{d^2 v}{ds^2}\right)^2 = \left(\frac{d^2 v}{ds^2}\right)\left(\frac{d^2 v}{ds^2}\right) = \underset{1\times 4}{\left\{\mathbf{q}^{(e)}\right\}^{\mathrm{T}}} \underset{4\times 1}{\{\mathbf{B}\}} \underset{1\times 4}{\{\mathbf{B}\}^{\mathrm{T}}} \underset{4\times 1}{\left\{\mathbf{q}^{(e)}\right\}}.$$

Note that for the first curvature term, we have used $\{\mathbf{q}\}^{\mathrm{T}}\{\mathbf{B}\}$, and for the second, $\{\mathbf{B}\}^{\mathrm{T}}\{\mathbf{q}\}$. Substituting the above relation in eq. (3.50), the strain energy of element e is derived as

$$\begin{aligned} U^{(e)} &= \frac{1}{2}\left\{\mathbf{q}^{(e)}\right\}^{\mathrm{T}}\left[\frac{EI}{L^3}\int_0^1 \{\mathbf{B}\}\{\mathbf{B}\}^{\mathrm{T}}ds\right]^{(e)}\left\{\mathbf{q}^{(e)}\right\} \\ &= \frac{1}{2}\left\{\mathbf{q}^{(e)}\right\}^{\mathrm{T}}\left[\mathbf{k}^{(e)}\right]\left\{\mathbf{q}^{(e)}\right\}, \end{aligned} \tag{3.51}$$

where $[\mathbf{k}^{(e)}]$ is the element stiffness matrix of the beam finite element. After integrating, the stiffness matrix can be derived as

$$\left[\mathbf{k}^{(e)}\right] = \frac{EI}{L^3}\int_0^1 \begin{Bmatrix} -6+12s \\ L(-4+6s) \\ 6-12s \\ L(-2+6s) \end{Bmatrix}\{-6+12s \quad L(-4+6s) \quad 6-12s \quad L(-2+6s)\}\,ds,$$

or

$$\left[\mathbf{k}^{(e)}\right] = \frac{EI}{L^3}\begin{bmatrix} 12 & 6L & -12 & 6L \\ 6L & 4L^2 & -6L & 2L^2 \\ -12 & -6L & 12 & -6L \\ 6L & 2L^2 & -6L & 4L^2 \end{bmatrix}. \tag{3.52}$$

which is a symmetric 4×4 matrix. Note that the element stiffness matrix is proportional to the flexural rigidity EI and inversely proportional to L^3. The stiffness matrices of all the elements can then be calculated from the element properties. The strain energy in the beam can then be obtained by summing strain energies of the individual elements as in eq. (3.49):

$$U = \sum_{e=1}^{NEL} U^{(e)} = \frac{1}{2}\sum_{e=1}^{NEL}\left\{\mathbf{q}^{(e)}\right\}^{\mathrm{T}}\left[\mathbf{k}^{(e)}\right]\left\{\mathbf{q}^{(e)}\right\},$$

or

$$U = \frac{1}{2}\{\mathbf{Q}_s\}^{\mathrm{T}}[\mathbf{K}_s]\{\mathbf{Q}_s\}, \tag{3.53}$$

where $\{\mathbf{Q}_s\}$ is the column matrix of all DOFs in the beam, and $[\mathbf{K}_s]$ is the *structural stiffness matrix* obtained by assembling the element stiffness matrices. The assembly procedure, which is similar to that of a uniaxial bar and truss elements, will be illustrated in the examples to follow.

EXAMPLE 3.4 *Assembly of two beam elements*

Consider a stepped beam structure modeled using two beam elements, as shown in figure 3.13. The flexural rigidity of elements 1 and 2 are, respectively, $2EI$ and EI. Construct element stiffness matrices and assemble them to build the structural stiffness matrix.

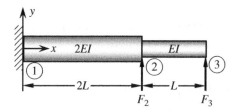

Figure 3.13 Finite element models of stepped cantilevered beam

SOLUTION   Since each node has two DOFs, the vector of total DOFs is defined by

$$\{\mathbf{Q}_s\} = \{v_1 \;\; \theta_1 \;\; v_2 \;\; \theta_2 \;\; v_3 \;\; \theta_3\}^{\mathrm{T}}.$$

From eq. (3.52), the element stiffness matrices of two beam elements can be obtained as

$$
\left[\mathbf{k}^{(1)}\right] = \frac{EI}{L^3}
\begin{array}{c}
\begin{array}{cccc} v_1 & \theta_1 & v_2 & \theta_2 \end{array} \\
\left[
\begin{array}{cccc}
3 & 3L & -3 & 3L \\
3L & 4L^2 & -3L & 2L^2 \\
-3 & -3L & 3 & -3L \\
3L & 2L^2 & -3L & 4L^2
\end{array}
\right]
\begin{array}{c} v_1 \\ \theta_1 \\ v_2 \\ \theta_2 \end{array}
\end{array}
,
$$

$$
\left[\mathbf{k}^{(2)}\right] = \frac{EI}{L^3}
\begin{array}{c}
\begin{array}{cccc} v_2 & \theta_2 & v_3 & \theta_3 \end{array} \\
\left[
\begin{array}{cccc}
12 & 6L & -12 & 6L \\
6L & 4L^2 & -6L & 2L^2 \\
-12 & -6L & 12 & -6L \\
6L & 2L^2 & -6L & 4L^2
\end{array}
\right]
\begin{array}{c} v_2 \\ \theta_2 \\ v_3 \\ \theta_3 \end{array}
\end{array}
.
$$

Corresponding row and column locations at the global DOFs are denoted in the above equation.

$$
[\mathbf{K}_s] = \frac{EI}{L^3}
\begin{bmatrix}
3 & 3L & -3 & 3L & 0 & 0 \\
3L & 4L^2 & -3L & 2L^2 & 0 & 0 \\
-3 & -3L & 15 & 3L & -12 & 6L \\
3L & 2L^2 & 3L & 8L^2 & -6L & 2L^2 \\
0 & 0 & -12 & -6L & 12 & -6L \\
0 & 0 & 6L & 2L^2 & -6L & 4L^2
\end{bmatrix}.
$$

Note that the structural stiffness matrix $[\mathbf{K}_s]$ is singular; more precisely, positive semi-definite. Applying the displacement boundary conditions is equivalent to deleting the first two rows and columns, to yield the global stiffness matrix $[\mathbf{K}]$. ∎

The next step is to derive an expression for the potential energy of external forces. If there are only concentrated forces and couples acting on the beam, the potential energy V can be written as

$$V = -\sum_{i=1}^{ND} (F_i v_i + C_i \theta_i), \tag{3.54}$$

where F_i and C_i, respectively, are the transverse force and couple acting at node i, and the total number of nodes in the beam model is denoted by ND. The above expression for V can be written in a matrix form as

$$V = -\{v_1\, \theta_1\, v_2 \dots \theta_{ND}\}\begin{Bmatrix} F_1 \\ C_1 \\ F_2 \\ \vdots \\ C_{ND} \end{Bmatrix} = -\{\mathbf{Q}_s\}^{\mathrm{T}}\{\mathbf{F}_s\}, \tag{3.55}$$

where $\{\mathbf{F}_s\}$ is the vector of nodal forces.

If there are distributed forces acting on the beam, then they have to be converted into equivalent nodal forces. Let us assume that the distributed load acting on the beam is given by $p(x)$. Then the potential energy of this load is given by

$$V = -\int_0^{L_T} p(x)v(x)\,dx, \tag{3.56}$$

where L_T is the total length of the beam. The integral can be broken down to integrals over each element as

$$V = -\sum_{e=1}^{NEL} \int_{x_1^{(e)}}^{x_2^{(e)}} p(x)v(x)\,dx = \sum_{e=1}^{NEL} V^{(e)}. \tag{3.57}$$

In the above equation, $V^{(e)}$ is the contribution to V by element e, which can be derived as

$$-V^{(e)} = \int_{x_1^{(e)}}^{x_2^{(e)}} p(x)v(x)\,dx = L\int_0^1 p(s)v(s)\,ds. \tag{3.58}$$

In the above equation, $p(s)$ should be expressed as a function of s using the change of variables given by eq. (3.37). Now we will use the shape functions to express $v(s)$ within an element in terms of nodal displacements using eq. (3.44) to obtain

$$\begin{aligned}
-V^{(e)} &= L^{(e)}\int_0^1 p(s)(v_1 N_1 + \theta_1 N_2 + v_2 N_3 + \theta_2 N_4)\,ds \\[2mm]
&= v_1\left(L^{(e)}\int_0^1 p(s)N_1(s)\,ds\right) + \theta_1\left(L^{(e)}\int_0^1 p(s)N_2(s)\,ds\right) \\[2mm]
&\quad + v_2\left(L^{(e)}\int_0^1 p(s)N_3(s)\,ds\right) + \theta_2\left(L^{(e)}\int_0^1 p(s)N_4(s)\,ds\right) \\[2mm]
&= v_1 F_1^{(e)} + \theta_1 C_1^{(e)} + v_2 F_2^{(e)} + \theta_2 C_2^{(e)}.
\end{aligned} \tag{3.59}$$

The terms inside the parentheses in the above equation are called *work equivalent loads* contributed by element e and are denoted by $F_1^{(e)}, C_1^{(e)}, F_2^{(e)}$, and $C_2^{(e)}$. These loads can be calculated from the distributed load p, the shape functions, and element length L. One can note that a transverse load p results not only in two concentrated forces at the nodes but also results in couples acting at the nodes. If a node belongs to more than one element, then the contributions from all elements at that node must be added together along with any applied concentrated forces and couples.

EXAMPLE 3.5 *Work-equivalent loads for a uniformly distributed load*

Calculate the work equivalent loads for a beam element of length L under a uniformly distributed load $p(x) = p$.

SOLUTION In the formula for the potential energy of the distributed load in eq. (3.59), $p(x) = p$ can be moved out from the integral because it is a constant. Thus, calculation of the work-equivalent nodal forces involves integral of the shape functions.

$$F_1 = pL \int_0^1 N_1(s)\,ds = pL \int_0^1 \left(1 - 3s^2 + 2s^3\right) ds = \frac{pL}{2},$$

$$C_1 = pL \int_0^1 N_2(s)\,ds = pL^2 \int_0^1 \left(s - 2s^2 + s^3\right) ds = \frac{pL^2}{12},$$

$$F_2 = pL \int_0^1 N_3(s)\,ds = pL \int_0^1 \left(3s^2 - 2s^3\right) ds = \frac{pL}{2},$$

$$C_2 = pL \int_0^1 N_4(s)\,ds = pL^2 \int_0^1 \left(-s^2 + s^3\right) ds = -\frac{pL^2}{12}.$$

Thus, the work-equivalent nodal forces for the uniformly distributed load becomes

$$\{\mathbf{F}\} = \left\{\frac{pL}{2} \quad \frac{pL^2}{12} \quad \frac{pL}{2} \quad -\frac{pL^2}{12}\right\}^{\mathrm{T}}.$$

Figure 3.14 illustrates the equivalent nodal forces. The equal transverse nodal force $pL/2$ is applied to the two nodes, and the couple of $pL^2/12$ is applied at the two nodes with the opposite signs (see figure 3.14). ■

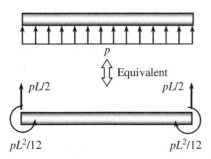

$pL^2/12$ $pL^2/12$ **Figure 3.14** Work equivalent nodal forces for the distributed load

---

Using the strain energy in eq. (3.53) and the potential energy of the applied loads in eq. (3.55), the total potential energy of the beam can be written as

$$\Pi = U + V = \frac{1}{2}\{\mathbf{Q}_s\}^{\mathrm{T}}[\mathbf{K}_s]\{\mathbf{Q}_s\} - \{\mathbf{Q}_s\}^{\mathrm{T}}\{\mathbf{F}_s\}. \tag{3.60}$$

The principle of minimum potential energy can be applied to determine the unknown DOFs $\{\mathbf{Q}_s\}$. One can recognize Π above as a quadratic function in $\{\mathbf{Q}_s\}$. Using the minimization principle derived in section A.6 of the appendix for a quadratic form, the stationary condition of Π with respect to $\{\mathbf{Q}_s\}$ yields

$$[\mathbf{K}_s]\{\mathbf{Q}_s\} = \{\mathbf{F}_s\}. \tag{3.61}$$

The above equations are the structural equations for the beam. In general, the structural stiffness matrix $[\mathbf{K}_s]$ will be singular. We can apply the boundary conditions by deleting the rows and columns corresponding to zero DOFs to obtain the global matrix equation as

$$[\mathbf{K}]\{\mathbf{Q}\} = \{\mathbf{F}\}.$$

This process is the same with that of the truss elements in chapter 1. The only difference is that the boundary conditions include not only the displacement but also the slope or rotation of the beam.

EXAMPLE 3.6   *Clamped-clamped beam element*

A beam of length 2 m is clamped at both ends and subjected to a downward transverse force of 240 N at the center, as shown in figure 3.15. Assume $EI = 1000 \text{ Nm}^2$. Use beam finite elements to determine the deflections and slopes at $x = 0.5$, 1.0, and 1.5 m.

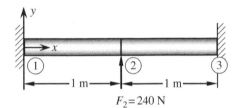

Figure 3.15 Finite element models of stepped cantilevered beam

SOLUTION   It is sufficient to use two elements of equal length for this problem. The element stiffness matrices of the two elements will be the same although the row and column addresses will be different. (See chapter 1 for row and column addresses.) Using the formula in eq. (3.52), we can write the element stiffness matrices as

$$\left[\mathbf{k}^{(1)}\right] = 1000 \begin{array}{c} \begin{array}{cccc} v_1 & \theta_1 & v_2 & \theta_2 \end{array} \\ \begin{bmatrix} 12 & 6 & -12 & 6 \\ 6 & 4 & -6 & 2 \\ -12 & -6 & 12 & -6 \\ 6 & 2 & -6 & 4 \end{bmatrix} \begin{array}{c} v_1 \\ \theta_1 \\ v_2 \\ \theta_2 \end{array} \end{array},$$

$$\left[\mathbf{k}^{(2)}\right] = 1000 \begin{array}{c} \begin{array}{cccc} v_2 & \theta_2 & v_3 & \theta_3 \end{array} \\ \begin{bmatrix} 12 & 6 & -12 & 6 \\ 6 & 4 & -6 & 2 \\ -12 & -6 & 12 & -6 \\ 6 & 2 & -6 & 4 \end{bmatrix} \begin{array}{c} v_2 \\ \theta_2 \\ v_3 \\ \theta_3 \end{array} \end{array}.$$

Assembling the element stiffness matrices, we obtain the 6×6 structural stiffness matrix $[\mathbf{K}_s]$. Also, the vector of applied nodal forces includes reactions at the clamped walls, as $\{\mathbf{F}_s\} = \{F_1 \quad C_1 \quad 240 \quad 0 \quad F_3 \quad C_3\}^{\mathrm{T}}$. Note that F_1 and

F_3 are reaction forces at the walls, whereas C_1 and C_3 are the reaction couples. The structural matrix equation is then obtained as

$$1000 \begin{bmatrix} 12 & 6 & -12 & 6 & 0 & 0 \\ 6 & 4 & -6 & 2 & 0 & 0 \\ -12 & -6 & 24 & 0 & -12 & 6 \\ 6 & 2 & 0 & 8 & -6 & 2 \\ 0 & 0 & -12 & -6 & 12 & -6 \\ 0 & 0 & 6 & 2 & -6 & 4 \end{bmatrix} \begin{Bmatrix} v_1 \\ \theta_1 \\ v_2 \\ \theta_2 \\ v_3 \\ \theta_3 \end{Bmatrix} = \begin{Bmatrix} F_1 \\ C_1 \\ 240 \\ 0 \\ F_3 \\ C_3 \end{Bmatrix}. \tag{3.62}$$

As discussed before, the structural stiffness matrix above is singular. Since both ends are clamped, the vertical deflection and rotation at these locations are zero. These boundary conditions can be applied by deleting the rows and columns corresponding to $v_1, \theta_1, v_3,$ and θ_3 (first, second, fifth, and sixth rows and columns) to obtain the global matrix equation as

$$1000 \begin{bmatrix} 24 & 0 \\ 0 & 8 \end{bmatrix} \begin{Bmatrix} v_2 \\ \theta_2 \end{Bmatrix} = \begin{Bmatrix} 240 \\ 0 \end{Bmatrix}. \tag{3.63}$$

Now, the global stiffness matrix above is positive definite, and thus a unique solution can be obtained by multiplying the inverse of the global stiffness matrix, to yield

$$v_2 = 0.01, \quad \theta_2 = 0.0.$$

As another approach, it is possible to assemble the global matrix equation (3.63) directly. This can be done by deleting the rows and columns corresponding to zero DOFs at the element level and only assemble those rows and columns corresponding to nonzero DOFs.

The deflection and slope at points in between nodes can be interpolated using the shape functions. The point $x = 0.5$ m is in the first element. Hence, the deflection and slope at this point is interpolated using the DOFs belonging to element 1. The local coordinate of this point is $s = 1/2$ (see eq. (3.37)). Then, the deflection and slope are obtained using the shape functions and their derivatives, as

$$v\left(\frac{1}{2}\right) = v_1 N_1\left(\frac{1}{2}\right) + \theta_1 N_2\left(\frac{1}{2}\right) + v_2 N_3\left(\frac{1}{2}\right) + \theta_2 N_4\left(\frac{1}{2}\right) = 0.01 \times N_3\left(\frac{1}{2}\right) = 0.005 \text{ m},$$

$$\theta\left(\frac{1}{2}\right) = \frac{1}{L^{(1)}} v_2 \frac{dN_3}{ds}\bigg|_{s=\frac{1}{2}} = 0.015 \text{ rad}.$$

Deflection and slope at $x = 1$ m are v_2 and θ_2. In this case, either element 1 or 2 can be used. If element 1 is used, the location $x = 1$ m corresponds to $s = 1$. Thus,

$$v(1) = v_1 N_1(1) + \theta_1 N_2(1) + v_2 N_3(1) + \theta_2 N_4(1) = 0.01 \times N_3(1) = 0.01 \text{ m},$$

$$\theta(1) = \frac{1}{L^{(1)}} v_2 \frac{dN_3}{ds}\bigg|_{s=1} = 0.0 \text{ rad}.$$

On the other hand, if element 2 is used, the location $x = 1$ m corresponds to $s = 0$. Thus,

$$v(1) = v_2 N_1(0) + \theta_2 N_2(0) + v_3 N_3(0) + \theta_3 N_4(0) = 0.01 \times N_1(0) = 0.01 \text{ m},$$

$$\theta(1) = \frac{1}{L^{(2)}} v_2 \frac{dN_1}{ds}\bigg|_{s=0} = 0.0 \text{ rad}.$$

Note that the two results are identical. This is true because in the beam element, the vertical deflection and the rotation are continuous at the connecting node.

To determine the deflection and slope at $x = 1.5$ m, we use the DOFs of element 2. Since the location is the center of the element, $s = 1/2$. Substituting in the shape functions, we obtain

$$v(1.5) = v_2 N_1\left(\frac{1}{2}\right) + \theta_2 N_2\left(\frac{1}{2}\right) + v_3 N_3\left(\frac{1}{2}\right) + \theta_3 N_4\left(\frac{1}{2}\right) = 0.01 \times N_1\left(\frac{1}{2}\right) = 0.005 \text{ m},$$

$$\theta(1.5) = \frac{1}{L^{(2)}} v_2 \frac{dN_1}{ds}\bigg|_{s=\frac{1}{2}} = -0.015 \text{ rad}.$$

As expected, the deflection is symmetric with respect to $x = 1$. Thus, the deflections at $x = 0.5$ and $x = 1.5$ are the same, whereas the rotations are equal and opposite. ▄▄

EXAMPLE 3.7   *Finite element analysis of a cantilevered beam*

A cantilever beam of length 1 m is subjected to a uniformly distributed load $p(x) = p_0 = 120$ N/m and a clockwise couple 50 N-m at the tip, as shown in figure 3.16. Use one element to determine the deflection curve $v(x)$ of the beam. What are the support reactions? Assume $EI = 1,000$ N-m^2.

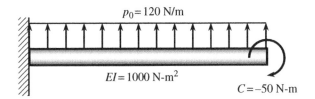

$p_0 = 120$ N/m

$EI = 1000$ N-m^2

$C = -50$ N-m

Figure 3.16 Cantilevered beam under uniformly distributed load and couple

SOLUTION   Since there is only one element, the structural stiffness matrix is the same as the element stiffness matrix:

$$[\mathbf{K}_s] = 1000 \begin{bmatrix} 12 & 6 & -12 & 6 \\ 6 & 4 & -6 & 2 \\ -12 & -6 & 12 & -6 \\ 6 & 2 & -6 & 4 \end{bmatrix} \begin{matrix} v_1 \\ \theta_1 \\ v_2 \\ \theta_2 \end{matrix}.$$

Next step is to calculate the work equivalent loads for the distributed force. Substituting $p(x) = p_0$ in eq. (3.59) we obtain

$$\begin{Bmatrix} F_{1e} \\ C_{1e} \\ F_{2e} \\ C_{2e} \end{Bmatrix} = p_0 L \int_0^1 \begin{Bmatrix} 1 - 3s^2 + 2s^3 \\ (s - 2s^2 + s^3)L \\ 3s^2 - 2s^3 \\ (-s^2 + s^3)L \end{Bmatrix} ds = p_0 L \begin{Bmatrix} 1/2 \\ L/12 \\ 1/2 \\ -L/12 \end{Bmatrix} = \begin{Bmatrix} 60 \\ 10 \\ 60 \\ -10 \end{Bmatrix}, \tag{3.64}$$

where the subscript e denotes the loads are equivalent loads. Note that the total force $p_0 L$ is equally divided between the two nodes. In addition, two equal and opposite couples of magnitude $p_0 L^2/12$ are also applied at the two nodes. These equivalent forces should be added to any concentrated forces and couples acting at the nodes. Thus, the force vector is written as

$$\{\mathbf{F}_s\} = \begin{Bmatrix} F_1 + 60 \\ C_1 + 10 \\ 60 \\ -10 - 50 \end{Bmatrix},$$

where F_1 and C_1 are the unknown reactions at the clamped end. The structural matrix equations becomes

$$1000 \begin{bmatrix} 12 & 6 & -12 & 6 \\ 6 & 4 & -6 & 2 \\ -12 & -6 & 12 & -6 \\ 6 & 2 & -6 & 4 \end{bmatrix} \begin{Bmatrix} v_1 \\ \theta_1 \\ v_2 \\ \theta_2 \end{Bmatrix} = \begin{Bmatrix} F_1 + 60 \\ C_1 + 10 \\ 60 \\ -10 - 50 \end{Bmatrix}.$$

Now the boundary conditions for the cantilevered beam is $v_1 = \theta_1 = 0$. Since the first two columns are going to be multiplied by zeros and the first two rows contain unknown reactions on the right-hand side, we delete these two rows and columns to obtain the global matrix equation, as

$$1000 \begin{bmatrix} 12 & -6 \\ -6 & 4 \end{bmatrix} \begin{Bmatrix} v_2 \\ \theta_2 \end{Bmatrix} = \begin{Bmatrix} 60 \\ -60 \end{Bmatrix}.$$

The global stiffness matrix in the above equation is positive definite and can be inverted to obtain the unknown DOFs, as

$$v_2 = -0.01 \, \text{m}, \quad \theta_2 = -0.03 \, \text{rad}.$$

The deflection curve is computed using the shape functions as

$$v(s) = -0.01 N_3(s) - 0.03 N_4(s) = -0.01 s^3,$$

where $s = x/L = x$. It is interesting to note that the tip deflection and rotations are exact. However, the deflection curve is approximate. The exact deflection curve is a quartic polynomial in x, whereas the approximate deflection is cubic. For comparison, the exact deflection curve is $v(x) = 0.005(x^4 - 4x^3 + x^2)$. Figure 3.17 compares the deflection and slope from the finite element and exact solutions.

The support reactions are determined from the first two of the four global equations:

$$1000(-12v_2 + 6\theta_2) = F_1 + 60,$$
$$1000(-6v_2 + 2\theta_2) = C_1 + 10.$$

Solving the above pair of equations, we obtain, $F_1 = -120\text{N}$, and $C_1 = -10$ as the reactions. The reader can verify that the reactions satisfy the force and moment equilibrium equations for the cantilever beam. ■

(a)

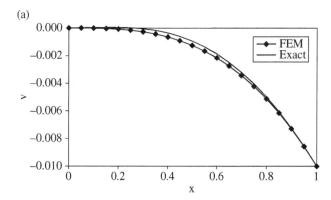

(b)

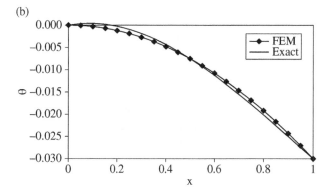

Figure 3.17 Comparison of beam deflection and rotation with exact solutions; (a) deflection, (b) slope

3.3.3   Bending Moment and Shear Force Distribution

After the nodal DOFs are determined, the bending moment $M(s)$ and shear force $V_y(s)$ distribution along the element can be calculated by substituting for the deflection curve into eqs. (3.6) and (3.8).
Bending moment:

$$M(s) = EI \frac{d^2 v}{dx^2} = \frac{EI}{L^2} \frac{d^2 v}{ds^2} = \frac{EI}{L^2} \{\mathbf{B}\}^T \{\mathbf{q}\}. \tag{3.65}$$

Shear force:

$$V_y(s) = -\frac{dM}{dx} = -EI \frac{d^3 v}{dx^3} = -\frac{EI}{L^3} \frac{d^3 v}{ds^3}$$

$$= \frac{EI}{L^3} \{-12 \quad -6L \quad 12 \quad -6L\} \begin{Bmatrix} v_1 \\ \theta_1 \\ v_2 \\ \theta_2 \end{Bmatrix}. \tag{3.66}$$

Note that the moment is a linear function of s, while the shear force is constant in an element. This is a limitation of the current beam element. If the loads on the beam are such that the shear force has a linear distribution, then several elements must be used so that the linear variation of the shear force can be better approximated by the piecewise constant shear force distribution.

We can use eq. (3.65) to calculate the bending moment at the two nodes of the beam element. We will denote the bending moments at nodes 1 and 2 by M_1 and M_2, respectively. They can be calculated as

$$M_1 = M(0) = \frac{EI}{L^2} \{-6 \quad -4L \quad 6 \quad -2L)\} \begin{Bmatrix} v_1 \\ \theta_1 \\ v_2 \\ \theta_2 \end{Bmatrix}, \tag{3.67}$$

$$M_2 = M(1) = \frac{EI}{L^2} \{6 \quad 2L \quad -6 \quad 4L\} \begin{Bmatrix} v_1 \\ \theta_1 \\ v_2 \\ \theta_2 \end{Bmatrix}. \tag{3.68}$$

From eqs. (3.66) through (3.68), we can write the shear force and bending moments at the two nodes in a matrix form as shown below:

$$\begin{Bmatrix} -V_{y1} \\ -M_1 \\ +V_{y2} \\ M_2 \end{Bmatrix} = \frac{EI}{L^3} \begin{bmatrix} 12 & 6L & -12 & 6L \\ 6L & 4L^2 & -6L & 2L^2 \\ -12 & -6L & 12 & -6L \\ 6L & 2L^2 & -6L & 4L^2 \end{bmatrix} \begin{Bmatrix} v_1 \\ \theta_1 \\ v_2 \\ \theta_2 \end{Bmatrix}. \tag{3.69}$$

One can note that the square matrix in the above equation is identical to the stiffness matrix of the beam element [see eq. (3.52)].

After calculating the bending moment, the axial stress in the beam at a cross section is calculated using:

$$\sigma_x = -\frac{My}{I}. \tag{3.70}$$

It is clear that the maximum stress appears either at the top or bottom of the beam element, which corresponds to $y = \pm h/2$ where the beam height is h.

The computation of the shear stress is more complicated and depends on the shape of the cross section. For a rectangular section of width b and height h, the shear stress distribution is given by

$$\tau_{xy}(y) = \frac{1.5V_y}{bh}\left(1 - \frac{4y^2}{h^2}\right). \tag{3.71}$$

The maximum shear stress appears at the neutral axis of the beam $(y = 0)$, and it is zero at the top and bottom of the cross section. The maximum shear stress is 1.5 times the average shear stress, which is equal to V_y/bh.

EXAMPLE 3.8   *Bending moment and shear force for a cantilevered beam*

Calculate the bending moment and the shear force along the cantilevered beam in example 3.7. Compare the results with exact solutions.

SOLUTION   From eq. (3.47), the bending moment can be written as

$$M(s) = \frac{EI}{L^2}\{\mathbf{B}\}^{\mathrm{T}}\{\mathbf{q}\}$$

$$= \frac{EI}{L^2}[(-6+12s)v_1 + L(-4+6s)\theta_1 + (6-12s)v_2 + L(-2+6s)\theta_2]$$

$$= 1000[-0.01(6-12s) - 0.03(-2+6s)]$$

$$= -60s \ \text{N·m}.$$

(a)

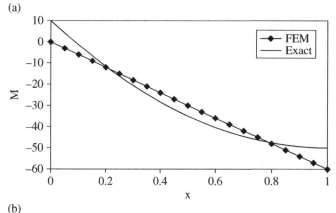

(b)

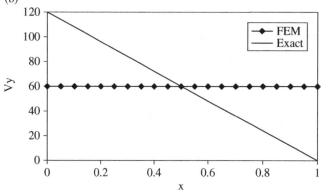

Figure 3.18 Comparison of bending moment and shear force with exact solutions; (a) bending moment, (b) shear force

From eq. (3.48), the shear force can be written as

$$V_y = \frac{EI}{L^3}[-12v_1 - 6L\theta_1 + 12v_2 - 6L\theta_2]$$

$$= 1000[12 \times (-0.01) - 6(-0.03)]$$

$$= 60\text{N}.$$

Since the deflection $v(s)$ is a cubic polynomial of s, the bending moment is linear, and the shear force is constant. Figure 3.18 shows the bending moment and shear force along with the exact values. The errors are in general larger than that of deflection and rotation. Thus, in the case of distributed load, more than one element is required to reduce the errors. ▬

EXAMPLE 3.9 *Finite element analysis of a simply supported beam*

Consider a simply supported beam under a uniformly distributed load p, as shown in figure 3.19. Using one beam element, calculate the deflection curve, the bending moment, and the shear force. Compare the finite element solutions with exact solutions.

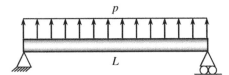

p

L

Figure 3.19 One element model with distributed force p

SOLUTION The whole beam is approximated using one beam finite element. The finite element equation can be written as

$$\frac{EI}{L^3}\begin{bmatrix} 12 & 6L & -12 & 6L \\ 6L & 4L^2 & -6L & 2L^2 \\ -12 & -6L & 12 & -6L \\ 6L & 2L^2 & -6L & 4L^2 \end{bmatrix}\begin{Bmatrix} v_1 \\ \theta_1 \\ v_2 \\ \theta_2 \end{Bmatrix} = \begin{Bmatrix} pL/2 \\ pL^2/12 \\ pL/2 \\ -pL^2/12 \end{Bmatrix} + \begin{Bmatrix} F_1 \\ 0 \\ F_2 \\ 0 \end{Bmatrix},$$

where F_1 and F_2 are unknown reactions at both nodes. No external couples are applied at the two ends of the beam. However, the work-equivalent nodal forces yield nonzero couples at the both nodes. Since there is only one element, the structural equations are the same as the local element equations. The displacement boundary conditions are $v_1 = v_2 = 0$ for the simply supported beam. These boundary conditions can be imposed using the procedure that is explained in section 1.1 (striking rows and columns that correspond to the zero displacement boundary conditions). Since v_1 and v_2 are fixed, the first and third rows and columns are removed. Then, the circled components in the equation below remain:

$$\frac{EI}{L^3}\begin{bmatrix} 12 & 6L & -12 & 6L \\ 6L & \boxed{4L^2} & -6L & \boxed{2L^2} \\ -12 & -6L & 12 & -6L \\ 6L & \boxed{2L^2} & -6L & \boxed{4L^2} \end{bmatrix}\begin{Bmatrix} 0 \\ \theta_1 \\ 0 \\ \theta_2 \end{Bmatrix} = \begin{Bmatrix} pL/2 + F_1 \\ pL^2/12 \\ pL/2 + F_2 \\ -pL^2/12 \end{Bmatrix}.$$

The global equations corresponding to unknown DOFs are

$$\frac{EI}{L^3}\begin{bmatrix} 4L^2 & 2L^2 \\ 2L^2 & 4L^2 \end{bmatrix}\begin{Bmatrix} \theta_1 \\ \theta_2 \end{Bmatrix} = p\begin{Bmatrix} L^2/12 \\ -L^2/12 \end{Bmatrix}.$$

Solving this matrix equation yields the solution:

$$\theta_1 = \frac{pL^3}{24EI}, \theta_2 = -\frac{pL^3}{24EI}.$$

Thus, the two ends of the beam rotate and the slopes are equal to θ_1 and θ_2. Using eq. (3.44), the displacement along the beam element can be approximated by

$$v(s) = \{N_1 \ N_2 \ N_3 \ N_4\}\begin{Bmatrix} 0 \\ \dfrac{pL^3}{24EI} \\ 0 \\ -\dfrac{pL^3}{24EI} \end{Bmatrix} = \frac{pL^4(s-s^2)}{24EI}. \tag{3.72}$$

The displacement $v(s)$ is a quadratic function of the parameter s. It may be noted that the exact deflection is a quartic function of x. The maximum deflection occurs at the center ($s = 0.5$). Substituting for s in eq. (3.72), we obtain the maximum deflection as $v_{max} = pL^4/96EI$. The exact solution for maximum deflection is $v_{max,exact} = 5pL^4/384EI$. In this example, the error in the maximum deflection is about 25%. It may be noted that the deflection from finite element analysis is smaller than the exact deflection; that is, the beam looks stiffer than it is. This is typical of approximate methods such as finite element analysis. As the number of elements is increased, the solution will approach the exact solution.

The support reactions can be calculated from the first and third of the four global equations as $F_1 = F_2 = -pL/2$. The element bending moment and shear force can be calculated as follows:

$$M(s) = \frac{EI}{L^2}\{-6+12s \ \ L(-4+6s) \ \ 6-12s \ \ L(-2+6s)\}\begin{Bmatrix} 0 \\ \dfrac{pL^3}{24EI} \\ 0 \\ -\dfrac{pL^3}{24EI} \end{Bmatrix} = -\frac{pL^2}{12},$$

$$V(s) = \frac{EI}{L^3}\{-12 \ \ -6L \ \ 12 \ \ -6L\}\begin{Bmatrix} 0 \\ \dfrac{pL^3}{24EI} \\ 0 \\ -\dfrac{pL^3}{24EI} \end{Bmatrix} = 0.$$

Since no shear force appears in the element, this loading condition produces a pure (constant) bending moment. The exact solution for bending moment is a quadratic in x, and the shear force has a linear variation along the length of the beam. Thus, the one-element solution is not enough to estimate the bending moment and shear force.

One of the biggest dangers in using finite element analysis without understanding the basic principles is to believe the accuracy of the solution without verification. Many people simply believe the output results from the computer have to be accurate. In the case of trusses, we have shown that the finite element solution is exactly the same as the analytical solution. That is not true for beam elements. The finite element solution for beams will be exact only when concentrated forces and couples act on the beam and the cross section is uniform between the nodes. In the following, we compare the finite element and analytical solutions for the problem shown in figure 3.19.

The analytical solution of the transverse displacement is given by

$$v(s)_{analytical} = \frac{pL^4}{24EI}(s - 2s^3 + s^4), \tag{3.73}$$

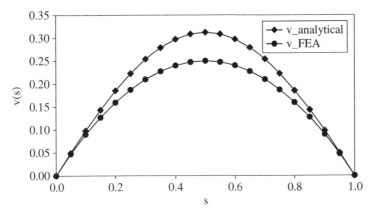

Figure 3.20 Transverse displacement of the beam element

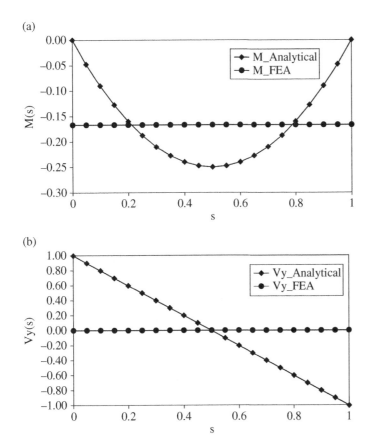

Figure 3.21 Comparison of FE and analytical solutions for the beam shown in figure 3.19; (a) bending moment, (b) shear force

which is a fourth-order function of s, while the finite element solution in eq. (3.72) is quadratic in s. Figure 3.20 compares the analytical and finite element solutions for transverse displacements. The center deflection from the finite element analysis is only 80% of the analytical solution.

The deviation of the finite element solution is more significant if the bending moment and shear force of the beam are compared. From the analytical solution, the bending moment and shear force of the beam can be calculated as

$$M(s)_{\text{analytical}} = \frac{qL^2}{2}\left(s^2 - s\right), \qquad (3.74)$$

$$V(s)_{\text{analytical}} = \frac{qL}{2}\left(1 - 2s\right). \qquad (3.75)$$

Notice that the bending moment of the beam finite element was a constant function, and the shear force was zero. Figure 3.21 compares the bending moment and shear force from the analytical and finite element solutions. One can note large differences between these solutions. However, the accuracy of the finite element solution can be improved by using a larger number of elements. ▄▄

3.4   PLANE FRAME ELEMENTS

Even if a beam element in the previous section can be useful, it is often not enough to model many practical applications. For example, each member of the portal frame in figure 3.22 needs to support not only transverse shear and bending moment but also axial load. That is, a structural member can play the role of both uniaxial bar and beam at the same time. Therefore, it would be necessary to combine the bar and beam elements for practical engineering applications.

3.4.1   Plane Frame Element Formulation

A plane frame is a structure similar to a truss, except the members can carry a transverse shear force and bending moment in addition to an axial force. Thus, a frame member combines the action of a uniaxial bar and a beam. This is accomplished by connecting the members by a rigid joint such as a gusset plate, which transmits the shear force and bending moment. Welding the ends of the members will also be sufficient in some cases. The cross sections of members connected to a joint (node) undergo the same rotation when the frame deforms. The nodes in a plane frame, which is in the x-y plane, have three DOFs, u, v, and θ, displacements in the x and y directions, and rotation about the z-axis. Consequently, one can apply two forces and one couple at each node corresponding to the three DOFs. An example of common frame is depicted in figure 3.22.

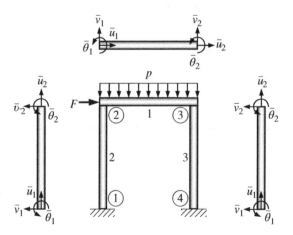

Figure 3.22 Frame structure and finite elements

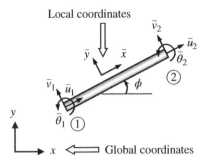

Local coordinates

Global coordinates **Figure 3.23** Local degrees of freedom of plane frame element

Consider the free-body diagram of a typical frame element shown in figure 3.23. It has two nodes and three DOFs at each node. Each element has a local coordinate system. The local or element coordinate system $\bar{x}-\bar{y}$ is such that the $\bar{x}$-axis is parallel to the element. The positive $\bar{x}$ direction is from the first node to the second node of the element. The $\bar{y}$-axis is such that the $\bar{z}$-axis is in the same direction as the z-axis. In the local coordinate system, the displacements in $\bar{x}$ and $\bar{y}$ directions are, respectively, $\bar{u}$ and $\bar{v}$, and the rotation in the $\bar{z}$ direction is $\bar{\theta}$. Each node has these three DOFs. The forces acting on the element, in local coordinates, are $f_{\bar{x}1}$, $f_{\bar{y}1}$, and $\bar{c}_1$ at node 1, and $f_{\bar{x}2}$, $f_{\bar{y}2}$, and $\bar{c}_2$ at node 2. Our goal is to derive a relation between the six element forces and the six DOFs. It will be convenient to use the local coordinate system to derive the force–displacement relation as the axial effects and bending effects are uncoupled in the local coordinates.

The element forces and nodal displacements are vectors, and they can be transformed to the $\bar{x}-\bar{y}$ coordinate system as follows:

$$
\begin{Bmatrix} f_{\bar{x}1} \\ f_{\bar{y}1} \\ \bar{c}_1 \\ f_{\bar{x}2} \\ f_{\bar{y}2} \\ \bar{c}_2 \end{Bmatrix} =
\begin{bmatrix}
\cos\phi & \sin\phi & 0 & 0 & 0 & 0 \\
-\sin\phi & \cos\phi & 0 & 0 & 0 & 0 \\
0 & 0 & 1 & 0 & 0 & 0 \\
0 & 0 & 0 & \cos\phi & \sin\phi & 0 \\
0 & 0 & 0 & -\sin\phi & \cos\phi & 0 \\
0 & 0 & 0 & 0 & 0 & 1
\end{bmatrix}
\begin{Bmatrix} f_{x1} \\ f_{y1} \\ c_1 \\ f_{x2} \\ f_{y2} \\ c_2 \end{Bmatrix},
$$

or

$$\{\bar{\mathbf{f}}\} = [\mathbf{T}]\{\mathbf{f}\}, \tag{3.76}$$

where the transformation matrix $[\mathbf{T}]$ is a function of the direction cosines of the element. Note that the size of the transformation matrix $[\mathbf{T}]$ in chapter 1 was 4×4, as a two-dimensional truss element has four DOFs. The transformation matrix in eq. (3.76) is basically the same, but the size is increased to 6×6 because of additional couples. However, the couple $\bar{c}$ is the same as the couple c, as the local $\bar{z}$-axis is parallel to the global z-axis. For the same reason, $\bar{\theta} = \theta$. Since the displacements are also vectors, a similar relation connects the DOFs in the local and global coordinates:

$$
\begin{Bmatrix} \bar{u}_1 \\ \bar{v}_1 \\ \bar{\theta}_1 \\ \bar{u}_2 \\ \bar{v}_2 \\ \bar{\theta}_2 \end{Bmatrix} =
\begin{bmatrix}
\cos\phi & \sin\phi & 0 & 0 & 0 & 0 \\
-\sin\phi & \cos\phi & 0 & 0 & 0 & 0 \\
0 & 0 & 1 & 0 & 0 & 0 \\
0 & 0 & 0 & \cos\phi & \sin\phi & 0 \\
0 & 0 & 0 & -\sin\phi & \cos\phi & 0 \\
0 & 0 & 0 & 0 & 0 & 1
\end{bmatrix}
\begin{Bmatrix} u_1 \\ v_1 \\ \theta_1 \\ u_2 \\ v_2 \\ \theta_2 \end{Bmatrix},
$$

or

$$\{\bar{\mathbf{q}}\} = [\mathbf{T}]\{\mathbf{q}\}. \tag{3.77}$$

In the local coordinate system, the axial deformation (uniaxial bar) and bending deformations (beam) are assumed uncoupled. When the beam is bent significantly, the axial load can affect the bending moment. Therefore, this assumption requires that bending deformation is infinitesimally small, which is the basic assumption in linear beam theory. Under such assumption, the axial forces and axial displacements are related by the uniaxial bar stiffness matrix, as

$$\frac{EA}{L} \begin{bmatrix} 1 & -1 \\ -1 & 1 \end{bmatrix} \begin{Bmatrix} \bar{u}_1 \\ \bar{u}_2 \end{Bmatrix} = \begin{Bmatrix} f_{\bar{x}1} \\ f_{\bar{x}2} \end{Bmatrix}. \tag{3.78}$$

On the other hand, the transverse force and couple are related to the transverse displacement and rotation by the bending stiffness matrix in eq. (3.52), as

$$\frac{EI}{L^3} \begin{bmatrix} 12 & 6L & -12 & 6L \\ 6L & 4L^2 & -6L & 2L^2 \\ -12 & -6L & 12 & -6L \\ 6L & 2L^2 & -6L & 4L^2 \end{bmatrix} \begin{Bmatrix} \bar{v}_1 \\ \bar{\theta}_1 \\ \bar{v}_2 \\ \bar{\theta}_2 \end{Bmatrix} = \begin{Bmatrix} f_{\bar{y}1} \\ \bar{c}_1 \\ f_{\bar{y}2} \\ \bar{c}_2 \end{Bmatrix}. \tag{3.79}$$

In a sense, the plane frame element is a combination of two-dimensional truss and beam elements. Combining eqs. (3.78) and (3.79), we obtain a relation between the element DOFs and forces in the local coordinate system:

$$\begin{bmatrix} a_1 & 0 & 0 & -a_1 & 0 & 0 \\ 0 & 12a_2 & 6La_2 & 0 & -12a_2 & 6La_2 \\ 0 & 6La_2 & 4L^2a_2 & 0 & -6La_2 & 2L^2a_2 \\ -a_1 & 0 & 0 & a_1 & 0 & 0 \\ 0 & -12a_2 & -6La_2 & 0 & 12a_2 & -6La_2 \\ 0 & 6La_2 & 2L^2a_2 & 0 & -6La_2 & 4L^2a_2 \end{bmatrix} \begin{Bmatrix} \bar{u}_1 \\ \bar{v}_1 \\ \bar{\theta}_1 \\ \bar{u}_2 \\ \bar{v}_2 \\ \bar{\theta}_2 \end{Bmatrix} = \begin{Bmatrix} \bar{f}_{x1} \\ \bar{f}_{y1} \\ \bar{c}_1 \\ \bar{f}_{x2} \\ \bar{f}_{y2} \\ \bar{c}_2 \end{Bmatrix}, \tag{3.80}$$

where

$$a_1 = \frac{EA}{L}, \ a_2 = \frac{EI}{L^3}. $$

It is clear that the equations at the first and fourth rows are the same as the uniaxial bar relation in eq. (3.78), whereas the remaining four equations are the beam relation. Equation (3.80) can be written in symbolic notation as

$$\boxed{[\bar{\mathbf{k}}]\{\bar{\mathbf{q}}\} = \{\bar{\mathbf{f}}\}}, \tag{3.81}$$

where $[\bar{\mathbf{k}}]$ is the element stiffness matrix in the local coordinate system.

As in the case of two-dimensional truss elements, the element matrix equation (3.81) cannot be used for assembly because different elements have different local coordinate systems. Thus, the element matrix equation needs to be transformed to the global coordinate system. Substituting for $\{\bar{\mathbf{f}}\}$ and $\{\bar{\mathbf{q}}\}$ from eqs. (3.76) and (3.77), we obtain

$$[\bar{\mathbf{k}}][\mathbf{T}]\{\mathbf{q}\} = [\mathbf{T}]\{\mathbf{f}\}. \tag{3.82}$$

Multiplying both sides of the above equation by $[\mathbf{T}]^{-1}$, we obtain

$$[\mathbf{T}]^{-1}[\bar{\mathbf{k}}][\mathbf{T}]\{\mathbf{q}\} = \{\mathbf{f}\}. \qquad (3.83)$$

It can be shown that for the transformation matrix $[\mathbf{T}]^{T} = [\mathbf{T}]^{-1}$. Hence, eq. (3.83) can be written as

$$[\mathbf{T}]^{T}[\bar{\mathbf{k}}][\mathbf{T}]\{\mathbf{q}\} = \{\mathbf{f}\},$$

or

$$[\mathbf{k}]\{\mathbf{q}\} = \{\mathbf{f}\}, \qquad (3.84)$$

where $[\mathbf{T}]^{T}[\bar{\mathbf{k}}][\mathbf{T}] = [\mathbf{k}]$ is the element stiffness matrix of a plane frame element in the global coordinate system. One can verify that $[\mathbf{k}]$ is symmetric and positive semi-definite.

Assembly of the element stiffness matrix to form the structural stiffness matrix $[\mathbf{K}_s]$ follows the same steps as for earlier elements. After deleting the rows and columns corresponding to zero DOFs, we obtain the global equations in the form $[\mathbf{K}]\{\mathbf{Q}\} = \{\mathbf{F}\}$.

3.4.2 Calculation of Element Forces

After solving the global equations, the DOFs of all elements are known in the frame structure. Let us represent the DOF of a typical element by $\{\mathbf{q}\}$ which is a 6×1 column matrix. We will transform the DOFs into local coordinates using eq. (3.77) to obtain $\{\bar{\mathbf{q}}\}$. Then, the axial force in the element can be obtained using

$$P = \frac{AE}{L}(\bar{u}_2 - \bar{u}_1). \qquad (3.85)$$

This relation is exactly the same as that of a uniaxial bar element. We note that the transverse shear force is constant throughout the element, and the bending moment varies linearly. Thus, it is enough to find the nodal values. From eq. (3.69), the nodal values of bending moments and shear forces can be found by

$$\begin{Bmatrix} -V_{\bar{y}1} \\ -\bar{M}_1 \\ +V_{\bar{y}2} \\ \bar{M}_2 \end{Bmatrix} = \frac{EI}{L^3} \begin{bmatrix} 12 & 6L & -12 & 6L \\ 6L & 4L^2 & -6L & 2L^2 \\ -12 & -6L & 12 & -6L \\ 6L & 2L^2 & -6L & 4L^2 \end{bmatrix} \begin{Bmatrix} \bar{v}_1 \\ \bar{\theta}_1 \\ \bar{v}_2 \\ \bar{\theta}_2 \end{Bmatrix}.$$

EXAMPLE 3.10 *Finite element analysis of a plane frame*

The two-member plane frame in figure 3.24 is subjected to a single horizontal force F. Both translations and rotational DOFs are fixed at nodes 1 and 3. Also, two members are welded together at node 2. Determine: (a) the displacements and rotation at the point of application of load; (b) axial force, shear force, and bending moment in each member; and (c) support reactions. Sketch the deformed shape of the frame. Assume $F = 1000$ N; $a = 0.3$ m. Element properties are: $E = 100$ GPa; $A = 10^{-4}$ m^2; $I = 10^{-8}$ m^4.

SOLUTION   Since a load acts only at a joint, each member can be modeled with one frame element. The first step is to calculate the element stiffness matrices. For element 1, node 1 is assigned as the first node and node 2 as the second node. Thus the angle of the element becomes $\phi = 45°$ and element length $L = 0.4243$ m. The local stiffness matrix $\left[\bar{\mathbf{k}}^{(1)} \right]$ can be calculated using eq. (3.80). Then it is transformed to the global coordinates using the transformation relations in eq. (3.84):

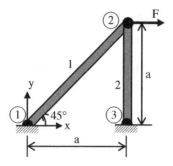

Figure 3.24 A two-member plane frame

$$\left[\mathbf{k}^{(1)} \right] = \left[\mathbf{T}(45) \right]^{\mathrm{T}} \left[\bar{\mathbf{k}}^{(1)} \right] \left[\mathbf{T}(45) \right]. \tag{3.86}$$

Note that $[\mathbf{k}^{(1)}]$ is a 6×6 symmetric matrix where the corresponding DOFs are $\{ \mathbf{q}^{(1)} \} = \{ u_1, v_1, \theta_1, u_2, v_2, \theta_2 \}^{\mathrm{T}}$. As discussed before, the boundary condition can be applied in the element level. Since the first three DOFs, u_1, v_1, and θ_1, are fixed, they can be removed in the element stiffness matrix. After removing the rows and columns corresponding to node 1, the element stiffness matrix in the global coordinates can be obtained as

$$\left[\mathbf{k}^{(1)} \right] = 10^6 \times \begin{bmatrix} 11.86 & 11.71 & 0.024 \\ 11.71 & 11.86 & -0.024 \\ 0.024 & -0.024 & 0.009 \end{bmatrix} \begin{matrix} u_2 \\ v_2 \\ \theta_2 \end{matrix}. \tag{3.87}$$

Using a similar procedure, the element stiffness matrix $[\mathbf{k}^{(2)}]$ of element 2 can also be obtained as

$$\left[\mathbf{k}^{(2)} \right] = 10^6 \times \begin{bmatrix} 0.444 & 0 & 0.067 \\ 0 & 33.33 & 0 \\ 0.067 & 0 & 0.013 \end{bmatrix} \begin{matrix} u_2 \\ v_2 \\ \theta_2 \end{matrix}. \tag{3.88}$$

Note that for element 2, fixed DOFs, u_3, v_3, and θ_3, are also removed. After assembly, the global matrix equations take the following form:

$$10^6 \times \begin{bmatrix} 12.31 & 11.71 & 0.090 \\ 11.71 & 45.2 & -0.024 \\ 0.090 & -0.024 & 0.023 \end{bmatrix} \begin{Bmatrix} u_2 \\ v_2 \\ \theta_2 \end{Bmatrix} = \begin{Bmatrix} 1000 \\ 0 \\ 0 \end{Bmatrix}. \tag{3.89}$$

Since boundary conditions are already applied, the global matrix in eq. (3.89) is positive definite and can be solved for unknown $\{ \mathbf{Q} \} = \{ u_2, v_2, \theta_2 \}^{\mathrm{T}}$ as:

$$u_2 = 0.113 \, \mathrm{m}; \; v_2 = -0.0295 \times 10^{-3} \, \mathrm{m}; \; \theta_2 = -0.478 \times 10^{-3} \, \mathrm{radians}.$$

Once nodal DOFs are given, it is possible to plot the deformed shape of the frame element using interpolation functions. We will plot the deflection shape of element 1. First we will transform the nodal displacements to the local coordinate system using eq. (3.77). It is obvious $\bar{u}_1 = \bar{v}_1 = \bar{\theta}_1 = 0$. Noting $\phi = 45°$ the DOFs at node 2 are transformed as:

$$\begin{Bmatrix} \bar{u}_2 \\ \bar{v}_2 \\ \bar{\theta}_2 \end{Bmatrix} = \begin{bmatrix} \cos 45 & \sin 45 & 0 \\ -\sin 45 & \cos 45 & 0 \\ 0 & 0 & 1 \end{bmatrix} \begin{Bmatrix} u_2 \\ v_2 \\ \theta_2 \end{Bmatrix} = \begin{Bmatrix} 0.0589 \\ -0.1006 \\ -0.4776 \end{Bmatrix} \times 10^{-3}. \tag{3.90}$$

In order to plot the deformed shape of the element, we calculate the displacements $\bar{u}(s)$ and $\bar{v}(s)$ at various points along the length of the element using shape functions:

$$\bar{u}(s) = \bar{u}_1(1-s) + \bar{u}_2 s,$$
$$\bar{v}(s) = \bar{v}_1 N_1(s) + \bar{\theta}_1 N_2(s) + \bar{v}_2 N_3(s) + \bar{\theta}_2 4_1(s), \tag{3.91}$$

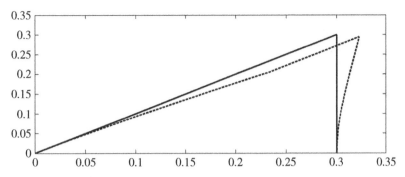

Figure 3.25 Deformed shape of the frame in figure 3.24. The displacements are magnified by a factor of 200

where N_i are the beam shape functions [see eq. (3.43)]. Then they have to be transformed to the global coordinates to obtain $u(s)$ and $v(s)$ using the inverse transform. Normally, the deformation is very small compared to the initial geometry of the frame. Therefore, in order to see the deformation, it is necessary to multiply the displacement by a magnification factor α. The magnification factor is normally determined such that the maximum deformation is about 10 ~ 20% of the size of the structure. Similar to displacements in eq. (3.91), the undeformed coordinate of the element can also be expressed in terms of the parameter s, that is, $x(s)$ and $y(s)$ for the undeformed coordinates of the structure. Then, the coordinates of various points along the length of the deformed element are obtained as $x'(s) = x(s) + \alpha u(s)$ and $y'(s) = y(s) + \alpha v(s)$. The deformed shape of the frame is plotted in figure 3.25. ■

Force and moment resultants: We will use eq. (3.81), $\{\bar{\mathbf{f}}\} = [\bar{\mathbf{k}}]\{\bar{\mathbf{q}}\}$, to obtain the force and moment resultants. Consider element 1 first. We already know the element stiffness matrix $[\bar{\mathbf{k}}]$ and the displacements $\{\bar{\mathbf{q}}\}$ for this element. Performing the matrix multiplication, we obtain the forces as:

$$\left\{\bar{\mathbf{f}}^{(1)}\right\} = \{-1390, -0.116, 1.11, 1390, 0.116, -1.15\}^T. \tag{3.92}$$

From the above result one can recognize the axial force P, shear force V, and moment resultants M_1 and M_2 in element 1 as:

$$
\begin{aligned}
P &= \bar{f}_4^{(1)} = 1390\,\text{N}, \\
V &= V_{\bar{y}2}^{(1)} = \bar{f}_5^{(1)} = 0.116\,\text{N}, \\
M_1 &= -M_1^{(1)} = -\bar{f}_3^{(1)} = -1.11\,\text{N·m}, \\
M_2 &= +M_2^{(1)} = \bar{f}_6^{(1)} = -1.15\,\text{N·m}.
\end{aligned}
\tag{3.93}
$$

The above forces are shown in the free-body diagram of element 1 in figure 3.26. Note the orientation of the local $\bar{x} - \bar{y}$ coordinate system. Force resultants P and V are constant along the length of the element. The bending moment varies linearly from M_1 at node 1 to M_2 at node 1.

The forces in element 2 were calculated using similar procedures to obtain:

$$
\begin{aligned}
\left\{\bar{\mathbf{f}}^{(2)}\right\} &= \{981, 18.3, 1.15, -981, -18.3, 4.33\}^T, \\
P &= -981\,\text{N}; V = -18.3\,\text{N}; M_2 = -1.15\,\text{N·m}; M_3 = 43.3\,\text{N·m}.
\end{aligned}
\tag{3.94}
$$

Note that nodes 2 and 3 were the first and second nodes for element 2. The free-body diagram of element 2 is shown in figure 3.26 along with that of element 1.

Support Reactions The reactions at node 1 are basically given by $\left\{\bar{f}_1^{(1)}, \bar{f}_2^{(1)}, \bar{f}_3^{(1)}\right\} = \{-1390, -0.116, 1.11\}$ (N-m units). Similarly reactions at node 3 are: $\left\{\bar{f}_4^{(2)}, \bar{f}_5^{(2)}, \bar{f}_6^{(2)}\right\} = \{-981, -18.3, 4.33\}$. But they are in respective local coordinate systems. They have to be resolved into the global x-y coordinates as shown in figure 3.27. Checking for global equilibrium is left to the student as an exercise in statics. Note there are roundoff errors in the reactions shown.

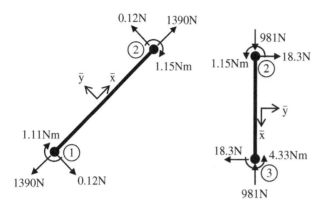

Figure 3.26 Free-body diagrams of elements 1 and 2 of the frame in example 3.10

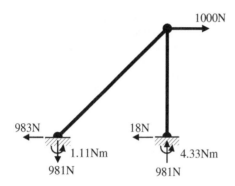

Figure 3.27 Support reactions for the frame in example 3.10

3.5   BUCKLING OF BEAMS

In engineering structures, buckling is instability that leads to a failure mode. It happens when the structure or elements within the structure are subjected to a compressive load or stress, especially when the structure has slender members such as a beam. For example, an eccentric compressive axial load may cause a small initial bending deformation to the beam. However, the bending deformation increases the eccentricity of the load and thus further bends the beam. This can possibly lead to a complete loss of the member's load-carrying capacity. Buckling may occur even though the stresses that develop in the structure are well below the failure strength of the material. Therefore, buckling is an important failure mode, and engineers should be aware of it when designing structures.

3.5.1   Review of Column Buckling

Consider a cantilevered beam subjected to an end couple C and an axial force P (figure 3.28). We are interested in the tip deflection of the beam. If we use the concept of linear superposition for elastic structures (see section 1.3.3), then the axial force P should have no effect on the flexural behavior due to the

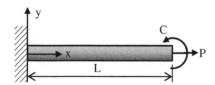

Figure 3.28 A beam subjected to axial force and an end couple

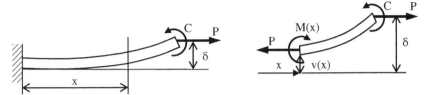

Figure 3.29 Beam subjected to an axial tension and an end couple with a free-body diagram to determine $M(x)$

couple C, and the tip deflection should be equal to $CL^2/2EI$. However, the engineering sense tells us that it should be difficult for the couple to bend the beam because the axial force is trying to straighten the beam. That is, the deflection of the beam should be less because of the axial force P. The opposite will be true if the axial force were compressive, that is, the deflection will be larger, because a compressive axial force will tend to aid the couple in bending the beam. The effect of the axial force can be easily understood, if we look at the free-body diagram of the beam in figure 3.29 and compute the bending moment at a cross section at a distance x.

The bending moment at x is given by

$$M(x) = C - P(\delta - v), \tag{3.95}$$

where $v(x)$ is the beam deflection, and δ is the tip deflection. In elementary beam theory, we do not consider the second term on the right-hand side of eq. (3.95) as it is assumed small compared to the applied couple C because either P is small or the deflection δ is small. This is the fundamental assumption for linear problems, and we used this assumption when we derive the equation for the frame element. From the constitutive relations of the beam [see eq. (3.6)], we know the curvature of the beam is proportional to the bending moment:

$$M(x) = EI\frac{\mathrm{d}^2 v}{\mathrm{d}x^2}. \tag{3.96}$$

By substituting eq. (3.96) in eq. (3.95), we obtain

$$EI\frac{\mathrm{d}^2 v}{\mathrm{d}x^2} = C - P(\delta - v) \quad \text{or} \quad EI\frac{\mathrm{d}^2 v}{\mathrm{d}x^2} - Pv = C - P\delta. \tag{3.97}$$

Equation (3.97) is an ordinary differential equation, whose solution consists of complementary functions and a particular integral given by:

$$v(x) = A\,\sinh\lambda x + B\,\cosh\lambda x - \left(\frac{C}{P} - \delta\right), \tag{3.98}$$

where A and B are constants to be determined from the boundary conditions, and the parameter λ is given by

$$\lambda^2 = \frac{|P|}{EI}. \tag{3.99}$$

The boundary conditions are similar to the cantilever beam such that at $x = 0$, deflection v and slope $\mathrm{d}v/\mathrm{d}x$ are equal to zero. Substituting the boundary conditions in eq. (3.98) and solving for A and B, we obtain the final solution as

$$v(x) = \left(\frac{C}{P} - \delta\right)(\cosh\lambda x - 1). \tag{3.100}$$

The tip deflection δ can be found by using $\delta = v(L)$ in eq. (3.100):

$$\delta = v(L) = \frac{C}{P}\left(1 - \frac{1}{\cosh\lambda L}\right), \quad P > 0. \tag{3.101}$$

It can be seen from eq. (3.101) that the effect of axial tension P is to reduce the tip deflection. As $P \to \infty$, $\lambda \to \infty$ and $\delta \to 0$. When $P \to 0$, $\lambda \to 0$, and the tip deflection approaches the beam theory deflection $CL^2/2EI$. Evaluation of this limit is left to the reader as an exercise in calculus.

If the force P is compressive, a similar procedure can be used to find the deflection $v(x)$. The only change will be in the sign of P in eq. (3.99). Since P is negative for a compressive load, the parameter λ becomes imaginary. This will lead to *sin* and *cos* terms in the solution instead of *sinh* and *cosh* terms. The solution after applying the boundary conditions can be derived as

$$v(x) = \left(\frac{C}{|P|} + \delta\right)(1 - \cos\lambda x), \quad \lambda = \sqrt{\frac{|P|}{EI}}, \quad P < 0. \tag{3.102}$$

Solving for delta using $v(L) = \delta$, we obtain

$$\delta = v(L) = \frac{C}{|P|}\left(\frac{1}{\cos\lambda L} - 1\right), \quad P < 0. \tag{3.103}$$

Again, one can show that as $P \to 0$ from the negative side, the tip deflection approaches the beam theory deflection $CL^2/2EI$. One can note that the tip deflection will become unbounded when $\lambda L \to \pi/2$. This corresponds to an axial compressive force given by

$$\lambda = \sqrt{\frac{|P|}{EI}} = \frac{\pi}{2L} \quad \Rightarrow \quad P = P_{cr} = \frac{\pi^2 EI}{4L^2} \approx 2.47\frac{EI}{L^2}, \tag{3.104}$$

where P_{cr} is called the *critical load* for buckling of the beam. One can note that P_{cr} does not depend on the external couple C that is applied but depends only on the flexural rigidity EI of the cross section and the beam length L. Thus, it is a structural property. When the axial load is tensile, the beam apparently becomes stiff, and this phenomenon is called stress stiffening. The effect of compressive axial load is called stress softening.

3.5.2 End Shortening Due to Compressive Load

In the previous section, we solved the differential equation of the beam to understand the effect of axial forces on the beam flexural behavior. In applying energy methods, we need to identify the additional energy terms due to coupling of P and the flexural deformation. We achieve this by accounting for the axial displacement of the beam due to bending. Let us assume that the axial rigidity of the beam, EA, is infinitely large. Then, the beam neutral surface does not undergo any stretching. From figure 3.30, one can note that as the beam undergoes a large deflection, the point of application of load P moves inwards in order to keep the beam length constant. Thus, the force P moves through a distance Δ as

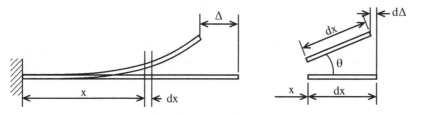

Figure 3.30 End shortening of a cantilever beam under a compressive load

shown in the figure. That means there is a work performed by P. We have to add a corresponding term in the expression for the total potential energy. The end shortening Δ depends only on the beam deflection, and its derivation is given below.

Consider the deformation of an infinitesimal length dx of the beam depicted in figure 3.30. The relative axial displacement of the two ends of this element can be derived as:

$$d\Delta = dx(1-\cos\theta) \approx 2dx\sin^2\frac{\theta}{2},$$

where θ is the rotation of the element. When the rotational angle is small, it can be approximated by $\sin\theta = dv/dx$. Therefore, we obtain

$$d\Delta \approx \frac{1}{2}\left(\frac{dv}{dx}\right)^2 dx. \tag{3.105}$$

The total axial displacement or the end shortening of the beam is obtained by integrating the differential end shortening to the entire beam, as

$$\Delta = \int_0^L d\Delta = \frac{1}{2}\int_0^L \left(\frac{dv}{dx}\right)^2 dx. \tag{3.106}$$

3.5.3   Rayleigh-Ritz Method

Consider the example given at the beginning of this section (i.e., figure 3.28). Let us assume the axial force P is tensile. We will solve this problem using the Rayleigh-Ritz method. First, the deflection of the beam is approximated as:

$$v(x) = cx^2, \tag{3.107}$$

which satisfies all kinematic boundary conditions at $x = 0$. The strain energy U in the beam can be computed in terms of c as [see eq. (3.17)]

$$U = \frac{1}{2}\int_0^L EI\left(\frac{d^2v}{dx^2}\right)^2 dx = 2EILc^2. \tag{3.108}$$

The next step is to express the potential energy of external forces in terms of c. If we use the small deflection theory, only the couple C would contribute to this term. However, in the present situation we need to account for the potential of the axial force P also. The energy term V can be expressed as

$$V = -C\frac{dv}{dx}\bigg|_{x=L} -(P)(-\Delta). \tag{3.109}$$

In the above equation, Δ has a negative sign because the beam end is moving in the negative x direction (end shortening), whereas the force P is in the positive direction. Substituting for Δ from eq. (3.106) and for $v(x)$ from eq. (3.107), we obtain

$$V = -2CLc + P\frac{1}{2}\int_0^L (2cx)^2 dx$$
$$= -2CLc + \frac{2}{3}PL^3c^2. \tag{3.110}$$

Now the total potential energy can be written as

$$\Pi = U + V = 2EILc^2 - 2CLc + \frac{2}{3}PL^3c^2. \tag{3.111}$$

The equilibrium of the structure occurs when the total potential energy is stationary in the Rayleigh-Ritz method. Since the only unknown variable is the coefficient c, the derivative of the total potential energy with respect to c has to be zero:

$$\frac{d\Pi}{dc} = 0 \quad \Rightarrow \quad \left(4EILc - 2CL + \frac{4}{3}PL^3c \right) = 0 \quad \Rightarrow \quad c = \frac{C}{2EI + \frac{2}{3}PL^2}. \tag{3.112}$$

Therefore, the beam deflection curve can be obtained using eq. (3.107). Using the same equation, the maximum deflection at the tip is given by

$$\delta = v(L) = cL^2 = \frac{CL^2}{2EI + \frac{2}{3}PL^2}, \quad P > 0. \tag{3.113}$$

From eq. (3.113) one can note that a tensile P will decrease the maximum deflection δ. The expression in eq. (3.113) can be used even if $P < 0$ by changing the sign of P. Then

$$\delta = \frac{CL^2}{2EI - \frac{2}{3}|P|L^2}, \quad P < 0. \tag{3.114}$$

From eq. (3.114) one can derive the approximate P_{cr} at which the deflection becomes unbounded by equating the denominator to zero:

$$P_{cr} = 3\frac{EI}{L^2}. \tag{3.115}$$

We note that the value of P_{cr} obtained by using the Rayleigh-Ritz method is about 20% larger than the exact buckling load for a cantilevered beam, which is given by $P_{cr} \approx 2.47EI/L^2$ in eq. (3.104). This is typical of approximate methods, as they tend to make the structure look stiffer.

The above results can be non-dimensionalized by dividing by the tip deflection of a beam:

$$\bar{\delta} = \frac{\delta}{CL^2/2EI}$$

$$= \frac{1}{1 + \frac{1}{3}(\lambda L)^2}, \quad P > 0 \tag{3.116}$$

$$= \frac{1}{1 - \frac{1}{3}(\lambda L)^2}, \quad P < 0.$$

The exact solution obtained in eqs. (3.101) and (3.103) can be written in non-dimensional forms as

$$\bar{\delta}_{exact} = \frac{2}{(\lambda L)^2} \left(1 - \frac{1}{\cosh \lambda L} \right), \quad P > 0$$

$$= \frac{2}{(\lambda L)^2} \left(\frac{1}{\cos \lambda L} - 1 \right), \quad P < 0. \tag{3.117}$$

The non-dimensional tip deflections are plotted as a function of the non-dimensional load term λL in figure 3.31. For the case of compressive axial load $P < 0$, one can note that the non-dimensional tip deflection will become unbounded when $\lambda L \to \pi/2$ for the exact solution in eq. (3.117) and when $\lambda L \to \sqrt{3}$ for the approximate solution given by eq. (3.116).

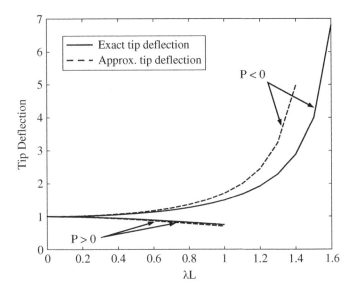

Figure 3.31 Non-dimensional tip deflection as a function of non-dimensional axial force λL for a given end couple in a cantilever beam

3.5.4 Finite Element Method for Buckling of Beams

The finite element formulation of buckling or stress stiffening follows a similar procedure as in the Rayleigh-Ritz method. The beam element discussed earlier can be modified for buckling analysis. Basically we divide the beam into a number of elements. Each element has two nodes, and each node has two DOFs, the transverse deflection v and rotation or slope θ. We assume that the axial force P is constant in each element. Then the potential energy of these forces can be given as [see eq. (3.109)]

$$V_{inc} = \sum_{e=1}^{NEL} P^{(e)} \Delta^{(e)}, \tag{3.118}$$

where V_{inc} denotes the potential energy of P, $P^{(e)}$ is the axial force in element e, $\Delta^{(e)}$ is the axial short-ening of node j with respect to node i of element e, and NEL is the total number of elements. The sub-script "inc" in V_{inc} refers to incremental energy because this is an additional energy as the beam undergoes large deflections. In fact, it will be shown later that V_{inc} will increase the stiffness for stress stiffening or decrease for buckling. The displacement $\Delta^{(e)}$ is the end-shortening of element e and is derived as:

$$\Delta^{(e)} = \frac{1}{2} \int_{x_i}^{x_j} \left(\frac{dv}{dx}\right)^2 dx, \tag{3.119}$$

where x_i and x_j are respectively the x-coordinates of the first and second nodes of the element e.

Before we proceed further, the sign conventions used herein are worth discussing. One may note the usual negative sign in the expression for potential is energy missing in eq. (3.118). In the derivation of potential energy, the axial force P is taken as tensile and hence positive. The term $\Delta^{(e)}$ is end shortening and hence is negative, which cancels the negative sign in the expression for potential energy.

By using the change of variable $s = (x - x_i)/L$ as defined in eq. (3.37), we obtain

$$\Delta^{(e)} = \frac{1}{2L} \int_0^1 \left(\frac{dv}{ds}\right)^2 ds. \tag{3.120}$$

The derivative dv/ds can be written in terms of the nodal DOFs and the shape functions $N_i(s)$ as

$$\frac{dv}{ds} = \{v_i \quad \theta_i \quad v_j \quad \theta_j\} \begin{Bmatrix} N'_1 \\ N'_2 \\ N'_3 \\ N'_4 \end{Bmatrix} = \{N'_1 \quad N'_2 \quad N'_3 \quad N'_4\} \begin{Bmatrix} v_i \\ \theta_i \\ v_j \\ \theta_j \end{Bmatrix}. \tag{3.121}$$

In the above equation, a prime ($'$) denotes the first derivative dN/ds with respect to s. The shape functions are given in eq. (3.43).

Substituting from (3.121) into (3.120) yields

$$\Delta^{(e)} = \frac{1}{2}\{\mathbf{q}\}^{\mathrm{T}}\left[\mathbf{k}^{(e)}_{inc}\right]\{\mathbf{q}\}, \tag{3.122}$$

where $\{\mathbf{q}\}$ is the column vector of element DOFs, and the components of symmetric incremental stiffness matrix $\left[\mathbf{k}^{(e)}_{inc}\right]$ can be calculated using the following formula:

$$\left(k^{(e)}_{inc}\right)_{ij} = \left(k^{(e)}_{inc}\right)_{ji} = \frac{1}{L}\int_0^1 N'_i N'_j ds. \tag{3.123}$$

After substituting four shape functions, $\left[\mathbf{k}^{(e)}_{inc}\right]$ can be obtained as

$$\left[\mathbf{k}^{(e)}_{inc}\right] = \frac{1}{30L}\begin{bmatrix} 36 & 3L & -36 & 3L \\ 3L & 4L^2 & -3L & -L^2 \\ -36 & -3L & 36 & -3L \\ 3L & -L^2 & -3L & 4L^2 \end{bmatrix} \begin{matrix} v_i \\ \theta_i \\ v_j \\ \theta_j \end{matrix}. \tag{3.124}$$

The row addresses of $\left[\mathbf{k}^{(e)}_{inc}\right]$ are shown next to the matrix in eq. (3.124).

Substituting for $\Delta^{(e)}$ from eq. (3.122) into eq. (3.118), the expression for V_{inc} is obtained as

$$V_{inc} = \sum_{e=1}^{NEL} P^{(e)}\Delta^{(e)} = \sum_{e=1}^{NEL} P^{(e)}\frac{1}{2}\{\mathbf{q}\}^{\mathrm{T}}\left[\mathbf{k}^{(e)}_{inc}\right]\{\mathbf{q}\}. \tag{3.125}$$

In general $P^{(e)}$ will be different in different elements, and hence it cannot be factored when assembling the incremental stiffness matrices. We will multiply and divide by a reference load P_r as shown below:

$$V_{inc} = \sum_{e=1}^{NEL} P_r \frac{P^{(e)}}{P_r}\frac{1}{2}\{\mathbf{q}\}^{\mathrm{T}}\left[\mathbf{k}^{(e)}_{inc}\right]\{\mathbf{q}\} = \frac{1}{2}\{\mathbf{Q}\}^{\mathrm{T}}[P_r\mathbf{K}_{inc}]\{\mathbf{Q}\}, \tag{3.126}$$

where $[P_r\mathbf{K}_{inc}]$ is the global incremental stiffness matrix obtained by assembling $\left[(P^{(e)}/P_r)\mathbf{k}^{(e)}_{inc}\right]$ and $\{\mathbf{Q}\}$ is the column vector of active DOFs. The total potential energy of the beam can be written as

$$\Pi = \frac{1}{2}\{\mathbf{Q}\}^{\mathrm{T}}[\mathbf{K}]\{\mathbf{Q}\} - \{\mathbf{Q}\}^{\mathrm{T}}\{\mathbf{F}\} + \frac{1}{2}\{\mathbf{Q}\}^{\mathrm{T}}[P_r\mathbf{K}_{inc}]\{\mathbf{Q}\}$$
$$= \frac{1}{2}\{\mathbf{Q}\}^{\mathrm{T}}[\mathbf{K} + P_r\mathbf{K}_{inc}]\{\mathbf{Q}\} - \{\mathbf{Q}\}^{\mathrm{T}}\{\mathbf{F}\}. \tag{3.127}$$

From eq. (3.127), one can see why $[\mathbf{K}_{inc}]$ is called the incremental stiffness matrix. The effect of axial tension is to add a positive definite matrix to the stiffness matrix, thus increasing the strain energy of the system, which makes the beam stiffer. By the same token, if the axial forces are negative (compressive), the beam will appear softer or more compliant. Minimization of Π with respect to $\{Q\}$ leads to the following standard finite element system of equations:

$$[\mathbf{K} + P_r\mathbf{K}_{inc}]\{\mathbf{Q}\} = \{\mathbf{F}\} \quad \text{or} \quad [\mathbf{K}_T]\{\mathbf{Q}\} = \{\mathbf{F}\}, \tag{3.128}$$

where $[\mathbf{K}_T]$ is the total stiffness matrix. If the inverse for $[\mathbf{K}_T]$ exists, then one can find it and determine the displacements $\{\mathbf{Q}\}$ for a given set of transverse forces and couples $\{\mathbf{F}\}$ acting on the beam. This will be the case when the beam is stable and the displacement $\{\mathbf{Q}\}$ determines the static deformation. The only difference from the previous beam analysis in section 3.3 is that the bending and axial deformation are coupled. The incremental stiffness matrix in eq. (3.124) has a constant of $1/30\,L$, while the beam stiffness matrix in eq. (3.52) has a constant of EI/L^3. Therefore, if the condition of $1/30\,L << EI/L^3$ is satisfied, then it is possible to ignore the effect of incremental stiffness, which is the case in the beam formulation in section 3.3. However, this is not the main interest in this section, where we look for a case when the effect of the incremental stiffness is significant.

If the axial forces P are such that the determinant of $[\mathbf{K}_T]$ is equal to zero, then it cannot be inverted, and the deflections $\{\mathbf{Q}\}$ are undefined. In physical terms, the deflections can become unbounded. Values of P_r that satisfy this condition are the *critical loads* (P_{cr}) for buckling of the beam, and they are solved using the polynomial equation for P_r obtained from

$$|\mathbf{K} - P_r\mathbf{K}_{inc}| = 0. \tag{3.129}$$

Usually we are concerned with the first or the lowest P_{cr} because the applied load increases gradually from zero. When the axial load reaches P_{cr}, the beam will buckle and collapse. Even if the zero-determinant equation in eq. (3.129) looks independent of the applied loads, it actually depends on the applied load because the element force is required in the form of $\left[(P^{(e)}/P_r)\mathbf{k}_{inc}^{(e)}\right]$ in the assembly of the incremental stiffness matrix. Therefore, when solving for buckling loads, the static analysis is solved first without considering buckling to calculate element forces, and then, eq. (3.129) is solved for the buckling loads. In fact, eq. (3.129) is the necessary condition for following the generalized eigenvalue problem (see eq. (A.66) in appendix):

$$[\mathbf{K} - P_r\mathbf{K}_{inc}]\{\mathbf{Q}\} = \{\mathbf{0}\} \quad \text{or} \quad [\mathbf{K}]\{\mathbf{Q}\} = P_r[\mathbf{K}_{inc}]\{\mathbf{Q}\}, \tag{3.130}$$

where P_r is the eigenvalue and $\{\mathbf{Q}\}$ is corresponding eigenvector. The eigenvectors for each eigenvalue define the corresponding mode shape or shape of the buckled beam. This will be illustrated in the following examples.

EXAMPLE 3.11 *Finite element analysis of a beam with axial force*

Consider a cantilever beam in figure 3.28 of length $L = 1$ m with flexural rigidity $EI = 1,000$ N-m^2. The beam is subjected to an axial force P and a couple $C = +1,000$ N-m at the tip. Find the deflection curve $v(x)$ for the following three cases: (a) $P = 0$; (b) $P = +2,000$ N; and (c) $P = -2,000$ N. Use one element for the entire beam.

Case a ($P = 0$): Since we are using one element, the element stiffness matrix will be the same as the global stiffness matrix. The stiffness matrix and the load vector are given by

$$[\mathbf{K}] = 1000\begin{bmatrix} 12 & -6 \\ -6 & 4 \end{bmatrix}\begin{matrix} v_2 \\ \theta_2 \end{matrix}, \quad \{\mathbf{F}\} = \begin{Bmatrix} 0 \\ 1,000 \end{Bmatrix}\begin{matrix} v_2 \\ \theta_2 \end{matrix}, \quad \{\mathbf{Q}\} = \begin{Bmatrix} v_2 \\ \theta_2 \end{Bmatrix}.$$

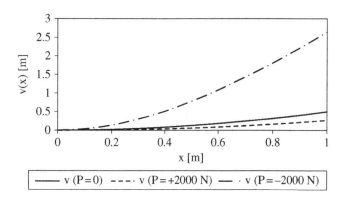

Figure 3.32 Deflection curve of a cantilever beam subjected to an end couple and different values of the axial force P

Solving $[\mathbf{K}]\{\mathbf{Q}\} = \{\mathbf{F}\}$, we obtain $v_2 = 0.5\,\mathrm{m}$, $\theta_2 = 1\,\mathrm{rad}$. The deflection curve can be obtained using the shape functions as shown below. Note that in the present example $x = s$ as $L = 1$ m. Furthermore, the following deflection curve is exact as the shape function is a cubic polynomial in x, as the exact solution:

$$v(s) = v_2 N_3(s) + \theta_2 N_4(s) = 0.5 s^2.$$

Case b ($P = +2,000$ N): The global stiffness matrix $[\mathbf{K}]$ and load vector $[\mathbf{Q}]$ will remain the same. In this simple example, it is unnecessary to solve for the static analysis because the element force is the same as the applied force $P_r = P = 2000$ N. Now we need to compute the incremental stiffness matrix from eq. (3.124), and then, the total stiffness $[\mathbf{K}_T]$ using eq. (3.128):

$$[\mathbf{K}_T] = [\mathbf{K}] + 2000[\mathbf{K}_{inc}]$$

$$= 1000 \begin{bmatrix} 12 & -6 \\ -6 & 4 \end{bmatrix} + 2000 \times \frac{1}{30} \begin{bmatrix} 36 & -3 \\ -3 & 4 \end{bmatrix}$$

$$= \begin{bmatrix} 14{,}400 & -6200 \\ -6200 & 4267 \end{bmatrix} \begin{matrix} v_2 \\ \theta_2 \end{matrix}.$$

Solving $[\mathbf{K}_T]\{\mathbf{Q}\} = \{\mathbf{F}\}$, we obtain $v_2 = 0.27\,\mathrm{m}$, $\theta_2 = 0.626\,\mathrm{rad}$. The deflection curve is obtained using the shape functions as before:

$$v(s) = v_2 N_3(s) + \theta_2 N_4(s) = 0.184 s^2 + 0.086 s^3.$$

The above deflection is approximate because the exact solution given in eq. (3.100) is a *cosh* function, whereas the finite element solution used polynomial shape functions. By using more elements, one can obtain a better approximation for the deflection.

Case c (P = –2000 N): The procedures are similar to case b above except P is negative.

$$K_T = K - 2000 K_{inc} = \begin{bmatrix} 9{,}600 & -5800 \\ -5800 & 3733 \end{bmatrix} \begin{matrix} v_2 \\ \theta_2 \end{matrix}.$$

Solving $[\mathbf{K}_T]\{\mathbf{Q}\} = \{\mathbf{F}\}$, we obtain $v_2 = 2.64\,\mathrm{m}$, $\theta_2 = 4.36\,\mathrm{rad}$. The deflection curve is obtained using the shape functions as before:

$$v(s) = v_2 N_3(s) + \theta_2 N_4(s) = 3.56 s^2 - 0.92 s^3.$$

In reality, such large deflection may not occur, as geometric and material nonlinearities will have to be taken into account. The beam may fracture or yield before such large deflections could occur. The deflection curves for all three cases are plotted in figure 3.32. As expected, the deflection is reduced when a positive axial load is applied, while the deflection is increased when a negative axial load is applied. ▄▄

EXAMPLE 3.12 *Finite element analysis of buckling of a beam*

Determine the critical values of P_{cr} for which buckling will occur in the beam given in example 3.11.

SOLUTION   Buckling will occur only for the case $P < 0$. The eigenvalue problem can be written as [see eq. (3.130)]

$$\begin{bmatrix} 12000 - 1.2P & -6000 + 0.1P \\ -6000 + 0.1P & 4000 - 0.133P \end{bmatrix} \begin{Bmatrix} v_2 \\ \theta_2 \end{Bmatrix} = \begin{Bmatrix} 0 \\ 0 \end{Bmatrix},$$

or

$$[\mathbf{K} - P\mathbf{K}_{inc}]\{\mathbf{Q}\} = \{\mathbf{0}\}.$$

Note we have used $P_r = P$, as only one element is used. By setting the determinant of the square matrix in the above equation to zero, we obtain a quadratic equation in P as

$$(12 - 1.2\beta)(4 - 0.133\beta) - (-6 + 0.1\beta)^2 = 0,$$

where $\beta = P/1000$ is introduced to make the expression simple. By solving the above equation, we can obtain the critical values of P as $P_{cr1} = 2{,}486$ N and $P_{cr2} = 32{,}180$ N. The exact values of the first two buckling loads for a cantilever beam are: $P_{cr1} = \pi^2 EI/4L^2 = 2{,}467$ N and $P_{cr2} = 9\pi^2 EI/4L^2 = 22{,}207$ N [5]. One can note that the one-element solution predicts the first buckling load within 1%, but there is an error of about 45% in the second buckling load. One has to use more elements to obtain an accurate estimate of higher buckling loads. Furthermore, the finite element solution for P_{cr} is larger than the exact solution. It is because the finite element models make the beam stiffer than it actually is.

In order to determine the mode shape of the beam, we substitute the critical load (eigenvalue) back into the eigenvalue problem above. Let us consider $P_{cr} = 2{,}467$ N. Then we obtain

$$\begin{bmatrix} 9017 & -5751 \\ -5751 & 3669 \end{bmatrix} \begin{Bmatrix} v2 \\ \theta 2 \end{Bmatrix} = \begin{Bmatrix} 0 \\ 0 \end{Bmatrix}.$$

One can verify that the square matrix in the above equation is nearly singular as its determinant is equal to zero. When the determinant of the coefficient matrix is zero, the solution of matrix equation may not exist, or there might be infinitely many solutions. Since the right-hand side is all zero, there will be infinitely many solutions in this case. In fact, it is possible to find the ratio between the two DOFs to be $\theta_2/v_2 = 1.57$. That means, any θ_2 and v_2 whose ratio is 1.57 can be a solution. Among infinitely many solutions, we choose $v_2 = 1$, $\theta_2 = 1.57$. It is noted that this choice is for convenience, and the magnitude of solution does not have any physical significance. In fact, different finite element programs use different ways of determining DOFs.

Using DOFs, the mode shape can be obtained by using the shape functions as:

$$v(s) = v_2 N_3(s) + \theta_2 N_4(s) = 1.43s^2 - 0.43s^3 \text{ (first mode)}.$$

Again, it is emphasized that $v(s)$ is different from the deflection curve of the beam. The mode shape shows the pattern of deformation due to the buckling load and does not represent the actual deformation amplitude.

Similarly, by using the second buckling load, we can obtain the relative values of DOFs as $v_2 = 1$, $\theta_2 = -9.57$. Then the mode shape for the second buckling load can be derived as

$$v(s) = v_2 N_3(s) + \theta_2 N_4(s) = 12.57s^2 - 11.57s^3 \text{ (second mode)}.$$

The two buckling mode shapes are plotted in figure 3.33. Note that both buckling mode shapes satisfy the displacement boundary condition (clamped at $x = 0$).

---

[5] Boresi, A. P. 2003. *Advanced Mechanics of Materials*, 6th Ed. John Wiley & Sons, New York, NY. (Page 439.)

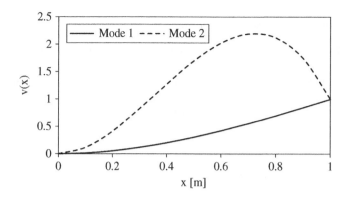

Figure 3.33 Buckling mode shapes of a cantilever beam obtained using one beam finite element

The above results can easily be obtained by using the MATLAB command $[\mathbf{Q},\mathbf{P}]=\text{eig}(\mathbf{K},\mathbf{K}_{inc})$. The outputs for $[\mathbf{Q}]$ and $[\mathbf{P}]$ are two square matrices:

$$[\mathbf{Q}] = \left[\begin{pmatrix} -0.91 \\ -1.4 \end{pmatrix} \begin{pmatrix} -0.26 \\ 2.4 \end{pmatrix} \right], [\mathbf{P}] = \begin{bmatrix} 2486 & 0 \\ 0 & 32181 \end{bmatrix}.$$

Matrix $[\mathbf{P}]$ is a diagonal matrix and has the eigenvalues or the P_{cr} values. The first column of $[\mathbf{Q}]$ is the mode shape corresponding to mode 1, and the second column corresponds to mode 2. One can note that MATLAB yields the same ratios for $\theta_2/v_2 = 1.57$ and -9.57 for mode 1 and mode 2, respectively, as obtained by hand calculations. ▄▄

EXAMPLE 3.13 *Finite element analysis of buckling of a clamped-hinged beam*

Calculate the buckling loads and corresponding mode shapes of the beam shown in figure 3.34 using: (a) one element and (b) two elements of equal length. Assume beam length $2L = 2$ m and flexural rigidity $EI = 1,000$ N-m^2.

SOLUTION   We will first solve the problem using two elements as the one-element solution is simple and straightforward. We first write down the element stiffness and incremental stiffness matrices. Since the elements are of equal length, the matrices are also the same for both elements.

$$\left[\mathbf{k}^{(1)}\right] = \frac{EI}{L^3} \begin{bmatrix} 12 & 6L & -12 & 6L \\ 6L & 4L^2 & -6L & 2L^2 \\ -12 & -6L & 12 & -6L \\ 6L & 2L^2 & -6L & 4L^2 \end{bmatrix} \begin{matrix} v_1 \\ \theta_1 \\ v_2 \\ \theta_2 \end{matrix},$$

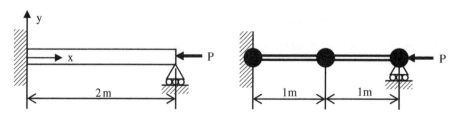

Figure 3.34 Clamped-hinged beam subjected to an axial force

$$\left[\mathbf{k}^{(2)}\right] = \frac{EI}{L^3}
\begin{bmatrix}
12 & 6L & -12 & 6L \\
6L & 4L^2 & -6L & 2L^2 \\
-12 & -6L & 12 & -6L \\
6L & 2L^2 & -6L & 4L^2
\end{bmatrix}
\begin{matrix} v_2 \\ \theta_2 \\ v_3 \\ \theta_3 \end{matrix},$$

$$\left[\mathbf{k}_{inc}^{(1)}\right] = \frac{1}{30L}
\begin{bmatrix}
36 & 3L & -36 & 3L \\
3L & 4L^2 & -3L & -L^2 \\
-36 & -3L & 36 & -3L \\
3L & -L^2 & -3L & 4L^2
\end{bmatrix}
\begin{matrix} v_1 \\ \theta_1 \\ v_2 \\ \theta_2 \end{matrix},$$

$$\left[\mathbf{k}_{inc}^{(2)}\right] = \frac{1}{30L}
\begin{bmatrix}
36 & 3L & -36 & 3L \\
3L & 4L^2 & -3L & -L^2 \\
-36 & -3L & 36 & -3L \\
3L & -L^2 & -3L & 4L^2
\end{bmatrix}
\begin{matrix} v_2 \\ \theta_2 \\ v_3 \\ \theta_3 \end{matrix}.$$

We will assemble the above stiffness matrices to obtain global matrices that contain only elements corresponding to nonzero DOFs. Note $L = 1$ m:

$$[\mathbf{K}] = 1000
\begin{bmatrix}
24 & 0 & 6 \\
0 & 8 & 2 \\
6 & 2 & 4
\end{bmatrix}
\begin{matrix} v_2 \\ \theta_2 \\ \theta_3 \end{matrix},$$

$$[\mathbf{K}_{inc}] = \frac{P}{30}
\begin{bmatrix}
72 & 0 & 3 \\
0 & 8 & -1 \\
3 & -1 & 4
\end{bmatrix}
\begin{matrix} v_2 \\ \theta_2 \\ \theta_3 \end{matrix}.$$

Since the axial force P is the same in both elements, we can use $P_r = P$. The global equations take the form:

$$[\mathbf{K} - P\mathbf{K}_{inc}]
\begin{Bmatrix} v_2 \\ \theta_2 \\ \theta_3 \end{Bmatrix} =
\begin{Bmatrix} 0 \\ 0 \\ 0 \end{Bmatrix}.$$

For the nontrivial solution of the above equation, the determinant of the total stiffness matrix must vanish, that is, $|\mathbf{K} - P\mathbf{K}_{inc}| = 0$. The eigenvalues and eigenvectors corresponding to the above global matrices can be obtained using the MATLAB command [Q,P]=eig(K,K$_{inc}$) :

$$[\mathbf{P}] =
\begin{bmatrix}
5{,}177 & 0 & 0 \\
0 & 18{,}775 & 0 \\
0 & 0 & 49{,}381
\end{bmatrix},$$

$$[\mathbf{Q}] =
\begin{bmatrix}
-0.5979 & 0.2693 & 0.0239 \\
-0.4143 & -1.2071 & 1.4987 \\
1.2623 & 1.3760 & 2.1244
\end{bmatrix}.$$

Thus, we obtain the first three buckling loads as: $P_{cr1} = 5{,}177$ N; $P_{cr2} = 18{,}775$ N; $P_{cr3} = 49{,}381$ N. The exact value of P_{cr1} is 5,036 N. The three eigenvectors are given by the three columns of $[\mathbf{Q}]$ above. The first column corresponds to the first eigenvalue and so on. The buckling mode shapes can be obtained using the beam shape functions. We will plot the mode shape for the first and second mode. Since there are two elements, there will be two expressions, one for each element: $v^{(1)}(s) = v_2 N_3(s) + L\theta_2 N_4(s)$ and $v^{(2)}(s) = v_2 N_1(s) + L\theta_2 N_2(s) + L\theta_3 N_4(s)$, where the superscripts

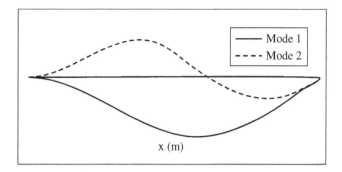

Figure 3.35 Buckling mode shapes for the beam in example 3.13 with two elements

denote the element number, and $N_i(s)$ are the beam shape functions given in eq. (3.43). For the first buckling load, we obtain the mode shape as:

$$v^{(1)}(s) = -0.5979(3s^2 - 2s^3) - 0.4143(-s^2 + s^3),$$
$$v^{(2)}(s) = -0.5979(1 - 3s^2 + 2s^3) - 0.4143(s - 2s^2 + s^3) + 1.2623(-s^2 + s^3).$$

In the above equations, the variable s is the normalized length such that $s = 0$ at the first node and $s = 1$ at the second node of the element. Similarly, we can derive the mode shape for the second buckling load as follows:

$$v^{(1)}(s) = 0.2693(3s^2 - 2s^3) - 1.2071(-s^2 + s^3),$$
$$v^{(2)}(s) = 0.2693(1 - 3s^2 + 2s^3) - 1.2071(s - 2s^2 + s^3) + 2.1244(-s^2 + s^3).$$

The buckling mode shapes are plotted in figure 3.35.

If we use just one element, the equations are greatly simplified as there is only one active DOF, the rotation at the second node, θ_2. The global stiffness matrices will just have one term and the buckling equation will be a scalar equation. Noting the element length will be equal to the beam length 2 m, we obtain: $[K] = 2{,}000$ and $[K_{inc}] = 0.2667$. The global equation takes the following form: $[2000 - 0.2667P]\theta_2 = 0$. This yields $P_{cr} = 7{,}500$ N. The eigenvector is given by $\theta_2 = 1$. Then the mode shape is simply the shape function $N_4(s)$, which is shown in figure 3.10. The error in buckling load is about 50% whereas the two-element solution above had an error of only 2.8%. ∎

3.6   BUCKLING OF FRAMES

Buckling analysis of plane frames follows essentially the same steps as buckling of beams. The stiffness matrix of a frame element was derived in section 3.4; for example, see eq. (3.80). For plane frames, each node has three DOFs, u, v, and θ. In deriving the stiffness matrix, we combined the stiffness matrices of a beam element and a uniaxial bar element. We will follow a similar procedure for deriving the incremental stiffness matrix $[\mathbf{k}_{inc}]$ also. It is noted that in the calculation of the stiffness matrix in eq. (3.80), we assumed that bending and tension are decoupled. On the other hand, in the derivation of buckling of the beam in section 3.5, it is assumed that the axial tension or compression can affect the bending deformation. However, the outcome of buckling analysis is the incremental stiffness matrix, which is solely a function of geometry. Therefore, it is still reasonable to assume that the bending and tension are decoupled in the calculation of the plane frame elements.

Consider a generic frame element shown in figure 3.36.

This is similar to the one in figure 3.23 except an axial compressive force $P^{(e)}$ is acting on the element. The axial force can be determined by a static analysis of the frame as described in section 3.4. The element stiffness matrix in the local $\bar{x} - \bar{y}$ coordinate system is given in eq. (3.80). The incremental

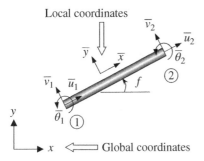

Figure 3.36 Degrees of freedom of plane portal frame

stiffness matrix of the frame element in local coordinates will be similar to that of a beam element in eq. (3.124) except extra zeros are added for the axial DOFs $\bar{u}_1$ and $\bar{u}_2$:

$$
[\bar{\mathbf{k}}_{inc}] = \frac{1}{30L}
\begin{bmatrix}
0 & 0 & 0 & 0 & 0 & 0 \\
0 & 36 & 3L & 0 & -36 & 3L \\
0 & 3L & 4L^2 & 0 & -3L & -L^2 \\
0 & 0 & 0 & 0 & 0 & 0 \\
0 & -36 & -3L & 0 & 36 & -3L \\
0 & 3L & -L^2 & 0 & -3L & 4L^2
\end{bmatrix}
\begin{matrix}
\bar{u}_1 \\ \bar{v}_1 \\ \bar{\theta}_1 \\ \bar{u}_2 \\ \bar{v}_2 \\ \bar{\theta}_2
\end{matrix}.
\tag{3.131}
$$

The above incremental stiffness matrix is transformed to global coordinates using the transformation in the same way as the stiffness matrix is transformed by $[\mathbf{k}] = [\mathbf{T}]^T[\bar{\mathbf{k}}][\mathbf{T}]$ (see section 3.4):

$$
[\mathbf{k}_{inc}] = [\mathbf{T}]^T[\bar{\mathbf{k}}_{inc}][\mathbf{T}].
\tag{3.132}
$$

Then the incremental stiffness matrices of various elements are assembled to obtain the global incremental stiffness matrix $[P_r\mathbf{K}_{inc}]$ as shown in eq. (3.126).

EXAMPLE 3.14 *Buckling of a portal frame*

A pair of loads P acts on a portal frame as shown in the figure 3.37. The length of all members is equal to 1 m, flexural rigidity $EI = 1,000$ N-m^2, and the axial rigidity $EA = 10^9$ N. Determine the critical values of P for buckling of the frame. Sketch the predicted buckled shape of the frame

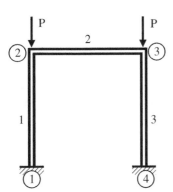

Figure 3.37 A portal frame subjected to two axial forces

SOLUTION We will use the simplest model and use one element per member. Thus, there will be six active DOFs, three each at nodes 2 and 3. Thus, we need to solve a 6 × 6 eigenvalue problem, and we will obtain six values of the critical loads. First, we need to solve the plane frame problem to determine the axial load carried by each member. By inspection, we can determine that elements 1 and 3 will carry a compressive load of P, and there will not be any axial force in element 2. Furthermore, we will assume that v_2 and v_3 are approximately equal to zero because of the large axial rigidity of elements 1 and 3. Thus, the active DOFs are u_2, θ_2, u_3, and θ_3.

The stiffness matrices of all three elements are as follows:

$$\left[\mathbf{k}^{(1)}\right] = 1000 \begin{bmatrix} 12 & 6 \\ 6 & 4 \end{bmatrix} \begin{matrix} u_2 \\ \theta_2 \end{matrix},$$

$$\left[\mathbf{k}^{(2)}\right] = 1000 \begin{bmatrix} 1 \times 10^6 & 0 & -1 \times 10^6 & 0 \\ 0 & 4 & 0 & 2 \\ -1 \times 10^6 & 0 & 1 \times 10^6 & 0 \\ 0 & 2 & 0 & 4 \end{bmatrix} \begin{matrix} u_2 \\ \theta_2 \\ u_3 \\ \theta_3 \end{matrix},$$

$$\left[\mathbf{k}^{(3)}\right] = 1000 \begin{bmatrix} 12 & 6 \\ 6 & 4 \end{bmatrix} \begin{matrix} u_3 \\ \theta_3 \end{matrix}.$$

The above matrices are assembled to obtain the global stiffness matrix as

$$[\mathbf{K}] = 1000 \begin{bmatrix} 12 + 10^6 & 6 & -1 \times 10^6 & 0 \\ 6 & 8 & 0 & 2 \\ -1 \times 10^6 & 0 & 12 + 10^6 & 6 \\ 0 & 2 & 6 & 8 \end{bmatrix} \begin{matrix} u_2 \\ \theta_2 \\ u_3 \\ \theta_3 \end{matrix}.$$

The next step is to calculate the incremental stiffness matrices for each element and transform them to the global coordinates using eq. (3.132). However, in hand calculations we can calculate $[\mathbf{k}_{inc}]$ by inspection. It is possible because the transformation involves angles such as 0 and 90 degrees. The element and global incremental stiffness matrices are calculated as follows:

$$\left[\mathbf{k}_{inc}^{(1)}\right] = \frac{1}{30} \begin{bmatrix} 36 & 3 \\ 3 & 4 \end{bmatrix} \begin{matrix} u_2 \\ \theta_2 \end{matrix},$$

$$\left[\mathbf{k}_{inc}^{(3)}\right] = \frac{1}{30} \begin{bmatrix} 36 & 3 \\ 3 & 4 \end{bmatrix} \begin{matrix} u_3 \\ \theta_3 \end{matrix},$$

$$[\mathbf{K}_{inc}] = \frac{1}{30} \begin{bmatrix} 36 & 3 & 0 & 0 \\ 3 & 4 & 0 & 0 \\ 0 & 0 & 36 & 3 \\ 0 & 0 & 3 & 4 \end{bmatrix} \begin{matrix} u_2 \\ \theta_2 \\ u_3 \\ \theta_3 \end{matrix}.$$

Note that we have used $P_r = P$. Since element 2 does not carry any axial force, the corresponding incremental stiffness matrix does not exist. The four eigenvalues and corresponding eigenvectors can be computed using the MATLAB command `eig(K, K_inc)`. Note that the actual system has six eigenvalues, but the simplified system has four eigenvalues. The two smallest eigenvalues are $P_{r1} = 7{,}445$ N and $P_{r2} = 45{,}000$ N, respectively. The corresponding eigenvectors $\{\mathbf{Q}_1\} = \{u_2, \theta_2, u_3, \theta_3\}^T = \{0.666, -0.388, 0.666, -0.388\}^T$ and $\{\mathbf{Q}_2\} = \{u_2, \theta_2, u_3, \theta_3\}^T = \{0, -1.94, 0, 1.94\}^T$, respectively. Note that $u_2 = u_3$ in both eigenvectors indicating the axial stiffness of element 2 is infinitely large.

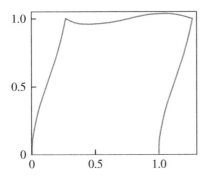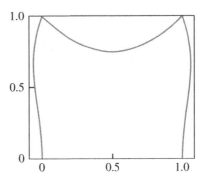

Figure 3.38 First mode (assymteric or swaying mode) and second mode (symmteric mode) buckling of the portal frame in example 3.14

It should be mentioned that the above buckling loads are approximate. In a static analysis of a frame, which has loads acting only at the joints or nodes, modeling each member using one element yields an exact solution. As we noted earlier, buckled shape of a beam cannot be represented using polynomial functions since they involve *sinh* and *cosh* functions. The cubic polynomial interpolation functions cannot represent *sinh* and *cosh* functions exactly. Each frame member should be subdivided into more beam elements if one desires more accurate buckling loads.

The mode shapes are plotted using procedures similar to those used for deflection shape for static analysis in section 3.4. Since we know the displacements and rotations at each node of an element, we can plot its deflection shape using beam shape functions. The mode shapes are depicted in figure 3.38. One can note that the first mode corresponds to swaying mode, and the second one is the symmetric mode. The other eigenvalues are much higher and of no practical significance as the structure is expected to buckle before those loads are reached.

3.7 FINITE ELEMENT MODELING PRACTICE FOR BEAMS

In this section, several analysis problems are used to discuss modeling issues as well as verifying the accuracy of the analysis results by comparing with literature. The examples are presented in such a way that any finite element analysis program can be used to solve the problems. However, the analysis results can be slightly different because of implementation differences between the finite element analysis programs.

3.7.1 Stress and Deflection Analysis of a Beam[6]

A standard 30 in. wide-flange beam, with a cross-sectional area $A = 50.65$ in^2 and the second moment of inertia $I_z = 7892$ in^4 is supported as shown in figure 3.39. A uniformly distributed load $w = 10,000$ lb/ft is applied on the two overhangs. Determine the maximum bending stress σ and the deflection δ at the middle portion of the beam. Use $E = 3.0 \times 10^7$ psi. $L = 20$ ft, $a = 10$ ft, $h = 30$ in.

---

[6] Timoshenko, S. 1995. *Strength of Material, Part I, Elementary Theory and Problems*, 3rd Edition. D. Van Nostrand Co., Inc., New York, NY. (Page 98, problem 4.)

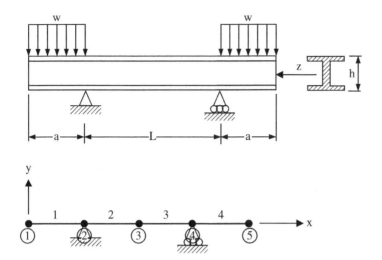

Figure 3.39 Beam bending with distributed loads

In order to solve for the stress and deflection, it is necessary to calculate the bending moment in the beam. Due to symmetry in geometry and loading, the portion between two supports will have a constant bending moment. From the moment equilibrium, the bending moment in the middle portion can be calculated as

$$M = \frac{wa^2}{2} = -6 \times 10^6 \text{ lb·in.}$$

Since the above bending moment is constant, stress will also be constant in the middle portion. At the top of the beam cross-section, the maximum stress can be calculated as

$$\sigma_{xx} = -\frac{M\frac{h}{2}}{I_z} = -\frac{-6 \times 10^6 \times 15}{7892} = 11,404 \text{ psi}$$

In order to calculate the deflection in the middle portion, the relation between bending moment and curvature $EIv'' = M$ can be used. For this purpose, let us assume that the origin is located at the left support. Since M is constant, v'' is also a constant; that is, the deflection curve is a quadratic function: $v(x) = a_0 + a_1 x + a_2 x^2$. Using the boundary conditions, $v(0) = v(240) = 0$, the following form of deflection curve can be obtained:

$$v(x) = a_2 x(x - 240), \quad v'' = 2a_2.$$

The unknown coefficient can be calculated from the relation $EIv'' = M$,

$$a_2 = \frac{M}{2EI_z} = -1.2671 \times 10^{-5}.$$

The maximum deflection occurs at the center of the middle portion at $x = 10$ ft,

$$v(120) = -1.2671 \times 10^{-5} \times 120 \times (120 - 240) = 0.1825 \text{ in.}$$

For finite element modeling, it is enough to use a single element in the middle portion because the bending moment is constant. However, since the stress and deflection at the center of the middle portion is of interest, two beam elements can be used. Including the two overhangs, a total of four equal-length beam elements are used to model the problem, as shown in figure 3.39.

The distributed load in the overhangs is constant in the negative y-coordinate direction. Different finite element analysis programs use different ways of applying distributed loads on a beam. The uniformly distributed load is the simplest one, and many programs allow to input a functional form of distributed load. Note that the distributed load is a force per unit length, not force per unit area.

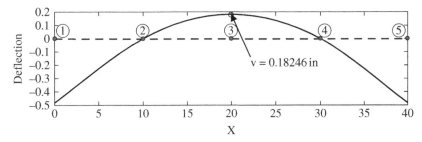

Figure 3.40 Deflection curve of the beam

In the case of a bar or truss, only the cross-sectional area is required to calculate stress. However, in the case of beam, the cross-sectional geometry is important as it affects the moment of inertia as well as stress calculation. Therefore, it is required to input cross-sectional geometry. Many finite element programs support a functionality of defining various cross-sectional geometries, such as circular, rectangular, hollow-cylinder, I-section, and so forth. As shown in figure 3.39, an I-section beam is used for this example.

The following table shows nodal displacements and rotations of the beam. As expected, the deformation is symmetric with respect to the center. A very small rotation $\theta = 0.24395E\text{-}17$ is calculated for node 3, but this is due to numerical error. Compared to other rotations, this can be considered as zero.

Node	v	θ
1	−0.48274	0.40547E-02
2	0.0	0.30411E-02
3	0.18246	0.24395E-17
4	0.0	−0.30411E-02
5	−0.48274	−0.40547E-02

Using the nodal displacements and rotations in the above table, the deflection curve of the beam can be obtained using the beam shape functions in eq. (3.43). Figure 3.40 shows the deflection curve of the beam. As expected, the deflections at nodes 2 and 4 are zero, and the maximum occurs at node 3.

In order to calculate stress, the bending moment should be calculated first. In element 2, the bending moment can be calculated from eq. (3.47) as

$$M(s) = \frac{EI}{L^2}\{\mathbf{B}\}^{\mathrm{T}}\{\mathbf{q}\} = -6 \times 10^6 \text{ lb·in.}$$

Note that the bending moment is constant in the element, and its value is identical to the analytical solution. Therefore, the stress will be the same as the analytical solution. The finite element solution happens to be accurate because the true deflection is a quadratic polynomial and the finite element shape functions are cubic polynomials.

3.7.2 Portal Frame Under Symmetric Loading[7]

A portal frame with I-beam sections is subjected to a uniformly distributed load $\omega = 500$ lb/in across the span as shown in figure 3.41. The length of the span is $L = 800$ in., while the height of the column

---

[7] Hoff, N. J. 1956. *The Analysis of Structures.* John Wiley and Sons, Inc., New York, NY. (Pages 115–119.)

is $a = 400$ in. The objective is to determine the maximum rotation and maximum bending moment. The moment of inertia for the span, I_{span} is five times the moment of inertia for the columns, I_{col}; $I_{col} = 20,300$ in^4 and $I_{span} = 101,500$ in^4. For material properties, Young's modulus $E = 30 \times 10^6$ psi and Poisson's ratio $\nu = 0.3$.

All the members of the frame are modeled using I-beam cross sections. The cross section for the columns is chosen to be a W 36×300 I-beam section, as shown in figure 3.42. The dimensions used in the horizontal span are scaled by a factor of 1.49535 to produce a moment of inertia that is five times the moment of inertia in the columns. The theoretical maximum rotation and maximum bending moment are, respectively,

$$\theta_{max} = \frac{wa^3}{27EI_{col}},$$
$$M_{max} = \frac{19}{54}wa^2.$$

The portal frames can be modeled using the plane beam element of most finite element software. If the plane beam element is not available, the space beam element can also be used after fixing all DOFs related to the out-of-plane motion. Since the distributed load is applied on the span, it would be necessary to use more than one element. In this case, 16 plane beam elements are used for the span, while four beam elements are used for each column as shown in figure 3.41.

In the beam modeling, it is important to understand the direction of coordinate systems. Some software uses the axial direction as local x-axis, and the cross section is defined in the y-z coordinate. In such a case, it is important to pay attention to the positive direction of each coordinate. Most finite element software support the standard cross-sections, such as I-beam section. Therefore, users can provide dimensions of the I-beam section. In the case of the vertical column, the cross-sectional dimensions are shown in the figure. In the case of the span, the dimensions are multiplied by 1.49535.

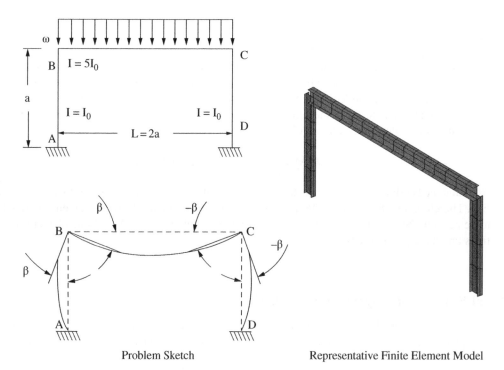

Problem Sketch

Representative Finite Element Model

Figure 3.41 Portal frame under symmetric loading

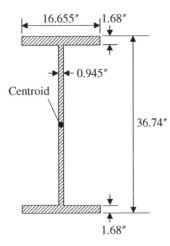

Figure 3.42 Cross-sectional dimensions for W 36×300 I-beam section

The following table shows the maximum rotation and the maximum bending moment of the portal frame. The finite element results show a close agreement with the theoretical solutions.

	Theoretical solution	Finite element results
Max. rotation (radian)	0.195E-2	0.213E-2
Max. bending moment (lb)	0.281E8	0.287E8

3.7.3   Buckling of a Bar with Hinged Ends[8]

Determine the critical buckling load of an axially loaded long slender bar of length $L = 200$ in. with hinged ends, as shown in figure 3.43. The bar has a square cross section with width and height set to 0.5 inches. For material property, Young's modulus $E = 30 \times 10^6$ psi. For the axial force, use $P = 1$ lb. Determine the critical buckling load of an axially loaded long slender bar of length L with hinged ends. The bar has a square cross section with width and height set to 0.5 inches.

Only the upper half of the bar is modeled because of symmetry. The upper half of the column is modeled using ten 2-node beam elements. The boundary conditions become free-fixed for the half symmetry model. The theoretical solution for buckling load is

$$P_{cr} = \frac{\pi^2 EI}{4L^2}.$$

It should be noted that $L = 100$ in. should be used in the above equation. The following table compares the finite element solution for the critical load with that of theoretical solution. Since we used 10 elements, the results are almost the same.

	Theoretical solution	Finite element solution
Lowest buckling load (lb)	38.553	38.553

---

[8] Timoshenko, S. 1956. *Strength of Material, Part II, Elementary Theory and Problems*, 3rd Edition. D. Van Nostrand Co., Inc., New York, NY. (Page 148, article 29.)

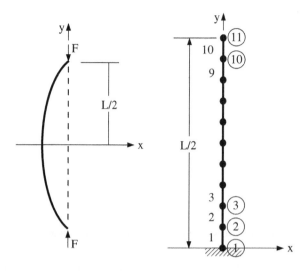

Figure 3.43 Buckling of a bar with hinged ends

3.8   PROJECT

(I)   This project is concerned with design of a bicycle frame using aluminum tubes. The schematic dimensions of the bicycle are shown in figure 3.44. The following two load cases should be considered.

(a)   Vertical loads: When an adult rides the bike, the nominal load is estimated as a downward load of 900 N at the seat position and a load of 300 N at the pedal crank location. When a dynamic environment is simulated using the static analysis, the static loads are often multiplied by a dynamic load factor G. In this design project, use $G = 2$. Use ball-joint boundary conditions for the front dropout (location 1) and sliding boundary conditions for the rear dropouts (locations 5 and 6).

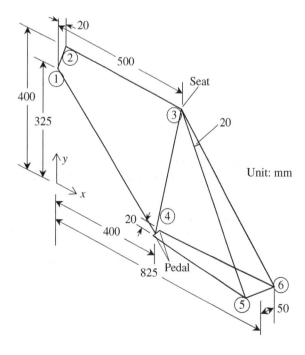

Figure 3.44 Bicycle frame structure

(b) Horizontal impact: The frame should be able to withstand a horizontal load of 1,000 N applied to the front dropout with rear dropouts constrained from any translational motion. For this load case, assume the front dropout can only move in the horizontal direction. Use $G = 2$.

Choose aluminum tubes of various diameters for the various members of the frame shown in figure 3.44 such that the bicycle is as light as possible. The minimum outside diameter is 12 mm and the wall thickness is 2 mm. Approximate the frame as a plane frame by giving the same (x, y) coordinates for nodes 5 and 6. Thus, all the nodes will be on the x-y plane. In addition to the dynamic load factor, use a safety factor of 1.5. Use the von Mises failure stress criterion for yielding. For compression members, include buckling as additional criterion. The buckling load of a member is approximated as $P_{cr} = 2\pi^2 EI/L^2$, where L is the member length. Use a safety factor of 1.5 for buckling also.

Properties of Aluminum

Material Property	Value
Young's Modulus (E)	70 GPa
Poisson's Ratio (ν)	0.33
Density (ρ)	2,580 kg/m^3
Yield Strength (σ_Y)	210 MPa

The report is supposed to be readable and complete by itself, for example, including introduction, approach, assumptions, results, conclusion, discussion, and references. In your report, you must include the following for each load case:

1. For each element the maximum normal stress, shear stress, and maximum von Mises stress and safety factor at each node.

2. For each element under compression ($P < 0$), buckling load, actual axial force P, and safety factor in buckling.

3. Nodal deflections at each node should be given. Calculate the weight of your frame.

3.9 EXERCISES

1. Answer the following descriptive questions.

   (a) Write the assumptions of the Euler-Bernoulli beam theory.

   (b) In an Euler-Bernoulli beam, what is the relationship between the vertical deflection and rotational angle?

   (c) For a fixed amount of cross-sectional area and a given bending moment, what is the best way to reduce the maximum stress for an Euler-Bernoulli beam?

   (d) What is the only nonzero stress component in an Euler-Bernoulli beam?

   (e) In an Euler-Bernoulli beam, how does the stress vary over the cross section?

   (f) How many Hermite beam elements do you need to get the exact solution (or analytical solution) for a cantilever beam that is subject only to a concentrated load at the tip? Explain.

   (g) For a simply supported beam element subjected to a uniformly distributed load, will you get the exact solution when it is modeled using a single beam element? Explain why or why not.

   (h) What is the difference between a truss-like structure and a frame-like structure?

   (i) List any three assumptions used in deriving the Euler-Bernoulli (Hermite) beam element.

   (j) If a beam element is clamped in both ends, when a uniformly distributed load is applied, what would be the deflection curve? Explain your answer.

(k) When a uniformly distributed load is applied to a cantilevered beam element, what would be the shear force diagram?

(l) What is the fundamental assumption of combining a beam element and a bar element to describe the behavior of a frame element?

(m) What is the best way of increasing the buckling load of a beam without increasing the structural weight?

2. Repeat example 3.1 with the approximate deflection in the following form: $v(x) = Cx(L-x)$.

   (a) Show that the above-assumed deflection satisfies the displacement boundary conditions of the beam.

   (b) Use the Rayleigh-Ritz method to determine the parameter C.

   (c) Compare the deflection curve with the exact solution.

3. Repeat example 3.2 with the approximate deflection in the following form: $v(x) = c_1x^2 + c_2x^3 + c_3x^4$. Compare the deflection curve with the exact solution.

4. The deflection of the simply supported beam shown in the figure is assumed as $v(x) = cx(x-1)$, where c is a constant. A force is applied at the center of the beam. Use the following properties: $EI = 2,000$ N-m^2 and $L = 1$ m. First, (a) show that the above approximate solution satisfies displacement boundary conditions, and (b) use the Rayleigh-Ritz method to determine c.

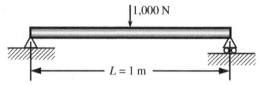

5. Use the Rayleigh-Ritz method to determine the deflection $v(x)$, bending moment $M(x)$, and shear force $V_y(x)$ for the beam shown in the figure. The bending moment and shear force are calculated from the deflection as: $M(x) = EId^2v/dx^2$ and $V_y(x) = -EId^3v/dx^3$. Assume the displacement as $v(x) = c_0 + c_1x + c_2x^2 + c_3x^3$, and $EI = 2,000$ N-m^2, $L = 1$ m, and $p_0 = 200$ N/m, and $C = 100$ N-m. Make sure the displacement boundary conditions are satisfied a priori.

   Hint: The potential energy of a couple is calculated as $V = -Cdv/dx$, where the rotation is calculated at the point of application of the couple.

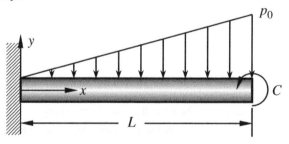

6. Repeat problem 5 with the linearly decreasing load as shown in the figure.

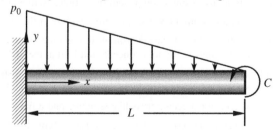

7. The right end of a cantilevered beam is resting on an elastic foundation that can be represented by as spring with spring constant $k = 2,000$ N/m. A force of 1,000 N acts at the center of the beam as shown. Use the Rayleigh-Ritz

method to determine the deflection $v(x)$ and the force in the spring. Assume $EI = 2,000$ N-m^2 and $v(x) = c_0 + c_1 x + c_2 x^2 + c_3 x^3$.

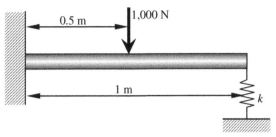

8. A cantilevered beam is modeled using one finite element. The nodal values of the beam element are given as

$$\{\mathbf{q}\} = \left\{ 0, \, 0, \, \frac{FL^3}{3EI}, \, \frac{FL^2}{2EI} \right\}^{\mathrm{T}}.$$

   Plot the deflection curve, bending moment, and shear force.

9. A simply supported beam with length L is under a concentrated vertical force $-F$ at the center. When two equal-length beam elements are used, the finite element analysis yields the following nodal DOFs:

$$\{\mathbf{Q}_s\}^{\mathrm{T}} = \{v_1, \theta_1, v_2, \theta_2, v_3, \theta_3\} = \left\{ 0, \, -\frac{FL^2}{16EI}, \, -\frac{FL^3}{48EI}, 0, 0, \, \frac{FL^2}{16EI} \right\}.$$

   Find the deflection curve $v(x)$ and compare it with the exact solution in a graph. Note: the exact deflection curve of the beam is, for $x \leq \frac{1}{2}L$, $v_{\mathrm{exact}}(x) = Fx(3L^2 - 4x^2)/48EI$ and symmetric for $x \geq \frac{1}{2}L$.

10. A simply supported beam with length L is under a uniformly distributed load $-p$. When two equal-length beam elements are used, the finite element analysis yields the following nodal DOFs:

$$\{\mathbf{Q}_s\}^{\mathrm{T}} = \{v_1, \theta_1, v_2, \theta_2, v_3, \theta_3\} = \left\{ 0, \, -\frac{pL^3}{24EI}, \, -\frac{5pL^4}{384EI}, 0, 0, \, \frac{pL^3}{24EI} \right\}.$$

   Find the deflection curve $v(x)$ and compare graphically with the exact solution. Note: the exact deflection curve of the beam is given as: $v_{\mathrm{exact}}(x) = -p(x^4 - 2Lx^3 + L^3x)/24EI$.

11. Consider a cantilevered beam with a Young's modulus E, moment of inertia I, height $2h$, and length L. A couple C is applied at the tip of the beam. One beam finite element is used to approximate the structure.

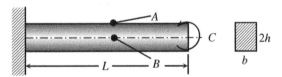

   (a) Calculate the tip displacement v and tip slope θ using the finite element equation.

   (b) Calculate the bending moment and shear force at the wall using the finite element equation.

   (c) Calculate the stress σ_{xx} at points A and B.

12. The cantilever beam shown is modeled using one finite element. If the deflection at the nodes of a beam element are $v_1 = \theta_1 = 0$, and $v_2 = 0.02$ m and slope $\theta_2 = 0$, write the equation of the deformed beam $v(s)$. In addition, compute the forces F_2 and C_2 acting on the beam to produce the above deformation in terms of E, I, and L. *Hint:* Use the equations $[\mathbf{k}]\{\mathbf{q}\} = \{\mathbf{f}\}$ for the beam element.

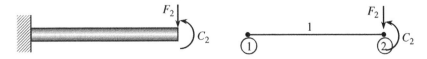

13. For the finite element model of a cantilever beam shown in problem 12, the beam has a concentrated force F_2 and couple C_2 acting at its end.

   (a) List three assumptions made in a beam element. Is one element sufficient to obtain an exact answer for this problem? Explain.

   (b) If the deflection at the end of the beam $w_2 = 0.01$ m and slope $\theta_2 = 0$, compute the applied loads (F_2 and C_2) and the reactions at the wall using one beam element. Assume $EI = 0.15$ N-m^2 and length of the beam $L = 1$ m.

14. Let a uniform cantilevered beam of length L be supported at the loaded end so that this end cannot rotate, as shown in the figure. For the given moment of inertia I, Young's modulus E, and applied tip load P, calculate the deflection curve $v(x)$ using one beam element.

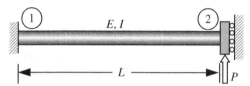

15. A cantilevered beam structure shown in the figure is under the distributed load. When $q = 1,000$ N/m, $L_T = 2$ m, $E = 207$ GPa, and the radius of the circular cross section $r = 0.1$ m, solve the displacement of the neutral axis and stress of the top surface. Use three equal-length beam finite elements in a commercial finite element analysis program. Compare the finite element solution with the exact solution. Provide the bending moment and shear force diagram from the finite element method and compare them with the exact solution. Explain why the finite element solutions are different from the exact solutions.

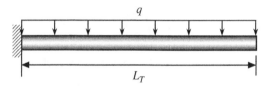

16. Solve the beam deflection problem shown in the figure using a two-node beam element. Note that node 1 is hinged.

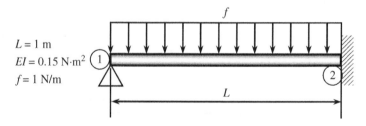

   (a) Using a beam stiffness matrix, set up the equation for this beam ($[\mathbf{K}]\{\mathbf{Q}\} = \{\mathbf{F}\}$).

   (b) Compute the angle of rotation at node 1, and write the equation of deformed shape of the beam using the shape functions.

   (c) Explain why the answer obtained above is not likely to be very accurate. If you want to obtain a better answer for this problem using the finite element method, what would you do?

17. The clamped-clamped beam shown below is subjected to uniformly distributed loads p_0 as well as a central concentrated force F. The beam is modeled using two beam elements. Use $EI = 1,000$ N-m^2, $L = 1$ m, $p_0 = 120$ N/m, and $F = 200$ N.

   (a) Calculate the consistent load vector for both elements.

   (b) Write down the 4×4 element stiffness matrices for both elements clearly showing the row addresses.

   (c) Write the assembled structural matrix equations in the form of $[\mathbf{K}_s]\{\mathbf{Q}_s\} = \{\mathbf{F}_s\}$.

(d) Obtain the global matrix equation $[\mathbf{K}]\{\mathbf{Q}\} = \{\mathbf{F}\}$ after applying boundary conditions.

(e) Solve the global matrix equation to calculate v_2 and θ_2.

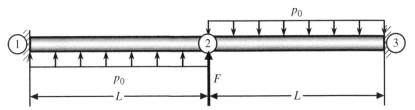

18. In this chapter, we derived the beam finite element equation using the principle of minimum potential energy. However, the same finite element equation can be derived from the Galerkin method, as in section 2.3. The governing differential equation of the beam is

$$EI\frac{d^4v}{dx^4} = f(x), \quad x \in [0,L],$$

where $f(x)$ is the distributed load. In the case of a clamped beam, the boundary conditions are given by

$$v(0) = v(L) = \frac{dv}{dx}(0) = \frac{dv}{dx}(L) = 0.$$

Using the Galerkin method and the interpolation scheme in eq. (3.44), derive the finite element matrix equation when a constant distributed load $f(x) = q$ is applied along the beam.

19. Repeat problem 18 for the case of a cantilevered beam whose boundary condition is given by

$$v(0) = \frac{dv}{dx}(0) = \frac{d^2v}{dx^2}(L) = \frac{d^3v}{dx^3}(L) = 0.$$

20. Solve the simply supported beam problem in example 3.9 using a commercial finite element analysis program. You can use either distributed load capability in the program or an equivalent nodal load. Plot the vertical displacement and rotation along the span of the beam. Compare them to the analytical solutions. Also, plot the bending moment and shear force along the span of the beam. Compare them to the analytical solutions.

21. Consider a cantilevered beam with spring support at the end, as shown in the figure. Assume $E = 200\,\text{ksi}, I = 1.0\,\text{in.}^4$, $L = 10\,\text{in.}, k = 300\,\text{lb./in.}$, beam height $h = 10\,\text{in.}$, and no gravity. The beam is subjected to a concentrated force $F = 200\,\text{lb.}$ at the tip.

(a) Using one beam element and one spring element, construct the structural matrix equation *before* applying boundary conditions. Clearly identify elements and nodes. Identify positive directions of all DOFs.

(b) Construct the global matrix equation after applying all the boundary conditions.

(c) Solve the matrix equation and calculate the tip deflection.

(d) Calculate the bending moment and shear force at the wall.

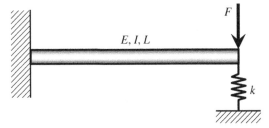

22. The right end of a cantilevered beam rests on a spring. A couple of magnitude 100 N-m acts at node 2. Calculate the deflection and rotation at node 2 of the beam. Assume Young's modulus of the beam material $E = 100\,\text{GPa}$, and the moment of inertia of beam cross section $I = 10^{-9}\,\text{m}^4$. The spring stiffness $= 1000\,\text{N/m}$. The stiffness matrix for the beam element is also given below.

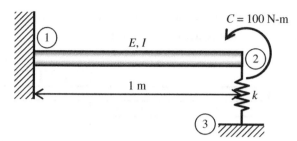

23. A beam is clamped at the left end and on a spring at the right end. The right support is such that the beam cannot rotate at that end. Thus, the only active DOF is v_2. A force $F = 5,000$ N acts downward at the right end as shown. The structure is modeled using two elements: one beam element and one spring element. The spring stiffness $k = 3,000$ N/m. The beam properties are $L = 1$ m, $EI = 2,000$ Nm^2.

   (a) Write the element stiffness matrices of both elements. Clearly show the DOFs.

   (b) Assemble the two element matrix equations and apply boundary conditions to obtain the global matrix equation.

   (c) Solve for unknown displacement v_2.

   (d) Calculate the deflection v and bending moment M at $x = 0.8$ m.

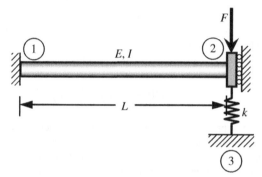

24. A beam of length $L = 0.1$ m is simply supported as shown in the figure. The beam is subjected to a rotation of $\theta = 1.75 \times 10^{-2}$ radians at node 1. Assume that the flexural rigidity of the cross section is 2.5 N-m^2.

   (a) Write the system equations $[K]\{Q\} = \{F\}$ for the beam (use one element). Apply the boundary conditions and solve for the unknown DOFs.

   (b) Derive the equation for the deformed shape of the beam element (deflection curve).

   (c) What is the couple applied at node 1 to produce the given rotation?

   (d) What are the bending moment and shear force at the midpoint of the element?

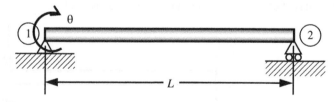

25. A straight shaft is modeled as a beam clamped at both ends. Due to the misalignment by an extraneous shim, the right bearing is located 1 mm above the straight position (see figure). The reactions at the right support are going to be calculated using two beam elements. Assume the flexural rigidity of the shaft $EI = 2,000$ N-m^2.

   (a) Build a global matrix equation *before* applying boundary conditions. Use the following nodal DOFs: $\{q\} = \{v_1, \theta_1, v_2, \theta_2, v_3, \theta_3\}^T$.

   (b) Reduce the global matrix equation by applying boundary conditions.

(c) Solve the global matrix equation for unknown nodal DOFs.

(d) Calculate the reactions at the right support.

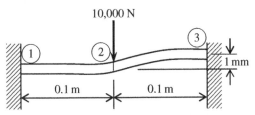

26. A linearly varying distributed load is applied to the beam finite element of length L. The maximum value of the load at the right side is q_0. Calculate "work equivalent" nodal forces and moment.

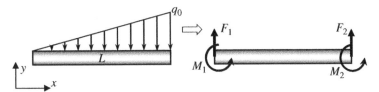

27. In general, a concentrated force can only be applied to the node. However, if we use the concept of "work-equivalent" load, we can convert the concentrated load within an element to corresponding nodal forces. A concentrated force P is applied at the center of one beam element of length L. Calculate "work equivalent" nodal forces and moments $\{F_1, M_1, F_2, M_2\}$ in terms of P and L.

Hint: the work done by a concentrated force can be obtained by multiplying the force by displacement at that point.

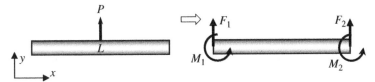

28. Use two equal-length beam elements to determine the deflection of the beam shown below. Estimate the deflection at point B, which is at 0.5 m from the left support. $EI = 1000 \ \text{N-m}^2$.

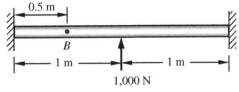

29. An external couple C_2 is applied at node 2 in the beam shown below. When $EI = 10^5 \ \text{N.m}^2$, the rotations in radians at the three nodes are determined to be $\theta_1 = -0.025$, $\theta_2 = +0.05$, $\theta_3 = -0.025$.

(a) Draw the shear force and bending moment diagrams for the entire beam.

(b) What is the magnitude of the couple C_2 applied at node 2?

(c) What is the support reaction F_{y1} at node 1?

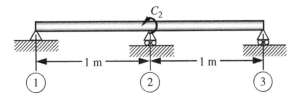

30. Two beam elements are used to model the structure shown in the figure. The beam is clamped to the wall at the left end (node 1), supports a load $P = 100$ N at the center (node 2), and is simply supported at the right end (node 3). Elements 1 and 2 are two-node beam element each of length 0.05 m, and the flexural rigidity $EI = 0.15$ N-m².

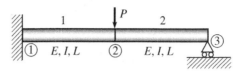

(a) Using the beam stiffness matrix, assemble the equations for the above model, apply boundary conditions, and solve for the deflection/slope at nodes 2 and 3.

(b) Write the equations and plot the deformed shape of each beam element showing defection and angles at all the nodes.

(c) What is the deflection v and slope θ at the midpoints of element 1 and element 2?

(d) What is the bending moments at the point of application of the load P?

(e) What are the reactions (bending moment and shear force) at the wall?

31. Consider the clamped-clamped beam shown below. Assume there are no axial forces acting on the beam. Use two elements to solve the problem. (a) Determine the deflection and slope at $x = 0.5$, 1, and 1.5 m; (b) Draw the bending moment and shear force diagrams for the entire beam; (c) What are the support reactions? (d) Use the beam element shape functions to plot the deflected shape of the beam. Use $EI = 2,000$ N.m, $L = 1$ m, and $F = 2,000$ N.

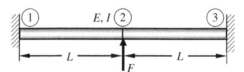

32. The structure shown in the figure consists of two cantilevered beams connected at the center by a riveted joint. Calculate the vertical deflection at the joint using two beam elements. Assume there is no friction at the riveted joints such that the beams can freely rotate relative to each other. Assume Young's modulus of the beam material $E = 100$ GPa and the moment of inertia of the beam cross section $I = 10^{-9}$ m⁴.

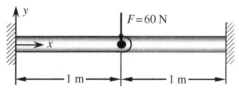

33. The frame shown in the figure is clamped at the left end and supported on a hinged roller at the right end. The radius of the circular cross section $r = 0.04$ m. An axial force P and a couple C act at the right end. Assume the following numerical values: $L = 1$ m, $E = 80$ GPa, $P = 10,000$ N, $C = 1,000$ Nm.

(a) Use one element to determine the rotation θ at the right support.

(b) What is the deflection of the beam at $x = L/2$?

(c) What is the maximum tensile stress? Where does it occur?

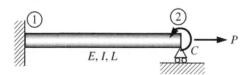

34. A frame structure is clamped on the left side, and there is an inclined roller supported on the right side, as shown in the figure. A uniformly distributed force q is applied, and the roller contact surface is assumed to be non-frictional. When one frame finite element is used to approximate the structure, the nodal displacement vector can be defined as $\{\mathbf{Q}\} = \{u_1, v_1, \theta_1, u_2, v_2, \theta_2\}^T$, where u and v are x- and y-directional displacements, respectively, and θ is the rotation with respect to z-direction.

(a) Construct the 6×6 structural matrix equation before applying boundary conditions.

(b) Reduce the structural matrix equation to a 2×2 global matrix equation by applying boundary conditions. You may need to use an appropriate transformation.

(c) Solve the global matrix equation and calculate the nodal displacement vector $\{\mathbf{Q}\}$.

(d) Write the expression of vertical displacement $v(x)$, $0 < x < L$, and sketch the vertical displacement $v(x)$ of the frame.

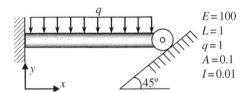

$E = 100$
$L = 1$
$q = 1$
$A = 0.1$
$I = 0.01$

35. A circular ring of square cross section is subjected to a pair of forces $F = 10,000$ N, as shown in the figure. Use a finite element analysis program to determine the compression of the ring, that is, relative displacements of the points where forces are applied. Assume $E = 70$ GPa. Determine the maximum values of the axial force P, bending moment M, and shear force V_r, and their respective locations.

Hints: Divide the ring into 40 elements. Use $x = R\cos\theta$ and $y = R\sin\theta$ to determine the coordinates. Fix all DOFs at the bottom-most point to constrain rigid body translation and rotation. Otherwise, the global stiffness matrix will be singular.

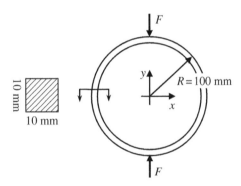

36. The ring in problem 35 can be solved using a smaller model considering the symmetry. Use the ¼ model to determine the deflection and maximum force resultants. What are the appropriate boundary conditions for this model? Show that both models yield the same results.

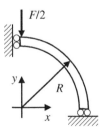

37. When nodal DOFs of a plane frame element with a rectangular cross section are given in the global coordinates as $\{\mathbf{q}\} = \{0, 0, 0, u_2, v_2, \theta_2\}^T$, calculate stress *at points A and B* using finite element method in terms of E, L, w, b, u_2, v_2, θ_2.

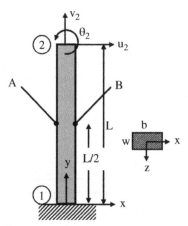

38. The frame shown in the figure is subjected to some forces at nodes 2 and 3. The resulting displacements are given in the table below. Sketch the axial force, shear force, and bending moment diagrams for element 3. What are the support reactions at Node 4? Assume $EI = 1000$ N-m^2 and $EA = 10^7$ m^2. Length of all elements = 1 m.

Node	u (mm)	v (mm)	θ (radian)
2	10	1	−0.1
3	11	−1	+0.1

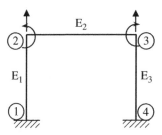

39. Solve the following frame structure using a commercial finite element analysis program. The frame structure is under a uniformly distributed load of $q = 1000$ N/m and has a circular cross section with radius $r = 0.1$ m. For material property, Young's modulus $E = 207$ GPa. Plot the deformed geometry with an appropriate magnification factor, and draw a bending moment and shear force diagrams.

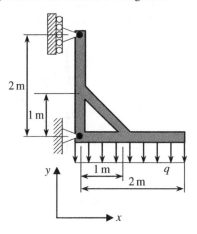

40. A cantilevered frame (element 1) and a uniaxial bar (element 2) are joined at node 2 using a bolted joint as shown in the figure. Assume there is no friction at the joint. The temperature of element 2 is raised by 200 °C above the reference temperature. Both elements have the same length $L = 1$ m, Young's modulus $E = 10^{11}$ Pa, cross-sectional area $A = 10^{-4}$ m^2. The frame has moment of inertia $I = 10^{-9}$ m^4, while the bar has the coefficient of thermal expansion $\alpha = 20 \times 10^{-6}/°$C. Using the finite element method, (a) determine displacements and rotation at node 2; (b) determine the axial force in both elements; (c) determine the shear forces and bending moments in element 1 at nodes 1 and 2; and (d) draw the free-body-diagram of node 2 and show the force equilibrium is satisfied. *Hint:* Treat element 1 as a plane frame element.

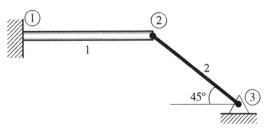

41. The figure shown below depicts a load cell made of aluminum. The ring and the stem both have square cross section: 0.1×0.1 m^2. Assume the Young's modulus is 72 GPa. The mean radius of the ring is 0.05 m. In a load cell, the axial load is measured from the average of strains at points P, Q, R, and S as shown in the figure. Points P and S are on the outside surface, and Q and R are on the inside surface of the ring. Model the load cell using plane frame elements. The step portions may be modeled using one element each. Use about 20 elements to model the entire ring. Compute the axial strain ε_{xx} at locations P, Q, R, and S for a load of 1,000 N. Draw the axial force, shear force and bending moment diagrams for one quarter of the ring. The strain can be computed using the beam formula:

$$\sigma_{xx} = E\varepsilon_{xx} = \frac{P}{A} \pm \frac{Mc}{I},$$

where P is the axial force, M the bending moment, A the cross-sectional area, I the moment of inertia, and c the distance from the midplane.

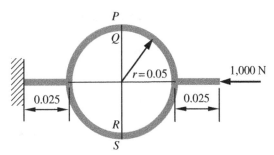

42. In problem 41, assume that load is applied eccentrically; that is, the distance between the line of action of the applied force and the centerline of load cell, e, is not equal to zero. Calculate the strain at P, Q, R, and S for $e = 0.002$ m. What is the average of these strains? Comment on the results. Note: The eccentric load can be replaced by a central load and a couple of $1000 \times e$ N·m.

43. A uniform bar clamped at both ends is subjected to an axial force P as shown in the figure. Calculate the critical value of P that will cause buckling. Assume $EI = 1,000$ N-m^2.

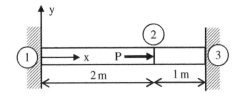

44. The composite beam in the figure is clamped at both ends and is subjected to an axial force P as shown. The cross sections of the two portions are such that $(EA)^{(2)} = 2(EA)^{(1)}$ and $(EI)^{(2)} = 2(EI)^{(1)}$. Calculate the critical value of P that will cause buckling. Assume $(EI)^{(1)} = 1{,}000\ \text{N-m}^2$.

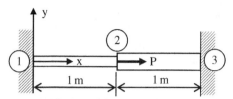

45. The left end of a shaft can be assumed to be perfectly clamped, but the right support has some play to accommodate misalignment. Its compliance is represented by a torsional spring and a linear spring. Determine the critical load for buckling of the shaft and corresponding mode shape. The length of the shaft is 1 m. The stiffnesses of the linear and torsional springs, respectively, are $k_L = 10{,}000\ \text{N/m}$ and $k_T = 1000\ \text{N-m/radian}$. The bending rigidity of the shaft cross section $EI = 1000\ \text{N-m}^2$.

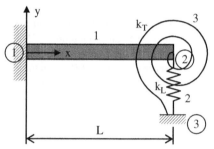

46. A pair of loads P acts on the horizontal member of the portal frame as shown in the figure. The length of all members is equal to 1 m, $EI = 1{,}000\ \text{N-m}^2$ and $EA = 10^9\ \text{N}$. Determine the critical value of P for buckling of the frame.

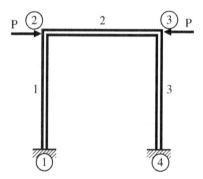

Chapter 4

Finite Elements for Heat Transfer Problems

4.1 INTRODUCTION

In this chapter, we will demonstrate the use of finite element analysis (FEA) in heat transfer problems, especially heat conduction in a solid. This is different from the thermal stress problem discussed in chapter 1. For simplicity, we will derive the equations for one-dimensional heat transfer. Although, on the surface, the heat transfer problem looks different from the structural mechanics problem, there are a number of similarities between the two. The thermal conductivity is the material property that plays the role of Young's modulus, and the temperature gradient is analogous to strain. Similarly, heat flow across the boundary of the solid is analogous to the surface traction in structural analysis, and internally generated heat is similar to the body force. In heat transfer problems, we solve for the temperature field instead of the displacement field. Table 4.1 compares the terms that are used in structural mechanics and their counterparts in heat conduction. Thus, as far as the finite element method is concerned, the two problems are similar, if these terms are interpreted appropriately.

It is possible to couple the structural and heat transfer problems together. This is required in some situations because the structure deforms due to thermal strains caused by temperature changes. In addition to conduction and convection, heat is also transferred through radiation, which makes the problem nonlinear. However, in this chapter we discuss only linear problems. Nonlinear and coupled problems are dealt with only in advanced finite element courses.

The matrix equation of the heat transfer problem is very similar to that of structural mechanics problems. In fact, the same finite element scheme can be used for both problems. Starting from conservation of energy, we will obtain a matrix equation similar to the structural finite element equations as shown below:

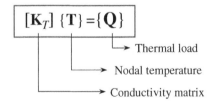

$$[\mathbf{K}_T]\,\{\mathbf{T}\} = \{\mathbf{Q}\}$$

Thermal load

Nodal temperature

Conductivity matrix

After applying the boundary conditions, the solution to the matrix equation will yield the nodal temperatures from which the temperature distribution within the solid can be calculated using the interpolation functions. Heat flow is calculated using the derivative of the temperature field.

Introduction to Finite Element Analysis and Design, Second Edition. Nam H. Kim, Bhavani V. Sankar, and Ashok V. Kumar.
© 2018 John Wiley & Sons Ltd. Published 2018 by John Wiley & Sons Ltd.
Companion website: www.wiley.com/go/kim/finite_element_analysis_design

Table 4.1 Analogy between structural and heat conduction problems

Structural Mechanics	Heat Transfer
Displacement (vector)	Temperature (scalar)
Stress (tensor)	Heat flux (vector)
Displacement boundary conditions	Temperature boundary conditions
Traction boundary conditions	Surface heat input boundary conditions
Body force	Internal heat generation

4.2 FOURIER HEAT CONDUCTION EQUATION

When there is a temperature gradient in a solid, heat flows from the high temperature region to the low temperature region. *Fourier's law* of heat conduction states that the magnitude of the *heat flux* (heat flow per unit time) is proportional to the temperature gradient:

$$q_x = -kA\frac{dT}{dx}, \tag{4.1}$$

where k, the *thermal conductivity*, is a material property, A is the area of cross section normal to the x-axis, and q_x is the *heat flux* in the x-direction. The unit of heat flux is watts, and that of thermal conductivity is W/m/°C. The negative sign in eq. (4.1) indicates that the direction of the heat flux is opposite to that of temperature gradient.

In eq. (4.1), we have assumed that the temperature varies only along the x-axis and is independent of y- and z- coordinates. This can happen in various situations. For example, in the case of heat transfer in a long wire, as shown in figure 4.1(a), the temperature variation in the lateral directions can be ignored because their dimensions are small compared to that in the axial direction. In this case, there could be heat transfer in the lateral directions, but there is no temperature gradient. On the other hand, say in the case of a furnace wall shown in figure 4.1(b), the temperature varies through the wall thickness, and hence there is heat flux in the thickness direction. However, there will not be any significant temperature gradient in the y- or z-directions, and hence the heat flow in those directions could be ignored. If the x-axis is parallel to the thickness direction, then the wall can also be modeled as a one-dimensional heat transfer problem.

We will derive the governing differential equation for one-dimensional heat conduction problems using the principle of *conservation of energy*. Consider an infinitesimal element (control volume) of the one-dimensional solid, as shown in figure 4.2. The heat flux through the cross section at x is given by q_x. The heat flux at the other end is then given by $q_x + (dq_x/dx)\Delta x$, where Δx is the length of the infinitesimal element. This is nothing but the first-order Taylor series expansion. Let us assume that heat energy is generated within the element at a rate Q_g per unit volume. Examples of such heat generation are chemical and nuclear reactions and electrical resistance heating. If the system absorbs energy due to an endothermic reaction, then Q_g will be negative. Let us also assume heat enters the control volume through the

(a)　　　　　　　(b)

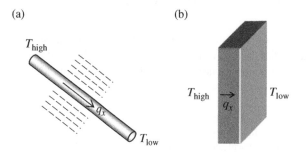

Figure 4.1 Examples of one-dimensional heat conduction problems; (a) heat conduction in a thin long rod; (b) a furnace wall with dimensions in the y- and z-directions much greater than the thickness in the x direction

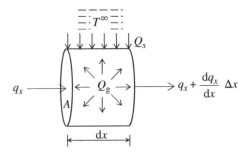

Figure 4.2 Energy balance in an infinitesimal volume

lateral surfaces. We will consider two modes of heat transfer at the lateral surface. In the first type, a surface heat flow given by Q_s per unit area enters the control volume. The second mode will be convective heat transfer given by the following equation:

$$Q_h = h(T^\infty - T),\tag{4.2}$$

where h is the *convection coefficient*, T^∞ is the temperature of the surrounding fluid, and T is the surface temperature of the solid. The unit of the heat transfer coefficient is $W/m^2/°C$. Convection can also occur at the end faces of the one-dimensional body.

Consider an infinitesimal element shown in figure 4.2. The principle of conservation of energy states that the change in the internal energy during a given time interval is equal to the sum of energy entering the element and the energy generated within the element minus the sum of the energy leaving the element. This relation can be written as

$$E_{in} + E_{gen} - E_{out} = \Delta U,\tag{4.3}$$

where E_{in} is the energy entering the system, E_{out} is the energy leaving the system, E_{gen} is the energy generated within the system, and ΔU is the increase of *internal energy*. In this text, we will derive the finite element equations only for steady-state problems wherein the temperature at a given cross section remains constant. That is, T is only a function of x and is independent of time. In that case the internal energy also remains constant, and, hence, $\Delta U = 0$. Referring to figure 4.2, eq. (4.3) can be written as

$$\underbrace{q_x + Q_s P \Delta x + h(T^\infty - T)P\Delta x}_{E_{in}} + \underbrace{Q_g A \Delta x}_{E_{gen}} = \underbrace{\left(q_x + \frac{dq_x}{dx}\Delta x\right)}_{E_{out}}.\tag{4.4}$$

In the above equation, A is the area of the cross section normal to the x-axis and P is the perimeter of the one-dimensional solid. The heat flux q_x can be cancelled as it appears on both sides of the above equation. Dividing by Δx throughout and letting Δx to approach zero, we obtain the following differential equation:

$$\frac{dq_x}{dx} = Q_g A + hP(T^\infty - T) + Q_s P, \quad 0 \le x \le L.\tag{4.5}$$

The term on the left-hand side (LHS) of the above equation represents the rate of change of heat flux along the length. The terms on the right-hand side (RHS) represent the heat generated and heat transferred into the system through the lateral surfaces. Our interest is in determining the temperature field $T(x)$. We use Fourier's law of heat conduction $q_x = -kA dT/dx$ in eq. (4.1) in the above equation to obtain

$$\boxed{\frac{d}{dx}\left(kA\frac{dT}{dx}\right) + Q_g A + hP(T^\infty - T) + Q_s P = 0,} \quad 0 \le x \le L.\tag{4.6}$$

Equation (4.6) is the governing differential equation for the steady-state one-dimensional heat transfer problem. The above differential equation can be uniquely solved when appropriate boundary conditions are provided. As has been discussed in chapter 2, there are two types of boundary conditions: the essential and the natural boundary conditions. The temperature at the boundary is prescribed in the former, while the derivative of the temperature or the heat flux is prescribed in the latter. Let $x = 0$ and $x = L$ represent the boundaries where the essential and natural boundary conditions are prescribed. Then, the boundary condition can be written as

$$
\begin{cases}
T(0) = T_0 \\
kA\dfrac{dT}{dx}\bigg|_{x=L} = q_L,
\end{cases}
\tag{4.7}
$$

where T_0 is the prescribed temperature at $x = 0$ and q_L is the prescribed heat flux at $x = L$. In eq. (4.7), the first one is the essential boundary condition, while the second is the natural boundary condition. In deriving the natural boundary conditions, we have used the sign convention that heat entering the body is considered positive, and that leaving is negative. Equations (4.6) and (4.7) together constitute the boundary value problem.

4.3   FINITE ELEMENT ANALYSIS – DIRECT METHOD

We will first use an engineering approach to derive the finite element equations for the heat conduction problem. This is similar to the direct stiffness method we used for uniaxial bar elements in chapter 1. In this case, we do not use the differential equation, but the principle of conservation of energy. Although the direct method is useful for discrete systems, it has limitations when multiple heat transfer modes, such as heat generation and convection, are present. Such cases will be treated later using the Galerkin method.

The one-dimensional problem we are interested in is depicted in figure 4.3. Consider a bar, which we refer to as a thermodynamic system or simply a system. Heat enters (or leaves) the system by various means. Heat can also be generated internally within the material volume. Examples of internal heat generation include heat generated in the core of a nuclear reactor due to nuclear fission, heat released due to a chemical reaction, electric resistance heating, and heat generated due to other forms of excitation using electromagnetic radiation. Heat can also enter through the lateral surface of the system by convection from the surrounding fluid. Another heat transfer mode is radiation of heat into the system wherein the system is heated by exposure to sun or a hot flame. In the following derivation, we will ignore the heat transfer by radiation.

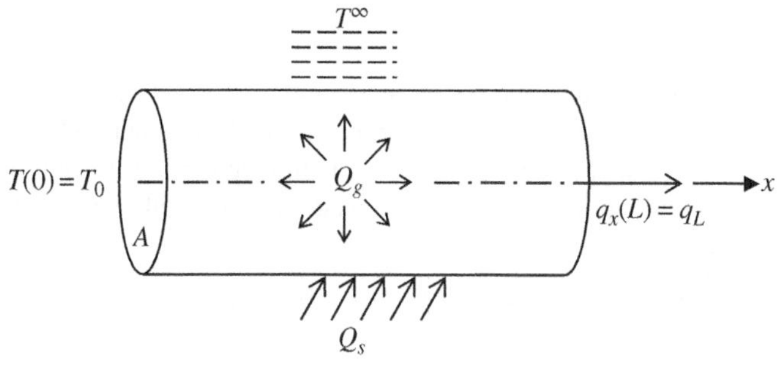

Figure 4.3 One-dimensional heat conduction of a long wire

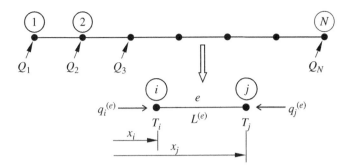

Figure 4.4 Finite elements for one-dimensional heat conduction problem

4.3.1   Element Conduction Equation

We divide the one-dimensional solid into a number of elements. The elements are connected at nodes. In this idealized model, we assume that the heat can enter the system only through the nodes. Thus, all of the aforementioned modes of heat input should be converted into equivalent nodal heat inputs. For that purpose, we can use a similar approach as the "work-equivalent" nodal forces in the structural elements. Let us assume that the heat input into the system at Node j is given by Q_j. The units of this heat input are Watts or Btu/s. The objective of finite element analysis is to determine the temperature distribution $T(x)$ along the length of the bar. Consider a typical element e of length $L^{(e)}$, as shown in figure 4.4. It has two nodes, which we refer to as the first and second nodes, and the temperatures at these nodes are denoted by T_i and T_j, respectively. This definition is consistent to the one-dimensional bar element in chapter 1. The only difference is that the nodal displacement u_i is replaced with nodal temperature T_i. The heat going into the system is considered positive and that leaving, negative.

The heat flow through the cross section of the bar at nodes i and j are denoted by $q_i^{(e)}$ and $q_j^{(e)}$. Using Fourier's law of heat conduction we can relate the heat flow to the temperature as

$$q_i^{(e)} = -kA\frac{dT}{dx} = -kA\frac{\left(T_j - T_i\right)}{L^{(e)}}. \tag{4.8}$$

In the above equation, the temperature gradient is approximated by $dT/dx = \left(T_j - T_i\right)/L^{(e)}$, which means that the temperature varies linearly along the x-axis.

Consider a typical element such as element e shown in figure 4.4. If the entire lateral surface is insulated, then from conservation of energy we have

$$q_i^{(e)} + q_j^{(e)} = 0. \tag{4.9}$$

Then, we can obtain the heat flow at node j from eqs. (4.8) and (4.9), as

$$q_j^{(e)} = +kA\frac{\left(T_j - T_i\right)}{L^{(e)}}. \tag{4.10}$$

Equations (4.8) and (4.10) can be combined to obtain

$$\left\{ \begin{matrix} q_i^{(e)} \\ q_j^{(e)} \end{matrix} \right\} = \frac{kA}{L^{(e)}} \begin{bmatrix} 1 & -1 \\ -1 & 1 \end{bmatrix} \left\{ \begin{matrix} T_i \\ T_j \end{matrix} \right\}, \tag{4.11}$$

where the square matrix on the RHS of the above equation is called the *element conductance matrix*. One can note the similarity between the above equation and the uniaxial bar element equation (1.16) in chapter 1. If we interpret the thermal conductivity as Young's modulus, nodal temperatures as nodal displacements, and heat flows as element forces, then the two equations are analogous to each other. An equation similar to eq. (4.11) can be derived for each element in the model.

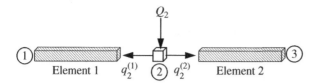

Figure 4.5 Balance in heat flow at node 2

4.3.2 Assembling Element Conduction Equations

Now consider the heat flow into node 2, which is connected to elements 1 and 2 (see figure 4.5). The heat input Q_2 at node 2 should be equal to the sum of the heat flow into elements 1 and 2 through node 2.

In general heat input at node i is equal to the sum of heat flows into all elements connected to node i. This can be written as

$$Q_i = \sum_{e=1}^{N_i} q_i^{(e)}, \tag{4.12}$$

where N_i is the number of elements connected to node i. Substituting for $q_i^{(e)}$ from element conduction equation (4.11) that includes node i, we obtain the global equations in the form

$$\underset{(N \times N)}{[\mathbf{K}_T]} \begin{Bmatrix} T_1 \\ T_2 \\ \vdots \\ T_N \end{Bmatrix} = \begin{Bmatrix} Q_1 \\ Q_2 \\ \vdots \\ Q_N \end{Bmatrix}, \tag{4.13}$$

where $[\mathbf{K}_T]$ is the global conductance matrix obtained by assembling the element matrices, and N is the number of nodes in the model. The assembly procedure of $[\mathbf{K}_T]$ is similar to that of uniaxial bar elements described in chapter 1. It must be mentioned that at each node either the temperature or the heat input should be known. Hence, the number of unknowns in eq. (4.13) is always equal to the number of equations N. Unlike the structural mechanics problems, often the rows and columns corresponding to prescribed temperature cannot be deleted. This is because in structural problems very often the prescribed displacement at a node is equal to zero. However, in heat conduction problems the prescribed temperature usually has a nonzero value. Thus, the action of "striking the rows" can still be used, but "striking the columns" cannot be used for heat conduction problems. Instead, the known temperature boundary condition has to be moved to the RHS of the global matrix equation. We will explain this operation in the following example.

EXAMPLE 4.1 *One-dimensional heat conduction using finite elements*

A heat conduction problem is modeled using four one-dimensional heat conduction elements, as shown in figure 4.6. All elements are of the same length, $L = 1$ m, cross-sectional area of $A = 1$ m^2, and thermal conductivity of $k = 10$ W/m·°C. The boundary conditions are given such that the temperature at node 1 is prescribed as $T_1 = 200\,°C$, while there is no heat flux at node 5, that is, $Q_5 = 0$. In addition, heat enters at nodes 2 with $Q_2 = 500$ watts, and heat leaves from node 4 with $Q_4 = -200$ watts. Determine all nodal temperatures and the heat input at node 1 in order to maintain the temperature distribution.

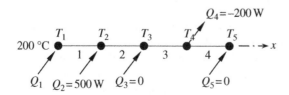

Figure 4.6 Finite elements for one-dimensional heat conduction problem

SOLUTION Since all elements are identical, the same element conduction equation can be used. The element matrix equations can be written as

element 1:

$$10 \begin{bmatrix} 1 & -1 \\ -1 & 1 \end{bmatrix} \begin{Bmatrix} T_1 \\ T_2 \end{Bmatrix} = \begin{Bmatrix} q_1^{(1)} \\ q_2^{(1)} \end{Bmatrix},$$

element 2:

$$10 \begin{bmatrix} 1 & -1 \\ -1 & 1 \end{bmatrix} \begin{Bmatrix} T_2 \\ T_3 \end{Bmatrix} = \begin{Bmatrix} q_2^{(2)} \\ q_3^{(2)} \end{Bmatrix},$$

element 3:

$$10 \begin{bmatrix} 1 & -1 \\ -1 & 1 \end{bmatrix} \begin{Bmatrix} T_3 \\ T_4 \end{Bmatrix} = \begin{Bmatrix} q_3^{(3)} \\ q_4^{(3)} \end{Bmatrix},$$

element 4:

$$10 \begin{bmatrix} 1 & -1 \\ -1 & 1 \end{bmatrix} \begin{Bmatrix} T_4 \\ T_5 \end{Bmatrix} = \begin{Bmatrix} q_4^{(4)} \\ q_5^{(4)} \end{Bmatrix}.$$

Now, the conservation of energy can be applied at each node using eq. (4.12). After replacing all element heat flows, q, by the element conduction equations, we can obtain the assembled global equations as

$$\begin{Bmatrix} Q_1 \\ Q_2 \\ Q_3 \\ Q_4 \\ Q_5 \end{Bmatrix} = \begin{Bmatrix} q_1^{(1)} \\ q_2^{(1)} + q_2^{(2)} \\ q_3^{(2)} + q_3^{(3)} \\ q_4^{(3)} + q_4^{(4)} \\ q_5^{(4)} \end{Bmatrix} = 10 \begin{bmatrix} 1 & -1 & 0 & 0 & 0 \\ -1 & 2 & -1 & 0 & 0 \\ 0 & -1 & 2 & -1 & 0 \\ 0 & 0 & -1 & 2 & -1 \\ 0 & 0 & 0 & -1 & 1 \end{bmatrix} \begin{Bmatrix} T_1 \\ T_2 \\ T_3 \\ T_4 \\ T_5 \end{Bmatrix}.$$

Since the temperature at node 1 is prescribed ($T_1 = 200\,°C$), the heat flow at node 1, Q_1, is unknown. However, external heat inputs at all other nodes are given. Thus, after substituting the known quantities in the above equations, we obtain

$$10 \begin{bmatrix} 1 & -1 & 0 & 0 & 0 \\ -1 & 2 & -1 & 0 & 0 \\ 0 & -1 & 2 & -1 & 0 \\ 0 & 0 & -1 & 2 & -1 \\ 0 & 0 & 0 & -1 & 1 \end{bmatrix} \begin{Bmatrix} 200 \\ T_2 \\ T_3 \\ T_4 \\ T_5 \end{Bmatrix} = \begin{Bmatrix} Q_1 \\ 500 \\ 0 \\ -200 \\ 0 \end{Bmatrix}.$$

Note that each row has only one unknown, either the nodal heat input or temperature. Since Q_1 is unknown, we want to remove the first row so that we only have temperatures as unknowns. In chapter 1, this was called the "striking the row" procedure. After that, we have

$$10 \begin{bmatrix} -1 & 2 & -1 & 0 & 0 \\ 0 & -1 & 2 & -1 & 0 \\ 0 & 0 & -1 & 2 & -1 \\ 0 & 0 & 0 & -1 & 1 \end{bmatrix} \begin{Bmatrix} 200 \\ T_2 \\ T_3 \\ T_4 \\ T_5 \end{Bmatrix} = \begin{Bmatrix} 500 \\ 0 \\ -200 \\ 0 \end{Bmatrix}.$$

Unlike in structural problems, we cannot perform the "striking the column" procedure because the prescribed temperature T_1 is not equal to zero. Observing that the first column will be multiplied by the prescribed temperature, we move this column, after multiplying by T_1, to the RHS to yield

$$10\begin{bmatrix} 2 & -1 & 0 & 0 \\ -1 & 2 & -1 & 0 \\ 0 & -1 & 2 & -1 \\ 0 & 0 & -1 & 1 \end{bmatrix}\begin{Bmatrix} T_2 \\ T_3 \\ T_4 \\ T_5 \end{Bmatrix} = \begin{Bmatrix} 500 \\ 0 \\ -200 \\ 0 \end{Bmatrix} + \begin{Bmatrix} 2000 \\ 0 \\ 0 \\ 0 \end{Bmatrix}.$$

Note the square matrix is not singular anymore, and we can solve for nodal temperatures:

$$\{\mathbf{T}\}^T = \{200 \quad 230 \quad 210 \quad 190 \quad 190\}°C.$$

The unknown heat input Q_1 at node 1 can be found from the deleted first equation of the global conduction equation as

$$Q_1 = 10T_1 - 10T_2 + 0T_3 + 0T_4 + 0T_5 = -300\text{W}.$$

Thus, in order to maintain constant temperature of $200\,°C$ at node 1, 300 W of heat must be removed from node 1. Note that the sum of all nodal heat inputs is equal to zero, which is consistent with the steady-state heat conduction problem. ▃▃

EXAMPLE 4.2 *Network of heat conduction elements*

Consider a network of one-dimensional heat conduction elements as shown in figure 4.7. The temperature at node 1 is fixed and heat inputs at other nodes are prescribed. Properties of the elements are listed in the table. Using the direct method, calculate the temperature at each node.

Element	k (W/m·°C)	A (m²)	L (m)	Node i	Node j
1	10	1	1	1	2
2	15	1	$\sqrt{2}$	2	3
3	20	1	2	2	4
4	15	1	$\sqrt{2}$	3	4
5	10	1	1	4	5

SOLUTION Using the element properties in the table, the element conduction equations can be written as

element 1:
$$10\begin{bmatrix} 1 & -1 \\ -1 & 1 \end{bmatrix}\begin{Bmatrix} T_1 \\ T_2 \end{Bmatrix} = \begin{Bmatrix} q_1^{(1)} \\ q_2^{(1)} \end{Bmatrix},$$

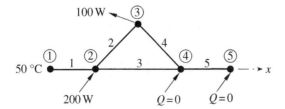

Figure 4.7 Network of heat conduction elements

element 2:
$$\frac{15}{\sqrt{2}}\begin{bmatrix} 1 & -1 \\ -1 & 1 \end{bmatrix}\begin{Bmatrix} T_2 \\ T_3 \end{Bmatrix} = \begin{Bmatrix} q_2^{(2)} \\ q_3^{(2)} \end{Bmatrix},$$

element 3:
$$10\begin{bmatrix} 1 & -1 \\ -1 & 1 \end{bmatrix}\begin{Bmatrix} T_2 \\ T_4 \end{Bmatrix} = \begin{Bmatrix} q_2^{(3)} \\ q_4^{(3)} \end{Bmatrix},$$

element 4:
$$\frac{15}{\sqrt{2}}\begin{bmatrix} 1 & -1 \\ -1 & 1 \end{bmatrix}\begin{Bmatrix} T_3 \\ T_4 \end{Bmatrix} = \begin{Bmatrix} q_3^{(4)} \\ q_4^{(4)} \end{Bmatrix},$$

element 5:
$$10\begin{bmatrix} 1 & -1 \\ -1 & 1 \end{bmatrix}\begin{Bmatrix} T_4 \\ T_5 \end{Bmatrix} = \begin{Bmatrix} q_4^{(5)} \\ q_5^{(5)} \end{Bmatrix}.$$

Equation (4.12) can be applied for the equilibrium of energy. Note that three elements are connected at nodes 2 and 4. After replacing all heat flows by the element conduction equation, we can obtain the assembled matrix equations as

$$\begin{Bmatrix} Q_1 \\ Q_2 \\ Q_3 \\ Q_4 \\ Q_5 \end{Bmatrix} = \begin{Bmatrix} q_1^{(1)} \\ q_2^{(1)}+q_2^{(2)}+q_2^{(3)} \\ q_3^{(2)}+q_3^{(4)} \\ q_4^{(3)}+q_4^{(4)}+q_4^{(5)} \\ q_5^{(5)} \end{Bmatrix} = \begin{bmatrix} 10 & -10 & 0 & 0 & 0 \\ -10 & 30.6 & -10.6 & -10 & 0 \\ 0 & -10.6 & 21.2 & -10.6 & 0 \\ 0 & -10 & -10.6 & 30.6 & -10 \\ 0 & 0 & 0 & -10 & 10 \end{bmatrix}\begin{Bmatrix} T_1 \\ T_2 \\ T_3 \\ T_4 \\ T_5 \end{Bmatrix}.$$

At each node, either temperature or heat flow, not both, should be known. For example, the temperature 50 °C is prescribed at node 1, while heat flows of 200, –100, 0, and 0 are prescribed at nodes 2–5. The following matrix equation shows all known boundary conditions:

$$\begin{bmatrix} 10 & -10 & 0 & 0 & 0 \\ -10 & 30.6 & -10.6 & -10 & 0 \\ 0 & -10.6 & 21.2 & -10.6 & 0 \\ 0 & -10 & -10.6 & 30.6 & -10 \\ 0 & 0 & 0 & -10 & 10 \end{bmatrix}\begin{Bmatrix} 50 \\ T_2 \\ T_3 \\ T_4 \\ T_5 \end{Bmatrix} = \begin{Bmatrix} Q_1 \\ 200 \\ -100 \\ 0 \\ 0 \end{Bmatrix}.$$

As with example 4.1, the first row is deleted as the heat flow is unknown, and the first column is moved to the RHS after being multiplied with $T_1 = 50$ °C. Then the global conduction equation can be obtained as

$$\begin{bmatrix} 30.6 & -10.6 & -10 & 0 \\ -10.6 & 21.2 & -10.6 & 0 \\ -10 & -10.6 & 30.6 & -10 \\ 0 & 0 & -10 & 10 \end{bmatrix}\begin{Bmatrix} T_2 \\ T_3 \\ T_4 \\ T_5 \end{Bmatrix} = \begin{Bmatrix} 700 \\ -100 \\ 0 \\ 0 \end{Bmatrix}.$$

Now, the global conductance matrix is positive definite, and hence unique nodal temperatures can be obtained by solving the above set of equations. The vector of nodal temperature becomes

$$\{\mathbf{T}\}^{\mathrm{T}} = \{50 \ 60 \ 53.6 \ 56.7 \ 56.7\}^{\circ}\text{C}.$$

After solving for the nodal temperature, the heat inputs at node 1 can be obtained from the first row of the global equation as

$$Q_1 = 10T_1 - 10T_2 + 0T_3 + 0T_4 + 0T_5 = -100\,\text{W}.$$

Thus, in order to maintain constant temperature of 70 °C at node 1, 100 W of heat must be removed. Note that the sum of all heat inputs Q_i is equal to zero as this is a steady-state problem. ■

4.4   GALERKIN'S METHOD FOR HEAT CONDUCTION PROBLEMS

The direct method described in the previous section used Fourier's law and the balance of heat flow. It does not require the differential equation and finite element interpolation. However, the direct method is not amenable when other modes of heat transfer such as heat generation and convection are present.

In this section, we use Galerkin's method discussed in section 2.4 in chapter 2 to derive the finite element equations for the one-dimensional heat conduction problem. Consider a typical element of length $L^{(e)}$, as shown in figure 4.4. The element is composed of two nodes: the first node i and the second node j. In the two-node element, the temperature within the element is interpolated in terms of nodal temperatures as

$$\widetilde{T}(x) = T_i N_i(x) + T_j N_j(x), \tag{4.14}$$

where the T_i and T_j are the temperatures of the first and second nodes of the element, respectively, and the tilde above the variable $\widetilde{T}$ indicates that we are seeking an approximate solution. The interpolation functions are given by

$$N_i(x) = \left(1 - \frac{x - x_i}{L^{(e)}}\right), \quad N_j(x) = \frac{x - x_i}{L^{(e)}}. \tag{4.15}$$

Note that $x_j - x_i = L^{(e)}$. If a parameter s is introduced such that $s = (x - x_i)/L^{(e)}$, then the two shape functions are $N_i(s) = 1 - s$ and $N_j(s) = s$. In the vector notation, the above interpolation can be written as

$$\widetilde{T}(x) = \{\mathbf{N}\}^{\mathrm{T}}\left\{\mathbf{T}^{(e)}\right\}, \tag{4.16}$$

where $\{\mathbf{N}\}^{\mathrm{T}} = \{N_i(x) \ \ N_j(x)\}$ is a row vector of shape functions, and $\{\mathbf{T}^{(e)}\}$ is the column vector of nodal temperatures. The above interpolation is valid within the element, that is, $x_i \leq x \leq x_j$. The above interpolation provides a linearly varying temperature field. In addition, since the heat flux is proportional to the temperature gradient, it is constant within an element.

$$\frac{d\widetilde{T}}{dx} = \left\{-\frac{1}{L^{(e)}} \ \ \frac{1}{L^{(e)}}\right\}\left\{\mathbf{T}^{(e)}\right\} = \{\mathbf{B}\}^{\mathrm{T}}\left\{\mathbf{T}^{(e)}\right\}, \tag{4.17}$$

where $\{\mathbf{B}\}^{\mathrm{T}}$ is a constant row vector (1×2). Note that the vectors $\{\mathbf{N}\}^{\mathrm{T}}$ and $\{\mathbf{B}\}^{\mathrm{T}}$ are identical to those of the uniaxial bar element. That is, the uniaxial bar element uses the same interpolation scheme with the one-dimensional heat transfer element.

In the following, we will derive the element conduction equations when a heat source, Q_g, exists within the element. Other heat transfer mechanisms will be discussed later. In the presence of heat source, the governing equation of the heat conduction problem in eq. (4.6) is modified as

$$\frac{d}{dx}\left(kA\frac{dT}{dx}\right) + Q_g A = 0, \quad 0 \leq x \leq L. \tag{4.18}$$

In the principle of virtual work in section 2.6 of chapter 2, the virtual displacement is introduced. Here we introduce a similar virtual temperature or variation in temperature $\delta T(x)$, which is any small arbitrary (virtual, not real) temperature whose values are zero on the essential boundary conditions. Then, the variation in energy can be obtained by multiplying the governing equilibrium equation in eq. (4.18) with the virtual temperature and integrate over the domain, as

$$\int_{x_i}^{x_j}\left(\frac{d}{dx}\left(kA\frac{d\widetilde{T}}{dx}\right)+AQ_g\right)\delta T(x)dx=0.\tag{4.19}$$

In eq. (4.19), we replaced the temperature field with the approximation in eq. (4.14). If the temperature satisfies the differential equation, then the differential equation would be zero at every point in the domain, and therefore eq. (4.19) will be valid regardless of the virtual temperature used. On the other hand, this equation can be valid even if the temperature has an error because by integrating over the domain, we are only requiring the result to be zero in an average sense. Therefore, the principle of virtual work in eq. (4.19) is called a "weak form" of the original boundary value problem. The virtual temperature is a weighting function that can be arbitrary, but in Galerkin's method it is interpolated using the same shape functions as the temperature field so that it has a similar form, which ensures that we will get a symmetric element conductance matrix. Therefore, we will interpolate the weighting function as

$$\delta T(x)=\{\mathbf{N}\}^{\mathrm{T}}\left\{\delta\mathbf{T}^{(e)}\right\},$$

where $\{\delta\mathbf{T}_e\}$ is the column vector of nodal values of the virtual temperature. We use integration by parts to obtain

$$kA\frac{d\widetilde{T}}{dx}\delta T(x)\Big|_{x_i}^{x_j}-\int_{x_i}^{x_j}kA\frac{d\widetilde{T}}{dx}\frac{d\delta T}{dx}dx=-\int_{x_i}^{x_j}AQ_g\delta T(x)dx.\tag{4.20}$$

Note that now the second term on the left-hand side is symmetric with respect to $\widetilde{T}$ and δT as they are both first derivatives. The derivative of the virtual temperature can be obtained by taking the derivative of the interpolation scheme as

$$\frac{d\delta T}{dx}=\left\{-\frac{1}{L^{(e)}}\ \ \frac{1}{L^{(e)}}\right\}\left\{\delta\mathbf{T}^{(e)}\right\}=\{\mathbf{B}\}^{\mathrm{T}}\left\{\delta\mathbf{T}^{(e)}\right\}.$$

Substituting for $\widetilde{T}$ and δT and their derivatives in the integrals of eq. (4.20), and rearranging the terms, we obtain

$$\int_{x_i}^{x_j}kA\left\{\delta\mathbf{T}^{(e)}\right\}^{\mathrm{T}}\{\mathbf{B}\}\{\mathbf{B}\}^{\mathrm{T}}\left\{\mathbf{T}^{(e)}\right\}dx=\int_{x_i}^{x_j}AQ_g\left\{\delta\mathbf{T}^{(e)}\right\}^{\mathrm{T}}\{\mathbf{N}\}dx-q\left(x_j\right)\delta T_j+q\left(x_i\right)\delta T_i.\tag{4.21}$$

We have substituted Fourier's law $q=-kAdT/dx$ in the boundary terms in the above equation. Note that the last term on the RHS becomes $q_i^{(e)}=q(x_i)$ because $N_i(x_i)=1$ and $N_i(x_j)=0$. On the other hand, for node j, $q_j^{(e)}=-q\left(x_j\right)$ because the heat flow is positive when the heat enters the element (see figure 4.4). The nodal values are not functions of x and therefore can be taken out of the integrals to rewrite eq. (4.21) as:

$$\left\{\delta\mathbf{T}^{(e)}\right\}^{\mathrm{T}}\left[\mathbf{k}_T^{(e)}\right]\left\{\mathbf{T}^{(e)}\right\}=\left\{\delta\mathbf{T}^{(e)}\right\}^{\mathrm{T}}\left\{\mathbf{Q}^{(e)}\right\}+\left\{\delta\mathbf{T}^{(e)}\right\}^{\mathrm{T}}\left\{\mathbf{q}^{(e)}\right\},\tag{4.22}$$

where

$$\left[\mathbf{k}_T^{(e)}\right] = \int_{x_i}^{x_j} kA\{\mathbf{B}\}\{\mathbf{B}\}^{\mathrm{T}}\mathrm{d}x = \frac{kA}{L^{(e)}}\begin{bmatrix} 1 & -1 \\ -1 & 1 \end{bmatrix}, \tag{4.23}$$

$$\left\{\mathbf{Q}^{(e)}\right\} = \int_{x_i}^{x_j} AQ_g\{\mathbf{N}\}\mathrm{d}x = \left\{ \begin{array}{c} Q_i^{(e)} \\ Q_j^{(e)} \end{array} \right\}, \tag{4.24}$$

$$\left\{\mathbf{q}^{(e)}\right\} = \left\{ \begin{array}{c} q(x_i) \\ -q(x_j) \end{array} \right\} = \left\{ \begin{array}{c} q_i^{(e)} \\ q_j^{(e)} \end{array} \right\}. \tag{4.25}$$

In eq. (4.23), we get an expression for the conductance matrix after performing the integration. Similarly, on the RHS, we get the load vector components or the equivalent heat input at node i due to the heat generation term Q_g as:

$$Q_i^{(e)} = \int_{x_i}^{x_j} AQ_gN_i(x)\mathrm{d}x,$$

which is the *thermal load* at node i corresponding to the distributed heat source, and $q_i^{(e)}$ is the heat flow across the cross section at node i into element e. In eq. (4.22), the nodal values of the virtual temperatures appeared on both sides of the equation and can be canceled out because the virtual temperature is arbitrary and this equation should be valid regardless of the nodal values of the virtual temperature. Therefore, using the principle of virtual work, we obtain the same equation as in the direct method:

$$\frac{kA}{L^{(e)}}\begin{bmatrix} 1 & -1 \\ -1 & 1 \end{bmatrix}\left\{ \begin{array}{c} T_i \\ T_j \end{array} \right\} = \left\{ \begin{array}{c} Q_i^{(e)} + q_i^{(e)} \\ Q_j^{(e)} + q_j^{(e)} \end{array} \right\},$$

or

$$\boxed{\left[\mathbf{k}_T^{(e)}\right]\left\{\mathbf{T}^{(e)}\right\} = \left\{\mathbf{Q}^{(e)}\right\} + \left\{\mathbf{q}^{(e)}\right\}}, \tag{4.26}$$

where $\left[\mathbf{k}_T^{(e)}\right]$ is the conductance matrix of element e, $\{\mathbf{Q}^{(e)}\}$ is the vector of thermal loads corresponding to the heat source, and $\{\mathbf{q}^{(e)}\}$ is the vector of nodal heat flows across the cross section.

When a uniform heat source exists in the element, eqs. (4.24) yield

$$\left\{\mathbf{Q}^{(e)}\right\} = \int_{x_i}^{x_j} AQ_g\begin{bmatrix} N_i(x) \\ N_j(x) \end{bmatrix}\mathrm{d}x = \frac{AQ_gL^{(e)}}{2}\left\{ \begin{array}{c} 1 \\ 1 \end{array} \right\}. \tag{4.27}$$

Note that $AQ_gL^{(e)}$ is the total heat generated in the element, and it is equally divided between the two nodes.

As mentioned before, the temperature is linear within an element. If the differential equation (4.18) is directly integrated with uniform heat source, the temperature will be a quadratic polynomial. Thus, the finite element solution is approximate, but the error can be reduced by using more elements.

EXAMPLE 4.3   *Temperature distribution in a heat chamber wall*

Consider a heat chamber in which the temperature inside is maintained at a constant value of 200 °C (see figure 4.8). The chamber is covered by a 1.0-meter-thick metal wall, and outside is insulated so that no heat flows to the outer

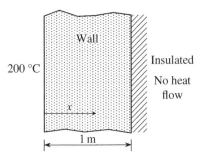

Figure 4.8 Heat transfer problem for insulated wall

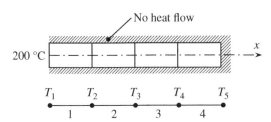

Figure 4.9 Finite element approximation of the wall

surface of the wall. There is a uniform heat source inside the wall generating $Q_g = 400 \, \text{W/m}^3$. The thermal conductivity of the wall is $k = 25 \, \text{W/m·°C}$. Assuming that the temperature only varies in the x–direction, find the temperature distribution in the wall.

SOLUTION From the assumption that the temperature only varies in the thickness direction, it is possible to model a unit slice of the wall as a one-dimensional problem (see figure 4.9). There is a uniform heat source inside the wall generating $Q_g = 400 \, \text{W/m}^3$. We will model the temperature variation in the thickness direction using four one-dimensional finite elements.

The boundary conditions are: temperature at the leftmost node is prescribed at $T_1 = 200 \, °\text{C}$, and the heat flux at the rightmost node vanishes, that is, $Q_5 = 0$.

Since the elements are identical, the same element matrix equation can be used for all elements. The element matrix equations can be written as

element 1:
$$100 \begin{bmatrix} 1 & -1 \\ -1 & 1 \end{bmatrix} \begin{Bmatrix} T_1 \\ T_2 \end{Bmatrix} = \begin{Bmatrix} 50 \\ 50 \end{Bmatrix} + \begin{Bmatrix} q_1^{(1)} \\ q_2^{(1)} \end{Bmatrix},$$

element 2:
$$100 \begin{bmatrix} 1 & -1 \\ -1 & 1 \end{bmatrix} \begin{Bmatrix} T_2 \\ T_3 \end{Bmatrix} = \begin{Bmatrix} 50 \\ 50 \end{Bmatrix} + \begin{Bmatrix} q_2^{(2)} \\ q_3^{(2)} \end{Bmatrix},$$

element 3:
$$100 \begin{bmatrix} 1 & -1 \\ -1 & 1 \end{bmatrix} \begin{Bmatrix} T_3 \\ T_4 \end{Bmatrix} = \begin{Bmatrix} 50 \\ 50 \end{Bmatrix} + \begin{Bmatrix} q_3^{(3)} \\ q_4^{(3)} \end{Bmatrix},$$

element 4:
$$100 \begin{bmatrix} 1 & -1 \\ -1 & 1 \end{bmatrix} \begin{Bmatrix} T_4 \\ T_5 \end{Bmatrix} = \begin{Bmatrix} 50 \\ 50 \end{Bmatrix} + \begin{Bmatrix} q_4^{(4)} \\ q_5^{(4)} \end{Bmatrix}.$$

Note that the first term on the RHS is the thermal load from a uniformly distributed heat source, which can be obtained using eq. (4.27). Now, the conservation of energy at each node can be applied using eq. (4.12). After replacing all heat flows by the element conduction equation, we obtain the assembled matrix equations as

$$
\begin{Bmatrix} Q_1 \\ Q_2 \\ Q_3 \\ Q_4 \\ Q_5 \end{Bmatrix} = \begin{Bmatrix} q_1^{(1)} \\ q_2^{(1)} + q_2^{(2)} \\ q_3^{(2)} + q_3^{(3)} \\ q_4^{(3)} + q_4^{(4)} \\ q_5^{(4)} \end{Bmatrix} = 100 \begin{bmatrix} 1 & -1 & 0 & 0 & 0 \\ -1 & 2 & -1 & 0 & 0 \\ 0 & -1 & 2 & -1 & 0 \\ 0 & 0 & -1 & 2 & -1 \\ 0 & 0 & 0 & -1 & 1 \end{bmatrix} \begin{Bmatrix} T_1 \\ T_2 \\ T_3 \\ T_4 \\ T_5 \end{Bmatrix} - \begin{Bmatrix} 50 \\ 100 \\ 100 \\ 100 \\ 50 \end{Bmatrix}.
$$

Two boundary conditions are given. The temperature at node 1 is fixed to 200 °C, while the heat flux at node 5 is zero. Since the temperature is given at node 1, we do not know how much heat needs to flow in order to maintain the constant temperature, that is, Q_1 is unknown. There is no heat input for other nodes except for the contribution from the heat source, that is, $Q_2 = Q_3 = Q_4 = Q_5 = 0$. Thus, after identifying known and unknown terms in the above equations, we obtain

$$
100 \begin{bmatrix} 1 & -1 & 0 & 0 & 0 \\ -1 & 2 & -1 & 0 & 0 \\ 0 & -1 & 2 & -1 & 0 \\ 0 & 0 & -1 & 2 & -1 \\ 0 & 0 & 0 & -1 & 1 \end{bmatrix} \begin{Bmatrix} 200 \\ T_2 \\ T_3 \\ T_4 \\ T_5 \end{Bmatrix} = \begin{Bmatrix} Q_1 + 50 \\ 100 \\ 100 \\ 100 \\ 50 \end{Bmatrix}.
$$

Note that each row has only one unknown, either heat input or temperature. Since Q_1 is unknown, we want to remove the first row so that we only have temperature unknowns. In chapter 2, this was called "striking the row" procedure. We have

$$
100 \begin{bmatrix} -1 & 2 & -1 & 0 & 0 \\ 0 & -1 & 2 & -1 & 0 \\ 0 & 0 & -1 & 2 & -1 \\ 0 & 0 & 0 & -1 & 1 \end{bmatrix} \begin{Bmatrix} 200 \\ T_2 \\ T_3 \\ T_4 \\ T_5 \end{Bmatrix} = \begin{Bmatrix} 100 \\ 100 \\ 100 \\ 50 \end{Bmatrix}.
$$

We cannot perform the "striking the column" procedure as we did in structural problems, because the prescribed temperature is not equal to zero. Observing that the first column will be multiplied by the prescribed temperature, we can move this column to the RHS to obtain

$$
100 \begin{bmatrix} 2 & -1 & 0 & 0 \\ -1 & 2 & -1 & 0 \\ 0 & -1 & 2 & -1 \\ 0 & 0 & -1 & 1 \end{bmatrix} \begin{Bmatrix} T_2 \\ T_3 \\ T_4 \\ T_5 \end{Bmatrix} = \begin{Bmatrix} 20100 \\ 100 \\ 100 \\ 50 \end{Bmatrix}.
$$

Note the square matrix is not singular, and we can solve for nodal temperatures:

$$
T_1 = 200 \text{ °C}, \ T_2 = 203.5 \text{ °C}, \ T_3 = 206 \text{ °C}, \ T_4 = 207.5 \text{ °C}, \ T_5 = 208 \text{ °C}.
$$

The unknown heat flow Q_1 at node 1 can be calculated from the first equation of the global conduction equations.

$$
Q_1 + 50 = 100 T_1 - 100 T_2 = -350
$$
$$
\Rightarrow Q_1 = -400 \text{ W}.
$$

Thus, in order to maintain a constant temperature at the inner wall surface, 400 W of heat should be removed from the inside wall. Note this is also equal to the total heat generated.

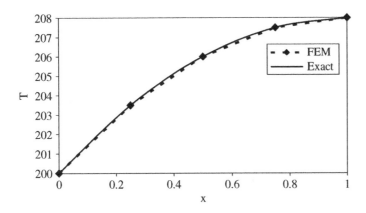

Figure 4.10 Temperature distribution along the wall thickness

Figure 4.10 plots the temperature variation in the wall. The exact solution is $T(x) = -8x^2 + 16x + 200$. Note that the finite element solution matches the exact solution at each node. However, due to linear interpolation, the temperature within an element deviates from the exact one. ▄▄▄

EXAMPLE 4.4 *Thermal protection system of a space vehicle*

A thermal protection system for a space vehicle consists of a 0.1-meter-thick outer ceramic foam and a 0.1-meter-thick inner metal foam, as shown in figure 4.11. The thermal conductivities of the ceramic and metal foams, respectively, are 0.05 and 0.025 W/m/K. A fluid is passed through the metal foam, which absorbs heat at the rate of 1,000 W/m³. The metal foam is attached to the vehicle structure and is insulated. The outside temperature of the ceramic foam is estimated to be 1,000 K. Use one-dimensional finite elements to determine the temperature distribution in the foams. What is the temperature of the vehicle structure? Calculate the heat entering the system due to aerodynamic heating.

SOLUTION From the assumption that the temperature only varies in the thickness direction, it is possible to model a unit slice of the wall as a one-dimensional problem. Since there is no distributed heat source in the ceramic foam, one finite element is enough. Due to heat absorption by fluid in the metal foam, more than one element is required for the metal foam. We will model the ceramic foam using one element, while two elements will be used for the metal foam as shown in figure 4.12.

The element conduction matrix equations can be written as

element 1:
$$\begin{bmatrix} 0.5 & -0.5 \\ -0.5 & 0.5 \end{bmatrix} \begin{Bmatrix} T_1 \\ T_2 \end{Bmatrix} = \begin{Bmatrix} q_1^{(1)} \\ q_2^{(1)} \end{Bmatrix},$$

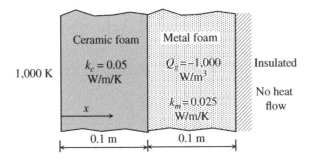

Figure 4.11 Heat transfer of a thermal protection system for a space vehicle

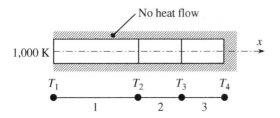

Figure 4.12 Finite element model of the thermal protection system

element 2 :
$$\begin{bmatrix} 0.5 & -0.5 \\ -0.5 & 0.5 \end{bmatrix} \begin{Bmatrix} T_2 \\ T_3 \end{Bmatrix} = \begin{Bmatrix} -25 \\ -25 \end{Bmatrix} + \begin{Bmatrix} q_2^{(2)} \\ q_3^{(2)} \end{Bmatrix},$$

element 3 :
$$\begin{bmatrix} 0.5 & -0.5 \\ -0.5 & 0.5 \end{bmatrix} \begin{Bmatrix} T_3 \\ T_4 \end{Bmatrix} = \begin{Bmatrix} -25 \\ -25 \end{Bmatrix} + \begin{Bmatrix} q_3^{(3)} \\ q_4^{(3)} \end{Bmatrix}.$$

Note that for elements 2 and 3, the first term on the RHS is the contribution from a uniformly distributed heat source, which can be obtained using eq. (4.24). For assembly, the conservation of energy is applied at each node using eq. (4.12). After replacing all heat flows by the element conduction equations, we can obtain the assembled matrix equations as

$$\begin{Bmatrix} Q_1 \\ Q_2 \\ Q_3 \\ Q_4 \end{Bmatrix} = \begin{Bmatrix} q_1^{(1)} \\ q_2^{(1)} + q_2^{(2)} \\ q_3^{(2)} + q_3^{(3)} \\ q_4^{(3)} \end{Bmatrix} = \begin{bmatrix} 0.5 & -0.5 & 0 & 0 \\ -0.5 & 1.0 & -0.5 & 0 \\ 0 & -0.5 & 1.0 & -0.5 \\ 0 & 0 & -0.5 & 0.5 \end{bmatrix} \begin{Bmatrix} T_1 \\ T_2 \\ T_3 \\ T_4 \end{Bmatrix} + \begin{Bmatrix} 0 \\ 25 \\ 50 \\ 25 \end{Bmatrix}.$$

Two boundary conditions are given. The temperature at node 1 is 1,000 K, while heat flux at node 4 is zero. Since the temperature is given at node 1, the heat flow is an unknown. There is no heat input for other nodes except for the contribution from the heat source. Thus, after identifying known and unknown terms from the above equation, we obtain

$$\begin{bmatrix} 0.5 & -0.5 & 0 & 0 \\ -0.5 & 1.0 & -0.5 & 0 \\ 0 & -0.5 & 1.0 & -0.5 \\ 0 & 0 & -0.5 & 0.5 \end{bmatrix} \begin{Bmatrix} 1000 \\ T_2 \\ T_3 \\ T_4 \end{Bmatrix} = \begin{Bmatrix} 0 \\ -25 \\ -50 \\ -25 \end{Bmatrix} + \begin{Bmatrix} Q_1 \\ 0 \\ 0 \\ 0 \end{Bmatrix}.$$

Since Q_1 is unknown, we want to remove the first row so that we only have temperature unknowns. As described in chapter 1, this was called the "striking the row" procedure. We have

$$\begin{bmatrix} -0.5 & 1.0 & -0.5 & 0 \\ 0 & -0.5 & 1.0 & -0.5 \\ 0 & 0 & -0.5 & 0.5 \end{bmatrix} \begin{Bmatrix} 1000 \\ T_2 \\ T_3 \\ T_4 \end{Bmatrix} = \begin{Bmatrix} -25 \\ -50 \\ -25 \end{Bmatrix}.$$

Observing that the first column will be multiplied by the prescribed temperature, we can move this column, after multiplying by the known temperature, to the RHS to yield

$$\begin{bmatrix} 1.0 & -0.5 & 0 \\ -0.5 & 1.0 & -0.5 \\ 0 & -0.5 & 0.5 \end{bmatrix} \begin{Bmatrix} T_2 \\ T_3 \\ T_4 \end{Bmatrix} = \begin{Bmatrix} 475 \\ -50 \\ -25 \end{Bmatrix}.$$

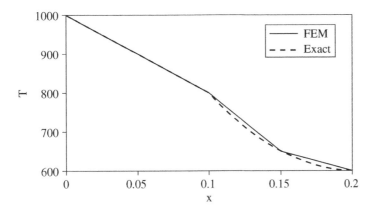

Figure 4.13 Temperature distribution in the thermal protection system

Note the matrix is not singular, and we can solve for nodal temperatures:

$$T_1 = 1{,}000 \text{ K}, \ T_2 = 800 \text{ K}, \ T_3 = 650 \text{ K}, \ T_4 = 600 \text{ K}.$$

The unknown heat flow Q_1 at node 1 can be calculated from the first equation of the global conduction equations:

$$Q_1 = 0.5T_1 - 0.5T_2 = 100 \text{ W}.$$

Q_1 is the heat entering the system due to aerodynamic heating.

The exact solution can be obtained by integrating the governing differential equation in the two separate regions:

$$T(x) = \begin{cases} -2000x + 1000 & 0 \le x \le 0.1 \\ 20000x^2 - 8000x + 1400 & 0.1 \le x \le 0.2. \end{cases}$$

Figure 4.13 shows the temperature distributions in the ceramic and metal foams obtained from FEA and an analytical method. The results from FEA are accurate in the ceramic foam. However, the FEA results for the metal foam are only approximate. The actual temperature variation is quadratic in x whereas in FEA the temperature is piecewise linear. ◼

4.5   CONVECTION BOUNDARY CONDITIONS

When a structure is surrounded by a fluid that has a different temperature from the structure, heat flow between the structure and fluid occurs and is called *convection*. Convection presents a special type of boundary condition called the mixed boundary condition. The closest analogy to this in structural mechanics problems is that of a beam resting on an elastic spring. The convection boundary condition contains an unknown temperature. For example, convective heat flow can be written as

$$q_h = hS(T^\infty - T), \tag{4.28}$$

where h is the convection coefficient, S is the area of the exposed surface, and T^∞ is the surrounding fluid temperature, which is assumed to be constant. Note that the amount of heat flow is not prescribed. Rather, it is a function of surface temperature, which is unknown.

We will consider two different types of convection heat transfer. In the first case, the convection occurs at the end faces of the one-dimensional body. For example, the inner surface of the heat chamber in example 4.3 can be exposed to a hot fluid, rather than prescribing a known temperature. The same applies to the outer surface of the ceramic foam in the thermal protection system in example 4.4. In this case, the exposed area S is the same as the cross-sectional area A. In the second case, the convection heat transfer occurs throughout the length of the one-dimensional body. For example, an electric wire immersed in water can transfer heat to the surrounding water. This type of convection is not a boundary condition in a strict sense. Rather it is treated as a distributed heat flow. Thus, it should be included in the governing differential equation as in eq. (4.6). In this case, the exposed area S is the perimeter times the length or the exposed area.

4.5.1 Convection on the Boundary

For illustration purpose, consider two one-dimensional heat transfer elements, as shown in figure 4.14. For simplicity, let us assume that both elements are identical in terms of geometry and material properties. Both ends of the system are under convective boundary conditions. An insulating wall can be an example for this type of problem.

Since both elements are identical, the element conduction equations can be written as

element 1:
$$\frac{kA}{L}\begin{bmatrix} 1 & -1 \\ -1 & 1 \end{bmatrix}\begin{Bmatrix} T_1 \\ T_2 \end{Bmatrix} = \begin{Bmatrix} q_1^{(1)} \\ q_2^{(1)} \end{Bmatrix},$$

element 2:
$$\frac{kA}{L}\begin{bmatrix} 1 & -1 \\ -1 & 1 \end{bmatrix}\begin{Bmatrix} T_2 \\ T_3 \end{Bmatrix} = \begin{Bmatrix} q_2^{(2)} \\ q_3^{(2)} \end{Bmatrix}.$$

The balance of heat flow in eq. (4.12) can be applied to all nodes in conjunction with convective heat flow in eq. (4.28) as

node 1:
$$q_1^{(1)} = h_1 A\left(T_1^\infty - T_1\right),$$

node 2:
$$q_2^{(1)} + q_2^{(2)} = 0,$$

node 3:
$$q_3^{(2)} = h_3 A\left(T_3^\infty - T_3\right).$$

At nodes 1 and 3, the amount of heat entering the element is equal to the heat transferred by convection, while at node 2 the sum of heat entering elements 1 and 2 is equal to zero.

After substituting the element conduction equation in the balance of heat flow equations, we can assemble the global heat conduction equation as

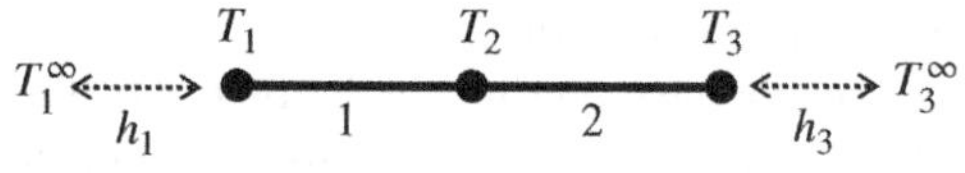

Figure 4.14 Finite element approximation of the furnace wall

$$\frac{kA}{L}\begin{bmatrix} 1 & -1 & 0 \\ -1 & 2 & -1 \\ 0 & -1 & 1 \end{bmatrix}\begin{Bmatrix} T_1 \\ T_2 \\ T_3 \end{Bmatrix} = \begin{Bmatrix} h_1 A\left(T_1^\infty - T_1\right) \\ 0 \\ h_3 A\left(T_3^\infty - T_3\right) \end{Bmatrix}.$$

The above equation cannot be solved as it is because the RHS includes unknown nodal temperatures, T_1 and T_3. This is due to the convection boundary conditions at nodes 1 and 3. In order to solve the above equations, we move the terms containing unknown nodal temperatures to the LHS. Then we obtain the following global equations:

$$\begin{bmatrix} \dfrac{kA}{L} + h_1 A & -\dfrac{kA}{L} & 0 \\ -\dfrac{kA}{L} & \dfrac{2kA}{L} & -\dfrac{kA}{L} \\ 0 & -\dfrac{kA}{L} & \dfrac{kA}{L} + h_3 A \end{bmatrix}\begin{Bmatrix} T_1 \\ T_2 \\ T_3 \end{Bmatrix} = \begin{Bmatrix} h_1 A T_1^\infty \\ 0 \\ h_3 A T_3^\infty \end{Bmatrix}. \tag{4.29}$$

The square matrix in the LHS is nonsingular due to the addition of terms to the diagonal, which makes the matrix positive definite.

EXAMPLE 4.5   *Convection in a furnace wall*

A furnace wall, as shown in figure 4.15, consists of two layers, firebrick inside and insulating brick on the outside. The thermal conductivities are $k_1 = 1.2$ W/m/°C for the firebrick and $k_2 = 0.2$ W/m/°C for the insulating brick. The temperature inside the furnace is $T_f = 1{,}500$ °C, and the convection coefficient at the inner surface is $h_i = 12$ W/m²/°C. The ambient temperature is $T_a = 25$ °C, and the convection coefficient at the outer surface is $h_o = 2.0$ W/m²/°C. The thermal resistance of the interface between firebrick and insulating brick can be neglected. Using two finite elements (one for each brick), determine the rate of heat loss through the outer wall, the temperature T_i at the inner surface, and the temperature T_o at the outer surface.

SOLUTION   Assuming that the furnace is tall enough, the heat flux along the vertical direction can be ignored, and we can assume that the heat flow is one-dimensional in the thickness direction. It is then possible to model a unit slice of the wall as a one-dimensional problem. We will model the temperature change in the thickness direction using two one-dimensional finite elements (see figure 4.16).

There are two convection boundaries: one at the inside of the wall and the other at the outer wall. Due to the high temperature inside, heat will be transferred from inside to the wall and from wall to the outside. Using eq. (4.29), we can construct the global matrix equations as

$$\begin{bmatrix} 16.8 & -4.8 & 0 \\ -4.8 & 6.47 & -1.67 \\ 0 & -1.67 & 3.67 \end{bmatrix}\begin{Bmatrix} T_1 \\ T_2 \\ T_3 \end{Bmatrix} = \begin{Bmatrix} 18{,}000 \\ 0 \\ 40 \end{Bmatrix}.$$

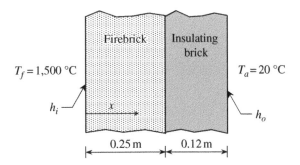

Figure 4.15 Heat transfer problem of an insulated wall

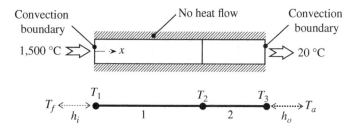

Figure 4.16 Finite element approximation of the furnace wall

Note that the lengths and conductivities of the two elements are different. The matrix in the above equation is positive definite, and a unique solution (nodal temperatures) can be obtained as

$$\{\mathbf{T}\}^{\mathrm{T}} = \{1{,}411 \quad 1{,}190 \quad 552\}\,^\circ\text{C}.$$

The surface temperature of the inner wall is 1,411 °C, and that of the outer wall is 552 °C. The rate of heat loss through the outer wall can be obtained from the convection boundary condition:

$$q_3^{(2)} = h_0(T_a - T_3) = -1054 \ \text{W/m}^2.$$

The negative sign means the heat leaves the system.

The exact solution of the problem can be obtained by integrating the governing differential equation and applying the convection boundary conditions:

$$T(x) = \begin{cases} -883x + 1411 & 0 \le x \le 0.25 \\ -5300x + 2513 & 0.25 \le x \le 0.37. \end{cases}$$

The finite element analysis results are identical to the exact solution. ∎

4.5.2 Convection along the Length of a Rod

When a long rod is submerged into a fluid, convection occurs across the entire surface. This is different from the previous convection boundary condition in which convection occurs only at the end faces. As shown in figure 4.17, the heat flow in convection is proportional to the perimeter of the cross section multiplied by the length of the rod. In such a case, it is convenient to think of the convection heat flow as a distributed thermal load.

The governing differential equation is (see eq. (4.6))

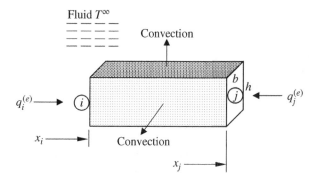

Figure 4.17 Heat conduction and convection in a long rod

$$\frac{\mathrm{d}}{\mathrm{d}x}\left(kA\frac{\mathrm{d}T}{\mathrm{d}x}\right)+AQ_g+hP(T^\infty-T)=0, \quad 0\le x\le L, \tag{4.30}$$

where P is the perimeter of the cross section, that is, $P=2(b+h)$. The weak form is obtained by multiplying the governing differential equation (4.30) with the virtual temperature $\delta T(x)$ and integrating over the element length. After replacing the temperature with the approximate temperature in eq. (4.14), we have

$$\int_{x_i}^{x_j}\left(\frac{\mathrm{d}}{\mathrm{d}x}\left(kA\frac{\mathrm{d}\widetilde{T}}{\mathrm{d}x}\right)+AQ_g+hP\left(T^\infty-\widetilde{T}\right)\right)\delta T(x)\mathrm{d}x=0, \tag{4.31}$$

We use integration by parts to obtain

$$kA\frac{\mathrm{d}\widetilde{T}}{\mathrm{d}x}\delta T(x)\Bigg|_{x_i}^{x_j}-\int_{x_i}^{x_j}kA\frac{\mathrm{d}\widetilde{T}}{\mathrm{d}x}\frac{\mathrm{d}\delta T}{\mathrm{d}x}dx-\int_{x_i}^{x_j}hP\widetilde{T}\delta Tdx$$
$$=-\int_{x_i}^{x_j}AQ_g\delta T(x)\mathrm{d}x-\int_{x_i}^{x_j}hPT^\infty\delta Tdx. \tag{4.32}$$

Substituting for $\widetilde{T}$, δT, and their derivatives in the integrals on the LHS of eq. (4.32), and rearranging the terms we obtain

$$\int_{x_i}^{x_j}kA\left\{\delta\mathbf{T}^{(e)}\right\}^{\mathrm{T}}\{\mathbf{B}\}\{\mathbf{B}\}^{\mathrm{T}}\left\{\mathbf{T}^{(e)}\right\}\mathrm{d}x+\int_{x_i}^{x_j}hP\left\{\delta\mathbf{T}^{(e)}\right\}^{\mathrm{T}}\{\mathbf{N}\}\{\mathbf{N}\}^{\mathrm{T}}\left\{\mathbf{T}^{(e)}\right\}\mathrm{d}x$$
$$=\int_{x_i}^{x_j}\left(AQ_g+hPT^\infty\right)\left\{\delta\mathbf{T}^{(e)}\right\}^{\mathrm{T}}\{\mathbf{N}\}\mathrm{d}x-q\left(x_j\right)\delta T_j+q(x_i)\delta T_i. \tag{4.33}$$

One may note that we have substituted Fourier's law $q=-kA\mathrm{d}T/\mathrm{d}x$ in the boundary terms in the above equation. Taking the constant nodal values outside the integral and performing the integration, the above equation takes the form

$$\left\{\delta\mathbf{T}^{(e)}\right\}^{\mathrm{T}}\left[\mathbf{k}_T^{(e)}\right]\left\{\mathbf{T}^{(e)}\right\}+\left\{\delta\mathbf{T}^{(e)}\right\}^{\mathrm{T}}\left[\mathbf{k}_h^{(e)}\right]\left\{\mathbf{T}^{(e)}\right\}=\left\{\delta\mathbf{T}^{(e)}\right\}^{\mathrm{T}}\left\{\mathbf{Q}^{(e)}\right\}+\left\{\delta\mathbf{T}^{(e)}\right\}^{\mathrm{T}}\left\{\mathbf{q}^{(e)}\right\}, \tag{4.34}$$

where

$$\left[\mathbf{k}_h^{(e)}\right]=\int_{x_i}^{x_j}hP\{\mathbf{N}\}\{\mathbf{N}\}^{\mathrm{T}}\mathrm{d}x.$$

The thermal load at node i due to the heat generation term Q_g and the convection term is given by

$$\left\{\mathbf{Q}^{(e)}\right\}=\int_{x_i}^{x_j}\left(AQ_g+hPT^\infty\right)\{\mathbf{N}\}\mathrm{d}x, \tag{4.35}$$

and $Q_i^{(e)}$ is the heat flow across the cross section at node i into element e. The nodal values of the virtual temperature, which is a factor on both sides of eq. (4.34), can be canceled out because that equation must be valid for any arbitrary value of $\{\delta \mathbf{T}^{(e)}\}$. This yields the following set of element equations:

$$\left[\frac{kA}{L^{(e)}}\begin{bmatrix} 1 & -1 \\ -1 & 1 \end{bmatrix} + \frac{hPL}{6}\begin{bmatrix} 2 & 1 \\ 1 & 2 \end{bmatrix}\right]\left\{\begin{matrix} T_i \\ T_j \end{matrix}\right\} = \left\{\begin{matrix} Q_i^{(e)} + q_i^{(e)} \\ Q_j^{(e)} + q_j^{(e)} \end{matrix}\right\},$$

or

$$\left[\left[\mathbf{k}_T^{(e)}\right] + \left[\mathbf{k}_h^{(e)}\right]\right]\{\mathbf{T}\} = \left\{\mathbf{Q}^{(e)}\right\} + \left\{\mathbf{q}^{(e)}\right\}. \tag{4.36}$$

Note that the first matrix on the LHS is the conductance matrix in eq. (4.26), and the equivalent conductance matrix due to convective heat transfer across the periphery is given by

$$\left[\mathbf{k}_h^{(e)}\right] = \frac{hPL}{6}\begin{bmatrix} 2 & 1 \\ 1 & 2 \end{bmatrix}. \tag{4.37}$$

Using the balance of heat flow in eq. (4.12), the above equations can be assembled to obtain the global matrix equations as

$$\boxed{[[\mathbf{K}_T] + [\mathbf{K}_h]]\{\mathbf{T}\} = \{\mathbf{Q}\}}. \tag{4.38}$$

Procedures for applying boundary conditions are identical to the previous section.

The thermal load vector on the RHS of eq. (4.36) includes the contribution from the heat source and the convection term. When there is a uniformly distributed heat source Q_g, the thermal load in eqs.(4.35) can be integrated to obtain

$$\left\{\mathbf{Q}^{(e)}\right\} = \left\{\begin{matrix} Q_i \\ Q_j \end{matrix}\right\} = \frac{AQ_gL^{(e)} + hPL^{(e)}T^\infty}{2}\left\{\begin{matrix} 1 \\ 1 \end{matrix}\right\}. \tag{4.39}$$

Note that the total amount of the thermal load is equally divided between the two nodes.

EXAMPLE 4.6   *Heat flow in a cooling fin*

Determine the steady-state temperature distribution in a thin rectangular fin shown in figure 4.18. The fin is 120 mm long, 160 mm wide, and 1.25 mm thick. The inside wall is at a temperature of 330 °C. Two side surfaces are insulated, while the top and bottom surfaces as well as the tip surface are exposed to the air. The ambient air temperature is 30 °C. Assume $k = 0.2$ W/mm/°C and $h = 2 \times 10^{-4}$ W/mm²/°C.

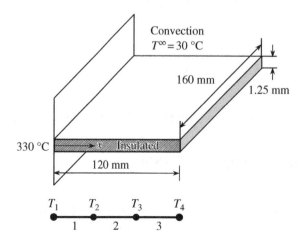

Figure 4.18 Heat flow through a thin fin and finite element model

SOLUTION   The problem can be treated as a one-dimensional problem because the thickness of the fin is small so that the temperature variation in the thickness direction can be neglected. A one-dimensional finite element model using three linear elements can be used to calculate the temperature distribution. The temperature at node 1 is prescribed, while the heat flux at node 4 is controlled by the convection boundary condition. In addition, there is convection heat transfer through the perimeter of the fin.

The conductance matrix and thermal load vector for all elements are identical and can be calculated from eq. (4.36) as

$$\left[\mathbf{k}_T^{(e)}\right] + \left[\mathbf{k}_h^{(e)}\right] = \frac{0.2 \times 200}{40}\begin{bmatrix} 1 & -1 \\ -1 & 1 \end{bmatrix} + \frac{2 \times 10^{-4} \times 320 \times 40}{6}\begin{bmatrix} 2 & 1 \\ 1 & 2 \end{bmatrix}.$$

and

$$\left\{\mathbf{Q}^{(e)}\right\} = \frac{2 \times 10^{-4} \times 320 \times 40 \times 30}{2}\begin{Bmatrix} 1 \\ 1 \end{Bmatrix}.$$

Thus, the element conduction equation becomes

element 1:
$$\begin{bmatrix} 1.8533 & -0.5733 \\ -0.5733 & 1.8533 \end{bmatrix}\begin{Bmatrix} T_1 \\ T_2 \end{Bmatrix} = \begin{Bmatrix} 38.4 \\ 38.4 \end{Bmatrix} + \begin{Bmatrix} q_1^{(1)} \\ q_2^{(1)} \end{Bmatrix},$$

element 2:
$$\begin{bmatrix} 1.8533 & -0.5733 \\ -0.5733 & 1.8533 \end{bmatrix}\begin{Bmatrix} T_2 \\ T_3 \end{Bmatrix} = \begin{Bmatrix} 38.4 \\ 38.4 \end{Bmatrix} + \begin{Bmatrix} q_2^{(2)} \\ q_3^{(2)} \end{Bmatrix},$$

element 3:
$$\begin{bmatrix} 1.8533 & -0.5733 \\ -0.5733 & 1.8533 \end{bmatrix}\begin{Bmatrix} T_3 \\ T_4 \end{Bmatrix} = \begin{Bmatrix} 38.4 \\ 38.4 \end{Bmatrix} + \begin{Bmatrix} q_3^{(3)} \\ q_4^{(3)} \end{Bmatrix}.$$

The balance of heat flow in eq. (4.12) can be applied for all nodes in conjunction with the convection heat flow in eq. (4.28) as

node 1:
$$q_1^{(1)} = Q_1,$$

node 2:
$$q_2^{(1)} + q_2^{(2)} = 0,$$

node 3:
$$q_3^{(2)} + q_3^{(2)} = 0,$$

node 4:
$$q_4^{(3)} = hA(T^\infty - T_4).$$

After substituting the element conduction equations, we obtain the global matrix equation as

$$\begin{bmatrix} 1.853 & -.573 & 0 & 0 \\ -.573 & 3.706 & -.573 & 0 \\ 0 & -.573 & 3.706 & -.573 \\ 0 & 0 & -.573 & 1.853 \end{bmatrix}\begin{Bmatrix} T_1 \\ T_2 \\ T_3 \\ T_4 \end{Bmatrix} = \begin{Bmatrix} 38.4 + Q_1 \\ 76.8 \\ 76.8 \\ 38.4 + hA(T^\infty - T_4) \end{Bmatrix}.$$

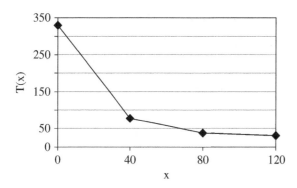

Figure 4.19 Temperature distribution in a thin fin

Note that the term containing the unknown temperature T_4 in the last equation needs to be moved to the LHS. After that, the following matrix equations is obtained:

$$\begin{bmatrix} 1.853 & -.573 & 0 & 0 \\ -.573 & 3.706 & -.573 & 0 \\ 0 & -.573 & 3.706 & -.573 \\ 0 & 0 & -.573 & 1.893 \end{bmatrix} \begin{Bmatrix} 330 \\ T_2 \\ T_3 \\ T_4 \end{Bmatrix} = \begin{Bmatrix} 38.4 + Q_1 \\ 76.8 \\ 76.8 \\ 39.6 \end{Bmatrix}.$$

In the above equation, the prescribed temperature $T_1 = 330\,°\text{C}$ is substituted. Now eliminating the first row and moving the first column, after multiplying with T_1, to the RHS we obtain

$$\begin{bmatrix} 3.706 & -.573 & 0 \\ -.573 & 3.706 & -.573 \\ 0 & -.573 & 1.893 \end{bmatrix} \begin{Bmatrix} T_2 \\ T_3 \\ T_4 \end{Bmatrix} = \begin{Bmatrix} 265.89 \\ 76.8 \\ 39.6 \end{Bmatrix}.$$

The solution of this matrix equation yields the three nodal temperatures. By combining prescribed temperature, the entire temperature of the fin is determined as

$$T_1 = 330°\text{C}, \quad T_2 = 77.57°\text{C}, \quad T_3 = 37.72°\text{C}, \quad T_4 = 32.34°\text{C}.$$

Since the temperature interpolation within the element is linear, the temperature distribution of the fin can be plotted as shown in figure 4.19. Due to convection across the lateral surfaces of the fin, the temperature drops dramatically along the x-axis. ■

4.6 TWO-DIMENSIONAL HEAT TRANSFER

A heat conduction problem can be considered two-dimensional if temperature is constant in one direction, the thickness direction, and all the heat flow occurs in a plane normal to the thickness direction. We will consider the plane where the heat flow occurs as the x-y plane as shown in figure 4.20 and assume that the temperature is constant in the z-direction (therefore there is no heat flow in that direction).

The three types of boundary conditions we described for the one-dimensional problem are also applicable to two-dimensional heat transfer problems. In some parts of the boundary S_T, the temperature may be known and fixed, while at other parts of the boundary, S_Q, heat flux may be known or the heat flux may be due to convection, S_h. The geometry is assumed to be of constant thickness t in the z-direction; therefore, we can consider a two-dimensional area as shown in the figure to be the domain of analysis where we need to compute temperature as a function of (x, y) coordinates. In this section, we will develop necessary equations for energy equilibrium using the Galerkin's method. In the following section, we would like to divide this domain into elements and describe the temperature field using nodal values while interpolating within each element to approximate the solution.

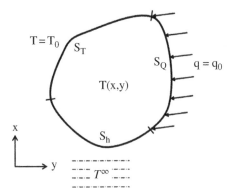

Figure 4.20 Two-dimensional heat transfer analysis domain

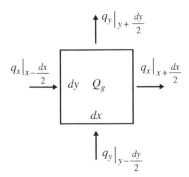

Figure 4.21 Energy balance in an infinitesimal element

4.6.1 Boundary Value Problem for Two-Dimensional Heat Transfer

Equilibrium equation for heat balance:

Similar to the one-dimensional heat transfer problem, the governing equations of two-dimensional heat transfer are the equilibrium equations, which can be easily derived from the energy balance condition. Consider a two-dimensional infinitesimal element shown in figure 4.21 that is in equilibrium. The heat flow into the element is equal to the heat flowing out of the element.

In two dimensions, the heat flux density is a vector that represents the rate of heat flow per unit area, and it is denoted here as $\mathbf{q} = q_x\mathbf{i} + q_y\mathbf{j}$. The energy balance is stated below as the sum of heat flowing in and out of the system in the x and y direction and the heat generated Q_g within the system should add up to zero:

$$\left(q_x|_{x-\frac{dx}{2}} - q_x|_{x+\frac{dx}{2}}\right)t\,dy + \left(q_y|_{y-\frac{dy}{2}} - q_y|_{y+\frac{dy}{2}}\right)t\,dx + Q_g\,t\,dx\,dy = 0. \tag{4.40}$$

As the heat flux density is per unit area, we multiply each term with the area across which the heat is flowing; that is, $t\,dy$ for the surface normal to x-axis and $t\,dx$ for the surface normal to y-axis. The thickness t can be canceled out from this equation as it occurs as a factor in all the terms, or equivalently, we can assume the equation is for unit thickness. Then, the area of the faces normal to the x- and y-directions are dy and dx, respectively, and the volume of the differential element is $dx\,dy$. The terms in parentheses represent the net rate of heat flux in the x- and y-directions. Expanding them using the first-order Taylor series expansion, we get

$$q_x|_{x-\frac{dx}{2}} - q_x|_{x+\frac{dx}{2}} = \left[q_x|_x + \frac{\partial q_x}{\partial x}\left(-\frac{dx}{2}\right)\right] - \left[q_x|_x + \frac{\partial q_x}{\partial x}\left(\frac{dx}{2}\right)\right]$$

$$= -\frac{\partial q_x}{\partial x}dx, \tag{4.41}$$

$$q_y\big|_{y-\frac{dy}{2}} - q_y\big|_{y+\frac{dy}{2}} = -\frac{\partial q_y}{\partial y}dy. \tag{4.42}$$

Substituting, these into the energy balance equation (4.40), we get the following partial differential equation as the equilibrium equation:

$$\left(-\frac{\partial q_x}{\partial x}dx\right)dy + \left(-\frac{\partial q_y}{\partial y}dy\right)dx + Q_g dxdy = 0. \tag{4.43}$$

Or, after canceling the volume term, we have

$$\frac{\partial q_x}{\partial x} + \frac{\partial q_y}{\partial y} = Q_g. \tag{4.44}$$

In would be informative if the above equation is compared with the governing equation for the one-dimensional problem in eq. (4.5). In the limit as the size of the differential element shrinks to zero, this equilibrium condition must apply at every point within the domain of analysis. Note that this equation does not yet involve temperature, which is the quantity we are interested in solving for. We need a relation between the heat flux and the temperature in order to rewrite this equation with temperature as the variable. This relation depends on the material through which heat transfer is occurring and is referred to as the constitutive equation. We need to develop a similar relationship given in eq. (4.1) for the case of the two-dimensional problem.

Constitutive equation (Fourier's law):
The constitutive equation for heat conduction is Fourier's law, which states that heat flux is proportional to the negative of temperature gradient. In other words, heat flows in the direction of decreasing temperature at a rate that is proportional to the rate at which temperature is decreasing. The proportionality constant is a material property, which is referred to as the thermal conductivity of the material. In the most general form, such a constitutive equation can be stated as:

$$\begin{aligned}
q_x &= -k_{xx}\frac{\partial T}{\partial x} - k_{xy}\frac{\partial T}{\partial y}, \\
q_y &= -k_{xy}\frac{\partial T}{\partial x} - k_{yy}\frac{\partial T}{\partial y}.
\end{aligned} \tag{4.45}$$

In the preceding equations, the heat flux densities in the x- and y-directions are written as proportional to the components of the temperature gradient. This equation can be restated as a matrix equation, where the material constants are all collected into a conductivity matrix, as

$$\begin{Bmatrix} q_x \\ q_y \end{Bmatrix} = -[\mathbf{k}]\{\nabla T\} = -\begin{bmatrix} k_{xx} & k_{xy} \\ k_{yx} & k_{yy} \end{bmatrix}\begin{Bmatrix} \dfrac{\partial T}{\partial x} \\ \dfrac{\partial T}{\partial y} \end{Bmatrix}, \tag{4.46}$$

where $\nabla = \{\partial/\partial x, \partial/\partial y\}^{\mathrm{T}}$ is called the gradient operator. When a function is multiplied with the gradient operator, it becomes the gradient vector of the function; for example, $\{\nabla T\}$ is the column vector of temperature gradient.

In practice, for most materials, the heat flux in any direction depends only on the component of temperature gradient in that direction[1]. In other words, heat conduction in the x-direction does not occur due to a temperature gradient in the y-direction. Therefore, the off-diagonal terms in

---

[1] Fiber reinforced composites can be tailored to have a non-zero k_{xy}. In that case k_{xy} will be equal to k_{yx} making the conductivity matrix symmetric.

the conductivity matrix are usually zero; that is, $k_{xy} = k_{yx} = 0$. Furthermore, if the material is isotropic, then the conductivity is the same in all directions. Therefore, for all isotropic materials $k_{xx} = k_{yy} = k$, and we can simplify Fourier's law to

$$\left\{ \begin{array}{c} q_x \\ q_y \end{array} \right\} = -k\{\nabla T\} = -k \left\{ \begin{array}{c} \dfrac{\partial T}{\partial x} \\ \dfrac{\partial T}{\partial y} \end{array} \right\}. \tag{4.47}$$

Governing differential equations:
The governing equation can be obtained by substituting the constitutive equation into the equilibrium equation (4.44). For two-dimensional heat conduction, the governing equation that must be satisfied at every point in the domain for equilibrium is

$$\frac{\partial}{\partial x}\left(k_{xx}\frac{\partial T}{\partial x} + k_{xy}\frac{\partial T}{\partial y} \right) + \frac{\partial}{\partial y}\left(k_{yx}\frac{\partial T}{\partial x} + k_{yy}\frac{\partial T}{\partial y} \right) + Q_g = 0. \tag{4.48}$$

This equation can be generalized to the three-dimensional problem by including the components in the z-direction as well. Using the notation of vector calculus, the governing equation can be stated more succinctly as

$$\nabla \cdot ([\mathbf{k}]\{\nabla T\}) + Q_g = 0. \tag{4.49}$$

In the case of isotropic materials, the above equations can be further simplified as

$$\frac{\partial}{\partial x}\left(k\frac{\partial T}{\partial x} \right) + \frac{\partial}{\partial y}\left(k\frac{\partial T}{\partial y} \right) + Q_g = 0, \tag{4.50}$$

and

$$\nabla \cdot (k\nabla T) + Q_g = 0. \tag{4.51}$$

These governing equations cannot be solved on their own to determine the temperature field. To fully define the problem, one needs to also define the domain of analysis and the boundary conditions that apply along its boundaries.

Boundary Value Problem: The domain of analysis for two-dimensional problems is an area as shown in figure 4.20. The boundary of this region can divided into different regions based on the type of boundary conditions that are applied. In the figure, regions subjected to temperature boundary conditions are labeled as S_T. Similarly, the parts of boundary subjected to heat flux boundary conditions and convection are named S_Q and S_h, respectively. It is assumed that these boundaries compose the entire boundary and do not overlap; that is, mathematically, $S = S_T \cup S_Q \cup S_h$ and $S_T \cap S_Q = S_T \cap S_h = S_Q \cap S_h = \varnothing$. Even though each of these regions on the boundary is shown as connected, it is not always be the case. For instance, it is possible that the heat flux is specified at two different parts of the boundary that are not adjacent to each other. In this case, S_Q is not a connected region but a union of all the regions where heat flux is specified on the boundary. Similarly, S_T and S_h could also be disconnected regions on the boundary. It should also be noted that if no boundary condition is specified in a particular region of the boundary, that region would be part of S_Q because it is implied that there is no heat flux on the boundary or, in other words, the heat flux at that boundary is known to be zero.

In two-dimensional heat transfer problems, the goal is to solve for the temperature field $T(x, y)$ in the given area of material, A, while satisfying all the specified boundary conditions. The boundary value problem to be solved for this purpose may be stated as follows:

Solve $T(x, y)$ in A such that

$$
\begin{vmatrix}
\nabla \cdot (k\nabla T) + Q_g = 0, & (x, y) \in A \\
T = T_0, & (x, y) \in S_T \\
q_n = -q_0, & (x, y) \in S_Q \\
q_n = h(T - T^\infty), & (x, y) \in S_h
\end{vmatrix}
, \tag{4.52}
$$

where $q_n = \mathbf{q} \cdot \mathbf{n} = -(k\nabla T) \cdot \mathbf{n}$ is the normal component of the heat flux on the boundary, and T^∞ is the temperature of the surrounding fluid. The negative sign in q_0 is because q_n is the heat flux out to the system, while by definition, q_0 is the heat flux coming into the system.

To keep our notations simple, we have assumed that the material is isotropic and used the scalar k for conductivity. We will use that assumption for the rest of the chapter with understanding that it can be replaced with the conductivity matrix if the material is anisotropic.

4.6.2 Galerkin's Method for Two-Dimensional Heat Conduction

The boundary value problem in eq. (4.52) states that the governing equation, which is a partial differential equation, must be satisfied at every point in the domain of interest while simultaneously satisfying all the boundary conditions. As we discussed in chapter 2, the solution to the boundary value problem is called the exact solution. In finite element analysis, we are looking for an approximate solution that satisfies the essential boundary conditions but not necessarily the natural boundary conditions. In this section, we use Galerkin's method in section 2.4 of chapter 2 to derive the governing equation that will be used for finite element analysis. In this approach, we first multiply the governing equation by a virtual temperature and integrate over the area, as

$$
\iint\limits_{A} \left[\nabla \cdot (k\nabla T) + Q_g \right] \delta T \, dA = 0. \tag{4.53}
$$

If the governing equation is satisfied at every point in the area A, then the above equation will be satisfied regardless of the nature of the virtual temperature δT, which is an arbitrary scalar field defined over the area A. Equation (4.53) is an approximation in the sense that it could be satisfied even if the governing equation is not strictly satisfied at every point and the residual is nonzero at some locations. Such error or residual is tolerated because the errors in different parts can cancel each other out, and the equation is satisfied in an average sense. This tolerance of error or nonzero residual is an important property that allows us to compute approximate solutions, and therefore we refer to this equation as a weak form. Then we use integration by parts to derive a more convenient form, as we did for one-dimensional problems in chapter 2. The equivalent of integration by parts for a two-dimensional problem is the Green's theorem. For convenience, we provide Green's theorem below:

Green's theorem: If u and v are continuous, differentiable scalar fields defined over an area A, then the following equality holds:

$$
\iint\limits_{A} v\nabla^2 u \, dA = \int\limits_{S} (\nabla u) \cdot \mathbf{n} v \, dS - \iint\limits_{A} \nabla u \cdot \nabla v \, dA, \tag{4.54}
$$

where, S is the boundary of the area, and $\mathbf{n}$ is the unit normal to the boundary pointing toward the outside of the area A.

Expanding eq. (4.53) using Green's theorem, we get a more convenient weak form that involves only the first derivatives of the temperature and the virtual temperature. Using this theorem, on the first part of eq. (4.53) we obtain

$$\iint_A \nabla \cdot (k\nabla T)\delta T \, dA = \int_S (k\nabla T)\cdot \mathbf{n}\delta T \, dS - \iint_A k\nabla \delta T \cdot \nabla T \, dA. \tag{4.55}$$

Substituting this into eq. (4.53), we obtain the following integral equation that is the starting point for formulating the discretized finite element equations:

$$\iint_A k\nabla \delta T \cdot \nabla T \, dA = \int_S (k\nabla T)\cdot \mathbf{n}\delta T \, dS + \iint_A Q_g \delta T \, dA. \tag{4.56}$$

The boundary integral over S can be split into three parts:

$$\int_S (k\nabla T)\cdot \mathbf{n}\delta T \, dS = -\int_S q_n \delta T \, dS$$
$$= -\int_{S_T} q_n \delta T \, dS - \int_{S_Q} q_n \delta T \, dS - \int_{S_h} q_n \delta T \, dS. \tag{4.57}$$

Here we have used the relationship $q_n = \mathbf{q}\cdot\mathbf{n} = -(k\nabla T)\cdot\mathbf{n}$. Along S_T, since the temperature is known, the virtual temperature vanishes, that is, $\delta T = 0$. Therefore, there is no contribution to the RHS from this part of the boundary. It is also important to note that the normal component of heat flux at this boundary is unknown. Indeed, to keep the temperature constant at this boundary, some heat has to flow through this boundary but it is an unknown quantity similar to the reaction at a fixed boundary for structural problems.

Along the S_Q boundary, since the heat flux is known, we can write the contribution to the RHS due to this part of the boundary as

$$\int_{S_Q} \mathbf{q}\cdot\mathbf{n}\delta T \, dS = \int_{S_Q} q_n \delta T \, dS = -\int_{S_Q} q_0 \delta T \, dS. \tag{4.58}$$

Note that $q_n = \mathbf{q}\cdot\mathbf{n}$ is the normal component of the heat flux flowing *out* of the area A because the unit normal $\mathbf{n}$ points to the outside of the analysis domain A. Therefore, if q_0 is the known heat flux coming *into* the system, then $q_n = -q_0$.

For convection boundary, the normal component of the heat flux flowing out of the volume is proportional to the difference between the surface temperature and the surrounding air temperature or the ambient temperature T^∞:

$$q_n = h(T - T_\infty). \tag{4.59}$$

The proportionality constant, h, is the convection coefficient. Using this relation, we can write the contribution from the convection boundary as

$$\int_{S_h} \mathbf{q}\cdot\mathbf{n}\delta T \, dS = \int_{S_h} q_n \delta T \, dS = \int_{S_h} h(T - T^\infty)\delta T \, dS. \tag{4.60}$$

Substituting the contribution from surface integrals along S_T, S_Q, and S_h, we can rewrite the weak form in eq. (4.55) in the final form as follows:

$$\iint_A k\nabla \delta T \cdot \nabla T \, dA + \int_{S_h} hT\delta T \, dS$$
$$= \int_{S_q} q_0 \delta T \, dS + \int_{S_h} hT^\infty \delta T \, dS + \iint_A Q_g \delta T \, dA. \tag{4.61}$$

It is apparent in the above equation that all the information from the original boundary value problem is incorporated, except for the temperature boundary condition on S_T. Therefore, the approximate temperature of finite element analysis should satisfy the temperature boundary condition, while other boundary conditions will be satisfied as a part of the analysis. The terms that involve the unknown temperature have been moved to the left-hand side while known quantities are on the right. As a result, the convection has a contribution to both sides of this equation.

Galerkin's method yields an integral equation where the integration must be carried out over any arbitrary structural geometry. To facilitate this integration, it is convenient to break up the geometry into simpler shapes called finite elements. The most commonly used elements for two dimensions are triangles and quadrilaterals. The process of subdividing the geometry into elements is called mesh generation wherein elements are created, and nodes are placed at the vertices, edges, or midpoints of these elements. The discretized geometry consisting of nodes and elements is referred to as the mesh. These nodes are shared between adjacent elements if two or more elements meet at a node. The mesh then serves as an approximation of the structural geometry. It is also used for approximating the solution by interpolating the nodal values within each element. In Galerkin's approach, we assume that the virtual temperature δT is interpolated within each element using the same shape functions as the temperature field. In this chapter, we will use a 3-node triangular element to illustrate the process of deriving the interpolation scheme as well as discretizing the weak form into a set of linear simultaneous equations. Elements that use higher-order interpolation will be introduced in chapters 6 and 7.

4.7 3-NODE TRIANGULAR ELEMENTS FOR TWO-DIMENSIONAL HEAT TRANSFER

The simplest way of dividing a planar region into elements is to use triangular elements. Figure 4.22 shows a planar region that is divided into triangular elements. Each element shares its edges and nodes with adjacent elements if any. The three vertices of a triangle are the nodes of that element as shown in figure 4.22. The first node of an element can be arbitrarily chosen. However, the sequence of the nodes 1, 2, and 3 should be in the counterclockwise direction.

The temperature within the element is interpolated using the nodal values of temperature using interpolation functions, better known as shape functions. The temperature field within the element is assumed to be a polynomial function whose coefficient must be determined by using the nodal values. For this element, the temperature field has to be a linear function in x and y because values of temperature are available only at three points (nodes), and a linear polynomial has three unknown coefficients. As the temperature is a linear function, its gradient is a constant within an element, and therefore, the heat flux density is a constant vector for each element in a mesh consisting of 3-node triangular elements. The heat flux density is of course not likely to be constant in the exact solution, and therefore, this element provides an approximate solution and should be used with caution. The elements in the mesh should be very

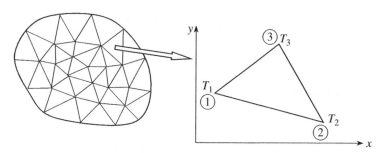

Figure 4.22 3-node triangular element

small so that the piecewise constant heat flux density can reasonably approximate the real heat flux density field. Later we will see higher-order elements that provide better approximation.

4.7.1 Temperature Interpolation

The first step in deriving the finite element matrix equation is to interpolate the temperature function in terms of the nodal temperatures. Clearly, the interpolated function must be a linear, three-term polynomial in x and y of the form:

$$T(x,y) = \alpha_1 + \alpha_2 x + \alpha_3 y, \tag{4.62}$$

where, the α's are constants to be determined. In finite element analysis, we would like to replace these constants by determining their value in terms of the nodal temperatures. At node 1, for example, x and y take the values of x_1 and y_1, respectively, and the nodal temperature is T_1. If we repeat this for the other two nodes, we obtain the following three simultaneous equations:

$$\begin{cases} T(x_1, y_1) \equiv T_1 = \alpha_1 + \alpha_2 x_1 + \alpha_3 y_1 \\ T(x_2, y_2) \equiv T_2 = \alpha_1 + \alpha_2 x_2 + \alpha_3 y_2 \\ T(x_3, y_3) \equiv T_3 = \alpha_1 + \alpha_2 x_3 + \alpha_3 y_3. \end{cases} \tag{4.63}$$

In matrix notation, the above equations can be written as

$$\begin{Bmatrix} T_1 \\ T_2 \\ T_3 \end{Bmatrix} = \begin{bmatrix} 1 & x_1 & y_1 \\ 1 & x_2 & y_2 \\ 1 & x_3 & y_3 \end{bmatrix} \begin{Bmatrix} \alpha_1 \\ \alpha_2 \\ \alpha_3 \end{Bmatrix}. \tag{4.64}$$

If the three points, (x_1, y_1), (x_2, y_2), and (x_3, y_3), are not on a straight line, then the inverse of the above coefficient matrix exists. Thus, we can calculate the unknown coefficients as

$$\begin{Bmatrix} \alpha_1 \\ \alpha_2 \\ \alpha_3 \end{Bmatrix} = \begin{bmatrix} 1 & x_1 & y_1 \\ 1 & x_2 & y_2 \\ 1 & x_3 & y_3 \end{bmatrix}^{-1} \begin{Bmatrix} T_1 \\ T_2 \\ T_3 \end{Bmatrix} = \frac{1}{2A_e} \begin{bmatrix} f_1 & f_2 & f_3 \\ b_1 & b_2 & b_3 \\ c_1 & c_2 & c_3 \end{bmatrix} \begin{Bmatrix} T_1 \\ T_2 \\ T_3 \end{Bmatrix}, \tag{4.65}$$

where A_e is the area of the triangle and

$$\begin{cases} f_1 = x_2 y_3 - x_3 y_2, & b_1 = y_2 - y_3, & c_1 = x_3 - x_2 \\ f_2 = x_3 y_1 - x_1 y_3, & b_2 = y_3 - y_1, & c_2 = x_1 - x_3 \\ f_3 = x_1 y_2 - x_2 y_1, & b_3 = y_1 - y_2, & c_3 = x_2 - x_1. \end{cases} \tag{4.66}$$

The area A_e of the triangle can be calculated from

$$A_e = \frac{1}{2} \det \begin{vmatrix} 1 & x_1 & y_1 \\ 1 & x_2 & y_2 \\ 1 & x_3 & y_3 \end{vmatrix}. \tag{4.67}$$

Note that the determinant in eq. (4.67) is zero when three nodes are collinear. In such a case, the area of the triangular element is zero, and we cannot uniquely determine the three coefficients. After calculating α_i the temperature interpolation can be written as

$$T(x,y) = \{N_1 \ N_2 \ N_3\} \begin{Bmatrix} T_1 \\ T_2 \\ T_3 \end{Bmatrix}, \tag{4.68}$$

where the shape functions are defined by

$$
\begin{cases}
N_1(x,y) = \dfrac{1}{2A_e}(f_1 + b_1 x + c_1 y) \\[2mm]
N_2(x,y) = \dfrac{1}{2A_e}(f_2 + b_2 x + c_2 y) \\[2mm]
N_3(x,y) = \dfrac{1}{2A_e}(f_3 + b_3 x + c_3 y).
\end{cases}
\tag{4.69}
$$

or

$$
\boxed{T(x,y) = \{\mathbf{N}(x,y)\}^{\mathrm{T}}\{\mathbf{T_e}\}}.
\tag{4.70}
$$

Note that $N_1(x,y)$, $N_2(x,y)$, and $N_3(x,y)$ are linear functions of x- and y-coordinates. Thus, interpolated temperature varies linearly in each coordinate direction.

After calculating the temperature field within an element, the gradient of temperature can be calculated by differentiating the temperature with respect to x and y. For example, the derivative with respect x can be written as

$$
\frac{\partial T}{\partial x} = \frac{\partial}{\partial x}\left(\sum_{i=1}^{3} N_i(x,y)T_i\right) = \sum_{i=1}^{3}\frac{\partial N_i}{\partial x}T_i = \sum_{i=1}^{3}\frac{b_i}{2A_e}T_i.
\tag{4.71}
$$

Note that T_1, T_2, and T_3 are nodal temperatures, and they are independent of the x-coordinate. Thus, only the shape function is differentiated with respect to x. Using the matrix notation, the temperature gradient can be written as

$$
\{\nabla T\} = \left\{\begin{array}{c} \dfrac{\partial T}{\partial x} \\[3mm] \dfrac{\partial T}{\partial y} \end{array}\right\} = \frac{1}{2A_e}\begin{bmatrix} b_1 & b_2 & b_3 \\ c_1 & c_2 & c_3 \end{bmatrix}\left\{\begin{array}{c} T_1 \\ T_2 \\ T_3 \end{array}\right\} \equiv [\mathbf{B}]^{\mathrm{T}}\{\mathbf{T_e}\}.
\tag{4.72}
$$

It may be noted that the $[\mathbf{B}]^{\mathrm{T}}$ matrix is constant and depends only on the coordinates of the three nodes of the triangular element. Thus, one can anticipate that if this element is used, then the temperature gradient will be constant over a given element and will depend only on nodal temperatures.

4.7.2   Conductance Matrix for 3-Node Triangular Element

The weak form in the previous section involves area integrals over the entire domain of analysis. This integral can now be split into integrals over individual elements in the mesh. The conductance term on the left-hand side (LHS) of eq. (4.61) can be written as the sum of the integration over all the elements in the mesh, as

$$
\iint\limits_{A} k\{\nabla\delta T\}^{\mathrm{T}}\{\nabla T\}\,\mathrm{d}A = \sum_{e=1}^{NEL}\iint\limits_{A_e} k\{\nabla\delta T\}^{\mathrm{T}}\{\nabla T\}\,\mathrm{d}A,
\tag{4.73}
$$

where NEL is the total number of elements in the system. The temperature field and its gradient were expressed in the matrix form using the shape functions of the three-node triangular element in the last section. These can now be used to write discretized equations for each element. In Galerkin's approach, we assume that the virtual temperature is also interpolated using the same shape functions as the temperature. Therefore, we can write these as,

$$
\{\delta T(x,y)\} = \{\mathbf{N}(x,y)\}^{\mathrm{T}}\{\delta \mathbf{T}^{(e)}\},
\tag{4.74}
$$

$$\{\nabla \delta T\} = \begin{Bmatrix} \dfrac{\partial \delta T}{\partial x} \\ \dfrac{\partial \delta T}{\partial y} \end{Bmatrix} = \frac{1}{2A_e} \begin{bmatrix} b_1 & b_2 & b_3 \\ c_1 & c_2 & c_3 \end{bmatrix} \begin{Bmatrix} \delta T_1 \\ \delta T_2 \\ \delta T_3 \end{Bmatrix} \equiv [\mathbf{B}]^{\mathrm{T}} \{\delta \mathbf{T}^{(e)}\}. \tag{4.75}$$

Substituting, these matrix expressions for temperature and the weighting function, we can rewrite the conductance term on the LHS of eq. (4.61) in the following matrix form:

$$\iint_{A_e} k\{\nabla \delta T\}^{\mathrm{T}} \{\nabla T\}\, \mathrm{d}A = \iint_{A_e} k\{\delta \mathbf{T}^{(e)}\}^{\mathrm{T}} [\mathbf{B}][\mathbf{B}]^{\mathrm{T}} \{\mathbf{T}^{(e)}\}\, \mathrm{d}A$$

$$= \{\delta \mathbf{T}^{(e)}\}^{\mathrm{T}} \left[\mathbf{k}_T^{(e)}\right] \{\mathbf{T}^{(e)}\}, \tag{4.76}$$

where $\left[\mathbf{k}_T^{(e)}\right]$ is the element conductance matrix for heat transfer and is analogous to the stiffness matrix computed for structural mechanics problems. Note that all the matrices within the integral are constant matrices for this element. Therefore, the conductance matrix can be computed as

$$\left[\mathbf{k}_T^{(e)}\right] = \iint_{A_e} k[\mathbf{B}][\mathbf{B}]^{\mathrm{T}}\, \mathrm{d}A = kA_e[\mathbf{B}][\mathbf{B}]^{\mathrm{T}}, \tag{4.77}$$

$$\left[\mathbf{k}_T^{(e)}\right] = \frac{k}{4A_e} \begin{bmatrix} b_1^2 + c_1^2 & b_1 b_2 + c_1 c_2 & b_1 b_3 + c_1 c_3 \\ b_1 b_2 + c_1 c_2 & b_2^2 + c_2^2 & b_2 b_3 + c_2 c_3 \\ b_1 b_3 + c_1 c_3 & b_2 b_3 + c_2 c_3 & b_3^2 + c_3^2 \end{bmatrix}.$$

4.7.3 Thermal Loads for a 3-Node Triangular Element

Thermal load due to distributed heat source:

The right-hand side (RHS) of eq. (4.61) contains an area integral, which corresponds to the thermal load due to the distributed heat source. An example of the distributed heat source is when there is heat being generated in the entire area of a solid due to electrical current flow. This term can also be discretized in a similar fashion to express it in a matrix form. The contribution of the distributed heat source or heat generation term for an element e can be expressed as:

$$\iint_{A_e} Q_g \delta T\, \mathrm{d}A = \iint_{A_e} Q_g \{\delta \mathbf{T}^{(e)}\}^{\mathrm{T}} \{\mathbf{N}\}\, \mathrm{d}A = \{\delta \mathbf{T}^{(e)}\}^{\mathrm{T}} \{\mathbf{Q}_g^{(e)}\}, \tag{4.78}$$

where

$$\{\mathbf{Q}_g^{(e)}\} = \iint_{A_e} Q_g \{\mathbf{N}\}\, \mathrm{d}A \tag{4.79}$$

is the thermal load due to the distributed heat source.

In later chapters, we will discuss numerical methods for integrating over triangles. For the 3-node triangles, the integration of the shape functions over the area of the triangle is rather straightforward as it is equivalent to finding the volume under the function $N_i(x, y)$. Figure 4.23 shows the plot of a typical linear shape function over a triangular element. At node 1, the value of the shape function $N_1 = 1$. The volume under the shape function is the tetrahedron whose height is one and its base is the area of the triangle, therefore,

$$\iint_{A_e} N_i(x, y)\, \mathrm{d}A = \frac{1}{3} A_e.$$

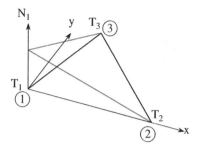

Figure 4.23 Plot of linear shape function for triangular element

Here, we use the well-known result that the volume of the tetrahedron is one-third of the product of its height and the area of the base.

Assuming the heat source is uniformly distributed within the element, the thermal load due to the distributed heat source can be calculated as

$$\left\{\mathbf{Q}_g^{(e)}\right\} = \iint_{A_e} Q_g \left\{\begin{matrix} N_1 \\ N_2 \\ N_3 \end{matrix}\right\} dA = \frac{1}{3} Q_g A_e \left\{\begin{matrix} 1 \\ 1 \\ 1 \end{matrix}\right\}. \tag{4.80}$$

Load due to applied heat flux:

The integral over S_Q on the right-hand side of eq. (4.61) is the contribution due to known heat flux entering the domain. This term contributes only to the thermal load vector of the elements that are on the boundary S_Q. To evaluate this term, we need to integrate along the edges of the elements that are on this boundary.

$$\int_{S_Q} q_0 \delta T \, dS = \int_{S_Q} q_0 \left\{\delta\mathbf{T}^{(e)'}\right\}^{\mathrm{T}} \{\mathbf{N}'\} dS = \left\{\delta\mathbf{T}^{(e)'}\right\}^{\mathrm{T}} \left\{\mathbf{Q}_q^{(e)}\right\}, \tag{4.81}$$

$$\left\{\mathbf{Q}_q^{(e)}\right\} = \int_{S_Q} q_0 \{\mathbf{N}'\} dS. \tag{4.82}$$

Here we have assumed that the virtual temperature δT is interpolated in the same fashion as the temperature. As δT is linearly interpolated over the element, we expect that δT is interpolated along the edge using linear shape functions. To interpolate nodal values of the edge nodes, we construct linear shape functions $N_i'(s)$ and use them to interpolate δT along the edge as:

$$\delta T = N_1'(s)\delta T_1 + N_2'(s)\delta T_2 = \{\mathbf{N}'\}^{\mathrm{T}}\left\{\delta\mathbf{T}^{(e)'}\right\}, \tag{4.83}$$

where the shape functions on the edge $\{\mathbf{N}'\}$ and nodal temperature on the edge $\{\delta\mathbf{T}^{(e)'}\}$ are defined as $\{\mathbf{N}'\} = \{N_1' \quad N_2'\}^{\mathrm{T}}$ and $\left\{\delta\mathbf{T}^{(e)'}\right\} = \{\delta T_1' \quad \delta T_2'\}^{\mathrm{T}}$, respectively.

The shape function vector $\{\mathbf{N}'\}$ consists of the linear shape functions that interpolate δT along the edge of the element and $\{\delta\mathbf{T}^{(e)'}\}$ containing the nodal values of δT. Figure 4.24 shows an element where a heat flux is applied to the edge between nodes 1 and 2. The local coordinate s has origin at node 1 and measures the distance from this node. The linear shape functions N_1' and N_2' are plotted along these edges. If we assume that the heat flux is constant along the edge, then we have

$$\left\{\mathbf{Q}_q^{(e)}\right\} = q_0 \int_{S_Q} \left\{\begin{matrix} N_1' \\ N_2' \end{matrix}\right\} dS. \tag{4.84}$$

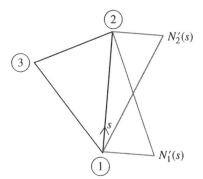

Figure 4.24 Linear shape functions for interpolation along edge

The linear shape functions derived earlier for one-dimensional elements in eq. (4.15) can be used here too after writing them in terms of the local coordinate system s shown in figure 4.24.

$$N_1'(s) = \left(1 - \frac{s}{L_e}\right), \quad N_2'(s) = \frac{s}{L_e}, \tag{4.85}$$

where L_e is the length of the edge 1-2 of the element. To evaluate the load due to heat flux, these shape functions can be integrated over s from 0 to L_e. We can evaluate the integral by making use of the fact that the area under the shape functions N_1' and N_2' is equal to half the length of the edge of the triangle L_e.

$$\left\{Q_q^{(e)}\right\} = \frac{q_0 L_e}{2} \begin{Bmatrix} 1 \\ 1 \end{Bmatrix}. \tag{4.86}$$

Convection boundary condition:
The contribution of the convection heat transfer is the integral over S_h, which was split into two parts in eq. (4.61). Again, we need to integrate along the edges of the elements that are on this boundary to evaluate this boundary integral, and as before we will interpolate δT linearly using the shape functions in (4.83). Consider the convection term on the LHS of eq. (4.61) first, as

$$\int_{S_h} h \delta T T \, dS = \int_{S_h} h \left\{\delta \mathbf{T}^{(e)\prime}\right\}^{\mathrm{T}} \{\mathbf{N}'\} \{\mathbf{N}'\}^{\mathrm{T}} \left\{\mathbf{T}^{(e)\prime}\right\} dS = \left\{\delta \mathbf{T}^{(e)\prime}\right\}^{\mathrm{T}} \left[\mathbf{k}_h^{(e)}\right] \left\{\mathbf{T}^{(e)\prime}\right\}, \tag{4.87}$$

where

$$\left[\mathbf{k}_h^{(e)}\right] = \int_{S_h} h \{\mathbf{N}'\} \{\mathbf{N}'\}^{\mathrm{T}} dS, \tag{4.88}$$

$$\left[\mathbf{k}_h^{(e)}\right] = h \int_{S_h} \begin{Bmatrix} N_1' \\ N_2' \end{Bmatrix} \{N_1' \quad N_2'\} dS = h \int_0^{L_e} \begin{Bmatrix} (N_1')^2 & N_1' N_2' \\ N_2' N_1' & (N_2')^2 \end{Bmatrix} ds, \tag{4.89}$$

$$\left[\mathbf{K}_h^{(e)}\right] = \frac{h L_e}{6} \begin{bmatrix} 2 & 1 \\ 1 & 2 \end{bmatrix}.$$

Along the edge of the element, both δT and T are interpolated linearly, and $\{\delta \mathbf{T}^{(e)\prime}\}$ is again a 2×1 column vector that contains that nodal values of δT for that edge. The effective conductance matrix $\left[\mathbf{k}_h^{(e)}\right]$ due to convection is a 2×2 matrix for elements on a convection boundary. This matrix is to be assembled into the global conductance matrix along with the conductance matrix of all the other elements.

On the right-hand side of eq. (4.61), the contribution due to convection is similar to a heat flux vector:

$$\int_{S_h} hT^\infty \delta T \, dS = \int_{S_h} hT^\infty \left\{\delta \mathbf{T}^{(e)\prime}\right\}^{\mathrm{T}} \{\mathbf{N}'\} \, dS = \left\{\delta \mathbf{T}^{(e)\prime}\right\}^{\mathrm{T}} \left\{\mathbf{Q}_h^{(e)}\right\}, \tag{4.90}$$

$$\left\{\mathbf{Q}_h^{(e)}\right\} = \int_{S_h} hT^\infty \{\mathbf{N}'\} \, dS = \frac{hT^\infty L_e}{2} \begin{Bmatrix} 1 \\ 1 \end{Bmatrix}. \tag{4.91}$$

So far, we have looked at how the contribution from each element in the mesh is computed in the form of element stiffness matrices and thermal load vectors. These contributions must be assembled into a global system of equations. This process can be best understood by looking at a simple example that shows the entire process. The essential boundary conditions are imposed after assembling the global equations.

EXAMPLE 4.7   *Heat transfer along a conducting block*

Consider heat transfer along a conducting block as shown in figure 4.25. The conductor, modeled here as a rectangular region, is insulated at the top and bottom, and there is no heat flow in the direction normal to the plane (z-direction). The thickness in the z-direction is $t = 5$ mm. The conductivity of the material is $k = 0.2 \, \mathrm{W/mm^oC}$. Heat is generated in the block due to electrical current flow, and the corresponding distributed heat source is $Q_g = 0.06 \, \mathrm{W/mm^3}$. On the left side, heat enters the block with a known heat flux of $q_0 = 0.04 \, \mathrm{W/mm^2}$. Heat is lost on the other side to the surrounding air, which has an ambient temperature of 25 °C. The convection coefficient for this heat transfer is $h = 1.2 \times 10^{-2} \, \mathrm{W/mm^{2o}C}$. Determine the temperature distribution in the domain using a 3-node triangular element.

SOLUTION   Even though the heat flow in this problem is only in the x-direction, and we could have modeled it as a one-dimensional problem, we model the problem as two-dimensional here mainly to illustrate the use of the 3-node triangular element. In figure 4.26, the rectangular domain is meshed using four triangular elements. In this figure, the numbers with circle around them are the node numbers while the numbers without circle are the element numbers. The edge between nodes 1-2 has a heat flux boundary condition, and a known quantity of heat is entering the domain through this edge. The edges between nodes 1-3, 3-5, 2-4, and 4-6 are insulated; therefore, they can be thought of as heat flux boundaries where the heat flux is zero. Finally, the edge 5-6 has a convection boundary condition applied on it.

The mesh for the analysis, shown in figure 4.26, consists of four triangular elements. The connectivity table for this mesh is shown in table 4.2 where, for each element, the columns show the global node numbers corresponding to the local node numbers (LN). Note that the connectivity for all the elements have nodes listed in the counterclockwise direction. If the order of nodes are not in counterclockwise direction, then the computed areas and stiffness will be negative.

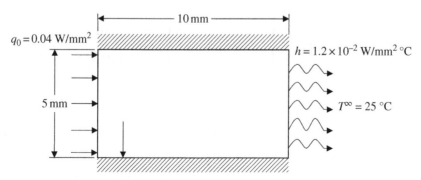

Figure 4.25 Heat conduction example

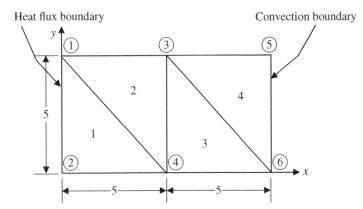

Figure 4.26 Finite element mesh for heat conduction analysis

Table 4.2   Connectivity table

Element	LN 1	LN 2	LN 3
1	1	2	4
2	1	4	3
3	3	4	6
4	3	6	5

As the first step in the process, we want to compute the stiffness of the elements. We expect the stiffness of elements 1 and 3 to be identical because they have the same size and shape. Similarly, the stiffness matrices for elements 2 and 4 will be identical. For element 1 and 3, we have

$$
\begin{aligned}
b_1 &= y_2 - y_3 = 0, & c_1 &= x_3 - x_2 = 5, \\
b_2 &= y_3 - y_1 = -5, & c_2 &= x_1 - x_3 = -5, \\
b_3 &= y_1 - y_2 = 5, & c_3 &= x_2 - x_1 = 0,
\end{aligned}
$$

$$
A_e = \frac{1}{2}\det\begin{vmatrix} 1 & x_1 & y_1 \\ 1 & x_2 & y_2 \\ 1 & x_3 & y_3 \end{vmatrix} = \frac{25}{2},
$$

$$
[\mathbf{B}]^{\mathrm{T}} = \frac{1}{2A_e}\begin{bmatrix} b_1 & b_2 & b_3 \\ c_1 & c_2 & c_3 \end{bmatrix} = \frac{1}{25}\begin{bmatrix} 0 & -5 & 5 \\ 5 & -5 & 0 \end{bmatrix},
$$

$$
\left[\mathbf{k}_T^{(1)}\right] = \left[\mathbf{k}_T^{(3)}\right] = [\mathbf{B}][\mathbf{B}]^{\mathrm{T}} kA_e t = 0.5\begin{bmatrix} 1 & -1 & 0 \\ -1 & 2 & -1 \\ 0 & -1 & 1 \end{bmatrix}.
$$

Note that the subscripts in the preceding equations are local node numbers whose corresponding global numbers are shown in the connectivity table. Similarly, for elements 2 and 4,

$$
b_1 = y_2 - y_3 = -5, \quad c_1 = x_3 - x_2 = 0,
$$

$$
b_2 = y_3 - y_1 = 0, \quad c_2 = x_1 - x_3 = -5,
$$

$$
b_3 = y_1 - y_2 = 5, \quad c_3 = x_2 - x_1 = 5,
$$

$$
A_e = \frac{1}{2}\det\begin{vmatrix} 1 & x_1 & y_1 \\ 1 & x_2 & y_2 \\ 1 & x_3 & y_3 \end{vmatrix} = \frac{25}{2},
$$

$$[\mathbf{B}]^{\mathrm{T}} = \frac{1}{2A_e}\begin{bmatrix} b_1 & b_2 & b_3 \\ c_1 & c_2 & c_3 \end{bmatrix} = \frac{1}{25}\begin{bmatrix} -5 & 0 & 5 \\ 0 & -5 & 5 \end{bmatrix},$$

$$\left[\mathbf{k}_T^{(2)}\right] = \left[\mathbf{k}_T^{(3)}\right] = [\mathbf{B}][\mathbf{B}]^{\mathrm{T}}kA_e t = 0.5\begin{bmatrix} 1 & 0 & -1 \\ 0 & 1 & -1 \\ -1 & -1 & 2 \end{bmatrix}.$$

Now the contribution from the other terms in Galerkin's method must be determined. Below we consider the three types of loads and the conductance that arises from the convection boundary condition.

Heat Generation: For each element in the mesh, the thermal load due to the distributed source or heat generation must be calculated. Using the expression in eq. (4.80), we have

$$\left\{\mathbf{Q}_g^{(e)}\right\} = \frac{1}{3}Q_g A_e t \begin{Bmatrix} 1 \\ 1 \\ 1 \end{Bmatrix} = \frac{1}{3}\cdot(0.06)\cdot\left(\frac{25}{2}\right)\cdot(5)\begin{Bmatrix} 1 \\ 1 \\ 1 \end{Bmatrix} = 1.25\begin{Bmatrix} 1 \\ 1 \\ 1 \end{Bmatrix}. \tag{4.92}$$

Heat Flux: The edge 1-2 is the only boundary on which a nonzero known heat flux is entering the domain. Therefore, we need to compute the load vector due to heat flux only for element 1, which contains this edge. Using eq. (4.84), we have

$$\left\{\mathbf{Q}_q^{(e)}\right\} = q_0 t \frac{L_e}{2}\begin{Bmatrix} 1 \\ 1 \end{Bmatrix} = 0.04\cdot5\cdot\frac{5}{2}\begin{Bmatrix} 1 \\ 1 \end{Bmatrix} = 0.5\begin{Bmatrix} 1 \\ 1 \end{Bmatrix}. \tag{4.93}$$

Convection: Convection on edge 5-6 has a conductance and load contribution. The conduction contribution on the left hand side is

$$\left[\mathbf{k}_h^{(e)}\right] = \frac{htL_e}{6}\begin{bmatrix} 2 & 1 \\ 1 & 2 \end{bmatrix} = \frac{(1.2\times10^{-2})\cdot5\cdot5}{6}\begin{bmatrix} 2 & 1 \\ 1 & 2 \end{bmatrix} = 0.05\begin{bmatrix} 2 & 1 \\ 1 & 2 \end{bmatrix}.$$

The load contribution on the right-hand side is

$$\left\{\mathbf{Q}_h^{(e)}\right\} = \int_{S_h} hT^\infty\{\mathbf{N}\}\,dS = \frac{htT^\infty L_e}{2}\begin{Bmatrix} 1 \\ 1 \end{Bmatrix} = \frac{1.2\times10^{-2}\cdot5\cdot25\cdot5}{2}\begin{Bmatrix} 1 \\ 1 \end{Bmatrix} = \begin{Bmatrix} 3.75 \\ 3.75 \end{Bmatrix}.$$

The global stiffness matrix can now be assembled as:

$$0.5\begin{bmatrix} 1+1 & -1 & -1 & 0 & 0 & 0 \\ -1 & 2 & 0 & -1 & 0 & 0 \\ -1 & 0 & 1+2+1 & -1-1 & -1 & 0 \\ 0 & -1 & -1-1 & 1+2+1 & 0 & -1 \\ 0 & 0 & -1 & 0 & 2+0.2 & -1+0.1 \\ 0 & 0 & 0 & -1 & -1+0.1 & 1+1+0.2 \end{bmatrix}\begin{Bmatrix} T_1 \\ T_2 \\ T_3 \\ T_4 \\ T_5 \\ T_6 \end{Bmatrix} = 1.25\begin{Bmatrix} 1+1+0.4 \\ 1+0.4 \\ 1+1+1 \\ 1+1+1 \\ 1+3 \\ 1+1+3 \end{Bmatrix}.$$

Solving this equation, we get the nodal temperatures as:

$$\{T_1\ T_2\ T_3\ T_4\ T_5\ T_6\} = \{95.75\ \ 94.92\ \ 90.59\ \ 90.58\ \ 77.93\ \ 78.75\}.$$

Heat flux computation: Now that we know the temperature at all the nodes, we can calculate other quantities such as gradient of temperature or heat flux at any point within the domain. To compute the heat flux, we make use of Fourier's law:

$$\{\mathbf{q}\} = \begin{Bmatrix} q_x \\ q_x \end{Bmatrix} = -k\nabla T = -k \begin{bmatrix} \dfrac{\partial T}{\partial x} \\ \dfrac{\partial T}{\partial y} \end{bmatrix} = -k[\mathbf{B}]^{\mathrm{T}}\{\mathbf{T}^{(e)}\}.$$

The heat flux at the midpoint of the edge between nodes 1-2 can be determined from element 1:

$$\{\mathbf{q}\} = \frac{-k}{2A_e}\begin{bmatrix} b_1 & b_2 & b_3 \\ c_1 & c_2 & c_3 \end{bmatrix}\begin{Bmatrix} T_1 \\ T_2 \\ T_3 \end{Bmatrix} = \frac{-0.2}{25}\begin{bmatrix} 0 & -5 & 5 \\ 5 & -5 & 0 \end{bmatrix}\begin{Bmatrix} 95.75 \\ 94.92 \\ 90.58 \end{Bmatrix} = \begin{Bmatrix} 0.174 \\ -0.0332 \end{Bmatrix}.$$

Note that the heat flux is constant within the element as per our assumption that the temperature is linear within the element. Due to this limitation, the heat flux will not be continuous on the element boundary with the adjacent element. This, of course, is not a very good assumption because we know that the heat flux should be linear in the x-direction for this problem. Therefore, we expect the solution we obtained to be approximate and not very accurate. The accuracy can be improved by increasing the mesh density if we use a finite element analysis software for this example.◼

4.8 FINITE ELEMENT MODELING PRACTICE FOR 2-D HEAT TRANSFER

The finite element method is most useful when applied to two- and three-dimensional problems with complex geometries for which no analytical solutions are available. For such problems, a large number of nodes and elements are typically needed, which makes it necessary to use software for the analysis. In this section, we look at some practical aspects to consider when constructing finite element models. These aspects include deciding types of boundary conditions and thermal load conditions that can represent the real physical system, as well as understanding periodicity and symmetry to simplify the physical model.

4.8.1 Heat Conduction in a Slab

A series of pipes carrying hot fluids are embedded in a slab. The spacing between pipes is 40 mm and the thickness of the slab is $t_h = 30$ mm. The surfaces of the block are exposed to air with ambient temperature of 30°C. The heat transfer takes place due to conduction between the pipes and the block and due to convection to the ambient air. The conductivity of the slab is $k = 1.6 \times 10^{-4}$ W/mm°C and the convection coefficient is assumed to be $h = 1.2 \times 10^{-5}$ W/mm²°C. We wish to determine the temperature distribution in the block.

Assuming that the slab extends much further in either direction and that the spacing between pipes is constant through the slab, we have a repeating pattern creating a periodic structure. We can make use of this periodicity to identify a rectangular region of the slab that contains just one single pipe as shown by the shaded region in figure 4.27(a). This is a *unit cell*, which repeats periodically in both directions. We need to model only one such unit cell, instead of modeling the entire slab, because we expect to find the same temperature distribution within any such cell within the slab. The unit cell is symmetric both vertically and horizontally. Therefore, making use of the symmetry, we can further simplify the model to a quadrant of the unit cell as shown in figure 4.27(b). For this two-dimensional model, we assume the depth of the model in the z-direction to be unity or in other words, we are modeling a unit depth of the slab. A temperature boundary condition ($T = 100$ °C) is applied to the circular edge, which represents the outside surface of the embedded pipe. One could argue that if the pipes contain hot fluid, then it may be more appropriate to use a convection boundary condition for the pipe surface. However, if the convection coefficient is sufficiently large, for example due to the high velocity of the fluid flow, then the temperature at the surface will be very close to the fluid temperature, and therefore a boundary condition specifying

(a)

(b)

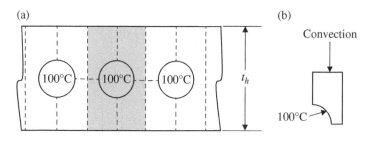

Figure 4.27 Finite element model for slab with pipes: (a) periodicity and symmetry, (b) model

(a) Continuous plot

(b) Fringe (or discrete) plot

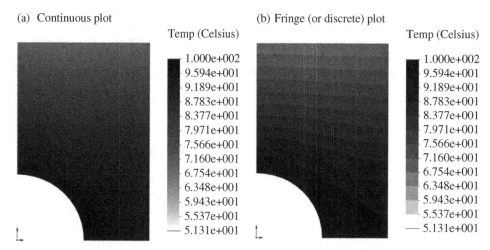

Figure 4.28 Temperature distribution in the slab

temperature at the boundary to be equal to the fluid temperature is justified. The correct boundary condition for any model depends on the real system being modeled. Therefore, it is very important to have a very good understanding of the system and the applicable material properties and parameters to arrive at the best model. Convection boundary condition is applied on the outer surface of the slab.

On the remaining edges, no boundary conditions or thermal loads are applied. This is equivalent to a zero heat flux condition on these boundaries, as if these boundaries were insulated. It is important to understand why such a boundary condition is appropriate. The reason has to do with the fact that these boundaries are symmetry boundaries, and therefore two neighboring points equidistant from these boundaries on its two sides will have the same temperature. This implies that the gradient of temperature normal to the boundary will be zero, and therefore the heat flux across the boundary will also be zero.

Figure 4.28 shows the temperature distribution in the slab computed using the finite element model described above. Figure 4.28(a) shows a continuous distribution plot while (b) shows discrete or fringe plots of the temperature. These are the two most common ways in which finite element analysis software programs display the solution. Some software programs are also able to plot contours of the variable. The contours of the temperature plot are also seen in the fringe plot. The contours provide useful visual information because the gradient of temperature and the heat flux are normal to these contours. It can be seen that the contours are horizontal near the vertical edges, which indicates that the heat flux is parallel to these vertical edges, and therefore the heat flux normal to these edges are zero as expected. Similarly, the heat flux component in the vertical direction is zero at the horizontal symmetric line.

The horizontal and vertical heat flux components are plotted in figure 4.29. Again, it can be seen that the horizontal component is nearly zero near the vertical edges of symmetry while the vertical component is zero at the horizontal symmetric edge.

(a) Horizontal heat flux component

(b) Vertical heat flux component

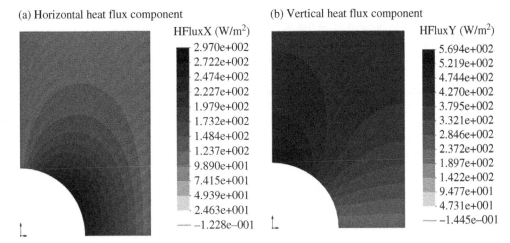

HFluxX (W/m^2)

2.970e+002
2.722e+002
2.474e+002
2.227e+002
1.979e+002
1.732e+002
1.484e+002
1.237e+002
9.890e+001
7.415e+001
4.939e+001
2.463e+001
−1.228e–001

HFluxY (W/m^2)

5.694e+002
5.219e+002
4.744e+002
4.270e+002
3.795e+002
3.321e+002
2.846e+002
2.372e+002
1.897e+002
1.422e+002
9.477e+001
4.731e+001
−1.445e–001

Figure 4.29 Heat flux components in the slab

The exact solution for the temperature distribution, its gradient, and the heat flux are all expected to be continuous functions. The heat flux shown in the figures suggests that this was true for the computed solution as well. However, we stated earlier that the triangular element computes a piecewise linear solution, and therefore, the heat flux will be constant within each element. In this case, the color within each element should be constant, and the heat flux distribution should be discontinuous. This raises the question about why the color plot of the heat flux appears continuous. This is because in most finite element analysis software, the heat flux (and also stresses and strains) are smoothed before plotting. There are many ways to perform the smoothing operation. A simple algorithm would compute the average value at the nodes by averaging the heat flux computed for each element that connects to that node. Then these averaged values are interpolated within each element to determine the heat flux values inside the triangular element.

Approximating the heat flux as constant within each element is not very desirable or accurate, and in fact, the smoothed results are closer to the real solution. The 3-node triangular element is not a very good element due to the piecewise linear approximation that it provides. A very dense mesh, that is, a mesh with very large number of small elements is needed to get even a reasonable or acceptable answer. Better elements can be derived if we assume that the temperature distribution is quadratic or even higher-order within each element. In later chapters, we will see other types of elements including higher-order triangular and quadrilateral elements that can be used to solve the heat conduction and solid mechanics problems.

4.9 EXERCISES

1. Answer the following descriptive questions.

    (a) In finite elements for solid mechanics, nodal forces are defined as positive when the force is applied in the positive coordinate direction. How is the positive heat flux defined for an element?

    (b) Two 1-D heat transfer elements have the same thermal conductivity and cross-sectional area but have different lengths. When the temperature differences between the two nodes of the two elements are the same, which element has a higher heat flow? A long element or a short one?

    (c) In solid mechanics, when a node is fixed, the column corresponding to the fixed DOF is deleted (strike the column). In heat transfer analysis, how can a prescribed temperature be imposed in the global matrix equation?

(d) In the weak form of Galerkin's method, when the temperature at node i, T_i, is prescribed, what is the value of the weighting function δT_i at that node?

(e) The wall of a heat chamber is modeled with three 1-D elements, where each element has different length. There is no heat generation within the wall. Will the heat flux for the three elements be the same or different? Explain why.

(f) Explain how a convection boundary condition makes the finite element matrix positive definite without striking the row or striking the column.

2. Consider heat conduction in a uniaxial rod surrounded by a fluid. The left end of the rod is at T_0. The free stream temperature is T^∞. There is convective heat transfer across the surface of the rod as well over the right end. The governing equation and boundary conditions are as follows:

$$\frac{d}{dx}\left(kA\frac{dT}{dx}\right) + hP(T^\infty - T) = 0, \quad 0 \le x \le L,$$

$$\begin{cases} T(0) = T_0, & x = 0, \\ k\dfrac{dT}{dx} = h(T^\infty - T), & x = L, \end{cases}$$

where k is the thermal conductivity, h is the convective heat transfer coefficient, and A and P are the area and circumference of the rod's cross section, respectively. The numerical values (SI units) are: $L = 0.5$, $k = 200$, $h = 10$, $T_0 = 700$, $T^\infty = 400$, $A = 10^{-4}$, $P = 4 \times 10^{-2}$. Use three finite elements to solve the problem. Use elements of lengths 0.1, 0.15, and 0.25, respectively.

3. Consider the heat conduction problem shown in the figure. Inside the bar, heat is generated from a uniform heat source $Q_g = 10 \text{ W/m}^3$, and the thermal conductivity of the material is $k = 0.2 \text{ W/m/°C}$. The cross-sectional area $A = 1 \text{ m}^2$. When the temperatures at both ends are fixed to 0 °C, calculate the temperature distribution using: (a) two equal-length elements and (b) three equal-length elements. Plot the temperature distribution along the bar and compare with the exact solution.

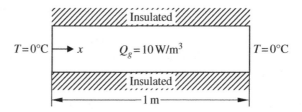

4. Repeat problem 3 with $Q_g = 20x$.

5. Determine the temperature distribution (nodal temperatures) of the bar shown in the figure using two equal-length finite elements with cross-sectional area of 1 m^2. The thermal conductivity is 15 W/m·°C. The left side is maintained at 300 °C. The right side is subjected to heat loss by convection with $h = 1.5 \text{ W/m}^2\cdot\text{°C}$ and $T_f = 30 \text{ °C}$. All other sides are insulated.

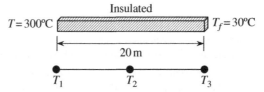

6. The nodal temperatures of a 1-D heat conduction problem are given in the figure. Calculate the temperature at $x = 0.2$ using: (a) two 2-node elements and (b) one 3-node element.

$$
\begin{array}{c}
\overset{0.5}{\longleftarrow\!\longrightarrow} \quad \overset{0.5}{\longleftarrow\!\longrightarrow} \\
\underset{T_1 = 0°C}{\bullet} \quad\quad \underset{T_2 = 100°C}{\bullet} \quad\quad \underset{T_3 = 50°C}{\bullet}
\end{array}
$$

7. In order to solve a 1-D steady-state heat transfer problem, one element with 3-nodes is used. The shape functions and the conductivity matrix before applying boundary conditions are given.

$$
\begin{cases}
N_1(x) = 1 - 3x + 2x^2 \\
N_2(x) = 4x - 4x^2 \\
N_3(x) = -x + 2x^2
\end{cases}
, [\mathbf{K}_T] =
\begin{bmatrix}
1 & -2 & 1 \\
-2 & 4 & -2 \\
1 & -2 & 2
\end{bmatrix}.
$$

   (a) When the temperature at node 1 is equal to 50 °C and a heat flux of 80 W is provided at node 3, calculate the temperature at $x = \frac{1}{4}$ m.

   (b) When the temperature at node 1 is equal to 50 °C and the convection boundary condition is applied at node 3 with $h = 4$ W/m²/°C, $T^\infty = 120$ °C, calculate the temperature at $x = \frac{1}{4}$ m.

   (c) Instead of the previous boundary conditions, heat fluxes at nodes 1 and 3 are given as Q_1 and Q_3, respectively. Can this problem be solved for the nodal temperatures? Explain your answer.

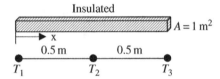

8. The one-dimensional wall in the figure is modeled using one element with three nodes. There is a uniform heat source inside the wall generating $Q = 200$ W/m³. The thermal conductivity of the wall is $k = 2$ W/(m·°C). Assume that $A = 1$ m² and $l = 1$ m. The left end has a fixed temperature of $T = 20$ °C, while the right end has zero heat flux.

   (a) Calculate the shape function $[\mathbf{N}] = [N_1, N_2, N_3]$ as a function of x.

   (b) Solve for the nodal temperature $\{\mathbf{T}\} = \{T_1, T_2, T_3\}^T$ using boundary conditions. Plot the temperature distribution in an x-T graph.

   (c) Calculate the heat fluxes at $x = 0$ and $x = 1$ when the solution is $\{\mathbf{T}\} = \{20, 58.25, 71\}$.

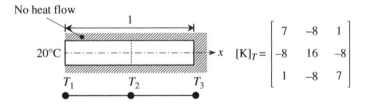

9. Consider heat conduction in a uniaxial rod surrounded by a fluid. The right end of the rod is attached to a wall and is at temperature T_R. One half of the bar is insulated as indicated. The free stream temperature is T_f. There is convective heat transfer across the un-insulated surface of the rod as well as over the left-end face. Use two equal-length elements to determine the temperature distribution in the rod. Use the following numerical values: $L = 0.2$ m, $k = 100$ W/m/°C, $h = 26$ W/m²/°C, $T_R = 600$ °C, $T_f = 300$ °C, $A = 10^{-4}$ m², and $P = 0.04$ m.

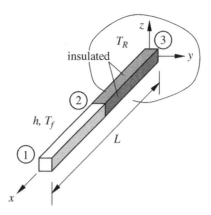

10. A well-mixed fluid is heated by a long iron plate of conductivity $k = 10$ W/m/°C and thickness $t = 0.12$ m. Heat is generated uniformly in the plate at the rate $Q_g = 5{,}000$ W/m^3. If the surface convection coefficient $h = 5$ W/m^2/°C and fluid temperature is $T_f = 45$ °C, determine the temperature at the center of the plate, T_c, and the heat flow rate to the fluid, q, using three one-dimensional elements.

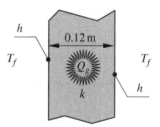

11. A cooling spine of square cross-sectional area $A = 0.25 \times 0.25$ m^2, length $L = 2$ m, and conductivity $k = 10$ W/m/°C extends from a wall maintained at temperature $T_w = 200$ °C. The surface convection coefficient between the spine and the surrounding air is $h = 0.6$ W/m^2/°C, and the air temperature is $T_a = 20$ °C. Determine the heat conducted by the spine and the temperature of the tip using five one-dimensional finite elements.

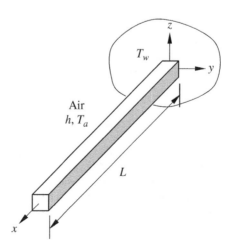

12. Find the heat transfer per unit area through the composite wall in the figure. Assume one-dimensional heat flow and that there is no heat flow between B and C. The thermal conductivities are $k_A = 0.04$ W/m/°C, $k_B = 0.2$ W/m/°C $k_C = 0.03$ W/m/°C, and $k_D = 0.06$ W/m/°C.

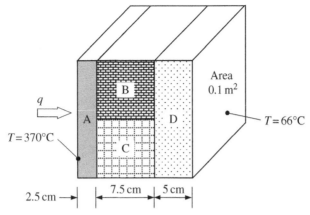

13. Consider a wall built up of concrete and thermal insulation. The outdoor temperature is $T_o = -20$ °C, and the temperature inside is $T_i = 20$ °C. The wall is subdivided into three elements. The thermal conductivity for concrete is $k_c = 2$ W/m/°C, and that of the insulator is $k_i = 0.05$ W/m/°C. Convection heat transfer is occurring at both surfaces with convection coefficients of $h_0 = 14$ W/m²/°C and $h_i = 5.5$ W/m²/°C. Calculate the temperature distribution of the wall. Also, calculate the amount of heat flow through the outdoor surface.

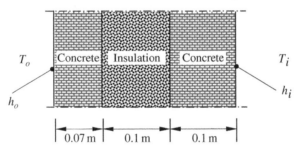

14. The heat conduction through the thickness of a plate can be modeled as a single 1-D quadratic (3-node) element assuming that the plate is large with uniform temperature on the same thickness. The temperature at x_1 is fixed at 100 °C while x_2 is at 25 °C. Heat is being generated within the plate at the rate of $Q_g = 1000$ W/m³. The conductivity of the plate is $k = 5.2 \times 10^{-2}$ W/(mm°C), and the thickness of the plate is 10 mm. When the shape functions and conductivity matrix are given,

    (a) compute the nodal equivalent heat source corresponding to the heat generation in the plate.
    Hint: $\left\{ \mathbf{Q}_g^{(e)} \right\} = \int_{V_e} Q_g \{\mathbf{N}\} dV$; and

    (b) determine the temperature distribution $T(r)$. Note k is conductivity, A = area, $L^{(e)}$ = length of the element.

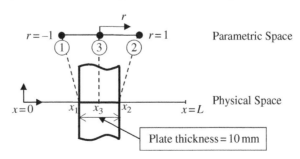

15. The shape functions of a linear 3-node element are also called barycentric or area coordinates because they uniquely define the location of a point in the triangle. Show that the shape functions in eq. (4.69) can also be computed as:

$$N_i = \frac{A_i}{A_e},$$

where A_1 is the area of triangle P23 in the figure, A_2 is the area of triangle P13, A_3 is the area of triangle P12, and $A_e = A_1 + A_2 + A_3$ is the area of triangle 123.

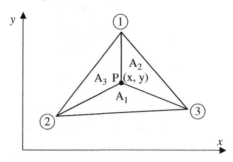

Chapter 5

Review of Solid Mechanics

5.1 INTRODUCTION

The finite element method is a powerful numerical method for solving partial differential equations. It has been applied to solve many physical problems whose governing equations are partial differential equations. The method has been implemented and is available as commercial software that can perform a variety of analysis including solids, structures, and thermal systems to mention a few. However, to use these programs effectively, one must understand the underlying physics of the problem being solved. This is important not only to be able to construct the right models for analysis but also to interpret the results and verify its accuracy. In this chapter, we review the main principles and the governing equations of solid mechanics. We explain the physical meaning behind the stress and strain tensors and the relation between them. Stress analysis is a major step, and in fact, it can be considered the most important one in the mechanical design process. There are many design considerations that influence the design of a machine element or structure. The most important design considerations are the following[1]: (i) the stress at every point should be below a certain limit for the material; (ii) the deflection should not exceed the maximum allowable for proper functioning of the system; (iii) the structure should be stable; and (iv) the structure or machine element should not fail due to fatigue. The failure mode corresponding to instability is also referred to as buckling. The failure due to excessive stress can take different forms such as brittle fracture, yielding of the material causing inelastic deformations, and fatigue failure. Stress analysis of structures plays a crucial role in predicting failure types (i), (ii), and (iv) above. The analysis of stability of a structure requires a slightly different approach but in general, uses most of the methods of stress analysis. Often, the results from the stress analysis can be used to predict buckling. Thus, stress is often used as a criterion for mechanical design.

In the elementary mechanics of materials or physics courses, stress is defined as force per unit area. While such a notion is useful and sufficient to analyze one-dimensional structures under a uniaxial state of stress, a complete understanding of the state of stress in a three-dimensional body requires a thorough understanding of the concept of stress at a point. Similarly, the strain is defined as the change in length per original length of a one-dimensional body. However, the concept of strain at a point in a three-dimensional body is quite interesting and is required for a complete understanding of the deformation a solid undergoes. While stresses and strains are concepts developed by engineers for a better understanding of the physics of deformation of a solid, the relation between stresses and strains is phenomenological in the sense that it is something observed and described as a simplified theory. Robert Hooke[2] was the first to establish the linear relation between stresses and strains in an elastic body. Although he

---

[1] Shigley, J. E., Mischke, C. R., and Budynas, R. G. 2004. *Mechanical Engineering Design*. McGraw-Hill.
[2] Hooke, R. 1678. *De Potentia Restitutiva*. London.

Introduction to Finite Element Analysis and Design, Second Edition. Nam H. Kim, Bhavani V. Sankar, and Ashok V. Kumar.
© 2018 John Wiley & Sons Ltd. Published 2018 by John Wiley & Sons Ltd.
Companion website: www.wiley.com/go/kim/finite_element_analysis_design

explained his theory for one-dimensional objects, later his theory became the generalized Hooke's law that relates the stresses and strains in three-dimensional elastic bodies.

In this chapter, we will introduce three-dimensional stresses and strains as second-order tensor quantities and study some of their properties. We will then present three-dimensional stress-strain relationship for isotropic, linear elastic materials and discuss three different types of two-dimensional solid mechanics problems. The solid mechanics boundary value problem and the principle of minimum potential energy will be presented as the basic equations to be solved to obtain the displacement field. Thereafter, we discuss various failure theories that are used while designing structures against failure.

5.2 STRESS

5.2.1 Surface Traction

Consider a solid subjected to external forces and in static equilibrium as shown in figure 5.1. We are interested in the state of stress at a point P in the interior of the solid. We cut the body into two halves by passing an imaginary plane through P. The unit vector normal to the plane is denoted by $\mathbf{n}$ [see figure 5.1(b)]. The portion of the body on the left side is in equilibrium because of the external forces $\mathbf{f}_1$, $\mathbf{f}_2$, and $\mathbf{f}_3$ and also the internal forces acting on the cut surface. "Surface traction" is defined as the internal force per unit area or the force intensity acting on the cut plane. In order to measure the intensity or traction specifically at P, we consider the force $\Delta\mathbf{F}$ acting over a small area ΔA that contains point P. Then the surface traction $\mathbf{T}^{(\mathbf{n})}$ acting at the point P is defined as

$$\mathbf{T}^{(\mathbf{n})} = \lim_{\Delta A \to 0} \frac{\Delta\mathbf{F}}{\Delta A}. \tag{5.1}$$

In eq. (5.1), the right superscript $(\mathbf{n})$ is used to denote the fact that this surface traction is defined on a plane whose normal is $\mathbf{n}$. It should be noted that at the same point P the traction vector $\mathbf{T}$ would be different on a different plane passing through P. It is clear from eq. (5.1) that the dimension of the traction vector is the same as that of pressure, or force per unit area.

Since $\mathbf{T}^{(\mathbf{n})}$ is a vector, one can resolve it into Cartesian components and write it as

$$\mathbf{T}^{(\mathbf{n})} = T_x\mathbf{i} + T_y\mathbf{j} + T_z\mathbf{k}, \tag{5.2}$$

and its magnitude can be computed from

$$\left\| \mathbf{T}^{(\mathbf{n})} \right\| = T = \sqrt{T_x^2 + T_y^2 + T_z^2}. \tag{5.3}$$

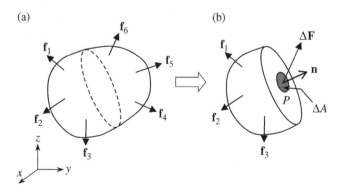

(a)   (b)

Figure 5.1 Surface traction acting on a plane at a point

EXAMPLE 5.1 *Stress in an inclined plane*

Consider a uniaxial bar with the cross-sectional area $A = 2 \times 10^{-4}$ m^2, as shown in figure 5.2. If an axial force $F = 100$ N is applied to the bar, determine the surface traction on the plane whose normal is at an angle θ from the axial direction.

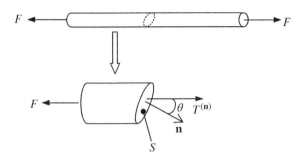

Figure 5.2 Equilibrium of a uniaxial bar under axial force

SOLUTION To simplify the analysis, let us assume that the traction on the plane is uniform, that is, the stresses are equally distributed over the cross section of the bar. In fact, this is the fundamental assumption in the analysis of bars. The force on the inclined plane S can be obtained by integrating the constant surface traction $T^{(\mathbf{n})}$ over the plane S. In this simple example, the direction of the surface traction $T^{(\mathbf{n})}$ must be opposite to that of the force F. Since the member is in static equilibrium, the integral of the surface traction must be equal to the magnitude of the force F.

$$F = \iint_S T^{(\mathbf{n})} \, dS = T \iint_S dS = T \frac{A}{\cos\theta},$$

$$\therefore T = \frac{F}{A}\cos\theta = 500 \cos\theta \; \frac{N}{m^2} = 500 \cos\theta \; \text{Pa}.$$

Note that the unit of traction is pascal (Pa or N/m^2). It is clear that the surface traction depends on the direction of the normal to the plane. ■

5.2.2 Normal Stresses and Shear Stresses

The surface traction $\mathbf{T}^{(\mathbf{n})}$ defined by eq. (5.1) does not act in general in the direction of $\mathbf{n}$, that is, $\mathbf{T}$ and $\mathbf{n}$ are not necessarily parallel to each other. Thus, we can decompose the surface traction into two components, one parallel to $\mathbf{n}$ and the other perpendicular to $\mathbf{n}$, which will lie on the plane. The component normal to the plane or parallel to $\mathbf{n}$ is called the normal stress and is denoted by σ_n. The other component, parallel to the plane or perpendicular to $\mathbf{n}$, is called the shear stress and is denoted by τ_n.

The normal stress can be obtained from the scalar product of $\mathbf{T}^{(\mathbf{n})}$ and $\mathbf{n}$, as (see figure 5.3)

$$\sigma_n = \mathbf{T}^{(\mathbf{n})} \cdot \mathbf{n}, \tag{5.4}$$

and the shear stress can be calculated from the relation

$$\tau_n = \sqrt{\left\| \mathbf{T}^{(\mathbf{n})} \right\|^2 - \sigma_n^2}. \tag{5.5}$$

The angle between $\mathbf{T}^{(\mathbf{n})}$ and $\mathbf{n}$ can be obtained from the definition of scalar product given in eq. (A.18) in the appendix.

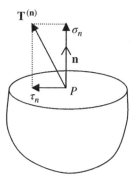

Figure 5.3 Normal and shear stresses at a point P

EXAMPLE 5.2   *Normal and shear stresses*

The surface traction at a point is $\mathbf{T}^{(\mathbf{n})} = \{3, \ 4, \ 5\}^{\mathrm{T}}$ on a plane whose normal vector is parallel to the z-axis. Calculate the normal and shear stresses on this plane. What is the angle between $\mathbf{T}^{(\mathbf{n})}$ and $\mathbf{n}$?

SOLUTION   Note that a direction parallel to the z-axis is given by $\mathbf{n} = \{0, \ 0, \ 1\}^{\mathrm{T}}$.

$$\sigma_n = \mathbf{T}^{(\mathbf{n})} \cdot \mathbf{n} = 3 \times 0 + 4 \times 0 + 5 \times 1 = 5,$$

$$\left\| \mathbf{T}^{(\mathbf{n})} \right\|^2 = 3^2 + 4^2 + 5^2 = 50,$$

$$\tau_n = \sqrt{50 - 5^2} = 5.$$

The angle θ between $\mathbf{T}^{(\mathbf{n})}$ and $\mathbf{n}$ can be calculated using the relation $\sigma_n = \mathbf{T}^{(\mathbf{n})} \cdot \mathbf{n} = T \cos \theta$. Thus,

$$\theta = \cos^{-1} \left(\sigma_n / T \right) = \cos^{-1} \left(5 / \sqrt{50} \right) = 45 \deg.$$

5.2.3   Rectangular or Cartesian Stress Components

The surface traction at a point varies depending on the direction of the normal to the plane, and therefore, one can obtain an infinite number of traction vectors $\mathbf{T}^{(\mathbf{n})}$ and corresponding normal and shear stresses for a given state of stress at a point. However, one might be interested in the maximum values of these stresses and the corresponding plane. Fortunately, the state of stress at a point can be completely characterized by defining the traction vectors on three mutually perpendicular planes passing through the point. That is, from the knowledge of $\mathbf{T}^{(\mathbf{n})}$ acting on three orthogonal planes, one can determine $\mathbf{T}^{(\mathbf{n})}$ on any arbitrary plane passing through the same point. For convenience, these planes are taken as the three planes that are normal to the x, y, and z axes.

Let us denote the traction vector on the yz–plane, which is normal to the x-axis, as $\mathbf{T}^{(x)}$. Instead of decomposing the traction vector into its normal and shear components, we will use its components parallel to the coordinate directions and denote them as $T_x^{(x)}, T_y^{(x)}$, and $T_z^{(x)}$. That is,

$$\mathbf{T}^{(x)} = T_x^{(x)} \mathbf{i} + T_y^{(x)} \mathbf{j} + T_z^{(x)} \mathbf{k}. \tag{5.6}$$

It may be noted that $T_x^{(x)}$ in eq. (5.6) is the normal stress, and $T_y^{(x)}$ and $T_z^{(x)}$ are the shear stresses in the y- and z-directions, respectively. In contemporary solid mechanics, the stress components in eq. (5.6) are denoted by σ_{xx}, τ_{xy}, and τ_{xz}, where σ_{xx} is the normal stress, and τ_{xy} and τ_{xz} are components of shear stress. In this notation, the first subscript denotes the plane on which the stress component acts—in this case, the plane normal to the *x*-axis or simply the *x*-plane—and the second subscript denotes the direction of the stress component. We can repeat this exercise by passing two more planes, normal to the *y*- and *z*-axes, respectively, through point *P*. Thus the surface tractions acting on the plane normal to *y*-plane will be τ_{yx}, σ_{yy}, and τ_{yz}. The stresses acting on the *z*-plane can be written as τ_{zx}, τ_{zy}, and σ_{zz}.

The stress components acting on the three planes can be depicted using a cube as shown in figure 5.4. It must be noted that this cube is not a physical cube and hence has no dimensions. The six faces of the cube represent the three pairs of planes normal to the coordinate axes. The top face, for example, is the + *z*-plane, and then the bottom face is the –*z*-plane, whose normal is in the –*z*-direction. Note that the three visible faces of the cube in figure 5.4 represent the three positive planes, that is, planes whose normal vectors are the positive *x*-, *y*- and *z*-axes. On these faces, all tractions are shown in the positive direction. For example, τ_{yz} is the traction on the *y*-plane acting in the positive *z*-direction. The description of all stress components is summarized in table 5.1.

Knowledge of the nine stress components is necessary in order to determine the components of the surface traction $\mathbf{T^{(n)}}$ acting on an arbitrary plane with normal **n**.

Stresses are second-order tensors, and their sign convention is different from that of regular force vectors. Stress components, in addition to the direction of the force, contain information of the surface on which they are defined. A stress component is positive when both the surface normal and the stress component are either in the positive or in the negative coordinate direction. If the surface normal is in the positive direction and the stress component is in the negative direction, then the stress component has a negative sign.

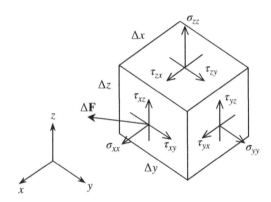

Figure 5.4 Stress components in a Cartesian coordinate system

Table 5.1 Description of stress components

Stress Component	Description
σ_{xx}	Normal stress on the *x* face in the *x* direction
σ_{yy}	Normal stress on the *y* face in the *y* direction
σ_{zz}	Normal stress on the *z* face in the *z* direction
τ_{xy}	Shear stress on the *x* face in the *y* direction
τ_{yx}	Shear stress on the *y* face in the *x* direction
τ_{yz}	Shear stress on the *y* face in the *z* direction
τ_{zy}	Shear stress on the *z* face in the *y* direction
τ_{xz}	Shear stress on the *x* face in the *z* direction
τ_{zx}	Shear stress on the *z* face in the *x* direction

Normal stress is positive when it is a tensile stress and negative when it is compressive. A shear stress acting on the positive face is positive when it is acting in the positive coordinate direction. The positive directions of all the stress components are shown in figure 5.4.

5.2.4   Traction on an Arbitrary Plane through a Point

If the components of stress at a point, say P, are known, it is possible to determine the surface traction acting on any plane passing through that point. Let $\mathbf{n}$ be the unit normal to the plane on which we want to determine the surface traction. The normal vector can be represented as,

$$\mathbf{n} = n_x\mathbf{i} + n_y\mathbf{j} + n_z\mathbf{k} = \left\{ \begin{array}{c} n_x \\ n_y \\ n_z \end{array} \right\}. \tag{5.7}$$

For convenience, we choose P as the origin of the coordinate system, as shown in figure 5.5, and consider a plane parallel to the intended plane but passing at an infinitesimally small distance h away from P. Note that the normal to the face ABC is also $\mathbf{n}$. We will calculate the tractions on this plane first and then take the limit as h tends to zero. We will consider the equilibrium of the tetrahedron PABC. If A is the area of the triangle ABC, then the areas of triangles PAB, PBC, and PAC are given by An_z, An_x, and An_y, respectively. Let $\mathbf{T}^{(\mathbf{n})} = T_x^{(\mathbf{n})}\mathbf{i} + T_y^{(\mathbf{n})}\mathbf{j} + T_z^{(\mathbf{n})}\mathbf{k}$ be the surface traction acting on the face ABC.

From the definition of surface traction in eq. (5.1), the force on the surface can be calculated by multiplying the stresses by the surface area. Since the tetrahedron should be in equilibrium, the sum of the forces acting on its surfaces should be equal to zero. Force balance in the x-direction yields

$$\sum F_x = T_x^{(\mathbf{n})}A - \sigma_{xx}An_x - \tau_{yx}An_y - \tau_{zx}An_z = 0.$$

In the above equation, we have assumed that the stresses acting on a surface are uniform. This will not be true if the size of the tetrahedron is not small. However, the tetrahedron is infinitesimally small, which is the case as h approaches zero. Dividing the above equation by A, we obtain the following relation:

$$T_x^{(\mathbf{n})} = \sigma_{xx}n_x + \tau_{yx}n_y + \tau_{zx}n_z. \tag{5.8}$$

Similarly, force balance in the y- and z-directions yield

$$T_y^{(\mathbf{n})} = \tau_{xy}n_x + \sigma_{yy}n_y + \tau_{zy}n_z,$$
$$T_z^{(\mathbf{n})} = \tau_{xz}n_x + \tau_{yz}n_y + \sigma_{zz}n_z. \tag{5.9}$$

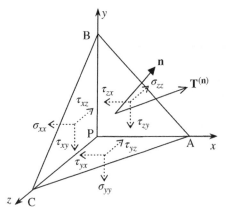

Figure 5.5 Surface traction and stress components acting on faces of an infinitesimal tetrahedron, at a given point P

From eqs. (5.8) and (5.9) it is clear that the surface traction acting on the surface whose normal is **n** can be determined if the nine stress components are available. Using matrix notation, we can write these equations as

$$\mathbf{T}^{(n)} = [\boldsymbol{\sigma}] \cdot \mathbf{n}, \tag{5.10}$$

where

$$[\boldsymbol{\sigma}] = \begin{bmatrix} \sigma_{xx} & \tau_{yx} & \tau_{zx} \\ \tau_{xy} & \sigma_{yy} & \tau_{zy} \\ \tau_{xz} & \tau_{yz} & \sigma_{zz} \end{bmatrix}. \tag{5.11}$$

$[\boldsymbol{\sigma}]$ is called the stress matrix and it completely characterizes the state of stress at a given point.

EXAMPLE 5.3 *Normal and shear stresses on a plane*

The state of stress at a particular point in the xyz coordinate system is given by the following stress matrix:

$$[\boldsymbol{\sigma}] = \begin{bmatrix} 3 & 7 & -7 \\ 7 & 4 & 0 \\ -7 & 0 & 2 \end{bmatrix}.$$

Determine the normal and shear stresses on a surface passing through the point and parallel to the plane given by the equation $4x - 4y + 2z = 2$.

SOLUTION   To determine the surface traction $\mathbf{T}^{(n)}$, it is necessary to determine the unit normal vector to the plane. From solid geometry, the normal to the plane is found to be in the direction $\mathbf{d} = \{4, -4, 2\}^T$ and $\|\mathbf{d}\| = 6$. Thus, the unit normal vector becomes

$$\mathbf{n} = \left\{ \frac{2}{3}, -\frac{2}{3}, \frac{1}{3} \right\}^T.$$

The surface traction can be obtained as

$$\mathbf{T}^{(n)} = [\boldsymbol{\sigma}] \cdot \mathbf{n} = \frac{1}{3} \begin{bmatrix} 3 & 7 & -7 \\ 7 & 4 & 0 \\ -7 & 0 & 2 \end{bmatrix} \cdot \left\{ \begin{array}{c} 2 \\ -2 \\ 1 \end{array} \right\} = \left\{ \begin{array}{c} -5 \\ 2 \\ -4 \end{array} \right\}.$$

The remaining calculations are similar to those of example 5.2:

$$\sigma_n = \mathbf{T}^{(n)} \cdot \mathbf{n} = -5 \times \frac{2}{3} - 2 \times \frac{2}{3} - 4 \times \frac{1}{3} = -6,$$

$$\|\mathbf{T}^{(n)}\|^2 = 5^2 + 2^2 + 4^2 = 45,$$

$$\tau_n = \sqrt{\|\mathbf{T}^{(n)}\|^2 - 6^2} = 3.$$

5.2.5 Symmetry of Stress Matrix and Vector Notation

The nine components of the stress matrix can be reduced to six using the symmetry property of the stress matrix. Figure 5.6 shows an infinitesimal portion ($\Delta l \ \times \ \Delta l$) of a solid with shear stresses acting on its surface. The dimension in the z-direction is taken as unity. The direction of the shear stress τ_{xy} on the surface BD is in the positive y-direction, while on the surface AC, it is in the negative y-direction. As the

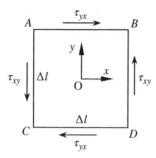

Figure 5.6 Equilibrium of a square element subjected to shear stresses

body is in static equilibrium, the sum of the moments about the z-axis must be equal to zero, which implies that the shear stresses τ_{xy} and τ_{yx} must be equal to each other as shown below.

$$\sum M_z = \Delta l \left(\tau_{xy} - \tau_{yx} \right) = 0$$
$$\Rightarrow \tau_{xy} = \tau_{yx}.$$

Similarly, we can derive the following relations:

$$\tau_{yz} = \tau_{zy},$$
$$\tau_{xz} = \tau_{zx}. \tag{5.12}$$

Thus, we need only six components to fully represent the stress at a point. In some occasions, stress at a point is written as a 6×1 pseudovector as shown below:

$$\{\boldsymbol{\sigma}\} = \left\{ \begin{array}{c} \sigma_{xx} \\ \sigma_{yy} \\ \sigma_{zz} \\ \tau_{yz} \\ \tau_{zx} \\ \tau_{xy} \end{array} \right\}. \tag{5.13}$$

Other textbooks often use a single subscript for normal stresses when the stress is written in a vector form.

5.2.6 Principal Stresses

As shown in the previous section, the normal and shear stresses acting on a plane passing through a given point in a solid change as the orientation of the plane is changed. Then a natural question is: Is there a plane on which the normal stress is the maximum? Similarly, we would also like to find the plane on which the shear stress attains a maximum. These questions are not only academic but also have significance in predicting the failure of the material at that point. In the following, we will provide some answers to the above questions without furnishing the proofs. The interested reader is referred to books on continuum mechanics, for instance, L. E. Malvern[3] or advanced solid mechanics,[4] for a more detailed treatment of the subject.

---

[3] Malvern, L. E. 1969. *Introduction to the Mechanics of a Continuous Medium*. Prentice-Hall, Englewood Cliffs, New Jersey.
[4] Boresi et al. 2003. *Advanced Mechanics of Materials*. John Wiley & Sons.

It can be shown that at every point in a solid there are at least three mutually perpendicular planes on which the normal stress attains extremum (maximum or minimum) values. On all these planes, the shear stresses vanish. Thus the traction vector $\mathbf{T^{(n)}}$ will be parallel to the normal vector $\mathbf{n}$, that is, $\mathbf{T^{(n)}} = \sigma_n \; \mathbf{n}$, on these planes. Of these three planes, one plane corresponds to the global maximum value of the normal stress and the other to the global minimum. The third plane will carry the intermediate normal stress. These special normal stresses are called the *principal stresses* at that point, the planes on which they act are the principal stress planes, and the corresponding normal vectors are the principal stress directions. The principal stresses are denoted by σ_1, σ_2, and σ_3 such that $\sigma_1 \geq \sigma_2 \geq \sigma_3$.

Based on the above observations, the principal stresses can be calculated as follows. When the normal direction to a plane is the principal direction, the surface normal and the surface traction are in the same direction ($\mathbf{T^{(n)}} \parallel \mathbf{n}$). Thus, the surface traction on a plane can be represented by the product of the normal stress σ_n and the normal vector $\mathbf{n}$, as

$$\mathbf{T^{(n)}} = \sigma_n \mathbf{n}. \tag{5.14}$$

Combining eq. (5.14) with eq. (5.10) for the surface traction, we obtain

$$[\boldsymbol{\sigma}] \cdot \mathbf{n} = \sigma_n \mathbf{n}. \tag{5.15}$$

Equation (5.15) represents the eigenvalue problem, where σ_n is the eigenvalue and $\mathbf{n}$ is the corresponding eigenvector (see section A.4 and eq. A.58). Equation (5.15) can be rearranged as

$$([\boldsymbol{\sigma}] - \sigma_n [\mathbf{I}]) \cdot \mathbf{n} = \mathbf{0}, \tag{5.16}$$

where $[\mathbf{I}]$ is a 3×3 identity matrix. In the component form, the above equation can be written as

$$\begin{bmatrix} \sigma_{xx} - \sigma_n & \tau_{yx} & \tau_{zx} \\ \tau_{xy} & \sigma_{yy} - \sigma_n & \tau_{zy} \\ \tau_{xz} & \tau_{yz} & \sigma_{zz} - \sigma_n \end{bmatrix} \begin{Bmatrix} n_x \\ n_y \\ n_z \end{Bmatrix} = \begin{Bmatrix} 0 \\ 0 \\ 0 \end{Bmatrix}. \tag{5.17}$$

Note that $\mathbf{n} = \mathbf{0}$ satisfies the above equation, which is not only a trivial solution but also physically not possible as $\|\mathbf{n}\|$ must be equal to unity. The above set of linear simultaneous equations will have a nontrivial, physically meaningful solution if and only if the determinant of the coefficient matrix is zero, that is,

$$\begin{vmatrix} \sigma_{xx} - \sigma_n & \tau_{yx} & \tau_{zx} \\ \tau_{xy} & \sigma_{yy} - \sigma_n & \tau_{zy} \\ \tau_{xz} & \tau_{yz} & \sigma_{zz} - \sigma_n \end{vmatrix} = 0. \tag{5.18}$$

Expanding this determinant, we obtain the following cubic equation in σ_n:

$$\sigma_n^3 - I_1 \sigma_n^2 + I_2 \sigma_n - I_3 = 0, \tag{5.19}$$

where

$$I_1 = \sigma_{xx} + \sigma_{yy} + \sigma_{zz},$$

$$I_2 = \begin{vmatrix} \sigma_{xx} & \tau_{xy} \\ \tau_{xy} & \sigma_{yy} \end{vmatrix} + \begin{vmatrix} \sigma_{yy} & \tau_{yz} \\ \tau_{yz} & \sigma_{zz} \end{vmatrix} + \begin{vmatrix} \sigma_{xx} & \tau_{zx} \\ \tau_{zx} & \sigma_{zz} \end{vmatrix} \tag{5.20}$$

$$= \sigma_{xx}\sigma_{yy} + \sigma_{yy}\sigma_{zz} + \sigma_{zz}\sigma_{xx} - \tau_{xy}^2 - \tau_{yz}^2 - \tau_{zx}^2,$$

$$I_3 = |[\sigma]| = \sigma_{xx}\sigma_{yy}\sigma_{zz} + 2\tau_{xy}\tau_{yz}\tau_{zx} - \sigma_{xx}\tau_{yz}^2 - \sigma_{yy}\tau_{zx}^2 - \sigma_{zz}\tau_{xy}^2.$$

In the above equation, I_1, I_2, and I_3 are the three invariants of the stress matrix $[\sigma]$, which can be shown to be independent of the coordinate system. The three roots of the cubic equation (5.19) correspond to the three principal stresses. We will denote them by σ_1, σ_2, and σ_3 in the order of $\sigma_1 \geq \sigma_2 \geq \sigma_3$. A method to solve the cubic equation is described in the appendix.

EXAMPLE 5.4   *Principal stresses*

For the stress matrix given below, determine the principal stresses.

$$[\sigma] = \begin{bmatrix} 3 & 1 & 1 \\ 1 & 0 & 2 \\ 1 & 2 & 0 \end{bmatrix}.$$

SOLUTION   Setting the determinant of the coefficient matrix to zero [see eq. (5.18)] yields

$$\begin{vmatrix} 3-\sigma_n & 1 & 1 \\ 1 & -\sigma_n & 2 \\ 1 & 2 & -\sigma_n \end{vmatrix} = 0.$$

By expanding the determinant, we obtain

$$(3-\sigma_n)(\sigma_n^2-4)-(-\sigma_n-2)+(2+\sigma_n)$$
$$= -(\sigma_n+2)(\sigma_n-1)(\sigma_n-4) = 0.$$

Three roots of the above equation are the principal stresses. They are

$$\sigma_1 = 4, \qquad \sigma_2 = 1, \qquad \sigma_3 = -2.$$

Principal Directions: Once the principal stresses have been computed, we can substitute them one at a time into eq. (5.17) to obtain the linear system of equations that have three unknowns, n_x, n_y, and n_z. Corresponding to each principal value, we will get a principal direction that will be denoted as $\mathbf{n}^1$, $\mathbf{n}^2$, and $\mathbf{n}^3$. For example, if σ_n is replaced by the first principal stress σ_1, then we obtain

$$\begin{bmatrix} \sigma_{xx}-\sigma_1 & \tau_{yx} & \tau_{zx} \\ \tau_{xy} & \sigma_{yy}-\sigma_1 & \tau_{zy} \\ \tau_{xz} & \tau_{yz} & \sigma_{zz}-\sigma_1 \end{bmatrix} \begin{Bmatrix} n_x^1 \\ n_y^1 \\ n_z^1 \end{Bmatrix} = \begin{Bmatrix} 0 \\ 0 \\ 0 \end{Bmatrix}. \tag{5.21}$$

In eq. (5.21), n_x^1, n_y^1, and n_z^1 are components of $\mathbf{n}^1$. The above equations are three linear simultaneous equations in three unknowns. However, since the determinant of the matrix is zero (i.e., the matrix is singular), they are not independent. Thus, an infinite number of solutions exist. We need one more equation to find a unique value for the principal directions $\mathbf{n}^i$. Note that $\mathbf{n}$ is a unit vector, and hence its components must satisfy the following relation:

$$\left\| \mathbf{n}^i \right\|^2 = \left(n_x^i\right)^2 + \left(n_y^i\right)^2 + \left(n_z^i\right)^2 = 1, \qquad i = 1,2,3. \tag{5.22}$$

It can be shown that the planes on which the principal stresses act are mutually perpendicular. Let us consider any two principal directions $\mathbf{n}^i$ and $\mathbf{n}^j$, with $i \neq j$. If σ_i and σ_j are the corresponding principal stresses, then they satisfy the following equations:

$$[\boldsymbol{\sigma}] \cdot \mathbf{n}^i = \sigma_i \mathbf{n}^i,$$
$$[\boldsymbol{\sigma}] \cdot \mathbf{n}^j = \sigma_j \mathbf{n}^j. \tag{5.23}$$

Multiplying the first equation by $\mathbf{n}^j$ and the second equation by $\mathbf{n}^i$, we obtain

$$\mathbf{n}^j \cdot [\boldsymbol{\sigma}] \cdot \mathbf{n}^i = \sigma_i \mathbf{n}^j \cdot \mathbf{n}^i,$$
$$\mathbf{n}^i \cdot [\boldsymbol{\sigma}] \cdot \mathbf{n}^j = \sigma_j \mathbf{n}^i \cdot \mathbf{n}^j. \tag{5.24}$$

Considering the symmetry of $[\boldsymbol{\sigma}]$, that is, $[\boldsymbol{\sigma}] = [\boldsymbol{\sigma}]^T$, and the rule for the transpose of matrix products (eq. (A.35) in the appendix), one can show that $\mathbf{n}^j \cdot [\boldsymbol{\sigma}] \cdot \mathbf{n}^i = \mathbf{n}^i \cdot [\boldsymbol{\sigma}] \cdot \mathbf{n}^j$. Then subtracting the first equation from the second in eq. (5.24), we obtain

$$(\sigma_i - \sigma_j) \mathbf{n}^i \cdot \mathbf{n}^j = 0. \tag{5.25}$$

This implies that if the principal stresses are distinct, that is, $\sigma_i \neq \sigma_j$, then

$$\mathbf{n}^i \cdot \mathbf{n}^j = 0, \tag{5.26}$$

which means that $\mathbf{n}^i$ and $\mathbf{n}^j$ are orthogonal. The three planes on which the principal stresses act are mutually perpendicular.

There are three different possibilities for principal stresses and directions:

(a) σ_1, σ_2, and σ_3 are distinct $\Rightarrow$ principal stress directions are three unique mutually orthogonal unit vectors.

(b) $\sigma_1 = \sigma_2 \neq \sigma_3 \Rightarrow \mathbf{n}^3$ is a unique principal stress direction, and any two orthogonal directions on the plane that is perpendicular to $\mathbf{n}^3$ are the other principal directions.

(c) $\sigma_1 = \sigma_2 = \sigma_3 \Rightarrow$ any three orthogonal directions are principal stress directions. This state of stress is called hydrostatic or isotropic state of stress.

EXAMPLE 5.5 *Principal directions*

Calculate the principal direction corresponding to $\sigma_3 = -2$ in example 5.4.

SOLUTION By substituting $\sigma_n = -2$ into eq. (5.17), we obtain the following simultaneous equations:

$$5n_x + n_y + n_z = 0,$$
$$n_x + 2n_y + 2n_z = 0,$$
$$n_x + 2n_y + 2n_z = 0.$$

We note that the three equations are not independent. In fact, the second and third equations are identical. From the first two equations we obtain the following ratios between components:

$$n_x : n_y : n_z = 0 : -1 : 1.$$

A unique solution can be obtained using eq. (5.22) as

$$\mathbf{n}^{(3)} = \frac{1}{\sqrt{2}} \left\{ \begin{array}{c} 0 \\ -1 \\ 1 \end{array} \right\}.$$

5.2.7   Transformation of Stress

Let the state of stress at a point be given by the stress matrix $[\boldsymbol{\sigma}]_{xyz}$, where the components are expressed with reference to the xyz coordinates as shown in figure 5.7. The question is what the components will look like in a different coordinate system, say $x'y'z'$, or how to determine $[\boldsymbol{\sigma}]_{x'y'z'}$. It will be shown that the stress matrix $[\boldsymbol{\sigma}]_{x'y'z'}$ can be obtained by rotating the stress matrix $[\boldsymbol{\sigma}]_{xyz}$ using a transformation matrix of direction cosines.

Let us define the $x'y'z'$ coordinate system whose unit vectors $\mathbf{b}^1$, $\mathbf{b}^2$, and $\mathbf{b}^3$, respectively, are given in the xyz coordinate system as

$$\mathbf{b}^1 = \left\{ \begin{array}{c} b_1^1 \\ b_2^1 \\ b_3^1 \end{array} \right\}, \quad \mathbf{b}^2 = \left\{ \begin{array}{c} b_1^2 \\ b_2^2 \\ b_3^2 \end{array} \right\}, \quad \mathbf{b}^3 = \left\{ \begin{array}{c} b_1^3 \\ b_2^3 \\ b_3^3 \end{array} \right\}.$$

For example, $\mathbf{b}^1 = \{1, 0, 0\}^T$ in the $x'y'z'$ coordinate system, while $\mathbf{b}^1 = \left\{ b_1^1, b_2^1, b_3^1 \right\}^T$ in the xyz coordinates. Using these vectors, let us define a transformation matrix as

$$[\mathbf{N}] = \begin{bmatrix} \mathbf{b}^1 & \mathbf{b}^2 & \mathbf{b}^3 \end{bmatrix} = \begin{bmatrix} b_1^1 & b_1^2 & b_1^3 \\ b_2^1 & b_2^2 & b_2^3 \\ b_3^1 & b_3^2 & b_3^3 \end{bmatrix}. \tag{5.27}$$

Matrix $[\mathbf{N}]$ transforms a vector in the $x'y'z'$ coordinates into a vector in the xyz coordinates, while its transpose $[\mathbf{N}]^T$ transforms a vector in the xyz coordinates into a vector in the $x'y'z'$ coordinates. For example, the unit vector $\mathbf{b}_{x'y'z'}^1 = \{1,0,0\}^T$ in $x'y'z'$ coordinates can be represented in the xyz coordinates as

$$\mathbf{b}_{xyz}^1 = [\mathbf{N}]^T \cdot \mathbf{b}_{x'y'z'}^1 = \left\{ \begin{array}{c} b_1^1 \\ b_2^1 \\ b_3^1 \end{array} \right\}. \tag{5.28}$$

The transformation of a stress matrix is more complicated than the transformation of a vector. We will perform the transformation in two steps. First, we will determine the traction vectors on the three planes normal to the x'-, y'- and z'-axes. This can be accomplished by multiplying the stress matrix and the corresponding direction cosine vector as described in eq. (5.10). The three sets of traction vectors are written as columns of a square matrix as shown below:

$$\begin{bmatrix} \mathbf{T}^{(\mathbf{b}^1)} & \mathbf{T}^{(\mathbf{b}^2)} & \mathbf{T}^{(\mathbf{b}^3)} \end{bmatrix}_{xyz} = [\boldsymbol{\sigma}]_{xyz} \begin{bmatrix} \mathbf{b}^1 & \mathbf{b}^2 & \mathbf{b}^3 \end{bmatrix} = [\boldsymbol{\sigma}]_{xyz}[\mathbf{N}]. \tag{5.29}$$

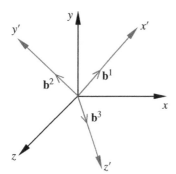

Figure 5.7 Coordinate transformation of stress

Equation (5.29) represents the three surface traction vectors on planes perpendicular to $\mathbf{b}^1$, $\mathbf{b}^2$, and $\mathbf{b}^3$ in the xyz coordinates system. In the next step, we would like to transform the traction vectors to the $x'y'z'$ coordinate system. This transformation can be accomplished by using eq. (5.28). It may be recognized that the traction vectors $\mathbf{T}^{(\mathbf{b}^1)}$, $\mathbf{T}^{(\mathbf{b}^2)}$, and $\mathbf{T}^{(\mathbf{b}^3)}$ represented in the $x'y'z'$ coordinate system will be the transformed stress matrix. Thus, the stress matrix in the new coordinate system can be obtained by

$$\boxed{[\boldsymbol{\sigma}]_{x'y'z'} = [\mathbf{N}]^{\mathrm{T}}[\boldsymbol{\sigma}]_{xyz}[\mathbf{N}].}$$

(5.30)

EXAMPLE 5.6 *Coordinate transformation*

The state of stress at a point in the xyz coordinates is

$$[\boldsymbol{\sigma}] = \begin{bmatrix} 2 & 1 & 0 \\ 1 & 2 & 0 \\ 0 & 0 & 2 \end{bmatrix}.$$

Determine the stress matrix relative to the $x'y'z'$ coordinates, which is obtained by rotating the xyz coordinates by $45°$ about the z-axis as shown in figure 5.8.

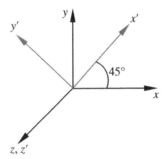

Figure 5.8 Coordinate transformation of example 5.6

SOLUTION Since the z'-direction is the same as the z-direction, the coordinate transformation is basically a two-dimensional rotation. Let $a = \cos 45° = \sin 45°$, then the transformation matrix becomes

$$[\mathbf{N}] = \begin{bmatrix} a & -a & 0 \\ a & a & 0 \\ 0 & 0 & 1 \end{bmatrix}.$$

Using eq. (5.30) the stress matrix in the transformed coordinates becomes

$$[\mathbf{N}]^{\mathrm{T}}[\boldsymbol{\sigma}][\mathbf{N}] = \begin{bmatrix} a & a & 0 \\ -a & a & 0 \\ 0 & 0 & 1 \end{bmatrix} \begin{bmatrix} 2 & 1 & 0 \\ 1 & 2 & 0 \\ 0 & 0 & 2 \end{bmatrix} \begin{bmatrix} a & -a & 0 \\ a & a & 0 \\ 0 & 0 & 1 \end{bmatrix} = \begin{bmatrix} 6a^2 & 0 & 0 \\ 0 & 2a^2 & 0 \\ 0 & 0 & 2 \end{bmatrix} = \begin{bmatrix} 3 & 0 & 0 \\ 0 & 1 & 0 \\ 0 & 0 & 2 \end{bmatrix}.$$

Note that the stress matrix is diagonal after the transformation, which means that the $x'y'z'$ coordinates are the principal stress directions and the diagonal terms are the principal stresses. ■

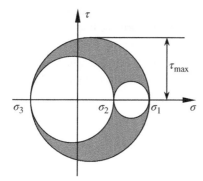

Figure 5.9 Maximum shear stress

5.2.8   Maximum Shear Stress

Maximum shear stress plays an important role in the failure of ductile materials. Let σ_1, σ_2, and σ_3 be the three principal stresses at a point such that $\sigma_1 \geq \sigma_2 \geq \sigma_3$. The stress state at this point can be described using three Mohr's circles as shown in figure 5.9.

In the diagram in figure 5.9, any point (σ, τ) located in the shaded area represents the normal and shear stresses, σ_n and τ_n, on a plane through the point. One can note that the maximum shear stress is given by the radius of the largest Mohr's circle as

$$\tau_{\max} = \frac{\sigma_1 - \sigma_3}{2}. \tag{5.31}$$

The plane on which the shear stress attains maximum bisects the first and third principal stress planes. The normal stress on this plane is given by

$$\sigma_n = \frac{\sigma_1 + \sigma_3}{2}. \tag{5.32}$$

5.3   STRAIN

When a solid is subjected to forces, it deforms. A quantitative measure of the deformation is provided by strains. Imagine an infinitesimal line segment in an arbitrary direction at a point in a solid. After deformation, the length of the line segment changes. Strain, specifically the normal strain, in the original direction of the line segment is defined as the change in length divided by the original length. However, this strain will be different in different directions at the same point. In the following, we develop the concept of strain in a three-dimensional body.

5.3.1   Definition of Strains

Figure 5.10 shows a body before and after deformation. Let the points P, Q, and R in the undeformed body moves to P', Q', and R', respectively, after deformation. The displacement of P can be represented by three displacement components, u, v, and w in the x-, y-, and z-directions. Thus the coordinates of P' are $(x + u, y + v, z + w)$. The functions $u(x,y,z)$, $v(x,y,z)$, and $w(x,y,z)$ are components of a vector field that

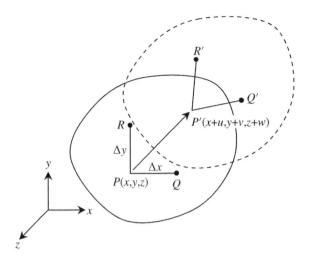

Figure 5.10 Deformation of line segments

is referred to as the deformation field or the displacement field. The displacements of point Q will be slightly different from that of P. They can be written as

$$u_Q = u + \frac{\partial u}{\partial x}\Delta x,$$

$$v_Q = v + \frac{\partial v}{\partial x}\Delta x, \qquad (5.33)$$

$$w_Q = w + \frac{\partial w}{\partial x}\Delta x.$$

Similarly, displacements of point R are

$$u_R = u + \frac{\partial u}{\partial y}\Delta y,$$

$$v_R = v + \frac{\partial v}{\partial y}\Delta y, \qquad (5.34)$$

$$w_R = w + \frac{\partial w}{\partial y}\Delta y.$$

The coordinates of P, Q, and R before and after deformation are as follows:

$$P : (x,y,z)$$
$$Q : (x+\Delta x, y, z)$$
$$R : (x, y+\Delta y, z)$$
$$P' : (x+u_P, y+v_P, z+w_P) = (x+u, y+v, z+w)$$
$$Q' : (x+\Delta x+u_Q, y+v_Q, z+w_Q) \qquad (5.35)$$
$$= \left(x+\Delta x+u+\frac{\partial u}{\partial x}\Delta x, y+v+\frac{\partial v}{\partial x}\Delta x, z+w+\frac{\partial w}{\partial x}\Delta x \right)$$
$$R' : (x+u_R, y+\Delta y+v_R, z+w_R)$$
$$= \left(x+u+\frac{\partial u}{\partial y}\Delta y, y+\Delta y+v+\frac{\partial v}{\partial y}\Delta y, z+w+\frac{\partial w}{\partial y}\Delta y \right)$$

Length of the line segment $P'Q'$ can be calculated as

$$P'Q' = \sqrt{(x_{P'} - x_{Q'})^2 + (y_{P'} - y_{Q'})^2 + (z_{P'} - z_{Q'})^2}. \tag{5.36}$$

Substituting for the coordinates of P' and Q' from eq. (5.35) we obtain

$$
\begin{aligned}
P'Q' &= \Delta x \sqrt{\left(1 + \frac{\partial u}{\partial x}\right)^2 + \left(\frac{\partial v}{\partial x}\right)^2 + \left(\frac{\partial w}{\partial x}\right)^2} \\
&= \Delta x \left(1 + 2\frac{\partial u}{\partial x} + \left(\frac{\partial u}{\partial x}\right)^2 + \left(\frac{\partial v}{\partial x}\right)^2 + \left(\frac{\partial w}{\partial x}\right)^2\right)^{1/2} \\
&\approx \Delta x \left(1 + \frac{\partial u}{\partial x} + \frac{1}{2}\left(\frac{\partial u}{\partial x}\right)^2 + \frac{1}{2}\left(\frac{\partial v}{\partial x}\right)^2 + \frac{1}{2}\left(\frac{\partial w}{\partial x}\right)^2\right).
\end{aligned} \tag{5.37}
$$

It may be noted that we have used a two-term binomial expansion in deriving an approximate expression for the change in length. In this book, we will consider only small deformations such that all deformation gradients are very small compared to unity, that is, $\partial u/\partial x \ll 1$, $\partial v/\partial x \ll 1$. Then we can neglect the higher-order terms in eq. (5.37) to obtain

$$P'Q' \approx \Delta x \left(1 + \frac{\partial u}{\partial x}\right). \tag{5.38}$$

Now we invoke the definition of normal strain as the ratio of change in length to original length to derive the expression for strain as

$$\varepsilon_{xx} = \frac{P'Q' - PQ}{PQ} = \frac{\partial u}{\partial x}. \tag{5.39}$$

Thus, the normal strain ε_{xx} at a point can be defined as the change in length per unit length of an infinitesimally long line segment originally parallel to the x-axis. Similarly, we can derive normal strains in the y- and z-directions as

$$\varepsilon_{yy} = \frac{\partial v}{\partial y}, \quad \varepsilon_{zz} = \frac{\partial w}{\partial z}. \tag{5.40}$$

The engineering shear strain, say γ_{xy}, is defined as the change in the angle between a pair of infinitesimal line segments that were originally parallel to the x- and y-axes. From figure 5.10, the angle between PQ and $P'Q'$ can be derived as

$$\theta_1 = \frac{y_{Q'} - y_Q}{\Delta x} = \frac{\partial v}{\partial x}. \tag{5.41}$$

Similarly, the angle between PR and $P'R'$ is

$$\theta_2 = \frac{x_{R'} - x_R}{\Delta y} = \frac{\partial u}{\partial y}. \tag{5.42}$$

Using the aforementioned definition of shear strain,

$$\gamma_{xy} = \theta_1 + \theta_2 = \frac{\partial u}{\partial y} + \frac{\partial v}{\partial x}. \tag{5.43}$$

Similarly, we can derive shear strains in the yz- and zx-planes as

$$\gamma_{yz} = \frac{\partial v}{\partial z} + \frac{\partial w}{\partial y},$$

$$\gamma_{zx} = \frac{\partial w}{\partial x} + \frac{\partial u}{\partial z}. \tag{5.44}$$

The shear strains, γ_{xy}, γ_{yz}, and γ_{zx}, are called engineering shear strains. We define tensorial shear strains as

$$\varepsilon_{xy} = \frac{1}{2}\left(\frac{\partial u}{\partial y} + \frac{\partial v}{\partial x}\right),$$

$$\varepsilon_{yz} = \frac{1}{2}\left(\frac{\partial v}{\partial z} + \frac{\partial w}{\partial y}\right), \tag{5.45}$$

$$\varepsilon_{zx} = \frac{1}{2}\left(\frac{\partial w}{\partial x} + \frac{\partial u}{\partial z}\right).$$

It may be noted that the tensorial shear strains are one half of the corresponding engineering shear strains. It can be shown that the normal strains and the tensorial shear strains transform from one coordinate system to another following tensor transformation rules.

In the general three-dimensional case, the strain matrix is defined as

$$[\varepsilon] = \begin{bmatrix} \varepsilon_{xx} & \varepsilon_{xy} & \varepsilon_{xz} \\ \varepsilon_{yx} & \varepsilon_{yy} & \varepsilon_{yz} \\ \varepsilon_{zx} & \varepsilon_{zy} & \varepsilon_{zz} \end{bmatrix}. \tag{5.46}$$

As is clear from the definition in eq. (5.45), the strain matrix is symmetric. Like the stress vector, the symmetric strain matrix can be represented as a pseudovector

$$\{\varepsilon\} = \begin{Bmatrix} \varepsilon_{xx} \\ \varepsilon_{yy} \\ \varepsilon_{zz} \\ \gamma_{yz} \\ \gamma_{zx} \\ \gamma_{xy} \end{Bmatrix}. \tag{5.47}$$

where γ_{yz}, γ_{zx}, and γ_{xy} are used instead of $\varepsilon_{yz}, \varepsilon_{zx}$, and ε_{xy}. The six components of strain completely define the deformation at a point. The normal strain in any arbitrary direction at that point and also the shear strain in any arbitrary plane passing through the point can be calculated using the above strain components, as explained in the following section. Strain is also a tensor and therefore it has properties similar to a stress tensor. For example, the transformation of strain, principal strains, and corresponding principal strain directions can be determined using the procedures we described for stresses.

5.3.2 Transformation of Strain

Let the state of strain at a point be given by the strain matrix $[\varepsilon]_{xyz}$, where the components are expressed in the xyz coordinates shown in figure 5.7. As for the case of stresses, the question again is: What are the components of strain at the same point in a different coordinate system, $x'y'z'$ (i.e., $[\varepsilon]_{x'y'z'}$)?

Let $\mathbf{b}^1$, $\mathbf{b}^2$, and $\mathbf{b}^3$ be unit vectors along the axes x', y', and z', respectively, represented in xyz coordinates. The strain matrix expressed with respect to the $x'y'z'$ coordinates system is

$$[\varepsilon]_{x'y'z'} = [\mathbf{N}]^{\mathrm{T}}[\varepsilon]_{xyz}[\mathbf{N}], \tag{5.48}$$

which is the same transformation used for stresses in eq. (5.30). It should be emphasized that tensorial shear strains must be used when using the transformation in eq. (5.48).

Let us consider the term $\varepsilon_{x'x'}$ in eq. (5.48). It can be written as

$$\varepsilon_{x'x'} = \left[\mathbf{b}^1\right]^{\mathrm{T}}[\varepsilon]_{xyz}\left[\mathbf{b}^1\right]. \tag{5.49}$$

We note that the normal strain in the x'-direction depends only on the strain tensor and the direction cosines of the x'-axis. In fact, this relation can be generalized to any arbitrary direction $\mathbf{n}$ and written as

$$\begin{aligned}
\varepsilon_{nn} &= \{\mathbf{n}\}^{\mathrm{T}}[\varepsilon]_{xyz}\{\mathbf{n}\} \\
&= \varepsilon_{xx}n_x^2 + \varepsilon_{yy}n_y^2 + \varepsilon_{zz}n_z^2 + 2\varepsilon_{xy}n_xn_y + 2\varepsilon_{yz}n_yn_z + 2\varepsilon_{zx}n_zn_x \\
&= \varepsilon_{xx}n_x^2 + \varepsilon_{yy}n_y^2 + \varepsilon_{zz}n_z^2 + \gamma_{xy}n_xn_y + \gamma_{yz}n_yn_z + \gamma_{zx}n_zn_x.
\end{aligned} \tag{5.50}$$

Similarly, one can derive the shear strain in a plane containing two mutually perpendicular vectors $\mathbf{m}$ and $\mathbf{n}$ as

$$\begin{aligned}
\varepsilon_{mn} &= \{\mathbf{m}\}^{\mathrm{T}}[\varepsilon]_{xyz}\{\mathbf{n}\} \\
&= \varepsilon_{xx}m_xn_x + \varepsilon_{yy}m_yn_y + \varepsilon_{zz}m_zn_z + \varepsilon_{xy}\left(m_xn_y + m_yn_x\right) \\
&\quad + \varepsilon_{yz}\left(m_yn_z + m_zn_y\right) + \varepsilon_{zx}\left(m_zn_x + m_yn_z\right).
\end{aligned} \tag{5.51}$$

5.3.3 Principal Strains

Strains, like stresses, are second-order tensors, and therefore one can compute eigenvalues and eigenvectors for a given strain matrix. The eigenvalues are the three principal strains, and the eigenvectors are the corresponding principal strain directions. The principal strain directions represent the directions in which the normal strain ε_{nn} in eq. (5.50) takes an extremum value. The maximum and minimum of the three principal strains are the global maximum and minimum, respectively. The three principal planes intersect along the principal strain directions. The shear strain vanishes on the principal strain planes. That is, a pair of infinitesimally small line segments parallel to any two principal directions remain perpendicular to each other after deformation, or, equivalently, the angle between them does not change after deformation. In other words, the shear strain on this plane is equal to zero.

The physical meaning of strains can be explained as follows. Consider an infinitesimal cube of size $\Delta a \times \Delta a \times \Delta a$ whose sides are parallel to the coordinate axes before deformation. After deformation, the rectangular parallelepiped will become an oblique parallelepiped of size $\Delta a(1+\varepsilon_{xx}) \times \Delta a\,(1+\varepsilon_{yy}) \times \Delta a(1+\varepsilon_{zz})$. The angles between the edges of the parallelepiped will reduce by γ_{xy}, γ_{yz}, and γ_{zx} (see figure 5.11). If the same parallelepiped is oriented such that its edges are parallel to the principal strain directions, then after deformation the parallelepiped will remain as a rectangular parallelepiped, and its dimensions will be $\Delta a(1+\varepsilon_1) \times \Delta a(1+\varepsilon_2) \times \Delta a(1+\varepsilon_3)$. Since the shear strains in the principal strain planes are zero, the angle between the edges will remain as 90°.

The principal strain is the extensional strain on a plane, where there is no shear strain. Principal strains and their directions are the eigenvalues and eigenvectors of the following eigenvalue problem:

$$[\boldsymbol{\varepsilon}]\cdot\mathbf{n} = \varepsilon_n\mathbf{n}, \tag{5.52}$$

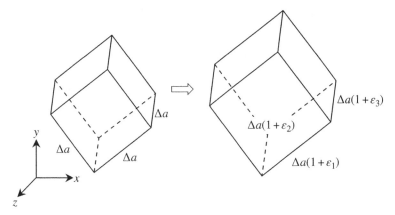

Figure 5.11 Deformation in the principal directions

where $\mathbf{n}$ is the eigenvector of matrix $[\varepsilon]$. The above eigenvalue problem has three solutions $\varepsilon_n = \varepsilon_1, \varepsilon_2,$ and ε_3, which are the principal strains.

If the principal strains ε_1, ε_2, and ε_3 are known, then the maximum engineering shear strain $\gamma_{\max}$ can be computed as

$$\frac{\gamma_{\max}}{2} = \frac{\varepsilon_1 - \varepsilon_3}{2}, \tag{5.53}$$

where ε_1 and ε_3 are the maximum and minimum principal strains, respectively.

5.3.4 Stress vs. Strain

Stresses and strains defined in the previous two sections are second-order tensors and hence share some common properties as shown in table 5.2. When the material is isotropic, which means the material properties are the same in all directions or independent of the coordinates system, the principal stress directions and principal strain directions coincide. If the material is anisotropic, the principal stress and strain directions are in general different. In this textbook, we will focus on isotropic materials only.

Table 5.2 Comparison of stress and strain

$[\boldsymbol{\sigma}]$ is a symmetric 3×3 matrix	$[\boldsymbol{\varepsilon}]$ is a symmetric 3×3 matrix
Normal stress in the direction $\mathbf{n}$ is $\sigma_{nn} = \mathbf{n}^{\mathrm{T}}[\boldsymbol{\sigma}]\mathbf{n}$	Normal strain in the direction $\mathbf{n}$ is $\varepsilon_{nn} = \mathbf{n}^{\mathrm{T}}[\boldsymbol{\varepsilon}]\mathbf{n}$
Shear stress in a plane containing two mutually perpendicular unit vectors $\mathbf{m}$ and $\mathbf{n}$ is $\tau_{mn} = \mathbf{m}^{\mathrm{T}}[\boldsymbol{\sigma}]\mathbf{n}$	Shear strain in a plane containing two mutually perpendicular unit vectors $\mathbf{m}$ and $\mathbf{n}$ is $\varepsilon_{mn} = \frac{\gamma_{mn}}{2} = \mathbf{m}^{\mathrm{T}}[\boldsymbol{\varepsilon}]\mathbf{n}$
Transformation of stresses $[\boldsymbol{\sigma}]_{x'y'z'} = [\mathbf{N}]^{\mathrm{T}}[\boldsymbol{\sigma}]_{xyz}[\mathbf{N}]$	Transformation of strains $[\boldsymbol{\varepsilon}]_{x'y'z'} = [\mathbf{N}]^{\mathrm{T}}[\boldsymbol{\varepsilon}]_{xyz}[\mathbf{N}]$
Three mutually perpendicular principal directions and principal stresses can be computed as eigenvectors and eigenvalues of the stress matrix: $[\boldsymbol{\sigma}]\mathbf{n} = \sigma_n \mathbf{n}$	Three mutually perpendicular principal directions and principal strains can be computed as eigenvectors and eigenvalues of the strain matrix: $[\boldsymbol{\varepsilon}]\mathbf{n} = \varepsilon_n \mathbf{n}$

5.4   STRESS–STRAIN RELATIONSHIP

Finding a relationship between the loads acting on a structure and its deflection has been of great interest to scientists since the seventeenth century.[5] Robert Hooke, Jacob Bernoulli, and Leonard Euler are some of the pioneers who developed various theories to explain the bending of beams and stretching of bars. Forces applied to a solid create stresses within the body in order to satisfy equilibrium. These stresses also cause deformation or strains. Accumulation of strains over the volume of a body manifests as deflections or gross deformation of the body. Hence, it is clear that a fundamental knowledge of the relationship between stresses and strains is necessary in order to understand the global behavior. Navier tried to explain deformations considering the forces between neighboring particles in a body, as they tend to separate and come closer. Later this approach was abandoned in favor of Cauchy's stresses and strains. Robert Hooke was the first one to propose the linear uniaxial stress–strain relationship, which states that the stress is proportional to strain. Later the general relationship between the six components of strains and stresses called the generalized Hooke's law was developed. The generalized Hooke's law states that each component of stress is a linear combination of strains. It should be mentioned that stress–strain relationships are called phenomenological models or theories as they are based on the commonly observed behavior of materials and verified by experiments. Only recently, with the advancement of computers and computational techniques, behavior of materials based on first principles or fundamental atomistic behavior is being developed. This new field of study is called computational materials and involves techniques such as molecular dynamics simulations and multiscale modeling. Stress–strain relationships are also called constitutive relationships as they describe the constitution of the material.

A cylindrical test specimen is loaded along its axis as shown figure 5.12. This type of loading ensures that the specimen is subjected to a uniaxial state of stress. If the stress–strain relation of the uniaxial tension test in figure 5.12 is plotted, then a typical ductile material may show a behavior as in figure 5.13. The explanation of terms in the figure is summarized in table 5.3.

After the material yields, the shape of the structure permanently changes. Hence, many engineering structures are designed such that the maximum stress is smaller than the yield stress of the material. Under this range of the stress, the stress–strain relation can be approximated by a linear relation.

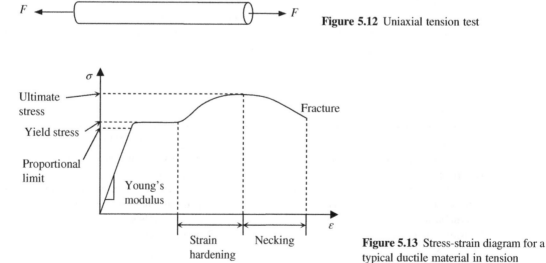

Figure 5.12 Uniaxial tension test

Figure 5.13 Stress-strain diagram for a typical ductile material in tension

---

[5] Timoshenko, S. P. 1983. *History of Strength of Materials*. Dover Publications, Inc., New York.

Table 5.3   Explanations of uniaxial tension test

Terms	Explanation
Proportional limit	The greatest stress for which the stress is still proportional to the strain
Elastic limit	The greatest stress that can be applied without resulting in any permanent strain upon unloading
Young's Modulus	Slope of the linear portion of the stress-strain curve
Yield stress	The stress required to produce 0.2% plastic strain
Strain hardening	A region where more stress is required to further plastically deform the material
Ultimate stress	The maximum stress the material can resist
Necking	Cross section of the specimen reduces during deformation

5.4.1   Linear Elastic Relationship (Generalized Hooke's Law)

The one-dimensional stress–strain relationship described in the previous section can be extended to the three-dimensional state of stress. When the stress–strain relationship is linear, it can be written as

$$\{\boldsymbol{\sigma}\} = [\mathbf{C}] \cdot \{\varepsilon\}, \tag{5.54}$$

where

$$\{\boldsymbol{\sigma}\} = \begin{Bmatrix} \sigma_{xx} \\ \sigma_{yy} \\ \sigma_{zz} \\ \tau_{yz} \\ \tau_{zx} \\ \tau_{xy} \end{Bmatrix}, \quad [\mathbf{C}] = \begin{bmatrix} C_{11} & C_{12} & C_{13} & C_{14} & C_{15} & C_{16} \\ C_{21} & C_{22} & C_{23} & C_{24} & C_{25} & C_{26} \\ C_{31} & C_{32} & C_{33} & C_{34} & C_{35} & C_{36} \\ C_{41} & C_{42} & C_{43} & C_{44} & C_{45} & C_{46} \\ C_{51} & C_{52} & C_{53} & C_{54} & C_{55} & C_{56} \\ C_{61} & C_{62} & C_{63} & C_{64} & C_{65} & C_{66} \end{bmatrix}, \quad \text{and} \quad \{\varepsilon\} = \begin{Bmatrix} \varepsilon_{xx} \\ \varepsilon_{yy} \\ \varepsilon_{zz} \\ \gamma_{yz} \\ \gamma_{zx} \\ \gamma_{xy} \end{Bmatrix}.$$

Matrix $[\mathbf{C}]$ is called the stress–strain matrix, or elasticity matrix. It can be shown that $[\mathbf{C}]$ must be a symmetric matrix and hence the number of independent coefficients or elastic constants for an anisotropic material is only 21. Many composite materials, naturally occurring composites such as wood or bone, and man-made materials such as fiber-reinforced composites can be modeled as an orthotropic material with nine independent elastic constants. Some composites are transversely isotropic and require only five independent elastic constants. Most materials are isotropic, and for such materials, the 21 constants in the symmetric matrix $[\mathbf{C}]$ can be expressed in terms of two independent constants called engineering elastic constants.

For isotropic materials, the relation between stress and strain can be written as:

$$\begin{Bmatrix} \varepsilon_{xx} \\ \varepsilon_{yy} \\ \varepsilon_{zz} \end{Bmatrix} = \frac{1}{E} \begin{bmatrix} 1 & -\nu & -\nu \\ -\nu & 1 & -\nu \\ -\nu & -\nu & 1 \end{bmatrix} \begin{Bmatrix} \sigma_{xx} \\ \sigma_{yy} \\ \sigma_{zz} \end{Bmatrix},$$
$$\gamma_{xy} = \frac{\tau_{xy}}{G}, \quad \gamma_{yz} = \frac{\tau_{yz}}{G}, \quad \gamma_{zx} = \frac{\tau_{zx}}{G}. \tag{5.55}$$

where E, Young's modulus, and ν, Poisson's ratio, are the two independent elastic constants, and G is the shear modulus defined by

$$G = \frac{E}{2(1+\nu)}. \tag{5.56}$$

Note that there are only two independent constants for isotropic materials.

Alternately, stresses can be written as a function of strains by inverting the relations in eq. (5.55), as

$$
\begin{Bmatrix} \sigma_{xx} \\ \sigma_{yy} \\ \sigma_{zz} \end{Bmatrix} = \frac{E}{(1+\nu)(1-2\nu)} \begin{bmatrix} 1-\nu & \nu & \nu \\ \nu & 1-\nu & \nu \\ \nu & \nu & 1-\nu \end{bmatrix} \begin{Bmatrix} \varepsilon_{xx} \\ \varepsilon_{yy} \\ \varepsilon_{zz} \end{Bmatrix},
$$

$$
\tau_{xy} = G\gamma_{xy}, \quad \tau_{yz} = G\gamma_{yz}, \quad \tau_{zx} = G\gamma_{zx}.
$$

(5.57)

The elasticity matrix [**C**] in eq. (5.54) can be written as

$$
[\mathbf{C}] = \frac{E}{(1+\nu)(1-2\nu)} \begin{bmatrix} 1-\nu & \nu & \nu & 0 & 0 & 0 \\ \nu & 1-\nu & \nu & 0 & 0 & 0 \\ \nu & \nu & 1-\nu & 0 & 0 & 0 \\ 0 & 0 & 0 & \frac{1}{2}-\nu & 0 & 0 \\ 0 & 0 & 0 & 0 & \frac{1}{2}-\nu & 0 \\ 0 & 0 & 0 & 0 & 0 & \frac{1}{2}-\nu \end{bmatrix}.
$$

(5.58)

EXAMPLE 5.7 *Stress–strain relationship*

The stress at a point in a body is given as

$$
[\boldsymbol{\sigma}] = \begin{bmatrix} 5 & 3 & 2 \\ 3 & -1 & 0 \\ 2 & 0 & 4 \end{bmatrix} \times 10^3 \quad \text{psi}.
$$

Determine the strain components for the isotropic material when $E = 10 \times 10^6$ psi, and $\nu = 0.3$.

SOLUTION   From eq. (5.55),

$$
\varepsilon_{xx} = \frac{1}{10 \times 10^6}[5 - 0.3(-1 + 4)] \times 10^3 = 4.1 \times 10^{-4},
$$

$$
\varepsilon_{yy} = \frac{1}{10 \times 10^6}[-1 - 0.3(5 + 4)] \times 10^3 = -3.7 \times 10^{-4},
$$

$$
\varepsilon_{zz} = \frac{1}{10 \times 10^6}[4 - 0.3(5 - 1)] \times 10^3 = 2.8 \times 10^{-4},
$$

$$
\gamma_{xy} = \frac{2(1 + 0.3)}{10 \times 10^6}3000 = 7.8 \times 10^{-4},
$$

$$
\gamma_{yz} = \frac{2(1 + 0.3)}{10 \times 10^6}0 = 0,
$$

$$
\gamma_{xz} = \frac{2(1 + 0.3)}{10 \times 10^6}2000 = 5.2 \times 10^{-4}.
$$

5.4.2 Simplified Laws for Two-Dimensional Analysis

The general three-dimensional stress–strain relationship in eq. (5.54) can be simplified for certain special situations that often occur in practice. The two-dimensional stress–strain relationship can be categorized into three cases: plane stress, plane strain, and axisymmetric.

Most practical structures consist of thin plate-like components in order to be efficient. Assume a thin plate that is parallel to the xy-plane. If we assume that the top and bottom surfaces of the plate are not subjected to any significant forces, that is, the plate is subjected to forces in its plane only, in the x- and y-directions, then the transverse stresses (stresses with a z subscript) vanish on the top and bottom surfaces, that is, $\sigma_{zz} = \tau_{xz} = \tau_{yz} = 0$ on the top and bottom surfaces. If the thickness is much smaller compared to the lateral dimensions of the plate, then we can assume that the aforementioned transverse stresses are approximately zero through the entire thickness. Then the plate is said to be in a state of plane stress parallel to the xy-plane or normal to the z-axis.

In order to derive the stress–strain relations for the state of plane stress we start with eq. (5.55). We set $\sigma_{zz} = \tau_{xz} = \tau_{yz} = 0$ on the right-hand side of the equations to obtain

$$
\begin{Bmatrix} \varepsilon_{xx} \\ \varepsilon_{yy} \end{Bmatrix} = \frac{1}{E} \begin{bmatrix} 1 & -\nu \\ -\nu & 1 \end{bmatrix} \begin{Bmatrix} \sigma_{xx} \\ \sigma_{yy} \end{Bmatrix},
$$
$$
\varepsilon_{zz} = -\frac{\nu}{E}\left(\sigma_{xx} + \sigma_{yy}\right),
$$
$$
\gamma_{xy} = \frac{\tau_{xy}}{G},
$$
$$
\gamma_{yz} = \gamma_{zx} = 0.
$$
(5.59)

Inverting the above relations, we obtain

$$
\{\sigma\} = \begin{Bmatrix} \sigma_{xx} \\ \sigma_{yy} \\ \tau_{xy} \end{Bmatrix} = \frac{E}{1-\nu^2} \begin{bmatrix} 1 & \nu & 0 \\ \nu & 1 & 0 \\ 0 & 0 & \frac{1}{2}(1-\nu) \end{bmatrix} \begin{Bmatrix} \varepsilon_{xx} \\ \varepsilon_{yy} \\ \gamma_{xy} \end{Bmatrix}.
$$
(5.60)

Similar to plane stress, one can define a state of plane strain in which strains with a z subscript are all equal to zero. This situation corresponds to a structure whose deformation in the z-direction is constrained (i.e., $w = 0$), so that the following relation holds:

$$
\varepsilon_{zz} = 0, \quad \varepsilon_{xz} = 0, \quad \varepsilon_{yz} = 0.
$$
(5.61)

Plane strain can also be used if the structure is infinitely long in the z-direction. The stress–strain relations for the case of plane strain can be derived by starting with eq. (5.54) with the stress–strain matrix $[\mathbf{C}]$ in eq. (5.58) and setting the strains with a z subscript, $\varepsilon_{zz}, \gamma_{xz},$ and $\gamma_{yz},$ equal to zero to obtain

$$
\{\sigma\} = \begin{Bmatrix} \sigma_{xx} \\ \sigma_{yy} \\ \tau_{xy} \end{Bmatrix} = \frac{E}{(1+\nu)(1-2\nu)} \begin{bmatrix} 1-\nu & \nu & 0 \\ \nu & 1-\nu & 0 \\ 0 & 0 & \frac{1}{2}-\nu \end{bmatrix} \begin{Bmatrix} \varepsilon_{xx} \\ \varepsilon_{yy} \\ \gamma_{xy} \end{Bmatrix}.
$$
(5.62)

The inverse relation is given by

$$
\begin{Bmatrix} \varepsilon_{xx} \\ \varepsilon_{yy} \\ \gamma_{xy} \end{Bmatrix} = \frac{(1+\nu)}{E} \begin{bmatrix} 1-\nu & \nu & 0 \\ \nu & 1-\nu & 0 \\ 0 & 0 & 2 \end{bmatrix} \begin{Bmatrix} \sigma_{xx} \\ \sigma_{yy} \\ \tau_{xy} \end{Bmatrix}.
$$
(5.63)

Note that the normal stress σ_{zz} is not zero in the plane strain problem but can be calculated from ε_{xx} and ε_{yy}:

$$\sigma_{zz} = \frac{E\nu}{(1+\nu)(1-2\nu)}\left(\varepsilon_{xx} + \varepsilon_{yy}\right). \tag{5.64}$$

Another class of problems that can be treated as two-dimensional is the axisymmetric problems where the geometry, the applied loads, and the boundary conditions are all symmetric about an axis. Examples of axisymmetric geometry are a cylinder, a cone, or any geometry that has rotational symmetry about an axis. If the loads and boundary conditions are symmetric about the same axis, then the deformation will be in the radial and axial direction only. That is, there will not be any twisting or rotation about that axis. Therefore, such problems can be treated as two-dimensional with the displacement vector having only two components. It is convenient to use a cylindrical coordinate system for such problems so that the displacement components, (u_r, u_z), are the radial and axial components. A two-dimensional section of this geometry is the domain of analysis. Even though there is no rotation or twisting, there will be strain and stress in the circumferential direction. This component of stress and strain are called hoop stress and hoop strain. The hoop strain occurs whenever there is radial expansion on contraction because any radial displacement would cause the circumference of the object to increase or decrease. In a cylindrical coordinate system, the stress and strain relation can be written as:

$$\{\sigma\} = \begin{Bmatrix} \sigma_{rr} \\ \sigma_{zz} \\ \sigma_{\theta\theta} \\ \tau_{rz} \end{Bmatrix} = \frac{E}{(1+\nu)(1-2\nu)} \begin{bmatrix} 1-\nu & \nu & \nu & 0 \\ \nu & 1-\nu & \nu & 0 \\ \nu & \nu & 1-\nu & 0 \\ 0 & 0 & 0 & \frac{1}{2}-\nu \end{bmatrix} \begin{Bmatrix} \varepsilon_{rr} \\ \varepsilon_{zz} \\ \varepsilon_{\theta\theta} \\ \varepsilon_{rz} \end{Bmatrix}. \tag{5.65}$$

This relation can be obtained by dropping the equations for the two out-of-plane shear components $\tau_{r\theta}$ and $\tau_{z\theta}$ from the 3D stress-strain relation in eq. (5.58).

5.5 BOUNDARY VALUE PROBLEMS

5.5.1 Equilibrium Equations

As we discussed earlier, the state of stress at a point is defined by the six stress components, three normal and three shear stresses. These components, in general, vary within the solid. In static problems the stresses can be represented by six functions of the spatial coordinates x, y, and z: $\sigma_{xx}(x,y,z)$, $\sigma_{yy}(x,y,z)$, and so forth. These six functions cannot be arbitrary and must satisfy certain relations called stress equilibrium equations. The functions are also called the stress field in the solid.

Consider the equilibrium of a differential element represented in two dimensions by a square (see figure 5.14) whose center is (x, y) and its sides are $\mathrm{d}x$ and $\mathrm{d}y$, respectively. The dimension in the z-direction is taken as unity. The stresses are assumed to be independent of z, and the solid is in a state of plane stress.

Equilibrium in the x-direction yields the following equation:

$$\left(\sigma_{xx}\big|_{x+\frac{\mathrm{d}x}{2}}\right)\mathrm{d}y - \left(\sigma_{xx}\big|_{x-\frac{\mathrm{d}x}{2}}\right)\mathrm{d}y + \left(\tau_{yx}\big|_{y+\frac{\mathrm{d}y}{2}}\right)\mathrm{d}x - \left(\tau_{yx}\big|_{y-\frac{\mathrm{d}y}{2}}\right)\mathrm{d}x = 0. \tag{5.66}$$

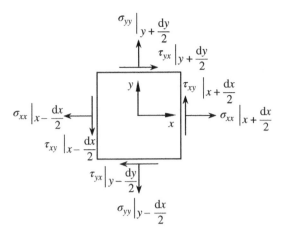

Figure 5.14 Stress variations in infinitesimal components

If the first-order Taylor series expansion is used to represent stresses on the surfaces of the rectangle in terms of stresses at the center, the first two terms in eq. (5.66) can be approximated by

$$\left(\sigma_{xx}\big|_{x+\frac{dx}{2}}\right)dy - \left(\sigma_{xx}\big|_{x-\frac{dx}{2}}\right)dy$$

$$= \left(\sigma_{xx}\big|_x + \frac{\partial\sigma_{xx}}{\partial x}\frac{dx}{2}\right)dy - \left(\sigma_{xx}\big|_x - \frac{\partial\sigma_{xx}}{\partial x}\frac{dx}{2}\right)dy = \frac{\partial\sigma_{xx}}{\partial x}dxdy.$$

Similarly, the last two terms can be approximated by

$$\left(\tau_{yx}\big|_{y+\frac{dy}{2}}\right)dx - \left(\tau_{yx}\big|_{y-\frac{dy}{2}}\right)dx$$

$$= \left(\tau_{yx}\big|_y + \frac{\partial\tau_{yx}}{\partial y}\frac{dy}{2}\right)dx - \left(\tau_{yx}\big|_y - \frac{\partial\tau_{yx}}{\partial y}\frac{dy}{2}\right)dx = \frac{\partial\tau_{yx}}{\partial y}dxdy.$$

By substituting these two equations into eq. (5.66), we obtain an equilibrium equation in the x-direction as

$$\frac{\partial\sigma_{xx}}{\partial x} + \frac{\partial\tau_{yx}}{\partial y} = 0. \tag{5.67}$$

Similarly, equilibrium in the y-direction yields the following equation:

$$\frac{\partial\tau_{xy}}{\partial x} + \frac{\partial\sigma_{yy}}{\partial y} = 0. \tag{5.68}$$

Equations (5.67) and (5.68) are the equilibrium equations for a solid subjected to a two-dimensional state of stress. We can similarly derive the equations for a three-dimensional state of stress, by considering the equilibrium of a three-dimensional differential element to obtain,

$$\begin{cases} \dfrac{\partial\sigma_{xx}}{\partial x} + \dfrac{\partial\tau_{yx}}{\partial y} + \dfrac{\partial\tau_{zx}}{\partial z} = 0, \\[2mm] \dfrac{\partial\tau_{xy}}{\partial x} + \dfrac{\partial\sigma_{yy}}{\partial y} + \dfrac{\partial\tau_{zy}}{\partial z} = 0, \\[2mm] \dfrac{\partial\tau_{xz}}{\partial x} + \dfrac{\partial\tau_{yz}}{\partial y} + \dfrac{\partial\sigma_{zz}}{\partial z} = 0. \end{cases} \tag{5.69}$$

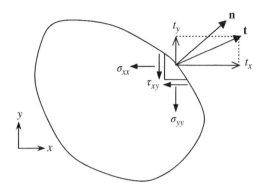

Figure 5.15 Traction boundary condition of a plane solid

Equation (5.69) is obtained by considering force equilibrium in the x-, y-, and z-directions. As has been shown in eq. (5.12), moment equilibrium yields symmetry of the stress matrix.

5.5.2 Traction or Stress Boundary Conditions

While the stress field must satisfy the differential equations of equilibrium in eq. (5.69), there are other conditions the stress field must satisfy on the boundaries of a solid. These are called traction or stress boundary conditions. Consider the surface of a solid subjected to distributed forces such that the tractions in the x-, y- and z-directions are, t_x, t_y, and t_z, respectively (see figure 5.15). Let the direction cosines of the normal to the surface be n_x, n_y, and n_z. Then the state of stress at a point on the surface of the body must satisfy the boundary conditions shown below:

$$\sigma_{xx}n_x + \tau_{yx}n_y + \tau_{zx}n_z = t_x,$$

$$\tau_{xy}n_x + \sigma_{yy}n_y + \tau_{zy}n_z = t_y, \qquad (5.70)$$

$$\tau_{xz}n_x + \tau_{yz}n_y + \sigma_{zz}n_z = t_z.$$

The derivation of the above boundary conditions is similar to those of eq. (5.10).

5.5.3 Boundary Value Problems

Consider an arbitrary body subjected to external forces and displacement constraints as shown in figure 5.16. Our goal is to determine the displacement field $\mathbf{u}(x,y,z)$. Once the displacements are determined, strains can be found using the strain displacement relations, that is in eqs. (5.39), (5.40), and (5.45), and then stress field can be determined using the constitutive relations, that is in eq. (5.54). This problem, typical of solid and structural mechanics, is called the elastostatic boundary value problem.

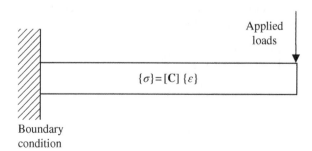

Boundary
condition

Figure 5.16 Boundary value problem

Techniques for solving the boundary value problems in two and three dimensions are considered in advanced courses in solid mechanics and theory of elasticity.[6] The elasticity problems can be simplified by making certain approximations in the displacement field, for instance, Euler-Bernoulli beam theory or Kirchhoff plate theory. The finite element method is a numerical method that can solve complex two- and three-dimensional elasticity problems and will be discussed in later chapters.

In general, solving an elasticity problem involves solving the three equilibrium equations in eq. (5.69) in conjunction with the strain–displacement relations in eqs. (5.39), (5.40), and (5.45) and constitutive relations in eq. (5.54). The boundary conditions also need to be provided in terms of given displacements or prescribed traction forces.

EXAMPLE 5.8 *Stress distribution of a cantilever beam*

The displacement field for the thin beam shown in figure 5.17 considering bending only is

$$u(x,y) = \frac{P}{EI}\left(Lx - \frac{x^2}{2}\right)y - \frac{\nu P}{6EI}y^3,$$

$$v(x,y) = \frac{-\nu P}{2EI}(L-x)y^2 - \frac{P}{EI}\left(\frac{Lx^2}{2} - \frac{x^3}{6}\right),$$

where P is the applied force at the tip, I is the area moment of inertia about the bending axis, and L is the length of the beam. Determine the entire stress field.

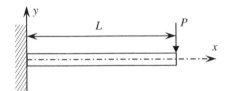

Figure 5.17 Cantilever beam bending problem

SOLUTION Since the thickness of the beam is small, we can assume the plane stress condition along the z-direction. From the definition of strain,

$$\varepsilon_{xx} = \frac{\partial u}{\partial x} = \frac{P}{EI}(L-x)y,$$

$$\varepsilon_{yy} = \frac{\partial v}{\partial y} = \frac{-\nu P}{EI}(L-x)y,$$

$$\gamma_{xy} = \frac{\partial v}{\partial x} + \frac{\partial u}{\partial y} = \left[\frac{\nu P y^2}{2EI} - \frac{P}{EI}\left(Lx - \frac{x^2}{2}\right)\right]$$

$$+ \left[\frac{P}{EI}\left(Lx - \frac{x^2}{2}\right)y - \frac{\nu P y^2}{2EI}\right] = 0.$$

---

[6] Boresi, A., Schmidt, R., and Sidebottom, O. J. 2003. *Advanced Mechanics of Materials*, 6th edition. Wiley.

Substituting into eq. (5.60) yields the stress field:

$$\sigma_{xx} = \frac{E}{1-\nu^2}\left[\frac{P}{EI}(L-x)y - \frac{\nu^2 P}{EI}(L-x)y\right] = \frac{P}{I}(L-x)y,$$

$$\sigma_{yy} = \frac{E}{1-\nu^2}\left[-\frac{\nu P}{EI}(L-x)y + \frac{\nu P}{EI}(L-x)y\right] = 0,$$

$$\tau_{xy} = 0.$$

Since the normal stress σ_{xx} changes linearly in the y-direction, the stress field represents the bending of a beam. ▬

5.5.4   Compatibility Conditions

In the previous section, we talked about the displacement field $\mathbf{u}(x,y,z)$. Usually, solutions to the boundary value problems are obtained by trial and error or inverse methods. We first assume a physically suitable displacement field and adjust the terms in the solution to satisfy the equilibrium equations and boundary conditions. More often one needs to start with a physically possible displacement field. In fact, a set of any three nonsingular functions will represent a possible displacement field. Of course, the forces required to produce such a displacement field may be complex and difficult in practice, but still physically possible. However, the same thing cannot be said about the strain field or stress field. That is, one cannot choose six arbitrary functions in spatial coordinates and claim the set to be a possible strain field. The particular strain field may not be physically possible. For example, a valid strain field should not create a discontinuous deformation or overlapping of certain points. The six strain functions must satisfy three relations called compatibility equations. The derivation of the three-dimensional compatibility equation is beyond the scope of this book. The interested reader is referred to more advanced elasticity books, such as Timoshenko.[7] For two-dimensional (plane) problems, the compatibility equation is given as

$$\frac{\partial^2 \gamma_{xy}}{\partial x \partial y} = \frac{\partial^2 \varepsilon_{xx}}{\partial y^2} + \frac{\partial^2 \varepsilon_{yy}}{\partial x^2}. \tag{5.71}$$

EXAMPLE 5.9   *Compatibility relation*

Choose two functions in x and y to represent the two displacements $u(x,y,z)$ and $v(x,y,z)$. Derive the strain field from the assumed displacement field. Show that the strains satisfy the compatibility equation in eq. (5.71).

SOLUTION   Let the displacement field be given by

$$u(x,y) = x^4 + y^4, \quad v(x,y) = x^2 y^2.$$

Then the strains can be derived as

$$\varepsilon_{xx} = 4x^3, \quad \varepsilon_{yy} = 2x^2 y, \quad \gamma_{xy} = 4y^3 + 2xy^2.$$

It is trivial to show that the above strains satisfy the compatibility requirement in eq. (5.71). ▬

---

[7] Timoshenko S. P., and Goodier, J. N. 1970. *Theory of Elasticity*, 3rd Ed. McGraw-Hill, New York.

5.6 PRINCIPLE OF MINIMUM POTENTIAL ENERGY FOR PLANE SOLIDS

There are different methods to solve for the boundary value problems in the previous section. In this section, the principle of minimum potential energy that is similar to the beam-bending problem in chapter 3 is derived for plane solids. The principle will be used to derive the finite element equations for different elements in the next chapter.

5.6.1 Strain Energy in a Plane Solid

Consider a plane elastic solid as illustrated in figure 5.18. The *strain energy* is a form of energy that is stored in the solid due to the elastic deformation. Formally, it can be defined as

$$
\begin{aligned}
U &= \frac{1}{2} \iiint_{\text{volume}} \{\varepsilon\}^{\mathrm{T}}\{\sigma\}\,\mathrm{d}V \\[4pt]
&= \frac{h}{2} \iint_{\text{area}} \{\varepsilon\}^{T}\{\sigma\}\,\mathrm{d}A \\[4pt]
&= \frac{h}{2} \iint_{\text{area}} \{\varepsilon\}^{\mathrm{T}}[C]\{\varepsilon\}\,\mathrm{d}A,
\end{aligned}
\tag{5.72}
$$

where h is the thickness of the plane solid ($h = 1$ for plane strain) and $[C] = [C_\sigma]$ for plane stress and $[C] = [C_\varepsilon]$ for plane strain. Since stress and strain are constant throughout the thickness, the volume integral is converted into the area integral by multiplying by the thickness in the second relation in eq. (5.72). The linear elastic relation in eq. (5.54) has been used in the last relation.

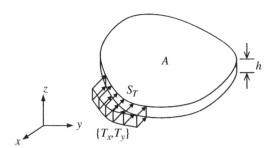

Figure 5.18 A plane solid under the distributed load $\{T_x, T_y\}$ on the traction boundary S_T

5.6.2 Potential Energy of Applied Loads

When a force acting on a body moves through a small distance, it loses its potential to do additional work, and hence its potential energy is given by the negative of the product of the force and corresponding displacement. For example, when concentrated forces are applied to the solid, the potential energy becomes

$$
V = -\sum_{i=1}^{ND} F_i q_i,
\tag{5.73}
$$

where F_i is the force in the i-th DOF, q_i is the displacement in the direction of the force, and ND is the total number of concentrated forces acting on the body. The negative sign indicates that the potential energy decreases as the force has expended some energy performing the work given by the product of force and corresponding displacement.

When distributed forces, such as a pressure load, act on the edge of a body, the summation sign in the above expression is replaced by integration over the edge of the body as shown below:

$$
\begin{aligned}
V &= -h \int_{S_T} \left(T_x u + T_y v \right) \mathrm{d}S \\
&= -h \int_{S_T} \{ u \quad v \} \begin{Bmatrix} T_x \\ T_y \end{Bmatrix} \mathrm{d}S \\
&\equiv -h \int_{S_T} \{ \mathbf{u} \}^{\mathrm{T}} \{ \mathbf{T} \} \mathrm{d}S,
\end{aligned}
\tag{5.74}
$$

where T_x and T_y are the components of applied surface forces in the x- and y-direction, respectively. In the finite element discretization, the vector of displacements, $\{\mathbf{u}\}$, is approximated by shape functions and nodal DOFs $\{\mathbf{q}\}$. Therefore, the potential energy of the applied loads is a function of nodal DOFs $\{\mathbf{q}\}$ after discretization by finite elements.

If body forces (forces distributed over the volume) are present, work done by these forces can be computed in a similar manner. The gravitational force is an example of body force. In this case, integration should be performed over the volume. We will discuss this further when we derive the finite element equations.

5.6.3 Principle of Minimum Potential Energy

As with the beam problem, the *potential energy* is defined as the sum of the strain energy and the potential energy of applied loads:

$$
\Pi = U + V,
\tag{5.75}
$$

where U is the strain energy and V is the potential energy of applied loads. The principle of minimum total potential energy states that of all possible displacement configurations of a solid/structure, the equilibrium configuration corresponds to the minimum total potential energy. That is, at equilibrium, we have

$$
\frac{\partial \Pi}{\partial \{\mathbf{Q}\}} = 0 \Rightarrow \frac{\partial \Pi}{\partial q_1} = 0, \; \frac{\partial \Pi}{\partial q_2} = 0, \cdots, \frac{\partial \Pi}{\partial q_N} = 0,
\tag{5.76}
$$

where $q_1, q_2, \ldots, q_N$ are nodal DOFs (i.e., displacements) that define the deformed configuration of the body. In finite element analysis, the deformation of the body is defined in terms of the displacements of the nodes. In the next chapter, we will use the principle of minimum total potential energy to derive finite element equations for different types of elements.

5.7 FAILURE THEORIES

In the previous section, we introduced the concept of stress, strain, and the relationship between stresses and strains. We also discussed the failure of materials under a uniaxial state of stress. Failure of engineering materials can be broadly classified into ductile and brittle failure. Most metals are ductile and fail due to yielding. Hence, the yield strength characterizes their failure. Ceramics and some polymers are brittle and rupture or fracture when the stress exceeds certain maximum value. Their stress–strain behavior is linear up to the point of failure and they fail abruptly.

Figure 5.19 Material failure due to relative sliding of atomic planes

The stress required to break the atomic bond and separate the atoms is called the theoretical strength of the material. It can be shown that the theoretical strength is approximately equal to $E/3$ where E is Young's modulus.[8] However, most materials fail at a stress about one-hundredth or even one-thousandth of the theoretical strength. For example, the theoretical strength of aluminum is about 22 GPa. However, the yield strength of aluminum is in the order of 100 MPa, which is 1/220 of the theoretical strength. This enormous discrepancy could be explained as follows.

In ductile materials, yielding occurs not due to separation of atoms but due to sliding of atoms (movement of dislocations) as depicted in figure 5.19. Thus, the stress or energy required for yielding is much less than that required for separating the atomic planes. Hence, in a ductile material, the maximum shear stress causes yielding of the material.

In brittle materials, the failure or rupture still occurs due to the separation of atomic planes. However, the high value of stress required is provided locally by stress concentration caused by small pre-existing cracks or flaws in the material. The stress concentration factors can be in the order of 100 to 1,000. That is, the applied stress is amplified by the enormous amount due to the presence of cracks, and it is sufficient to separate the atoms. When this process becomes unstable, the material separates over a large area causing brittle failure of the material.

Although research is underway not only to explain but also quantify the strength of materials in terms of its atomic structure and properties, it is still not practical to design machines and structures based on such atomistic models. Hence, we resort to phenomenological failure theories, which are based on observations and testing over a period. The purpose of failure theories is to extend the strength values obtained from uniaxial tests to multiaxial states of stress that exist in practical structures. It is not practical to test a material under all possible combinations of stress states. In the following, we describe some well-established phenomenological failure theories for both ductile and brittle materials.

5.7.1 Strain Energy

When a force is applied to a solid, it deforms. Then, we can say that work is done on the solid, which is proportional to the force and deformation. The work done by the applied force is stored in the solid as potential energy, which is called the *strain energy*. The strain energy in the solid may not be distributed uniformly throughout the solid. We introduce the concept of strain energy density, which is strain energy per unit volume, and we denote it by U_0. Then the strain energy in the body can be obtained by integration as follows:

$$U = \iiint_V U_o(x,y,z)\mathrm{d}V, \tag{5.77}$$

where the integration is performed over the volume V of the solid. In the case of uniaxial stress state, the strain energy density is equal to the area under the stress–strain curve (see figure 5.20). Thus, it can be written as

$$U_0 = \frac{1}{2}\sigma\varepsilon. \tag{5.78}$$

---

[8] Anderson, T. L. 2006. *Fracture Mechanics – Fundamentals and Applications*, Third Edition. CRC Press, Boca Raton, FL.

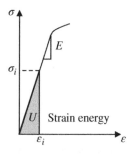

Figure 5.20 Stress–strain curve and the strain energy

For the general 3-D case the strain energy density is expressed as

$$U_0 = \frac{1}{2}\left(\sigma_x \varepsilon_x + \sigma_y \varepsilon_y + \sigma_z \varepsilon_z + \tau_{yz}\gamma_{yz} + \tau_{zx}\gamma_{zx} + \tau_{xy}\gamma_{xy}\right). \tag{5.79}$$

If the material is elastic, then the strain energy can be completely recovered by unloading the body.

The strain energy density in eq. (5.79) can be further simplified. Consider a coordinate system that is parallel to the principal stress directions. In this coordinate system, no shear components exist. Extending eq. (5.79) to this stress states yields

$$U_0 = \frac{1}{2}\left(\sigma_1 \varepsilon_1 + \sigma_2 \varepsilon_2 + \sigma_3 \varepsilon_3\right). \tag{5.80}$$

From section 5.3, we know that stresses and strains are related through the linear elastic relationship. For example, in case of principal stresses and strains,

$$\begin{cases} \varepsilon_1 = \dfrac{1}{E}(\sigma_1 - \nu\sigma_2 - \nu\sigma_3), \\[2mm] \varepsilon_2 = \dfrac{1}{E}(\sigma_2 - \nu\sigma_1 - \nu\sigma_3), \\[2mm] \varepsilon_3 = \dfrac{1}{E}(\sigma_3 - \nu\sigma_1 - \nu\sigma_2). \end{cases} \tag{5.81}$$

Substituting from eq. (5.81) into eq. (5.80), we can write the strain energy density in terms of principal stresses as

$$U_0 = \frac{1}{2E}\left[\sigma_1^2 + \sigma_2^2 + \sigma_3^2 - 2\nu(\sigma_1\sigma_2 + \sigma_2\sigma_3 + \sigma_1\sigma_3)\right]. \tag{5.82}$$

The strain energy density can be thought of as consisting of two components: one due to dilation or change in volume and the other due to distortion or change in shape. The former is called dilatational strain energy and the latter, distortional energy. Many experiments have shown that ductile materials can be hydrostatically stressed to levels beyond their ultimate strength in compression without failure. This is because the hydrostatic state of stress reduces the volume of the specimen without changing its shape.

5.7.2   Decomposition of Strain Energy

The strain energy density at a point in a solid can be divided into two parts: dilatational strain energy density, U_h, that is due to change in volume, and distortional strain energy density, U_d, that is responsible for the change in shape. In order to compute these components, we divide the stress matrix also into similar components, dilatational stress matrix, σ_h, and deviatoric stress matrix, σ_d. For convenience, we will use the stress components with respect to the principal stress coordinates. Then the aforementioned stress components can be derived as follows:

$$
\begin{bmatrix} \sigma_1 & 0 & 0 \\ 0 & \sigma_2 & 0 \\ 0 & 0 & \sigma_3 \end{bmatrix} = \begin{bmatrix} \sigma_h & 0 & 0 \\ 0 & \sigma_h & 0 \\ 0 & 0 & \sigma_h \end{bmatrix} + \begin{bmatrix} \sigma_{1d} & 0 & 0 \\ 0 & \sigma_{2d} & 0 \\ 0 & 0 & \sigma_{3d} \end{bmatrix}.
\tag{5.83}
$$

The dilatational component σ_h is defined as

$$
\sigma_h = \frac{\sigma_1 + \sigma_2 + \sigma_3}{3} = \frac{\sigma_{xx} + \sigma_{yy} + \sigma_{zz}}{3},
\tag{5.84}
$$

which is also called the volumetric stress. Note that $3\sigma_h$ is the first invariant I_1 of the stress matrix in eq. (5.20). Thus, it is independent of the coordinate system. Note that σ_h is a state of hydrostatic stress and hence the subscript h is used to denote the dilatational stress component as well as dilatational energy density

The dilatational energy density can be obtained by substituting the stress components of the hydrostatic stress state in eq. (5.84) into the expression for strain energy density in eq. (5.82),

$$
\begin{aligned}
U_h &= \frac{1}{2E} \left[\sigma_h^2 + \sigma_h^2 + \sigma_h^2 - 2\nu \left(\sigma_h \sigma_h + \sigma_h \sigma_h + \sigma_h \sigma_h \right) \right] \\
&= \frac{3(1-2\nu)}{2} \frac{\sigma_h^2}{E},
\end{aligned}
\tag{5.85}
$$

and using the relation in eq. (5.84),

$$
\begin{aligned}
U_h &= \frac{3(1-2\nu)}{2E} \left(\frac{\sigma_1 + \sigma_2 + \sigma_3}{3} \right)^2 \\
&= \frac{1-2\nu}{6E} \left[\sigma_1^2 + \sigma_2^2 + \sigma_3^2 + 2(\sigma_1\sigma_2 + \sigma_2\sigma_3 + \sigma_1\sigma_3) \right].
\end{aligned}
\tag{5.86}
$$

The distortion part of the strain energy is now found by subtracting eq. (5.86) from eq. (5.82), as

$$
\begin{aligned}
U_d &= U_0 - U_h \\
&= \frac{1+\nu}{3E} \left[\sigma_1^2 + \sigma_2^2 + \sigma_3^2 - \sigma_1\sigma_2 - \sigma_2\sigma_3 - \sigma_1\sigma_3 \right] \\
&= \frac{1+\nu}{3E} \frac{(\sigma_1 - \sigma_2)^2 + (\sigma_2 - \sigma_3)^2 + (\sigma_3 - \sigma_1)^2}{2}.
\end{aligned}
\tag{5.87}
$$

It is customary to write U_d in terms of an equivalent stress called von Mises stress, σ_{VM}, as

$$
U_d = \frac{1+\nu}{3E} \sigma_{VM}^2.
\tag{5.88}
$$

The von Mises stress is defined in terms of principal stresses as

$$
\sigma_{VM} = \sqrt{\frac{(\sigma_1 - \sigma_2)^2 + (\sigma_2 - \sigma_3)^2 + (\sigma_3 - \sigma_1)^2}{2}}.
\tag{5.89}
$$

5.7.3 Distortion Energy Theory (von Mises)

According to von Mises's theory, a ductile solid will yield when the distortion energy density reaches a critical value for that material. Since this should be true for the uniaxial stress state also, the critical value of the distortional energy can be estimated from the uniaxial test. At the instance of yielding in a uniaxial tensile test, the state of stress in terms of principal stress is given by $\sigma_1 = \sigma_Y$ (yield stress) and $\sigma_2 = \sigma_3 = 0$. The distortion energy density associated with yielding is

$$U_d = \frac{1+\nu}{3E}\sigma_Y^2. \tag{5.90}$$

Thus, the energy density given in eq. (5.90) is the critical value of the distortional energy density for the material. Then according to von Mises's failure criterion, the material under multiaxial loading will yield when the distortional energy is equal to or greater than the critical value for the material:

$$\frac{1+\nu}{3E}\sigma_{VM}^2 \geq \frac{1+\nu}{3E}\sigma_Y^2,$$
$$\therefore \quad \sigma_{VM} \geq \sigma_Y. \tag{5.91}$$

Thus, the distortion energy theory states that material yields when the von Mises stress exceeds the yield stress obtained in a uniaxial tensile test.

The von Mises stress in eq. (5.87) can be rewritten in terms of stress components as

$$\sigma_{VM} = \sqrt{\frac{\left(\sigma_{xx}-\sigma_{yy}\right)^2 + \left(\sigma_{yy}-\sigma_{zz}\right)^2 + \left(\sigma_{zz}-\sigma_{xx}\right)^2 + 6\left(\tau_{xy}^2 + \tau_{yz}^2 + \tau_{zx}^2\right)}{2}}. \tag{5.92}$$

For a two-dimensional plane stress state, $\sigma_3 = 0$, the von Mises stress can be defined in terms of principal stresses as

$$\sigma_{VM} = \sqrt{\sigma_1^2 - \sigma_1\sigma_2 + \sigma_2^2}, \tag{5.93}$$

and in terms of general stress components as

$$\sigma_{VM} = \sqrt{\sigma_{xx}^2 + \sigma_{yy}^2 - \sigma_{xx}\sigma_{yy} + 3\tau_{xy}^2}. \tag{5.94}$$

The two-dimensional distortion energy equation in eq. (5.94) becomes an ellipse when plotted on the σ_1-σ_2 plane as shown in figure 5.21. The interior of this ellipse defines the region of combined biaxial stress where the material is safe against yielding under static loading.

Consider a situation in which only a shear stress exists, such that $\sigma_x = \sigma_y = 0$, and $\tau_{xy} = \tau$. For this stress state, the principal stresses are $\sigma_1 = -\sigma_3 = \tau$ and $\sigma_2 = 0$. On the σ_1-σ_3 plane, this pure shear state is represented as a straight line through the origin at $-45°$ as shown in figure 5.21. The line intersects the

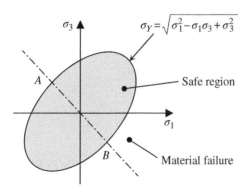

Figure 5.21 Failure envelope of the distortion energy theory

von Mises failure envelope at two points, A and B. The magnitude of σ_1 and σ_2 at these points can be found from eq. (5.93) as

$$\sigma_Y^2 = \sigma_1^2 + \sigma_1\sigma_1 + \sigma_1^2 = 3\sigma_1^2 = 3\tau_{\max}^2,$$

$$\tau_{\max} = \sigma_1 = \frac{\sigma_Y}{\sqrt{3}} = 0.577\sigma_Y. \tag{5.95}$$

Thus, in a pure shear stress state, the material yields when the shear stress reaches $0.577\sigma_Y$. This value will be compared to the maximum shear stress theory described below.

5.7.4 Maximum Shear Stress Theory (Tresca)

According to the maximum shear stress theory, the material yields when the maximum shear stress at a point equals the critical shear stress value for that material. Since this should be true for a uniaxial stress state, we can use the results from uniaxial tension test to determine the maximum allowable shear stress. The stress state in a tensile specimen at the point of yielding is given by $\sigma_1 = \sigma_Y$, $\sigma_2 = \sigma_3 = 0$. The maximum shear stress is calculated as

$$\tau_{\max} = \frac{\sigma_1 - \sigma_3}{2} \geq \tau_Y = \frac{\sigma_Y}{2}. \tag{5.96}$$

This value of maximum shear stress is also called the yield shear stress of the material and is denoted by τ_Y. Note that $\tau_Y = \sigma_Y/2$. Thus, Tresca's yield criterion is that yielding will occur in a material when the maximum shear stress equals the yield shear strength, τ_Y, of the material.

The hexagon in figure 5.22 represents the two-dimensional failure envelope according to maximum shear stress theory. The ellipse corresponding to von Mises's theory is also shown in the same figure. The hexagon is inscribed within the ellipse and contacts it at six vertices. Combinations of principal stresses σ_1 and σ_3 that lie within this hexagon are considered safe based on the maximum shear stress theory, and failure is considered to occur when the combined stress state reaches the hexagonal boundary. This is obviously more conservative failure theory than distortion energy theory as it is contained within the latter. In the pure shear stress state, the shear stress at points C and D correspond to $0.5\sigma_Y$, which is smaller than $0.577\sigma_Y$, the failure point according to the distortion energy theory.

5.7.5 Maximum Principal Stress Theory (Rankine)

According to the maximum principal stress theory, a brittle material ruptures when the maximum principal stress in the specimen reaches some limiting value for the material. Again, this critical value can be inferred as the tensile strength measured using a uniaxial tension test. In practice, this theory is simple but

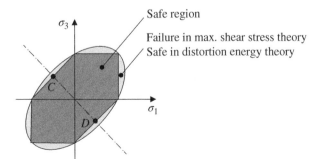

Figure 5.22 Failure envelope of the maximum shear stress theory

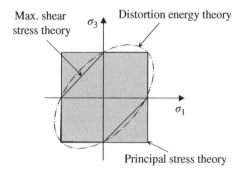

Figure 5.23 Failure envelope of the maximum principal stress theory

can only be used for brittle materials. Some practitioners have modified this theory for ductile materials as

$$\sigma_1 \geq \sigma_U, \tag{5.97}$$

where σ_1 is the maximum principal stress, and σ_U is the ultimate strength described in table 5.3. Figure 5.23 shows the failure envelope based on the maximum principal stress theory. Note that the failure envelopes in the first and third quadrants are coincident with those of the maximum shear stress theory and contained within the distortion energy theory. However, the envelopes in the second and fourth quadrants are well outside of the other two theories. Hence, the maximum principal stress theory is not considered suitable for ductile materials. However, it can be used to predict failure in brittle materials.

5.8 SAFETY FACTOR

One can notice that the all aforementioned failure theories are of the form:

$$A \text{ function of stress} \geq \text{strength.} \tag{5.98}$$

The term on the left-hand side of a failure criterion depends on the state of stress at a point. Various stress analysis methods, including the finite element method discussed in this book, are used to evaluate the stress term. The right-hand side of the equation is a material property usually determined from material tests. There are many uncertainties in calculating the state of stress at a point. These include uncertainties in the loads, material properties such as Young's modulus, dimensions and geometry of the solid, and so forth. Similarly, there are uncertainties in the strength of a material depending on the tests used to measure the strength, manufacturing process, and so forth. In order to account for the uncertainties, engineers use a factor of safety in the design of a solid or structural component. Thus, the failure criteria are modified as:

$$N \times \text{stress} = \text{strength}, \tag{5.99}$$

where N is the safety factor. That is, we assume the stress is N times the calculated or estimated state of stress. Another interpretation is that the strength is reduced by a factor N to account for uncertainties. Thus, the space of allowable stresses is contained well within the failure envelope shown in figure 5.23. For example, the safety factor in the von Mises theory is defined as

$$N_{VM} = \frac{\sigma_Y}{\sigma_{VM}}. \tag{5.100}$$

In many engineering applications, N is in the range of 1.1–1.5.

The safety factor in the maximum shear stress theory is defined as

$$N_\tau = \frac{\tau_Y}{\tau_{max}} = \frac{\sigma_Y/2}{\tau_{max}}.$$ (5.101)

It can be shown that for any two-dimensional loading, $N_{VM} \geq N_\tau$.

$$N_{VM} \geq N_\tau,$$
$$\frac{\sigma_Y}{\sigma_{VM}} \geq \frac{\tau_Y}{\tau_{max}}.$$ (5.102)

EXAMPLE 5.10   *Yield criteria of a shaft*

Estimate the torque on a 10-mm-diameter steel shaft when yielding begins using (a) the maximum shear stress theory and (b) the maximum distortion energy theory. The yield stress of the steel is 140 MPa.

SOLUTION

(a) For torsion, the maximum shear stress occurs on the outside surface of the shaft:

$$T_{max} = \frac{\tau_{max} I_p}{d/2},$$ (5.103)

where $I_p = \pi d^4/32$ is the polar moment of inertia for a solid circular shaft. Using the relation that the maximum shear stress is half of the yield stress, the shear yield stress for torsion can be obtained from eq. (5.103), as

$$\tau_{max} = \frac{16T}{\pi d^3} = \frac{1}{2}\sigma_Y.$$

Solving for the torque yields

$$T = \frac{\pi d^3}{32}\sigma_Y = \frac{\pi(0.01)^3}{32}140 \times 10^6 = 13.74\,N \cdot m.$$

(b) The principal stresses for the torsion problem can be obtained from the Mohr's circle,

$$\sigma_1 = \tau, \qquad \sigma_2 = 0, \qquad \sigma_3 = -\tau.$$

From eq. (5.95),

$$\tau = 0.577\sigma_Y.$$

Substituting into eq. (5.103) and solving for the torque results in

$$T = \frac{0.577\pi d^3}{16}\sigma_Y = \frac{0.577\pi(0.01)^3}{16}140 \times 10^6 = 15.86\,N \cdot m.$$

Thus, it can be seen that for yielding in pure torsion, the distortion energy theory predicts a torque that is 15% greater than the prediction of the maximum shear stress theory. Tests on ductile materials have shown that the distortion energy theory is much more accurate for predicting yield, but in design work, the more conservative answer predicted by shear stress is commonly used. ■

EXAMPLE 5.11   *Safety factor of a bracket*

The bracket shown in figure 5.24 consists of a rigid arm and a flexible rod. The latter has the following properties: moment of inertia $I = 1.0$, polar moment of inertia $J = 0.5$, radius $r = 0.1$, length $l = 10.0$, and yield stress $\sigma_Y = 2.8$.

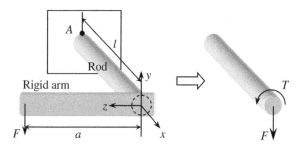

Figure 5.24 Bracket structure

When a vertical force of $F = 1.0$ is applied at the end of the rigid arm ($a = 5\sqrt{2}$), calculate the stress matrix at point A, and evaluate the safety factors at that point using both the distortion energy theory and maximum shear stress theory.

SOLUTION   As shown in figure 5.24, the torque and the transverse shear force are applied to the rod. Since the transverse shear stress caused by F is zero at point A, only the bending moment and the torque contribute to stress on the rod. Thus, the stress components are

$$\sigma_{xx} = \frac{M \cdot r}{I} = \frac{F \cdot l \cdot r}{I} = 1,$$

$$\tau_{xz} = \frac{T \cdot r}{J} = \frac{5\sqrt{2} \cdot 0.1}{0.5} = \sqrt{2}.$$

$$\sigma_{yy} = \sigma_{zz} = \tau_{xy} = \tau_{yz} = 0.0.$$

Thus, the symmetric stress matrix can be written as

$$[\sigma] = \begin{bmatrix} 1 & 0 & \sqrt{2} \\ 0 & 0 & 0 \\ \sqrt{2} & 0 & 0 \end{bmatrix}.$$

The principal stresses are found by solving the following eigenvalue problem:

$$([\sigma] - \lambda[\mathbf{I}]) \cdot \mathbf{n} = \mathbf{0}.$$

The solution to the above eigenvalue problem is nontrivial only if the coefficient matrix is singular or, equivalently, its determinant is equal to zero.

$$\begin{vmatrix} 1-\lambda & 0 & \sqrt{2} \\ 0 & -\lambda & 0 \\ \sqrt{2} & 0 & -\lambda \end{vmatrix} = 0.$$

After expanding the Jacobian, the following three solutions are obtained:

$$-\lambda[-\lambda(1-\lambda)-2] = -\lambda(\lambda^2 - \lambda - 2) = 0$$
$$\Rightarrow -\lambda(\lambda-2)(\lambda+1) = 0, \quad \therefore \lambda = 2, 0, -1,$$

which are the three principal stresses. Thus, we obtain

$$\sigma_1 = 2, \quad \sigma_2 = 0, \quad \sigma_3 = -1.$$

In order to apply failure theory, the maximum shear stress and von Mises stress are calculated by

$$\tau_{max} = \frac{\sigma_1 - \sigma_3}{2} = 1.5,$$

$$\sigma_{VM} = \sqrt{4 + 2 + 1} = \sqrt{7}.$$

The safety factor from distortion energy theory is

$$N = \frac{\sigma_Y}{\sigma_{VM}} = \frac{2.8}{\sqrt{7}} = 1.0583,$$

which means that the structure is safe.

The safety factor from the maximum shear stress theory is

$$N = \frac{\tau_Y}{\tau_{max}} = \frac{1.4}{1.5} = 0.9333,$$

which means that the structure will yield. ■

5.9 EXERCISES

1. Answer the following descriptive questions.

   (a) How can the sign of Cartesian components of stress be determined?

   (b) When the three principal stresses are identical, that is, $\sigma_1 = \sigma_2 = \sigma_3 = \sigma$, what are the principal directions?

   (c) When the only nonzero stress components are $\sigma_{xx} = \sigma_{yy} = \sigma_{zz} = \sigma$, will $\mathbf{n} = \{1, 2, 3\}^T$ be a principal direction?

   (d) When $\gamma_{xy} = 150$ MPa is the only nonzero strain component, what are the three principal strains?

   (e) For a 2D model, when is it more appropriate to use plane stress elements rather than plane strain elements? Give examples of both.

   (f) When can a structure be modeled using axisymmetric elements? Give two examples.

   (g) If a thin rectangular plate is rigidly held along all its edges and a transverse load is applied normal to the plate, is it appropriate to model it as 2D (plane stress or plane strain)? Explain why or why not.

   (h) In the real world, all structures are 3D. Why is it then not appropriate to always use 3D elements for all structures? Explain this for the case of a frame-like structure.

2. A vertical force F is applied to a two-bar truss as shown in the figure. Let the cross-sectional areas of trusses 1 and 2 be A_1 and A_2, respectively. Determine the area ratio A_1/A_2 in order to have the same magnitude of stresses in both members.

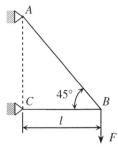

3. The stress matrix at a point P is given below. The direction cosines of the normal $\mathbf{n}$ to a plane that passes through P have the ratio $n_x:n_y:n_z = 3:4:12$. Determine: (a) the traction vector $\mathbf{T}^{(n)}$; (b) the magnitude T of $\mathbf{T}^{(n)}$; (c) the normal stress σ_n; (d) the shear stress τ_n; and (e) the angle between $\mathbf{T}^{(n)}$ and $\mathbf{n}$. *Hint:* Use $n_x^2 + n_y^2 + n_z^2 = 1$.

$$[\sigma] = \begin{bmatrix} 13 & 13 & 0 \\ 13 & 26 & -13 \\ 0 & -13 & -39 \end{bmatrix}.$$

4. At a point P in a body, Cartesian stress components are given by $\sigma_{xx} = 80$ MPa, $\sigma_{yy} = -40$ MPa, $\sigma_{zz} = -40$ MPa, and $\tau_{xy} = \tau_{yz} = \tau_{zx} = 80$ MPa. Determine the traction vector and its normal component and shear component on a plane that is equally inclined to all three axes.
   Hint: When a plane is equally inclined to all the three coordinate axes, the direction cosines of the normal are equal to each other.

5. If $\sigma_{xx} = 90$ MPa, $\sigma_{yy} = -45$ MPa, $\tau_{xy} = 30$ MPa, and $\sigma_{zz} = \tau_{xz} = \tau_{yz} = 0$, compute the surface traction $\mathbf{T}^{(n)}$ on the plane shown in the figure, which makes an angle of $\theta = 40°$ with the vertical axis. What are the normal and shear components of stress on this plane?

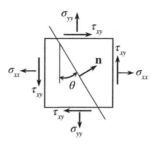

6. Find the principal stresses and the orientation of the principal axes of stresses for the following cases of plane stress.

(a) $\sigma_{xx} = 40$ MPa, $\sigma_{yy} = 0$ MPa, $\tau_{xy} = 80$ MPa

(b) $\sigma_{xx} = 140$ MPa, $\sigma_{yy} = 20$ MPa, $\tau_{xy} = -60$ MPa

(c) $\sigma_{xx} = -120$ MPa, $\sigma_{yy} = 50$ MPa, $\tau_{xy} = 100$ MPa

7. If the minimum principal stress is -7 MPa, find σ_{xx} and the angle that the principal stress axes make with the xy axes for the case of plane stress illustrated.

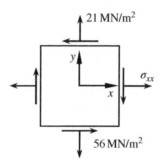

8. Determine the principal stresses and their associated directions, when the stress matrix at a point is given by

$$[\sigma] = \begin{bmatrix} 1 & 1 & 1 \\ 1 & 1 & 2 \\ 1 & 2 & 1 \end{bmatrix} \text{ MPa.}$$

9. Let an $x'y'z'$ coordinate system be defined using the three principal directions obtained from problem 8. Determine the transformed stress matrix $[\sigma]_{x'y'z'}$ in the new coordinate system.

10. For the stress matrix below, the two principal stresses are given as $\sigma_3 = -3$ and $\sigma_1 = 2$, respectively. In addition, two principal directions corresponding to the two principal stresses are also given below.

$$[\sigma] = \begin{bmatrix} 1 & 0 & 2 \\ 0 & 1 & 0 \\ 2 & 0 & -2 \end{bmatrix}, \mathbf{n}^1 = \begin{bmatrix} \dfrac{2}{\sqrt{5}} \\ 0 \\ \dfrac{1}{\sqrt{5}} \end{bmatrix}, \text{ and } \mathbf{n}^3 = \begin{bmatrix} \dfrac{1}{\sqrt{5}} \\ 0 \\ \dfrac{-2}{\sqrt{5}} \end{bmatrix}.$$

(a) What are the normal and shear stresses on the plane whose normal vector is parallel to (2, 1, 2)?

(b) Calculate the principal stress σ_2 and the principal direction $\mathbf{n}^2$.

(c) Write the stress matrix in the new coordinate system that is aligned with $\mathbf{n}^1$, $\mathbf{n}^2$, and $\mathbf{n}^3$.

11. With respect to the coordinate system xyz, the state of stress at a point P in a solid is:

$$[\sigma] = \begin{bmatrix} -20 & 0 & 0 \\ 0 & 50 & 0 \\ 0 & 0 & 50 \end{bmatrix} \text{MPa.}$$

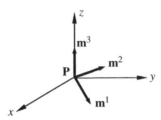

   (a) $\mathbf{m}^1$, $\mathbf{m}^2$ and $\mathbf{m}^3$ are three mutually perpendicular vectors such that $\mathbf{m}^1$ makes $45°$ with both the x- and y-axis and $\mathbf{m}^3$ is aligned with the z-axis. Compute the normal stresses on planes normal to $\mathbf{m}^1$, $\mathbf{m}^2$, and $\mathbf{m}^3$.

   (b) Compute two components of shear stress on the plane normal to $\mathbf{m}^1$ in the directions $\mathbf{m}^2$ and $\mathbf{m}^3$.

   (c) Is the vector $\mathbf{n} = \{0, 1, 1\}^T$ a principal direction of stress? Explain. What is the normal stress in the direction $\mathbf{n}$?

   (d) Draw an infinitesimal cube with faces normal to $\mathbf{m}^1$, $\mathbf{m}^2$, and $\mathbf{m}^3$ and display the stresses on the positive faces of this cube.

   (e) Express the state of stress at the point P with respect to the $x'y'z'$ coordinates system that is aligned with the vectors $\mathbf{m}^1$, $\mathbf{m}^2$ and $\mathbf{m}^3$?

   (f) What are the principal stress and principal directions of stress at the point P with respect to the $x'y'z'$ coordinate system? Explain.

   (g) Compute the maximum shear stress at point P. Which plane(s) does this maximum shear stress act on?

12. A solid shaft of diameter $d = 5$ cm as shown in the figure is subjected to tensile force $P = 13,000$ N and a torque $T = 6,000$ N·cm. At point A on the surface, what is the state of stress (write in matrix form), the principal stresses, and the maximum shear stress? Show the coordinate system you are using.

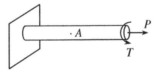

13. The solid shaft in problem 12 has diameter $d = 10$ cm and is subjected to tensile force $P = 80,000$ N and a torque $T = 5,000$ N·cm. Assume that the yield stress of the material of the shaft is $\sigma_y = 200$ MPa.

   (a) Compute the state of stress at a potential point of failure.

   (b) Based on the Maximum Shear stress theory (Tresca's law) will this shaft fail? What is the safety factor?

14. If the displacement field is given by

$$\begin{cases} u = x^2 + 2y^2, \\ v = -y^2 - 2x(y-z), \\ w = -z^2 - 2xy. \end{cases}$$

   (a) Write down the 3×3 strain matrix.

   (b) What is the normal strain component in the direction of $(1,1,1)$ at point $(1,-3,1)$?

15. If the displacement field is given by

$$\begin{cases} u = kx^2, \\ v = 2kxy^2, \\ w = k(x+y)z. \end{cases}$$

where k is a constant,

(a) Write down the strain matrix.

(b) What is the normal strain in the direction of $\mathbf{n} = \{1, 1, 1\}^T$?

16. Consider the following displacement field in a plane solid:

$$u(x,y) = 0.4 - 0.09x + 0.06y, \quad v(x,y) = 0.6 + 0.04x + 0.15y.$$

(a) Compute the infinitesimal strain components ε_x, ε_y, and γ_{xy}. Is this a state of uniform strain?

(b) Determine the principal strains and their corresponding directions. Express the principal strain directions in terms of the *angles* (in degrees) the directions make with the x-axis.

(c) What is the normal strain at (x,y) in a direction $45°$ to the x-axis?

17. Draw a 2-inch $\times$ 2-inch square OABC on the engineering paper. The coordinates of O are $(0, 0)$ and those of B are $(2, 2)$. Using the displacement field in problem 16, determine the u and v displacements of the corners of the square. Let the deformed square be denoted as O'A'B'C'.

(a) Determine the change in lengths of OA and OC. Relate the changes to the strain components.

(b) Determine the change in $\angle AOC$. Relate the change to the shear strain.

(c) Determine the change in length in the diagonal OB. How is it related to the strain(s)?

(d) Show that the relative change in the area of the square (change in area/original area) is given by $\Delta A/A = \varepsilon_{xx} + \varepsilon_{yy} = \varepsilon_1 + \varepsilon_2$.

Hint: You can use the old-fashioned method of using set-squares (triangles) and protractor or use a spreadsheet to do the calculations. Place the origin somewhere in the bottom middle of the paper so that you have enough room to the left of the origin.

18. Draw a 2-inch $\times$ 2-inch square OPQR such that OP makes $+73°$ to the x-axis. Repeat questions (a) through (d) in problem 17 for OPQR. Give physical interpretations to your results.

Note: The principal strains and the principal strain directions are given by:

$$\varepsilon_{1,2} = \frac{(\varepsilon_{xx} + \varepsilon_{yy})}{2} \pm \sqrt{\left(\frac{\varepsilon_{xx} - \varepsilon_{yy}}{2}\right)^2 + \left(\frac{\gamma_{xy}}{2}\right)^2}$$

$$\tan 2\theta = \frac{\gamma_{xy}}{\varepsilon_{xx} - \varepsilon_{yy}}$$

19. For steel, the following material data are applicable: Young's modulus $E = 207$ GPa and shear modulus $G = 80$ GPa. For the strain matrix at a point shown below, determine the symmetric 3×3 stress matrix.

$$[\varepsilon] = \begin{bmatrix} 0.003 & 0 & -0.006 \\ 0 & -0.001 & 0.003 \\ -0.006 & 0.003 & 0.0015 \end{bmatrix}.$$

20. Strain at a point is such that $\varepsilon_{xx} = \varepsilon_{yy} = 0$, $\varepsilon_{zz} = -0.001$, $\varepsilon_{xy} = 0.006$, and $\varepsilon_{xz} = \varepsilon_{yz} = 0$. Note: You need not solve the eigenvalue problem for this question.

(a) Show that $\mathbf{n}^1 = \mathbf{i} + \mathbf{j}$ and $\mathbf{n}^2 = -\mathbf{i} + \mathbf{j}$ are principal directions of strain at this point.

(b) What is the third principal direction?

(c) Compute the three principal strains.

21. Derive the stress–strain relationship in eq. (5.60) from eq. (5.55) and the plane stress conditions.

22. A thin plate of width b, thickness t, and length L is placed between two frictionless rigid walls a distance b apart and is acted on by an axial force P. The material properties are Young's modulus E and Poisson's ratio ν.

   (a) Find the stress and strain components in the xyz coordinate system.

   (b) Find the displacement field.

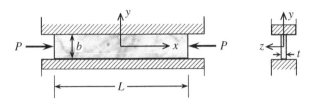

23. A solid with Young's modulus $E = 70$ GPa and Poisson's ratio $= 0.3$ is in a state of ***plane strain*** parallel to the xy-plane. The in-plane strain components are measured as: $\varepsilon_{xx} = 0.007$, $\varepsilon_{yy} = -0.008$, and $\gamma_{xy} = 0.02$.

   (a) Compute the principal strains and corresponding principal strain directions.

   (b) Compute the stresses, including σ_{zz}, corresponding to the above strains.

   (c) Determine the principal stresses and corresponding principal stress directions. Are the principal stress and principal strain directions the same?

   (d) Show the principal stresses could have been obtained from the principal strains using the stress-strain relations.

   (e) Compute the strain energy density using the stress and strain components in the xy-coordinate system.

   (f) Compute the strain energy density using the principal stresses and principal strains.

24. Assume that the solid in problem 23 is under a state of plane stress. Repeat (b) through (f).

25. A strain rosette consisting of three strain gauges was used to measure the strains at a point in a thin-walled plate. The measured strains in the three gauges are: $\varepsilon_A = 0.001$, $\varepsilon_B = -0.0006$, and $\varepsilon_C = 0.0007$. Note that gauge C is at 45° with respect to the x-axis. Assume the plane stress state.

   (a) Determine the complete state of strains and stresses (all 6 components) at that point. Assume $E = 70$ GPa, and $\nu = 0.3$.

   (b) What are the principal strains and their directions?

   (c) What are the principal stresses and their directions?

   (d) Show that the principal strains and stresses satisfy the stress-strain relations.

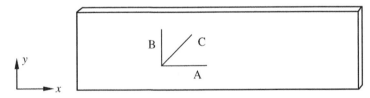

26. A strain rosette consisting of three strain gauges was used to measure the strains at a point in a thin-walled plate. The measured strains in the three gauges are: $\varepsilon_A = 0.016$, $\varepsilon_B = 0.004$, and $\varepsilon_C = 0.016$. Determine the complete state of strains and stresses (all six components) at that point. Assume $E = 100$ GPa, and $\nu = 0.3$.

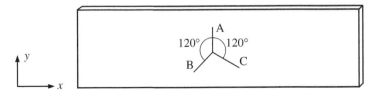

27. A strain rosette consisting of three strain gauges was used to measure the strains at a point in a thin plate of dimensions $100 \times 20 \times 1$ mm. The measured strains in the three gauges are: $\varepsilon_A = 0.008$, $\varepsilon_B = 0.002$, and $\varepsilon_C = 0.008$. Assume $E = 100$ GPa, and $\nu = 0.3$.

    (a) Assuming the plate is in a state of plane stress, determine the complete state of strains and stresses (all six components) at that point.

    (b) Assuming the state of stress is uniform, calculate the strain energy stored in the plate.

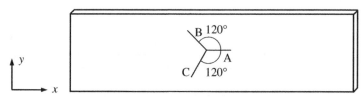

28. The figure below illustrates a thin plate of thickness t. An approximate displacement field, which accounts for displacements due to the weight of the plate, is given by

$$u(x,y) = \frac{\rho}{2E}\left(2bx - x^2 - \nu y^2\right),$$

$$v(x,y) = -\frac{\nu\rho}{E}y(b-x).$$

    (a) Determine the corresponding plane stress field.

    (b) Qualitatively draw the deformed shape of the plate.

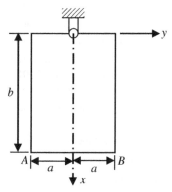

29. The stress matrix at a particular point in a body is

$$[\boldsymbol{\sigma}] = \begin{bmatrix} -2 & 1 & -3 \\ 1 & 0 & 4 \\ -3 & 4 & 5 \end{bmatrix} \times 10^7 \ \text{Pa}.$$

Determine the corresponding strain if $E = 20 \times 10^{10}$ Pa and $\nu = 0.3$.

30. For a *plane stress* problem, the strain components in the *xy*-plane at a point P are computed as:

$$\varepsilon_{xx} = \varepsilon_{yy} = .125 \times 10^{-2}, \quad \varepsilon_{xy} = .25 \times 10^{-2}.$$

    (a) Compute the state of stress at this point if Young's modulus $E = 2 \times 10^{11}$ Pa and Poisson's ratio $\nu = 0.3$.

    (b) What is the normal strain in the z-direction?

    (c) Compute the normal strain in the direction of $\mathbf{n} = \{1, 1, 1\}^T$.

31. The state of strain at a point P in a structure is

$$[\varepsilon] = \begin{bmatrix} 0.125 & .25 & 0 \\ .25 & 0.125 & 0 \\ 0 & 0 & .1 \end{bmatrix} \times 10^{-2}.$$

(a) Compute the extensional strain in the direction $\mathbf{n} = \{1 \ \ 1 \ \ 1\}^{T}$.

(b) If $\hat{\mathbf{n}}_1 = \{1 \ \ 1 \ \ 0\}^{T}/\sqrt{2}$ and $\hat{\mathbf{n}}_2 = \{-1 \ \ 1 \ \ 0\}^{T}/\sqrt{2}$ are principal directions, compute the three principal strains at this point.

(c) Compute the principal *stresses* at this point if Young's Modulus $E = 2\times10^{11}$ Pa and Poisson's ratio $\nu = 0.3$. (Use principal strains from part (b)).

(d) Can the state of stress at this point be described as *plane stress*? Explain.

(e) Compute the strain energy density at the point P.

(f) If the yield stress of the material of the plate is $\sigma_y = 2 \times 10^8$ Pa, has this plate undergone plastic deformation due to the applied strain? Use maximum shear stress criterion (Tresca's law) to answer this question.

32. The state of stress at a point is given by

$$[\sigma] = \begin{bmatrix} 80 & 20 & 40 \\ 20 & 60 & 10 \\ 40 & 10 & 20 \end{bmatrix} \text{MPa}.$$

(a) Determine the strains using Young's modulus of 100 GPa and Poisson's ratio of 0.25.

(b) Compute the strain energy density using these stresses and strains.

(c) Calculate the principal stresses.

(d) Calculate the principal strains from the strains calculated in (a).

(e) Show that the principal stresses and principal strains satisfy the constitutive relations.

(f) Calculate the strain energy density using the principal stresses and strains.

33. Consider the state of stress in problem 32. The yield strength of the material is 100 MPa. Determine the safety factors according to: (a) maximum principal stress criterion, (b) Tresca criterion, and (c) von Mises criterion.

34. A thin-walled tube is subject to a torque T. The only nonzero stress component is the shear stress τ_{xy}, which is given by $\tau_{xy} = 10,000 \ T$ (Pa), where T is the torque in N.m. When the yield strength $\sigma_Y = 300$ MPa and the safety factor $N = 2$, calculate the maximum torque that can be applied using:

(a) maximum principal stress criterion (Rankine);

(b) maximum shear stress criterion (Tresca); and

(c) distortion energy criterion (Von Mises).

35. A thin-walled cylindrical pressure vessel with closed ends is subjected to an internal pressure $p = 100$ psi and also a torque T about its axis of symmetry. Determine T that will cause yielding according to von Mises' yield criterion. The design requires a safety factor of 2. The nominal diameter D of the pressure vessel = 20 inch, wall thickness $t = 0.1$ in., the yield strength of the material = 30 ksi. (1 ksi = 1000 psi). Stresses in a thin-walled cylinder are: longitudinal stress σ_l, hoop stress σ_h, and shear stress τ due to torsion. They are given by

$$\sigma_l = \frac{pD}{4t}, \sigma_h = \frac{pD}{2t}, \tau = \frac{2T}{\pi D^2 t}.$$

36. A cold-rolled steel shaft is used to transmit 60 kW at 500 rpm from a motor. What should be the diameter of the shaft, if the shaft is 6 m long and is simply supported at both ends? The shaft also experiences bending due to a distributed transverse load of 200 N/m. Ignore bending due to the weight of the shaft. Use a safety factor of 2. The tensile yield limit is 280 MPa. Find the diameter using both maximum shear stress theory and von Mises' criterion for failure.

37. For the stress matrix below, the two principal stresses are given as $\sigma_1 = 2$ and $\sigma_3 = -3$, respectively. In addition, two principal directions corresponding to the two principal stresses are also given below. The yield stress of the structure is given as $\sigma_Y = 4.5$.

$$[\sigma] = \begin{bmatrix} 1 & 0 & 2 \\ 0 & 1 & 0 \\ 2 & 0 & -2 \end{bmatrix}, \mathbf{n}^1 = \begin{bmatrix} \dfrac{2}{\sqrt{5}} \\ 0 \\ \dfrac{1}{\sqrt{5}} \end{bmatrix} \text{ and } \mathbf{n}^3 = \begin{bmatrix} \dfrac{1}{\sqrt{5}} \\ 0 \\ -\dfrac{2}{\sqrt{5}} \end{bmatrix}.$$

(a) Calculate the safety factor based on the maximum shear stress theory and determine whether the structure is safe.

(b) Calculate the safety factor based on the distortion energy theory and determine whether the structure is safe or not.

38. The figure below shows a round shaft of diameter 1.5 in. loaded by a bending moment $M_z = 5,000$ lb.·in., a torque $T = 8,000$ lb.·in., and an axial tensile force $N = 6,000$ lb. If the material is ductile with the yielding stress $\sigma_Y = 40,000$ psi, determine the safety factor corresponding to yield using (a) the maximum shear stress theory and (b) the maximum distortion energy theory.

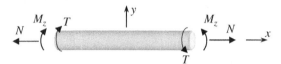

39. A 20-mm diameter rod made of a ductile material with a yield strength of 350 MPa is subject to a torque of $T = 100$ N·m and a bending moment of $M = 150$ N·m. An axial tensile force P is then gradually applied. What is the value of the axial force when yielding of the rod occurs? Solve the problem two ways using (a) the maximum shear stress theory and (b) the maximum distortional energy theory.

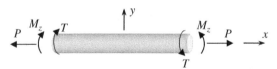

40. A circular shaft of radius r in the figure has a moment of inertia I and polar moment of inertia J. The shaft is under torsion T_z in the positive z-axis and bending moment M_x in the positive x-axis. The material is mild steel with a yield strength of 2.8 MPa. Use only the given coordinate system for your calculations.

(a) If T_z and M_x are gradually increased, which point (or points) will fail first among four points (A, B, C, and D)? Identify all.

(b) Construct the stress matrix $[\sigma]_A$ at point A in xyz-coordinates in terms of the given parameters (i.e., T_z, M_x, I, J, and r).

(c) Calculate three principal stresses at point B in terms of the given parameters.

(d) When the principal stresses at point C are $\sigma_1 = 1$, $\sigma_2 = 0$, and $\sigma_3 = -2$ MPa, calculate safety factors (1) from maximum shear stress theory and (2) from distortion energy theory.

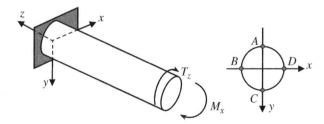

41. A rectangular plastic specimen of size $100 \times 100 \times 10 \text{ mm}^3$ is placed in a rectangular metal cavity. The dimensions of the cavity are $101 \times 101 \times 9 \text{ mm}^3$. The plastic is compressed by a rigid punch until it is completely inside the cavity. Due to Poisson effect, the plastic also expands in the x- and y-direction and fill the cavity. Calculate all stress and strain components and the force exerted by the punch. Assume there is no friction between all contacting surfaces. The metal cavity is rigid. Elastic constants of the plastic are: $E = 10 \text{ GPa}$, $\nu = 0.3$.

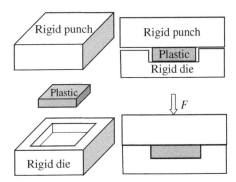

42. Repeat problem 41 with elastic constants of the plastic as $E = 10 \text{ GPa}$ and $\nu = 0.485$.

43. Repeat problem 41 with the specimen of size $100 \times 100 \times 10 \text{ mm}^3$ and the dimensions of the cavity $104 \times 104 \times 9 \text{ mm}^3$. Elastic constants of the plastic are: $E = 10 \text{ GPa}$, $\nu = 0.3$.

Chapter **6**

Finite Elements for Two-Dimensional Solid Mechanics

6.1 INTRODUCTION

All real-life structures are three–dimensional. It is engineers who make the approximation as a one-dimensional (e.g., beam) or a two-dimensional structure (e.g., plate or plane solid). In chapter 5 we explained in detail the conditions under which such approximation could be made. When the stresses on a plane normal to one of the axes are approximately zero, then we say that the solid is in the state of plane stress. Similarly, when the corresponding strains are zero, the solid is in the state of plane strain. A two-dimensional solid is also called a plane solid. Some examples of plane solids are: (1) a thin plate subjected to in-plane forces; or (2) a very thick solid with a constant cross-section in the thickness direction. In this chapter, we will discuss when an engineering problem can be assumed to be two-dimensional and how to solve such a problem using two-dimensional finite elements. We will introduce three different types of two-dimensional problems and corresponding two-dimensional elements. Every element has its own characteristics. In order to use the finite element method appropriately, a thorough understanding of the capabilities and limitations of each element is required.

In general, two-dimensional elasticity problem can be expressed by a system of coupled second-order partial differential equations. Based on the constraints imposed in the thickness direction, a two-dimensional problem can be a plane stress or plane strain problem. The main difference is in the stress-strain relations. Yet another type of problem that can be classified as two-dimensional is the axisymmetric problem where the structure has symmetry about an axis and it deforms only in the radial and axial directions.

In an elasticity problem, the main variables are the displacements in the coordinate directions. After solving for the displacements, stresses and strains can be calculated from the derivatives of displacements. The displacements are calculated using the fact that the structure is in equilibrium when the total potential energy has its minimum value. This will yield a matrix equation similar to the beam problem in chapter 3.

6.2 TYPES OF TWO-DIMENSIONAL PROBLEMS

6.2.1 Governing Differential Equations

In two-dimensional problems, the stresses and strains are independent of the coordinate in the thickness direction, (usually the z-axis). By setting all the derivatives with respect to the z-coordinate in eq. (5.69) in chapter 5 to zero, we obtain the governing differential equations for plane problems as

---

Introduction to Finite Element Analysis and Design, Second Edition. Nam H. Kim, Bhavani V. Sankar, and Ashok V. Kumar.
© 2018 John Wiley & Sons Ltd. Published 2018 by John Wiley & Sons Ltd.
Companion website: www.wiley.com/go/kim/finite_element_analysis_design

$$\begin{cases} \dfrac{\partial \sigma_{xx}}{\partial x} + \dfrac{\partial \tau_{xy}}{\partial y} + b_x = 0 \\[3mm] \dfrac{\partial \tau_{xy}}{\partial x} + \dfrac{\partial \sigma_{yy}}{\partial y} + b_y = 0, \end{cases} \tag{6.1}$$

where b_x and b_y are the body forces.

Let u and v be the displacement in the x- and y-direction, respectively. From eqs. (5.39)–(5.45), the strain components in a plane solid are defined as

$$\varepsilon_{xx} = \frac{\partial u}{\partial x}, \qquad \varepsilon_{yy} = \frac{\partial v}{\partial y}, \qquad \gamma_{xy} = \left(\frac{\partial u}{\partial y} + \frac{\partial v}{\partial x} \right). \tag{6.2}$$

In addition, the stresses and strains are related by the following constitutive relation:

$$\begin{Bmatrix} \sigma_{xx} \\ \sigma_{yy} \\ \tau_{xy} \end{Bmatrix} = \begin{bmatrix} C_{11} & C_{12} & C_{13} \\ C_{21} & C_{22} & C_{23} \\ C_{31} & C_{32} & C_{33} \end{bmatrix} \begin{Bmatrix} \varepsilon_{xx} \\ \varepsilon_{yy} \\ \gamma_{xy} \end{Bmatrix} \quad \Leftrightarrow \quad \{\boldsymbol{\sigma}\} = [\mathbf{C}]\{\boldsymbol{\varepsilon}\}. \tag{6.3}$$

Substituting the stress-strain relations in eq. (6.3) and strain-displacement relations in eq. (6.2) in the equilibrium equations in eq. (6.1), we obtain a pair of second-order partial differential equations in two variables $u(x,y)$ and $v(x,y)$. The explicit form of the equations is available in textbooks on elasticity, for instance, Timoshenko and Goodier[1].

The differential equation must be accompanied by boundary conditions. Two types of boundary conditions can be defined. The first one is the boundary in which the values of displacements are prescribed (*essential boundary condition*). The other is the boundary in which the tractions are prescribed (*natural boundary condition*). The boundary conditions can be formally stated as

$$\mathbf{u} = \mathbf{g}, \qquad \text{on } S_g$$
$$\boldsymbol{\sigma}\mathbf{n} = \mathbf{T}, \qquad \text{on } S_T, \tag{6.4}$$

where S_g and S_T, respectively, are the boundaries where the displacement and traction boundary conditions are prescribed. The objective is to determine the displacement field $u(x,y)$ and $v(x,y)$ that satisfies the differential equation (6.1) and the boundary conditions in eq. (6.4). Now, we will discuss the stress-strain relations in eq. (6.3) for the two different plane problems—plane stress and plane strain. The third type of 2D problem, namely axisymmetric problems, will be discussed later in the chapter.

6.2.2   Plane Stress Problems

Plane stress conditions exist when the thickness dimension (usually the z-direction) is much smaller than the length and width dimensions of a solid. Since stresses at the two surfaces normal to the z-axis are zero, it is assumed that stresses in the normal direction are zero throughout the body, that is, $\sigma_{zz} = \tau_{xz} = \tau_{yz} = 0$. In such a case, the structure can be modeled in two dimensions. An example of the plane stress problem is a thin plate or disk with applied in-plane forces (see figure 6.1).

– Nonzero stress components: $\sigma_{xx}, \sigma_{yy}, \tau_{xy}$.

– Nonzero strain components: $\varepsilon_{xx}, \varepsilon_{yy}, \gamma_{xy}, \varepsilon_{zz}$.

---

[1] Timoshenko, S.P., and Goodier, J.N. *Theory of Elasticity*. McGraw-Hill.

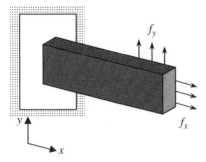

Figure 6.1 Thin plate with in-plane applied forces

Under plane stress conditions, for linear isotropic materials, the stress-strain relation can be written as (see section 5.3)

$$
\begin{Bmatrix} \sigma_{xx} \\ \sigma_{yy} \\ \tau_{xy} \end{Bmatrix} = \frac{E}{1-\nu^2} \begin{bmatrix} 1 & \nu & 0 \\ \nu & 1 & 0 \\ 0 & 0 & \frac{1-\nu}{2} \end{bmatrix} \begin{Bmatrix} \varepsilon_{xx} \\ \varepsilon_{yy} \\ \gamma_{xy} \end{Bmatrix}
$$

(6.5)

$$\Leftrightarrow \quad \{\boldsymbol{\sigma}\} = [\mathbf{C}_\sigma]\{\boldsymbol{\varepsilon}\},$$

where $[\mathbf{C}_\sigma]$ is the stress-strain matrix or elasticity matrix for the plane stress problem. It should be noted that the normal strain ε_{zz} in the thickness direction is not zero; it can be calculated from the following relation:

$$
\varepsilon_{zz} = -\frac{\nu}{E}\left(\sigma_{xx} + \sigma_{yy}\right).
$$

(6.6)

6.2.3   Plane Strain Problems

A state of plane strain will exist in a solid when the thickness dimension is much larger than other two dimensions. When the deformation in the thickness direction is constrained, the solid is assumed to be in a state of plane strain even if the thickness dimension is small. A proper assumption is that strain components with z subscript are zero, that is, $\varepsilon_{zz} = \gamma_{xz} = \gamma_{yz} = 0$. In such a case, it is sufficient to model a slice of the solid with unit thickness. Some examples of plane strain problems are the retaining wall of a dam and long cylinder such as a gun barrel (see figure 6.2).

– Nonzero stress components: σ_{xx}, σ_{yy}, τ_{xy}, σ_{zz}.

– Nonzero strain components: ε_{xx}, ε_{yy}, γ_{xy}.

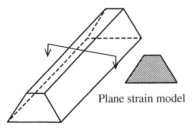

Plane strain model

Figure 6.2 Dam structure with plane strain assumption

For linear isotropic materials, the stress-strain relations under plane strain conditions can be written as (see section 5.3)

$$\begin{Bmatrix} \sigma_{xx} \\ \sigma_{yy} \\ \tau_{xy} \end{Bmatrix} = \frac{E}{(1+\nu)(1-2\nu)} \begin{bmatrix} 1-\nu & \nu & 0 \\ \nu & 1-\nu & 0 \\ 0 & 0 & \frac{1}{2}-\nu \end{bmatrix} \begin{Bmatrix} \varepsilon_{xx} \\ \varepsilon_{yy} \\ \gamma_{xy} \end{Bmatrix} \tag{6.7}$$

$$\Leftrightarrow \{\sigma\} = [\mathbf{C}_\varepsilon]\{\varepsilon\},$$

where $[\mathbf{C}_\varepsilon]$ is the stress-strain matrix for the plane strain problem. It should be noted that the transverse stress σ_{zz} is not equal to zero in the plane strain case, and it can be calculated from the following relation:

$$\sigma_{zz} = \frac{E\nu}{(1+\nu)(1-2\nu)}\left(\varepsilon_{xx} + \varepsilon_{yy}\right). \tag{6.8}$$

6.2.4 Equivalence between Plane Stress and Plane Strain Problems

Although plane stress and plane strain problems are different by definition, they are quite similar from the computational viewpoint. Thus, it is possible to use the plane strain formulation and solve the plane stress problem. In such case, two material properties, E and ν, need to be modified. Similarly, it is also possible to convert the plane stress formulation into the plane strain formulation. Table 6.1 summarizes the conversion relations.

Table 6.1 Material property conversion between plane strain and plane stress problems

From → To	E	ν
Plane strain → Plane stress	$E\left[1-\left(\dfrac{\nu}{1+\nu}\right)^2\right]$	$\dfrac{\nu}{1+\nu}$
Plane stress → Plane strain	$\dfrac{E}{1-\left(\dfrac{\nu}{1-\nu}\right)^2}$	$\dfrac{\nu}{1-\nu}$

6.3 CONSTANT STRAIN TRIANGULAR (CST) ELEMENT

In finite element analysis, a plane solid can be divided into a number of contiguous elements. In chapter 4, we introduced the 3-node triangular element for heat transfer problems. In this section, we will use the same element for plane solid problems. Figure 6.3 shows a typical mesh consisting of triangular elements. The three nodes of the element, as shown in figure 6.3, have two displacement components, u and v, respectively in the x- and y-direction for two-dimensional solid mechanics problems. The first node of an element can arbitrarily be chosen. However, the sequence of the nodes 1, 2, and 3 should be in the counterclockwise direction.

The nodal values of the displacement components are interpolated within the element using shape functions. In the polynomial approximation, the displacement has to be a linear function in x and y because displacement information is available only at three points (nodes), and a linear polynomial has three unknown coefficients. Since displacement is a linear function, strain and stress are constant within an element, and that is why a triangular element is called a *constant strain triangular element*.

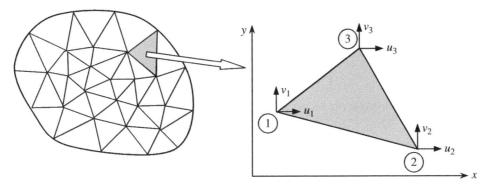

Figure 6.3 Constant strain triangular (CST) element

6.3.1 Displacement and Strain Interpolation

The first step in deriving the finite element matrix equation is to interpolate the displacement function within the element using the nodal values of displacements. Let the x- and y-directional displacements be $u(x,y)$ and $v(x,y)$, respectively. Since the two coordinates are perpendicular to each other (orthogonal), $u(x,y)$ and $v(x,y)$ are independent of each other. Hence, $u(x,y)$ needs to be interpolated in terms of u_1, u_2, and u_3, and $v(x,y)$ in terms of v_1, v_2, and v_3. It is obvious that the interpolation function must be a three-term polynomial in x and y. Since we must have rigid body displacements (constant displacements) and constant strain terms in the interpolation function, the displacement interpolation must be of the following form:

$$\begin{cases} u(x,y) = \alpha_1 + \alpha_2 x + \alpha_3 y \\ v(x,y) = \beta_1 + \beta_2 x + \beta_3 y, \end{cases} \tag{6.9}$$

where α's and β's are constants. In finite element analysis, we would like to replace the constants by the nodal displacements. Let us consider x-directional displacements, which are u_1, u_2, and u_3. At node 1, for example, x and y take the values of x_1 and y_1, respectively, and the nodal displacement is u_1. If we repeat this for other two nodes, we obtain the following three simultaneous equations:

$$\begin{cases} u(x_1,y_1) \equiv u_1 = \alpha_1 + \alpha_2 x_1 + \alpha_3 y_1 \\ u(x_2,y_2) \equiv u_2 = \alpha_1 + \alpha_2 x_2 + \alpha_3 y_2 \\ u(x_3,y_3) \equiv u_3 = \alpha_1 + \alpha_2 x_3 + \alpha_3 y_3. \end{cases} \tag{6.10}$$

In matrix notation, the above equations can be written as

$$\begin{Bmatrix} u_1 \\ u_2 \\ u_3 \end{Bmatrix} = \begin{bmatrix} 1 & x_1 & y_1 \\ 1 & x_2 & y_2 \\ 1 & x_3 & y_3 \end{bmatrix} \begin{Bmatrix} \alpha_1 \\ \alpha_2 \\ \alpha_3 \end{Bmatrix}. \tag{6.11}$$

If the three points (x_1, y_1), (x_2, y_2), and (x_3, y_3) are not on a straight line, then the inverse of the above coefficient matrix exists. Thus, we can calculate the unknown coefficients as

$$\begin{Bmatrix} \alpha_1 \\ \alpha_2 \\ \alpha_3 \end{Bmatrix} = \begin{bmatrix} 1 & x_1 & y_1 \\ 1 & x_2 & y_2 \\ 1 & x_3 & y_3 \end{bmatrix}^{-1} \begin{Bmatrix} u_1 \\ u_2 \\ u_3 \end{Bmatrix} = \frac{1}{2A} \begin{bmatrix} f_1 & f_2 & f_3 \\ b_1 & b_2 & b_3 \\ c_1 & c_2 & c_3 \end{bmatrix} \begin{Bmatrix} u_1 \\ u_2 \\ u_3 \end{Bmatrix}. \tag{6.12}$$

where A is the area of the triangle and

$$
\begin{cases}
f_1 = x_2 y_3 - x_3 y_2, & b_1 = y_2 - y_3, & c_1 = x_3 - x_2 \\
f_2 = x_3 y_1 - x_1 y_3, & b_2 = y_3 - y_1, & c_2 = x_1 - x_3 \\
f_3 = x_1 y_2 - x_2 y_1, & b_3 = y_1 - y_2, & c_3 = x_2 - x_1.
\end{cases}
\tag{6.13}
$$

The area A of the triangle can be calculated from

$$
A = \frac{1}{2} \det \begin{vmatrix} 1 & x_1 & y_1 \\ 1 & x_2 & y_2 \\ 1 & x_3 & y_3 \end{vmatrix}.
\tag{6.14}
$$

Note that the determinant in eq. (6.14) is zero when three nodes are collinear. In such a case, the area of the triangular element is zero, and we cannot uniquely determine the three coefficients.

A similar procedure can be applied to the y-directional displacement $v(x, y)$, and the unknown coefficients β_i, $(i = 1, 2, 3)$ are determined using the following equation:

$$
\begin{Bmatrix} \beta_1 \\ \beta_2 \\ \beta_3 \end{Bmatrix} = \frac{1}{2A} \begin{bmatrix} f_1 & f_2 & f_3 \\ b_1 & b_2 & b_3 \\ c_1 & c_2 & c_3 \end{bmatrix} \begin{Bmatrix} v_1 \\ v_2 \\ v_3 \end{Bmatrix}.
\tag{6.15}
$$

After calculating α_i and β_i, the displacement interpolation can be written as

$$
u(x,y) = \begin{bmatrix} N_1 & N_2 & N_3 \end{bmatrix} \begin{Bmatrix} u_1 \\ u_2 \\ u_3 \end{Bmatrix} \quad \text{and} \quad v(x,y) = \begin{bmatrix} N_1 & N_2 & N_3 \end{bmatrix} \begin{Bmatrix} v_1 \\ v_2 \\ v_3 \end{Bmatrix},
\tag{6.16}
$$

where the shape functions are defined by

$$
\begin{cases}
N_1(x,y) = \dfrac{1}{2A}(f_1 + b_1 x + c_1 y) \\[2mm]
N_2(x,y) = \dfrac{1}{2A}(f_2 + b_2 x + c_2 y) \\[2mm]
N_3(x,y) = \dfrac{1}{2A}(f_3 + b_3 x + c_3 y).
\end{cases}
\tag{6.17}
$$

Note that N_1, N_2, and N_3 are linear functions of x- and y-coordinates. Thus, interpolated displacement varies linearly in each coordinate direction.

In order to make the derivations simple, we will rewrite the interpolation relation in eq. (6.16) in matrix form. Let $\{\mathbf{u}\} = \{u, v\}^\mathrm{T}$ be the displacement vector at any point (x, y). The interpolation can be written in the matrix notation by

$$
\{\mathbf{u}\} \equiv \begin{Bmatrix} u \\ v \end{Bmatrix} = \begin{bmatrix} N_1 & 0 & N_2 & 0 & N_3 & 0 \\ 0 & N_1 & 0 & N_2 & 0 & N_3 \end{bmatrix} \begin{Bmatrix} u_1 \\ v_1 \\ u_2 \\ v_2 \\ u_3 \\ v_3 \end{Bmatrix},
$$

or

$$
\boxed{\{\mathbf{u}(x,y)\} = [\mathbf{N}(x,y)]\{\mathbf{q}\}}.
\tag{6.18}
$$

Equation (6.18) is the critical relationship in finite element approximation. When a point (x, y) within a triangular element is given, the shape function $[\mathbf{N}]$ is calculated at this point. Then, the displacement at this point can be calculated by multiplying this shape function matrix with the nodal displacement vector $\{\mathbf{q}\}$. Thus, if we solve for the nodal displacements, we can calculate the displacement everywhere in the element. Note that the nodal displacements will be evaluated using the principle of minimum total potential energy in the following section.

After calculating displacement within an element, it is possible to calculate the strain by differentiating the displacement with respect to x and y. From the expression in eq. (6.18), it can be noted that the nodal displacements are constant, but the shape functions are functions of x and y. Thus, the strain can be calculated by differentiating the shape function with respect to the coordinates. For example, ε_{xx} can be written as

$$\varepsilon_{xx} \equiv \frac{\partial u}{\partial x} = \frac{\partial}{\partial x}\left(\sum_{i=1}^{3} N_i(x,y)u_i\right) = \sum_{i=1}^{3}\frac{\partial N_i}{\partial x}u_i = \sum_{i=1}^{3}\frac{b_i}{2A}u_i. \tag{6.19}$$

Note that u_1, u_2, and u_3 are nodal displacements, and they are independent of coordinate x. Thus, only the shape function is differentiated with respect to x. A similar calculation can be carried out for ε_{yy} and γ_{xy}. Using the matrix notation, we have

$$\{\varepsilon\} = \left\{\begin{array}{c} \dfrac{\partial u}{\partial x} \\[2mm] \dfrac{\partial v}{\partial y} \\[2mm] \dfrac{\partial u}{\partial y} + \dfrac{\partial v}{\partial x} \end{array}\right\} = \frac{1}{2A}\begin{bmatrix} b_1 & 0 & b_2 & 0 & b_3 & 0 \\ 0 & c_1 & 0 & c_2 & 0 & c_3 \\ c_1 & b_1 & c_2 & b_2 & c_3 & b_3 \end{bmatrix}\left\{\begin{array}{c} u_1 \\ v_1 \\ u_2 \\ v_2 \\ u_3 \\ v_3 \end{array}\right\} \equiv [\mathbf{B}]\{\mathbf{q}\}. \tag{6.20}$$

It may be noted that the $[\mathbf{B}]$ matrix is constant and depends only on the coordinates of the three nodes of the triangular element. Thus, one can anticipate that if this element is used, then the computed strains will be constant over a given element and will depend only on nodal displacements. Hence, this element is called the **Constant Strain Triangular Element** or **CST** element. The interpolation of displacement in eq. (6.18) and the expression for strain in eq. (6.20) are used for approximating the strain energy and potential energy of applied loads.

6.3.2 Properties of the CST Element

Before we derive the strain energy, it may be useful to study some interesting aspects of the CST element. Since the displacement field is assumed to be a linear function in x and y, one can show that the triangular element deforms into another triangle when forces are applied. Furthermore, an imaginary straight line drawn within an element before deformation becomes another straight line after deformation.

Let us consider the displacements of points along one of the edges of the triangle. Consider the points along the edge 1–2 in figure 6.4. These points can be conveniently represented by a coordinate ξ. The coordinate $\xi = 0$ at node 1 and $\xi = a$ at node 2. Along this edge, x and y are related to ξ. By substituting this relation in the displacement functions, one can express the displacements of points on the edge 1–2 as a function of ξ. It can be easily shown that the displacement functions, for both u and v, must be linear in ξ, that is,

$$\begin{cases} u(\xi) = \gamma_1 + \gamma_2\xi \\ v(\xi) = \gamma_3 + \gamma_4\xi, \end{cases}$$

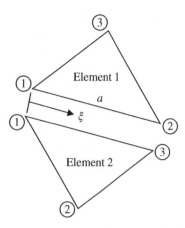

Figure 6.4 Inter-element displacement compatibility of constant strain triangular element

where γ's are constants to be determined. Since the variation of displacement is linear, it might be argued that the displacements should depend only on u_1 and u_2, and not on u_3. Then, the displacement field along the edge 1-2 takes the form:

$$\begin{cases} u(\xi) = \left(1 - \dfrac{\xi}{a}\right)u_1 + \dfrac{\xi}{a}u_2 = H_1(\xi)u_1 + H_2(\xi)u_2 \\ v(\xi) = \left(1 - \dfrac{\xi}{a}\right)v_1 + \dfrac{\xi}{a}v_2 = H_1(\xi)v_1 + H_2(\xi)v_2, \end{cases} \tag{6.21}$$

where H_1 and H_2 are shape functions defined along the edge 1-2, and a is the length of edge 1-2. One can also note that a condition called inter-element displacement compatibility is satisfied by triangular elements. This condition can be described as follows: any point should have a unique displacement including along shared edges between elements. After the loads are applied and the solid is deformed, the displacements at any point in an element can be computed from the nodal displacements of that particular element and the interpolation functions in eq. (6.18). Consider a point on a common edge of two adjacent elements. This point can be considered as belonging to either of the elements. Then the nodes of either triangle can be used in interpolating the displacements of this point. However, one must obtain a unique set of displacements independent of the choice of the element. This can be true only if the displacements of the points depend only on the nodes common to both elements. In fact, this will be satisfied because of eq. (6.21). Thus, the CST element satisfies the inter-element displacement compatibility.

EXAMPLE 6.1 *Interpolation in a triangular element*

Consider the two triangular elements shown in figure 6.5. The nodal displacements are given as $\{u_1, v_1, u_2, v_2, u_3, v_3, u_4, v_4\} = \{-0.1, 0, 0.1, 0, -0.1, 0, 0.1, 0\}$. Calculate displacements and strains in both elements.

SOLUTION Element 1 has nodes 1, 2, and 4. Then, using the nodal coordinates, we can derive the shape functions as shown below:

$$\begin{array}{lll} x_1 = 0 & x_2 = 1 & x_3 = 0 \\ y_1 = 0 & y_2 = 0 & y_3 = 1 \\ f_1 = 1 & f_2 = 0 & f_3 = 0 \\ b_1 = -1 & b_2 = 1 & b_3 = 0 \\ c_1 = -1 & c_2 = 0 & c_3 = 1. \end{array}$$

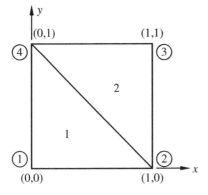

Figure 6.5 Interpolation of displacements in triangular elements

In addition, the area of the element is 0.5. Thus, from eq. (6.17) the shape functions can be derived as

$$N_1(x,y) = 1-x-y,$$
$$N_2(x,y) = x,$$
$$N_3(x,y) = y.$$

Then, the displacements in element 1 can be interpolated as

$$u^{(1)}(x,y) = \sum_{I=1}^{3} N_I(x,y)u_I = 0.1(2x+2y-1),$$
$$v^{(1)}(x,y) = \sum_{I=1}^{3} N_I(x,y)v_I = 0.0.$$

Strains can be calculated from eq. (6.20), or directly differentiating the above expressions for displacements, as

$$\varepsilon_{xx}^{(1)} = \frac{\partial u^{(1)}}{\partial x} = 0.2,$$
$$\varepsilon_{yy}^{(1)} = \frac{\partial v^{(1)}}{\partial y} = 0.0,$$
$$\gamma_{xy}^{(1)} = \frac{\partial u^{(1)}}{\partial y} + \frac{\partial v^{(1)}}{\partial x} = 0.2.$$

Element 2 connects nodes $2-3-4$. Thus, using the nodal coordinates, we can build the shape functions, as

$$
\begin{array}{lll}
x_1 = 1 & x_2 = 1 & x_3 = 0 \\
y_1 = 0 & y_2 = 1 & y_3 = 1 \\
f_1 = 1 & f_2 = -1 & f_3 = 1 \\
b_1 = 0 & b_2 = 1 & b_3 = -1 \\
c_1 = -1 & c_2 = 1 & c_3 = 0.
\end{array}
$$

In addition, the area of the element is 0.5. Thus, from eq. (6.17), the shape functions can be obtained as

$$N_1(x,y) = 1-y,$$
$$N_2(x,y) = x+y-1,$$
$$N_3(x,y) = 1-x.$$

Then, the displacements of element 2 can be interpolated as

$$u^{(2)}(x,y) = \sum_{i=1}^{3} N_i(x,y) u_i = 0.1(3 - 2x - 2y),$$

$$v^{(2)}(x,y) = \sum_{i=1}^{3} N_i(x,y) v_i = 0.0.$$

Strains can be calculated from eq. (6.20), or directly differentiating the above expressions for displacements, as

$$\varepsilon_{xx}^{(2)} = \frac{\partial u^{(2)}}{\partial x} = -0.2,$$

$$\varepsilon_{yy}^{(2)} = \frac{\partial v^{(2)}}{\partial y} = 0.0,$$

$$\gamma_{xy}^{(2)} = \frac{\partial u^{(2)}}{\partial y} + \frac{\partial v^{(2)}}{\partial x} = -0.2.$$

Note that the displacements are linear and the strains are constant in each element. From the given nodal displacements, it is clear that the top edge has strain $\varepsilon_{xx} = -0.2$, while the bottom edge has $\varepsilon_{xx} = 0.2$. The strain varies linearly along the y-coordinate. However, the triangular element cannot represent this change, and provides constant values of $\varepsilon_{xx} = 0.2$ for element 1 and $\varepsilon_{xx} = -0.2$ for element 2. In general, if a plane solid is under the constant strain states, the CST element will provide accurate solutions. However, if the strain varies in the solid, then the CST element cannot represent it accurately. In such a case, a dense mesh with many elements should be used to approximate it as a series of step functions. Note that the strains along the interface between two elements are discontinuous.

6.3.3   Strain Energy

Let us calculate the strain energy in a typical triangular element, say element e. In eq. (5.72), the strain energy of the plane solid was derived in terms of strains and the elasticity matrix $[\mathbf{C}]$. Substituting for strains from eq. (6.20), we obtain

$$
U^{(e)} = \frac{h}{2} \iint_A \{\varepsilon\}^{\mathrm{T}} [\mathbf{C}] \{\varepsilon\} \, dA^{(e)}
$$

$$
= \frac{h}{2} \left\{ \mathbf{q}^{(e)} \right\}^{\mathrm{T}} \left[\iint_A [\mathbf{B}]_{6\times3}^{\mathrm{T}} [\mathbf{C}]_{3\times3} [\mathbf{B}]_{3\times6} \, dA^{(e)} \right] \left\{ \mathbf{q}^{(e)} \right\}
\tag{6.22}
$$

$$
\equiv \frac{1}{2} \left\{ \mathbf{q}^{(e)} \right\}^{\mathrm{T}} \left[\mathbf{k}^{(e)} \right]_{6\times6} \left\{ \mathbf{q}^{(e)} \right\},
$$

where $[\mathbf{k}^{(e)}]$ is the *element stiffness matrix* of the triangular element. The column vector $\{\mathbf{q}^{(e)}\}$ contains the displacements of the three nodes that belong to the element. The dimension of $[\mathbf{k}^{(e)}]$ is 6×6. In the case of the triangular element, all entries in matrices $[\mathbf{B}]$ and $[\mathbf{C}]$ are constant and can be integrated easily. After integration, the element stiffness matrix takes the form

$$
\boxed{\left[\mathbf{k}^{(e)} \right] = h A^{(e)} [\mathbf{B}]^{\mathrm{T}} [\mathbf{C}] [\mathbf{B}]},
\tag{6.23}
$$

where $A^{(e)}$ is the area of the plane element. Using the expression for $[\mathbf{B}]$ in eq. (6.20) and stress–strain relation in eq. (6.5), the element stiffness matrix can be calculated.

One may note that in the case of truss and frame elements, we used a transformation matrix $[\mathbf{T}]$ in deriving the element stiffness matrix. However, in the present case, we have used the global coordinate system in the derivation of $[\mathbf{k}^{(e)}]$, and that is the reason for not using a transformation matrix. In some cases, however, it is required to define the element in a local coordinate system. For example, if the material is not isotropic, it will have a specific directional property. In such a case, $[\mathbf{k}^{(e)}]$ is first derived in a local coordinate system and transformed to the global coordinates by multiplying by appropriate transformation matrices.

The strain energy of the entire solid is simply the sum of the element strain energies. That is,

$$U = \sum_{e=1}^{NEL} U^{(e)} = \frac{1}{2}\sum_{e=1}^{NEL}\left\{\mathbf{q}^{(e)}\right\}^{\mathrm{T}}\left[\mathbf{k}^{(e)}\right]\left\{\mathbf{q}^{(e)}\right\}, \tag{6.24}$$

where NEL is the number of elements in the model. The superscript (e) in $\mathbf{q}^{(e)}$ implies that it is the vector of displacements or DOFs of element e. The summation in the above equation leads to the assembling of the element stiffness matrices into the global stiffness matrix.

$$U = \frac{1}{2}\{\mathbf{Q}_s\}^{\mathrm{T}}[\mathbf{K}_s]\{\mathbf{Q}_s\}, \tag{6.25}$$

where $\{\mathbf{Q}_s\}$ is the column vector of all displacements in the model and $[\mathbf{K}_s]$ is the structural stiffness matrix obtained by assembling the element stiffness matrices.

6.3.4   Potential Energy of Applied Loads

Concentrated forces at nodes: The next step is to calculate the potential energy of external forces. We will consider three different types of applied forces. The first type is concentrated forces at nodes. It may be noted that the model expects two forces, one in the x-direction and the other in the y-direction, at each node. In general, the potential energy of concentrated nodal forces can be written as

$$V = -\sum_{i=1}^{ND}\left(F_{ix}u_i + F_{iy}v_i\right) \equiv -\{\mathbf{Q}_s\}^{\mathrm{T}}\{\mathbf{F}_N\}, \tag{6.26}$$

where $\{\mathbf{F}_N\} = \begin{bmatrix} F_{1x} & F_{1y} & \cdots & F_{NDx} & F_{NDy}\end{bmatrix}^{\mathrm{T}}$ is the vector of applied nodal forces, and ND is the number of nodes in the solid. The contribution of a particular node to the potential energy will be zero if no force is applied at the node, or the displacement of the node becomes zero. The above potential energy of concentrated forces does not include a supporting reaction because the displacement at those nodes will be zero.

Distributed forces along element edges: The second type of applied force is the distributed force (traction) on the side surface of the plane solid. In the plane solid, the traction is assumed to be a constant through the thickness. Let the surface traction force $\{\mathbf{T}\} = \{T_x, T_y\}^{\mathrm{T}}$ is applied on the element edge 1-2 as shown in figure 6.6. The unit of the surface traction is Pa (N/m^2) or psi. Since the force is distributed along the edge, the potential energy of the surface traction force must be defined in the form of integral as

$$V^{(e)} = -h\int_{S_T}\{\mathbf{u}(s)\}^{\mathrm{T}}\{\mathbf{T}(s)\}\,\mathrm{d}s = -\{\mathbf{d}\}^{\mathrm{T}}h\int_{S_T}[\mathbf{H}(s)]^{\mathrm{T}}\{\mathbf{T}(s)\}\,\mathrm{d}s, \tag{6.27}$$

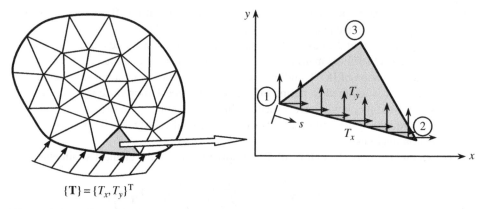

$$\{\mathbf{T}\} = \{T_x, T_y\}^T$$

Figure 6.6 Applied surface traction along edge 1-2

where $\{\mathbf{u}\} = \{u(s) \quad v(s)\}^T$ is the vector of displacements along the edge 1-2, $\{\mathbf{T}\} = \{T_x(s) \quad T_y(s)\}^T$ is the vector of applied tractions along the edge 1-2, $\{\mathbf{d}\} = \{u_1 \quad v_1 \quad u_2 \quad v_2\}^T$ is the vector of displacements of nodes 1 and 2, and

$$[\mathbf{H}] = \begin{bmatrix} H_1 & 0 & H_2 & 0 \\ 0 & H_1 & 0 & H_2 \end{bmatrix}$$

is the matrix of shape functions defined in eq. (6.21). The integration can be performed in a closed form if the specified surface tractions (T_x and T_y) are simple functions of s. We will modify eq. (6.27) to include all the six DOFs of the element and rewrite as

$$V^{(e)} = -\left\{\mathbf{q}^{(e)}\right\}^T h \int_{S_T} [\mathbf{N}(s)]^T \{\mathbf{T}(s)\} ds = -\left\{\mathbf{q}^{(e)}\right\}^T \left\{\mathbf{f}_T^{(e)}\right\}. \tag{6.28}$$

We have used the complete shape function matrix in eq. (6.28):

$$[\mathbf{N}] = \begin{bmatrix} \dfrac{l-s}{l} & 0 & \dfrac{s}{l} & 0 & 0 & 0 \\ 0 & \dfrac{l-s}{l} & 0 & \dfrac{s}{l} & 0 & 0 \end{bmatrix}. \tag{6.29}$$

If the last expression in eq. (6.28) is examined carefully, it is possible to note that the force vector $\{\mathbf{f}_T\}$ is nodal force vector that is equivalent to the distributed force applied on the edge of the element. This is also called the work-equivalent nodal force vector. For a constant surface traction T_x and T_y, we can calculate the equivalent nodal force, as

$$\left\{\mathbf{f}_T^{(e)}\right\} = h \int_0^l [\mathbf{N}]^T \{\mathbf{T}\} ds = h \int_0^l \begin{bmatrix} (l-s)/l & 0 \\ 0 & (l-s)/l \\ s/l & 0 \\ 0 & s/l \\ 0 & 0 \\ 0 & 0 \end{bmatrix} \begin{Bmatrix} T_x \\ T_y \end{Bmatrix} ds = \frac{hl}{2} \begin{Bmatrix} T_x \\ T_y \\ T_x \\ T_y \\ 0 \\ 0 \end{Bmatrix}. \tag{6.30}$$

For the uniform surface traction force, the equivalent nodal forces are obtained by simply dividing the total force equally between the two nodes on the edge.

The potential energy of distributed forces of all elements whose edge belongs to the traction boundary S_T must be assembled to build the global force vector of distributed forces:

$$V = -\sum_{e=1}^{NS} \left\{ \mathbf{q}^{(e)} \right\}^{\mathrm{T}} \left\{ \mathbf{f}_T^{(e)} \right\} = -\{\mathbf{Q}_s\}^{\mathrm{T}} \{\mathbf{F}_T\}, \tag{6.31}$$

where NS is the number of elements whose edge belongs to S_T.

Body forces: The body forces are distributed over the entire element (e.g., centrifugal forces, gravitational forces, inertia forces, or magnetic forces). For simplification of the derivation, let us assume that a constant body force $\mathbf{b} = \{b_x, b_y\}^{\mathrm{T}}$ is applied to the whole element. The potential energy of body force becomes

$$V^{(e)} = -h \iint_A \{u \;\; v\} \begin{Bmatrix} b_x \\ b_y \end{Bmatrix} dA = -\left\{ \mathbf{q}^{(e)} \right\}^{\mathrm{T}} h \iint_A [\mathbf{N}]^{\mathrm{T}} dA \begin{Bmatrix} b_x \\ b_y \end{Bmatrix}$$

$$\equiv -\left\{ \mathbf{q}^{(e)} \right\}^{\mathrm{T}} \left\{ \mathbf{f}_b^{(e)} \right\}, \tag{6.32}$$

where

$$\left\{ \mathbf{f}_b^{(e)} \right\} = \frac{hA}{3} \begin{bmatrix} 1 & 0 \\ 0 & 1 \\ 1 & 0 \\ 0 & 1 \\ 1 & 0 \\ 0 & 1 \end{bmatrix} \begin{Bmatrix} b_x \\ b_y \end{Bmatrix} = \frac{hA}{3} \begin{Bmatrix} b_x \\ b_y \\ b_x \\ b_y \\ b_x \\ b_y \end{Bmatrix}. \tag{6.33}$$

The resultant of body forces is hAb_x in the x-direction and hAb_y in the y-direction. Equation (6.33) equally distributes these forces to the three nodes. Similar to the distributed force, $\{\mathbf{f}_B\}$ is the equivalent nodal force corresponding to the constant body force.

The potential energy of body forces of all elements must be assembled to build the global force vector of body forces:

$$V = -\sum_{e=1}^{NEL} \left\{ \mathbf{q}^{(e)} \right\}^{\mathrm{T}} \left\{ \mathbf{f}_b^{(e)} \right\} = -\{\mathbf{Q}_s\}^{\mathrm{T}} \{\mathbf{F}_B\}, \tag{6.34}$$

where NEL is the number of elements.

6.3.5 Global Finite Element Equations

Since the strain energy and potential energy of applied forces are now available, let us go back to the potential energy of the triangular element. The discrete version of the potential energy becomes

$$\Pi = U + V = \frac{1}{2}\{\mathbf{Q}_s\}^{\mathrm{T}}[\mathbf{K}_s]\{\mathbf{Q}_s\} - \{\mathbf{Q}_s\}^{\mathrm{T}} \{\mathbf{F}_N + \mathbf{F}_T + \mathbf{F}_B\}. \tag{6.35}$$

The principle of minimum potential energy in chapter 2 states that the structure is in equilibrium when the potential energy is minimum. Since the potential energy in eq. (6.35) is a quadratic form, the displacement vector $\{\mathbf{Q}_s\}$, we can differentiate Π to obtain

$$\frac{\partial \Pi}{\partial \{\mathbf{Q}_s\}} = 0 \quad \Rightarrow \quad [\mathbf{K}_s]\{\mathbf{Q}_s\} = \{\mathbf{F}_N + \mathbf{F}_T + \mathbf{F}_B\}. \tag{6.36}$$

The stationary condition of the potential energy yields the global finite element matrix equations.

The assembled structural stiffness matrix $[\mathbf{K}_s]$ is singular due to the rigid body motion. After constructing the global matrix equation, the boundary conditions are applied by removing those DOFs that are fixed or prescribed. After imposing the boundary condition, the global stiffness matrix becomes nonsingular and it can be inverted to solve for the nodal displacements.

6.3.6 Calculation of Strains and Stresses

Once the nodal displacements are calculated, strains and stresses in individual elements can be calculated. First, the nodal displacement vector $\{\mathbf{q}^{(e)}\}$ for the element of interest needs to be extracted from the global displacement vector. Then, the strains and stresses in the element can be obtained from

$$\{\varepsilon\} = [\mathbf{B}]\left\{\mathbf{q}^{(e)}\right\}, \tag{6.37}$$

and

$$\{\sigma\} = [\mathbf{C}]\{\varepsilon\} = [\mathbf{C}][\mathbf{B}]\left\{\mathbf{q}^{(e)}\right\}, \tag{6.38}$$

where $[\mathbf{C}] = [\mathbf{C}_\sigma]$ for the plane stress problems and $[\mathbf{C}] = [\mathbf{C}_\varepsilon]$ for the plane strain problems.

As discussed before, stress and strain are constant within an element because matrices $[\mathbf{B}]$ and $[\mathbf{C}]$ are constant. This property can cause difficulties in interpreting the results of the finite element analysis. When two adjacent elements have different stress values, it is difficult to determine the stress value at the interface. Such discontinuity is not caused by the physics of the problem but by the inability of the triangular element in describing the continuous change of stresses across element boundary. In fact, most finite elements cannot maintain continuity of stresses across the element boundary. Most programs average the stress at the element boundaries in order to make the stress look continuous. However, as we refine the model using smaller-size elements, this discontinuity can be reduced. The following example illustrates discontinuity of stress and strain between two adjacent CST elements.

EXAMPLE 6.2 *Cantilevered plate*

Consider a cantilevered plate as shown in figure 6.7. The plate has the following properties: $h = 0.1$ in., $E = 30 \times 10^6$ psi and $\nu = 0.3$. Model the plate using two CST elements to determine the displacements and stresses.

SOLUTION This problem can be modeled as plane stress because the thickness of the plate is small compared to the other dimensions.

1. Element 1: Nodes 1–2–3

   Using nodal coordinates, we can calculate the constants defined in eq. (6.13) as

$$x_1 = 0, \ y_1 = 0 \qquad x_2 = 10, \ y_2 = 5 \qquad x_3 = 10, \ y_3 = 15$$
$$b_1 = y_2 - y_3 = -10 \qquad b_2 = y_3 - y_1 = 15 \qquad b_3 = y_1 - y_2 = -5$$
$$c_1 = x_3 - x_2 = 0 \qquad c_2 = x_1 - x_3 = -10 \qquad c_3 = x_2 - x_1 = 10.$$

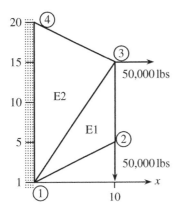

Figure 6.7 Cantilevered plate

In addition, from the geometry of the element, the area of the triangle $A_1 = 0.5 \times 10 \times 10 = 50$. Matrix $[\mathbf{B}]$ in eq. (6.20) and $[\mathbf{C}_\sigma]$ can be written as

$$[\mathbf{B}] = \frac{1}{2A} \begin{bmatrix} b_1 & 0 & b_2 & 0 & b_3 & 0 \\ 0 & c_1 & 0 & c_2 & 0 & c_3 \\ c_1 & b_1 & c_2 & b_2 & c_3 & b_3 \end{bmatrix}$$

$$= \frac{1}{100} \begin{bmatrix} -10 & 0 & 15 & 0 & -5 & 0 \\ 0 & 0 & 0 & -10 & 0 & 10 \\ 0 & -10 & -10 & 15 & 10 & -5 \end{bmatrix},$$

and

$$[\mathbf{C}_\sigma] = \frac{E}{1-\nu^2} \begin{bmatrix} 1 & \nu & 0 \\ \nu & 1 & 0 \\ 0 & 0 & \frac{1}{2}(1-\nu) \end{bmatrix} = 3.297 \times 10^7 \begin{bmatrix} 1 & .3 & 0 \\ .3 & 1 & 0 \\ 0 & 0 & .35 \end{bmatrix}.$$

Using the above two matrices, the element stiffness matrix can be obtained as

$$[\mathbf{k}^{(1)}] = hA [\mathbf{B}]^{\mathrm{T}} [\mathbf{C}_\sigma][\mathbf{B}]$$

$$= 3.297 \times 10^6 \begin{bmatrix} .5 & 0. & -.75 & .15 & .25 & -.15 \\ & .175 & .175 & -.263 & -.175 & .088 \\ & & 1.3 & -.488 & -.55 & .313 \\ & & & .894 & .338 & -.631 \\ & & & & .3 & -.163 \\ \text{Symmetric} & & & & & .544 \end{bmatrix}.$$

2. Element 2: Nodes 1–3–4

By following similar procedures as for element 1, the constants in eq. (6.13) for element 2 can be written as

$$x_1 = 0, \; y_1 = 0 \qquad x_2 = 10, \; y_2 = 15 \qquad x_3 = 0, \; y_3 = 20$$

$$b_1 = y_2 - y_3 = -5 \qquad b_2 = y_3 - y_1 = 20 \qquad b_3 = y_1 - y_2 = -15$$

$$c_1 = x_3 - x_2 = -10 \qquad c_2 = x_1 - x_3 = 0 \qquad c_3 = x_2 - x_1 = 10.$$

The area of element 2 is twice of element 1: $A_2 = 0.5 \times 20 \times 10 = 100$. The strain–displacement matrix $[\mathbf{B}]$ can be obtained as

$$[\mathbf{B}] = \frac{1}{200} \begin{bmatrix} -5 & 0 & 20 & 0 & -15 & 0 \\ 0 & -10 & 0 & 0 & 0 & 10 \\ -10 & -5 & 0 & 20 & 10 & -15 \end{bmatrix}.$$

By following the same procedure, the stiffness matrix for element 2 can be computed as

$$[\mathbf{k}^{(2)}] = 3.297 \times 10^6 \begin{bmatrix} .15 & .081 & -.25 & -.175 & .1 & .094 \\ & .272 & -.15 & -.088 & .069 & -.184 \\ & & 1. & 0. & -.75 & .15 \\ & & & .35 & .175 & -.263 \\ & \text{Symmetric} & & & .65 & -.244 \\ & & & & & .447 \end{bmatrix}.$$

3. Global finite element matrix equations

The two element stiffness matrices are assembled to form the global stiffness matrix. Since there are four nodes, the model has eight DOFs: each node has two DOFs. Thus, the global matrix has a dimension of 8×8. After assembly, the global matrix equation can be written as

$$3.297 \times 10^6 \begin{bmatrix} .65 & .081 & -.75 & .15 & .0 & -.325 & .1 & .094 \\ & .447 & .175 & -.263 & -.325 & .0 & .069 & -.184 \\ & & 1.3 & -.488 & -.55 & .313 & .0 & .0 \\ & & & .894 & .338 & -.631 & .0 & .0 \\ & & & & 1.3 & -.163 & -.75 & .15 \\ & & & & & .894 & .175 & -.263 \\ & & & & & & .65 & -.244 \\ & \text{Symmetric} & & & & & & .447 \end{bmatrix} \begin{Bmatrix} u_1 \\ v_1 \\ u_2 \\ v_2 \\ u_3 \\ v_3 \\ u_4 \\ v_4 \end{Bmatrix} = \begin{Bmatrix} R_{x1} \\ R_{y1} \\ 0 \\ -50,000 \\ 50,000 \\ 0 \\ R_{x4} \\ R_{y4} \end{Bmatrix},$$

where R_{x1}, R_{y1}, R_{x4}, and R_{y4} are unknown reaction forces at nodes 1 and 4.

4. Applying boundary conditions

The displacement boundary conditions are: $u_1 = v_1 = u_4 = v_4 = 0$. Thus, we remove the first, second, seventh, and eighth rows and columns. After removing those rows and columns, we obtain the following reduced matrix equation:

$$3.297 \times 10^6 \begin{bmatrix} 1.3 & -.488 & -.55 & .313 \\ & .894 & .338 & -.631 \\ & & 1.3 & -.163 \\ \text{Symmetric} & & & .894 \end{bmatrix} \begin{Bmatrix} u_2 \\ v_2 \\ u_3 \\ v_3 \end{Bmatrix} = \begin{Bmatrix} 0 \\ -50,000 \\ 50,000 \\ 0 \end{Bmatrix}.$$

Note that the stiffness matrix of the above equation is nonsingular and, therefore, the unique solution can be obtained.

5. Solution

The above matrix equation can be solved for unknown nodal displacements:

$$u_2 = -2.147 \times 10^{-3}$$
$$v_2 = -4.455 \times 10^{-2}$$
$$u_3 = 1.891 \times 10^{-2}$$
$$v_3 = -2.727 \times 10^{-2}.$$

6. Strain and stress in element 1:

After calculating the nodal displacements, strain and stress can be calculated at the element level. First, the displacements for those nodes that belong to the element need to be extracted from the global nodal displacement vector. Since nodes 1, 2, and 3 belong to element 1, the nodal displacements will be $\{q\} = \{u_1, v_1, u_2, v_2, u_3, v_3\}^T = \{0, 0, -2.147 \times 10^{-3}, -4.455 \times 10^{-2}, 1.891 \times 10^{-2}, -2.727 \times 10^{-2}\}^T$. Then, strain in eq. (6.20) can be calculated using $\{\varepsilon\} = [B]\{q\}$

$$\begin{Bmatrix} \varepsilon_{xx} \\ \varepsilon_{yy} \\ \gamma_{xy} \end{Bmatrix} = \frac{1}{100} \begin{bmatrix} -10 & 0 & 15 & 0 & -5 & 0 \\ 0 & 0 & 0 & -10 & 0 & 10 \\ 0 & -10 & -10 & 15 & 10 & -5 \end{bmatrix} \begin{Bmatrix} 0 \\ 0 \\ -2.147 \times 10^{-3} \\ -4.455 \times 10^{-2} \\ 1.891 \times 10^{-2} \\ -2.727 \times 10^{-2} \end{Bmatrix}$$

$$= \begin{Bmatrix} -1.268 \times 10^{-3} \\ 1.727 \times 10^{-3} \\ -3.212 \times 10^{-3} \end{Bmatrix}.$$

The stresses in the element are obtained from eq. (6.5)

$$\begin{Bmatrix} \sigma_{xx} \\ \sigma_{yy} \\ \tau_{xy} \end{Bmatrix} = 3.297 \times 10^7 \begin{bmatrix} 1 & .3 & 0 \\ .3 & 1 & 0 \\ 0 & 0 & .35 \end{bmatrix} \begin{Bmatrix} -1.268 \times 10^{-3} \\ 1.727 \times 10^{-3} \\ -3.212 \times 10^{-3} \end{Bmatrix} = \begin{Bmatrix} -24,709 \\ 44,406 \\ -37,063 \end{Bmatrix} \text{psi}.$$

7. Strains and stresses in element 2:

Element 2 has nodes 1, 3, and 4. Thus, the nodal displacements will be $\{\mathbf{q}\} = \{u_1, v_1, u_3, v_3, u_4, v_4\}^T = \{0, 0, 1.891 \times 10^{-2}, -2.727 \times 10^{-2}, 0, 0\}^T$. Using the element displacements, the strains and stresses in the element can be obtained as

$$
\begin{Bmatrix} \varepsilon_{xx} \\ \varepsilon_{yy} \\ \gamma_{xy} \end{Bmatrix} = \frac{1}{200} \begin{bmatrix} -5 & 0 & 20 & 0 & -15 & 0 \\ 0 & -10 & 0 & 0 & 0 & 10 \\ -10 & -5 & 0 & 20 & 10 & -15 \end{bmatrix} \begin{Bmatrix} 0 \\ 0 \\ 1.891 \times 10^{-2} \\ -2.727 \times 10^{-2} \\ 0 \\ 0 \end{Bmatrix}
$$

$$
= \begin{Bmatrix} 1.891 \times 10^{-3} \\ 0 \\ -2.727 \times 10^{-3} \end{Bmatrix},
$$

and

$$
\begin{Bmatrix} \sigma_{xx} \\ \sigma_{yy} \\ \tau_{xy} \end{Bmatrix} = 3.297 \times 10^7 \begin{bmatrix} 1 & .3 & 0 \\ .3 & 1 & 0 \\ 0 & 0 & .35 \end{bmatrix} \begin{Bmatrix} 1.891 \times 10^{-3} \\ 0 \\ -2.727 \times 10^{-3} \end{Bmatrix} = \begin{Bmatrix} 62,354 \\ 18,706 \\ -31,469 \end{Bmatrix} \text{psi.}
$$

If the stresses in the two elements are examined, one can note that the stress value changes suddenly across the element boundary. For example, σ_{xx} in element 1 is $-24,709$ psi, whereas in element 2, it is 62,354 psi. Such a drastic change in stresses is an indicator that the finite element analysis results from the current model are not accurate and further mesh refinement is necessary. ▪

From example 6.2, we can conclude the following:

– Stresses are constant over the individual element.

– The solution is not accurate because there are large discontinuities in stresses across element boundaries.

– With only two elements, the mesh is very coarse, and we obviously cannot expect very good results.

6.4 FOUR–NODE RECTANGULAR ELEMENT

6.4.1 Lagrange Interpolation for a Rectangular Element

A rectangular element is composed of four nodes and eight DOFs (see figure 6.8). It is a part of a plane solid that is composed of many rectangular elements. Each element shares its edge and two corner nodes with an adjacent element, except for those on the boundary. The four vertices of a rectangle are the nodes of that element as shown in figure 6.8. The first node of an element can arbitrarily be chosen. However, the sequence of the nodes 1, 2, 3, and 4 should be in the counterclockwise direction. Each node has two displacements, u and v, respectively in the x- and y-direction.

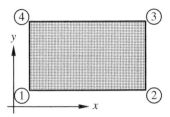

Figure 6.8 Four–node rectangular element

Since all edges are parallel to the coordinate directions, this element is not practical, but it is useful, as it is the basis for the quadrilateral element discussed in the following chapter. In addition, the behavior of the rectangular element is similar to that of the quadrilateral element. Shape functions can be calculated using procedures similar to that of CST element, but it is more instructive to use the Lagrange interpolation functions in the x- and y-direction.

Consider the rectangular element in figure 6.8. From the geometry, it is clear that $x_3 = x_2$, $y_4 = y_3$, $x_4 = x_1$, and $y_2 = y_1$. We will use a polynomial in x and y as the interpolation function. Since there are four nodes, we can apply four boundary conditions, and hence the polynomial should have four terms as follows:

$$u = \alpha_1 + \alpha_2 x + \alpha_3 y + \alpha_4 xy,$$
$$v = \beta_1 + \beta_2 x + \beta_3 y + \beta_4 xy. \tag{6.39}$$

Let us calculate unknown coefficients α_i using the x-directional displacement u:

$$\begin{cases} u_1 = \alpha_1 + \alpha_2 x_1 + \alpha_3 y_1 + \alpha_4 x_1 y_1, \\ u_2 = \alpha_1 + \alpha_2 x_2 + \alpha_3 y_2 + \alpha_4 x_2 y_2, \\ u_3 = \alpha_1 + \alpha_2 x_3 + \alpha_3 y_3 + \alpha_4 x_3 y_3, \\ u_4 = \alpha_1 + \alpha_2 x_4 + \alpha_3 y_4 + \alpha_4 x_4 y_4. \end{cases}$$

It is obvious that we need to invert the 4×4 matrix in order to calculate the interpolation coefficients.

Instead of matrix inversion method, we use the *Lagrange interpolation* method to interpolate u and v. The goal is to obtain the following expression:

$$u(x,y) = [N_1 \ \ N_2 \ \ N_3 \ \ N_4] \begin{Bmatrix} u_1 \\ u_2 \\ u_3 \\ u_4 \end{Bmatrix}, \tag{6.40}$$

where $N_1, \ldots, N_4$ are the interpolation functions. In order to do that, let us first consider displacement along edge 1-2 in figure 6.8. Along edge 1-2, $y = y_1$ (constant); therefore, shape functions must be functions of x only as shown below:

$$u_I(x,y_1) = [n_1(x) \ \ n_2(x)] \begin{Bmatrix} u_1 \\ u_2 \end{Bmatrix}. \tag{6.41}$$

Using the one-dimensional Lagrange interpolation formula, the shape functions can be obtained as

$$n_1(x) = \frac{x - x_2}{x_1 - x_2}, \quad n_2(x) = \frac{x - x_1}{x_2 - x_1}. \tag{6.42}$$

This is the same procedure that was used in chapter 2. Next, since $y = y_3 = y_4$ along edge 4-3 in figure 6.8, the displacement can be interpolated as

$$u_{II}(x, y_3) = [n_4(x) \quad n_3(x)] \begin{Bmatrix} u_4 \\ u_3 \end{Bmatrix}. \tag{6.43}$$

Again from the one-dimensional Lagrange interpolation formula, we have

$$n_4(x) = \frac{x - x_3}{x_4 - x_3}, \quad n_3(x) = \frac{x - x_4}{x_3 - x_4}. \tag{6.44}$$

Equations (6.41) and (6.43) represent interpolation of displacements at the top and bottom of the element, respectively. So far, we have interpolated displacements in the x-direction only. Now, we can extend the interpolation in the y-direction between $u_I(x, y_1)$ and $u_{II}(x, y_3)$ using the same Lagrange interpolation method. By considering $u_I(x, y_1)$ and $u_{II}(x, y_3)$ as nodal displacements, we have the following interpolation formula:

$$u(x, y) = [n_1(y) \quad n_4(y)] \begin{Bmatrix} u_I(x, y_1) \\ u_{II}(x, y_3) \end{Bmatrix}, \tag{6.45}$$

where

$$n_1(y) = \frac{y - y_4}{y_1 - y_4}, \quad n_4(y) = \frac{y - y_1}{y_4 - y_1} \tag{6.46}$$

are the Lagrange interpolations in the y-direction. By substituting eqs. (6.41) and (6.43) into eq. (6.45), we have the following formula:

$$u(x, y) = [n_1(y) \quad n_4(y)] \begin{Bmatrix} [n_1(x) \quad n_2(x)] \begin{Bmatrix} u_1 \\ u_2 \end{Bmatrix} \\ [n_4(x) \quad n_3(x)] \begin{Bmatrix} u_4 \\ u_3 \end{Bmatrix} \end{Bmatrix}. \tag{6.47}$$

Thus,

$$u(x, y) = [n_1(x)n_1(y) \quad n_2(x)n_1(y) \quad n_3(x)n_4(y) \quad n_4(x)n_4(y)] \begin{Bmatrix} u_1 \\ u_2 \\ u_3 \\ u_4 \end{Bmatrix}. \tag{6.48}$$

Comparing the above expression with eq. (6.40) we can define shape functions. $N_1, \ldots, N_4$. In the rectangular element, it is enough to use the coordinates of two nodes, because $x_1 = x_4$, $y_1 = y_2$, and so forth. We will use the coordinates of nodes 1 and 3. Using the property that the area of the element is $A = (x_3 - x_1)(y_3 - y_1)$, we obtain

$$\begin{cases} N_1 \equiv n_1(x)n_1(y) = \dfrac{1}{A}(x_3 - x)(y_3 - y), \\[2mm] N_2 \equiv n_2(x)n_1(y) = -\dfrac{1}{A}(x_1 - x)(y_3 - y), \\[2mm] N_3 \equiv n_3(x)n_4(y) = \dfrac{1}{A}(x_1 - x)(y_1 - y), \\[2mm] N_4 \equiv n_4(x)n_4(y) = -\dfrac{1}{A}(x_3 - x)(y_1 - y). \end{cases} \tag{6.49}$$

Note that the *shape functions* for rectangular elements are the product of Lagrange interpolations in the two coordinate directions. Let us discuss the properties of the shape functions. It can be easily verified that $N_1(x, y)$ is:

– 1 at node 1 and 0 at other nodes

– a linear function of x along edge 1-2 and linear function of y along edge 1-4 (bilinear interpolation)

– zero along edges 2-3 and 3-4

Other shape functions have similar behavior. Because of these characteristics, the i-th shape function is considered associated with node i of the element.

In order to make the derivations simple, we rewrite the interpolation relation in eq. (6.40) in matrix form. Let $\{\mathbf{u}\} = \{u, v\}^{\mathrm{T}}$ be the displacement vector at any point (x, y). The interpolation can be written using the matrix notation by

$$\{\mathbf{u}\} \equiv \begin{Bmatrix} u \\ v \end{Bmatrix} = \begin{bmatrix} N_1 & 0 & N_2 & 0 & N_3 & 0 & N_4 & 0 \\ 0 & N_1 & 0 & N_2 & 0 & N_3 & 0 & N_4 \end{bmatrix} \begin{Bmatrix} u_1 \\ v_1 \\ u_2 \\ v_2 \\ u_3 \\ v_3 \\ u_4 \\ v_4 \end{Bmatrix},$$

or

$$\boxed{\{\mathbf{u}\} = [\mathbf{N}]_{2 \times 8} \{\mathbf{q}\}_{8 \times 1}}. \tag{6.50}$$

Note that the dimension of the shape function matrix is 2×8.

EXAMPLE 6.3 *Shape functions of a rectangular element*

A rectangular element is shown in figure 6.9. By substituting the numerical values of nodal coordinates into the above shape function formulas, the explicit expressions for shape functions for this rectangular element can be obtained as

$$N_1 = \frac{(3-x)(2-y)}{6} \quad N_2 = \frac{x(2-y)}{6}$$

$$N_3 = \frac{xy}{6} \quad N_4 = \frac{y(3-x)}{6}.$$

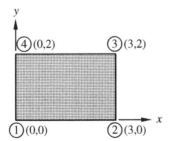

Figure 6.9 Four-node rectangular element

(a)

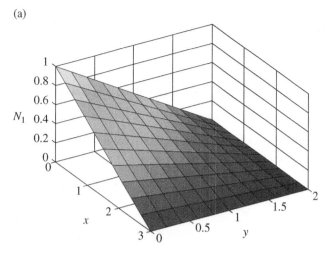

(b)

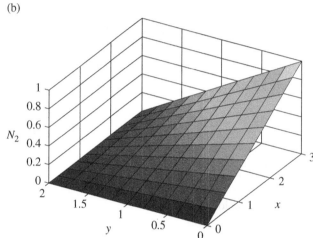

Figure 6.10 Three-dimensional surface plots of shape functions for a rectangular element; (a) $N_1(x, y)$, (b) $N_2(x, y)$

Three-dimensional plots of N_1 and N_2 are shown in figure 6.10(a) and figure 6.10(b).

Since the shape functions are given as a function of x- and y-coordinates, we can use an approach similar to that of CST element to obtain the strain–displacement relations. Thus, the strain can be calculated by differentiating the shape function with respect to the coordinates. For example, ε_{xx} can be written as

$$\varepsilon_{xx} \equiv \frac{\partial u}{\partial x} = \frac{\partial}{\partial x}\left(\sum_{i=1}^{4} N_i(x,y)u_i\right) = \sum_{i=1}^{4}\frac{\partial N_i}{\partial x}u_i. \tag{6.51}$$

Note that u_1, u_2, u_3, and u_4 are nodal displacements, and they are independent of coordinate x. Thus, only the shape function is differentiated with respect to x. Similar calculation can be carried out for ε_{yy} and γ_{xy}. Then, we have

$$\{\varepsilon\} = \frac{1}{A} \begin{bmatrix} y-y_3 & 0 & y_3-y & 0 & y-y_1 & 0 & y_1-y & 0 \\ 0 & x-x_3 & 0 & x_1-x & 0 & x-x_1 & 0 & x_3-x \\ x-x_3 & y-y_3 & x_1-x & y_3-y & x-x_1 & y-y_1 & x_3-x & y_1-y \end{bmatrix} \begin{Bmatrix} u_1 \\ v_1 \\ u_2 \\ v_2 \\ u_3 \\ v_3 \\ u_4 \\ v_4 \end{Bmatrix} \tag{6.52}$$

$$\equiv [\mathbf{B}]\{\mathbf{q}\}.$$

Note that matrix $[\mathbf{B}]$ is a linear function of x and y. Thus, the strain will change linearly within the element. For example, ε_{xx} will vary linearly in the y-direction, while it is constant with respect to x. Thus, the element will have approximation error, if the actual strains vary in the x-direction.

6.4.2 Element Stiffness Matrix

The element stiffness matrix can be calculated from the strain energy of the element. By substituting for strains from eq. (6.52) into the expression for strain energy in eq. (6.52) we have

$$U^{(e)} = \frac{h}{2} \iint_A \{\varepsilon\}^{\mathrm{T}} [\mathbf{C}] \{\varepsilon\} \, dA^{(e)}$$

$$= \frac{h}{2} \{\mathbf{q}^{(e)}\}^{\mathrm{T}} \left[\iint_A [\mathbf{B}]_{8\times3}^{\mathrm{T}} [\mathbf{C}]_{3\times3} [\mathbf{B}]_{3\times8} \, dA^{(e)} \right] \{\mathbf{q}^{(e)}\} \tag{6.53}$$

$$\equiv \frac{1}{2} \{\mathbf{q}^{(e)}\}^{\mathrm{T}} \left[\mathbf{k}^{(e)} \right]_{8\times8} \{\mathbf{q}^{(e)}\},$$

where $[\mathbf{k}^{(e)}]$ is the element stiffness matrix. Calculation of the element stiffness matrix requires two-dimensional integration. We will discuss numerical integration in the next chapter. When the element is square and the problem is plane stress, analytical integration of the strain energy yields the following form of element stiffness matrix:

$$\left[\mathbf{k}^{(e)}\right] = \frac{Eh}{1-\nu^2} \begin{bmatrix} \dfrac{3-\nu}{6} & \dfrac{1+\nu}{8} & -\dfrac{3+\nu}{12} & \dfrac{-1+3\nu}{8} & \dfrac{-3+\nu}{12} & -\dfrac{1+\nu}{8} & \dfrac{\nu}{6} & \dfrac{1-3\nu}{8} \\[2mm] \dfrac{1+\nu}{8} & \dfrac{3-\nu}{6} & \dfrac{1-3\nu}{8} & \dfrac{\nu}{6} & -\dfrac{1+\nu}{8} & \dfrac{-3+\nu}{12} & \dfrac{-1+3\nu}{8} & -\dfrac{3+\nu}{12} \\[2mm] -\dfrac{3+\nu}{12} & \dfrac{1-3\nu}{8} & \dfrac{3-\nu}{6} & -\dfrac{1+\nu}{8} & \dfrac{\nu}{6} & \dfrac{-1+3\nu}{8} & \dfrac{-3+\nu}{12} & \dfrac{1+\nu}{8} \\[2mm] \dfrac{-1+3\nu}{8} & \dfrac{\nu}{6} & -\dfrac{1+\nu}{8} & \dfrac{3-\nu}{6} & \dfrac{1-3\nu}{8} & \dfrac{3+\nu}{12} & \dfrac{1+\nu}{8} & \dfrac{-3+\nu}{12} \\[2mm] \dfrac{-3+\nu}{12} & -\dfrac{1+\nu}{8} & \dfrac{\nu}{6} & \dfrac{1-3\nu}{8} & \dfrac{3-\nu}{6} & \dfrac{1+\nu}{8} & \dfrac{3+\nu}{12} & \dfrac{-1+3\nu}{8} \\[2mm] -\dfrac{1+\nu}{8} & \dfrac{-3+\nu}{12} & \dfrac{-1+3\nu}{8} & \dfrac{3+\nu}{12} & \dfrac{1+\nu}{8} & \dfrac{3-\nu}{6} & \dfrac{1-3\nu}{8} & \dfrac{\nu}{6} \\[2mm] \dfrac{\nu}{6} & \dfrac{-1+3\nu}{8} & \dfrac{-3+\nu}{12} & \dfrac{1+\nu}{8} & \dfrac{3+\nu}{12} & \dfrac{1-3\nu}{8} & \dfrac{3-\nu}{6} & -\dfrac{1+\nu}{8} \\[2mm] \dfrac{1-3\nu}{8} & -\dfrac{3+\nu}{12} & \dfrac{1+\nu}{8} & \dfrac{-3+\nu}{12} & \dfrac{-1+3\nu}{8} & \dfrac{\nu}{6} & -\dfrac{1+\nu}{8} & \dfrac{3-\nu}{6} \end{bmatrix}. \tag{6.54}$$

It is interesting to note that the element stiffness matrix does not depend on the actual element dimensions, but it is a function only of material properties (E and ν) and thickness h.

The strain energy of the entire solid can be obtained using eq. (6.24), which involves the assembly process.

6.4.3 Potential Energy of Applied Loads

In the CST element, we discussed three different types of applied loads: concentrated forces at nodes, distributed forces along element edges, and body force. The first two types are independent of the element used. Thus, the same forms in eqs. (6.26) and (6.31) can be used for the potential energies of concentrated force and distributed force, respectively.

In the case of the body force, the element shape functions are used to calculate equivalent nodal forces. When a constant body force $\mathbf{b} = \{b_x, b_y\}^T$ acts on a rectangular element, the potential energy of body force becomes

$$
V^{(e)} = -h \iint_A [u \quad v] \begin{Bmatrix} b_x \\ b_y \end{Bmatrix} dA = -\left\{ \mathbf{q}^{(e)} \right\}^T h \iint_A [\mathbf{N}]^T dA \begin{Bmatrix} b_x \\ b_y \end{Bmatrix}
$$

$$
\equiv \left\{ \mathbf{q}^{(e)} \right\}^T \left\{ \mathbf{f}_b^{(e)} \right\},
$$

(6.55)

where

$$
\left\{ \mathbf{f}_b^{(e)} \right\} = \frac{hA}{4} \begin{bmatrix} 1 & 0 \\ 0 & 1 \\ 1 & 0 \\ 0 & 1 \\ 1 & 0 \\ 0 & 1 \\ 1 & 0 \\ 0 & 1 \end{bmatrix} \begin{Bmatrix} b_x \\ b_y \end{Bmatrix} = \frac{hA}{4} \begin{Bmatrix} b_x \\ b_y \\ b_x \\ b_y \\ b_x \\ b_y \\ b_x \\ b_y \end{Bmatrix}.
$$

(6.56)

Equation (6.56) equally divides the total magnitude of the body force to the four nodes. $\{\mathbf{f}_b\}$ is the equivalent nodal force corresponding to the constant body force. The potential energy of body forces of all elements must be assembled to build the global force vector of body forces as in eq. (6.34).

Using the principle of minimum total potential energy in eq. (6.36), a similar global matrix equation for the rectangular elements can be obtained. Applying boundary conditions and solving the matrix equations are identical to those of the CST element. After solving for nodal displacements, strains and stresses in each element can be calculated using eqs. (6.37) and (6.38), respectively.

EXAMPLE 6.4 *Simple shear deformation of a square element*

A square element shown in figure 6.11 is under a simple shear deformation. Material properties are given as $E = 10$ GPa, $\nu = 0.25$, and thickness is $h = 0.1$ m. When a distributed force $f = 100$ kN/m^2 is applied at the top edge, calculate stress and strain components. Compare the results with the exact solution.

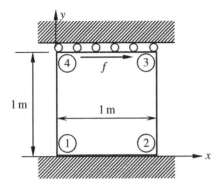

Figure 6.11 A square element under a simple shear condition

SOLUTION Since the problem consists of one element, we do not need assembly process. The element has eight DOFs: $\{Q_s\} = \{u_1, v_1, u_2, v_2, u_3, v_3, u_4, v_4\}^T$. From the boundary condition given in figure 6.11, only two DOFs are nonzero: u_3 and u_4. Thus, from the element stiffness matrix given in eq. (6.54), all fixed DOFs are deleted to obtain

$$[K] = \frac{Eh}{1-\nu^2} \begin{bmatrix} \dfrac{3-\nu}{6} & -\dfrac{3+\nu}{12} \\ -\dfrac{3+\nu}{12} & \dfrac{3-\nu}{6} \end{bmatrix} \begin{matrix} u_3 \\ u_4 \end{matrix} = 10^8 \begin{bmatrix} 4.88 & -2.88 \\ -2.88 & 4.88 \end{bmatrix} \begin{matrix} u_3 \\ u_4 \end{matrix}.$$

The total distributed load of 10,000 N at the top edge will be equally divided into two nodes: 4 and 3. Thus, the global matrix equation becomes

$$10^8 \begin{bmatrix} 4.88 & -2.88 \\ -2.88 & 4.88 \end{bmatrix} \begin{Bmatrix} u_3 \\ u_4 \end{Bmatrix} = \begin{Bmatrix} 5,000 \\ 5,000 \end{Bmatrix}.$$

The above equation can be solved for unknown nodal displacements, as $u_3 = u_4 = 0.025$ mm. Then, from eq. (6.52), the strain components can be obtained, as

$$\{\varepsilon\} = \begin{bmatrix} y-1 & 0 & 1-y & 0 & y & 0 & -y & 0 \\ 0 & x-1 & 0 & -x & 0 & x & 0 & 1-x \\ x-1 & y-1 & -x & 1-y & x & y & 1-x & -y \end{bmatrix} \begin{Bmatrix} 0 \\ 0 \\ 0 \\ 0 \\ 2.5 \times 10^{-5} \\ 0 \\ 2.5 \times 10^{-5} \\ 0 \end{Bmatrix}$$

$$= \begin{Bmatrix} 0 \\ 0 \\ 2.5 \times 10^{-5} \end{Bmatrix}.$$

Note that the shear strain is the only nonzero strain. Thus, the rectangular element can accurately represent the simple shear condition. Figure 6.12 shows the deformed shape of the solid. Note that the deformation is magnified for illustration purposes.

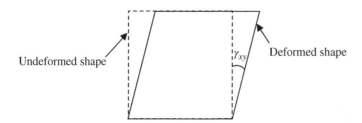

Figure 6.12 Simple shear deformation of a square element

Using the stress–strain relation in eq. (6.5) for plane stress, the stress components can be obtained, as

$$\begin{Bmatrix} \sigma_{xx} \\ \sigma_{yy} \\ \tau_{xy} \end{Bmatrix} = \frac{10^{10}}{1-0.25^2} \begin{bmatrix} 1 & 0.25 & 0 \\ 0.25 & 1 & 0 \\ 0 & 0 & 0.375 \end{bmatrix} \begin{Bmatrix} 0 \\ 0 \\ 2.5 \times 10^{-5} \end{Bmatrix} = \begin{Bmatrix} 0 \\ 0 \\ 10^5 \end{Bmatrix} \text{Pa.}$$

Since the distributed force $f = 10$ kN/m^2 is applied at the top edge, the above shear stress is exact. ■

EXAMPLE 6.5   *Pure bending deformation of a square element*

A pure bending condition can be achieved by applying a couple in the case of a beam (see chapter 3). For the plane solid, the effect of a couple can be achieved by applying equal forces in opposite directions. A square element shown in figure 6.13 is under a pure bending condition. Material properties are given as $E = 10$ GPa, $\nu = 0.25$, and thickness is $h = 0.1$ m. When an equal and opposite force $f = 100$ kN is applied at nodes 2 and 3, calculate stress and strain components. Compare the results with the exact solutions from the beam theory.

SOLUTION

(a) Analytical solution: If we consider the above plane solid as a cantilevered beam, the moment of inertia $I = 8.333 \times 10^{-3}$ m^4 and the applied couple $M = 100$ kN·m. Thus, the maximum stress will occur at the bottom edge with the magnitude of

$$(\sigma_{xx})_{max} = -\frac{M\left(-\dfrac{h}{2}\right)}{I} = 6.0 \text{MPa.}$$

The minimum stress will occur at the top edge with the same magnitude but in compression. Since the stress varies linearly along the y-coordinate, we have

$$\sigma_{xx} = 6.0(1-2y) \text{ MPa.}$$

All other stress components are zero.

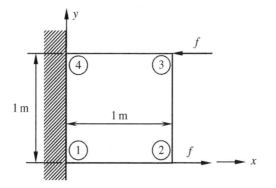

Figure 6.13 A square element under pure bending condition

(b) Numerical solution: Since we use only one element, we do not need the assembly process. The element has eight DOFs: $\{\mathbf{Q}_s\} = \{u_1, v_1, u_2, v_2, u_3, v_3, u_4, v_4\}^{\mathrm{T}}$. From the boundary condition given in figure 6.13, only four DOFs are nonzero: u_2, v_2, u_3, and v_3. Thus, from the element stiffness matrix given in eq. (6.54), all fixed DOFs are deleted to obtain

$$[\mathbf{K}] = \frac{Eh}{1-\nu^2} \begin{bmatrix} \dfrac{3-\nu}{6} & -\dfrac{1+\nu}{8} & \dfrac{\nu}{6} & \dfrac{-1+3\nu}{8} \\ -\dfrac{1+\nu}{8} & \dfrac{3-\nu}{6} & \dfrac{1-3\nu}{8} & -\dfrac{3+\nu}{12} \\ \dfrac{\nu}{6} & \dfrac{1-3\nu}{8} & \dfrac{3-\nu}{6} & \dfrac{1+\nu}{8} \\ \dfrac{-1+3\nu}{8} & -\dfrac{3+\nu}{12} & \dfrac{1+\nu}{8} & \dfrac{3-\nu}{6} \end{bmatrix} = 10^8 \begin{bmatrix} 4.89 & -1.67 & 0.44 & -0.33 \\ -1.67 & 4.89 & 0.33 & -2.89 \\ 0.44 & 0.33 & 4.89 & 1.67 \\ -0.33 & -2.89 & 1.67 & 4.89 \end{bmatrix} \begin{matrix} u_2 \\ v_2 \\ u_3 \\ v_3 \end{matrix}.$$

Using the applied nodal forces, the global matrix equation becomes

$$10^8 \begin{bmatrix} 4.89 & -1.67 & 0.44 & -0.33 \\ -1.67 & 4.89 & 0.33 & -2.89 \\ 0.44 & 0.33 & 4.89 & 1.67 \\ -0.33 & -2.89 & 1.67 & 4.89 \end{bmatrix} \begin{Bmatrix} u_2 \\ v_2 \\ u_3 \\ v_3 \end{Bmatrix} = \begin{Bmatrix} 100{,}000 \\ 0 \\ -100{,}000 \\ 0 \end{Bmatrix}.$$

The above equation can be solved for unknown nodal displacements as

$$u_2 = 0.4091 \text{ mm}, \quad v_2 = 0.4091 \text{ mm}$$
$$u_3 = -0.4091 \text{ mm}, \quad v_3 = 0.4091 \text{ mm}.$$

Then, from eq. (6.52), the strain components can be obtained as

$$\{\varepsilon\} = \begin{bmatrix} y-1 & 0 & 1-y & 0 & y & 0 & -y & 0 \\ 0 & x-1 & 0 & -x & 0 & x & 0 & 1-x \\ x-1 & y-1 & -x & 1-y & x & y & 1-x & -y \end{bmatrix} \begin{Bmatrix} 0 \\ 0 \\ 0.4091 \\ 0.4091 \\ -0.4091 \\ 0.4091 \\ 0 \\ 0 \end{Bmatrix} \times 10^{-3}$$

$$= \begin{Bmatrix} 0.4091 \times 10^{-3}(1-2y) \\ 0 \\ 0.4091 \times 10^{-3}(1-2x) \end{Bmatrix}.$$

Using the stress–strain relation in eq. (6.5) for plane stress, the stress components can be obtained as

$$\begin{Bmatrix} \sigma_{xx} \\ \sigma_{yy} \\ \tau_{xy} \end{Bmatrix} = \frac{10^{10}}{1-0.25^2} \begin{bmatrix} 1 & 0.25 & 0 \\ 0.25 & 1 & 0 \\ 0 & 0 & 0.375 \end{bmatrix} \begin{Bmatrix} 0.4091 \times 10^{-3}(1-2y) \\ 0 \\ 0.4091 \times 10^{-3}(1-2x) \end{Bmatrix}$$

$$= \begin{Bmatrix} 4.364(1-2y) \\ 1.091(1-2y) \\ 1.636(1-2x) \end{Bmatrix} \text{MPa}.$$

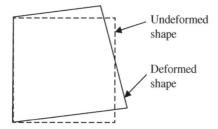

Undeformed
shape

Deformed
shape

Figure 6.14 Pure bending deformation of a square element

The deformed shape of the element is shown in figure 6.14. In a plane solid, the applied couple produces a curvature, but the rectangular element is unable to produce deformation corresponding to the curvature because the displacement can only change linearly within the element. The rectangular shape deforms to the trapezoidal shape, and as a result, nonzero shear stress is produced. Note that the maximum stress $(\sigma_{xx})_{\max}$ is only 73% (4.364/6.0) of the exact solution. In addition, σ_{yy} and τ_{xy} have nonzero values. The applied couple is supported by other stress components, σ_{yy} and τ_{xy}, and as a result, the element shows smaller $(\sigma_{xx})_{\max}$. In a sense, the element shows a *stiff* behavior. ▄▄

6.5   AXISYMMETRIC ELEMENT

Axisymmetric problems are also classified as two-dimensional solid mechanics problems. For such problems, the structure has a geometry that is symmetric about an axis. For instance, cylinders and cones are axisymmetric geometries. Any structure where the geometry can be described as the volume swept by a two-dimensional profile or section revolved about an axis is a structure with symmetry about that axis. If the applied loads and constraints on such geometry are also symmetric about the same axis, then the structure would deform in an axisymmetric manner. Even though all real structures are three-dimensional, we can state the problem as two-dimensional only if all the quantities of interest are constant in the direction normal to a plane and the displacement field has only two components that are parallel to this plane. In this case, the domain of the analysis is an area on this plane. For plane stress and plane strain problems, the thickness in the z-direction is constant and we are able to easily convert volume integrals into area integrals because the field variables of interest are constant in the z-direction.

Axisymmetric problems can also be stated as two-dimensional problems when viewed with respect to a cylindrical coordinate system as shown in figure 6.15 where the geometry in (a) is obtained by

(a)   (b)

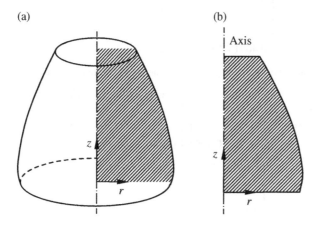

Figure 6.15 Axisymmetric geometry; (a) revolved geometry, (b) section – plane of deformation

revolving the section in (b). The section plane is on the r-z plane and is the plane on which the displacement vector must lie. If the z-axis is the axis of symmetry, then points on the r-z plane must remain on this plane as the structure deforms in order for it to be considered a two-dimensional deformation. Any twisting about the z-axis will result in a three-dimensional deformation.

To summarize, the conditions under which we have axisymmetric deformation are:

(a) The geometry of the structure must be symmetric about an axis.

(b) The applied load and boundary conditions must also be symmetric about the same axis.

(c) The structure should not be subjected to loads that produce a torque about the axis of symmetry. In other words, the structure should not twist about the axis.

For an axisymmetric problem, the displacement field has two components (u_r, u_z) which are the radial and axial components in a cylindrical coordinate system. The component of displacement in the tangential or circumferential direction must be zero ($u_\theta = 0$). Any displacement in the circumferential direction would cause twisting about the axis, which is the reason for assuming that there is no torque applied about the axis. Even though there is no displacement in the θ-direction, there is a stress in this direction, often referred to as the hoop stress. To understand the reason for this stress, imagine a circle in the structure on a plane normal to the axis with its center at the axis of symmetry of the structure as shown in figure 6.16. During the deformation, if there is any displacement u_r in the radial direction, this will cause the radius of the circle to increase, which implies that the circumference will increase resulting in a strain in the circumferential/tangential direction, which is the hoop strain, $\varepsilon_{\theta\theta}$. This strain is an extensional strain and can be defined as the change in circumference over the original circumference of this circle as shown in the following equation.

$$\varepsilon_{\theta\theta} = \frac{2\pi(r + u_r) - 2\pi r}{2\pi r} = \frac{u_r}{r}. \tag{6.57}$$

From the above discussion, it is clear that there are four strain components for axisymmetric problems. Similarly, there are four stress components: σ_{rr} the radial stress, σ_{zz} the axial stress, $\sigma_{\theta\theta}$ or the hoop stress, and finally τ_{rz} the shear stress. The strain components, placed in a column matrix, can be defined as:

$$\{\varepsilon\} = \begin{Bmatrix} \varepsilon_{rr} \\ \varepsilon_{zz} \\ \varepsilon_{\theta\theta} \\ \varepsilon_{rz} \end{Bmatrix} = \begin{Bmatrix} \dfrac{\partial u_r}{\partial r} \\[6pt] \dfrac{\partial u_z}{\partial z} \\[6pt] \dfrac{u_r}{r} \\[6pt] \dfrac{\partial u_r}{\partial z} + \dfrac{\partial u_z}{\partial r} \end{Bmatrix}. \tag{6.58}$$

For finite element analysis, within each element, we can compute the strains using the interpolated displacement components. For convenience we will use the notation: $u_r \equiv u$ and $u_z \equiv v$. The nodal values of

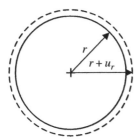

Figure 6.16 Circumferential strain due to radial displacement

the displacement components can then be denoted as $\{u_i, v_i\}$. We will assume that when modeling the two-dimensional cross section, we will create the model on the xy-plane such that the r- and z-axis of the cylindrical coordinate system align with the x- and y-axis of the Cartesian coordinates. Therefore, in the subsequent discussion, we will assume that $r \equiv x$ and $z \equiv y$. The strain components can be numerically computed within each element as:

$$\{\varepsilon\} = \begin{Bmatrix} \varepsilon_{rr} \\ \varepsilon_{zz} \\ \varepsilon_{\theta\theta} \\ \varepsilon_{rz} \end{Bmatrix} = \begin{Bmatrix} \dfrac{\partial u}{\partial x} \\ \dfrac{\partial v}{\partial y} \\ \dfrac{u}{x} \\ \dfrac{\partial u}{\partial y} + \dfrac{\partial v}{\partial x} \end{Bmatrix} = \begin{bmatrix} \dfrac{\partial N_1}{\partial x} & 0 & \cdot & \cdot & \dfrac{\partial N_n}{\partial x} & 0 \\ 0 & \dfrac{\partial N_1}{\partial y} & \cdot & \cdot & 0 & \dfrac{\partial N_n}{\partial y} \\ \dfrac{N_1}{x} & 0 & \cdot & \cdot & \dfrac{N_n}{x} & 0 \\ \dfrac{\partial N_1}{\partial y} & \dfrac{\partial N_1}{\partial x} & \cdot & \cdot & \dfrac{\partial N_n}{\partial y} & \dfrac{\partial N_n}{\partial x} \end{bmatrix} \begin{Bmatrix} u_1 \\ v_1 \\ \cdot \\ \cdot \\ u_n \\ v_n \end{Bmatrix} \equiv [\mathbf{B}]\{\mathbf{q}\}.$$

The preceding equation is valid for any two-dimensional element with n nodes. The first two columns of the $[\mathbf{B}]$ matrix are multiplied by (u_1, v_1) or the degrees of freedom of the first node, and the same pattern repeats for every node. The third row corresponds to the hoop strain, where the denominator is the distance of the point at which the strain is being computed from the axis of revolution. It appears as though the hoop strain could be undefined at the axis because the distance from the axis is zero along the axis. Indeed it would be infinite unless the numerator is also zero or in other words, points along the axis cannot have any displacement in the radial direction regardless of the applied loads. This makes sense when you consider that displacement of points at the axis in the radial direction is equivalent to a hole being created at the axis. Therefore, for axisymmetric problems, it is not necessary to add displacement boundary conditions in the radial direction on any edge of the domain that is along the axis of symmetry.

The stress-strain relation for axisymmetric problems can be obtained by eliminating from the three-dimensional stress-strain relations the two rows and columns that correspond to the two shear strains that are zero. For linear isotropic materials, the stress-strain relations under axisymmetric conditions can be written as:

$$\{\boldsymbol{\sigma}\} = \begin{Bmatrix} \sigma_{rr} \\ \sigma_{zz} \\ \sigma_{\theta\theta} \\ \tau_{rz} \end{Bmatrix} = \frac{E}{(1+\nu)(1-2\nu)} \begin{bmatrix} 1-\nu & \nu & \nu & 0 \\ \nu & 1-\nu & \nu & 0 \\ \nu & \nu & 1-\nu & 0 \\ 0 & 0 & 0 & \frac{1}{2}-\nu \end{bmatrix} \begin{Bmatrix} \varepsilon_{rr} \\ \varepsilon_{zz} \\ \varepsilon_{\theta\theta} \\ \varepsilon_{rz} \end{Bmatrix}. \tag{6.59}$$

The stiffness matrix computation for axisymmetric elements is very similar to other two-dimensional elements with the main difference being that the integration is carried out in a cylindrical coordinate system. The real geometry of a triangular axisymmetric element is the triangle revolved around the axis of symmetry. For any element, the stiffness matrix is computed as:

$$\left[\mathbf{k}^{(e)}\right] = \iiint_{V^{(e)}} [\mathbf{B}]^{\mathrm{T}}[\mathbf{C}][\mathbf{B}] \mathrm{d}V.$$

The integration in the preceding equation is over the volume of an element $V^{(e)}$. Performing this volume integration in cylindrical coordinates, we can restate the stiffness matrix computation as:

$$\left[\mathbf{k}^{(e)}\right] = \iint \int_0^{2\pi} [\mathbf{B}]^{\mathrm{T}}[\mathbf{C}][\mathbf{B}] \, r \, \mathrm{d}\theta \mathrm{d}r \mathrm{d}z = \iint_{A^{(e)}} [\mathbf{B}]^{\mathrm{T}}[\mathbf{C}][\mathbf{B}] \, 2\pi r \, \mathrm{d}r \mathrm{d}z. \tag{6.60}$$

The volume integral is converted to an area integral by first integrating in the θ direction, noting that the integrand does not vary in the θ direction and therefore can be treated as a constant while integrating with respect to θ. Here we have integrated from 0 to 2π, though we could have integrated from 0 to 1 radian instead. Either way, the resultant equations will be the same as long as all the volume integrals in the energy equation are computed in the same way because the constant 2π will then occur in all the terms and will cancel out. For axisymmetric problems, we do not need to specify a thickness as we did earlier for plane stress problems.

EXAMPLE 6.6   *Triangular axisymmetric element*

Consider an annular ring with a rectangular cross section expanded by radial pressure. The radial pressure p is applied uniformly along its entire inner radius. For a triangular element in the mesh that is along the inner boundary as shown in figure 6.17: (i) Compute the equivalent nodal forces to be applied at the nodes of the element. (ii) Assuming that the computed nodal displacement vector for this element is: $\{\mathbf{q}\} = \{u_1 \ v_1 \ u_2 \ v_2 \ u_3 \ v_3\}^{\mathrm{T}}$, compute the components of strain in this element.

SOLUTION   The geometry and the applied pressure is symmetric about an axis, and therefore we can use an axisymmetric model for this problem. The model of the annular ring is shown in figure 6.17(a), where the rectangular region is the domain of the analysis and it represents the cross section of the ring. The axis of symmetry is the y-axis, so the distance of the left edge of the rectangle from the axis must be equal to the inner radius (R_i) of the ring, and the right edge corresponds to the outer surface of the ring. The real geometry of the ring can be obtained by revolving this rectangle around the axis. The height of the rectangle corresponds to the thickness of the ring, h. The figure shows one element labeled (e) in the mesh with the local node numbering shown in figure 6.17(b).

(i) The equivalent nodal forces acting on the edge can be computed as:

$$\left\{\mathbf{f}_T^{(e)}\right\} = \int\limits_{S_e} [\mathbf{N}]^{\mathrm{T}}\{\mathbf{T}\}\,dS = \int\limits_{0}^{l_e} [\mathbf{N}]^{\mathrm{T}}\{\mathbf{T}\}(2\pi r)\,ds.$$

In the preceding equation, S_e is the surface of the element on which the load is applied. The integral over this surface can be decomposed into integrals in the circumferential direction and the axial direction. The integrand is constant in the circumferential direction and therefore can be carried out analytically resulting in the term $2\pi r$, and for the axial direction, we use a local coordinate s attached to node 1 and along the edge of the element as shown in the figure.

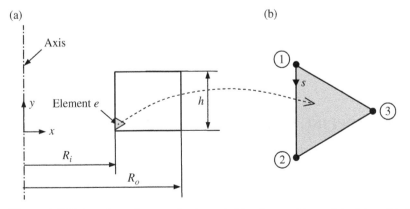

Figure 6.17 Triangular axisymmetric element; (a) axisymmetric model of ring, (b) element e

$$\left\{\mathbf{f}_T^{(e)}\right\} = 2\pi R_i \int_0^{l_e} \begin{bmatrix} N_1' & 0 \\ 0 & N_1' \\ N_2' & 0 \\ 0 & N_2' \end{bmatrix} \left\{\begin{array}{c} p \\ 0 \end{array}\right\} ds = 2\pi p R_i \int_0^{l_e} \left\{\begin{array}{c} N_1' \\ 0 \\ N_2' \\ 0 \end{array}\right\} ds.$$

As the triangular element is linear, we can use linear shape functions derived earlier for one-dimensional elements

$$N_1'(s) = \left(1 - \frac{s}{l_e}\right), \quad N_2'(s) = \frac{s}{l_e}.$$

Evaluating the integral on the length of the edge of the triangle l_e we get,

$$\left\{\mathbf{f}_T^{(e)}\right\} = \frac{2\pi p R_i l_e}{2} \left\{\begin{array}{c} 1 \\ 0 \\ 1 \\ 0 \end{array}\right\}.$$

(ii) The strain components in the element can be computed by first computing the displacement field within the element and then computing the required derivatives. In the finite element method, the strain components are typically computed using the [**B**] matrix as follows:

$$\{\varepsilon\} = \left\{\begin{array}{c} \varepsilon_{rr} \\ \varepsilon_{zz} \\ \varepsilon_{\theta\theta} \\ \varepsilon_{rz} \end{array}\right\} = [\mathbf{B}]\{\mathbf{q}\} = \begin{bmatrix} \dfrac{\partial N_1}{\partial x} & 0 & \dfrac{\partial N_2}{\partial x} & 0 & \dfrac{\partial N_3}{\partial x} & 0 \\ 0 & \dfrac{\partial N_1}{\partial y} & 0 & \dfrac{\partial N_2}{\partial y} & 0 & \dfrac{\partial N_3}{\partial y} \\ \dfrac{N_1}{x} & 0 & \dfrac{N_2}{x} & 0 & \dfrac{N_3}{x} & 0 \\ \dfrac{\partial N_1}{\partial y} & \dfrac{\partial N_1}{\partial x} & \dfrac{\partial N_2}{\partial x} & \dfrac{\partial N_2}{\partial x} & \dfrac{\partial N_3}{\partial y} & \dfrac{\partial N_3}{\partial x} \end{bmatrix} \left\{\begin{array}{c} u_1 \\ v_1 \\ u_2 \\ v_2 \\ u_3 \\ v_3 \end{array}\right\},$$

$$\{\varepsilon\} = \frac{1}{2A} \begin{bmatrix} b_1 & 0 & b_2 & 0 & b_3 & 0 \\ 0 & c_1 & 0 & c_2 & 0 & c_3 \\ \dfrac{f_1 + b_1 x + c_1 y}{x} & 0 & \dfrac{f_2 + b_2 x + c_2 y}{x} & 0 & \dfrac{f_3 + b_3 x + c_3 y}{x} & 0 \\ c_1 & b_1 & c_2 & b_2 & c_3 & b_3 \end{bmatrix} \left\{\begin{array}{c} u_1 \\ v_1 \\ u_2 \\ v_2 \\ u_3 \\ v_3 \end{array}\right\},$$

where, f_i, b_i, and c_i are the coefficients of the shape function defined in eq. (6.13) and A is the area of the triangle. Note that for this element, the hoop strain is not constant within the element while all the other components are constant. ▪

6.6 FINITE ELEMENT MODELING PRACTICE FOR SOLIDS

The most important step in finite element analysis is selecting the right model to use for solving a given problem. There may be more than one way to correctly model a problem. For example, a beam bending problem can be solved using beam elements, but it is also a plane stress problem. The geometry of the structure plays a big role in determining the appropriate model, but as we have seen, the applied load and boundary conditions also can change the nature of the problem and therefore the type of analysis. Very often the right model also depends on the purpose of the analysis, that is, the question you are trying to answer through the analysis. For example, to find the maximum deflection of a frame-like structure, a model using beam/frame elements is often the most appropriate, but if one needs to calculate stress concentration at a joint of the frame, a different type of model is needed. To make the right decision, it is important to understand the underlying theory and the assumptions used for various types of analysis and elements. In this sections, we will study the nature of the solution obtained by the finite element method using some examples solved using commercial software. We have selected examples that can be modeled correctly in more than one way.

EXAMPLE 6.7   *Cantilevered beam*

Model a cantilever beam with a concentrated tip load whose length-to-height ratio is equal to 10 using plane stress elements, and compare the solution to analytical and beam element solutions. Assume that the length is $L = 100$ mm, the cross section is 10×10 mm^2, the applied concentrated load is $P = 100$ N, the material properties are Young's modulus $E = 2 \times 10^{11}$ Pa and Poisson's ratio $\nu = 0.3$.

SOLUTION   A cantilever beam is best modeled using beam elements if it is slender, that is, if the length-to-height ratio is greater than 10. Otherwise, if this ratio is less than 10, then the shear strain energy is no longer negligible, and therefore a plane stress model is more appropriate. Using Euler-Bernoulli beam theory, the analytical solution for a cantilever beam with a concentrated load at the tip is:

$$\delta = \frac{PL^3}{3EI} = \frac{100 \times (0.1)^3}{3 \times 2 \times 10^{11} \times 8.333 \times 10^{-10}} = 2 \times 10^{-4} \text{ m.}$$

When the beam is subjected to concentrated forces, the analytical solution for the beam deflection is a cubic function. Beam elements use cubic shape functions and therefore, a model using a single beam element is sufficient to yield the exact solution for the beam deflection in this example. The normal stresses in beam elements are computed using beam theory (see chapter 3). The normal stress on any section of the beam is a linear function of the distance from the neutral axis. When using software for analysis, one has to provide the height of the cross section so that the maximum stress can be computed.

In contrast, the two-dimensional plane stress model requires a planar rectangular geometry that is meshed using triangular or quadrilateral elements. In this chapter, we described the 3-node constant strain triangular (CST) elements. The strain is approximated to be constant within this element, and therefore it is not a very accurate element. A relatively high-density mesh is needed to get acceptable results since the normal stress and strain vary linearly through the height of the cross section of the beam. To reasonably approximate this linear variation using elements with constant stress and strain, we will need a lot of elements through the height of the beam. Better elements will be discussed in a subsequent chapter that use higher-order polynomials for the interpolation.

Figure 6.18 shows the plane stress model of the beam constructed using triangular elements. To clamp the beam at the left end, a sliding boundary condition has been applied along the left edge so that the displacement is fixed only in the x-direction while the y-component of the displacement is not fixed. If the y-component were also fixed, then shrinkage or expansion in that direction that occurs due to Poisson's effect will be prevented causing some artificial stresses. It is important to apply sufficient boundary conditions to prevent any rigid body motion. To prevent rigid

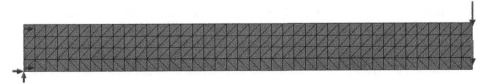

Figure 6.18 Beam model using plane stress CST elements

UV (mm)

——	3.776e–005
·-	–1.675e–002
·-	–3.353e–002
·-	–5.032e–002
·-	–6.710e–002
·-	–8.389e–002
·-	–1.007e–001
·-	–1.175e–001
·-	–1.342e–001
·-	–1.510e–001
·-	–1.678e–001
·-	–1.846e–001
	–2.014e–001

Figure 6.19 Beam deflection computed using CST elements

body motion in the *y*-direction, a point along the left edge is fixed as shown in the figure. Ideally, if a node is available at the exact midpoint of the left edge, then this node should be fixed. However, with automatic mesh generation, this is not guaranteed. In some software packages, a node can be placed there during model definition, but these details vary widely between programs. The load, which is typically modeled as a concentrated force in theoretical and beam models, is better defined as a distributed load in a plane stress model. So the load is applied at the right end as a traction load that is distributed evenly over the right face of the beam. This option is better than applying a concentrated force at one of the vertices, which could create stress concentration at the point of application that is unrealistic. Figure 6.19 shows the deformed shape of the beam computed using the plane stress model. As seen in the figure, the maximum deflection computed is very close to the analytical solution indicating that the mesh density was sufficient for this example if computing the deflection of the beam was the main intent. Compared to beam elements, however, this is clearly a less efficient model that requires a lot more elements and computation.

The normal strain computed using CST elements is displayed in figure 6.20. The linear variation through the beam height is approximated by constant values within each triangle. Notice that the strain is not perfectly symmetrical about the neutral axis as one would expect. The scale indicates that the maximum and minimum do not have exactly the same magnitude. This is due to the fact that the mesh is not symmetric about the neutral axis, and unlike displacements, the strains computed using this model is not as accurate. In fact, the analytical solution for the maximum strain is 3×10^{-4}.

The plot for stresses and strains can be improved by smoothing the computed result. A very simple method for smoothing is to average the computed values from all the elements attached to a node and then using these averaged nodal values to interpolate the stresses or strains. Figure 6.21 shows the normal stress plot after smoothing the results. Now the stress variation through the thickness appears more realistic because the stress is interpolated using average nodal values and is therefore not constant within elements. The maximum normal stress should be 6×10^7 Pa. The results can be improved by further refining the mesh using much smaller elements.

Even though the results do not exactly match the analytical solution, the CST element models do give a reasonable approximation of the solution and are able to tell us where the maximum and minimum stresses would occur in a structure. As the beam height decreases, the stiffness of the beam is overestimated for this type of 2D elements leading to a phenomenon called shear locking for very slender beams if modeled using plane stress elements. This happens because the typical finite element solution has errors due to the fact that the shear does not reduce to zero as

Figure 6.20 Computed normal strain component without smoothing

Figure 6.21 Normal stress component after smoothing

the beam height decreases, and this leads to overestimation of stiffness. For such slender beams, the beam element is more appropriate and is, in fact, able to predict the theoretical solution if the applied loads are all concentrated forces. On the other hand, for short beams where the shear strain is not negligible, plane stress models are preferred. In general, the accuracy of the solution improves with increasing mesh density. With increasing mesh density, the solutions should converge, meaning that the change in the solution should decrease with increasing mesh density. The solution is said to have converged when the solution does not change significantly with further increase in mesh density. ▬

EXAMPLE 6.8   *Thick-walled cylinder*

A thick-walled tube subjected to internal pressure is studied here as illustrated in figure 6.22(a) where its cross section is shown. The inner and outer radii are 0.3 and 0.5 in., respectively. The inner surface is subjected to a pressure of 500 psi. The material is assumed to be made of steel with a modulus of elasticity, $E = 30 \times 10^6$ psi and a Poisson's ratio of 0.3. Using two-dimensional models, compute the radial displacement and the stress distribution in the tube.

SOLUTION   This problem can be modeled in a finite element software in two different ways. As the tube has a cylindrical geometry and the load is symmetric about its axis, it is a problem that can be modeled as axisymmetric. On the other hand, if the tube is very long compared to the diameter of its cross section, then we can assume that the internal pressure will not cause a change in length (or axial strain). Therefore, it can also be modeled as a plane strain problem. Obviously, the geometry of the analysis domain will be very different depending on the model.

To model the cylinder as a plane strain problem, we model only a quarter of the cross section of the cylinder as shown in figure 6.22(b). The displacement boundary conditions are applied on the straight edges of the quarter such that only radial displacement is allowed while tangential displacement is restricted. For the axisymmetric model, as shown in figure 6.22(c), the domain of analysis is a rectangle placed such that its inner (left-side) boundary is at a distance equal to the inner radius from the axis. An arbitrary length of the cylinder can be modeled if in fact the tube is very long and we expect no axial displacements/strains. In this case, since the model length is significantly less than the real length, it is important to also apply displacement boundary conditions such that there is no axial displacement. The internal pressure is applied along the inner radius as shown in the figure.

Results obtained from the two models are compared in figure 6.23. When the thick-walled cylinder is modeled using plain strain elements, the maximum displacement was computed to be 1.165×10^{-5} *in.* whereas the same cylinder (same dimensions and loads) when modeled using axisymmetric elements yields displacement equal to 1.162×10^{-5} *in.*, which is approximately the same with a difference of less than 0.3%. The two models will converge toward the analytical solution as the element size is decreased. Similarly, the von Mises stress computed by the two models are also approximately equal. Triangular elements linearly interpolate nodal values of displacement, and therefore most of the strain and stress components are computed to be constant within each element as noted in the previous example. The stress plot in figure 6.23 does not show a discontinuity at the edges of the elements because stresses are smoothed before plotting. Smoothing the results in this manner improves the accuracy of the solutions since the exact solution for stresses is indeed continuous throughout the domain.

(a)   (b)   (c)

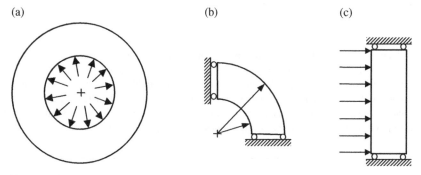

Figure 6.22 Thick-walled cylinder; (a) cross-section, (b) plane strain model, (c) axisymmetric model

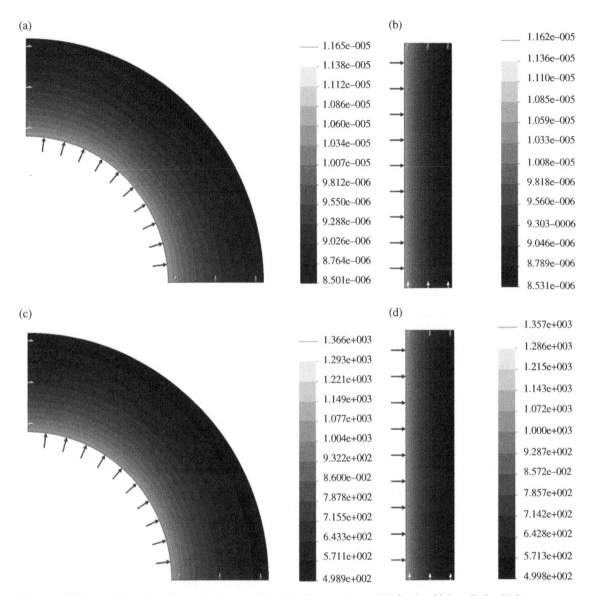

Figure 6.23 Comparison of results using plane strain and axisymmetric models for the thick-walled cylinder; (a) displacement magnitude with plane strain model, (b) displacement magnitude with axisymmetric model, (c) von Mises stress with plane strain model, (d) von Mises stress with axisymmetric model

6.7 PROJECT

Project 6.1 Accuracy and convergence analysis of a cantilever beam

In this project, we want to compare the finite element results of plane solid elements with that of uniaxial bar and beam elements. Consider a cantilever beam shown in figure 6.24 under horizontal and transverse

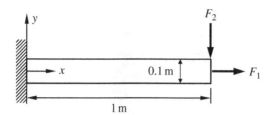

Figure 6.24 Cantilever beam model

forces at the tip. The beam has a square cross section of $0.1\,\mathrm{m} \times 0.1\,\mathrm{m}$, length of $L = 1\,\mathrm{m}$, Young's modulus $E = 207\,\mathrm{GPa}$, and Poisson's ratio $\nu = 0.3$.

PART I

   (a) Consider the case of $F_1 = 100\,\mathrm{N}$ and $F_2 = 0$. Solve the problem using a uniaxial bar element to find the elongation $u(x)$. Calculate ε_{xx} and σ_{xx}. Assume that $\sigma_{yy} = \sigma_{zz} = \tau_{xy} = \tau_{yz} = \tau_{xz} = 0$. Compare the results with analytical solution.

   (b) Consider the case of $F_1 = 0$ and $F_2 = 500\,\mathrm{N}$. Solve the problem using a beam element to find the deflection $w(x)$. Calculate ε_{xx} and σ_{xx}. Assume that $\sigma_{yy} = \sigma_{zz} = \tau_{xy} = \tau_{yz} = \tau_{xz} = 0$. Compare the results with analytical solution. Plot σ_{xx} as a function of y at $x = L/2$.

PART II

   (a) Consider the case of $F_1 = 100\,\mathrm{N}$ and $F_2 = 0$. Solve the problem using: (i) 20 CST elements and (ii) 10 rectangular elements to find the elongation $u(x)$. Calculate ε_{xx} and σ_{xx}. Compare the results with those from part I. Explain the results using an interpolation scheme.

   (b) Consider the case of $F_1 = 0$ and $F_2 = 500\,\mathrm{N}$. Solve the problem using: (i) 20 CST elements and (ii) 10 rectangular elements to find the deflection $w(x)$. Calculate ε_{xx} and σ_{xx}. Compare the results with those of part I. Explain the results using an interpolation scheme.

   (c) Consider the case of $F_1 = 0$ and $F_2 = 500\,\mathrm{N}$. Perform convergence study by gradually decreasing the element size, and show the deflection and stress converge to the exact solution.

6.8   EXERCISES

   1. Answer the following descriptive questions.

   (a) What are nonzero stress and strain components for plane stress problems?

   (b) What are nonzero stress and strain components for plane strain problems?

   (c) What are nonzero stress and strain components for axisymmetric problems?

   (d) When would a 3-node triangular element be invalid?

   (e) If only nonzero displacements of a 3-node triangular element are u_3 and v_3, what would be the displacement along the edge of nodes 1 and 2?

   (f) How do the strains, ε_{xx}, ε_{yy}, and γ_{yx}, vary within a 3-node triangular element?

   (g) How do the strains, ε_{xx}, ε_{yy}, and γ_{xy}, vary within a 4-node rectangular element?

   (h) If a gravitational force is applied in the y-coordinate direction for a 3-node triangular element, what would the equivalent nodal forces be? Assume that the total gravitational force is $\rho h A g$, where ρ is density, h is thickness, A is area, and g is gravitational acceleration.

   (i) For a four-node rectangular plane element, we use $u(x,y) = a_0 + a_1 x + a_2 y + a_3 xy$ as the form of solution. What would be the problem if we use $u(x,y) = a_0 x^2 + a_1 xy + a_2 y^2 + a_3 x^2 y^2$ instead?

   (j) For a rectangular element, plot the shape function $N_1(x,y)$.

   (k) Define inter-element displacement compatibility.

2. Repeat example 6.2 with the following element connectivity:

   Element 1: 1–2–4

   Element 2: 2–3–4
   Does the different element connectivity change the results?

3. Solve example 6.2 using a commercial finite element analysis program.

4. Using two CST elements, solve the simple shear problem depicted in the figure and determine whether the CST elements can represent the simple shear condition accurately or not. Material properties are given as $E = 10$ GPa, $\nu = 0.25$, and thickness is $h = 0.1$ m. The distributed force $f = 100$ kN/m^2 is applied at the top edge.

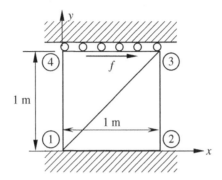

5. Solve problem 4 using a commercial finite element analysis program.

6. A structure shown in the figure is modeled using one triangular element. Plane strain assumption is used.

   (a) Calculate the strain–displacement matrix $[\mathbf{B}]$.

   (b) When nodal displacements are given by $\{u_1, v_1, u_2, v_2, u_3, v_3\} = \{0, 0, 2, 0, 0, 1\}$, calculate element strains.

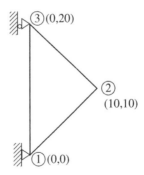

7. Calculate the shape function matrix $[\mathbf{N}]$ and strain–displacement matrix $[\mathbf{B}]$ of the triangular element shown in the figure

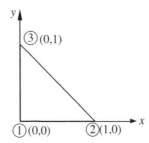

8. The nodal coordinates and corresponding displacements in a plane triangular element are given in the table below. Calculate the u-displacement at a point given by $(1,1)$.

Node number	(x,y) (mm)	u (mm)	v (mm)
1	(0,0)	0	0
2	(3,0)	1	1
3	(0,3)	2	0

9. The nodal displacements of the triangular element are given in the table below.

Node	(x, y)	u	v
1	(0, 0)	0	6
2	(3, 0)	3	5
3	(0, 4)	8	4

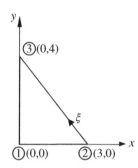

(a) It is intended to use the polynomial $u(x,y) = a_1 + a_2 x + a_3 y$ to interpolate the u displacements in the triangular element. Calculate the constants a_1, a_2, and a_3.

(b) What is the strain ε_{xx} at the centroid of the triangle?

(c) Derive an expression for u displacements of points on the edge 2–3 as a function of ξ. That is, derive an expression for $u(\xi)$.

Hint: Along the edge 2–3, $x = \left(3 - \frac{3}{5}\xi\right), y = \frac{4}{5}\xi$.

10. The coordinate of the nodes and corresponding displacements in a triangular element are given in the table. Calculate the displacement u and v and strains $\varepsilon_{xx}, \varepsilon_{yy}$, and γ_{xy} at the centroid of the element given by the coordinates $(1/3, 1/3)$

Node	x (m)	y (m)	u (m)	v (m)
1	0	0	0	0
2	1	0	0.1	0.2
3	0	1	0	0.1

11. A $2 \times 2 \times 1$ mm^3 square plate with $E = 70$ GPa and $\nu = 0.3$ is subjected to a uniformly distributed load as shown in the left figure. Due to symmetry, it is sufficient to model one-quarter of the plate with artificial boundary conditions as shown in the right figure. Use two triangular elements to find the displacements, strains, and stresses in the plate. Check the answers using simple calculations from mechanics of materials.

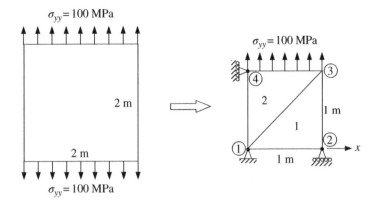

12. A beam problem under the pure bending moment is solved using CST finite elements, as shown in the figure. Assume $E = 200$ GPa and $\nu = 0.3$. The thickness of the beam is 0.01 m. In order to simulate the pure bending moment, two opposite forces $F = \pm 100,000$ N are applied at the end of the beam. Using any available finite element program, calculate the stresses in the beam along the neutral axis and top and bottom surfaces. Compare the numerical results with the elementary beam theory. Provide an element stress contour plot for σ_{xx}.

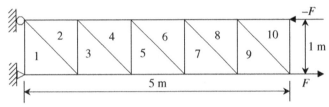

13. For a rectangular element shown in the figure, displacements at four nodes are given by $\{u_1, v_1, u_2, v_2, u_3, v_3, u_4, v_4\} = \{0.0, 0.0, 1.0, 0.0, 2.0, 1.0, 0.0, 2.0\}$. Calculate displacement (u, v) and strain ε_{xx} at point $(x, y) = (2, 1)$.

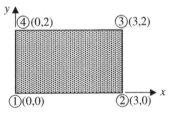

14. The figure shows the assemblage of two square elements, elements 23 and 24. The coordinates and displacements of all six nodes are given in the table below.

   (a) Consider point P given by the global coordinates $(2, 1)$. Calculate the displacements u and v at this point. As this point belongs to both elements, one can calculate the displacements u and v at this point using the nodal displacements of either element 23 or element 24. Will they be same? Explain.

   (b) Repeat the above question (a) including discussion for calculating the strain ε_{xx} at point P.

Node	1	2	3	4	5	6
x	0	2	4	0	2	4
y	0	0	0	2	2	2
u	0	0.1	0.3	0	0.2	0.4
v	0	-0.05	-0.2	0	-0.5	-0.2

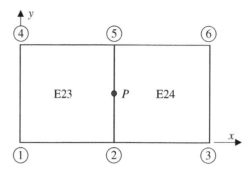

15. The FE model of a plane solid consists of two square elements of dimensions 1×1 as shown in the figure below. The nodal coordinates and corresponding displacements are given in the table.

    (a) Calculate the displacement u at $(x,y) = (1, 0.5)$;

    (b) Estimate the strain ε_{xx} at $(x,y) = (1, 0.5)$ using element 1 and element 2.

Node	(x, y)	u	v
1	(0,0)	0	0
2	(0,1)	0.1	−0.1
3	(1,0)	0.1	0.1
4	(1,1)	−0.2	0.2
5	(2,0)	0.15	−0.15
6	(2,1)	0.2	0

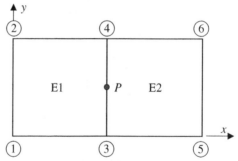

16. For the 4-node plane stress finite element shown in the figure, finite element computation yields the displacements at the nodes as follows.

u_1	v_1	u_2	v_2	u_3	v_3	u_4	v_4
0.005	0.003	0.006	0.005	0.0	0.0	0.0	0.0

(a) What is the displacement field $u(x,y)$ and $v(x,y)$ within the element?

(b) Compute the strain displacement matrix $[\mathbf{B}]$ for this element to express strain as $\{\varepsilon\} = [\mathbf{B}]\{\mathbf{q}\}$?

(c) Compute the displacements (u,v) and the strain components at the origin.

(d) Show that you can use this element to represent a constant state of strain.

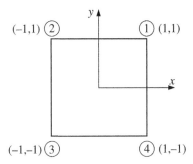

17. A rectangular element with thickness 1.0 m is under gravity. Express strain ε_{yy} in terms of vertical nodal displacement $-v$ at the top. ($v_1 = v_2 = 0$, $v_3 = v_4 = -v$). Explain if the calculated strain ε_{yy} is exact or not. (If you need, use g: gravitational acceleration, ρ: density, E: elastic modulus).

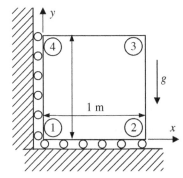

18. For the 4-node element shown in the figure, a linearly varying pressure p is applied along the edge. The finite element method converts the distributed force into an equivalent set of nodal forces $\{\mathbf{F}_e\}$ such that

$$\int_S [\mathbf{N}]^T \{\mathbf{T}\} dS = \{\mathbf{F}_e\},$$

where $\{\mathbf{T}\}$ is the traction (force per unit area) on the surface S and $[\mathrm{N}]$ is a 2×8 matrix of shape functions. The applied pressure in the above figure is normal to the surface (in the x-direction), therefore the traction can be expressed as $\{\mathbf{T}\} = \{p \ \ 0\}^T$ where p can be expressed as $p = p_0 y / L_e$. L_e is the length of the edge. Integrate the left-hand side of the equation above to compute the work-equivalent nodal forces $\{\mathbf{F}_e\}$ (8×1 vector).

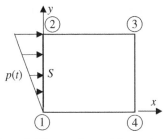

19. Six rectangular elements are used to model the cantilevered beam shown in the figure. Sketch the graph of σ_{xx} along the top surface that a finite element analysis would yield. There is no need to actually solve the problem, but use your knowledge of shape functions for rectangular elements.

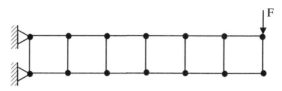

20. A rectangular element as shown in the figure is used to represent a pure bending problem. Due to the bending moment M, the element is deformed as shown in the figure with displacement $\{q\} = \{u_1, v_1, u_2, v_2, u_3, v_3, u_4, v_4\}^T = \{-1, 0, 1, 0, -1, 0, 1, 0\}^T$.

   (a) Write the mathematical expressions of strain component ε_{xx}, ε_{yy}, and γ_{xy}, as functions of x and y.

   (b) Does the element satisfy pure bending condition? Explain your answer.

   (c) If two CST elements are used by connecting nodes 1–2–4 and 4–2–3, what will be ε_{xx} along line A–B?

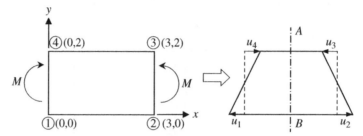

21. Five rectangular elements are used to model a plane beam under pure bending. The element in the middle has nodal displacements as shown in the figure. Using the bilinear interpolation scheme, calculate the shear strain along the edge AB and compare it with the exact shear strain.

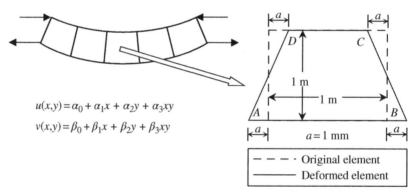

$$u(x,y) = \alpha_0 + \alpha_1 x + \alpha_2 y + \alpha_3 xy$$
$$v(x,y) = \beta_0 + \beta_1 x + \beta_2 y + \beta_3 xy$$

- - - - · Original element
———— Deformed element

22. A uniform beam is modeled by two rectangular elements with thickness b. Qualitatively, and without performing calculations, plot σ_{xx} and τ_{xy} along the top edge from A to C, as predicted by FEA. Also, plot the exact stresses according to beam theory.

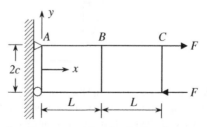

23. A beam problem under the pure bending moment is solved using five rectangular finite elements, as shown in the figure. Assume $E = 200$ GPa and $\nu = 0.3$ are used. The thickness of the beam is 0.01 m. In order to simulate pure bending moment, two opposite forces $F = \pm 100,000$ N are applied at the end of the beam. Using a commercial FE program, calculate strains in the beam along the bottom surface. Draw graphs of ε_{xx} and γ_{xy} with the x–axis being the beam length. Compare the numerical results with the elementary beam theory. Provide an explanation for the differences, if any. Is the rectangular element stiff or soft compared to the CST element?

Normally, a commercial finite element program provides stress and strain at the nodes of the element by averaging the stresses computed in the adjacent elements. Thus, you may use nodal displacement data from FE code to calculate strains along the bottom surface of the element. Calculate the strains at about ten points in each element for plotting purposes. Make sure that the commercial program uses the standard Lagrange shape function.

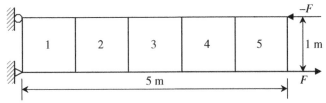

Repeat the above procedure when an upward vertical force of 200,000 N is applied at the tip of the beam. Use boundary conditions similar to the clamped boundary conditions of a cantilevered beam.

Chapter 7

Isoparametric Finite Elements

7.1 INTRODUCTION

The shape functions for triangular and rectangular elements in chapter 6 are derived in the global coordinates and are dependent on the nodal coordinates of the element. Therefore, different elements have different shape functions. Knowing that hundreds of thousands of elements are often used in solving practical problems, evaluating the shape functions for individual elements might be laborious and computationally inefficient. In addition, deriving the shape functions for quadrilaterals in global coordinates is difficult as compared to rectangular elements. It is often convenient to use local coordinate systems for interpolating the field variables such as displacement or temperature fields because the shape functions are easier to derive and have simpler expressions. This involves a change of variables from the physical (x,y) or (x,y,z) coordinates to parametric (s,t) or (r,s,t) coordinates. Isoparametric elements use a parametric coordinate system that transforms the elements by scaling and deforming it. The domain in the parametric coordinate system is often referred to as the *parametric space*, as opposed to the space occupied by the element in the global coordinate system, which is the *physical space*. A mapping is established between the physical and parametric spaces. The advantage of using this approach is that more complex-shaped elements can be constructed such as quadrilateral and hexahedral elements. In addition, since shape functions are calculated in the parametric space, and since all elements are mapped into the same parametric space, all elements in different geometry share the same shape functions. Only the mapping from individual elements to the parametric space is different for different elements.

In this chapter, we introduce the concept of isoparametric elements first with one-dimensional elements. The linear one-dimensional elements are very similar to the elements we have already seen in the previous chapters. For higher-order elements such as quadratic or even cubic elements, the isoparametric element formulation has several advantages. The shape functions of these elements are much easier to derive when parametric coordinates are used. Systematic methods for deriving these shape functions using Lagrange interpolation techniques are available that are also presented in section 7.2 for one-dimensional elements and in section 7.3 for two-dimensional elements. Even though the isoparametric formulation simplifies the element geometry in the parametric space, it is still difficult to analytically integrate over the geometry to compute a stiffness matrix or load vector, and therefore numerical integration is needed. Another advantage of isoparametric element is that it is convenient to apply numerical integration methods. In finite element analysis, the method of Gauss quadrature is predominantly used for numerical integration to evaluate the stiffness matrix or load vector, which will be discussed in 7.4. Using the isoparametric formulation, higher-order quadrilateral elements and triangular elements will be derived and discussed in sections 7.5 and 7.6. Although most important theoretical and numerical aspects of the finite element method can be illustrated using one- and two-dimensional problems, some

Introduction to Finite Element Analysis and Design, Second Edition. Nam H. Kim, Bhavani V. Sankar, and Ashok V. Kumar.
© 2018 John Wiley & Sons Ltd. Published 2018 by John Wiley & Sons Ltd.
Companion website: www.wiley.com/go/kim/finite_element_analysis_design

three-dimensional elements are introduced in section 7.7 to demonstrate the complexity of the formulation. Section 7.8 discusses some modeling issue for practical problems, followed by three projects in section 7.9.

7.2 ONE-DIMENSIONAL ISOPARAMETRIC ELEMENTS

7.2.1 2-Node Linear Isoparametric Element

A one-dimensional 2-node linear element is used here to introduce the concept of an isoparametric element. Figure 7.1 shows a one-dimensional domain of length L where the physical coordinate x has a domain of $0 \le x \le L$. This domain can be the one-dimensional bar in chapter 1 or one-dimensional heat transfer in a rod in chapter 4. The domain is discretized by *NEL* number of elements. An element e in this domain is shown, which has nodal coordinate of x_i and x_j at nodes i and j respectively. This element is mapped to a reference element, which is defined using the reference coordinate s. The reference element has a domain of $-1 \le s \le 1$. The reference element has two nodes. Node i in physical element is mapped to node 1 in the reference element, and node j to node 2. Note that the physical element has a length $L^{(e)} = x_j - x_i$, while the reference element has a fixed length of 2.

We need a mapping function $x(s)$, which defines the relation between these two coordinates such that $x(s = -1) = x_i$ and $x(s = +1) = x_j$. For isoparametric elements, the shape functions used for interpolation are also used for defining this mapping function. The shape functions are first derived in the reference element as a function of the parameter s. For this 2-node linear element, we can assume that the shape functions are in the form of $N_i(s) = a_i + b_i s$.

The shape functions must satisfy the Kronecker's delta condition, that is,

$$N_i(s) = \begin{cases} 1 & \text{at node } i \\ 0 & \text{at all other nodes.} \end{cases} \tag{7.1}$$

Using this condition for node 1, we get the following two equations to determine $N_1(s)$

$$N_1(-1) = a_1 - b_1 = 1,$$
$$N_1(1) = a_1 + b_1 = 0.$$

Solving these two equations, we get $a_1 = 1/2$ and $b_1 = -1/2$. Substituting this into $N_1(s)$, we get,

$$N_1(s) = \frac{1-s}{2}. \tag{7.2}$$

Similarly, using eq. (7.1) for node 2, we can solve for the second shape function as,

$$N_2(s) = \frac{1+s}{2}. \tag{7.3}$$

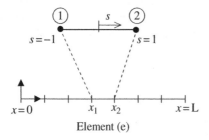

Parametric space

Physical space

Element (e)

Figure 7.1 One-dimensional 2-node linear isoparametric element

These shape functions can be used for interpolation as we have done in the previous chapters, but this yields an interpolation in parametric space. For example, if we use this element for heat conduction analysis, the temperature within the element can be interpolated using nodal temperature as

$$T(s) = \frac{(1-s)}{2}T_1 + \frac{(1+s)}{2}T_2$$

$$= N_1(s)T_1 + N_2(s)T_2$$

$$= \{N_1 \quad N_2\}\begin{Bmatrix} T_1 \\ T_2 \end{Bmatrix}$$

$$= \{\mathbf{N}\}^{\mathrm{T}}\{\mathbf{T}^{(e)}\}.$$

Note that the temperature is now a function of the parameter s rather than the physical coordinate x. Therefore we need a mapping function that establishes the relation between x and s. In isoparametric elements, this mapping function is constructed using the same shape functions as

$$x(s) = x_i N_1(s) + x_j N_2(s). \tag{7.4}$$

It can be easily verified that this mapping satisfies the condition that at node 1, $s = -1$, $x(-1) = x_i$ and similarly, at node 2, $x(1) = x_j$.

EXAMPLE 7.1   *One-dimensional isoparametric heat conduction element*

Using the 2-node isoparametric element in the heat conduction equations in eq. (4.20) in chapter 4, derive the conductivity matrix.

SOLUTION   A weak form for one-dimensional heat transfer, similar to that of the principle of virtual work for solid elements, can be derived from the governing equation using the weighted residual method as:

$$\int_{x_i}^{x_j} kA\frac{d\delta T}{dx}\frac{dT}{dx}dx = \int_{x_i}^{x_j} AQ_g\delta T dx - kA\frac{dT}{dx}\delta T\bigg|_{x_i}^{x_j}. \tag{7.5}$$

We want to use the shape functions derived for the isoparametric element on the left-hand side of this equation to determine the conductivity matrix. The integration is with respect to the physical coordinate x, but the shape functions of the isoparametric element is a function of the parameter s. Therefore a change of variables is needed for which we will use the mapping function.

$$\int_{x_1}^{x_2} kA\frac{d\delta T}{dx}\frac{dT}{dx}dx = \int_{-1}^{1} kA\frac{d\delta T}{ds}\frac{ds}{dx}\frac{dT}{ds}\frac{ds}{dx}\frac{dx}{ds}ds = \int_{-1}^{1} kA\frac{d\delta T}{ds}\frac{dT}{ds}\frac{ds}{dx}ds. \tag{7.6}$$

Now the mapping function can be used to express the integrand entirely as a function of s. The rate of change of x with respect to s is referred to as the Jacobian.

$$x(s) = N_1(s)x_1 + N_2(s)x_2$$

$$\frac{dx}{ds} = \frac{dN_1}{ds}x_1 + \frac{dN_2}{ds}x_2$$

$$= -\frac{1}{2}x_1 + \frac{1}{2}x_2 \tag{7.7}$$

$$= \frac{L_e}{2}.$$

The inverse of the Jacobian is its reciprocal or the rate of change of s with respect to x.

$$\frac{ds}{dx} = \frac{1}{dx/ds} = \frac{2}{x_2 - x_1} = \frac{2}{L_e}. \tag{7.8}$$

Note that the Jacobian is a constant equal to the ratio between the lengths of the element in the physical coordinate system to that of the element in reference (or parametric) coordinate system. Later we will see that it is not always as simple as this for higher-order elements.

As described earlier, in the weak form, we use the same interpolation for the virtual temperature to get

$$\delta T(s) = N_1(s)\delta T_1 + N_2(s)\delta T_2.$$

Using these relations, we can reduce eq. (7.6) to the discretized matrix form as:

$$\int_{-1}^{1} kA \frac{d\delta T}{ds} \frac{dT}{ds} \frac{ds}{dx} ds = 2\frac{kA}{L_e} \int_{-1}^{1} \{\delta T_1 \quad \delta T_2\} \begin{Bmatrix} \dfrac{dN_1}{ds} \\[2mm] \dfrac{dN_2}{ds} \end{Bmatrix} \left\{ \dfrac{dN_1}{ds} \quad \dfrac{dN_2}{ds} \right\} \begin{Bmatrix} T_1 \\[2mm] T_2 \end{Bmatrix} ds$$

$$= \{\delta T_1 \quad \delta T_2\} \left[\mathbf{k}^{(e)}\right] \begin{Bmatrix} T_1 \\[2mm] T_2 \end{Bmatrix}.$$

From this, we get the conductivity matrix as

$$\left[\mathbf{k}^{(e)}\right] = 2\frac{kA}{L_e} \int_{-1}^{1} \begin{Bmatrix} \dfrac{dN_1}{ds} \\[2mm] \dfrac{dN_2}{ds} \end{Bmatrix} \left\{ \dfrac{dN_1}{ds} \quad \dfrac{dN_2}{ds} \right\} ds$$

$$= 2\frac{kA}{L_e} \int_{-1}^{1} \begin{bmatrix} 1/4 & -1/4 \\ -1/4 & 1/4 \end{bmatrix} ds = \frac{kA}{L_e} \begin{bmatrix} 1 & -1 \\ -1 & 1 \end{bmatrix}.$$

As expected we get the same stiffness matrix that we obtained earlier when we were not using the isoparametric formulation. Clearly, the element is the same whether we use an isoparametric formulation or not. But by changing the variable from the physical coordinate to the reference coordinate with a more convenient domain, it becomes a little easier to derive the shape functions and to carry out the integration for evaluating the stiffness matrix. This advantage is all the more critical for more complex-shaped elements such as quadrilaterals as well as higher-order elements, where it is much more difficult to derive the shape functions in terms of the global coordinates.

7.2.2   Quadratic 3-Node Isoparametric Element

In the previous section, we saw that if the element has 2-nodes, then we can fit a straight line between the values at the two nodes to obtain a linear interpolation. This is because we had two nodal values to fit a function, and a linear function has two unknown coefficients. If we need a higher-order element, we need more nodes in the element or have more variables per node. A quadratic element can be constructed if the element has three nodes so that we can fit a parabolic curve between the three nodal values.

For the physical element, the element connectivity is given as i-j-k, where node i and j are two end nodes and node k is the one in between. In the finite element community, it is customary that node numbers are given to the corner nodes first followed by nodes on the edges. Node k may not be located at the

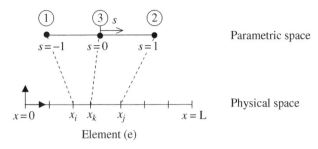

Parametric space

Physical space

Figure 7.2 3-node quadratic isoparametric element

Element (e)

center of the element in the global coordinates even though it is at the center in the parametric space. However, if the node is not centrally placed in the global coordinates, then the mapping is distorted as we shall see in later discussion. For the reference element, the element connectivity is given as 1-2-3, where node 1 is at $s = -1$, node at $s = +1$, and node 3 at $s = 0$.

We will assume that the quantity we are interpolating (say, the temperature) and the shape functions are quadratic polynomials, as

$$T(r) = a + bs + cs^2. \tag{7.9}$$

That means, the shape functions are also a quadratic polynomial, as

$$N_i(s) = a_i + b_i s + c_i s^2 \quad i = 1,2,3. \tag{7.10}$$

Again, one could use the Kronecker's delta condition to derive the shape functions. For the first shape function

$$N_1(-1) = a_1 - b_1 + c_1 = 1,$$
$$N_1(1) = a_1 + b_1 + c_1 = 0,$$
$$N_1(0) = a_1 = 0.$$

Solving these equations, we get $a_1 = 0$, $b_1 = -1/2$, and $c = 1/2$. Similarly, we can solve for the other shape functions. Therefore, three shape functions for a quadratic element can be obtained as

$$\begin{cases} N_1(s) = -\dfrac{1}{2}s(1-s), \\ N_2(s) = \dfrac{1}{2}s(1+s), \\ N_3(s) = 1 - s^2. \end{cases} \tag{7.11}$$

The mapping function for the element is again constructed using the same shape functions in all isoparametric elements.

$$x(s) = N_1(s)x_1 + N_2(s)x_2 + N_3(s)x_3. \tag{7.12}$$

We could have used the mapping function we derived for the 2-node linear element here. That is, the displacement is interpolated using quadratic shape functions, while the geometry is interpolated using linear shape functions. But then the element would not be isoparametric. Instead, it would be a sub-parametric element because the mapping function is of a lower order than the interpolation. Similarly, one can construct super-parametric elements that use a higher-order mapping function than the order of the interpolation used for the element. So the term isoparametric comes from the notion of using the same shape functions to create both the mapping and the interpolation.

EXAMPLE 7.2   *Mapping using a quadratic element*

Consider regular and irregular 3-node one-dimensional elements in figure 7.3. Both elements are mapped into the same reference element. When three different displacement conditions are imposed:

(a) zero strain: $u(x) = $ constant, $du/dx = 0$

(b) constant strain: $u(x) = x$, $du/dx = 1$

(c) linear strain: $u(x) = x^2$, $du/dx = 2x$

Plot the mapping relation $x(s)$, Jacobian dx/ds, displacement gradient du/ds in the reference element, and strain du/dx for each condition and check whether the interpolation yields accurate results or not.

SOLUTION   Regular Element:

The length of the physical element is the same as that of the reference element. Therefore, for the regular element it is expected that the Jacobian is unity, that is, $dx/ds = 1$. This can be shown using eq. (7.12) but also can be shown be generating the relationship based on observation. Since the physical element is in the range of $0 \leq x \leq 2$ and the reference element $-1 \leq s \leq 1$, the relation between x and s can be obtained as $x(s) = s + 1$. Therefore, the Jacobian can be obtained by differenting x with respect to s as $dx/ds = 1$. As shown in figure 7.4, the mapping relationship is linear and the Jacobian is constant for the regular element.

(a) In the case of constant displacement, all nodal displacements are the same, and therefore, $du/ds = du/dx = 0$.

(b) In the case of constant strain, the nodal displacements can be obtained using $u(x_i) = x_i$, as $u_1 = 0$, $u_2 = 2$, $u_3 = 1$. The interpolation relation can be used to obtained $u(s)$, as

$$u(s) = N_1(s)u_1 + N_2(s)u_2 + N_3(s)u_3$$
$$= -\frac{1}{2}s(1-s) \times 0 + \frac{1}{2}s(1+s) \times 2 + (1-s^2) \times 1$$
$$= 1 + s.$$

Then $du/ds = 1$. From the fact that Jacobian is unity, it is obvious that $du/dx = 1$.

(c) In the case of linear strain, the nodal displacements can be obtained using $u(x_i) = x_i^2$, as $u_1 = 0$, $u_2 = 4$, $u_3 = 1$. The interpolation relation can be used to obtained $u(s)$, as

$$u(s) = N_1(s)u_1 + N_2(s)u_2 + N_3(s)u_3$$
$$= -\frac{1}{2}s(1-s) \times 0 + \frac{1}{2}s(1+s) \times 4 + (1-s^2) \times 1$$
$$= (1+s)^2.$$

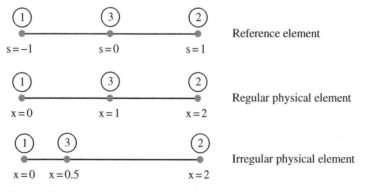

Figure 7.3 Regular versus irregular quadratic element

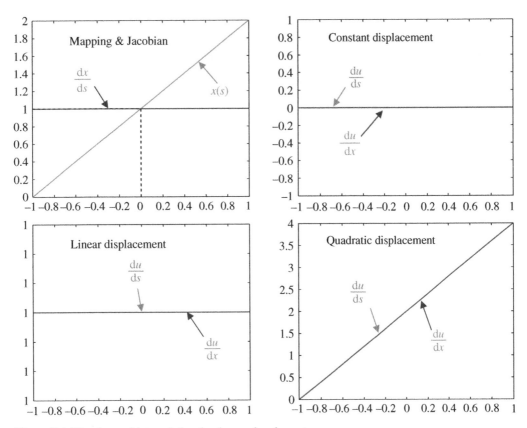

Figure 7.4 Mapping and interpolation for the regular element

Then $du/ds = 2(1 + s)$. From the fact that the Jacobian is unity and the relation $x(s) = s + 1$, it is obvious that $du/dx = 2x$. Figure 7.4 shows the plots for all these cases. These plots show that the quadratic element can exactly represent constraint displacement, constraint strain, and linear strain.

Irregular Element:

In the case of the irregular element, even if the length of the physical element is the same as that of the reference element, the Jacobian is not identity because of the biased mid-side node. Using eq. (7.12), the relationship between x and s can be obtained as

$$x(s) = N_1(s)x_1 + N_2(s)x_2 + N_3(s)x_3$$

$$= -\frac{1}{2}s(1-s) \times 0 + \frac{1}{2}s(1+s) \times 2 + (1-s^2) \times 0.5$$

$$= \frac{1}{2}(1+s)^2.$$

Therefore, the Jacobian becomes $dx/ds = 1 + s$. As shown in figure 7.5, the mapping relationship is quadratic, and the Jacobian is linear for the irregular element. A technical difficulty occurs at $s = -1$, where the Jacobian is singular, and so its inverse $ds/dx = 1/(1 + s)$ becomes infinite. This singularity is in general undesirable because the mapping is invalid and strain or heat flux cannot be calculated at this point. However, when a physical problem has a singularity, such as a crack tip in fracture mechanics, this type of irregular element is intentionally used.

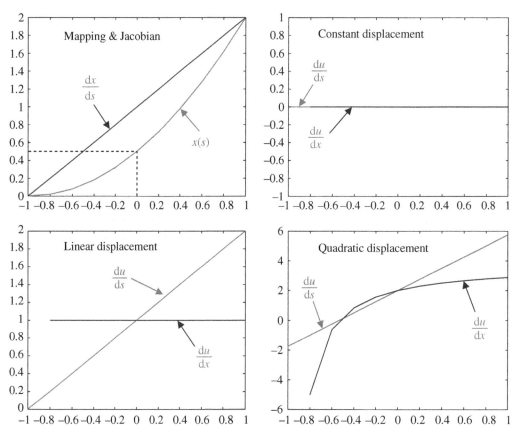

Figure 7.5 Irregular versus irregular quadratic element

(a) In the case of constant displacement, all nodal displacements are the same, and therefore, $du/ds = du/dx = 0$, except at $s = -1$. At $s = -1$, du/ds can be defined, but du/dx cannot because of a singularity at this point.

(b) In the case of constant strain, the nodal displacements can be obtained using $u(x_i) = x_i$, as $u_1 = 0$, $u_2 = 2$, $u_3 = 0.5$. The interpolation relation can be used to obtain $u(s)$, as

$$u(s) = N_1(s)u_1 + N_2(s)u_2 + N_3(s)u_3$$
$$= -\frac{1}{2}s(1-s) \times 0 + \frac{1}{2}s(1+s) \times 2 + (1-s^2) \times 0.5$$
$$= \frac{1}{2}(1+s)^2.$$

Then $du/ds = 1 + s$ is a linear function, but $du/dx = (du/ds)(ds/dx) = 1$ is a constant. Therefore, the strain is accurate at all points except the singular point.

(c) In the case of linear strain, the nodal displacements can be obtained using $u(x_i) = x_i^2$, as $u_1 = 0$, $u_2 = 4$, $u_3 = 0.25$. The interpolation relation can be used to obtain $u(s)$, as

$$u(s) = N_1(s)u_1 + N_2(s)u_2 + N_3(s)u_3$$
$$= -\frac{1}{2}s(1-s) \times 0 + \frac{1}{2}s(1+s) \times 4 + (1-s^2) \times 0.25$$
$$= \frac{1}{4} + 2s + \frac{7}{4}s^2.$$

Then $du/ds = 2 + 7 s/2$. In this case, the linear strain is expected, but the strain becomes

$$\frac{du}{dx} = \frac{4+7s}{2(1+s)},$$

which is far from a linear function.

Figure 7.5 shows the plots for all these cases. These plots show that the irregular quadratic element is exact up to constant strain, but has an error in representing linear strain. Therefore, it is important to keep the mid-side node at the center of the edge unless a different location is chosen intentionally.▪

EXAMPLE 7.3   *One-dimensional 3-node isoparametric heat conduction element*

Derive the element conductivity matrix of the 3-node quadratic isoparametric element for one-dimensional heat transfer analysis.

SOLUTION   As in example 7.1, we modify the left-hand side of the weak form by changing the variable.

$$\int_{x_1}^{x_2} kA \frac{dT}{dx} \frac{d\delta T}{dx} dx = \int_{-1}^{1} kA \frac{dT}{dx} \frac{d\delta T}{dx} \frac{dx}{ds} ds.$$

Using the mapping function, equation (7.12), we can determine the Jacobian or the rate of change of x with respect to s.

$$\frac{dx}{ds} = \frac{dN_1}{ds} x_1 + \frac{dN_2}{ds} x_2 + \frac{dN_3}{ds} x_3$$

$$= \left(-\frac{1}{2} + s \right) x_1 + \left(\frac{1}{2} + s \right) x_2 + (-2s) x_3 \tag{7.13}$$

$$= \frac{x_2 - x_1}{2} + (x_1 + x_2 - 2x_3) s.$$

If we assume that the mid-node of the element is exactly at the midpoint of the element, then $(x_1 + x_2 - 2x_3) = 0$. Therefore, the above Jacobian becomes identical to that of linear element. The implication is that when we create a mesh in the physical space, we need to ensure that the middle node has coordinates that place it exactly at the center of the element. In the previous example, we studied the consequences of not placing the mid-node at the center of the element. For some special applications such as fracture mechanics, the mid-node is deliberately placed off-center to create singularities within the element. But for most applications, this is not desired. So the assumption that the mid-node is at the center makes the Jacobian a constant equal to the ratio of the lengths of the element in real and parametric space respectively.

$$\frac{dx}{ds} = \frac{x_2 - x_1}{2} = \frac{L^{(e)}}{2}.$$

The inverse of the Jacobian is, therefore,

$$\frac{ds}{dx} = \frac{2}{x_2 - x_1} = \frac{2}{L_e}.$$

The derivatives of the temperature and the virtual temperature within the element can be obtained from the temperature interpolation and expressed in matrix form as:

$$\frac{dT}{dx} = \frac{dT}{ds}\frac{ds}{dx} = \frac{2}{L_e}\left\{\begin{array}{ccc}\frac{dN_1}{ds} & \frac{dN_2}{ds} & \frac{dN_3}{ds}\end{array}\right\}\left\{\begin{array}{c}T_1 \\ T_2 \\ T_3\end{array}\right\} = \{\mathbf{B}\}^{\mathrm{T}}\left\{\mathbf{T}^{(e)}\right\},$$

$$\frac{d\delta T}{dx} = \frac{d\delta T}{ds}\frac{ds}{dx} = \frac{2}{L_e}\left\{\begin{array}{ccc}\frac{dN_1}{ds} & \frac{dN_2}{ds} & \frac{dN_3}{ds}\end{array}\right\}\left\{\begin{array}{c}\delta T_1 \\ \delta T_2 \\ \delta T_3\end{array}\right\} = \{\mathbf{B}\}^{\mathrm{T}}\left\{\delta\mathbf{T}^{(e)}\right\}.$$

Substituting these into the left-hand side of the weak form, we get,

$$\int_{x_1}^{x_2} kA\frac{dT}{dx}\frac{d\delta T}{dx}dx = \int_{-1}^{1} kA\frac{dT}{dx}\frac{d\delta T}{dx}\frac{dx}{ds}ds$$

$$= \left\{\delta\mathbf{T}^{(e)}\right\}^{\mathrm{T}}\left[\int_{-1}^{1}kA\{\mathbf{B}\}\{\mathbf{B}\}^{\mathrm{T}}\frac{L^{(e)}}{2}ds\right]\left\{\mathbf{T}^{(e)}\right\}$$

$$= \left\{\delta\mathbf{T}^{(e)}\right\}^{\mathrm{T}}\left[\mathbf{k}^{(e)}\right]\left\{\mathbf{T}^{(e)}\right\}.$$

The stiffness matrix $[\mathbf{k}^{(e)}]$ is a 3×3 matrix in this case and can be obtained by integrating over the length of the element in parametric space. The integration is always from -1 to +1 for isoparametric elements, which makes it very convenient especially when using numerical algorithms for integration. For two-dimensional and three-dimensional elements, numerical integration is necessary, but for this element, it is not difficult to carry out the integration analytically, as

$$\left[\mathbf{k}^{(e)}\right] = 2\frac{kA^{(e)}}{L^{(e)}}\int_{-1}^{1}\left[\begin{array}{ccc}\left(s-\frac{1}{2}\right)^2 & s^2-\frac{1}{4} & s(1-2s) \\ s^2-\frac{1}{4} & \left(s+\frac{1}{2}\right)^2 & -s(1+2s) \\ s(1-2s) & -s(1+2s) & 4s^2\end{array}\right]ds.$$

The integration yields the following 3×3 matrix as the conductivity matrix for this 3-node quadratic element.

$$\left[\mathbf{k}^{(e)}\right] = \frac{kA^{(e)}}{3L^{(e)}}\left[\begin{array}{ccc}7 & 1 & -8 \\ 1 & 7 & -8 \\ -8 & -8 & 16\end{array}\right].$$

EXAMPLE 7.4 *One-dimensional heat transfer*

Model the heat conduction through the thickness of a concrete slab using one-dimensional quadratic elements. The temperature on one-side is fixed at 100 °C while the other side it is 25 °C. Heat is being generated within the plate at the rate of 10^{-3} W/mm^3. If the thermal conductivity of the slab is 5.2×10^{-4} W/mm/°C and its thickness is 10 mm, determine the temperature distribution through the thickness of the slab.

SOLUTION   The heat transfer across the thickness of a slab can be modeled as one-dimensional if the temperature variation is only in the thickness direction and not along the length and breadth of the slab. A region of the slab is shown in figure 7.6 where we assume that the heat transfer is only in the *x*-direction. To illustrate the assembly process, we will

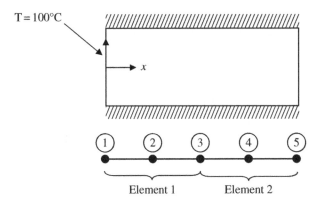

Figure 7.6 1-D heat conduction model using 3-node elements

Table 7.1 Element connectivity

Element	Local Node 1	Local Node 2	Local Node 3
1	1	3	2
2	3	5	4

use two elements for this model even though one element would suffice for this example. Note that in this case, element 1 extends from node 1 to node 3. Node 2 is the node at the mid-point of the element as shown figure 7.6.

The connectivity table for these elements is shown in table 7.1, and we see that there is a difference in how nodes are numbered in the local coordinates versus how they are numbered in global coordinates. Again, the convention for node numbering in a quadratic element is the corner nodes first and then mid-side node for the reference element.

The conductivity matrix is the same for both elements in the model since they both have the same length and material properties.

$$\left[\mathbf{k}^{(1)}\right] = \left[\mathbf{k}^{(2)}\right] = \frac{kA^{(e)}}{3L^{(e)}} \begin{bmatrix} 7 & 1 & -8 \\ 1 & 7 & -8 \\ -8 & -8 & 16 \end{bmatrix}.$$

The equivalent nodal heat source corresponding the heat generation in the plate can be found, as described in a previous chapter, in the following manner.

$$\left\{\mathbf{Q}_g^{(1)}\right\} = \left\{\mathbf{Q}_g^{(2)}\right\} = \int_0^{L_e} Q_g \begin{Bmatrix} N_1 \\ N_2 \\ N_3 \end{Bmatrix} A^{(e)} dx = Q_g A^{(e)} \int_{-1}^{1} \begin{Bmatrix} -\frac{s}{2}(1-s) \\ \frac{s}{2}(1+s) \\ 1-s^2 \end{Bmatrix} \frac{dx}{ds} ds = \frac{1}{6} Q_g A^{(e)} L^{(e)} \begin{Bmatrix} 1 \\ 1 \\ 4 \end{Bmatrix}.$$

It is interesting to note that in the case of linear elements, the uniformly generated heat source is equally contributed to the two nodes. In the case of a quadratic element, however, the mid-side node has four times larger contribution than the two end nodes.

Assembling the conductivity matrices on the left-hand side and the contribution for the heat source and boundary heat flux on the right-hand side, we get the following global system of equations:

$$\frac{kA^{(e)}}{3L^{(e)}} \begin{bmatrix} 7 & -8 & 1 & 0 & 0 \\ -8 & 16 & -8 & 0 & 0 \\ 1 & -8 & 7+7 & -8 & 1 \\ 0 & 0 & -8 & 16 & -8 \\ 0 & 0 & 1 & -8 & 7 \end{bmatrix} \begin{Bmatrix} T_1 \\ T_2 \\ T_3 \\ T_4 \\ T_5 \end{Bmatrix} = \begin{Bmatrix} q_1 A^{(e)} \\ 0 \\ 0 \\ 0 \\ q_5 A^{(e)} \end{Bmatrix} + \frac{1}{6} Q_g A^{(e)} L^{(e)} \begin{Bmatrix} 1 \\ 4 \\ 2 \\ 4 \\ 1 \end{Bmatrix}.$$

We discussed how to apply nonzero boundary conditions in chapter 4. Here since the temperature is specified at both boundaries, we know that $T_1 = 100°C$ and $T_5 = 25°C$. At these boundary nodes, we do not know the heat fluxes q_1 and q_5, but since the temperatures are known, we do not need those two equations to solve for the known temperatures. Therefore, we strike out the first and fifth rows, and move the first and fifth columns to the right-hand side after multiplying with the known nodal temperatures. Then we can rewrite the remaining equations as:

$$\frac{k}{3L^{(e)}} \begin{bmatrix} 16 & -8 & 0 \\ -8 & 14 & -8 \\ 0 & -8 & 16 \end{bmatrix} \begin{Bmatrix} T_2 \\ T_3 \\ T_4 \end{Bmatrix} = \frac{1}{6}Q_g L^{(e)} \begin{Bmatrix} 4 \\ 2 \\ 4 \end{Bmatrix} - T_1 \frac{k}{3L^{(e)}} \begin{Bmatrix} -8 \\ 1 \\ 0 \end{Bmatrix} - T_5 \frac{k}{3L^{(e)}} \begin{Bmatrix} 0 \\ 1 \\ -8 \end{Bmatrix}.$$

Substituting the numerical values of the known constants, we get the following linear system of equations that can be solved for the unknown temperatures:

$$3.467 \times 10^{-5} \begin{bmatrix} 16 & -8 & 0 \\ -8 & 14 & -8 \\ 0 & -8 & 16 \end{bmatrix} \begin{Bmatrix} T_2 \\ T_3 \\ T_4 \end{Bmatrix} = \begin{Bmatrix} 0.0311 \\ -0.0027 \\ 0.0103 \end{Bmatrix}.$$

Upon solving these equations, we get the nodal temperatures as: $T_2 = 99.28°C$, $T_3 = 86.54°C$, and $T_4 = 61.78°C$. The temperature distribution within element 1 is:

$$T(s) = N_1(s)T_1 + N_2(s)T_2 + N_3(s)T_3$$
$$= -\frac{s(1-s)}{2} \times 100 + \frac{s(1+s)}{2} \times 86.54 + (1-s^2) \times 99.28$$
$$= 99.28 - 6.73s - 6.01s^2.$$

Note that the interpolation equation uses a local node numbering scheme where the third node is at the center of the element, the left node is node 1, and the right node is node 2. The temperature distribution obtained by the interpolation is a function of the reference coordinate s. To express the temperature distribution as a function of the physical coordinate x, the mapping between function relating x and s must be used. The mapping function is:

$$x(s) = N_1(s)x_1 + N_2(s)x_2 + N_3(s)x_3$$
$$= -\frac{1}{2}s(1-s)x_1 + \frac{1}{2}s(1+s)x_2 + (1-s^2)x_3$$
$$= x_3 + \frac{(x_2-x_1)}{2}s + \left(\frac{x_1+x_2}{2} - x_3\right)s^2$$
$$= \frac{5}{2}(1+s).$$

The mapping is linear because the middle node is exactly at the center of the element in real space, and therefore, the coefficient of the quadratic term goes to zero. One could solve for s as function of x and substitute in temperature interpolation to express temperature as a function of x. Similarly, one can interpolate the temperature within the second element to obtain the temperature distribution in the second half of the domain. In practice, it is not necessary to express temperature as a function of the physical coordinate x because it is possible to plot the temperature distribution for graphical display even if the temperature is expressed as a function of s. ■

7.3   TWO-DIMENSIONAL ISOPARAMETRIC QUADRILATERAL ELEMENT

Four–node quadratic isoparametric finite element is one of the most commonly used elements in engineering applications. A mesh consisting of quadrilateral elements can be used to approximate any arbitrary shape in two dimensions, unlike rectangular elements. Since the geometry of the element is

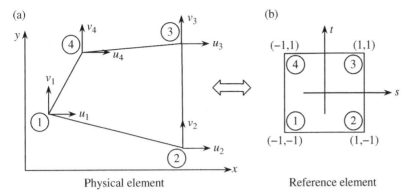

Figure 7.7 Four–node quadrilateral element for plane solids

irregular, it is convenient to introduce a *reference element* in the parametric coordinate system and use a mapping relation between the physical element and the reference element. The terms "isoparametric" comes from the fact that the same interpolation scheme is used for interpolating both the field variable (e.g., displacement or temperature) and geometry.

7.3.1 Isoparametric Mapping

The physical element in figure 7.7 is a general quadrilateral shape. However, all interior angles should be less than 180 degrees. The order of node numbers is the same as that of the rectangular element: starting from one corner and moving in the counterclockwise direction. For two-dimensional solid mechanics, each node has two DOFs: u and v. Thus, the element has a total of eight DOFs.

Since different elements have different shapes, it would not be a trivial task if the interpolation functions need to be developed for an individual element. The interpolation functions must satisfy the inter-element displacement compatibility condition discussed earlier in the context of triangular elements. Instead, the concept of mapping to the reference element will be used. The physical element in figure 7.7(a) will be mapped into the reference element shown in figure 7.7(b). The physical element is defined in x-y physical coordinates, while the reference element is defined in s-t or parametric coordinates. The reference element is a square element and has the origin at the center. Although the physical element can have the first node at any corner, the reference element always has the first node at the lower–left corner $(-1,-1)$.

The interpolation functions are defined for the reference element in the parametric coordinates so that all the elements have the same interpolation functions. But the mapping relation will be unique to each element. Since the reference element is of square shape, it is easy to derive Lagrange interpolation functions. The interpolation or shape functions can be written in s-t coordinates as

$$\begin{cases} N_1(s,t) = \dfrac{1}{4}(1-s)(1-t), \\[2mm] N_2(s,t) = \dfrac{1}{4}(1+s)(1-t), \\[2mm] N_3(s,t) = \dfrac{1}{4}(1+s)(1+t), \\[2mm] N_4(s,t) = \dfrac{1}{4}(1-s)(1+t). \end{cases} \tag{7.14}$$

Since the above shape functions are Lagrange interpolation functions, they satisfy the property that a shape function N_I is equal to unity at node-I and zero at other nodes.

In an isoparametric element, the shape functions are used for also mapping between the physical element and the reference element. The quadrilateral element is defined by the coordinates of four corner nodes. These four corner nodes are mapped into the four corner nodes of the reference element. In addition, every point in the physical element is also mapped into a point in the reference element. The mapping relation is one–to–one so that every point in the reference element is mapped to a point in the physical element. Thus a physical point (x, y) is a function of the reference point (s, t). The relation between (x, y) and (s, t) is the mapping function that is be derived using the same shape functions as

$$x(s,t) = [N_1(s,t) \ N_2(s,t) \ N_3(s,t) \ N_4(s,t)] \begin{Bmatrix} x_1 \\ x_2 \\ x_3 \\ x_4 \end{Bmatrix},$$

$$\qquad\qquad (7.15)$$

$$y(s,t) = [N_1(s,t) \ N_2(s,t) \ N_3(s,t) \ N_4(s,t)] \begin{Bmatrix} y_1 \\ y_2 \\ y_3 \\ y_4 \end{Bmatrix}.$$

It can be easily checked that at node 1, for example, $(s, t) = (-1, -1)$ and $N_1 = 1$, $N_2 = N_3 = N_4 = 0$. Thus, we have $x(-1, -1) = x_1$ and $y(-1, -1) = y_1$, that is, node 1 in the physical element is mapped into node 1 in the reference element. The above mapping relation is called *isoparametric mapping* because the same shape functions are used for interpolating geometry as well as displacements.

The above mapping relation is explicit in terms of x and y, which means that when s and t are given, x and y can be calculated explicitly. The reverse relation is not straightforward. However, the following example explains how s and t can be calculated for a given x and y.

EXAMPLE 7.5 *Isoparametric mapping of two-dimensional quadrilateral element*

Consider a quadrilateral element of the trapezoidal shape shown in figure 7.8. Using the isoparametric mapping method calculate: (a) the physical coordinates of point A (0.5, 0.5), and (b) the reference coordinate of point $B(1, 2)$.

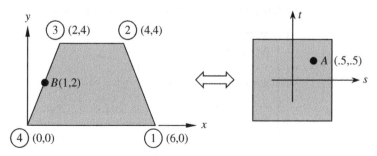

Figure 7.8 Mapping of a quadrilateral element

SOLUTION

(a) At point A, $(s, t) = (0.5, 0.5)$. The values of the shape functions at A are

$$N_1\left(\frac{1}{2}, \frac{1}{2}\right) = \frac{1}{16}, \quad N_2\left(\frac{1}{2}, \frac{1}{2}\right) = \frac{3}{16}, \quad N_3\left(\frac{1}{2}, \frac{1}{2}\right) = \frac{9}{16}, \quad N_4\left(\frac{1}{2}, \frac{1}{2}\right) = \frac{3}{16}.$$

Thus, the physical coordinate becomes

$$x\left(\frac{1}{2}, \frac{1}{2}\right) = \sum_{I=1}^{4} N_I\left(\frac{1}{2}, \frac{1}{2}\right) x_I = \frac{1}{16}\cdot 6 + \frac{3}{16}\cdot 4 + \frac{9}{16}\cdot 2 + \frac{3}{16}\cdot 0 = 2.25,$$

$$y\left(\frac{1}{2}, \frac{1}{2}\right) = \sum_{I=1}^{4} N_I\left(\frac{1}{2}, \frac{1}{2}\right) y_I = \frac{1}{16}\cdot 0 + \frac{3}{16}\cdot 4 + \frac{9}{16}\cdot 4 + \frac{3}{16}\cdot 0 = 3.$$

Thus, the reference point $(s, t) = (0.5, 0.5)$ is mapped into the physical point $(x, y) = (2.25, 3.0)$.

(b) At point B, $(x, y) = (1, 2)$. From the isoparametric mapping relation, we have

$$x = 1 = \sum_{I=1}^{4} N_I(s, t) x_I = \frac{1}{4}(1-s)(1-t)\cdot 6 + \frac{1}{4}(1+s)(1-t)\cdot 4$$

$$+ \frac{1}{4}(1+s)(1+t)\cdot 2 + \frac{1}{4}(1-s)(1+t)\cdot 0$$

$$= st - 2t + 3,$$

$$y = 2 = \sum_{I=1}^{4} N_I(s, t) y_I = \frac{1}{4}(1-s)(1-t)\cdot 0 + \frac{1}{4}(1+s)(1-t)\cdot 4$$

$$+ \frac{1}{4}(1+s)(1+t)\cdot 4 + \frac{1}{4}(1-s)(1+t)\cdot 0$$

$$= 2 + 2s.$$

From the above two relations, we obtain $(s, t) = (0, 1)$. Note that the above results will not be the same if the sequence of node numbers in the physical element is changed. ■

7.3.2   Jacobian of Mapping

The idea of using the reference element is convenient because it is unnecessary to build different shape functions for different elements. The same shape functions can be used for all elements. However, it has its own drawbacks. The strain energy in the plane solid element requires the derivative of displacement, that is, strains. As we know, the strains are defined as derivatives of displacements. In the case of CST and rectangular elements, the shape functions could be differentiated directly because the nodal displacements are explicit functions of x and y. For those elements, the derivatives of the shape functions can be easily obtained because they are defined as a function of physical coordinates (x, y). However, in the case of the isoparametric quadrilateral element, the shape functions are defined in the reference coordinates. Thus, differentiation with respect to the physical coordinates is not straightforward. In this case, we use a Jacobian relation and the chain rule of differentiation. From the fact that $s = s(x, y)$ and $t = t(x, y)$, we can write the derivatives of N_I as follows:

$$\frac{\partial N_I}{\partial s} = \frac{\partial N_I}{\partial x}\frac{\partial x}{\partial s} + \frac{\partial N_I}{\partial y}\frac{\partial y}{\partial s},$$

$$\frac{\partial N_I}{\partial t} = \frac{\partial N_I}{\partial x}\frac{\partial x}{\partial t} + \frac{\partial N_I}{\partial y}\frac{\partial y}{\partial t}.$$

Using the matrix form, the above equation can be written as

$$
\left\{ \begin{array}{c} \dfrac{\partial N_I}{\partial s} \\[2mm] \dfrac{\partial N_I}{\partial t} \end{array} \right\} = \left[\begin{array}{cc} \dfrac{\partial x}{\partial s} & \dfrac{\partial y}{\partial s} \\[2mm] \dfrac{\partial x}{\partial t} & \dfrac{\partial y}{\partial t} \end{array} \right] \left\{ \begin{array}{c} \dfrac{\partial N_I}{\partial x} \\[2mm] \dfrac{\partial N_I}{\partial y} \end{array} \right\} = [\mathbf{J}] \left\{ \begin{array}{c} \dfrac{\partial N_I}{\partial x} \\[2mm] \dfrac{\partial N_I}{\partial y} \end{array} \right\}, \tag{7.16}
$$

where $[\mathbf{J}]$ is the *Jacobian matrix* and its determinant is called the Jacobian. By inverting the Jacobian matrix, the desired derivatives with respect to x and y can be obtained:

$$
\left\{ \begin{array}{c} \dfrac{\partial N_I}{\partial x} \\[2mm] \dfrac{\partial N_I}{\partial y} \end{array} \right\} = [\mathbf{J}]^{-1} \left\{ \begin{array}{c} \dfrac{\partial N_I}{\partial s} \\[2mm] \dfrac{\partial N_I}{\partial t} \end{array} \right\} = \dfrac{1}{|\mathbf{J}|} \left[\begin{array}{cc} \dfrac{\partial y}{\partial t} & -\dfrac{\partial y}{\partial s} \\[2mm] -\dfrac{\partial x}{\partial t} & \dfrac{\partial x}{\partial s} \end{array} \right] \left\{ \begin{array}{c} \dfrac{\partial N_I}{\partial s} \\[2mm] \dfrac{\partial N_I}{\partial t} \end{array} \right\}, \tag{7.17}
$$

where $|\mathbf{J}|$ is the *Jacobian* and is the determinant of $[\mathbf{J}]$.

$$
|\mathbf{J}| = \frac{\partial x}{\partial s}\frac{\partial y}{\partial t} - \frac{\partial x}{\partial t}\frac{\partial y}{\partial s}. \tag{7.18}
$$

Since isoparametric mapping is used, the above Jacobian can be obtained by differentiating the relation in eq. (7.15) with respect to s and t. For example,

$$
\frac{\partial x}{\partial s} = \sum_{I=1}^{4} \frac{\partial N_I}{\partial s} x_I = \frac{1}{4}(-x_1 + x_2 + x_3 - x_4) + \frac{t}{4}(x_1 - x_2 + x_3 - x_4),
$$

$$
\frac{\partial x}{\partial t} = \sum_{I=1}^{4} \frac{\partial N_I}{\partial t} x_I = \frac{1}{4}(-x_1 - x_2 + x_3 + x_4) + \frac{s}{4}(x_1 - x_2 + x_3 - x_4).
$$

A similar expression can be obtained for $\partial y/\partial s$ and $\partial y/\partial t$ by replacing x_i with y_i. Note that $\partial x/\partial s$ is the function of t only, while $\partial x/\partial t$ is the function of s only.

As seen from eq. (7.17), the derivative of the shape function cannot be obtained if the Jacobian is zero anywhere in the element. In fact, the mapping relation between (x, y) and (s, t) is not valid if the Jacobian is zero or negative anywhere in the element $(-1 \leq s, t \leq 1)$.

The Jacobian plays an important role in evaluating the validity of mapping as well as the quality of the quadrilateral element. The fundamental requirement is that every point in the reference element should be mapped into the interior of the physical element, and vice versa. When an interior point in (s, t) coordinates is mapped into an exterior point in the (x, y) coordinates, the Jacobian becomes negative. If multiple points in (s, t) coordinates are mapped into a single point in (x, y) coordinates, the Jacobian becomes zero at that point. Thus, it is important to maintain the element shape so that the Jacobian is positive everywhere in the element.

EXAMPLE 7.6 *Jacobian of mapping.*

Check the validity of isoparametric mapping for the two elements shown in figure 7.9

SOLUTION

(a) Nodal coordinates:

$$
x_1 = 0, \; x_2 = 1, \; x_3 = 2, \; x_4 = 0
$$
$$
y_1 = 0, \; y_2 = 0, \; y_3 = 2, \; y_4 = 1.
$$

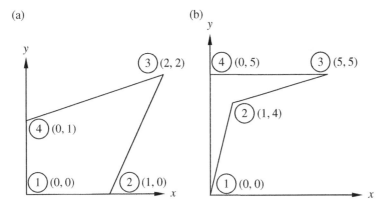

Figure 7.9 Four–node quadrilateral element

– Isoparametric mapping:

$$x = \sum_{I=1}^{4} N_I x_I = N_2 + 2N_3 = \frac{1}{4}(3 + 3s + t + st),$$

$$y = \sum_{I=1}^{4} N_I y_I = 2N_3 + N_4 = \frac{1}{4}(3 + s + 3t + st).$$

– Jacobian:

$$[\mathbf{J}] = \begin{bmatrix} \dfrac{\partial x}{\partial s} & \dfrac{\partial y}{\partial s} \\[2mm] \dfrac{\partial x}{\partial t} & \dfrac{\partial y}{\partial t} \end{bmatrix} = \frac{1}{4}\begin{bmatrix} 3+t & 1+t \\ 1+s & 3+s \end{bmatrix},$$

$$|\mathbf{J}| = \frac{1}{4}[(3+t)(3+s) - (1+t)(1+s)] = \frac{1}{2} + \frac{1}{8}s + \frac{1}{8}t.$$

Thus, it is clear that $|\mathbf{J}| > 0$ for $-1 \leq s \leq 1$ and $-1 \leq t \leq 1$. Figure 7.10 shows constant s and t lines. Since all lines are within the element boundary, the mapping is valid.

(b) Nodal coordinates:

$$x_1 = 0,\ x_2 = 1,\ x_3 = 5,\ x_4 = 0,$$
$$y_1 = 0,\ y_2 = 4,\ y_3 = 5,\ y_4 = 5.$$

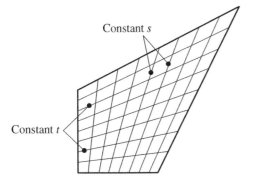

Figure 7.10 Isoparametric lines of a quadrilateral element

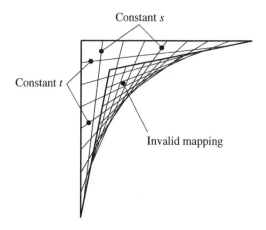

Constant s

Constant t

Invalid mapping

Figure 7.11 An example of invalid mapping

– Isoparametric mapping:

$$x = \sum_{I=1}^{4} N_I x_I = \frac{1}{2}(1+s)(3+2t),$$

$$y = \sum_{I=1}^{4} N_I y_I = \frac{1}{2}(7+2s+3t-2st).$$

– Jacobian:

$$|\mathbf{J}| = \frac{1}{4}(5-10s+10t).$$

Note that $|\mathbf{J}| = 0$ at $5 - 10s + 10\,t = 0$, that is, $s - t = 1/2$. The mapping illustrated in figure 7.11 clearly shows that the mapping is invalid. Some points in the reference element are mapped into the outside of the physical element. ■

In practice, maintaining a positive Jacobian is not enough because of other potential numerical problems. For example, when the Jacobian is small, that is, $|\mathbf{J}| \ll 1$, calculation of stress and strain is not accurate, and the integration of the strain energy will lose its accuracy. A small value of the Jacobian occurs when the element shape is far from a rectangle. To avoid problems due to badly shaped elements, it is recommended that the inside angles in quadrilateral elements be $> 15°$ and $< 165°$ as illustrated in figure 7.12.

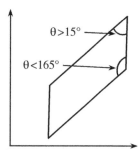

$\theta > 15°$

$\theta < 165°$

Figure 7.12 Recommended ranges of internal angles in a quadrilateral element

7.3.3 Interpolation of Displacements and Computation of Strains

As we explained earlier, in an isoparametric element, the same shape functions are used for mapping and for interpolating displacements. For two-dimensional solid mechanics problems, the quadrilateral element has eight DOFs. Then the displacements within the element can be interpolated as

$$
\begin{Bmatrix} u \\ v \end{Bmatrix} = \begin{bmatrix} N_1 & 0 & N_2 & 0 & N_3 & 0 & N_4 & 0 \\ 0 & N_1 & 0 & N_2 & 0 & N_3 & 0 & N_4 \end{bmatrix} \begin{Bmatrix} u_1 \\ v_1 \\ u_2 \\ v_2 \\ u_3 \\ v_3 \\ u_4 \\ v_4 \end{Bmatrix} = [\mathbf{N}]\{\mathbf{q}\},
\tag{7.19}
$$

where the shape functions in eq. (7.14) are used for interpolation. The difference between the previously described elements in chapter 6 (CST and rectangular elements) and the isoparametric quadrilateral element is that the interpolation is done in the reference coordinates (s, t). However, the behavior of the element is similar to that of the rectangular element because both of them are based on the bilinear Lagrange interpolation.

Now, we derive the strain-displacement relationship for the quadrilateral element. In order make the following matrix operation convenient, we first reorder the strain components into the derivatives of displacements, as

$$
\{\boldsymbol{\varepsilon}\} = \begin{Bmatrix} \varepsilon_{xx} \\ \varepsilon_{yy} \\ \gamma_{xy} \end{Bmatrix} = \begin{Bmatrix} \partial u/\partial x \\ \partial v/\partial y \\ \partial u/\partial y + \partial v/\partial x \end{Bmatrix} = \begin{bmatrix} 1 & 0 & 0 & 0 \\ 0 & 0 & 0 & 1 \\ 0 & 1 & 1 & 0 \end{bmatrix} \begin{Bmatrix} \partial u/\partial x \\ \partial u/\partial y \\ \partial v/\partial x \\ \partial v/\partial y \end{Bmatrix}.
$$

As we discussed above, the derivatives of displacements cannot be obtained directly. Instead, we use the inverse Jacobian relation so that the derivatives of displacements are written in terms of the reference coordinates. Thus, we have

$$
\begin{Bmatrix} \dfrac{\partial u}{\partial x} \\[2mm] \dfrac{\partial u}{\partial y} \end{Bmatrix} = \frac{1}{|\mathbf{J}|} \begin{bmatrix} \dfrac{\partial y}{\partial t} & -\dfrac{\partial y}{\partial s} \\[2mm] -\dfrac{\partial x}{\partial t} & \dfrac{\partial x}{\partial s} \end{bmatrix} \begin{Bmatrix} \dfrac{\partial u}{\partial s} \\[2mm] \dfrac{\partial u}{\partial t} \end{Bmatrix},
$$

$$
\begin{Bmatrix} \dfrac{\partial v}{\partial x} \\[2mm] \dfrac{\partial v}{\partial y} \end{Bmatrix} = \frac{1}{|\mathbf{J}|} \begin{bmatrix} \dfrac{\partial y}{\partial t} & -\dfrac{\partial y}{\partial s} \\[2mm] -\dfrac{\partial x}{\partial t} & \dfrac{\partial x}{\partial s} \end{bmatrix} \begin{Bmatrix} \dfrac{\partial v}{\partial s} \\[2mm] \dfrac{\partial v}{\partial t} \end{Bmatrix}.
$$

Writing the two equations together, we have

$$
\begin{Bmatrix} \partial u/\partial x \\ \partial u/\partial y \\ \partial v/\partial x \\ \partial v/\partial y \end{Bmatrix} = \frac{1}{|\mathbf{J}|} \begin{bmatrix} \partial y/\partial t & -\partial y/\partial s & 0 & 0 \\ -\partial x/\partial t & \partial x/\partial s & 0 & 0 \\ 0 & 0 & \partial y/\partial t & -\partial y/\partial s \\ 0 & 0 & -\partial x/\partial t & \partial x/\partial s \end{bmatrix} \begin{Bmatrix} \partial u/\partial s \\ \partial u/\partial t \\ \partial v/\partial s \\ \partial v/\partial t \end{Bmatrix}.
$$

The strains can now be expressed as

$$
\begin{Bmatrix} \varepsilon_{xx} \\ \varepsilon_{yy} \\ \gamma_{xy} \end{Bmatrix} = \frac{1}{|\mathbf{J}|} \begin{bmatrix} 1 & 0 & 0 & 0 \\ 0 & 0 & 0 & 1 \\ 0 & 1 & 1 & 0 \end{bmatrix} \begin{bmatrix} \partial y/\partial t & -\partial y/\partial s & 0 & 0 \\ -\partial x/\partial t & \partial x/\partial s & 0 & 0 \\ 0 & 0 & \partial y/\partial t & -\partial y/\partial s \\ 0 & 0 & -\partial x/\partial t & \partial x/\partial s \end{bmatrix} \begin{Bmatrix} \partial u/\partial s \\ \partial u/\partial t \\ \partial v/\partial s \\ \partial v/\partial t \end{Bmatrix}
$$

$$
\equiv [\mathbf{A}] \begin{Bmatrix} \partial u/\partial s \\ \partial u/\partial t \\ \partial v/\partial s \\ \partial v/\partial t \end{Bmatrix},
$$

where $[\mathbf{A}]$ is a 3×4 matrix. The derivatives of the displacements with respect to s and t can be obtained by differentiating $u(s,t)$ and $v(s,t)$ in eq. (7.19), which involves the derivatives of the shape functions:

$$
\begin{Bmatrix} \partial u/\partial s \\ \partial u/\partial t \\ \partial v/\partial s \\ \partial v/\partial t \end{Bmatrix} = \frac{1}{4} \begin{bmatrix} -1+t & 0 & 1-t & 0 & 1+t & 0 & -1-t & 0 \\ -1+s & 0 & -1-s & 0 & 1+s & 0 & 1-s & 0 \\ 0 & -1+t & 0 & 1-t & 0 & 1+t & 0 & -1-t \\ 0 & -1+s & 0 & -1-s & 0 & 1+s & 0 & 1-s \end{bmatrix} \begin{Bmatrix} u_1 \\ v_1 \\ u_2 \\ v_2 \\ u_3 \\ v_3 \\ u_4 \\ v_4 \end{Bmatrix}
$$

$$
\equiv [\mathbf{G}]\{\mathbf{q}\},
$$

where the dimension of matrix $[\mathbf{G}]$ is 4×8. The strain-displacement matrix $[\mathbf{B}]$ can now be written as follows:

$$
\begin{Bmatrix} \varepsilon_{xx} \\ \varepsilon_{yy} \\ \gamma_{xy} \end{Bmatrix} = [\mathbf{A}] \begin{Bmatrix} \partial u/\partial s \\ \partial u/\partial t \\ \partial v/\partial s \\ \partial v/\partial t \end{Bmatrix} = [\mathbf{A}][\mathbf{G}]\{\mathbf{q}\} \equiv [\mathbf{B}]\{\mathbf{q}\}. \tag{7.20}
$$

where $[\mathbf{B}]$ is a 3×8 matrix. The explicit expression of $[\mathbf{B}]$ is not readily available because the matrix $[\mathbf{A}]$ involves the inverse of the Jacobian matrix. However, for a given reference coordinate (s, t), it can be calculated using eq. (7.20). Note that the strain-displacement matrix $[\mathbf{B}]$ is not constant as in CST elements. Thus, the strains and stresses within an element vary as a function of s and t coordinates.

EXAMPLE 7.7   *Interpolation using quadrilateral element*

For a rectangular element shown in figure 7.13, displacements at four nodes are given by $\{u_1, v_1, u_2, v_2, u_3, v_3, u_4, v_4\} = \{0.0, 0.0, 1.0, 0.0, 2.0, 1.0, 0.0, 2.0\}$. Calculate displacement and strain at point $(s, t) = (1/3, 0)$.

SOLUTION   When the reference coordinate $(s, t) = (1/3, 0)$, the shape functions become

$$N_1 = \frac{1}{6}, \quad N_2 = \frac{1}{3}, \quad N_3 = \frac{1}{3}, \quad N_4 = \frac{1}{6}.$$

Using eq. (7.19), the displacements can be interpolated, as

$$\begin{cases} u = \displaystyle\sum_{I=1}^{4} N_I u_I = \frac{1}{6}\cdot 0 + \frac{1}{3}\cdot 1 + \frac{1}{3}\cdot 2 + \frac{1}{6}\cdot 0 = 1, \\[3mm] v = \displaystyle\sum_{I=1}^{4} N_I v_I = \frac{1}{6}\cdot 0 + \frac{1}{3}\cdot 0 + \frac{1}{3}\cdot 1 + \frac{1}{6}\cdot 2 = \frac{2}{3} \end{cases}.$$

In order to calculate strains, we need the derivatives of the shape functions. First, we calculate the derivatives with respect to the reference coordinates, as

$$\begin{cases} \dfrac{\partial N_1}{\partial s} = -\dfrac{1}{4}(1-t) = -\dfrac{1}{4} \\[2mm] \dfrac{\partial N_2}{\partial s} = \dfrac{1}{4}(1-t) = \dfrac{1}{4} \\[2mm] \dfrac{\partial N_3}{\partial s} = \dfrac{1}{4}(1+t) = \dfrac{1}{4} \\[2mm] \dfrac{\partial N_4}{\partial s} = -\dfrac{1}{4}(1+t) = -\dfrac{1}{4} \end{cases} \quad \begin{cases} \dfrac{\partial N_1}{\partial t} = -\dfrac{1}{4}(1-s) = -\dfrac{1}{6} \\[2mm] \dfrac{\partial N_2}{\partial t} = -\dfrac{1}{4}(1+s) = -\dfrac{1}{3} \\[2mm] \dfrac{\partial N_3}{\partial t} = \dfrac{1}{4}(1+s) = \dfrac{1}{3} \\[2mm] \dfrac{\partial N_4}{\partial t} = \dfrac{1}{4}(1-s) = \dfrac{1}{6} \end{cases}$$

In addition, the Jacobian matrix can be calculated using eq. (7.17), as

$$\begin{cases} \dfrac{\partial x}{\partial s} = -\dfrac{1}{4}\cdot 0 + \dfrac{1}{4}\cdot 3 + \dfrac{1}{4}\cdot 3 - \dfrac{1}{4}\cdot 0 = \dfrac{3}{2} \\[2mm] \dfrac{\partial y}{\partial s} = -\dfrac{1}{4}\cdot 0 + \dfrac{1}{4}\cdot 0 + \dfrac{1}{4}\cdot 2 - \dfrac{1}{4}\cdot 2 = 0 \\[2mm] \dfrac{\partial x}{\partial t} = -\dfrac{1}{6}\cdot 0 - \dfrac{1}{3}\cdot 3 + \dfrac{1}{3}\cdot 3 + \dfrac{1}{6}\cdot 0 = 0 \\[2mm] \dfrac{\partial y}{\partial t} = -\dfrac{1}{6}\cdot 0 - \dfrac{1}{3}\cdot 0 + \dfrac{1}{3}\cdot 2 + \dfrac{1}{6}\cdot 2 = 1 \end{cases}$$

$$[\mathbf{J}] = \begin{bmatrix} \dfrac{\partial x}{\partial s} & \dfrac{\partial y}{\partial s} \\[2mm] \dfrac{\partial x}{\partial t} & \dfrac{\partial y}{\partial t} \end{bmatrix} = \begin{bmatrix} \dfrac{3}{2} & 0 \\[1mm] 0 & 1 \end{bmatrix}, \; [\mathbf{J}]^{-1} = \begin{bmatrix} \dfrac{2}{3} & 0 \\[1mm] 0 & 1 \end{bmatrix}.$$

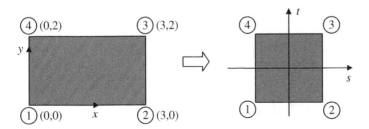

Figure 7.13 Mapping of a rectangular element

The Jacobian is positive, and the mapping is valid at this point. In fact, the Jacobian matrix is constant throughout the element. Note that the Jacobian matrix only has diagonal components, which means that the physical element is a rectangle. The horizontal dimension of the physical element is 1.5 times that of the reference element, and the vertical dimension is the same.

Using the inverse Jacobian matrix and the derivatives of the shape functions, we can calculate the following:

$$
\begin{cases}
\dfrac{\partial N_I}{\partial s} = \dfrac{\partial N_I}{\partial x}\dfrac{\partial x}{\partial s} + \dfrac{\partial N_I}{\partial y}\dfrac{\partial y}{\partial s} \\[2mm]
\dfrac{\partial N_I}{\partial t} = \dfrac{\partial N_I}{\partial x}\dfrac{\partial x}{\partial t} + \dfrac{\partial N_I}{\partial y}\dfrac{\partial y}{\partial t}
\end{cases}
\Rightarrow
\begin{Bmatrix}
\dfrac{\partial N_I}{\partial s} \\[2mm]
\dfrac{\partial N_I}{\partial t}
\end{Bmatrix}
= [\mathbf{J}]
\begin{Bmatrix}
\dfrac{\partial N_I}{\partial x} \\[2mm]
\dfrac{\partial N_I}{\partial y}
\end{Bmatrix},
$$

$$
\begin{Bmatrix}
\dfrac{\partial N_I}{\partial x} \\[2mm]
\dfrac{\partial N_I}{\partial y}
\end{Bmatrix}
= [\mathbf{J}]^{-1}
\begin{Bmatrix}
\dfrac{\partial N_I}{\partial s} \\[2mm]
\dfrac{\partial N_I}{\partial t}
\end{Bmatrix}
=
\begin{bmatrix}
\dfrac{2}{3} & 0 \\[2mm]
0 & 1
\end{bmatrix}
\begin{Bmatrix}
\dfrac{\partial N_I}{\partial s} \\[2mm]
\dfrac{\partial N_I}{\partial t}
\end{Bmatrix}
=
\begin{Bmatrix}
\dfrac{2}{3}\dfrac{\partial N_I}{\partial s} \\[2mm]
\dfrac{\partial N_I}{\partial t}
\end{Bmatrix}.
$$

Using the derivatives of the shape functions, the strains can be calculated using eq. (7.20) as

$$
\varepsilon_{xx} = \frac{\partial u}{\partial x} = \sum_{I=1}^{4}\frac{\partial N_I}{\partial x}u_I = \sum_{I=1}^{4}\frac{2}{3}\frac{\partial N_I}{\partial s}u_I
$$

$$
= \frac{2}{3}\left(-\frac{1}{4}\cdot 0 + \frac{1}{4}\cdot 1 + \frac{1}{4}\cdot 2 - \frac{1}{4}\cdot 0\right) = \frac{1}{2},
$$

$$
\varepsilon_{yy} = \frac{\partial v}{\partial y} = \sum_{I=1}^{4}\frac{\partial N_I}{\partial y}v_I = \sum_{I=1}^{4}\frac{\partial N_I}{\partial t}v_I
$$

$$
= -\frac{1}{6}\cdot 0 - \frac{1}{3}\cdot 0 + \frac{1}{3}\cdot 1 + \frac{1}{6}\cdot 2 = \frac{2}{3},
$$

$$
\gamma_{xy} = \frac{\partial u}{\partial y} + \frac{\partial v}{\partial x} = \sum_{I=1}^{4}\left(\frac{\partial N_I}{\partial y}u_I + \frac{\partial N_I}{\partial x}v_I\right) = \sum_{I=1}^{4}\left(\frac{\partial N_I}{\partial t}u_I + \frac{2}{3}\frac{\partial N_I}{\partial s}v_I\right)
$$

$$
= -\frac{1}{6}\cdot 0 - \frac{1}{3}\cdot 1 + \frac{1}{3}\cdot 2 + \frac{1}{6}\cdot 0 + \frac{2}{3}\left(-\frac{1}{4}\cdot 0 + \frac{1}{4}\cdot 0 + \frac{1}{4}\cdot 1 - \frac{1}{4}\cdot 2\right) = \frac{1}{6}.
$$

7.3.4   Finite Element Matrix Equation

As in the case of the CST element, the element stiffness matrix can be calculated from the strain energy of the element. By substituting for strains from eq. (7.20) into the strain energy in eq. (5.72) we have

$$
U^{(e)} = \frac{h}{2}\iint_A \{\boldsymbol{\varepsilon}\}^{\mathrm{T}}[\mathbf{C}]\{\boldsymbol{\varepsilon}\}\,\mathrm{d}A^{(e)}
$$

$$
= \frac{h}{2}\left\{\mathbf{q}^{(e)}\right\}^{\mathrm{T}}\iint_A [\mathbf{B}]_{8\times 3}^{\mathrm{T}}[\mathbf{C}]_{3\times 3}[\mathbf{B}]_{3\times 8}\,\mathrm{d}A\left\{\mathbf{q}^{(e)}\right\} \qquad (7.21)
$$

$$
\equiv \frac{1}{2}\left\{\mathbf{q}^{(e)}\right\}^{\mathrm{T}}\left[\mathbf{k}^{(e)}\right]_{8\times 8}\left\{\mathbf{q}^{(e)}\right\},
$$

where $[\mathbf{k}^{(e)}]$ is the element stiffness matrix. Calculation of the element stiffness matrix has two challenges. First, the integration domain is a general quadrilateral shape, and second, the strain–displacement matrix $[\mathbf{B}]$ is written in (s, t) coordinates. Thus, the integration in eq. (7.21) is not trivial. Using the idea of mapping the physical element into the reference element, we can perform the integration in eq. (7.21) in the reference element. Since the reference element is a square and it is defined in (s, t) coordinates, the above two challenges can be resolved simultaneously. Again, the Jacobian plays an important role in transforming the integral to the reference element. Let us consider an infinitesimal area dA of the physical element that is mapped into an infinitesimal rectangle ds·dt in the reference element. Then, the relation between the two areas becomes

$$dA = |\mathbf{J}|dsdt. \tag{7.22}$$

Thus, the element stiffness matrix in the reference element can be written as

$$\left[\mathbf{k}^{(e)}\right] = h\iint_A [\mathbf{B}]^T[\mathbf{C}][\mathbf{B}]\,dA \equiv h\int_{-1}^{1}\int_{-1}^{1}[\mathbf{B}]^T[\mathbf{C}][\mathbf{B}]|\mathbf{J}|\,dsdt. \tag{7.23}$$

Although the integration has been transformed to the reference element, still the integration in eq. (7.23) is not trivial because the integrand cannot be written down as an explicit function of s and t. Note that the matrix $[\mathbf{B}]$ includes the inverse of the Jacobian matrix. Thus, it is going to be extremely difficult, if not impossible, to integrate eq. (7.23) analytically. However, since the integral domain is a square, numerical integration can be used to calculate the element stiffness matrix. Numerical integration methods using Gauss quadrature, which is the most popular method, will be discussed in the following section. Similar to the other elements, the strain energy of entire solid can be obtained using eq. (6.24) in chapter 6, which involves the assembly process.

The potentials of applied loads can be obtained by following a similar procedure as the CST and rectangular elements. The potential energy of concentrated forces and distributed forces will be the same with that of the CST element. The potential energy of the body force can be calculated using eq. (6.32), except that the transformation in eq. (7.22) should be used so that the integration be performed in the reference element. For rectangular elements, the uniform body force yields the equally divided nodal forces. In the case of the quadrilateral element, however, the work–equivalent nodal forces will not divide the body force equally because the Jacobian is not constant within the element. The numerical integration can be used for integrating eq. (6.32).

Using the principle of minimum total potential energy in eq. (6.36), a similar global matrix equation for the quadrilateral elements can be obtained. Applying boundary conditions and solving the matrix equations are identical to those of the CST element. After solving for nodal displacements, strain and stress of the element can be calculated using eqs. (7.20) and (6.38), respectively.

7.4   NUMERICAL INTEGRATION

7.4.1   Gauss Quadrature

As discussed before, it is not trivial to analytically integrate the element stiffness matrix and body force for the quadrilateral element. Although there are many numerical integration methods available, Gauss quadrature is the preferred method in the finite element analysis because it requires fewer function evaluations compared to other methods. We will explain the one–dimensional Gauss quadrature first.

Table 7.2 Gauss quadrature points and weights

NG	Integration Points (s_i)	Weights (w_i)	Exact polynomial degree
1	0.0	2.0	1
2	±0.5773502692	1.0	3
3	±0.7745966692	0.5555555556	5
	0.0	0.8888888889	
4	±0.8611363116	0.3478546451	7
	±0.3399810436	0.6521451549	
5	±0.9061798459	0.2369268851	9
	±0.5384693101	0.4786286705	
	0.0	0.5688888889	

In Gauss quadrature, the integrand is evaluated at predefined points (called Gauss points). The sum of these integrand values, multiplied by integration weights (called Gauss weight) provides an approximation to the integral:

$$I = \int_{-1}^{1} f(s)\,ds \approx \sum_{i=1}^{n} w_i f(s_i), \tag{7.24}$$

where n is the number of Gauss points, s_i are the Gauss points, w_i are the Gauss weights, and $f(s_i)$ is the function value at the Gauss point s_i. The locations of Gauss points and weights are derived in such a way that with n points, a polynomial of degree $2n-1$ can be integrated exactly. Note that the integral domain is normalized, that is, [−1, 1], which is why the range of the one-dimensional reference element in section 7.2 is defined by [−1, 1]. The Gauss quadrature performs well when the integrand is a smooth function. Table 7.2 shows the locations of the Gauss points and corresponding weights.

EXAMPLE 7.8 *Numerical integration*

Evaluate the following integral using Gauss quadrature with 1–4 integration points. Compare the integral results with the analytical integration.

$$I = \int_{-1}^{1} \left(8x^7 + 7x^6\right) dx.$$

SOLUTION It can be easily verified that the exact integral will yield $I = 2$. Now we assume that the exact integral is unknown and calculate its approximate value using Gauss quadrature.

(a) 1–point integral:

$$s_1 = 0, \quad f(s_1) = 0, \quad w_1 = 2,$$
$$I = w_1 f(s_1) = 2 \times 0 = 0.$$

Obviously, the one–point integral is not accurate.

(b) 2–point integral:

$$s_1 = -.577, \quad f(s_1) = 8(-.577)^7 + 7(-.577)^6 = .0882, \quad w_1 = 1,$$
$$s_2 = .577, \quad f(s_2) = 8(.577)^7 + 7(.577)^6 = .4303, \quad w_2 = 1.$$
$$I = w_1 f(x_1) + w_2 f(x_2) = .0882 + .4303 = .5185.$$

The 2–point integral still has a large error because it is accurate only up to the third-order polynomial.

(c) 3–point integral:

$$s_1 = -.7746, \quad f(s_1) = .17350, \quad w_1 = .5556,$$
$$x_2 = 0.0, \qquad f(s_2) = 0.0 \qquad w_2 = .8889.$$
$$x_3 = \ \ .7746, \quad f(s_3) = 2.8505, \quad w_3 = .5556.$$
$$I = w_1 f(s_1) + w_2 f(s_2) + w_3 f(s_3) = .5556(.17350 + 2.8505) = 1.6800.$$

(d) 4–point integral:

$$s_1 = -.8611, \quad f(s_1) = .0452, \quad w_1 = .3479,$$
$$s_2 = -.3400, \quad f(s_2) = .0066, \quad w_2 = .6521,$$
$$s_3 = \ \ .3400, \quad f(s_3) = .0150, \quad w_3 = .6521,$$
$$s_3 = \ \ .8611, \quad f(s_3) = 5.6638, \quad w_3 = .3479,$$
$$I = w_1 f(s_1) + w_2 f(s_2) + w_3 f(s_3) + w_4 f(s_4) = 2.0.$$

Note that the 4-point integral is exact up to 7th-order polynomials. Since the given problem is 7th-order polynomial, the numerical integration is exact. ■

Two-dimensional Gauss integration formulas can be obtained by combining two one-dimensional Gauss quadrature formulas as shown below:

$$I = \int_{-1}^{1}\int_{-1}^{1} f(s,t)\,ds\,dt$$

$$\approx \int_{-1}^{1} \sum_{i=1}^{m} w_i f(s_i, t)\,dt \tag{7.25}$$

$$= \sum_{j=1}^{n}\sum_{i=1}^{m} w_i w_j f(s_i, t_j),$$

where s_i and t_j are Gauss points, m is the number of Gauss points in s direction, n is the number of Gauss points in the t-direction, and w_i and w_j are Gauss weights. The total number of Gauss points becomes $m \times n$. Figure 7.14 shows few commonly used integration points.

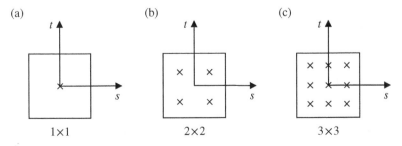

Figure 7.14 Gauss integration points in two-dimensional parent elements

The element stiffness matrix in eq. (7.23) can be evaluated using 2×2 Gauss integration formulas:

$$\left[\mathbf{k}^{(e)}\right] = h \int\limits_{-1}^{1}\int\limits_{-1}^{1} [\mathbf{B}]^T[\mathbf{C}][\mathbf{B}]\,|\mathbf{J}|\,dsdt$$

$$\approx h\sum_{i=1}^{2}\sum_{j=1}^{2} w_i w_j \left[\mathbf{B}(s_i, t_j)\right]^T [\mathbf{C}]\left[\mathbf{B}(s_i, t_j)\right]\left|\mathbf{J}(s_i, t_j)\right|. \tag{7.26}$$

EXAMPLE 7.9 *Numerical integration of element stiffness matrix*

Calculate the element stiffness matrix of the square element shown in figure 7.15 using (a) 1×1 Gauss quadrature and (b) 2×2 Gauss quadrature. Compare the numerically integrated element stiffness matrix with the exact one calculated using eq. (6.54) in chapter 6. Assume plane stress with thickness $h = 0.1$ m, Young's modulus $E = 10$ GPa and Poisson's ratio $\nu = 0.25$.

SOLUTION   Since the element size is the same as that of the reference element, the Jacobian matrix becomes the identity matrix. Thus, from eq. (7.20), the displacement-strain matrix [**B**] becomes

$$[\mathbf{B}] = \frac{1}{4}\begin{bmatrix} -1+t & 0 & 1-t & 0 & 1+t & 0 & -1-t & 0 \\ 0 & -1+s & 0 & -1-s & 0 & 1+s & 0 & 1-s \\ -1+s & -1+t & -1-s & 1-t & 1+s & 1+t & 1-s & -1-t \end{bmatrix}.$$

(a) 1×1 Gauss quadrature use one-point integration at $(s, t) = (0, 0)$ with weight four. The [**B**] matrix at this point becomes

$$[\mathbf{B}] = \frac{1}{4}\begin{bmatrix} -1 & 0 & 1 & 0 & 1 & 0 & -1 & 0 \\ 0 & -1 & 0 & -1 & 0 & 1 & 0 & 1 \\ -1 & -1 & -1 & 1 & 1 & 1 & 1 & -1 \end{bmatrix}.$$

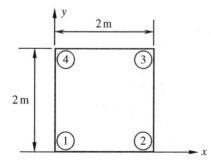

Figure 7.15 Numerical integration of a square element

Then, the numerical integration of the element stiffness matrix becomes

$$[\mathbf{k}_1] \approx h w_1 w_1 \left[\mathbf{B}(0,0) \right]^{\mathrm{T}} [\mathbf{C}][\mathbf{B}(0,0)].$$

$$= 10^9 \begin{bmatrix} .367 & .167 & -.167 & -.033 & -.367 & -.167 & .167 & .033 \\ & .367 & .033 & .167 & -.167 & -.367 & -.033 & -.167 \\ & & .367 & -.167 & .167 & -.033 & -.367 & .167 \\ & & & .367 & .033 & -.167 & .167 & -.367 \\ & & & & .367 & .167 & -.167 & -.033 \\ & & & & & .367 & .033 & .167 \\ & \text{Symmetric} & & & & & .367 & -.167 \\ & & & & & & & .367 \end{bmatrix}$$

(b) For 2×2 Gauss quadrature, we need four integration points and weights are a unit.

Integration point	s	t
1	−0.5773502692	−0.5773502692
2	+0.5773502692	−0.5773502692
3	+0.5773502692	+0.5773502692
4	−0.5773502692	+0.5773502692

Then, the numerical integration of the element stiffness matrix in eq. (7.26) becomes

$$[\mathbf{k}_2] \approx h \sum_{i=1}^{2} \sum_{j=1}^{2} w_i w_j [\mathbf{B}(s_i,t_j)]^{\mathrm{T}} [\mathbf{C}][\mathbf{B}(s_i,t_j)] |\mathbf{J}(s_i,t_j)|.$$

$$= 10^9 \begin{bmatrix} .489 & .167 & -.289 & -.033 & -.244 & -.167 & .044 & .033 \\ & .489 & .033 & .044 & -.167 & -.244 & -.033 & -.289 \\ & & .489 & -.167 & .044 & -.033 & -.244 & .167 \\ & & & .489 & .033 & -.289 & .167 & -.244 \\ & & & & .489 & .167 & -.289 & -.033 \\ & \text{Symmetric} & & & & .489 & .033 & .044 \\ & & & & & & .489 & -.167 \\ & & & & & & & .489 \end{bmatrix}$$

Using the exact stiffness in eq. (6.54), we can find that the element stiffness matrix obtained from 2×2 Gauss quadrature is exact. ■

In general, the 2×2 Gauss quadrature is not exact for quadrilateral elements. The exact results in the above example occur because the element shape is a square.

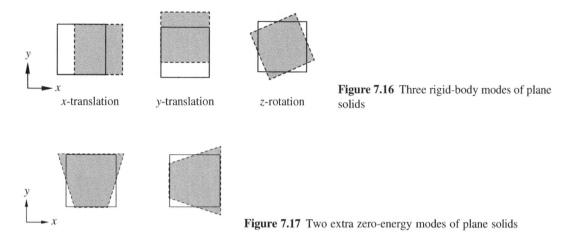

Figure 7.16 Three rigid-body modes of plane solids

x-translation   y-translation   z-rotation

Figure 7.17 Two extra zero-energy modes of plane solids

7.4.2 Lower–Order Integration and Extra Zero–Energy Modes

It is important that the proper order of Gauss quadrature should be used. Otherwise, the element may show undesirable behavior. One of the well-known phenomena of lower–order integration is *extra zero energy modes*. The zero-energy mode is the deformation of an element without changing its strain energy. In plane solids, there are three types of deformations (more precisely, motions) that do not change the strain energy: x-translation, y-translation, and z-rotation. Figure 7.16 illustrates these modes. Since the relative locations of nodes do not change, the stress and strain of the elements are zero, and the strain energy remains constant. In finite element analysis, these modes should be fixed by applying displacement boundary conditions. Otherwise, the stiffness matrix will be singular, and there will be no unique solution.

While the zero-energy modes in figure 7.16 are proper modes, there are improper modes, called extra zero-energy modes, which often occur when an element is under-integrated. For example, if a square element is integrated using 1×1 Gauss quadrature, there will be two extra zero-energy modes in addition to the three rigid-body modes. Figure 7.17 illustrates the two extra zero-energy modes of plane solids. It is clear that the element is being deformed, but the centroid (the quadrature point) of the element does not experience any deformation and hence the strain energy remains constant. In other words, the element will deform without having externally applied forces, which is a numerical artifact. Thus, the extra zero-energy modes must be removed in order to obtain meaningful deformation.

The most common way of checking whether extra zero-energy modes exists is by computing the eigenvalues of the stiffness matrix. For a plane solid, the number of zero eigenvalues must be equal to three, which represents the three rigid-body motion. However, the element stiffness matrix with 1×1 integration will have five zero eigenvalues corresponding to five zero energy modes shown in figure 7.16 and figure 7.17. In the following example, we will show another method of checking for extra zero-energy modes.

EXAMPLE 7.10 *Extra zero-energy modes*

Consider two stiffness matrices of the square element in example 7.9: $[\mathbf{k}_1]$ for 1×1 integration and $[\mathbf{k}_2]$ for 2×2 integration. When nodal displacements are given as $\{\mathbf{q}\}^T = \{0.1, 0, -0.1, 0, 0.1, 0, -0.1, 0\}$, check the reaction forces and determine if the stiffness matrix has extra zero-energy modes.

SOLUTION

(a) For $[\mathbf{k}_1]$ (1×1 integration), the reaction force can be calculated by multiplying the stiffness matrix with the nodal displacements as

$$
[\mathbf{k}_1]\{\mathbf{q}\} = 10^9
\begin{bmatrix}
.367 & .167 & -.167 & -.033 & -.367 & -.167 & .167 & .033 \\
 & .367 & .033 & .167 & -.167 & -.367 & -.033 & -.167 \\
 & & .367 & -.167 & .167 & -.033 & -.367 & .167 \\
 & & & .367 & .033 & -.167 & .167 & -.367 \\
 & & & & .367 & .167 & -.167 & -.033 \\
 & & & & & .367 & .033 & .167 \\
 & \text{Symmetric} & & & & & .367 & -.167 \\
 & & & & & & & .367
\end{bmatrix}
\begin{Bmatrix}
0.1 \\ 0.0 \\ -0.1 \\ 0.0 \\ 0.1 \\ 0.0 \\ -0.1 \\ 0.0
\end{Bmatrix}
=
\begin{Bmatrix}
0 \\ 0 \\ 0 \\ 0 \\ 0 \\ 0 \\ 0 \\ 0
\end{Bmatrix}.
$$

No force is required to deform the element. Thus, the $[\mathbf{k}_1]$ matrix has an extra zero-energy mode.

(b) For $[\mathbf{k}_2]$ (2×2 integration), the reaction force can be calculated by multiplying the stiffness matrix with the nodal displacements as

$$
[\mathbf{k}_2]\{\mathbf{q}\} = 10^9
\begin{bmatrix}
.489 & .167 & -.289 & -.033 & -.244 & -.167 & .044 & .033 \\
 & .489 & .033 & .044 & -.167 & -.244 & -.033 & -.289 \\
 & & .489 & -.167 & .044 & -.033 & -.244 & .167 \\
 & & & .489 & .033 & -.289 & .167 & -.244 \\
 & & & & .489 & .167 & -.289 & -.033 \\
 & & & & & .489 & .033 & .044 \\
 & \text{Symmetric} & & & & & .489 & -.167 \\
 & & & & & & & .489
\end{bmatrix}
\begin{Bmatrix}
0.1 \\ 0 \\ -0.1 \\ 0 \\ 0.1 \\ 0 \\ -0.1 \\ 0
\end{Bmatrix}
= 10^7
\begin{bmatrix}
4.89 \\ 0 \\ -4.89 \\ 0 \\ 4.89 \\ 0 \\ -4.89 \\ 0
\end{bmatrix}.
$$

Nonzero nodal forces are required to deform the element. Thus, the $[\mathbf{k}_2]$ matrix does not have an extra zero–energy mode corresponding to the given deformation. ■

7.5   HIGHER-ORDER QUADRILATERAL ELEMENTS

The quadrilateral element described in the previous section is a bilinear element that is widely used but often requires a high mesh density for the solution to converge toward the exact solution. The four-node quadrilateral interpolates the variable linearly along the edge. Higher-order elements can similarly be derived, which are quadratic or cubic along the edges. These elements have more nodes per element and therefore are more computationally expensive than bilinear elements. But on the other hand, one can obtain a better quality solution with fewer elements in the mesh. Since the geometry is also interpolated using the same shape functions, the geometry is also better approximated by these elements. For all isoparametric quadrilateral elements, the shape functions are expressed in terms of the parametric coordinates. Here we will look at several higher-order quadrilateral elements.

When the order of element goes up, the element has more nodes and DOFs. Therefore, it is possible to choose the interpolation using higher-order polynomials. The question is how to choose an appropriate order of polynomials for a given element. A useful strategy is based on the polynomial triangle,

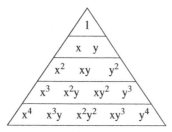

Figure 7.18 Polynomial triangle

similar to Pascal's triangle, shown in figure 7.18. In the case of three-dimensional space, a similar polynomial pyramid can be defined. In the figure, polynomials are chosen from the top to bottom. The chosen polynomials are also called the basis. For example, the linear triangular element has three nodes, so the interpolation can choose $\{1, s, t\}$. In the case of a quadrilateral element, the first three bases are the same as a triangular element. Additional bases can be chosen from next level. Among s^2, st, and t^2, st is chosen because the interpolation should perform equally for both the s- and t-direction.

In the polynomial triangle, it is important for the interpolation functions to include the first level, that is, 1, and the second level, s and t. The first level is required for the element to represent rigid body motion properly, while the second level is required for the constant strain field. If an interpolation scheme misses these two levels, it is possible that the element can generate fictitious strain under rigid body motion.

7.5.1 Nine-Node Lagrange Element

The 9-node Lagrange element is a bi-quadratic element whose shape functions can be derived using the Lagrangian interpolation approach and hence is a part of the family of Lagrange elements. Figure 7.19 shows the reference element in the parametric space, which is again a 2×2 square with its centroid at the origin. While the nodes of the physical element can be numbered arbitrarily, the element connectivity must be defined in the sequence shown in figure 7.19. Here we use a node-numbering scheme where the first four nodes are the corners going in the counterclockwise direction. The next four nodes are the mid-edge nodes, and finally, the last node is at the centroid of the element.

In the case of a heat conduction problem, the temperature $T(s, t)$ is the scalar field to interpolate. In the case of solid mechanics, displacements, $u(s, t)$ and $v(s, t)$, are the vector field to interpolate. In this case, since each node has two DOFs, u_I and v_I, $u(s, t)$ and $v(s, t)$ are interpolated independently. Therefore, we look for an interpolation scheme that uses nine nodal DOFs. From the polynomial triangle in

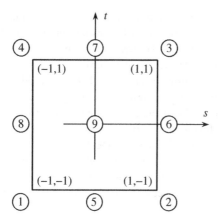

Figure 7.19 9-node Lagrange element in parametric space

figure 7.18, the first three levels include six terms, and we need additional three terms. Among s^3, s^2t, st^2, t^3, there is no way that we can maintain symmetry by choosing three out of four. Instead, we choose two terms, s^2t and st^2, from this level, and choose the last term from next level, s^2t^2. Therefore, the interpolation of a scalar field, for example, displacement $u(s, t)$ within this element can be expressed as a polynomial in s and t in the following form:

$$u(s,t) = a_1 + a_2s + a_3t + a_4st + a_5s^2 + a_6t^2 + a_7s^2t + a_8t^2s + a_9s^2t^2. \tag{7.27}$$

Using the condition that displacement at node i is u_i, we could solve for the constants a_i and substitute it back to express the interpolation in the standard form using shape functions.

$$u(s,t) = \sum_{i=1}^{9} N_i(s,t)u_i. \tag{7.28}$$

Alternatively, we can derive the shape functions using the requirement that $N_i(s_i, t_i) = 1$ at node i and zero at all other nodes. In order for $N_1(s, t)$ to be zero at all other nodes, it must have the form

$$N_1(s,t) = k(1-s)(1-t)st. \tag{7.29}$$

This ensures that N_1 is zero along the lines $s = 1$, $t = 1$, $s = 0$, and $t = 0$ along which all the other nodes are located. Furthermore, we can determine k such that $N_1(-1, -1) = 1$. This yields $N_1(-1, -1) = k(1+1)(1+1)1 = 1$ or $k = 1/4$. The shape functions for the other corner nodes can also be derived in a similar fashion yielding the following shape functions for the corner nodes.

$$N_1(s,t) = \frac{1}{4}st(1-s)(1-t), \tag{7.30}$$

$$N_2(s,t) = -\frac{1}{4}st(1+s)(1-t), \tag{7.31}$$

$$N_2(s,t) = \frac{1}{4}st(1+s)(1+t), \tag{7.32}$$

$$N_4(s,t) = -\frac{1}{4}st(1-s)(1+t). \tag{7.33}$$

The shape functions of the mid-edge nodes such as node 5 should be zero along all the other edges as well as at the origin; therefore, it must be of the form:

$$N_5 = k(1-s)(1+s)t(1-t). \tag{7.34}$$

Again the constant k can be derived such that N_5 is unity at node 5 or $N_5(0, -1) = 1$ and $k = 1/2$. Similarly, we can derive the other mid-edge node shape functions as

$$N_5(s,t) = \frac{1}{2}t(1-s^2)(1-t), \tag{7.35}$$

$$N_6(s,t) = \frac{1}{2}s(1-t^2)(1+s), \tag{7.36}$$

$$N_7(s,t) = \frac{1}{2}t(1-s^2)(1+t), \tag{7.37}$$

$$N_8(s,t) = \frac{1}{2}s(1-t^2)(1-s). \tag{7.38}$$

Finally, the last node, which is at the centroid of the element has a shape function that is zero along all the edges of the element and is unity at the centroid.

$$N_9(s,t) = (1-s^2)(1-t^2). \tag{7.39}$$

These shape functions are quadratic along the edges of the element, and therefore, the interpolation of the field variable will also be quadratic. For this element to be isoparametric, the mapping between the reference element and the physical element must be constructed using the same shape functions that we have derived above for interpolation. The mapping or the geometry interpolation is, therefore, bi-quadratic, which implies that the edges of the element will be parabolic allowing this element to better approximate curved boundaries than the bilinear element. As discussed earlier for the one-dimensional quadratic element, the placement of the nodes in the real element is critical to ensure that the mapping is uniform. The mid-edge nodes of the real elements in the mesh must be at the midpoint of the edges and likewise, the last node must be at the centroid of the element.

7.5.2   Eight-Node Serendipity Elements

A two-dimensional quadratic element with just 8 nodes is possible that has only corner nodes and mid-edge nodes as shown in figure 7.20. The first eight shape functions of the 9-node element cannot be used here as the shape functions for this element because all those shape functions are zero at the centroid, and therefore any interpolation within the element will also be zero. We need to derive shape functions that satisfy Kronecker's delta condition in eq. (7.1) but are not simultaneously zero at the centroid.

It is easy to derive such shape functions for the mid-edge nodes. For example, the shape function for node 5 must be zero at the edges $s = \pm 1$ and $t = 1$ and will be of the form $N_5(s,t) = k(1-s^2)(1-t)$. The constant k must equal one half for this shape function to have a unit value at the node 5. Following this procedure, the shape functions of the mid-edge nodes can be derived as

$$N_5(s,t) = \frac{1}{2}\left(1-s^2\right)(1-t), \tag{7.40}$$

$$N_6(s,t) = \frac{1}{2}\left(1-t^2\right)(1+s), \tag{7.41}$$

$$N_7(s,t) = \frac{1}{2}\left(1-s^2\right)(1+t), \tag{7.42}$$

$$N_8(s,t) = \frac{1}{2}\left(1-t^2\right)(1-s). \tag{7.43}$$

The shape functions of the corner nodes must be zero at the opposite edges. These conditions are satisfied also by the shape functions of the 4-node elements. However, for the 8-node element, we want the shape functions to be zero at the midpoint of the adjacent edges as well. Figure 7.21 shows the plot of the shape function for node 1 of a 4-node isoparametric element. At the midpoint of the adjacent edges of node 1, this shape function has a value of one half.

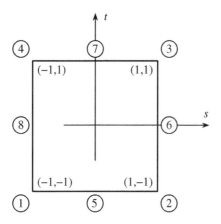

Figure 7.20 8-node serendipity element in parametric space

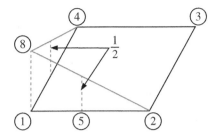

Figure 7.21 Shape function for node 1 of a 4-node element

To derive shape functions for the 8-node element, we need to reduce the value of the 4-node element shape function to zero at node 5 and node 8. This can be accomplished using the shape functions for nodes 5 and 8 as

$$N_1(s,t) = \frac{1}{4}(1-s)(1-t) - \frac{1}{2}N_5(s,t) - \frac{1}{2}N_8(s,t). \tag{7.44}$$

A similar procedure can be employed to derive the shape functions of all the corner nodes as

$$N_1(s,t) = -\frac{1}{4}(1-s)(1-t)(s+t+1), \tag{7.45}$$

$$N_2(s,t) = \frac{1}{4}(1+s)(1-t)(s-t-1), \tag{7.46}$$

$$N_3(s,t) = \frac{1}{4}(1+s)(1+t)(s+t-1), \tag{7.47}$$

$$N_4(s,t) = \frac{1}{4}(1-s)(1+t)(t-s-1). \tag{7.48}$$

The 8-node element is also quadratic along the edges of the element and therefore has parabolic edges, as the 9-node element. Therefore it is able to represent curved boundaries with the same accuracy as the 9-node element but with one fewer node. For this reason, the 8-node element is the more popular quadratic element.

The technique employed here to derive the shape functions of an 8-node element can also be used to derive shape functions for 5-node, 6-node, or 7-node elements. Such elements are needed for adaptive mesh regeneration techniques where instead of increasing the number of elements, the order of the elements is increased in regions where there is a stress concentration. In order to transition from a higher-order element, such as an 8-node element, to a lower-order element, such as a 4-node element, one needs transition elements in between that have 3 nodes along their edges on one side and only 2 nodes on the opposite edge. This can be achieved by a 5-node element. Such elements are therefore often referred to as transition elements.

7.5.3 Practical Considerations

Higher-order elements are able to approximate the solution better than lower-order elements and, in general, for the same number of elements, they would provide a better quality solution. Another way to compare elements is to compare the number of elements needed to reach the same level of accuracy or to obtain a converged solution. Again, the higher-order elements are better in this regard because you can use a much lower mesh density to reach the same level of accuracy. However, higher order elements have more nodes per element, and they also require higher-order Gauss quadrature for accurate integration when computing the stiffness matrix. Therefore, from the viewpoint of the cost of computation, higher-order elements can be more expensive because the cost of computation to obtain a converged result is often higher even though the number of elements in the mesh is lower.

The elements discussed in this chapter interpolate the field variables such that it is continuous between elements so that the interpolation is said to satisfy C^0 continuity. But the derivatives of these variables are not guaranteed to be continuous or in other words, the interpolation does not satisfy C^1 continuity across boundaries. This implies that quantities such as stress /strain and heat flux are not continuous between elements. The field variables such as displacement (or temperature) also converge faster with increasing mesh density than derived quantities like stress/strain (or heat flux). If the stress is computed at a node or edge, the value obtained will be different for each element adjacent to that node. To plot such quantities, finite element analysis software programs smooth the results. One simple approach to smooth the solution is to compute the stresses at a node for each adjacent element and then to use the average value as the nodal value, which can be interpolated to obtain a continuous solution. The results for stresses and other similar derived quantities are most accurate at the integration (Gauss) points. Therefore, average nodal values are often computed by extrapolating from the nearest integration points. Another popular approach involves fitting a patch (curve for two-dimension and surface for three-dimension) that best approximates the values at the nearest integration point to estimate the value at the node.

If the structure being analyzed has stress concentration or the solution has large gradients, then smaller elements are needed in such regions. In regions where the solution is nearly constant and the gradients are small, larger elements would suffice, so it is not beneficial to globally reduce the size of the elements. To reduce computation, the mesh must be locally refined in regions that need smaller-sized elements. Adaptive mesh refinement capability is available in most commercial software to automatically refine the initial uniform mesh in regions where the solution has large errors. This can be achieved by sub-dividing the elements in such regions to create smaller-sized elements. This method is referred to as h-adaptive mesh refinement. This technique is used in an example later in this chapter to show why such refinement is necessary to compute stresses accurately at stress concentrations.

Another approach for adaptive mesh refinement is p-adaptive mesh refinement where the order of the elements is raised in regions where the solution is not accurate. This is illustrated in figure 7.22 where the left-most element has been converted to an 8-node quadratic element to improve accuracy. In order to do so, it is necessary to have transition elements like the 5-node element shown in the figure because 4-node elements and 8-node elements are not compatible and cannot be next to each other. Along the edges, displacement is interpolated linearly for 4-node elements while it is interpolated as a quadratic for an 8-node element. This means that if these two types of elements are side by side, then along the shared edge, the displacement will not be continuous.

The 5-node element shown in figure 7.22 is a transition element that uses quadratic interpolation along the edge that it shares with the 8-node element, and these two elements share the 3 nodes along this edge. Along the opposite edge where it is adjacent to the 4-node element it has linear interpolation and only two nodes. Shape functions for such an element can be derived using the same technique that we used earlier for the 8-node serendipity element. Other types of transition elements may be needed in a mesh during p-adaptive refinement such as elements with 6 nodes and 7 nodes depending on the type of element it is adjacent to at each edge.

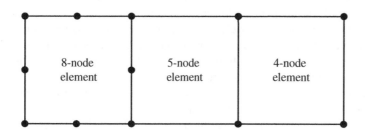

| 8-node element | 5-node element | 4-node element |

Figure 7.22 5-node transition element

7.6 ISOPARAMETRIC TRIANGULAR ELEMENTS

7.6.1 Collapsed 4-Node Quadrilateral Element

In the previous chapters, we have seen the classical triangular element, which is not formulated as an isoparametric element. Since we explained the benefits of isoparametric elements, it would be good to derive the isoparametric formulation for a triangular element. There are a couple of ways to derive the isoparametric triangular element. First, the 4-node quadrilateral isoparametric element in section 7.3 can be converted to a triangular element by collapsing or merging two of its nodes together into one node. By doing so, we get a mapping relation between a square in the parametric space to a triangle in the physical coordinates. This is illustrated in figure 7.23 where nodes 3 and 4 are merged and mapped to the same physical node.

In practice, this can be easily achieved by creating a connectivity table for the element in which two of the nodes are identical. This implies that if only the 4-node isoparametric element is available in a finite element analysis software program, then one can still use a mesh that contains triangular elements. This also facilitates the use a mesh that contains both quadrilateral and triangular elements. For the examples shown in figure 7.23, the row in the connectivity table corresponding to this element will look as follows:

Element #	1	2	3	4
45	51	57	61	61

The interpolation of a scalar field $u(s, t)$ in parametric space for the 4-node element is simplified to obtain the three shape functions of the collapsed element where we assume that two of the nodes are the same.

$$\begin{aligned} T(s,t) &= N_1(s,t)T_1 + N_2(s,t)T_2 + (N_3(s,t) + N_4(s,t))T_3 \\ &= N_1(s,t)T_1 + N_2(s,t)T_2 + N_3'(s,t)T_3. \end{aligned} \tag{7.49}$$

The three shape functions of the triangular element thus obtained are

$$N_1 = \frac{1}{4}(1-s)(1-t), \tag{7.50}$$

$$N_2 = \frac{1}{4}(1+s)(1-t), \tag{7.51}$$

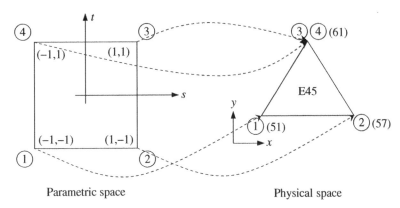

Figure 7.23 Triangular element by collapsing a 4-node quadrilateral

$$N_3' = N_3 + N_4 = \frac{1}{2}(1+t). \tag{7.52}$$

The mapping can also be now expressed using these three shape functions for the collapsed 4-node element. This provides a mapping from the square in parametric space to a triangle in real space.

$$x(s,t) = N_1(s,t)x_1 + N_2(s,t)x_2 + N_3'(s,t)x_3,$$
$$y(s,t) = N_1(s,t)y_1 + N_2(s,t)y_2 + N_3'(s,t)y_3. \tag{7.53}$$

It is interesting to note that this element is similar to the classical triangular elements in terms of the quality of the interpolation within the element. Since it interpolates between three nodal values, we expect that the interpolation in the real space is linear because three points define a plane. Therefore we also expect that the derivatives should be constant. The following example illustrates this idea and explores some numerical aspects.

EXAMPLE 7.11 *Collapsed 4-node triangular element*

Figure 7.24 shows a triangular element in the physical space whose nodal coordinates are known. If the 4-node isoparametric element is used for the analysis, compute the heat flux in the element assuming that the nodal temperatures have already been computed as $\{\mathbf{T}^{(e)}\} = \{T_1 \quad T_2 \quad T_3\}^{\mathrm{T}}$.

SOLUTION It is illustrative to study the heat flux in this element because heat flux is proportional to the gradient of temperature $\mathbf{q} = -k\nabla T$. The gradient of temperature in the element can be computed as

$$\nabla T = \left\{ \begin{array}{c} \dfrac{\partial T}{\partial x} \\[2mm] \dfrac{\partial T}{\partial y} \end{array} \right\} = [\mathbf{B}]\left\{ \mathbf{T}^{(e)} \right\}.$$

The [**B**] matrix contains derivatives of the shape functions with respect to the global coordinates. We use the inverse of the Jacobian matrix to transform derivatives with respect to the parametric coordinates to derivatives with respect to global coordinates as:

$$[\mathbf{B}] = [\mathbf{J}]^{-1} \begin{bmatrix} \dfrac{\partial N_1}{\partial s} & \dfrac{\partial N_2}{\partial s} & \dfrac{\partial N_3'}{\partial s} \\[3mm] \dfrac{\partial N_1}{\partial t} & \dfrac{\partial N_2}{\partial t} & \dfrac{\partial N_3'}{\partial t} \end{bmatrix}.$$

In order to determine the Jacobian, we need to derive the mapping relation first. Using the given nodal coordinates of the element and the shape functions of the collapse element we get,

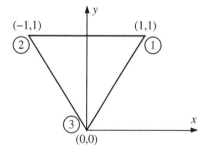

Figure 7.24 Triangular element in physical space

$$x(s,t) = N_1(s,t) \times 1 - N_2(s,t) \times 1 + N_3'(s,t) \times 0$$

$$= N_1(s,t) - N_2(s,t)$$

$$= -\frac{1}{2}s(1-t),$$

$$y(s,t) = N_1(s,t) \times 1 + N_2(s,t) \times 1 + N_3'(s,t) \times 0$$

$$= N_1(s,t) + N_2(s,t)$$

$$= \frac{1}{2}(1-t).$$

Now the Jacobian and its inverse can be computed using these mapping functions. Note that the Jacobian is not diagonal or constant because the element has been distorted from square to a triangle.

$$[\mathbf{J}] = \begin{bmatrix} \dfrac{\partial x}{\partial s} & \dfrac{\partial y}{\partial s} \\ \dfrac{\partial x}{\partial t} & \dfrac{\partial y}{\partial t} \end{bmatrix} = \begin{bmatrix} -\dfrac{1}{2}(1-t) & 0 \\ \dfrac{s}{2} & -\dfrac{1}{2} \end{bmatrix}.$$

$$[\mathbf{J}]^{-1} = \frac{4}{1-t} \begin{bmatrix} -\dfrac{1}{2} & 0 \\ -\dfrac{s}{2} & -\dfrac{1}{2}(1-t) \end{bmatrix}.$$

Now the [**B**] matrix can be determined, and the heat flux can be computed.

$$[\mathbf{B}] = \frac{4}{1-t} \begin{bmatrix} -\dfrac{1}{2} & 0 \\ -\dfrac{s}{2} & -\dfrac{1}{2}(1-t) \end{bmatrix} \begin{bmatrix} -\dfrac{1}{4}(1-t) & \dfrac{1}{4}(1-t) & 0 \\ -\dfrac{1}{4}(1-s) & -\dfrac{1}{4}(1+s) & \dfrac{1}{2} \end{bmatrix} = \frac{1}{2} \begin{bmatrix} 1 & -1 & 0 \\ 1 & 1 & -2 \end{bmatrix}.$$

$$\mathbf{q} = -k[\mathbf{B}]\left\{\mathbf{T}^{(e)}\right\} = -\frac{k}{2} \begin{bmatrix} 1 & -1 & 0 \\ 1 & 1 & -2 \end{bmatrix} \begin{Bmatrix} T_1 \\ T_2 \\ T_3 \end{Bmatrix} = -\frac{k}{2} \begin{Bmatrix} T_1 - T_2 \\ T_1 + T_2 - 2T_3 \end{Bmatrix}.$$

As expected the [**B**] matrix and the heat flux are constant within this element even though the Jacobian inverse, as well as the derivatives of the shape functions with respect to the parametric coordinates, were not constants.

7.6.2 Three-Node Linear Triangular Element

Another way of formulating the isoparametric triangular element is to use a mapping where the reference element is assumed to be a right triangle that is mapped to an arbitrary triangle in the physical space as shown in figure 7.25. This approach has the same advantages as the quadrilateral isoparametric elements. The shape functions are derived with respect to the parametric coordinates to perform interpolation and are therefore very simple functions. The stiffness matrix can be computed by integrating over a right triangle of fixed size and shape in the parametric space.

The shape functions for a 3-node linear element can be expressed simply as:

$$N_1 = r, \tag{7.54}$$

$$N_2 = s, \tag{7.55}$$

$$N_3 = 1 - r - s = t. \tag{7.56}$$

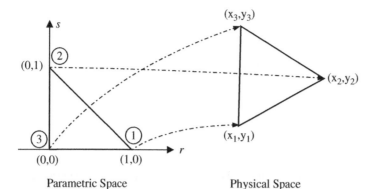

Figure 7.25 3-node isoparametric triangular element

The shape functions for this element are identical to the coordinates in the parametric space. The third shape function can also be thought of as a coordinate, even though only any two of (r, s, t) are sufficient to identify the location of a point in the triangle. Note that the third coordinate t is unity at node 3 and is zero along the line between nodes 1 and 2. This three-node element is identical in behavior to the classic 3-node triangle discussed earlier in terms of the quality of the results obtained. Therefore, when used for two-dimensional solid mechanics, this element is a constant-strain triangle. As strain is approximated as constant within each element, these elements are not very accurate and must be used only to obtain rough, qualitative results for displacement. Stresses and strains computed using this element are likely to be inaccurate unless the mesh is very dense. Despite these shortcomings, these elements are popular and available in most commercial software since these elements are computationally very inexpensive and often the user may only need a qualitative answer, for example, to visualize the deformation pattern. Mesh generation is also easier for triangular elements and automated mesh generation is available for triangular elements in many commercial finite element analysis software programs.

7.6.3   Six-Node Quadratic Element

As we saw earlier for the isoparametric quadrilateral elements, it is fairly easy to derive the shape functions of the higher-order triangular elements using the Kronecker's delta conditions. Consider a 6-node quadratic element with 3 corner nodes and 3 mid-edge nodes as shown in figure 7.26. In the parametric

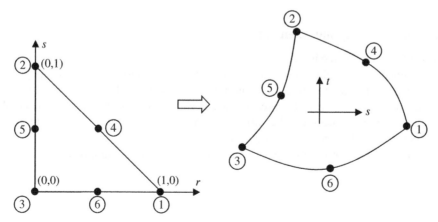

Figure 7.26 Six-node isoparametric triangular element

coordinates, we will assume that the six nodes are located at: (1, 0), (0, 1), (0, 0), (0.5, 0.5), (0, 0.5) and (0.5, 0) and numbered in that order.

For node 1, we want the shape function to be zero along the lines $r = 0$ and $r = 0.5$ and equal to unity at (1, 0). Similarly, N_4 should be zero at $r = 0$ and $s = 0$ and have a unit value at (0.5, 0.5). Using such conditions, the shape functions for this element can be derived as:

$$\begin{cases} N_1(r,s,t) = r(2r-1) \\ N_2(r,s,t) = s(2s-1) \\ N_3(r,s,t) = t(2t-1) \\ N_4(r,s,t) = 4rs \\ N_5(r,s,t) = 4st \\ N_6(r,s,t) = 4rt \end{cases} \qquad (7.57)$$

In the preceding equations, $t = 1 - r - s$, and it can be easily verified that these shape functions satisfy the Kronecker's delta condition. The interpolation of the field variables for these triangular elements is done the same way as all other isoparametric elements yielding an interpolation in the parametric coordinate system.

$$T(r,s,t) = \sum_{i=1}^{6} N_i(r,s,t) T_i. \qquad (7.58)$$

The mapping between parametric and real space is again constructed using the shape functions so that the edges of a linear element are straight lines while the edges of a quadratic element are parabolic.

$$x(r,s,t) = \sum_{i=1}^{6} N_i(r,s,t) x_i,$$

$$y(r,s,t) = \sum_{i=1}^{6} N_i(r,s,t) y_i. \qquad (7.59)$$

EXAMPLE 7.12 *Three-node isoparametric triangular element*

Assuming that the triangular element in example 7.11 was formulated as an isoparametric element, compute the heat flux in the element as a function of the nodal temperatures.

SOLUTION The temperature within the element can be interpolated in the parametric coordinates as:

$$T(r,s,t) = \sum_{i=1}^{3} N_i T_i = rT_1 + sT_2 + tT_3 = T_3 + r(T_1 - T_3) + s(T_2 - T_3).$$

The nodal coordinates of the triangles are: $(x_1,y_1) = (1,1)$, $(x_2,y_2) = (-1,1)$, and $(x_3,y_3) = (0,0)$. Using the shape functions, the mapping relation between the real coordinates and parametric coordinates can be derived to be:

$$x(r,s) = \sum_{i=1}^{3} N_i x_i = r - s,$$

$$y(r,s) = \sum_{i=1}^{3} N_i y_i = r + s.$$

The Jacobian matrix for this element is:

$$[\mathbf{J}] = \begin{bmatrix} \dfrac{\partial x}{\partial r} & \dfrac{\partial y}{\partial r} \\[2mm] \dfrac{\partial x}{\partial s} & \dfrac{\partial y}{\partial s} \end{bmatrix} = \begin{bmatrix} 1 & 1 \\ -1 & 1 \end{bmatrix}.$$

As with any isoparametric element, since the [**B**] matrix contains derivatives of the shape functions with respect to the physical coordinates, we use the inverse of the Jacobian matrix to transform derivatives with respect to the parametric coordinates to determine derivatives with respect to global coordinates as:

$$[\mathbf{B}] = [\mathbf{J}]^{-1} \begin{bmatrix} \dfrac{\partial N_1}{\partial r} & \dfrac{\partial N_2}{\partial r} & \dfrac{\partial N_3}{\partial r} \\[2mm] \dfrac{\partial N_1}{\partial s} & \dfrac{\partial N_2}{\partial s} & \dfrac{\partial N_3}{\partial s} \end{bmatrix} = \frac{1}{2} \begin{bmatrix} 1 & -1 \\ 1 & 1 \end{bmatrix} \begin{bmatrix} 1 & 0 & -1 \\ 0 & 1 & -1 \end{bmatrix} = \frac{1}{2} \begin{bmatrix} 1 & -1 & 0 \\ 1 & 1 & -2 \end{bmatrix},$$

$$\mathbf{q} = -k[\mathbf{B}]\left\{\mathbf{T}^{(e)}\right\} = -\frac{k}{2}\begin{bmatrix} 1 & -1 & 0 \\ 1 & 1 & -2 \end{bmatrix}\begin{Bmatrix} T_1 \\ T_2 \\ T_3 \end{Bmatrix} = -\frac{k}{2}\begin{Bmatrix} T_1 - T_2 \\ T_1 + T_2 - 2T_3 \end{Bmatrix}.$$

The [**B**] matrix and the heat flux are constant within this element and identical to what we obtained using the collapsed 4-node element showing that these are essentially identical elements even though the formulations of these elements are different. ■

7.6.4 Numerical Integration for Isoparametric Triangular Elements

As in the case of quadrilateral isoparametric elements (see section 7.4), Gauss quadrature is used for integrating over isoparametric triangular elements as well. Again, the integral is evaluated as the sum of the values of the integrand at Gauss points multiplied by Gauss weighting values:

$$I = \int_{-1}^{1} \int_{-1}^{1-r} f(r,s)\,\mathrm{d}r\mathrm{d}s \approx \sum_{i=1}^{n} w_i f(r_i, s_i), \tag{7.60}$$

Table 7.3 Gauss quadrature points and weights for triangles

| n | Integration Points | | Weights |
	r_i	s_i	w_i
1	1/3	1/3	1.0
3	1/6	2/3	1/6
	1/6	1/6	1/6
	2/3	1/6	1/6
4	1/3	1/3	−9/32
	1/5	1/5	25/96
	1/5	3/5	25/96
	3/5	1/5	25/96
6	0.445948490915965	0.445948490915965	0.111690794839005
	0.445948490915965	0.108103018168070	0.111690794839005
	0.108103018168070	0.445948490915965	0.111690794839005
	0.091576213509771	0.091576213509771	0.054975871827661
	0.091576213509771	0.816847572980458	0.054975871827661
	0.816847572980458	0.091576213509771	0.054975871827661

where n is the number of Gauss points, (r_i, s_i) are the Gauss points, w_i are the Gauss weights, and $f(r_i, s_i)$ the function value at the Gauss points. A polynomial of degree $2n-1$ can be integrated exactly using n-point integration. The domain of the integral is normalized to $[-1, 1]$ for convenience. Even if the integrand is not a polynomial, Gauss quadrature performs well when the integrand is a smooth function that can be well-approximated by a polynomial. Table 7.3 shows the locations of the Gauss points and the corresponding weights for quadrature over triangles.

7.7   THREE-DIMENSIONAL ISOPARAMETRIC ELEMENTS

Three-dimensional finite elements are rather straightforward extensions of similar two-dimensional elements, and no new concepts or methodology is needed. In general, three-dimensional elements are far more computationally expensive than two-dimensional elements not only due to the larger number of nodes and equations but also because numerical integration is more expensive in three dimensions. So if it is possible to model a structure as two-dimensional, then it is better to avoid three-dimensional models. Using simplified models that use half or a quarter of the structure, if the symmetry of the structure allows it, can also help to reduce the cost of computation. Popular elements in three dimensions are tetrahedral and hexahedral elements even though specialized elements shaped like pyramids or wedges are also available in some software packages. Below we present the shape functions of a few isoparametric elements that are commonly found in finite element software.

7.7.1   Four-Node Tetrahedral Element

The 4-node tetrahedral element is similar to the 3-node triangular element. The shape functions and the displacement interpolation are linear while the strain components are constant within the element. The shape of the element in parametric coordinates and the mapping to the real space are shown in figure 7.27. For this mapping, the shape functions are identical to the parametric coordinates.

$$\begin{cases} N_1 = r \\ N_2 = s \\ N_3 = t \\ N_4 = 1 - r - s - t \end{cases} \tag{7.61}$$

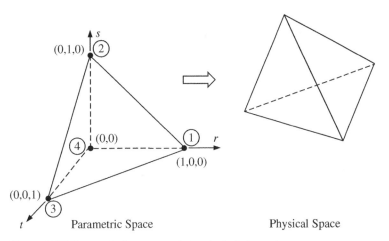

Figure 7.27 Four-node isoparametric tetrahedral element

For this element, the interpolation of the variable is linear within the element. For a field variable such as temperature, the interpolation for the three-dimensional element is no different than for two-dimensional elements.

$$T(r,s,t) = \sum_{i=1}^{4} N_i(r,s,t)T_i. \tag{7.62}$$

As the shape functions are linear, the physical quantities that are proportional to the gradient of the field variable such as strains, stresses, or heat flux are constant within the element. Therefore, the quality of the solution is not likely to be very good using this element especially when the strain is varying rapidly, for example, near regions with stress concentration. This element should be used with caution, and it is important to be aware that the solution can have large errors in stress/strain when mesh density is inadequate. As with the other parametric elements, the mapping from the parametric to the real space is accomplished using shape functions as

$$x(r,s,t) = \sum_{i=1}^{4} N_i(r,s,t)x_i,$$
$$y(r,s,t) = \sum_{i=1}^{4} N_i(r,s,t)y_i, \tag{7.63}$$
$$z(r,s,t) = \sum_{i=1}^{4} N_i(r,s,t)z_i.$$

As with the 3-node triangular element, the 4-node tetrahedron has constant strain within the element and is, therefore, a constant strain tetrahedron (CST). Tetrahedral elements are popular in finite element analysis software due to the ease of mesh generation compared to hexahedral elements. Automated mesh generation is available in many finite element analysis programs for tetrahedral elements. But the 4-node element is not very accurate, so the 10-node quadratic tetrahedral element should be used when more accurate results are needed especially for stresses or heat flux.

7.7.2   10-Node (Quadratic) Tetrahedral Element

The quadratic tetrahedral element is shown in figure 7.28. As shown in the figure, it has nodes at the midpoint of all the edges in addition to the nodes at the vertices. The shape functions for this element can be defined using the shape functions of the 4-node tetrahedral element: $r,s,t,u = 1-r-s-t$.

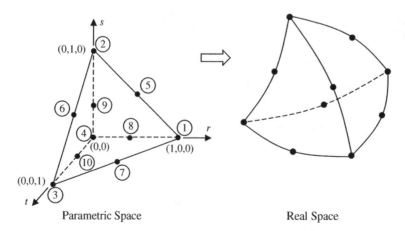

Parametric Space   Real Space

Figure 7.28 Ten-node isoparametric tetrahedral element

As in the previous elements, the shape functions can be derived easily using the Kronecker's delta properties. For example, the shape function for node 1 is zero along the planes $r = 0$ and $r = 1/2$, and it is unity at node 1. The shape functions are:

$$
\begin{cases}
N_1 = r(2r-1) \\
N_2 = s(2s-1) \\
N_3 = t(2t-1) \\
N_4 = u(2u-1) \\
N_5 = 4rs \\
N_6 = 4st \\
N_7 = 4rt \\
N_8 = 4ru \\
N_9 = 4su \\
N_{10} = 4tu
\end{cases}
\tag{7.64}
$$

The interpolation of field variables for this element will yield a quadratic function within the element. Therefore, we expect the gradients and quantities related to gradients such as strains and stresses to be linearly varying within the element. The shape of the element in the parametric space is its ideal shape, and any distortion of this shape in the physical space will reduce the accuracy of the numerical integration.

7.7.3   Eight-Node Hexahedral (Brick Element)

The 8-node hexahedral element is the three-dimensional analog of the two-dimensional 4-node quadrilateral element. The hexahedral elements are often referred to as brick elements. This 8-node element has shape functions that are trilinear, that is, linear along its edges but not internally. All eight shape functions of this element can be expressed using the following equation in index notation for $i = 1$ to 8.

$$
N_i(r,s,t) = \frac{1}{8}(1 + r_i r)(1 + s_i s)(1 + t_i t).
\tag{7.65}
$$

In the preceding equations (r_i, s_i, t_i) are the parametric coordinates of the nodes of the elements. As shown in figure 7.29, in the parametric coordinates (r, s, t) the element is a cube of size $2 \times 2 \times 2$ with its centroid at the origin while in the real space it is a hexahedron where the opposite faces and edges are

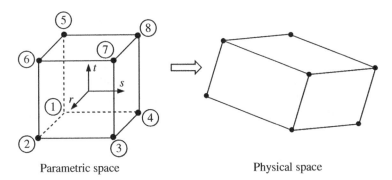

Parametric space   Physical space

Figure 7.29 Eight-node isoparametric hexahedral element

not necessarily parallel to each other. For the element numbering shown in the parametric space, the coordinate of node-1 in parametric space is $(-1, -1, -1)$ and the coordinates of the diagonally opposite node-7 are $(1,1,1)$.

The interpolation of the field variables for all the 3D elements described in this section is similar to other isoparametric elements yielding an interpolation in the r-s-t coordinate system. For solid mechanics, the three components of the displacement field are interpolated in the parametric space as:

$$u(r,s,t) = \sum_{i=1}^{8} N_i(r,s,t)u_i,$$

$$v(r,s,t) = \sum_{i=1}^{8} N_i(r,s,t)v_i, \tag{7.66}$$

$$w(r,s,t) = \sum_{i=1}^{8} N_i(r,s,t)w_i.$$

Here again, the summation is over the number of nodes per element (n_e). The mapping between parametric and real space is again constructed using the shape functions so that the edges of a linear element are straight lines while the edges of a quadratic element are parabolic.

$$x(r,s,t) = \sum_{i=1}^{8} N_i(r,s,t)x_i,$$

$$y(r,s,t) = \sum_{i=1}^{8} N_i(r,s,t)y_i, \tag{7.67}$$

$$z(r,s,t) = \sum_{i=1}^{8} N_i(r,s,t)z_i.$$

The 8-node hexahedral element is superior to the 4-node tetrahedral element due to the fact that strains are not constant within this element, and therefore better-quality results are obtained. However, automated mesh generation is more difficult for hexahedral elements. Earlier we discussed how a triangular element can be created by collapsing two nodes of a 4-node quadrilateral element. Similarly, it is possible to create a pyramid element by collapsing the four nodes on any face of an 8-node hexahedron into a single node. A wedge-shaped element can be created by collapsing the two nodes on the opposite edges of a face.

Higher-order hexahedral elements can be derived using the same techniques that we discussed earlier for quadrilateral elements. Popular elements include 20-node serendipity element and 27-node triquadratic elements analogous to the 8-node serendipity and 9-node bi-quadratic elements that we discussed for two dimensions. For such elements, it is important that the mesh generator places the nodes on the edges of the element at the midpoint of the edge to avoid distorting the mapping of the element from parametric to real space. For all isoparametric elements, the order of the nodes in the connectivity table should also match the order in which the nodes are numbered in parametric space.

Mesh generation is more difficult for hexahedral elements because fully automated mesh generators are rarely available. Traditional mesh generation programs require the user to subdivide the shape into simpler shapes and specify the number of nodes along the edges of the sub-divided regions. The quality of the solution deteriorates if the element has to be severely distorted to fit the geometry. Adaptive mesh refinement is also more difficult to implement for hexahedral elements. For these reasons, tetrahedral elements are still more popular than hexahedral elements for three-dimensional analysis.

7.7.4 Numerical Integration in Three-Dimensional Hexahedral Elements

For three-dimensional elements, we need to extend Gauss quadrature to three dimensions, which involves combining three one-dimensional Gauss quadrature formulas as shown below:

$$I = \int_{-1}^{1} \int_{-1}^{1} \int_{-1}^{1} f(r,s,t) \, dr \, ds \, dt \approx \sum_{k=1}^{n} \sum_{j=1}^{m} \sum_{i=1}^{l} w_i w_j w_k f(r_i, s_j, t_k), \qquad (7.68)$$

where r_i, s_j, and t_k are Gauss points, l is the number of Gauss points in the r direction, m is the number of Gauss points in s direction, and n is the number of Gauss points in the t direction, and w_i, w_j, and w_k are the Gauss weights. The total number of Gauss points is therefore $l \times m \times n$.

The element stiffness matrix for an 8-node hexahedral element can be evaluated using the $2 \times 2 \times 2$ Gauss quadrature formula:

$$\left[\mathbf{k}^{(e)} \right] = \int_{-1}^{1} \int_{-1}^{1} \int_{-1}^{1} [\mathbf{B}]^T [\mathbf{C}] [\mathbf{B}] |\mathbf{J}| \, dr \, ds \, dt$$

$$\approx \sum_{i=1}^{2} \sum_{j=1}^{2} \sum_{k=1}^{2} w_i w_j w_k \left[\mathbf{B}(r_i, s_i, t_j) \right]^T [\mathbf{C}] \left[\mathbf{B}(r_i, s_i, t_j) \right] |\mathbf{J}(r_i, s_i, t_j)|. \qquad (7.69)$$

The Jacobian matrix for 3D elements is defined as:

$$[\mathbf{J}] = \begin{bmatrix} \dfrac{\partial x}{\partial r} & \dfrac{\partial y}{\partial r} & \dfrac{\partial z}{\partial r} \\[2mm] \dfrac{\partial x}{\partial s} & \dfrac{\partial y}{\partial s} & \dfrac{\partial z}{\partial s} \\[2mm] \dfrac{\partial x}{\partial t} & \dfrac{\partial y}{\partial t} & \dfrac{\partial z}{\partial t} \end{bmatrix}. \qquad (7.70)$$

As in the case of 2D problems, the determinant of the Jacobian is the ratio of the volumes in real space and parametric space. The Jacobian matrix is, in general, a function of the coordinates and is not constant within the element. However, if the element is regular shaped (cubes or cuboid) and the element edges are aligned with the coordinate system, then this matrix will be constant and diagonal. The numerical integration is exact only if the integrand is a polynomial. If the element is distorted and the Jacobian is not constant, then the integrand is not a polynomial because the inverse of the Jacobian will be a rational function (a ratio of polynomials). The more distorted the element is, the less accurate the numerical integration is. This is the reason why the quality of the mesh is very important to ensure the accuracy of the results. Ideally, the elements should be as close to cuboids as possible for a three-dimensional hexahedral mesh.

7.8 FINITE ELEMENT MODELING PRACTICE FOR ISOPARAMETRIC ELEMENTS

In this section, we consider several examples where we explore some of the practical applications of the elements discussed in this chapter to model solids and structures. Many commonly used practices for selecting the right model and boundary conditions, as well as method for simplifying the model, are discussed.

7.8.1 U-Shaped Beam

The U-shaped aluminum frame shown in figure 7.30 has a $20 \times 20 \text{ mm}^2$ square cross section. A load P equal to a magnitude of 5 kN is located at a distance of 40 mm from the center of curvature of the curved

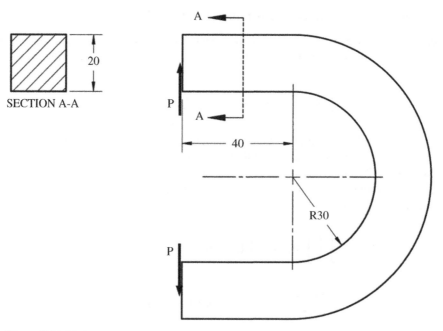

Figure 7.30 U-shaped beam

portion of the frame. The inner radius of the curved beam is 30 mm. The Young's modulus of the material is 69 GPa, and Poisson's ratio is 0.33. Describe the model you would use for this structure. Compute the deflection due to the applied load and the maximum/minimum stresses in the structure.

The structure has a uniform thickness in the z-direction and the loads are in the x-y plane, so clearly, this structure can be modeled using 2D elements as a plane stress problem. The structure and the loads are symmetric about the mid-plane of the structure, so a plane stress model of half the structure can be used as shown in figure 7.31.

The load is applied as a shear stress (traction parallel to the face) at the tip of the structure, and a sliding boundary condition is used along the symmetry face. The sliding boundary condition is necessary to allow the width to change due to Poisson's effect. In this model, sliding implies that the nodes on the edge are fixed in the vertical direction and free in the horizontal direction. In addition to this sliding boundary condition, we need some constraint to prevent rigid body motion in the horizontal direction.

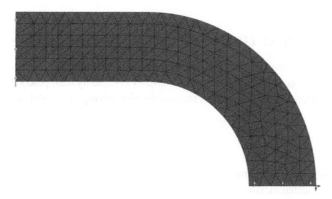

Figure 7.31 Plane stress model of U-shaped beam

(a)

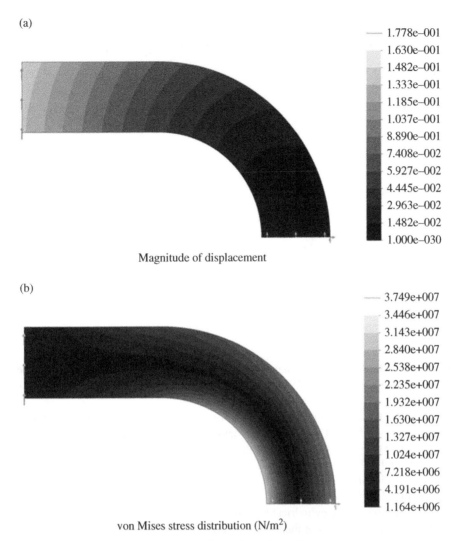

	1.778e–001
	1.630e–001
	1.482e–001
	1.333e–001
	1.185e–001
	1.037e–001
	8.890e–001
	7.408e–002
	5.927e–002
	4.445e–002
	2.963e–002
	1.482e–002
	1.000e–030

Magnitude of displacement

(b)

	3.749e+007
	3.446e+007
	3.143e+007
	2.840e+007
	2.538e+007
	2.235e+007
	1.932e+007
	1.630e+007
	1.327e+007
	1.024e+007
	7.218e+006
	4.191e+006
	1.164e+006

von Mises stress distribution (N/m^2)

Figure 7.32 Deflection and stress distribution in U-shaped beam

This can be achieved by fixing a point along the symmetry plane. In this model, we have fixed the vertex at the right end of the sliding edge.

Figure 7.32 shows the deflection and von Mises stress distribution in the structure. As expected, the maximum deflection is at the tip where the load is applied, and the maximum stress is at the symmetry plane where the vertical displacement is constrained. The stress varies through the thickness of the beam, so it is important to have sufficient elements through the thickness to capture this stress distribution accurately.

7.8.2 Plate with Holes

A steel plate with holes as shown in figure 7.33 serves as a bracket and is bolted at the top two holes and supports a load via a pin that goes through the lower hole. The dimensions of the plate are as shown in figure 7.33, and its material properties are Young's modulus = 200 GPa and Poisson's ratio = 0.3.

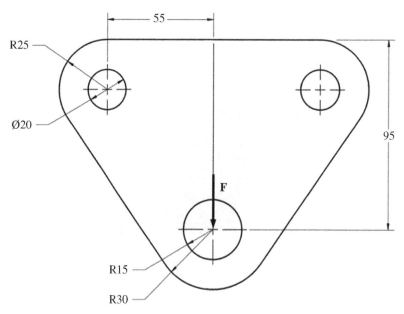

Figure 7.33 Plate with holes

The vertical load applied through the pin is 5000 N. Compute the maximum deflection and peak stress in the structure due to this applied load. What would be the appropriate model if the load were normal to the plate at the pin?

The plate geometry and the applied loads are symmetric; therefore, half of this structure can be modeled as a plane stress problem. The two bolted holes can be treated as complete fixed if it is in fact very tightly bolted so that the plate is not likely to rotate about the bolts during its deformation when the load is applied. On the other hand, if it is likely to rotate, then they need to be modeled as hinges. In some finite element analysis software programs, the fixed hinge is a predefined boundary condition that can be applied to cylindrical surfaces. This is equivalent to using a cylindrical coordinate system with its origin and z-axis along the axis of the cylindrical hole and then fixing the radial and axial components of displacement with respect to this coordinate system while allowing tangential displacement that causes rotation about the z-axis. In this example, we will study both options to see how it affects the solution.

The load that is applied to the plate is clearly not going to be uniformly distributed along the surface of the hole. The pin will transmit the load through contact over the lower half of the hole only since it cannot pull on the upper half and furthermore the contact pressure distribution is not uniform over this region. The contact load distribution is often assumed to be parabolic, and some commercial finite element analysis software programs provide a "bearing load" option that will automatically apply this type of load distribution on cylindrical holes in the specified direction. Figure 7.34 shows the pressure

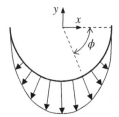

$$p(\phi) = p_0 \sin^2 \phi \, \cos\phi \hat{\mathbf{i}} - p_0 \sin^3\phi \hat{\mathbf{j}}$$ **Figure 7.34** Pressure distribution $p(\phi)$ for bearing load

distribution typically assumed for the bearing load over the lower half of the hole that supports the pin. In this example, we will explore how much the deformation of the plate would differ if the load is applied as a bearing load versus applying it as a uniform traction along the entire cylindrical surface.

Figure 7.35 shows the finite element model that can be used for this analysis where one half of the symmetric plate is modeled. Along the symmetry plane, the sliding boundary condition is applied so that the nodes are able to slide in the vertical direction while being fixed in the horizontal or x-direction. As discussed above, the bolted hole was modeled both as fixed and later as hinged to compare the resulting deflection of the plate. Similarly, the load was applied both as uniformly distributed traction and as the parabolic bearing load in the vertical direction for comparison. As we are modeling one half of the plate, a load of 2500 N is applied, which is half the total load on the plate.

Figure 7.36(a) and (b) shows the plot of the deflection of the plate for uniform and bearing loads respectively. In both cases, the bolt hole is allowed to rotate about its axis. The von Mises stress distributions for these two models are shown in figure 7.37.

As can be seen in these figures, the results are significantly different when the load is applied differently even though the total magnitude of the applied load is the same. There is more deflection at the hole where the load is applied. The location where the highest stress occurs is also different. The highest stress is mostly localized near the load-application hole when the bearing load is applied whereas it is highest near the bolted hole when the load is applied as uniformly distributed.

The results are summarized in table 7.4 where the maximum displacement and the maximum von Mises stresses are listed for the various options for boundary conditions and load application. As discussed earlier, there is a significant difference in displacement and stresses for different load distributions, but the results do not differ much between fixing the bolt hole versus allowing it to rotate. As expected, we get a slightly larger maximum displacement when the bolt hole is allowed to rotate compared to when it is fixed.

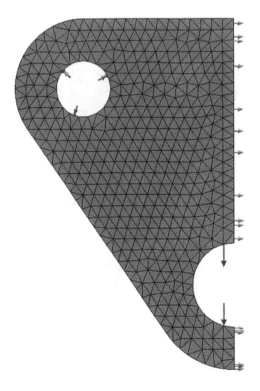

Figure 7.35 Finite element model of plate with holes

(a)

(b)

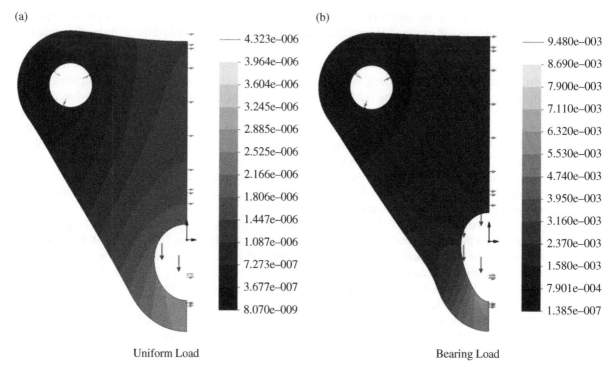

Uniform Load

Bearing Load

Figure 7.36 Deflection (mm) due to uniform load versus bearing load

(a)

(b)

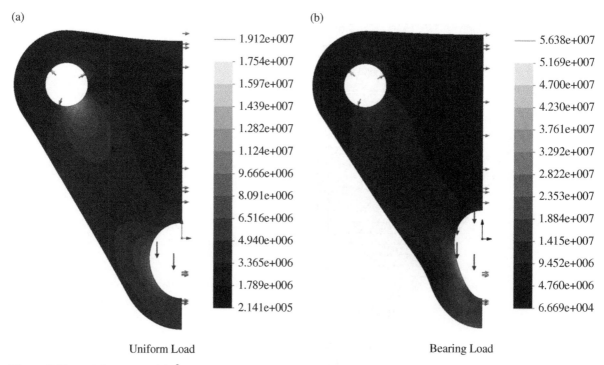

Uniform Load

Bearing Load

Figure 7.37 von Mises stress (N/m^2) due to uniform load versus bearing load

Table 7.4   Results for the plate with holes

Result	Uniform load and Rotating at bolt	Bearing load and Rotating at bolt	Bearing load and Fixed at bolt
Max. Displacement (mm)	4.323×10^{-3}	9.479×10^{-3}	9.277×10^{-3}
Max. Stress (Von-Mises, MPa)	19.12	56.40	56.49

(a)   (b)

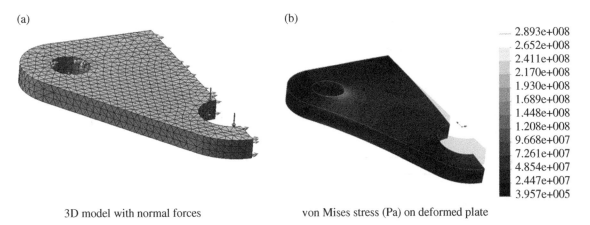

2.893e+008
2.652e+008
2.411e+008
2.170e+008
1.930e+008
1.689e+008
1.448e+008
1.208e+008
9.668e+007
7.261e+007
4.854e+007
2.447e+007
3.957e+005

3D model with normal forces   von Mises stress (Pa) on deformed plate

Figure 7.38 Plate with normal forces

If the load were normal to the plate, then it can no longer be modeled as a two-dimensional plane stress problem. The geometry alone does not determine the type of model that is needed. Obviously, the applied load plays an important role since the deformation due to this load will now be normal to the plate and no longer in the plane of the plate. In this case, this plate could be modeled using a three-dimensional finite element model as shown in figure 7.38 where the load is applied as a shear force along the surface of the hole assuming that the pin is press fitted or has a flange to transmit the load.

As the plate bends, the stresses vary linearly through the thickness of the plate (as in a beam), and therefore several elements are needed through the thickness. In this case, ideally quadratic tetrahedrons or even higher-order elements must be used to ensure accurate results. If this plate were very thin, as in a sheet metal where thickness may be several orders of magnitude smaller than the overall dimensions, then it becomes impractical to use three-dimensional elements. In that case, one should use plate or shell elements that are based on plate/shell theory. These elements are beyond the scope of this text book.

7.8.3   The 3D Bracket

An alloy steel L-shaped bracket, as shown in figure 7.39, is bolted to a rigid wall through the holes on its upright side. A stepped shaft going through the 140-mm hole rests on this bracket and applies a 10,000-N vertical load on the bracket. Assume that the load is transmitted through an annular region around the hole that has a radial width of 10 mm. Compute the maximum stress due to this load assuming that the material properties are as follows: $E = 2.1 \times 10^{11}$ Pa and $\nu = 0.28$.

The geometry of this bracket is non-planar, and therefore a 2D model is obviously not applicable for this example. We will assume that the bolt joints rigidly fix the holes, and since it is against a rigid wall, we will fix the back side of the bracket as well. The geometry and the loads are symmetric, so one could create a model using half the geometry, but the whole geometry is modeled here for better visualization of the results. To apply the load, the face supporting the load is split to create an annular region as shown

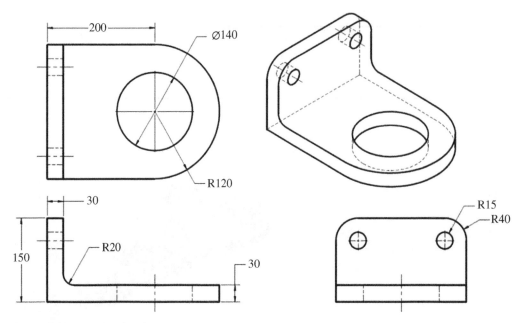

Figure 7.39 A 3D bracket drawing

in figure 7.40. Most CAD software programs provide the capability to split a face by sketching a region on it. The mesh generator then automatically creates a mesh that includes the edges of the faces as shown in the mesh in figure 7.41(a). This allows a normal pressure to be applied on the face.

The initial mesh and the computed von Mises stresses are shown in figure 7.41(a) and (c), respectively. There is a stress concentration at the rounded edge, which is not accurately calculated by this default mesh. Increasing the mesh density will cause the computed stress at this corner to increase. One way to increase the mesh density is to specify a smaller element size, but doing so for the entire mesh is inefficient. Therefore, most FEA programs provide various methods to locally refine the mesh

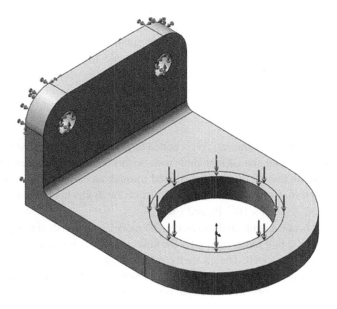

Figure 7.40 A 3D bracket loads and boundary conditions

(a)

(b)

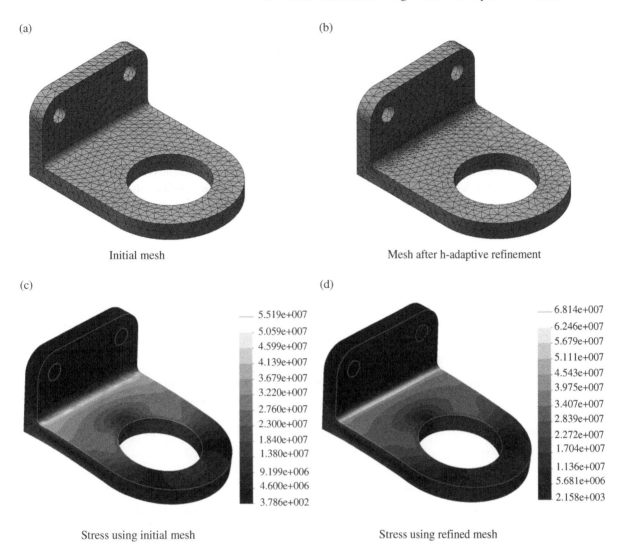

Initial mesh

Mesh after h-adaptive refinement

(c)

5.519e+007
5.059e+007
4.599e+007
4.139e+007
3.679e+007
3.220e+007
2.760e+007
2.300e+007
1.840e+007
1.380e+007
9.199e+006
4.600e+006
3.786e+002

(d)

6.814e+007
6.246e+007
5.679e+007
5.111e+007
4.543e+007
3.975e+007
3.407e+007
2.839e+007
2.272e+007
1.704e+007
1.136e+007
5.681e+006
2.158e+003

Stress using initial mesh

Stress using refined mesh

Figure 7.41 h-adaptive mesh refinement for 3D bracket

where needed. For example, a smaller element size can be specified on the rounded face at the corner so that the mesh generator will create a mesh where the element size is small near the corner and gradually increases to the default size away from this rounded corner. Alternatively, one could use adaptive mesh generation schemes where the software automatically determines regions with high error and refines the mesh only in these regions. The error can be estimated in several ways such as checking how well the solution satisfies the equilibrium equations or by the magnitude of the discontinuity in the derivatives of the displacement between elements. As stated earlier, there are two approaches that are popular for adaptive refinement: p-adaptive and h-adaptive. In the p-adaptive refinement, the order of the elements is raised in regions with high error. Here we will use h-adaptive refinement, where the elements are subdivided into smaller elements in regions where the error is high. Figure 7.41(b) and (d) shows the locally refined mesh obtained by h-adaptive refinement and the corresponding computed von Mises stress distribution, respectively. The refined mesh is able to more accurately determine the stress at the rounded edge, and the computed stresses are therefore much higher than those computed with the default mesh. Notice that the elements at the round are extremely small, and it would not be efficient or practical to use such small elements through the entire structure. The h-adaptive refinement determines that the solution is not accurate at the corner with the initial mesh and automatically subdivides the elements in this region until the desired accuracy is achieved.

7.9   PROJECTS

Project 7.1   Accuracy and convergence analysis of a plate with a hole

The aluminum plate with a central hole is subjected to uniform uniaxial stress as shown in figure 7.42. Use a commercial finite element analysis software to determine the deformed shape, and calculate the maximum von Mises stress and stress concentration factor. Properties of aluminum are: $E = 70$GPa and $\nu = 0.33$. Dimensions of the plate is: $200\,\text{mm} \times 100\,\text{mm} \times 5\,\text{mm}$. The diameter of the hole $= 40\,\text{mm}$. Remote uniaxial stress $= 50\,\text{MPa}$.

PART I

(a) Use triangular elements only. Repeat your calculations with the number of nodes $N = K$, $2\,K$, and $4\,K$, where $4\,K$ is approximately the maximum number of nodes you can use. Plot the von Mises stress contour for each case. Plot the deformed shape. You may like to use a magnification factor so that the deformed shape is distinct from the original model.

(b) Repeat (a) using quadrilateral elements.

Plot the following graphs: (i) stress concentration factor vs. $\log N$; (ii) maximum von Mises stress vs. $\log N$; (iii) u_A vs. $\log N$, where u_A is the displacement in the loading direction at point A with respect to the hole center. Compare your stress concentration factors with the theoretical value. What is the factor of safety for the plate according to von Mises yield criterion? The yield stress of aluminum $= 280\,\text{MPa}$.

Project 7.2   Design of a torque–arm

A torque arm shown in figure 7.43 is under horizontal and vertical loads transmitted from a shaft at the right hole, while the left hole is fixed. Assume: Young's modulus $= 206.8$ GPa, Poisson's ratio $= 0.29$, and thickness $= 1.0$ cm.

1. Provide a preliminary analysis result that can estimate the maximum von Mises stress.

2. Using plane stress elements, carry out finite element analysis for the given loads. Clearly state all assumptions and simplifications that you adopted in modeling. Carry out convergence study and determine the size of elements for a reasonably accurate solution.

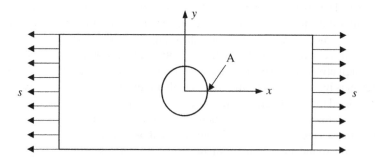

Figure 7.42 Cantilever beam model

All dimensions in cm

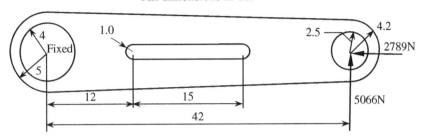

Figure 7.43 Dimensions of torque arm model

7.10 EXERCISES

1. Answer the following descriptive questions.

   (a) Explain what the "isoparametric" mapping means.

   (b) For a 1D 3-node quadratic element, what is the condition to make the Jacobian constant?

   (c) When quadratic elements are used, will displacements be continutous across the element boundary with the adjacent element? Will strains be continuous?

   (d) The stress in a bar varies linearly. When three nodes are equally spaced in the bar, which result is more accurate, two linear elements or one quadratic element?

   (e) What will happen if the shape of an element is distorted in a such a way that there is a concave region within the element?

   (f) List possible problems when the Jacobian of an element is negative.

   (g) How many Gauss quadrature points are required to accurately integrate a polynomial of order 7?

   (h) How many zero eigenvalues are expected for the properly integrated stiffness matrix of a quadrilateral element?

   (i) When the stiffness matrix of a quadrilateral element has four zero eigenvalues, explain the possible cause of the problem.

   (j) Explain why it is not acceptable to put a 4-node quadrilateral element adjacent to an 8-node quadrilateral element in a finite element mesh.

   (k) Why is numerical quadrature needed for 2D and 3D elements?

   (l) What are the main sources of error in the solutions obtained using isoparametric elements?

2. The nodal coordinates and corresponding displacements in a plane triangular element shown in the figure are given in the table below. Use isoparametric formulation where $N_1 = s$, $N_2 = t$, and $N_3 = r = 1 - s - t$

Node number	(x,y) (mm)	u (mm)	v (mm)
1	(3,0)	1	1
2	(0,3)	2	0
3	(0,0)	0	0

   (a) What is the parametric coordinates of the point P whose real coordinates are $(x,y) = (1,1)$?

   (b) Calculate the displacement vector at this point.

   (c) Calculate the strain at this point.

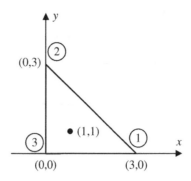

3. The quadrilateral element shown in the figure has the nodal displacement of $\{u_1, v_1, u_2, v_2, u_3, v_3, u_4, v_4\} = \{-1, 0, -1, 0, 0, 1, 0, 1\}$.

   (a) Find the (s, t) reference coordinates of point A $(0.5, 0)$ using the isoparametric mapping method.

   (b) Calculate the displacement at point B whose reference coordinate is $(s,t) = (0,-0.5)$

   (c) Calculate the Jacobian matrix $[\mathbf{J}]$ at point B.

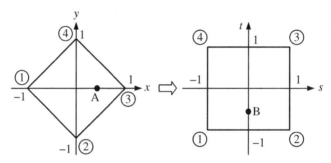

Physical element   Reference element

4. A four–node quadrilateral element is defined as shown in the figure.

   (a) Find the coordinates in the reference element corresponding to $(x, y) = (0, 0.5)$.

   (b) Calculate the Jacobian matrix as a function of s and t.

   (c) Is the mapping valid? Explain your answer.

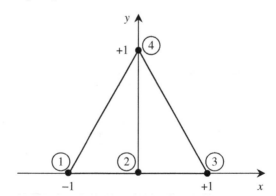

5. Consider the isoparametric quadrilateral element shown in figure 7.9(a).

   (a) Calculate the (s,t) coordinates of the point $(x,y) = (1.3, 0.6)$, which lies on the 2-3 edge.

   (b) Given that the displacements at all the nodes are zero except $v_3 = 0.02$ at node 3, calculate the value of v at the point $(x,y) = (1.3, 0.6)$.

6. For the rectangular element shown determine the strain ε_{xx} at the centroid of the element using the isoparametric formulation. The nodal coordinates in the figure are in meters. The nodal displacements given in the table are also in meters.

Node Number	1	2	3	4
u (m)	0	1	2	0
v (m)	0	1	2	0

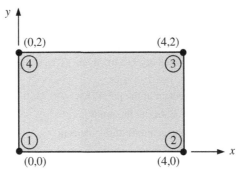

7. For the 4-node *plane strain* finite element, finite element computation yields the displacements at the nodes as follows: $\{u_1, v_1, u_2, v_2, u_3, v_3, u_4, v_4\} = \{-1, 0, -1, 0, 0, 1, 0, 1\} \times 10^{-4}$. The connectivity table is as shown in the table.

Element	Local node 1	Local node 2	Local node 3	Local node 4
i	5	7	16	14

The local and global node numbering are shown in the physical and parametric spaces, respectively, in the figure above.

(a) Where is the point $(s, t) = (0.0, 0.0)$, located in the global (x, y) coordinates?

(b) Determine the Jacobian matrix for this element.

(c) Find the derivate of u with respect to x and y, $(\partial u/\partial x, \partial u/\partial y)$ at the centroid.

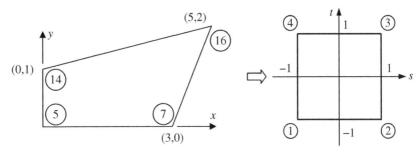

Physical element   Reference element

8. A quadrilateral element in the figure is mapped into the parent element.

(a) A point P has a coordinate $(x, y) = (\tfrac{1}{2}, y)$ in the physical element and $(s, t) = (-\tfrac{1}{2}, t)$ in the parent element. Find the y and t coordinates of the point using isoparametric mapping.

(b) Calculate the Jacobian matrix *at the center* of the element.

(c) Is the mapping valid? Explain your answer.

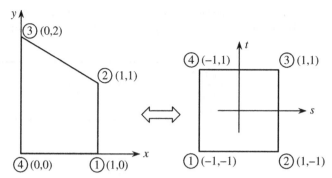

9. Consider the plane stress 4–node element shown in the figure. Its global node numbers are shown in the figure with the element connectivity in the table. The coordinates of the nodes in the global x-y coordinate system is shown next to each node. Nodal displacements are given as $\{\mathbf{q}\}^{\mathrm{T}} = \{u_{51}, v_{51}, u_{52}, v_{52}, u_{63}, v_{63}, u_{64}, v_{64}\} = \{0, 0, 0.1, 0, 0.1, 0.1, 0, 0\}$.

   (a) Determine the displacement at the point $(x, y) = (0.75, 0.75)$ by interpolating the nodal displacements.

   (b) Compute the Jacobian matrix at the point in (a).

   (c) Compute strain ε_{yy} at the centroid of the element.

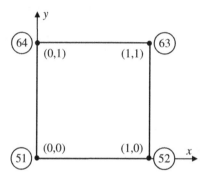

Element #	Local node 1	Local node 2	Local node 3	Local node 4
27	51	52	63	64

10. A linearly varying pressure p is applied along the edge of the 4–node element shown in the figure. The finite element method converts the distributed force into an equivalent set of nodal forces $\{\mathbf{F}^e\}$ such that

$$\int_S \mathbf{u}^{\mathrm{T}}\mathbf{T}\,dS = \left\{\mathbf{q}^{(e)}\right\}^{\mathrm{T}}\left\{\mathbf{F}^{(e)}\right\},$$

where $\mathbf{T}$ is the applied traction (force per unit area), and $\mathbf{u}$ is the vector of displacements. Since the applied pressure is normal to the surface (in the x-direction), the traction can be expressed as $\mathbf{T} = \{p, 0\}^T$, where p can be expressed as $p = p_0(t + 1)/2$, where $t = -1$ at node 1 and $t = +1$ at node 4. The length of the edge is L. Integrate the left-hand side of the above equation to compute the work–equivalent nodal forces $\{\mathbf{F}^{(e)}\}$ when $\{\mathbf{q}^{(e)}\}^{\mathrm{T}} = \{u_1, v_1, u_2, v_2, u_3, v_3, u_4, v_4\}$.

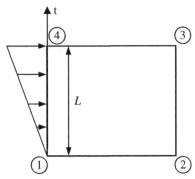

11. Determine the Jacobian matrix for the following isoparametric elements. If the temperatures at the nodes of both above elements are $\{T_1, T_2, T_3, T_4\} = \{100, 90, 80, 90\}$, compute the temperature at the midpoint of the element and at the mid point of the edge between connecting nodes 1 and 4.

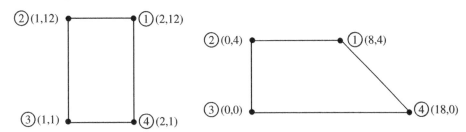

12. Integrate the following function using one-point and two-point numerical integration (Gauss quadrature). The exact integral is equal to 2. Compare the accuracy of the numerical integration with the exact solution.

$$I = \int_0^{\pi} \sin(x) \, dx.$$

13. Use 1×2 Gauss rule to evaluate the integral $I = \iint xy^2 dxdy$ over the rectangular region shown in the figure. Explain why the above estimate is or is not exact.

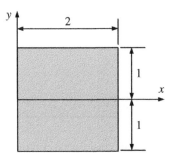

14. A six-node finite element as shown in the figure is used for approximating the beam problem.

    (a) Write the expressions of displacements $u(x,y)$ and $v(x,y)$ in terms of polynomials with unknown coefficients. For example, $u(x,y) = a_0 + a_1 x + \cdots$.

    (b) Can this element represent the pure bending problem accurately? Explain your answer. A bending moment M is applied to the edge 2–3.

    (c) Can this element represent a uniformly distributed load problem accurately? The distributed load q is applied along the edge 4–6–3.

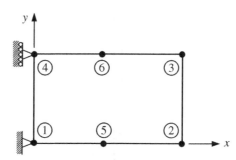

15. Consider a quadrilateral element shown in the figure below. The nodal temperatures of the element are given as $\{T_1, T_2, T_3, T_4\} = \{80, 40, 40, 80\}$.

    (a) Compute the expression of the temperature T along the line ξ that connects nodes 3 and 1. For example, $T = 3 + 5\xi + 3\xi^2 + \cdots$. You can assume that $\xi = 0$ at node 3 and $\xi = 1$ at node 1. Plot the graph of $T(\xi)$ with respect to ξ.

    (b) Compute the temperature gradient $\partial T/\partial x$ at the center of the element.

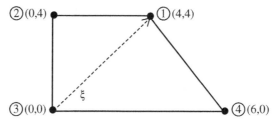

16. Compute the work (or energy) equivalent nodal forces that correspond to the uniformly distributed load of $T_y = 20$ kN/m assuming that the element is a 9-node quadrilateral plane strain element. Use Gauss quadrature for the integration.

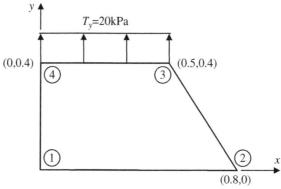

17. The figure below shows a 4-node quadrilateral element in parametric space and in real space where two of the nodes are collapsed or merged to form a triangle.

    (a) Determine the mapping functions $x(s,t)$ and $y(s,t)$.

    (b) The displacement at the nodes of the element are found to be: $\{u_1, v_1, u_2, v_2, u_3, v_3\} = \{1, 0, 0, 2, 0, 0\}$ $\times 10^{-3}$ in. Compute the displacement field $u(s,t)$ and $v(s,t)$.

    (c) Compute the Jacobian matrix.

    (d) Compute the strain in the element. Is this is a constant strain element?

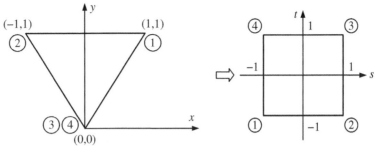

Physical element   Reference element

18. An 8-node element is shown in the figure in the parametric space.

   (a) Derive the shape functions for this element.

   (b) If in the real (x,y) space, an element of this type is square and has side length of 5 m, write the Jacobian matrix for this element. Assume that the edges of the element are parallel or perpendicular to the x- and y-axis and that the mid-edge nodes are at the midpoint of the edge. *Hint:* You do not need the shape functions.

   (c) What is the order of Gauss quadrature that is appropriate to evaluate the stiffness matrix of this element? Will this integration be exact for the element described in (b) using this order of quadrature. Explain.

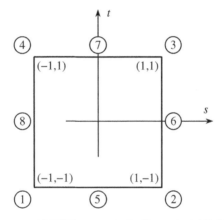

Chapter 8

Finite Element Analysis for Dynamic Problems

8.1 INTRODUCTION

In the previous chapters, we have discussed the static analysis of solids and structures, in which the loads are assumed to be applied slowly such that the structure is in equilibrium at every instant until the full load is applied. When the effect of velocity and acceleration can be ignored due to slow application of loads, it is called a quasi-static loading. The structure may be considered to be under *quasi-static* deformation when the inertial forces are orders of magnitude smaller than the internal forces caused by the deformation. The structure gradually deforms to produce internal forces (i.e., stresses) so that the internal forces are in equilibrium with the externally applied load. In structural mechanics, this is called *static equilibrium*. We did not consider the time involved, that is, how quickly the load is applied or how quickly the structure reaches equilibrium with the applied loads. However, when the loads vary rapidly with time, the inertial effects cannot be neglected. In such cases, the quasi-static assumption is not valid and we need to compute the *dynamic equilibrium*, which is the main topic of this chapter. Dynamic problems are concerned with the study of motion of structures under external loads. When the external loads are time varying, or steady but applied suddenly, it will trigger dynamic effects that cause structural vibration immediately upon the application of the load. If the load is steady, then due to the presence of intrinsic damping in most materials, these vibrations will eventually die down and the structure will settle at the equilibrium position where the displacements and stresses are as calculated in static analysis. However, before reaching this equilibrium, the maximum displacements and stresses during the vibration can be significantly higher than the corresponding static values and could cause catastrophic failure.

Dynamic problems can be classified into three broader groups: Natural vibration, forced vibration, and wave propagation. The natural vibration of a structure is characterized by its natural frequencies and mode shape of vibration. These natural modes of vibration characterize the dynamic behavior of a structure and are of great value in dynamic analysis. Some forced vibration analysis techniques use the natural modes and frequencies to compute the forced response of the system. Forced vibration problems can be further divided into impact loading problems and periodic loading problems. When intense loads are suddenly applied for a short period, as in an impact problem, the response during a short initial period immediately after the impact is of importance. Whereas, when a structure is subjected to periodically fluctuating loads, we are interested in the *steady-state response* which is the steady periodic response of the structure after the transient effects have dissipated due to damping. Different analysis procedures are needed for these two cases in order to make the computation efficient.

Introduction to Finite Element Analysis and Design, Second Edition. Nam H. Kim, Bhavani V. Sankar, and Ashok V. Kumar.
© 2018 John Wiley & Sons Ltd. Published 2018 by John Wiley & Sons Ltd.
Companion website: www.wiley.com/go/kim/finite_element_analysis_design

Among many different dynamic problems, we will focus on problems where the overall dynamic response of a structure is sought rather than a relatively local response, such as shock propagation. That is, we are interested in inertial problems, where wave effects such as focusing, reflection, and diffraction are not important. In this chapter, we will discuss only the computation of the natural modes of vibration and forced response of structures. In all vibration problems, damping is an important practical consideration, and therefore we will also indicate approaches to include damping in the dynamic analysis of structures.

8.2 DYNAMIC EQUATION OF MOTION AND MASS MATRIX

8.2.1 Equation of Motion of Uniaxial Bars

In static analysis, since the applied loads are independent of time, the displacements in equilibrium do not vary with time. In dynamic analysis, however, the displacement at each point within a structure is a function of time. Thus, in dynamic finite element analysis, it is necessary to introduce the concept of nodal velocity or the rate of change of nodal displacements with respect to time, and nodal acceleration or the second time derivative of the nodal displacements. In the global level, the vectors of nodal displacements, velocities, an accelerations are denoted by $\{\mathbf{Q}\}$, $\{\dot{\mathbf{Q}}\}$, and $\{\ddot{\mathbf{Q}}\}$, respectively.

As shown in the previous chapters, static analysis involves two energy terms—the strain energy of the solid and the potential energy of the external loads. On the other hand, dynamic analysis involves another energy term—the kinetic energy of the solid. The kinetic energy of a structure is the energy that it possesses due to its motion. It is defined as the work needed to accelerate a structure of a given mass. Like the other two energy terms, the kinetic energy also has to be expressed in terms of nodal DOFs, which in this case are the nodal velocities.

Consider a uniaxial bar element with two nodes, as shown in figure 8.1. The two DOFs are $u_1(t)$ and $u_2(t)$, which are functions of time. In the rest of the chapter, it will be implied that the nodal DOFs such as displacements and velocities are functions of time. Even if the nodal DOFs are functions of time, it is assumed that their magnitudes are small enough so that the infinitesimal deformation assumption is still valid. Under this assumption, the expression of the strain energy and hence the stiffness matrix will remain the same as static analysis, which was shown in eq. (1.16) of chapter 1. The displacement field is interpolated using the same shape functions as in static analysis (eq. (2.34) in chapter 2):

$$u(x,t) = u_1(t)N_1(x) + u_2(t)N_2(x). \tag{8.1}$$

The displacement field is now a function of position x as well as time t. Note that the *interpolation functions* or the *shape functions* are independent of time; only the nodal DOFs vary with time. Therefore, the velocity of a cross section at position x can then be derived as

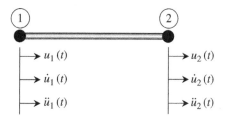

Figure 8.1 Uniaxial bar element in dynamic analysis

$$\dot{u}(t) = \frac{\partial u}{\partial t} = \dot{u}_1(t)N_1(x) + \dot{u}_2(t)N_2(x)$$

$$= \{\dot{u}_1 \ \ \dot{u}_2\} \left\{ \begin{array}{c} N_1(x) \\ N_2(x) \end{array} \right\} \tag{8.2}$$

$$= \{N_1(x) \ \ N_2(x)\} \left\{ \begin{array}{c} \dot{u}_1 \\ \dot{u}_2 \end{array} \right\},$$

where, the dot "." over the variable name denotes differentiation with respect to time t. Equation (8.2) implies that both displacement and velocity are interpolated using the same shape functions. If the velocity is differentiated one more time, it can be shown that the acceleration is also interpolated using the same shape functions. We have written the velocity expression as two different matrix products in eq. (8.2), which will be useful in the next step.

The *kinetic energy* of the element is given by

$$T^{(e)} = \frac{1}{2}\int_0^L \rho A \dot{u}^2 \mathrm{d}x. \tag{8.3}$$

In this equation, ρ is the mass *density* and A is the cross-sectional area of the element. We can replace the velocity term in the integral using the matrix products in eq. (8.2), to obtain

$$T^{(e)} = \frac{1}{2}\int_0^L \rho A \{\dot{u}_1 \ \ \dot{u}_2\} \left\{ \begin{array}{c} N_1(x) \\ N_2(x) \end{array} \right\} \{N_1(x) \ \ N_2(x)\} \left\{ \begin{array}{c} \dot{u}_1 \\ \dot{u}_2 \end{array} \right\} \mathrm{d}x. \tag{8.4}$$

Note that from the expression of $\dot{u}^2 = \dot{u} \cdot \dot{u}$, the two different expressions of $\dot{u}$ are used from eq. (8.2). This is because the vector cannot be squared, i.e., $\{\mathbf{u}\}^2$ is not a valid operation. Instead, $\{\mathbf{u}\}^T\{\mathbf{u}\}$ should be used.

In the expression of the kinetic energy, since the nodal velocities are independent of the spatial variable x, they can be moved out of the integral, resulting in

$$T^{(e)} = \frac{1}{2}\{\dot{u}_1 \ \ \dot{u}_2\} \int_0^L \rho A \left\{ \begin{array}{c} N_1(x) \\ N_2(x) \end{array} \right\} \{N_1(x) \ \ N_2(x)\} \mathrm{d}x \left\{ \begin{array}{c} \dot{u}_1 \\ \dot{u}_2 \end{array} \right\}. \tag{8.5}$$

The integral above can be evaluated using the shape functions $N_1(x) = (1 - x/L)$ and $N_2(x) = x/L$ to yield

$$T^{(e)} = \frac{1}{2}\{\dot{u}_1 \ \ \dot{u}_2\} \left[\mathbf{m}^{(e)}\right] \left\{ \begin{array}{c} \dot{u}_1 \\ \dot{u}_2 \end{array} \right\}, \tag{8.6}$$

where the 2×2 element mass matrix is derived as

$$\left[\mathbf{m}^{(e)}\right] = \rho A \int_0^L \left\{ \begin{array}{c} N_1(x) \\ N_2(x) \end{array} \right\} \{N_1(x) \ \ N_2(x)\} \mathrm{d}x$$

$$= \rho A \int_0^L \left[\begin{array}{cc} N_1^2 & N_1 N_2 \\ N_1 N_2 & N_2^2 \end{array} \right] \mathrm{d}x \tag{8.7}$$

$$= \frac{\rho A L}{6} \left[\begin{array}{cc} 2 & 1 \\ 1 & 2 \end{array} \right].$$

The above mass matrix is called the *consistent mass matrix* as it is derived using the principle of equivalent kinetic energy. Note that the total mass of the element is $\rho A L$. Therefore, the sum of all components in the mass matrix is the same as the mass of the element. Unlike the element stiffness matrix, the mass matrix of the element is positive definite. This is because the kinetic energy of the bar element is always positive for any nonzero velocities.

The kinetic energy of the bar is the sum of the kinetic energies of the elements:

$$T = \sum_{e=1}^{NEL} T^e = \sum_{e=1}^{NEL} \frac{1}{2} \{ \dot{u}_i \quad \dot{u}_j \} \left[\mathbf{m}^{(e)} \right] \begin{Bmatrix} \dot{u}_i \\ \dot{u}_j \end{Bmatrix} = \frac{1}{2} \{ \dot{\mathbf{Q}} \}^{\mathrm{T}} [\mathbf{M}] \{ \dot{\mathbf{Q}} \}, \tag{8.8}$$

where i and j are the first and second nodes of element e, NEL is the number of elements in the model, $\{ \dot{\mathbf{Q}} \}$ denotes the column vector of the global nodal velocities, and $[\mathbf{M}]$ is the global mass matrix obtained by assembling the element mass matrices. The assembly of mass matrix follows exactly the same steps as the assembly of the global stiffness matrix discussed in chapter 1.

Now we have all three energy terms given by[1]

$$U = \frac{1}{2} \{ \mathbf{Q} \}^{\mathrm{T}} [\mathbf{K}] \{ \mathbf{Q} \},$$

$$V = -\{ \mathbf{Q} \}^{\mathrm{T}} \{ \mathbf{F} \}, \tag{8.9}$$

$$T = \frac{1}{2} \{ \dot{\mathbf{Q}} \}^{\mathrm{T}} [\mathbf{M}] \{ \dot{\mathbf{Q}} \}.$$

As mentioned before, the stiffness and mass matrices are independent of time. Therefore, these matrices need to be constructed once. However, the applied load vector is given as a function of time, and accordingly, the vectors of nodal DOFs and velocities are also a function of time.

We use the *Lagrange equations*[2] to derive the finite element equations of motion. Similar to the total potential energy in chapter 2, the Lagrangian is defined for a dynamic problem as $L = T - \Pi = T - (U + V)$. For an N DOF system, the Lagrange equations are given by

$$\frac{\mathrm{d}}{\mathrm{d}t} \left(\frac{\partial L}{\partial \dot{Q}_i} \right) - \frac{\partial L}{\partial Q_i} = 0, \quad i = 1, \ldots, N. \tag{8.10}$$

Note that the Lagrange equations consider that $\{ \mathbf{Q} \}$ and $\{ \dot{\mathbf{Q}} \}$ are independent. Therefore, only the kinetic energy term is involved in $\partial L / \partial \dot{Q}_i$, while the strain energy and the potential energy of the external loads are involved in $\partial L / \partial Q_i$. By substituting for all three energy terms of L from eq. (8.9) into eq. (8.10), we obtain the following dynamic finite element equations of motion:

$$[\mathbf{M}] \{ \ddot{\mathbf{Q}} \} + [\mathbf{K}] \{ \mathbf{Q} \} = \{ \mathbf{F} \}. \tag{8.11}$$

In the above equation, $\{ \ddot{\mathbf{Q}} \}$ is the vector of global nodal accelerations. Since the acceleration is the second-order derivative of the displacement, the above equation is the system of second-order differential equations with respect to time. Methods for solving the above equations will be discussed in the rest of this chapter.

The equation of motion in eq. (8.11) is the second-order differential equation with respect to time. In order to solve for the differential equation, it is necessary to provide the initial conditions for the vectors of nodal displacements and velocities:

---

[1] In earlier chapters, we used $[\mathbf{K}_s]$ and $\{\mathbf{F}_s\}$ for structural matrices before applying boundary conditions, and $[\mathbf{K}]$ and $\{\mathbf{F}\}$ for global matrices after applying boundary conditions. However, here we do not distinguish these two cases for the simplicity of explanation.
[2] Virgin, Lawrence N. 2007. *Vibration of Axially Loaded Structures*. Cambridge University Press, New York.

$$\{\mathbf{Q}(t=0)\} = \{\mathbf{Q}_0\},$$
$$\{\dot{\mathbf{Q}}(t=0)\} = \{\dot{\mathbf{Q}}_0\}. \tag{8.12}$$

In addition to the initial conditions, the differential equations also need boundary conditions as with the static problems. In general, the boundary conditions are assumed independent of time. It is also assumed that the boundary conditions are given only for displacements. When the displacement is fixed, then the velocity and acceleration are automatically vanish. It is possible to have a prescribed motion, that is, the displacement, velocity or acceleration on the boundary are given as a function of time. However, the prescribed motion will not be considered here.

Unlike static problems, dynamic problems may or may not have displacement boundary conditions. If there is no displacement boundary conditions, the structure will have a rigid-body motion with deformation, such as a flying airplane. In such a case, both the rigid-body motion and deformation are a function of time.

In general, the stiffness matrix, [**K**], and mass matrix, [**M**], are independent of time when the deformation is small. Therefore, they are calculated at the initial time and repeatedly used. The force, $\{\mathbf{F}(t)\}$, is given as a function of time.

8.2.2 Lumped Mass Matrix of a Uniaxial Bar Element

As mentioned in the previous section, the consistent mass matrix is derived by equating the kinetic energy of the bar, a continuous system, to that of the finite element model, a discrete system, using the shape functions. Therefore, the derived mass matrix is consistent with the finite element interpolation scheme. It is noted that the consistent mass matrix is fully populated like the element stiffness matrix. There is another simpler heuristic way of deriving the mass matrix. In this method, the mass of the element is distributed equally between the two nodes. That is, all mass is placed at the nodes, and the bar is supposed to possess only elasticity. This is similar to a discrete spring-mass system discussed in chapter 1 (see figure 8.2).

In the lumped mass assumption, the kinetic energy of the element is the sum of the kinetic energies of the two concentrated masses placed at the nodes:

$$T^{(e)} = \frac{1}{2} \frac{\rho AL}{2} \left(\dot{u}_1^2 + \dot{u}_2^2 \right). \tag{8.13}$$

Note that the term ρAL is the mass of the element. The above expression can be written in a matrix form as

$$T^{(e)} = \frac{1}{2} \{ \dot{u}_1 \quad \dot{u}_2 \} \left(\frac{\rho AL}{2} \right) \begin{bmatrix} 1 & 0 \\ 0 & 1 \end{bmatrix} \begin{Bmatrix} \dot{u}_1 \\ \dot{u}_2 \end{Bmatrix}$$
$$= \frac{1}{2} \{ \dot{u}_1 \quad \dot{u}_2 \} \left[\mathbf{m}_L^{(e)} \right] \begin{Bmatrix} \dot{u}_1 \\ \dot{u}_2 \end{Bmatrix}, \tag{8.14}$$

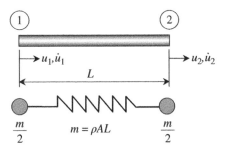

Figure 8.2 Lumped mass idealization of a uniaxial bar element

where $\left[\mathbf{m}_L^{(e)}\right]$ is called the *lumped mass matrix*. Note that the sum of all components of the lumped mass matrix is equal to the mass of the bar element, which is the same as the consistent mass matrix. The lumped mass matrix is also positive definite. The consistent mass matrix assumes that the mass is distributed throughout the element, while the lumped mass matrix assumes that the mass is concentrated at the nodes.

As can be seen in eq. (8.14), the lumped mass matrix of an element is a diagonal matrix. If the element mass matrix is diagonal, the global mass matrix after assembly will also be diagonal. Many numerical advantages exist when the global mass matrix is diagonal. Therefore, it would be important to understand how much accuracy may be lost and how much numerical advantage may be achieved by using the lumped mass matrix.

8.2.3 Mass Matrix of a Beam Element

As the first step in the dynamic analysis of beam elements, we will derive the mass matrix for a beam element. Consider a generic beam element discussed in section 3.3. The deflection of the beam in terms of nodal DOFs and shape functions is given in eq. (3.44). The kinetic energy of the beam can be written as

$$T^{(e)} = \frac{1}{2}\int_0^L \rho A \dot{v}^2 \, dx, \qquad (8.15)$$

where $\dot{v}$ is the transverse velocity of the vibrating beam. The above expression for kinetic energy is consistent with the Euler-Bernoulli beam theory described in chapter 3. Even though the rotation of the cross section of the beam also contributes to the rotational kinetic energy, it is neglected in the current formulation as it is considered small compared to the kinetic energy due to transverse motion. The velocity can be derived in terms of nodal DOFs as

$$
\begin{aligned}
\dot{v}(t) = \frac{\partial v}{\partial t} &= \dot{v}_1(t)N_1(x) + \dot{\theta}_1(t)N_2(x) + \dot{v}_2(t)N_3(x) + \dot{\theta}_2(t)N_4(x) \\
&= \underset{(1\times 4)}{\left\{ \dot{v}_1 \;\; \dot{\theta}_1 \;\; \dot{v}_2 \;\; \dot{\theta}_2 \right\}} \underset{(4\times 1)}{\left\{ \mathbf{N}(x) \right\}} \\
&= \underset{(1\times 4)}{\left\{ N_1(x) \;\; N_2(x) \;\; N_3(x) \;\; N_4(x) \right\}} \underset{(4\times 1)}{\left\{ \dot{\mathbf{q}} \right\}},
\end{aligned}
\qquad (8.16)
$$

where $\{\mathbf{N}(x)\}$ and $\{\dot{\mathbf{q}}\}$, respectively, are the column vectors of shape functions and the nodal velocities corresponding to the four DOFs of beam element. Even if the shape functions are written as a function of coordinate x, the shape functions of beam elements in eq. (3.43) were given in terms of parametric coordinate s, as

$$
\begin{aligned}
N_1(s) &= 1 - 3s^2 + 2s^3 \\
N_2(s) &= L\left(s - 2s^2 + s^3\right) \\
N_3(s) &= 3s^2 - 2s^3 \\
N_4(s) &= L\left(-s^2 + s^3\right),
\end{aligned}
\qquad (8.17)
$$

with the relationship of $s = x/L$. Substituting for $\dot{v}$ from eq. (8.16) into eq. (8.15) and following the steps used in deriving the mass matrix for uniaxial bar elements, the components of the consistent mass matrix can be defined as

$$m_{ij} = \rho A \int_0^L N_i(s)N_j(s)\,dx$$

$$= \rho AL \int_0^1 N_i(s)N_j(s)\,ds, \quad (i,j = 1,4). \tag{8.18}$$

Substituting for $N_i(s)$ from eq. (8.17) and evaluating the integrals, we obtain the consistent mass matrix of a beam element as

$$[\mathbf{m}] = \frac{\rho AL}{420}
\begin{bmatrix}
156 & 22L & 54 & -13L \\
22L & 4L^2 & 13L & -3L^2 \\
54 & 13L & 156 & -22L \\
-13L & -3L^2 & -22L & 4L^2
\end{bmatrix}
\begin{matrix}
v_1 \\ \theta_1 \\ v_2 \\ \theta_2
\end{matrix}. \tag{8.19}$$

Unlike the consistent mass matrix of the bar element, it is not straightforward to interpret the consistent mass matrix of the beam element. Some components even have a negative value. This is because the beam element has both translational DOFs (v_1 and v_2) and rotational DOFs (θ_1 and θ_2). However, if the components of translational DOFs are added, it is the same as the mass of the beam element: $(156 + 54 + 54 + 156) \times \rho AL/420 = \rho AL$. Again, like the bar element, the consistent mass matrix of the beam element is also positive definite.

The kinetic energy of a beam element is then given by

$$T^{(e)} = \frac{1}{2}\{\dot{\mathbf{q}}\}^T \left[\mathbf{m}^{(e)}\right]\{\dot{\mathbf{q}}\}. \tag{8.20}$$

Assembling of element mass matrices to obtain the global mass matrix follows similar procedures as before.

8.2.4   Lumped Mass Matrix for a Beam

In the case of the bar element, the lumped mass matrix is obtained by assuming that the mass of the element is distributed equally between the two nodes. In the case of beam element, however, this is not straightforward, as the beam element has rotational DOFs. In deriving the lumped mass matrix of the beam element, we assume the same lumped mass distribution in figure 8.2. Therefore, the components of mass matrix corresponding to the translation DOFs (v_1 and v_2) are the same as that of the bar element that is, a half of mass to each DOF. For rotational DOFs (θ_1 and θ_2), the mass moment of inertia is used. When a mass of $\rho AL/2$ is located at the distance of $L/2$, the mass moment of inertia becomes $I = (\rho AL/2) \times (L/2)^2/3 = \rho AL^3/24$. Therefore, the lumped mass matrix of the beam element can be defined as

$$[\mathbf{m}_L] = \frac{\rho AL}{2}
\begin{bmatrix}
1 & 0 & 0 & 0 \\
0 & L^2/12 & 0 & 0 \\
0 & 0 & 1 & 0 \\
0 & 0 & 0 & L^2/12
\end{bmatrix}
\begin{matrix}
v_1 \\ \theta_1 \\ v_2 \\ \theta_2
\end{matrix}. \tag{8.21}$$

EXAMPLE 8.1   *Consistent mass versus lumped mass of clamped-sliding beam*

A clamped-sliding beam is modeled using a beam element. When the tip of the beam is moved with the velocity of $\dot{v}_2$, compare the kinetic energy using (a) consistent mass matrix and (b) lumped mass matrix. The beam has density ρ, cross section A, and length L.

SOLUTION The kinetic energy of a beam element can be calculated using eq. (8.20). Since the beam is clamped on the wall and does not allow to rotate at the tip, $\dot{v}_1 = \dot{\theta}_1 = \dot{\theta}_2 = 0$. Therefore, we can strike out all rows and columns of the mass matrix except for the row and column corresponding to $\dot{v}_2$. With the consistent mass matrix in eq. (8.19), the kinetic energy becomes

$$
\begin{aligned}
T^{(e)} &= \frac{1}{2}\{\dot{\mathbf{q}}\}^{\mathrm{T}}\left[\mathbf{m}^{(e)}\right]\{\dot{\mathbf{q}}\} \\
&= \frac{1}{2}\frac{\rho AL}{420}\{\dot{v}_2\}[156]\{\dot{v}_2\} \\
&= \frac{39\rho AL}{210}\dot{v}_2^2 \\
&\approx 0.186\rho AL\dot{v}_2^2.
\end{aligned}
$$

On the other hand, with the lumped mass matrix in eq. (8.21), the kinetic energy becomes

$$
\begin{aligned}
T^{(e)} &= \frac{1}{2}\{\dot{\mathbf{q}}\}^{\mathrm{T}}\left[\mathbf{m}_L^{(e)}\right]\{\dot{\mathbf{q}}\} \\
&= \frac{1}{2}\frac{\rho AL}{2}\{\dot{v}_2\}[1]\{\dot{v}_2\} \\
&= \frac{\rho AL}{4}\dot{v}_2^2 \\
&= 0.25\rho AL\dot{v}_2^2.
\end{aligned}
$$

It is noted that the kinetic energy using the consistent mass matrix is about 20% smaller than the kinetic energy using the lumped mass matrix. Even if the consistent mass matrix used the finite element shape functions, it does not mean that it is more accurate than the lumped mass matrix. In fact, based on rigorous study, the lumped mass underestimates the natural frequency, while the consistent mass overestimates it. Therefore, it is possible to average these two mass matrices to obtain a better estimate of natural frequency[3]. ▮

8.3 NATURAL VIBRATION: NATURAL FREQUENCIES AND MODE SHAPES

A system is said to undergo natural vibration when it oscillates in one of its natural or fundamental modes of vibration. For example, when a mass-spring system is stretched from its undeformed position and released, it will oscillate at its natural frequency. All dynamics systems, such as a vibrating structure, have certain preferred or fundamental modes of vibration. For discrete systems, the number of fundamental modes is equal to the number of degrees of freedom of the system. In this section, we will study *modal analysis* approach (also known as *frequency analysis*) for computing the natural modes of vibration of structures. When a structure is vibrating in one of its fundamental modes, it vibrates at a frequency that is unique to that mode of vibration. The frequency of a fundamental mode of vibration is called its *natural frequency*, which is an important dynamic characteristic of the system. To understand the dynamic characteristics of a system, we need to know its natural frequencies and mode shapes that together describe the natural vibration modes of the system. In the study of natural vibrations, the effect of damping is often ignored to understand the dynamic characteristics of the system. In this section, a simple one-dimensional mass-spring system is introduced first, in order to derive the dynamic equation of free vibration, followed by the system of matrix equations for finite element models.

---

[3] Kim, K. O. 1993. "A review of mass matrices for eigenproblems." *Computers & Structures*, 46(6): 1041–1048.

Figure 8.3 Free vibration of 1D spring-mass system

8.3.1   Natural Vibration of One-Dimensional Mass-Spring System

In this section, a simple mass-spring system is considered to derive the dynamic equation of free vibration. For free vibration, it is assumed that the damping is negligible and that there is no external force applied to the mass. Referring to figure 8.3, the only force applied to the mass is from the deformation of spring, as

$$F_s = -ku, \tag{8.22}$$

where k is the spring constant in the units of force/displacement (e.g., lb/in or N/m). The negative sign indicates that the force is always opposing the motion of the mass attached to it. Based on Newton's second law of motion, the force applied to the mass is proportional to the acceleration of the mass as given by

$$F_s = ma = m\ddot{u}. \tag{8.23}$$

By substituting from eq. (8.22), the following ordinary differential equation can be obtained:

$$m\ddot{u} + ku = 0. \tag{8.24}$$

This is the dynamic equation for free vibration for one-dimensional mass-spring system. Note that this equation has a similar form as eq. (8.11) when the external force is zero. It will be shown in the next section that a similar equation but with a higher dimension can be obtained for the finite element model with multiple degrees of freedom.

If the spring is initially stretched by magnitude A and then released, the solution to the above free vibration equation that describes the motion of mass becomes (see figure 8.4)

$$u(t) = A \cos(2\pi f_n t). \tag{8.25}$$

This solution says that it will oscillate with simple harmonic motion that has an amplitude of A and a frequency of f_n (units of Hz). The value f_n is called the undamped natural frequency. For the simple mass–spring system, f_n is defined as

$$f_n = \frac{1}{T} = \frac{1}{2\pi}\sqrt{\frac{k}{m}}. \tag{8.26}$$

where T is the period of oscillation (seconds). The natural frequency can also be represented using the angular velocity (units of rad/sec) $\omega_n = 2\pi f_n$.

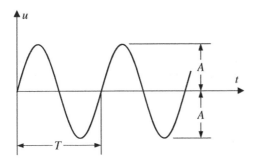

Figure 8.4 Free vibration of 1D mass-spring system

If the mass and stiffness of the system is known, the formula above can determine the frequency at which the system vibrates once set in motion by an initial disturbance. The system will vibrate slowly if the mass is increased or the stiffness is decreased. Every vibrating system has one or more natural frequencies that it vibrates at once disturbed. This simple relation can be used to understand in general what happens to a more complex system once we add mass or stiffness.

An interesting observation is that the natural frequency is independent of the initial perturbation, but the amplitude depends on it. That is, no matter what amplitude is applied for the initial perturbation, the system will oscillate in the same natural frequency, which means that the natural frequency is a system-related characteristic. In fact, in the free vibration analysis, the initial perturbation is assumed arbitrary, and the natural frequency is the main interest.

The natural frequency is important when designing a system under dynamic loads. If the excitation frequency of the applied load is the same as the natural frequency of the system, then a phenomenon called resonance occurs. At resonant frequencies, small periodically applied forces have the ability to produce large amplitude oscillations, which can cause a catastrophic failure of the system. Therefore, it is important for engineers to identify all natural frequencies of the system and to make sure that the excitation frequencies of applied loads are not close to the natural frequencies.

8.3.2 Natural Vibration Analysis Using Finite Element Models

As explained in the previous section, the dynamic equation of natural vibration of a finite element model can be obtained from eq. (8.11) by removing the externally applied forces. Therefore, the dynamic equation of natural vibration for finite element model can be written as

$$[\mathbf{M}]\{\ddot{\mathbf{Q}}\} + [\mathbf{K}]\{\mathbf{Q}\} = \{\mathbf{0}\}. \tag{8.27}$$

The detailed expression of this equation for a bar, a beam, and the system of springs will be given in examples. Since the nodal DOFs of finite elements will undergo a harmonic motion under free vibration, it is possible to assume the following form of nodal DOFs:

$$\{\mathbf{Q}(t)\} = \{\mathbf{A}\}e^{i\omega t}, \tag{8.28}$$

where $\{\mathbf{A}\} = \{A_1, A_2, \ldots A_N\}^{\mathrm{T}}$ is a vector of unknown constants, and ω is the angular velocity of periodic motion. Physically A_i is the amplitude of vibration of the i-th DOF.

An interesting observation from eq. (8.28) is that the vibration response $\{\mathbf{Q}(t)\}$ is decomposed into a constant deformation mode and a time-varying part. If the vector $\{\mathbf{A}\}$ is considered as nodal displacements, it represents the shape of deformation (or mode shape) during the vibration. On the other hand, the time-varying term, $e^{i\omega t}$, varies between +1 and −1 to represent the oscillatory motion. Therefore, the finite element model will vibrate with the shape of $\{\mathbf{A}\}$ and the frequency of $f_n = 2\pi\omega$ Hz. The vector $\{\mathbf{A}\}$ is called the mode shape and f_n is called the natural frequency. Since the mode shape depends on the initial perturbation, its magnitude is not of interest; it is often normalized. Note that even though $\{\mathbf{A}\} = \{\mathbf{0}\}$ satisfies the dynamics equation in eq. (8.27), it is called a trivial solution because there is no vibration. So we are looking for a nontrivial solution with $\{\mathbf{A}\} \neq \{\mathbf{0}\}$.

Although we have defined mode shapes and natural frequencies, they are still unknown quantities to be determined. We will now describe a standard approach called modal analysis for computing these quantities where the assumed solution in eq. (8.28) is substituted into the equations of motion in eq. (8.27) to yield an eigenvalue problem.

$$-\omega^2[\mathbf{M}]\{\mathbf{A}\}e^{i\omega t} + [\mathbf{K}]\{\mathbf{A}\}e^{i\omega t} = \{\mathbf{0}\}$$
$$\Rightarrow ([\mathbf{K}] - \omega^2[\mathbf{M}])\{\mathbf{A}\} = \{\mathbf{0}\}. \tag{8.29}$$

One can recognize the above as the generalized eigenvalue problem in which ω^2 is the eigenvalue and $\{A\}$ is the corresponding eigenvector (see section A.4 of the appendix). The physical meaning of the above equation is as follows. The equations of motion have nontrivial solutions for only certain values of ω for which the determinant of the matrix vanishes; that is,

$$\left|[K] - \omega^2[M]\right| = 0. \tag{8.30}$$

Otherwise, the solution of the set of equations is $\{A\} = \{0\}$. That means there is no motion or vibration, which is a trivial solution. In general, the number of natural frequencies is the same as the number of DOFs of the system. Therefore, theoretically, there are N number of natural frequencies and mode shapes: $(\omega_1, \{A^{(1)}\}), (\omega_2, \{A^{(2)}\}), \ldots, (\omega_N, \{A^{(N)}\})$.

For a given eigenvalue ω_i^2, eq. (8.29) is solved for the corresponding eigenvector $\{A^{(i)}\}$. However, since the determinant of the coefficient matrix is zero, eq. (8.29) may yield infinitely many solutions. That is, if $\{A^{(i)}\}$ is a solution to eq. (8.29), then $\alpha\{A^{(i)}\}$ can also be a solution for an arbitrary α. Therefore, among infinitely many solutions, the one that satisfies a condition is used as an eigenvector. For example, it is possible to choose the eigenvector whose magnitude is one; that is, $\left\|A^{(i)}\right\| = 1$. Imposing the condition that the eigenvector has unit magnitude is called normalization. In fact, normalization is commonly used in solving standard eigenvalue problems. In the case of generalized eigenvalue problems, as in free vibration problems, the eigenvectors are normalized by using the mass matrix; that is, eigenvectors are chosen such that $\{A^{(i)}\}^{\mathrm{T}}[M]\{A^{(i)}\} = 1$. This choice of normalization does not have any physical meaning; rather it is more of computational convenience.

If the structure is excited by an applied load with frequency of ω_1, it will vibrate in the shape of $\{A^{(1)}\}$, and so on. In general, however, excitation frequencies are much lower than the natural frequencies of the structure. Therefore, a handful of the lowest natural frequencies are of interest. Often, only the lowest natural frequencies are required in design. Therefore, many numerical algorithms in solving the eigenvalue problem only calculate the lowest n (n is much smaller than N) number of eigenvalues instead of calculating all eigenvalues, to reduce the computational cost.

EXAMPLE 8.2 *Natural vibration of a clamped-free bar*

The bar shown in figure 8.5 is fixed at one end and free at the other end. Use two bar elements to estimate the natural frequencies and corresponding mode shapes. The length of the bar is 2 m, the axial rigidity $EA = 10^6$ N, and the mass per unit length of the bar $\rho A = 0.042$ kg/m. Determine the natural frequencies of the bar using: (a) consistent mass matrix; and (b) lumped mass matrix.

SOLUTION   Since the system has two DOFs after applying the boundary condition at $x = 0$, it will have two natural frequencies and corresponding mode shapes.

(a) Using the consistent mass matrix:
   Natural frequencies. We will use two elements of equal length, $L = 1$ m. The element stiffness matrices and mass matrices can be written as follows:

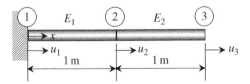

Figure 8.5 Vibration of a clamped-free bar modeled using two elements

$$\left[\mathbf{k}^{(1)}\right] = 10^6 \begin{bmatrix} 1 & -1 \\ -1 & 1 \end{bmatrix} \begin{matrix} u_1 \\ u_2 \end{matrix} \quad \left[\mathbf{m}^{(1)}\right] = 0.007 \begin{bmatrix} 2 & 1 \\ 1 & 2 \end{bmatrix} \begin{matrix} u_1 \\ u_2 \end{matrix}$$

$$\left[\mathbf{k}^{(2)}\right] = 10^6 \begin{bmatrix} 1 & -1 \\ -1 & 1 \end{bmatrix} \begin{matrix} u_2 \\ u_3 \end{matrix} \quad \left[\mathbf{m}^{(2)}\right] = 0.007 \begin{bmatrix} 2 & 1 \\ 1 & 2 \end{bmatrix} \begin{matrix} u_2 \\ u_3 \end{matrix}.$$

After assembly, the row and column corresponding to u_1 can be deleted in order to apply the displacement boundary condition. After that, the global matrices are obtained as:

$$[\mathbf{K}] = 10^6 \begin{bmatrix} 2 & -1 \\ -1 & 1 \end{bmatrix} \begin{matrix} u_2 \\ u_3 \end{matrix}; \quad [\mathbf{M}] = 0.007 \begin{bmatrix} 4 & 1 \\ 1 & 2 \end{bmatrix} \begin{matrix} u_2 \\ u_3 \end{matrix}.$$

Using the MATLAB command [Q,D]=eig(K,M), we obtain the two eigenvalues as $\omega_1 = \sqrt{D_{11}} = 3{,}931 \text{s}^{-1}$ and $\omega_2 = \sqrt{D_{22}} = 13{,}734 \text{s}^{-1}$. Note that the 2×2 diagonal matrix [D] contains eigenvalues ω_i^2, while the column vectors of the 2×2 matrix [Q] are the corresponding eigenvectors. The exact natural frequencies of a fixed-free bar[4] are given by $\omega_n = ((2n-1)\pi/2L)\sqrt{E/\rho}$. The first two frequencies are obtained by using mode numbers $n = 1$ and $n = 2$. The exact frequencies are: $\omega_1 = 3{,}832 \text{s}^{-1}$ and $\omega_2 = 11{,}496 \text{s}^{-1}$. The error in ω_1 is only 2.6%, and error in ω_2 is about 19%. As discussed before, the consistent mass matrix overestimates the natural frequencies.

From the results, we can make the following observations. The natural frequencies from finite element analysis are higher than the exact values, or in the other words, the finite element model of the structure is stiffer than the real structure, as we have seen before. When we calculate the first two natural frequencies, the first mode is more accurate than the second mode. If we would like to calculate the second frequency also accurately, then we have to use a larger number of elements.

Mode shapes. The two columns of the square matrix [Q] from the MATLAB command represent the two eigenvectors; they are

$$[\mathbf{Q}] = \begin{bmatrix} 3.6322 & 5.2558 \\ 5.1367 & -7.4328 \end{bmatrix}.$$

Each column of [Q] represents the nodal mode shape of the corresponding natural frequency. That is, $[\mathbf{Q}] = [\mathbf{A}^{(1)}, \mathbf{A}^{(2)}]$. We will demonstrate the procedure for determining mode shapes using the first mode. The values of nodal DOFs corresponding to the first eigenvector are $u_2 = 3.6322$ and $u_3 = 5.1367$. Also, from the boundary condition, $u_1 = 0$. These values can be considered as nodal displacements for the mode shape, although their absolute magnitudes do not have any physical meaning. Similar to recovering an element displacement function, the shape functions can be used to calculate the mode shapes of elements 1 and 2, as

$$u^{(1)}(s) = \mathbf{x}_1(1-s) + u_2 s = 3.6322s,$$
$$u^{(2)}(s) = u_2(1-s) + u_3 s = 3.6322(1-s) + 5.1367s.$$

In the above expressions, superscripts denote the element number and s is the local coordinate defined by $s = (x - x_i)/L$, where x_i is the x-coordinate of the first node, and L is the length of the element.

Similarly, the mode shape of the second mode can be derived as

$$u^{(1)}(s) = 5.2558s,$$
$$u^{(2)}(s) = 5.2558(1-s) - 7.4328s.$$

The first mode shape shows that both nodes 2 and 3 move in the same direction and both elements are in elongation. On the other hand, the second mode shape shows that when element 1 is in elongation, element 2 is in contraction. Therefore, their motions are in opposite directions.

---

[4] Refer to page 599 of S. S. Rao. 2004. *Mechanical Vibrations*, Fourth Ed. Pearson Prentice Hall, Upper Saddle River, NJ.

(b) Using the lumped mass matrix:
Since the lumped mass matrix is independent of the stiffness, the stiffness matrix will remain the same. The lumped mass matrices are evaluated as follows:

$$\left[\mathbf{m}_L^{(1)}\right] = \frac{0.042}{2}\begin{bmatrix} 1 & 0 \\ 0 & 1 \end{bmatrix}\begin{matrix} u_1 \\ u_2 \end{matrix},$$

$$\left[\mathbf{m}_L^{(2)}\right] = \frac{0.042}{2}\begin{bmatrix} 1 & 0 \\ 0 & 1 \end{bmatrix}\begin{matrix} u_2 \\ u_3 \end{matrix},$$

$$[\mathbf{M}_L] = 0.021\begin{bmatrix} 2 & 0 \\ 0 & 1 \end{bmatrix}\begin{matrix} u_2 \\ u_3 \end{matrix}.$$

By solving the generalized eigenvalue problem in eq. (8.29) with the lumped mass matrix $[\mathbf{M}_L]$, we obtain the two natural frequencies as: $\omega_1 = 3{,}735\,\mathrm{s}^{-1}$ and $\omega_2 = 9{,}016\,\mathrm{s}^{-1}$. First, we note that the frequencies are lower than the exact values. This is because the lumped matrix method assigns all the masses at the nodes, which overestimates the kinetic energy of the system. When the mass of a system is concentrated at the nodes instead of being distributed evenly over the length of the element, the natural frequency is underestimated. Recall the natural frequency of a simple one-dimensional mass-spring system is given by $\omega = \sqrt{k/m}$. A larger m at the nodes reduces the natural frequency. The error in ω_1 is −2.5%, while the error in ω_2 is −22%. ■

EXAMPLE 8.3 *Natural vibration of a clamped-clamped beam*

Use two elements of equal length to determine the natural frequencies and mode shapes of a clamped-clamped beam shown in figure 8.6. The length of the beam is 2 m, $E = 100\,\mathrm{GPa}$, $A = 10^{-5}\,\mathrm{m}^2$, $I = 10^{-8}\,\mathrm{m}^4$, and $\rho = 4000\,\mathrm{kg/m}^3$. Determine the natural frequencies of the beam using: (a) consistent mass matrix; and (b) lumped mass matrix.

SOLUTION Since each node of a beam element has two DOFs, v and θ, the system has a total of six DOFs. However, since nodes 1 and 3 are clamped, only two DOFs, v_2 and θ_2, remain after applying displacement boundary conditions. Therefore, in the assembly of the global matrices, we only need to assemble these two DOFs.

(a) Using the consistent mass matrix:
Natural frequency. The element stiffness and consistent mass matrices are derived as follows. Only terms corresponding to the nonzero (active) DOFs are shown below:

$$\left[\mathbf{k}^{(1)}\right] = 1000\begin{bmatrix} 12 & -6 \\ -6 & 4 \end{bmatrix}\begin{matrix} v_2 \\ \theta_2 \end{matrix} \quad \left[\mathbf{m}^{(1)}\right] = \frac{0.04}{420}\begin{bmatrix} 156 & -22 \\ -22 & 4 \end{bmatrix}\begin{matrix} v_2 \\ \theta_2 \end{matrix}$$

$$\left[\mathbf{k}^{(2)}\right] = 1000\begin{bmatrix} 12 & 6 \\ 6 & 4 \end{bmatrix}\begin{matrix} v_2 \\ \theta_2 \end{matrix} \quad \left[\mathbf{m}^{(2)}\right] = \frac{0.04}{420}\begin{bmatrix} 156 & 22 \\ 22 & 4 \end{bmatrix}\begin{matrix} v_2 \\ \theta_2 \end{matrix}.$$

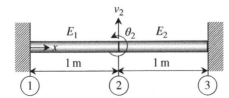

Figure 8.6 Free vibration of a clamped-clamped beam using two beam elements

The global stiffness and mass matrix can be obtained by assembling the two-element matrices as

$$[\mathbf{K}] = 1000 \begin{bmatrix} 24 & 0 \\ 0 & 8 \end{bmatrix} \begin{matrix} v_2 \\ \theta_2 \end{matrix} \quad [\mathbf{M}] = \frac{0.04}{420} \begin{bmatrix} 312 & 0 \\ 0 & 8 \end{bmatrix} \begin{matrix} v_2 \\ \theta_2 \end{matrix}.$$

Since both the stiffness and mass matrices are diagonal, it is possible to calculate the eigenvalues using $\omega_1^2 = \sqrt{K_{11}/M_{11}}$ and $\omega_2^2 = \sqrt{K_{22}/M_{22}}$. It is also possible to use the MATLAB command [Q,D]=eig(K,M) to solve the generalized eigenvalue problem shown in eq. (8.29). The resulting eigenvalues and eigenvectors are obtained as

$$[\mathbf{Q}] = \begin{bmatrix} 5.8 & 0 \\ 0 & 36 \end{bmatrix} \text{ and } [\mathbf{D}] = 10^7 \begin{bmatrix} 0.0808 & 0 \\ 0 & 1.0500 \end{bmatrix}.$$

Noting that $[\mathbf{D}]$ is the diagonal matrix of squares of the natural frequencies, we obtain $\omega_1 = \sqrt{D_{11}} = 899\,\mathrm{s}^{-1}$ and $\omega_2 = \sqrt{D_{22}} = 3{,}240\,\mathrm{s}^{-1}$. From the analytical solution[5], the exact values of the first two natural frequencies are: $\omega_1 = 885\,\mathrm{s}^{-1}$ and $\omega_2 = 2{,}439\,\mathrm{s}^{-1}$. The errors in the first and second natural frequencies are 1.6% and 33%, respectively. Same with the bar elements, the finite element model with consistent mass matrix overestimates the frequencies.

Mode shapes. We note the first column of $[\mathbf{Q}]$ from the MATLAB output is the first eigenvector, which defines the first mode. Similarly, the second column defines the second mode. These are the nodal mode shapes. The mode shape of the beam element can be derived by using the shape functions given in eq. (8.17). The first mode shape is given by the first eigenvector $\{\mathbf{A}^{(1)}\} = \{v_2, \theta_2\}^{\mathrm{T}} = \{5.8, 0\}^{\mathrm{T}}$:

$$v^{(1)}(s) = v_2 N_3(s) + \theta_2 N_4(s) = 5.8(3s^2 - 2s^3),$$
$$v^{(2)}(s) = v_2 N_1(s) + \theta_2 N_2(s) = 5.8(1 - 3s^2 + 2s^3).$$

Note that the shape is written as separate equations, $v^{(1)}(s)$ and $v^{(2)}(s)$, for each element. For the first element, $v^{(1)}(s)$, the boundary conditions $v_1 = \theta_1 = 0$ are used. Similarly, for $v^{(2)}(s)$, the boundary conditions $v_3 = \theta_3 = 0$ are used. The variable s is not the same for the two equations because it is the local coordinate of each element. Similarly, the second mode shape is derived using the second eigenvector $\{\mathbf{A}^{(2)}\} = \{v_2, \theta_2\}^{\mathrm{T}} = \{0, 36\}^{\mathrm{T}}$:

$$v^{(1)}(s) = v_2 N_3(s) + \theta_2 N_4(s) = 36(-s^2 + s^3),$$
$$v^{(2)}(s) = v_2 N_1(s) + \theta_2 N_2(s) = 36(s - 2s^2 + s^3).$$

The mode shapes are plotted in figure 8.7. The first mode is the symmetric deflection of the beam, while the second mode is the antisymmetric deflection of the beam.

(b) Using the lumped mass matrix:

The lumped mass matrix will be a diagonal matrix as shown below:

$$\left[\mathbf{m}^{(1)}\right] = \frac{0.04}{2} \begin{bmatrix} 1 & 0 \\ 0 & 1/12 \end{bmatrix} \begin{matrix} v_2 \\ \theta_2 \end{matrix},$$

$$\left[\mathbf{m}^{(2)}\right] = \frac{0.04}{2} \begin{bmatrix} 1 & 0 \\ 0 & 1/12 \end{bmatrix} \begin{matrix} v_2 \\ \theta_2 \end{matrix},$$

$$[\mathbf{M}] = \begin{bmatrix} 0.04 & 0 \\ 0 & 0.0033 \end{bmatrix} \begin{matrix} v_2 \\ \theta_2 \end{matrix}.$$

One can easily calculate the two natural frequencies as $\omega_1 = \sqrt{K_{11}/M_{11}} = 775\,\mathrm{s}^{-1}$ and $\omega_2 = \sqrt{K_{22}/M_{22}} = 1{,}557\,\mathrm{s}^{-1}$. The errors in the first and second natural frequencies are −12.4% and −36.2%,

---

[5] Refer to page 613 of S. S. Rao. 2004. *Mechanical Vibrations*, Fourth Ed. Pearson Prentice Hall, Upper Saddle River, NJ.

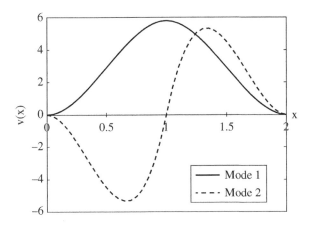

Figure 8.7 Mode shapes of a clamped-clamped beam of length 2 m

respectively. Thus, the lumped mass matrix underestimates the frequencies. Furthermore, the error is also large compared to consistent mass matrix method. ▪

EXAMPLE 8.4 *Natural frequencies of a discrete spring-mass system*

Consider the rigid body-spring system in example 1.1 in chapter 1, which is shown again in figure 8.8. Let the mass of bodies 2, 3, and 4 be equal to 20, 30, and 40 kg, respectively. Determine the natural frequencies of the system.

SOLUTION   The global stiffness matrix has already been derived in chapter 1. It is given below in N-m units:

$$[\mathbf{K}] = 10^5 \begin{bmatrix} 15 & -6 & -4 \\ -6 & 12 & -4 \\ -4 & -4 & 11 \end{bmatrix} \begin{matrix} u_2 \\ u_3 \\ u_4 \end{matrix}.$$

Note that the rows and columns corresponding to nodes 1 and 5 are struck out when the displacement boundary conditions are imposed. In a spring-mass system, the masses are already lumped, and the springs do not have any mass. Hence, the lumped mass matrix is indeed the exact representation of the mass matrix:

$$[\mathbf{M}] = \begin{bmatrix} 20 & 0 & 0 \\ 0 & 30 & 0 \\ 0 & 0 & 40 \end{bmatrix} \begin{matrix} u_2 \\ u_3 \\ u_4 \end{matrix}.$$

Using the MATLAB command [Q,D]=eig(K,M), we obtain three eigenvalues of the system as

$$[\mathbf{D}] = 10^4 \begin{bmatrix} 1.0363 & 0 & 0 \\ 0 & 4.3591 & 0 \\ 0 & 0 & 8.8546 \end{bmatrix}$$

$$\Rightarrow \begin{Bmatrix} \omega_1 \\ \omega_2 \\ \omega_3 \end{Bmatrix} = \begin{Bmatrix} 101.8 \\ 208.8 \\ 297.6 \end{Bmatrix} \text{rad/s} \Rightarrow \begin{Bmatrix} f_1 \\ f_2 \\ f_3 \end{Bmatrix} = \begin{Bmatrix} 16.2 \\ 33.2 \\ 47.4 \end{Bmatrix} \text{Hz}.$$

In the above equation, we have used $\omega_i = \sqrt{D_{ii}}$ and $f_i = \omega_i/2\pi$. ▪

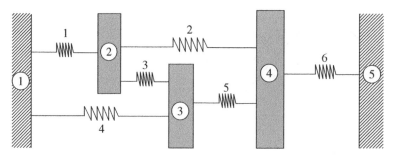

Figure 8.8 Rigid bodies connected by springs

So far, we only consider the cases when the displacement boundary conditions are given such that the rigid body motions are eliminated. However, it is possible that the generalized eigenvalue problem in eq. (8.29) can be solved in the structural matrix level; that is, before applying boundary conditions. In such a case, the structural stiffness matrix is positive semi-definite and will have zero eigenvalues. The number of zero eigenvalues is the same as the number of rigid body motions. For example, in the case of a uniaxial bar, there is one rigid body motion, while there are two rigid body motions for plane beams. Therefore, when the structure has rigid body motions, the zero eigenvalues should not be used in calculating the natural frequencies of the structure.

As a closing remark, it is emphasized again that the natural vibration analysis does not require any externally applied loads. Natural vibration analysis is related to the vibration characteristics of the system, not to the applied loads or initial conditions. The magnitude of mode shapes (eigenvectors) do not have any physical meaning; their shapes are the mode of vibration when the system is excited at the corresponding frequency. When the applied load is composed of the combination of multiple frequencies, the system will also respond as the combination of multiple mode shapes, which requires a method called mode superposition. This will be discussed further in section 8.5.

8.4 FORCED VIBRATION: DIRECT INTEGRATION APPROACH

When time-dependent loads and/or boundary conditions are applied to the structure, the dynamic response of the structure becomes important, and inertial effects and wave propagation are dominant throughout the structure compared with the quasi-static response. Time history analysis is concerned with computing the response of the structure under time-dependent loads. In general, the methods of analysis can be classified into two broad categories: the direct integration method and the mode superposition method. In the direct integration method, the ordinary differential equations are integrated using a step-by-step numerical procedure. There is no transformation of the finite element matrix equation. In the mode superposition method, on the other hand, the finite element matrix equation is transformed so that the solution space is spanned by the eigenvectors of the system. The unknown displacements are represented as a linear combination of the mode shapes. The direct integration method will be explained in this section, while the mode superposition method will be discussed in section 8.5.

In chapter 2, we showed that the static equilibrium of a structure could be written as partial differential equations whose weighted residual becomes the finite element matrix equation after discretization. This is in fact the partial differential equation over the structural space. The dynamic equation is an ordinary differential equation over time. Unlike partial differential equations, the dynamic equation is often solved using finite difference methods. While partial differential equations have boundary conditions, ordinary differential equations have initial conditions at $t = 0$.

In the finite difference methods, the continuous time interval $[0, T]$ is discretized by $[t_0 = 0, t_1, t_2, \ldots, t_n, \ldots, t_N = T]$ first, and then, the equilibrium equation in eq. (8.11) is imposed at each discrete time.

It is possible that the time increment can be variable, but it is assumed that a constant time increment is used; that is, $\Delta t = t_n - t_{n-1}$. Let the current time be t_n. Then, the dynamic equilibrium equation can be written as

$$[\mathbf{M}]\{\ddot{\mathbf{Q}}_n\} + [\mathbf{K}]\{\mathbf{Q}_n\} = \{\mathbf{F}_n\}, \quad n = 1, \ldots, N, \tag{8.31}$$

where $\{\mathbf{Q}_n\} = \{\mathbf{Q}(t_n)\}$. Our goal is to determine $\{\mathbf{Q}(t)\}$ and thus $\{\ddot{\mathbf{Q}}(t)\}$ given the initial conditions of the nodal DOFs and the forcing functions $\{\mathbf{F}(t)\}$. In the discretized time intervals, it is equivalent to determining $\{\mathbf{Q}_n\}$ and $\{\ddot{\mathbf{Q}}_n\}$. In the following derivations, we assume that the dynamic problem has been solved up to time t_n, and the solution at t_{n+1} is required.

The typical solution procedure of the direct integration method is calculating the unknown nodal displacements, velocities, and acceleration at time t_{n+1} based on known information from the previous times, t_n, t_{n-1}, and so forth. If the integration method only requires the information from previous time t_n, it is called a one-step method; if it requires more than one previous times, it is called a multi-step method. In addition, if the displacements, velocities, and accelerations at t_{n+1} are expressed in terms of those at t_n or previous times, then it is called an explicit method. On the other hand, if the method requires information at t_{n+1}, then it is called an implicit method. The implicit method requires solving a system of equations to calculate the information at time t_{n+1}, and therefore, it is more expensive than the explicit method. However, in general, larger time-step size can be used for an implicit method than for an explicit method in order to satisfy the stability condition.

The key concept of the direct integration methods is to approximate the time derivatives, $\{\ddot{\mathbf{Q}}\}$ and $\{\dot{\mathbf{Q}}\}$ using finite differences. Different methods have been proposed depending on how the time derivatives are approximated. Since the accuracy and stability of the solution depend on the method of choice, it is important for the users to fully understand the characteristics of different time integration methods. In this section, some commonly used methods are presented. The advantages and limitations of various methods will be discussed.

8.4.1 Implicit versus Explicit Time Integration Methods

In general, various time integration methods for ordinary differential equations can be categorized either as explicit methods or implicit methods. When a direct computation of the field variables can be made in terms of known quantities, the computational method is called an explicit method. When the field variables are defined by coupled sets of equations, and either a matrix or an iterative technique is needed to calculate the variable, the numerical method is called an implicit method. In general, the explicit method is computationally less expensive during each step but requires many more time steps because it requires a very small time step in order to make the numerical integration stable. On the other hand, many implicit methods are stable no matter what time step is used. However, if the time step is too large, the numerical method may not be accurate. Therefore, an appropriate time-step size should be used by considering both accuracy and stability.

Explicit method: The simplest method of explicit time integration is the forward Euler method, which is used to solve first-order ordinary differential equations. Although we are interested in the second-order ordinary differential equation in eq. (8.31), we will use the first-order differential equation to explain the difference between the implicit and explicit methods. Consider that we want to solve the following differential equation with the initial condition $y(t_0) = y_0$ in the time interval $[0, T]$:

$$\dot{y} = f(t, y(t)), \quad y(t_0) = y_0. \tag{8.32}$$

In the numerical time integration approach, the above differential equation is solved at a discrete set of times. Let the continuous time interval be discretized by $[t_0 = 0, t_1, t_2, \ldots, t_n, \ldots, t_N = T]$. As the initial condition is given at t_0, we want to use the information at t_0 to calculate the information at t_1. If we repeat this process, we can solve for information at all discrete times.

The forward Euler method is based on the first-order Taylor series expansion with respect to t_n. To determine $y(t_{n+1}) = y_{n+1}$, the following integration rule can be obtained:

$$y_{n+1} = y_n + \Delta t \cdot f(t_n, y_n).$$

(8.33)

Since $y(t_n) = y_n$ is already calculated at the previous time, and the function $f(t_n, y_n)$ is available, all the terms on the right-hand side of eq. (8.33) are known. Therefore, the unknown variable y_{n+1} can be explicitly calculated using the information from the previous time step.

The numerical integration formula in eq. (8.33) can be viewed as an approximation of the time derivative $\dot{y}(t)$. That is, in the forward Euler method, the time derivative is approximated by

$$\dot{y}(t_n) = f(t_n, y_n) \approx \frac{y_{n+1} - y_n}{\Delta t}.$$

(8.34)

The explicit method calculates the system response at the next time from the state of the system at the current time, which is why it is called a "forward" difference method.

Implicit method: Unlike the explicit method, the implicit method calculates the system response at the next time by solving an equation involving both the current and next times. For example, the counter-part of the forward Euler method is the backward Euler method, which can be written as

$$y_{n+1} = y_n + \Delta t \cdot f(t_{n+1}, y_{n+1}).$$

(8.35)

If $f(t, y)$ is a nonlinear function of y, an iterative approach is required to solve the implicit time integration. Even if we showed only two cases of evaluating $f(t, y)$ at t_n and t_{n+1} to approximate the time derivative, it is possible that it can be generalized to $f(t_{n+\alpha}, y_{n+\alpha})$, where $0 \leq \alpha \leq 1$. Many different numerical methods are based on this generalization.

Even if the implicit method requires iteration, it has an advantage of stability. That is, eq. (8.35) is unconditionally stable no matter what size of time step Δt is used. On the other hand, eq. (8.33) is conditionally stable, which requires a relatively small time step. However, if a large time step is used for the implicit method, the accuracy of the approximation will be sacrificed.

8.4.2 Central Difference Method: Explicit Time Integration

In the central difference method (CDM), the velocity and acceleration are approximated as

$$\dot{\mathbf{Q}}_n = \frac{\mathbf{Q}_{n+1} - \mathbf{Q}_{n-1}}{2\Delta t},$$

(8.36)

$$\ddot{\mathbf{Q}}_n = \frac{\mathbf{Q}_{n+1} - 2\mathbf{Q}_n + \mathbf{Q}_{n-1}}{(\Delta t)^2},$$

(8.37)

where the subscript n denotes time step $t_n = n\Delta t$ and $\mathbf{Q}_n = \mathbf{Q}(t_n)$. Note that the curly brackets for a vector are omitted for the simplification of notation. In approximating acceleration in eq. (8.37), the velocity is calculated using the forward Euler method in eq. (8.34). Therefore, the velocity and acceleration are calculated using different schemes in CDM. Then, the equilibrium is applied at t_n as

$$\mathbf{M}\ddot{\mathbf{Q}}_n + \mathbf{K}\mathbf{Q}_n = \mathbf{F}_n.$$

(8.38)

Substituting for $\ddot{\mathbf{Q}}_n$ from eq. (8.37) into the above equation and rearranging the terms, we obtain

$$\mathbf{Q}_{n+1} = 2\mathbf{Q}_n - \mathbf{Q}_{n-1} + (\Delta t)^2 \mathbf{M}^{-1}(\mathbf{F}_n - \mathbf{K}\mathbf{Q}_n).$$

(8.39)

Note that we use the displacements at the n-th and $(n-1)$-th time steps in order to solve for displacements at the $(n+1)$-th step. That is, in order to calculate the displacement at a new time step, only information at the previous time steps is required. Since information at the previous time steps is already available, the unknown displacement at a new step is an explicit function of information from the previous steps. That is why the central difference method is called an explicit method.

It is noted that the inverse of the mass matrix is required to calculate the displacement in eq. (8.39). If the lumped mass matrix is used, then the inverse is trivial. Therefore, the CDM is computationally efficient with the lumped mass matrix. By careful implementation, it is unnecessary to build the global stiffness matrix [K]. The calculation can be done in the element level.

The central difference method is also called a two-step method because the displacements at two consecutive time steps, $(n-1)$ and n, are used to determine displacements at $(n+1)$-th step. This can cause an issue to calculate the displacement at t_1 because in order to determine $\mathbf{Q}_1$ we need the initial condition $\mathbf{Q}_0$, and also $\mathbf{Q}_{-1}$. The steps involved in the CDM are listed below:

1. The initial acceleration can be calculated from the equations of motion in eq. (8.38) as

$$\ddot{\mathbf{Q}}_0 = \mathbf{M}^{-1}(\mathbf{F}_0 - \mathbf{K}\mathbf{Q}_0). \tag{8.40}$$

2. By using Taylor series expansion, we obtain

$$\mathbf{Q}_{-1} = \mathbf{Q}_0 - \Delta t\dot{\mathbf{Q}}_0 + \frac{\Delta t^2}{2}\ddot{\mathbf{Q}}_0. \tag{8.41}$$

3. By substituting for initial conditions in the above equation, $\mathbf{Q}_{-1}$ can be calculated. Knowing the displacements at $n=-1$ and $n=0$, one can calculate $\mathbf{Q}_1$ using eq. (8.39) as:

$$\mathbf{Q}_1 = 2\mathbf{Q}_0 - \mathbf{Q}_{-1} + (\Delta t)^2\mathbf{M}^{-1}(\mathbf{F}_0 - \mathbf{K}\mathbf{Q}_0). \tag{8.42}$$

4. Knowing $\mathbf{Q}_0$ and $\mathbf{Q}_1$, use eq. (8.39) to calculate $\mathbf{Q}_n$ for $n = 2, 3, \ldots$, iteratively.

Once displacement Q_{n+1} at t_{n+1} is calculated, the velocity and acceleration at time t_n can be calculated using eq. (8.36) and (8.37). Note that the velocity and acceleration are calculated one step behind than the displacement. That is, after calculating displacement $\mathbf{Q}_{n+1}$, the velocity and acceleration at t_n are updated. Another method of starting the CDM is to use one of the implicit methods described in the next section to calculate $\mathbf{Q}_1$ as accurately as possible and proceed to step 4 above to use the CDM.

The CDM is the simplest of all direct integration methods. First, it is unnecessary to factorize the stiffness matrix as in the quasi-static problem. Only the mass matrix needs to be factorized. If the lumped mass matrix is used, then $\mathbf{M}$ will be diagonal, and calculating $\mathbf{M}^{-1}$ in eq. (8.39) is straightforward and computationally efficient. However, CDM is only conditionally stable. To ensure stability, the time step Δt must be less than a critical value Δt_{cr}. The critical value of Δt is given as[6]

$$\Delta t_{cr} = \frac{T_{\min}}{\pi} = \frac{1}{\pi}\frac{2\pi}{\omega_{\max}} = \frac{2}{\omega_{\max}}, \tag{8.43}$$

where $\omega_{\max}$ is the highest natural frequency of the finite element model. In general, $\omega_{\max}$ increases as the element size decreases or the stiffness of the material increases. Therefore, when small-sized elements are used with a highly stiff material, a very small time step must be used to make the explicit time integration stable.

EXAMPLE 8.5 *Response of a uniaxial bar for a ramp loading*

Consider a clamped uniaxial bar subjected to a tip load as shown in figure 8.9. Use three elements of equal length to determine the tip displacement as a function of time. Use the central difference method for time integration and the consistent mass matrix. The properties of the bar are: $E = 100$ GPa, $L = 1$ m, $A = 10^{-5}$ m^2, $\rho = 4000$ kg/m^3. The tip

---

[6] Bathe, K. J. 1982. *Finite Element Procedures in Engineering Analysis.* Prentice-Hall, New Jersey.

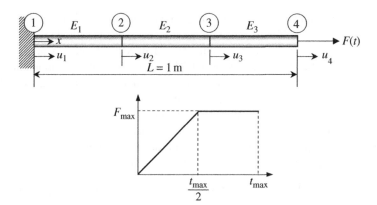

Figure 8.9 A clamped-free uniaxial bar subjected to a tip force $F(t)$

force is given by: $F(t) = 2F_{max}t/t_{max}$ for $0 < t < t_{max}/2$, and $F(t) = F_{max}$ for $t > t_{max}/2$, where $F_{max} = 0.001\ EA$ and $t_{max} = 3$ ms. Calculate the tip displacement $u(t)$ for $0 < t < t_{max}$.

SOLUTION First, the element stiffness and mass matrices are calculated using eqs. (1.16) and (8.7), respectively. Note that these matrices are identical for all three elements. Since node 1 is fixed, the assembly can be done without node 1. The assembled global stiffness and (consistent) mass matrices are

$$
\mathbf{K} = 10^7 \begin{bmatrix} 6 & -3 & 0 \\ -3 & 6 & -3.03 \\ 0 & -3 & 3 \end{bmatrix} \begin{matrix} u_2 \\ u_3 \\ u_4 \end{matrix} \quad \mathbf{M} = \begin{bmatrix} 0.089 & 0.022 & 0 \\ 0.022 & 0.089 & 0.022 \\ 0 & 0.022 & 0.044 \end{bmatrix} \begin{matrix} u_2 \\ u_3 \\ u_4 \end{matrix}.
$$

By solving the generalized eigenvalue problem, the highest frequency was calculated as $\omega_{max} = 47,133\,\mathrm{s}^{-1}$. Then the critical time step can be derived as $\Delta t_{cr} = 2/\omega_{max} = 42\,\mu\mathrm{s}$. We choose the time step as half of the critical value, that is, $\Delta t = 21\,\mu\mathrm{s}$. The following MATLAB script was used to perform the time integration.

```
%Central Difference Method
%Problem parameters
Area=10e-5;E=100e9;rho=4000;L=1/3;Fmax=0.001*Area*E;tmax=3e-3;
Dt=21e-6; Nmax=round(tmax/Dt)+1;
t = linspace(0, tmax, Nmax);
%Global matrices
K=(Area*E/L)*[2 -1 0;-1 2 -1;0 -1 1];
M=(rho*Area*L/6)*[4 1 0;1 4 1;0 1 2];
F=[zeros(2,Nmax);
    linspace(0,Fmax,round(Nmax/2)) Fmax*ones(1,Nmax-round(Nmax/2))];
Q=zeros(3,Nmax);
%Time integration
Minv=(Dt^2)*inv(M);
Q(:,2)=Q(:,1)+Dt*zeros(3,1);
for n=2:Nmax-1
 Q(:,n+1) = 2*Q(:,n) - Q(:,n-1) + Minv*(F(:,n)-K*Q(:,n));
end
%Static solution
Qs=K\F;
plot (t,Q(3,:), t,Qs(3,:));
```

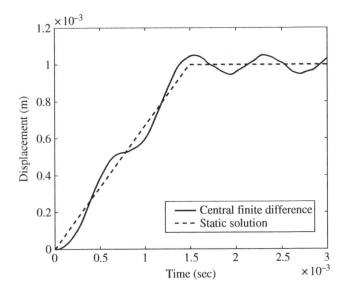

Figure 8.10 Tip-displacement of a uniaxial bar in figure 8.9 using the central difference method. The quasi-static response is shown in a dashed line.

The resulting tip displacement history is shown in a solid line in figure 8.10. In the same figure, a quasi-static solution is shown in a dashed line. The quasi-static solution can be thought of as the response when there are no inertial effects. It essentially follows the force history. In the dynamic response, there are oscillations about the static response.

The bar has the lowest natural frequency of 7,944 s^{-1}. Therefore, the period of vibration can be calculated from eq. (8.26), as

$$T = \frac{2\pi}{\omega} = 0.791 \times 10^{-3} \text{ sec.}$$

The figure shows that the period of vibration is about the same as the theoretical prediction. That is, the bar is oscillating according to its lowest natural frequency, while following the trend of static deformation. ▪

As mentioned before, the central difference method is easy to implement because the variables at time t_{n+1} are an explicit function of variables at time t_n. However, the method is conditionally stable; that is, it requires a small time step to be stable. In order to show the issue related to instability, the same problem in example 8.5 is solved with a larger time step. As we calculated in example 8.5, the critical time step is $\Delta t_{cr} = 42 \mu s$. Therefore, we solved the same problem with $\Delta t = 42.5 \mu s$, whose results are shown in figure 8.11. In the figure, a small oscillatory error starts in early time, and then, the error exponentially increases in the following time and eventually diverged. Therefore, it is important to keep the time step smaller than the critical time step. In order to be conservative, often the time step is chosen much smaller than the critical time step.

It would be also interesting to solve the same problem in example 8.5, but this time with a slow application of the load. In order to do that, $t_{max} = 30$ ms is used. Note that even if the ending time is increased by 10 times, the time step needs to remain the same as $\Delta t = 21 \mu s$ in order to maintain stability. Figure 8.12 shows the results from the central difference method and from the quasi-static solution. It is interesting to note that there is no significant difference between the dynamic and quasi-static analysis results. As shown in example 8.5, the period of vibration for the bar is about 0.8 ms; that is, the bar can vibrate about 1.26 cycles in a millisecond. Therefore, if the applied load is in the order of 10 ms, the load is relatively slow compared to the structure's natural vibration frequency. Therefore, the structure behaves as if it were under a quasi-static load.

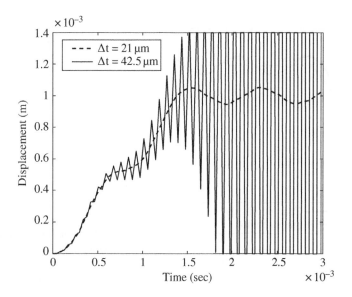

Figure 8.11 Instability of the central finite difference method due to a large time step

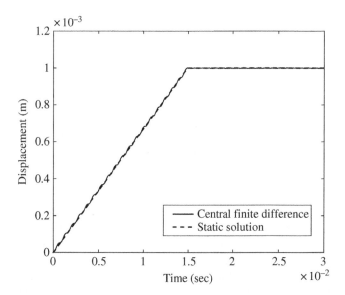

Figure 8.12 Equivalence of dynamic and static solution under slowly applied load

8.4.3 Newmark Method: Implicit Time Integration

In the implicit methods, we apply the equations of motion in eq. (8.38) at time step $(n + 1)$ in order to calculate the displacements $\mathbf{Q}_{n+1}$. This is different from the explicit CDM wherein the equations were applied at the previous time step n. The Newmark family of time integration algorithms is one of the most popular time integration methods as a single-step algorithm. Consider the following Taylor series expansions for velocity:

$$\dot{\mathbf{Q}}_{n+1} = \dot{\mathbf{Q}}_n + (1-\gamma)\Delta t\ddot{\mathbf{Q}}_n + \gamma\Delta t\ddot{\mathbf{Q}}_{n+1}, \tag{8.44}$$

and displacement:

$$\mathbf{Q}_{n+1} = \mathbf{Q}_n + \Delta t\dot{\mathbf{Q}}_n + \left(\frac{1}{2}-\beta\right)(\Delta t)^2\ddot{\mathbf{Q}}_n + \beta(\Delta t)^2\ddot{\mathbf{Q}}_{n+1}, \tag{8.45}$$

where γ and β are two parameters that determine the accuracy and stability of integration. Note that when $\beta = 0$, the Newmark integration method becomes an explicit method. Therefore, in the following derivations, we assume that $\beta > 0$. The Newmark family of algorithms are based on different choices of these parameters. Some common algorithms will be presented later.

The Newmark algorithm can be split into two steps: predictor and correction steps. In the predictor step, the velocity and displacement are predicted using known information at time t_n. After calculating the acceleration at time t_{n+1}, the corrector step updates the velocity and displacement with the acceleration. From eqs. (8.44) and (8.45), the predictors for velocity and displacement are defined as

$$\dot{\mathbf{Q}}_{n+1}^{pr} = \dot{\mathbf{Q}}_n + (1-\gamma)\Delta t \ddot{\mathbf{Q}}_n, \tag{8.46}$$

$$\mathbf{Q}_{n+1}^{pr} = \mathbf{Q}_n + \Delta t \dot{\mathbf{Q}}_n + \left(\frac{1}{2}-\beta\right)(\Delta t)^2 \ddot{\mathbf{Q}}_n. \tag{8.47}$$

That is, the predictors include those terms at time t_n. Once the acceleration at time t_{n+1} is given, they are corrected as

$$\dot{\mathbf{Q}}_{n+1} = \dot{\mathbf{Q}}_{n+1}^{pr} + \gamma\Delta t \ddot{\mathbf{Q}}_{n+1}, \tag{8.48}$$

$$\mathbf{Q}_{n+1} = \mathbf{Q}_{n+1}^{pr} + \beta(\Delta t)^2 \ddot{\mathbf{Q}}_{n+1}. \tag{8.49}$$

Therefore, the only remaining task is to calculate the acceleration at time t_{n+1}, which can be achieved by substituting eq. (8.49) into the equations of motion in eq. (8.38) at t_{n+1}:

$$\mathbf{M}\ddot{\mathbf{Q}}_{n+1} + \mathbf{K}\mathbf{Q}_{n+1} = \mathbf{F}_{n+1}. \tag{8.50}$$

Substituting for $\mathbf{Q}_{n+1}$ from eq. (8.49) into the above equation and rearranging the terms we obtain

$$\hat{\mathbf{M}}\ddot{\mathbf{Q}}_{n+1} = \hat{\mathbf{F}}_{n+1}, \tag{8.51}$$

where the effective mass matrix and load vector are derived as

$$\hat{\mathbf{M}} = \mathbf{M} + \beta(\Delta t)^2 \mathbf{K},$$
$$\hat{\mathbf{F}}_{n+1} = \mathbf{F}_{n+1} - \mathbf{K}\mathbf{Q}_{n+1}^{pr}. \tag{8.52}$$

Note that the Newmark method solves for acceleration in eq. (8.51), while the central difference method in section 8.4.2 solves for displacement. In general, it is possible that the Newmark method can be formulated to solve for displacement, which is left as an exercise problem. Once the acceleration is calculated, the velocity and displacement can be corrected using eqs. (8.48) and (8.49).

The steps involved in implementing Newmark's method are summarized below. We note that there is no need for any special starting procedure, which was required in the case of the central difference method.

1. Calculate the initial acceleration using

$$\ddot{\mathbf{Q}}_0 = \mathbf{M}^{-1}(\mathbf{F}_0 - \mathbf{K}\mathbf{Q}_0).$$

2. Construct the effective mass matrix

$$\hat{\mathbf{M}} = \mathbf{M} + \beta(\Delta t)^2 \mathbf{K}.$$

3. Set $n = 0$, and calculate predictors:

$$\dot{\mathbf{Q}}_{n+1}^{pr} = \dot{\mathbf{Q}}_n + (1-\gamma)\Delta t \ddot{\mathbf{Q}}_n,$$
$$\mathbf{Q}_{n+1}^{pr} = \mathbf{Q}_n + \Delta t \dot{\mathbf{Q}}_n + \left(\frac{1}{2}-\beta\right)(\Delta t)^2 \ddot{\mathbf{Q}}_n.$$

4. Construct the effective force vector

$$\hat{\mathbf{F}}_{n+1} = \mathbf{F}_{n+1} - \mathbf{K}\mathbf{Q}_{n+1}^{pr}.$$

5. Solve for $\ddot{\mathbf{Q}}_{n+1}$

$$\hat{\mathbf{M}}\ddot{\mathbf{Q}}_{n+1} = \hat{\mathbf{F}}_{n+1}.$$

6. Correct velocity and displacement

$$\dot{\mathbf{Q}}_{n+1} = \dot{\mathbf{Q}}_{n+1}^{pr} + \gamma\Delta t\ddot{\mathbf{Q}}_{n+1},$$
$$\mathbf{Q}_{n+1} = \mathbf{Q}_{n+1}^{pr} + \beta(\Delta t)^2\ddot{\mathbf{Q}}_{n+1}.$$

7. If $n = N$, stop. Otherwise, set $n = n + 1$ and go to step 3

As mentioned before, the performance of Newmark integration algorithms depend on the choice of the two parameters, γ and β. The most important factors in choosing appropriate parameters are accuracy, stability, and dissipation. Accuracy and stability of a numerical integration were explained when we discussed the central difference method. When a numerical solution is approximated, the energy of the initial wave may be reduced in a way analogous to a diffusional process, which is called numerical dissipation. In some cases, "artificial dissipation" is intentionally added to improve numerical stability of the solution. Some combinations of γ and β may or may not include the numerical dissipation. Table 8.1 shows some examples of the Newmark family time integration algorithms.

In conditionally stable time integration algorithms, such as the central difference method, stability is affected by the size of the time step. Normally, if the chosen time step that can provide stability is small enough, accuracy is not a major issue for the conditionally stable algorithms. In unconditionally stable time integration algorithms, however, a time step size can be chosen independent of stability considerations. Therefore, the size of the time step in unconditionally stable algorithms can be much larger than that of the conditionally stable algorithms. However, the size of the time step must be chosen to provide the required level of accuracy because the accuracy decreases as the size of time step increases.

Advanced studies on stability and accuracy using eigenvalue analysis show that the Newmark family methods show second-order accuracy with $\gamma = 1/2$, while they show first-order accuracy when $\gamma \neq 1/2$. Since accuracy is an important criterion, many Newmark family algorithms use $\gamma = 1/2$. However, numerical damping does not provide stability for $\gamma = 1/2$. Therefore, for Newmark methods, $\gamma > 1/2$ is necessary to introduce high-frequency dissipation.

For stability, the same eigenvalue analysis shows that the Newmark method has the following stability conditions:

$$\text{Unconditionally stable}: \gamma \geq \frac{1}{2}, \quad \beta \geq \frac{1}{2}\gamma. \tag{8.53}$$

$$\text{Conditionally stable}: \gamma \geq \frac{1}{2}, \quad \beta < \frac{1}{2}\gamma. \tag{8.54}$$

For conditionally stable algorithms, the time step should be less than the critical time step. For the undamped system, the critical time step, Δt_{cr}, can be calculated by

$$\omega_{\max}\Delta t_{cr} = \frac{1}{\sqrt{\frac{1}{2}\gamma - \beta}}, \tag{8.55}$$

Table 8.1   Newmark family of time integration algorithms

Method	γ	β	Properties
Average acceleration	1/2	1/4	Implicit, unconditionally stable
Linear acceleration	1/2	1/6	Implicit, conditionally stable
Fox-Goodwin	1/2	1/12	Implicit, conditionally stable. 4th-order accuracy
Central difference	1/2	0	Explicit, conditionally stable

where ω_{max} is the largest eigenvalue of the system. In general, the element eigenvalue is larger than that of the system. Since the maximum eigenvalue of the system is difficult to calculate, the maximum element eigenvalue is often used as a conservative estimate. Let the element size be L. It is also beneficial to note that the maximum element eigenvalue is $\omega_{max} \propto 1/L^2$. Therefore, as the element size is decreased, the critical time step also decreases.

EXAMPLE 8.6 *Newmark's time integration method*

Solve example 8.5 using Newmark's method with two different time steps $\Delta t = 42\mu$ sec and $\Delta t = 75\mu$ sec. Compare the tip displacement using the average acceleration method (unconditionally stable) and the linear acceleration method (conditionally stable).

SOLUTION   The Newmark integration parameters are $\gamma = 2\beta = 1/2$ for the average acceleration method, and $\gamma = 1/2$ and $\beta = 1/6$ for linear acceleration. The following MATLAB script was used to perform the time integration. Different methods and different time steps can be tested by changing the variables gamma, beta, and Dt.

```
%Newmark method
%Problem parameters
Area=10e-5;E=100e9;rho=4000;L=1/3;Fmax=0.001*Area*E;tmax=3e-3;
gamma=0.5; beta=0.25;Dt=42e-6; Nmax=round(tmax/Dt)+1;
t = linspace(0, tmax, Nmax);
%Global matrices
K=(Area*E/L)*[2 -1 0;-1 2 -1;0 -1 1];
M=(rho*Area*L/6)*[4 1 0;1 4 1;0 1 2];
F=[zeros(2,Nmax);
    linspace(0,Fmax,round(Nmax/2)) Fmax*ones(1,Nmax-round(Nmax/2))];
Minv=inv(M + (beta*Dt^2)*K);
Q=zeros(3,Nmax);V=zeros(3,Nmax);A=zeros(3,Nmax);
%Time integration
for n=1:Nmax-1
 %Predictor
 Vpr = V(:,n)+(1- gamma)*Dt*A(:,n);
 Qpr = Q(:,n)+Dt*V(:,n)+(0.5-beta)*Dt^2*A(:,n);
 A(:,n+1)=Minv*(F(:,n+1)-K*Qpr);
 %Corrector
 V(:,n+1)=Vpr + gamma*Dt*A(:,n+1);
 Q(:,n+1)=Qpr + (beta*Dt^2)*A(:,n+1);
end
%Static solution
Qs=K\F;
plot (t,Q(3,:),t,Qs(3,:));
```

From the stability condition in eq. (8.55), the stability condition for the linear acceleration can be given as

$$\omega_{max}\Delta t_{cr} = \frac{1}{\sqrt{\frac{1}{2}\gamma - \beta}} = 3.464.$$

Using the maximum natural frequency from example 8.5, the critical time step for the linear acceleration method can be obtained as

$$\Delta t_{cr} = \frac{3.464}{\omega_{max}} = 72.7\mu \text{ sec.}$$

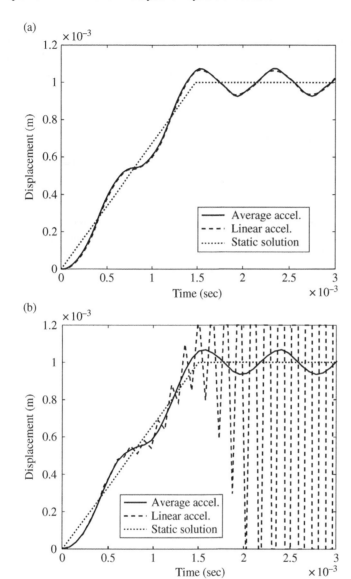

Figure 8.13 Tip displacement of the uniaxial bar in figure 8.9 using the Newmark method; (a) $\Delta t = 42$ μsec, and (b) $\Delta t = 75$ μsec

While it is expected that the average acceleration method will always be stable, the linear acceleration method will be stable at $\Delta t = 42\mu$ sec but not at $\Delta t = 75\mu$ sec. Figure 8.13(a) shows the time history of displacements with $\Delta t = 42\mu$ sec. As expected, both methods yield a similar trend by oscillating around the static solution. However, when $\Delta t = 75\mu$ sec as shown in figure 8.13(b), the linear acceleration method diverges as the time step size is larger than the critical time step for stability. ∎

EXAMPLE 8.7 *Impact on an aluminum beam*

A rigid mass of 1 kg drops from a height of 1 m onto a simply supported aluminum beam at the center (figure 8.14). Use Newmark time integration to calculate the maximum impact force and impact duration. Plot the center deflection of the beam as a function of time. The dimensions of the beam are: $1 \times 1 \times 30$ cm; $E = 70$ GPa, and density $\rho = 2{,}700$ kg/m^3.

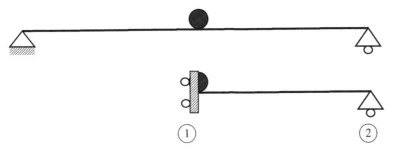

Figure 8.14 (Top) Impact of a mass on a simply supported beam; (bottom) one-element FE model of one half of the beam.

SOLUTION   A realistic approach to this problem should include Hertizian contact[7] so that the mass and beam locally deform during impact. However, this is beyond the scope of this book. We will assume that the impact mass attaches itself to the beam until it rebounds. This is a good approximation considering the fact that the impact mass is much greater than that of the beam, which is about 81 g.

Before modeling the problem using the finite element method, we will obtain a quick but approximate analytical solution. If we ignore the mass effects, the beam can be considered as a spring with stiffness given by $k_b = 48EI/L^3 = 103.7$ kN/m. We can model the problem as a single DOF spring-mass system with the mass equal to 1 kg and initial velocity $v_0 = \sqrt{2g} = 4.43$ m/s. The maximum deflection of the spring (beam) w_{max}, maximum impact force F_{max}, and impact duration T_i, can be calculated as $w_{max} = 13.8$ mm, $F_{max} = 1,426$ N, and $T_i = 9.8$ ms. The impact force history will take a simple form $F(t) = F_{max} \sin \pi t/T_i$.

We will now use Newmark method to solve the same problem. Due to symmetry, we will use one half of the beam and one half of the impact mass as shown in figure 8.14. Furthermore, one half of the beam will be modeled using a one-beam element for the purpose of illustration. In practice, a large number of elements must be used to get a more accurate solution. The element stiffness and mass matrices are derived as follows:

$$[K] = \frac{EI}{l^3}\begin{bmatrix} 12 & 6l \\ 6l & 4l^2 \end{bmatrix} = 10^5 \begin{bmatrix} 2.07 & 0.156 \\ 0.156 & 0.0156 \end{bmatrix} \begin{matrix} v_1 \\ \theta_2 \end{matrix},$$

$$[M] = \frac{m}{420}\begin{bmatrix} 156 & -13l \\ -13l & 4l^2 \end{bmatrix} + \begin{bmatrix} 0.5m_i & 0 \\ 0 & 0 \end{bmatrix} = \begin{bmatrix} 0.515 & -0.188 \times 10^{-3} \\ -0.188 \times 10^{-3} & 0.868 \times 10^{-5} \end{bmatrix} \begin{matrix} v_1 \\ \theta_2 \end{matrix}.$$

In the above calculation, l and m are the length and mass of one half of the beam, and m_i is the impacting mass. Note that the mass $m_i/2$ is added to M_{11} of the beam mass matrix. The natural frequencies are calculated using the MATLAB command $[\mathbf{q},\mathbf{D}] = \text{eig}(\mathbf{K},\mathbf{M})$. The two natural frequencies are: $\omega_1 = 316$ rad/s and $\omega_2 = 13,502$ rad/s. The corresponding periods are $T_1 = 2\pi/\omega_1 = 0.0199$ s and $T_2 = 2\pi/\omega_2 = 465\,\mu$s. We will use $\Delta t = T_2/10$ and perform the calculations up to $t = T_1/2$. The procedure for the Newmark method were implemented in a MATLAB script. The impact force can be calculated using the expression $F_i(t) = -m_i\ddot{v}_1$, where $\ddot{v}_1$ is the acceleration of node 1. The impact force history is plotted in the figure below. In the same figure, the result corresponding to the single DOF approximation is also shown in dashed lines.

In the very beginning, the impact force becomes negative (less than zero) for a brief time, which is not physically possible. It means the mass loses contact with the beam briefly, but it will hit the beam again and regain full contact for the rest of the period. We will choose to ignore this episode in the present calculations.

One can observe that the impact force above has two harmonics. The single DOF solution (dashed line) is a single sinusoidal variation. In the finite element solution, a high-frequency content is superposed over the single DOF solution. This is because the finite element has two DOFs, and thus, two natural frequencies. Therefore, the beam will vibrate according to $T_1 = 2\pi/\omega_1 = 0.0199$ s, which is the dashed curve in figure 8.15, and

---

[7] Sankar, B. V., and Sun, C. T. 1985. "An Efficient Numerical Algorithm for Transverse Impact Problems." *Computers & Structures*, 20(6):1009–1012.

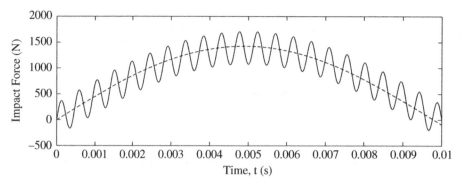

Figure 8.15 Impact response of a simply supported beam subjected to central impact. The figure shows the impact force history. The dotted line represents the approximate single DOF solution.

$T_2 = 2\pi/\omega_2 = 465\,\mu s$, which is the period of the high frequency. The amplitude of the vibration depends on the magnitude of impact velocity, but the beam vibrates according to its natural frequencies. If a large number of elements were used, then the solution would include several harmonics. The maximum impact force is 1,706 N, and the impact duration is about 0.01 s. ▄▄

8.5 METHOD OF MODE SUPERPOSITION

The direct integration method discussed in section 8.4 is most suitable when the loads act over a short duration, and only the response during the same short duration is of crucial interest. Since the time step required is very small, direct integration methods are not suitable when the response is required for a long period. In such a case, we propose to identify a transformation of displacements that will enable us to make the solution technique very efficient for a long-duration response.

As explained in section 8.3, structures vibrate at their natural frequencies when their natural modes of vibration are excited. If the excitation frequency of the applied force is much smaller than the lowest natural frequencies of the structure, it will behave as if a quasi-static load is applied. On the other hand, if the excitation frequency is the same as one of the natural frequencies, a phenomenon called resonance occurs, and the amplitude of vibration greatly increases and can cause structural failure. Since the structure vibrates in its mode shapes, it is possible to express the forced vibration response of a structure as a linear combination of its mode shapes, which allows us to decouple the equation of motion. This approach for computing the forced response of structure is called the method of mode superposition. In fact, the method of mode superposition is commonly used in forced vibrations of linear systems because it can provide insight into the structure's behavior that is not otherwise available and because it is usually significantly more cost effective than the direct time integration methods.

Before we proceed further, we will derive certain properties of eigenvectors. Consider the generalized eigenvalue problem in eq. (8.29), which can be written as

$$[\mathbf{K}]\{\mathbf{A}\} = \omega^2[\mathbf{M}]\{\mathbf{A}\}. \tag{8.56}$$

As we noted earlier, for an N-DOFs system, the above eigenvalue problem leads to N natural frequencies, $\omega_1, \omega_2, \ldots, \omega_N$, and corresponding eigenvectors $\mathbf{A}_1, \mathbf{A}_2,\ldots, \mathbf{A}_N$. Each eigenvalue and its eigenvector must satisfy the above equation. Let us consider such equations for two arbitrary modes i and j:

$$\mathbf{KA}_i = \omega_i^2\mathbf{MA}_i,$$
$$\mathbf{KA}_j = \omega_j^2\mathbf{MA}_j. \tag{8.57}$$

We multiply the first equation by $\mathbf{A}_i^T$ and the second by $\mathbf{A}_j^T$ and subtract to obtain:

$$0 = \left(\omega_i^2 - \omega_j^2\right)\mathbf{A}_j^T\mathbf{M}\mathbf{A}_i. \tag{8.58}$$

In arriving at the above equation, we have used the identities $\mathbf{A}_j^T\mathbf{K}\mathbf{A}_i = \mathbf{A}_i^T\mathbf{K}\mathbf{A}_j$ and $\mathbf{A}_j^T\mathbf{M}\mathbf{A}_i = \mathbf{A}_i^T\mathbf{M}\mathbf{A}_j$ because $\mathbf{K}$ and $\mathbf{M}$ are symmetric matrices. From eq. (8.58) we find that when $i \neq j$, that is, $\omega_i^2 \neq \omega_j^2$, $\mathbf{A}_j^T\mathbf{M}\mathbf{A}_i$ must be equal to zero. That is, the eigenvectors are orthogonal in some sense, and we call them M-orthogonal. Now consider the case $i = j$. The product $\mathbf{A}_i^T\mathbf{M}\mathbf{A}_i$ must be positive if it is nonzero, because $\mathbf{M}$ is positive definite. Let us denote the product by r_i^2. Then we can normalize the eigenvector by dividing by r_i such that $\mathbf{A}_i^T\mathbf{M}\mathbf{A}_i = 1$. Then from eq. (8.57), it follows that $\mathbf{A}_i^T\mathbf{K}\mathbf{A}_i = \omega_i^2$. Thus, the properties of the eigenvectors can be summarized as:

$$\left.\begin{aligned}\mathbf{A}_i^T\mathbf{M}\mathbf{A}_j &= 1, \\ \mathbf{A}_i^T\mathbf{K}\mathbf{A}_i &= \omega_i^2\end{aligned}\right\} \text{ if } i = j,$$

$$\left.\begin{aligned}\mathbf{A}_i^T\mathbf{M}\mathbf{A}_j &= 0, \\ \mathbf{A}_i^T\mathbf{K}\mathbf{A}_i &= 0\end{aligned}\right\} \text{ if } i \neq j. \tag{8.59}$$

Let us define a square matrix $\mathbf{\Lambda}$ whose columns are the eigenvectors $\mathbf{A}_j$. That is

$$\mathbf{\Lambda} = [\mathbf{A}_1\,\mathbf{A}_2\ldots\mathbf{A}_N] = \begin{bmatrix} A_1^{(1)} & A_1^{(2)} & \cdots & A_1^{(N)} \\ A_2^{(1)} & A_2^{(2)} & \ddots & \vdots \\ \vdots & \vdots & \ddots & \vdots \\ A_N^{(1)} & A_2^{(N)} & \cdots & A_N^{(N)} \end{bmatrix}, \tag{8.60}$$

where $A_m^{(n)}$ is the m-th DOF of n-th eigenvector. Then from eq. (8.59), we can deduce the following:

$$\mathbf{\Lambda}^T\mathbf{M}\mathbf{\Lambda} = \mathbf{I}_N,$$

$$\mathbf{\Lambda}^T\mathbf{K}\mathbf{\Lambda} = \mathbf{\Omega} = \begin{bmatrix} \omega_{11}^2 & 0 & \cdots & 0 \\ 0 & \omega_{22}^2 & \ddots & \vdots \\ \vdots & \vdots & \ddots & \vdots \\ 0 & 0 & \cdots & \omega_{NN}^2 \end{bmatrix}, \tag{8.61}$$

where $\mathbf{I}_N$ is the $N \times N$ identity matrix. Note that $\mathbf{\Omega} = \mathbf{\Lambda}^T\mathbf{K}\mathbf{\Lambda}$ is a diagonal matrix of eigenvalues.

8.5.1 Modal Decomposition

Returning to forced vibration problems, our goal is to simplify the time integration procedures. As we mentioned before, the response of forced vibration can be represented using a linear combination of mode shapes. That is, the dynamic response can be written as

$$\{\mathbf{Q}(t)\} = x_1(t)\{\mathbf{A}_1\} + x_2(t)\{\mathbf{A}_2\} + \cdots + x_N(t)\{\mathbf{A}_N\},$$

where the mode shapes are independent of time, while the vector of coefficients, $\{\mathbf{X}(t)\} = \{x_1(t), x_2(t), ..., x_N(t)\}^T$, are a function of time. Therefore, the shape of deformation is determined by mode shapes, while the time-varying magnitude of deformation is represented by $\{\mathbf{X}(t)\}$.

Since eigenvectors are M-orthogonal based on the discussion in eq. (8.59), they can be considered as modal basis vectors. Then, the above modal decomposition can be considered as a transformation from physical DOFs $\{\mathbf{Q}(t)\}$ to modal DOFs $\{\mathbf{X}(t)\}$, as

$$\{\mathbf{Q}(t)\} = [\mathbf{\Lambda}]\{\mathbf{X}(t)\}. \tag{8.62}$$

Substituting the transformation in eq. (8.31), we obtain the equation of motion as:

$$[\mathbf{M}][\mathbf{\Lambda}]\{\ddot{\mathbf{X}}\} + [\mathbf{K}][\mathbf{\Lambda}]\{\mathbf{X}\} = \{\mathbf{F}\}. \tag{8.63}$$

That is, instead of solving physical DOFs, the transformed equation solves for the modal DOFs. This transformation to modal DOFs can provide significant computational efficiency. In practice, only a small number of modes are enough to accurately approximate the physical DOFs in eq. (8.62). That is, the number of modes N can be much smaller than the size of matrix.

The matrix equation in eq. (8.63) can be further simplified by pre-multiplying the above equation by $[\mathbf{\Lambda}]^T$ to obtain

$$[\mathbf{\Lambda}]^T[\mathbf{M}][\mathbf{\Lambda}]\{\ddot{\mathbf{X}}\} + [\mathbf{\Lambda}]^T[\mathbf{K}][\mathbf{\Lambda}]\{\mathbf{X}\} = [\mathbf{\Lambda}]^T\{\mathbf{F}\}, \tag{8.64}$$

which can be written as

$$[\hat{\mathbf{M}}]\{\ddot{\mathbf{X}}\} + [\hat{\mathbf{K}}]\{\mathbf{X}\} = \{\mathbf{R}\}, \tag{8.65}$$

where $\{\mathbf{R}\} = [\mathbf{\Lambda}]^T\{\mathbf{F}\}$ is the generalized force vector given. From eq. (8.61), we note that $[\hat{\mathbf{M}}] = [\mathbf{\Lambda}]^T[\mathbf{M}][\mathbf{\Lambda}]$ is an identity matrix, and $[\hat{\mathbf{K}}] = [\mathbf{\Lambda}]^T[\mathbf{K}][\mathbf{\Lambda}]$ is a diagonal matrix containing the squares of the natural frequencies, that is, $\hat{\mathbf{K}}_{ii} = \omega_i^2$. Then the equations are completely decoupled and we obtain N number of differential equations as

$$\ddot{X}_i + \omega_i^2 X_i = R_i(t), \quad i = 1, ..., N. \tag{8.66}$$

This is a significant simplification from the N-dimensional system of coupled equations to N decoupled equations.

An analytical solution can be derived for eq. (8.66) and can be written as

$$X_i(t) = A_i \sin\omega t + B_i \cos\omega t + \frac{1}{\omega_i}\int_0^t \sin\omega_i(t-\tau)R_i(\tau)\mathrm{d}\tau. \tag{8.67}$$

The first two terms in this solution are referred to as the *complementary solution*, and it corresponds to the response if no external loads are acting. Therefore, the complementary solution is also called the *free vibration response*. The second part of the solution involves a convolution integral and is called the *particular solution*, which is the response of the structure to the external load. When damping is ignored, both of these components will last forever, but in reality there will be damping and the complementary part will be damped out so that only the particular part of the solution will continue as the *steady-state solution*. The constants, A_i and B_i, are evaluated using the initial conditions of $\mathbf{X}$, that is, $\mathbf{X}(0)$ and $\dot{\mathbf{X}}(0)$. The initial conditions are usually known for the nodal DOFs. Let the initial value of nodal variables and corresponding nodal velocities be represented as $\mathbf{Q}(0)$ and $\dot{\mathbf{Q}}(0)$. The inverse transformation corresponding to eq. (8.62) can be derived by pre-multiplying the equation by $\mathbf{\Lambda}^T\mathbf{M}$ and using the orthogonality condition in eq. (8.59), as

$$\mathbf{X} = \mathbf{\Lambda}^T\mathbf{M}\mathbf{Q}. \tag{8.68}$$

Then, the initial conditions of $\mathbf{X}$ can be written as

$$\mathbf{X}(0) = \mathbf{\Lambda}^{\mathrm{T}}\mathbf{M}\mathbf{Q}(0),$$
$$\dot{\mathbf{X}}(0) = \mathbf{\Lambda}^{\mathrm{T}}\mathbf{M}\dot{\mathbf{Q}}(0). \tag{8.69}$$

Putting $t = 0$ in eq. (8.67), we obtain

$$B_i = X_i(0). \tag{8.70}$$

Differentiating the solution for $X_i(t)$ in eq. (8.67) and putting $t = 0$, we have

$$\dot{X}_i(0) = A_i\omega_i \quad \Rightarrow \quad A_i = \frac{\dot{X}_i(0)}{\omega_i}. \tag{8.71}$$

If the forcing function $F(t)$ and hence $R(t)$ are available in a closed form, then the integration above can also be performed in closed form. If the forcing function is known only in a discrete form, then numerical integration is used to evaluate the above integral. Once the solution is obtained for $\mathbf{X}(t)$, the solution in terms of nodal variables is obtained using the transformation $\mathbf{Q} = \mathbf{\Lambda}\mathbf{X}$ (see eq. (8.62)).

EXAMPLE 8.8   *Response of a uniaxial bar to a ramp loading*

Consider a clamped uniaxial bar subjected to a tip load as shown in figure 8.9. Use three elements of equal length to determine the tip displacement as a function of time. The properties of the bar are: $E = 100$ GPa, $L = 1$ m, $A = 10^{-5}$ m^2, $\rho = 4000$ kg/m^3. The tip force is given by: $f(t) = 2f_{max}t/t_{max}$ for $0 < t < t_{max}/2$, and $f(t) = f_{max}$ for $t > t_{max}/2$, where $f_{max} = 0.001EA$ and $t_{max} = 3$ ms. Calculate the tip displacement $u(t)$ for $0 < t < t_{max}$.

SOLUTION   The global stiffness and mass matrices are derived as:

$$[\mathbf{K}] = 10^7 \begin{bmatrix} 6 & -3 & 0 \\ -3 & 6 & -3 \\ 0 & -3 & 3 \end{bmatrix} \begin{matrix} u_2 \\ u_3, \\ u_4 \end{matrix}$$

$$[\mathbf{M}] = \begin{bmatrix} 0.089 & 0.022 & 0 \\ 0.022 & 0.089 & 0.022 \\ 0 & 0.022 & 0.044 \end{bmatrix} \begin{matrix} u_2 \\ u_3. \\ u_4 \end{matrix}$$

Using the MATLAB command $[\Lambda,\mathbf{D}]$ =eig($\mathbf{K},\mathbf{M}$), we obtain the natural frequencies (s^{-1}): $\omega_1 = \sqrt{D_{11}} = 8{,}025$, $\omega_2 = \sqrt{D_{22}} = 26{,}243$, and $\omega_3 = \sqrt{D_{33}} = 47{,}609$. In addition, the M-normalized matrix of eigenvectors Λ can be obtained as

$$[\Lambda] = \begin{bmatrix} 1.1496 & -2.7524 & 1.8277 \\ 1.9912 & 0 & -3.1656 \\ 2.2993 & 2.7524 & 3.6553 \end{bmatrix}.$$

The generalized forces are calculated as

$$\mathbf{R}(t) = \mathbf{\Lambda}^{\mathrm{T}}\mathbf{F}(t) = [\Lambda] \begin{Bmatrix} 0 \\ 0 \\ F(t) \end{Bmatrix} = \begin{Bmatrix} 1.8277 \\ -3.1656 \\ 3.6553 \end{Bmatrix} f(t),$$

where with $t_0 = t_{max}/2$

$$f(t) = \begin{cases} f_{max}t/t_0, & 0 < t < t_0 \\ f_{max}, & t_0 < t < t_{max} \end{cases}.$$

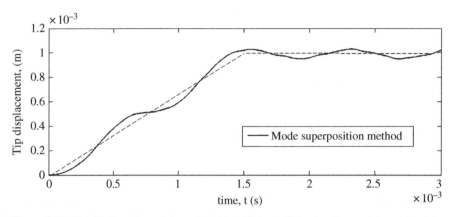

Figure 8.16 Tip displacement of the uniaxial bar in figure 8.9 using the modal superposition method

Since the initial conditions are given as $\mathbf{Q}(0) = \mathbf{0}$ and $\dot{\mathbf{Q}}(0) = \mathbf{0}$, we obtain $\mathbf{X}(0) = \mathbf{0}$ and $\dot{\mathbf{X}}(0) = \mathbf{0}$. Thus the constants $A_i = B_i = 0$ in eq. (8.71). Let $g_i(t)$ be defined as

$$g_i(t) = \frac{1}{\omega_i} \int_0^t \sin \omega_i (t - \tau) f(\tau) d\tau.$$

Performing the integration, we obtain

$$g_i(t) = \begin{cases} \dfrac{f_{\max}}{\omega_i^2} \left(\dfrac{t}{t_0} - \dfrac{\sin \omega_i t}{\omega_i t} \right), & 0 < t < t_0 \\[4mm] \dfrac{f_{\max}}{\omega_i^2} \cos \omega_i (t - t_0) + \dfrac{f_{\max}}{\omega_i^3 t_0} (\sin \omega_i (t - t_0) - \sin \omega_i t) \\[4mm] \quad + \dfrac{f_{\max}}{\omega_i^2} (1 - \cos \omega_i (t - t_0)) & t_0 < t < t_{\max} \end{cases}$$

Using eq. (8.67), we obtain the solution for $\mathbf{X}(t)$ as

$$\begin{Bmatrix} X_1(t) \\ X_2(t) \\ X_3(t) \end{Bmatrix} = \begin{Bmatrix} 1.8277 g_1(t) \\ -3.1656 g_2(t) \\ 3.6553 g_3(t) \end{Bmatrix}.$$

Finally, the solution for nodal displacements is obtained using the transformation $\mathbf{Q} = \mathbf{\Lambda}\mathbf{X}$:

$$\begin{Bmatrix} Q_1(t) \\ Q_2(t) \\ Q_3(t) \end{Bmatrix} = [\mathbf{\Lambda}] \begin{Bmatrix} 1.8277 g_1(t) \\ -3.1656 g_2(t) \\ 3.6553 g_3(t) \end{Bmatrix}.$$

The tip deflection Q_3 is plotted as a function of time t in figure 8.16. It is instructive to compare the solutions to the uniaxial bar problem obtained using different methods.■

EXAMPLE 8.9 *Impact response of a beam*

Use the mode superposition method to solve example 8.7: A rigid mass of 1 kg drops from a height of 1 m onto a simply supported aluminum beam at the center (figure 8.14). Use finite element analysis to calculate the maximum impact force and impact duration. Plot the center deflection of the beam as a function of time. The dimensions of the beam are: $1 \times 1 \times 30$cm; $E = 70$ GPa and density $\rho = 2{,}700$kg/m^3.

SOLUTION   The global stiffness and mass matrices and natural frequencies are calculated as:

$$[\mathbf{K}] = 10^5 \begin{bmatrix} 2.07 & 0.156 \\ 0.156 & 0.0156 \end{bmatrix} \begin{matrix} v_1 \\ \theta_2 \end{matrix},$$

$$[\mathbf{M}] = \begin{bmatrix} 0.515 & -0.188 \times 10^{-3} \\ -0.188 \times 10^{-3} & 0.868 \times 10^{-5} \end{bmatrix} \begin{matrix} v_1 \\ \theta_2 \end{matrix}.$$

The generalized eigenvalue problem is solved for two natural frequencies as $\omega_1 = 316 \text{rad/s}$ and $\omega_2 = 13,502 \text{rad/s}$. The corresponding eigenvectors are calculated as

$$\mathbf{\Lambda} = \begin{bmatrix} -1.34 & 0.181 \\ 0.139 & 341 \end{bmatrix}.$$

It should be noted that there are no external forces acting on the system. However, the initial velocities are nonzero. The initial conditions are derived as

$$\mathbf{X}(0) = \mathbf{\Lambda}^{\mathrm{T}} \mathbf{M} \mathbf{Q}(0) = \begin{Bmatrix} 0 \\ 0 \end{Bmatrix},$$

$$\dot{\mathbf{X}}(0) = \mathbf{\Lambda}^{\mathrm{T}} \mathbf{M} \dot{\mathbf{Q}}(0) = \mathbf{\Lambda}^{\mathrm{T}} \mathbf{M} \begin{Bmatrix} v_i \\ 0 \end{Bmatrix} = \begin{Bmatrix} \Lambda_{11} M_{11} + \Lambda_{21} M_{12} \\ \Lambda_{12} M_{11} + \Lambda_{22} M_{12} \end{Bmatrix} v_i.$$

The two differential equations along with corresponding initial conditions are:

$$\ddot{X}_1 + \omega_1^2 X_1 = 0, \ X_1(0) = 0, \ \dot{X}_1(0) = C_1 v_i, \ (C_1 = \Lambda_{11} M_{11} + \Lambda_{21} M_{12}),$$
$$\ddot{X}_2 + \omega_2^2 X_2 = 0, \ X_2(0) = 0, \ \dot{X}_2(0) = C_2 v_i, \ (C_2 = \Lambda_{12} M_{11} + \Lambda_{22} M_{12}).$$

The solution consists of only the complementary part because there is no load applied on the system. The complementary solution is given below:

$$X_1(t) = \frac{C_1 v_i}{\omega_1} \sin \omega_1 t, \quad X_2(t) = \frac{C_2 v_i}{\omega_2} \sin \omega_2 t.$$

The solution in terms of nodal variables is given by $\mathbf{Q} = \mathbf{\Lambda} \mathbf{X}$. We are interested in the response of v_1 at the impact point. The beam deflection and corresponding velocity and acceleration are obtained as

$$v_1(t) = \Lambda_{11} X_1 + \Lambda_{12} X_2 = \frac{C_1 v_i}{\omega_1} \Lambda_{11} \sin \omega_1 t + \frac{C_2 v_i}{\omega_2} \Lambda_{12} \sin \omega_2 t,$$

$$\dot{v}_1(t) = C_1 v_i \Lambda_{11} \cos \omega_1 t + C_2 v_i \Lambda_{12} \cos \omega_2 t,$$

$$\ddot{v}_1(t) = -v_i (C_1 \Lambda_{11} \omega_1 \sin \omega_1 t + C_2 \Lambda_{12} \omega_2 \sin \omega_2 t).$$

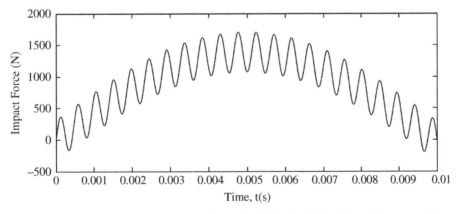

Figure 8.17 Impact response of a beam in figure 8.15 obtained using the mode superposition method

The impact force history can be calculated as

$$F_i(t) = -m_i \ddot{v}_1(t) = m_i v_i (C_1 \Lambda_{11} \omega_1 \sin \omega_1 t + C_2 \Lambda_{12} \omega_2 \sin \omega_2 t).$$

The impact force history is plotted in figure 8.17. It is identical to the one obtained using the Newmark method, but the mode superposition solution was obtained in a closed-form solution with less computational effort.

8.6 DYNAMIC ANALYSIS WITH STRUCTURAL DAMPING

In the previous sections, we considered systems that have no damping. Now we will consider the implementation of damping in a spring-mass-dashpot (viscous damping) system shown in figure 8.18. The procedures are similar to the direct method of formulating the finite element equations described in section 1.1. Consider the free-body diagram of a dashpot element in figure 8.18. The forces $f_i^{(e)}$ and $f_j^{(e)}$ can be related to the nodal velocities as

$$\begin{bmatrix} c & -c \\ -c & c \end{bmatrix} \begin{Bmatrix} \dot{u}_i \\ \dot{u}_j \end{Bmatrix} = \begin{Bmatrix} f_i^{(e)} \\ f_j^{(e)} \end{Bmatrix},$$

or

$$\left[\mathbf{c}^{(e)} \right] \left\{ \dot{\mathbf{q}}^{(e)} \right\} = \left\{ \mathbf{f}^{(e)} \right\}. \tag{8.72}$$

Recall that we have used a similar equation for spring elements of the form: $\left[\mathbf{k}^{(e)} \right] \left\{ \mathbf{q}^{(e)} \right\} = \left\{ \mathbf{f}^{(e)} \right\}$ in eq. (1.6). Now consider the equation of motion of a rigid mass at node i:

$$m_i \ddot{u}_i = F_i - \sum_{e=1}^{i_e} f_i^{(e)}, \tag{8.73}$$

where i_e is the number of elements connected to node i. Note that the nodal forces $f_i^{(e)}$ can arise from a spring or damper (dashpot) or both. Substituting for $f_i^{(e)}$ as a combination from eqs. (1.5) and (8.72), we obtain

$$[\mathbf{M}_s] \begin{Bmatrix} \ddot{u}_1 \\ \ddot{u}_2 \\ \vdots \\ \ddot{u}_{ND} \end{Bmatrix} + [\mathbf{C}_s] \begin{Bmatrix} \dot{u}_1 \\ \dot{u}_2 \\ \vdots \\ \dot{u}_{ND} \end{Bmatrix} + [\mathbf{K}_s] \begin{Bmatrix} u_1 \\ u_2 \\ \vdots \\ u_{ND} \end{Bmatrix} = \begin{Bmatrix} F_1 \\ F_2 \\ \vdots \\ F_{ND} \end{Bmatrix}. \tag{8.74}$$

After deleting the rows and columns corresponding to zero DOFs, we obtain the equations of motion as:

$$[\mathbf{M}] \{ \ddot{\mathbf{Q}}(t) \} + [\mathbf{C}] \{ \dot{\mathbf{Q}}(t) \} + [\mathbf{K}] \{ \mathbf{Q}(t) \} = \{ \mathbf{F}(t) \}. \tag{8.75}$$

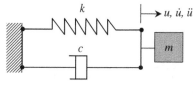

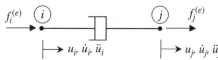

Figure 8.18 One-dimensional spring-mass-dashpot element

The above equation can be solved using one of the time integration methods or the mode superposition method.

Damping in a continuous system is difficult to decouple. Hence, a heuristic method called proportional damping, also called Rayleigh damping, is used in practice. The damping matrix [C] is assumed to be a linear combination of stiffness and mass matrices as

$$[\mathbf{C}] = c_1 [\mathbf{K}] + c_2 [\mathbf{M}], \tag{8.76}$$

where c_1 and c_2 are constants of proportionality. We make the proportional damping assumption mainly because it provides computational convenience.

8.6.1   Central Difference Method with Damping

Recall that in the CDM, the equilibrium condition is stated at time t_n to calculate the displacements at time t_{n+1}. We use central difference to approximate the velocity term in eq. (8.75):

$$\dot{\mathbf{Q}}_n = \frac{1}{2 \Delta t} (\mathbf{Q}_{n+1} - \mathbf{Q}_{n-1}). \tag{8.77}$$

Substituting for $\ddot{\mathbf{Q}}_n$ from eq. (8.37) and for $\dot{\mathbf{Q}}_n$ from eq. (8.76) into (8.75) at time t_n, and solving for $\mathbf{Q}_{n+1}$, we obtain

$$\left[\frac{1}{\Delta t^2} \mathbf{M} + \frac{1}{2 \Delta t} \mathbf{C} \right] \mathbf{Q}_{n+1} = \mathbf{F}_n - \left(\mathbf{K} - \frac{2}{\Delta t^2} \mathbf{M} \right) \mathbf{Q}_n - \left(\frac{1}{\Delta t^2} \mathbf{M} - \frac{1}{2 \Delta t} \mathbf{C} \right) \mathbf{Q}_{n-1}. \tag{8.78}$$

Since we know the right-hand side, the above equation can be solved for $\mathbf{Q}_{n+1}$. Note that the expression for $\ddot{\mathbf{Q}}_0$ has to be modified as

$$\ddot{\mathbf{Q}}_0 = \mathbf{M}^{-1} \left(\mathbf{F}_0 - \mathbf{C} \dot{\mathbf{Q}}_0 - \mathbf{K} \mathbf{Q}_0 \right). \tag{8.79}$$

EXAMPLE 8.10   *Explicit time integration of uniaxial bar with damping*

Repeat the uniaxial bar problem in example 8.5 including damping in the bar. Assume $c_1 = 10^{-4}$ and $c_2 = 5 \times 10^{-3}$.

SOLUTION   Equations (8.77) were implemented in MATLAB. The tip displacement is shown in figure 8.19.

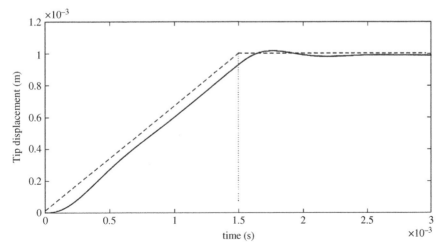

Figure 8.19 Tip displacement of a uniaxial bar in figure 8.9 using the central difference method when structural damping is included

8.6.2 Newmark Method with Damping

As discussed earlier, the equilibrium is applied at time t_{n+1} in order to determine $\mathbf{Q}_{n+1}$:

$$\mathbf{M}\ddot{\mathbf{Q}}_{n+1} + \mathbf{C}\dot{\mathbf{Q}}_{n+1} + \mathbf{K}\mathbf{Q}_{n+1} = \mathbf{F}_{n+1}. \tag{8.80}$$

When damping is included, the effective mass matrix and force vector in eq. (8.52) can be defined as

$$\hat{\mathbf{M}} = \mathbf{M} + \gamma\Delta t\mathbf{C} + \beta(\Delta t)^2\mathbf{K},$$

$$\hat{\mathbf{F}}_{n+1} = \mathbf{F}_{n+1} - \mathbf{C}\dot{\mathbf{Q}}_{n+1}^{pr} - \mathbf{K}\mathbf{Q}_{n+1}^{pr}. \tag{8.81}$$

EXAMPLE 8.11 *Implicit time integration of uniaxial bar with damping*

Consider again the uniaxial bar problem in example 8.5. Solve this problem using the Newmark method assuming that damping is present with $c_1 = 10^{-4}$ and $c_2 = 5 \times 10^{-3}$.

SOLUTION   The equations of motion were implemented in a MATLAB script using the effective stiffness matrix and force vector in eq. (8.80). The response of the bar is shown in figure 8.20.

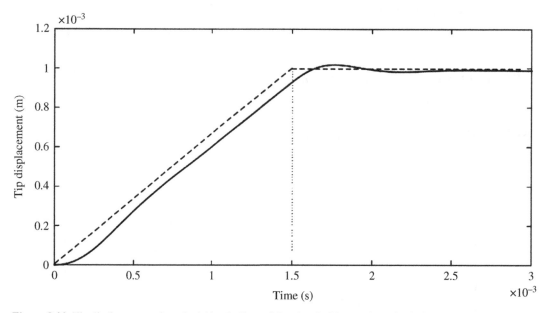

Figure 8.20 Tip displacement of a uniaxial bar in figure 8.9 using the Newmark method when structural damping is included

8.6.3 Modal Superposition with Damping

In the modal superposition method, we apply the transformation $\mathbf{Q} = \mathbf{\Lambda X}$ in eq. (8.62). Substituting the transformation in the equations of motion in eq. (8.75) and pre-multiplying the equation by $\mathbf{\Lambda}^T$, we obtain

$$\mathbf{\Lambda}^T \mathbf{M\Lambda}\ddot{\mathbf{X}} + \mathbf{\Lambda}^T \mathbf{C\Lambda}\dot{\mathbf{X}} + \mathbf{\Lambda}^T \mathbf{K\Lambda X} = \mathbf{\Lambda}^T \mathbf{F}. \tag{8.82}$$

We have already shown that $\mathbf{\Lambda}^T \mathbf{M\Lambda} = \mathbf{I}$, and $\mathbf{\Lambda}^T \mathbf{K\Lambda}$ is a diagonal matrix consisting of squares of natural frequencies. Proportional damping allows the decoupling of the equations of motion in the modal superposition method as it did for the undamped cases.

$$\begin{aligned}\mathbf{\Lambda}^T \mathbf{C\Lambda} &= \mathbf{\Lambda}^T [c_1 \mathbf{M} + c_2 \mathbf{K}]\mathbf{\Lambda} \\ &= c_1 \mathbf{I} + c_2 \mathbf{\Omega},\end{aligned} \tag{8.83}$$

where $\mathbf{I}$ is an identity matrix of size $N \times N$ and $\mathbf{\Omega}$ is defined in eq. (8.61). Then we note that eq. (8.81) will decompose into N number of ordinary differential equations similar to eq. (8.66) but with an extra term for damping:

$$\ddot{X}_i + 2\xi_i\omega_i\dot{X}_i + \omega_i^2 X_i = R_i(t), \quad i = 1, \ldots, N, \tag{8.84}$$

where the damping ratio ξ_i is defined as (see (8.83))

$$2\xi_i\omega_i = \left(c_1 + c_2\omega_i^2\right) \text{ or } \xi_i = \frac{1}{2}\left(\frac{c_1}{\omega_i} + c_2\omega_i\right). \tag{8.85}$$

The nature of the solution of equations (8.84) depends on the value of the damping ratios as shown below:

Underdamped $(\xi_i < 1)$

$$X_i(t) = \frac{1}{\omega_{di}}\int_0^t R_i(\tau)e^{-\xi_i\omega_i(t-\tau)}\sin\omega_{di}(t-\tau)d\tau + e^{-\xi_i\omega_i t}(a_i\sin\omega_{di}t + b_i\cos\omega_{di}t), \tag{8.86}$$

where $\omega_{di} = \omega_i\sqrt{1-\xi_i^2}$. Here the first part of the solution is a convolution integral that corresponds to the forced response, and the second part is the complementary solution, which will die out due to damping, and is a *transient response*. After the transient solution dies out, only the particular solution exists as the *steady-state response*. Note the frequency of the steady state response is slightly different from the natural frequency due to damping.

Critically damped $(\xi_i = 1)$

$$X_i(t) = \int_0^t R_i(\tau)e^{-\xi_i\omega_i(t-\tau)}(t-\tau)d\tau + e^{-\xi_i\omega_i t}(a_i t + b_i). \tag{8.87}$$

When critical damping is present, the transient response or the complementary part does not have any oscillations. The steady-state response, which is the particular solution, depends on the external load and will be periodic if the load is periodic.

Overdamped $(\xi_i > 1)$

$$X_i(t) = \frac{1}{\varpi_i} \int_0^t R_i(\tau) e^{-\xi_i \omega_i (t-\tau)} \sinh \varpi_i (t-\tau) d\tau + e^{-\xi_i \omega_i t} (a_i \sinh \varpi_i t + b_i \cosh \varpi_i t), \qquad (8.88)$$

where $\varpi_i = \omega_i \sqrt{\xi_i^2 - 1}$. In overdamped systems also, transient oscillations are suppressed entirely, and the system reaches steady state faster than critically damped systems. In the above solutions, a_i and b_i are constants to be determined from the initial conditions $X_i(0)$ and $\dot{X}_i(0)$. When the forcing function $\mathbf{F}(t)$ is not available in a closed form, numerical integration can be used. Procedures for numerical integration of eq. (8.84) can be found in many elementary books on vibration.[8]

8.7 FINITE ELEMENT MODELING PRACTICE FOR DYNAMIC PROBLEMS

EXAMPLE 8.12 *Natural vibration of a cantilever beam*

Model a cantilever beam, whose length is 1 m, $E = 100$ GPa, $\rho = 4000$ kg/m^3, and the cross section is a square of size 20×20 mm^2, to compute the first six natural frequencies and mode shapes. Compare solutions obtained by (i) beam elements, (ii) plane stress elements, and (iii) 3D elements.

SOLUTION   Figure 8.21 shows three different models for the cantilever beam. The model using plane beam elements has only 10 elements in it, the plane solid model uses 60×3 quadratic rectangular elements, and the three-dimensional model uses 9108 quadratic tetrahedral elements. As the beam is expected to have many bending modes of vibration, it is important to have a mesh that uses multiple quadratic elements through the thickness so that the linearly varying strain can be accurately modeled. When beam elements are used, they are better able to model the bending modes, and therefore 10 elements are sufficient though more elements would be needed if more frequencies are to be computed. Additional elements can also improve the accuracy of the computed natural frequencies.

The eigenvalue problem is computed using numerical algorithms that can either compute a specified number of the lowest natural frequencies or the highest natural frequencies. Some software can also compute frequencies within a range. Here we have computed the lowest six natural frequencies as: $f = \omega/2\pi$ and shown in table 8.2. The frequencies computed using beam elements and plane stress elements match very closely, with the first two frequencies

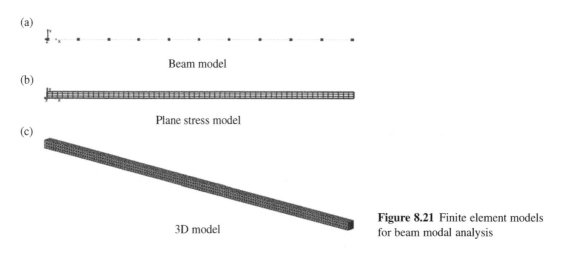

(a)

Beam model

(b)

Plane stress model

(c)

3D model

Figure 8.21 Finite element models for beam modal analysis

---

[8] Craig, R.R., and Kurdila, A.J. 2006. *Fundamentals of Structural Dynamics*. John Wiley & Sons.

Table 8.2 First six natural frequencies of cantilever beam

Mode	Beam model	Plane Stress model	3D model
1	$f = 16.2\,\text{Hz}$	$f = 16.2\,\text{Hz}$	$f = 16.2\,\text{Hz}$
2	$f = 101\,\text{Hz}$	$f = 101\,\text{Hz}$	$f = 16.2\,\text{Hz}$
3	$f = 284\,\text{Hz}$	$f = 282\,\text{Hz}$	$f = 101\,\text{Hz}$
4	$f = 556\,\text{Hz}$	$f = 551\,\text{Hz}$	$f = 101\,\text{Hz}$
5	$f = 920\,\text{Hz}$	$f = 905\,\text{Hz}$	$f = 282\,\text{Hz}$
6	$f = 1251\,\text{Hz}$	$f = 1250\,\text{Hz}$	$f = 282\,\text{Hz}$

Table 8.3 Modes shapes of vibration

Mode	Beam / 2D model	3D model

matching exactly. With the three-dimensional model, repeating natural frequencies are obtained with the first two mode frequencies matching the first frequency obtained by the beam and plane solid models. Similarly, the third and fourth frequencies of the three-dimensional model match the second frequency of the plane solid model, and so on. The reason for this is clear when the mode shapes are examined.

Table 8.3 shows the mode shapes of vibration obtained by the plane solid and three-dimensional models. The beam model and plane solid model have identical mode shapes and so one column is used for both. The number of waves in the mode shapes increase with increasing frequency but the sixth mode obtained using the plane solid model is an extensional vibration. The mode shapes predicted by the three-dimensional model have repeating mode shapes just as there were repeating natural frequencies.

The first two mode shapes of the three-dimensional model seem identical to the first mode of the plane solid model, but they are in different planes. For the three-dimensional model, the first mode is a vibration in the x-y plane while the second mode is a similar mode of vibration in the x-z plane. As the cross section of the beam is square, the mode shapes and the frequencies are the same for the modes of vibration in these two planes. The plane solid model is restricted to vibrate in the x-y plane and therefore is not able to predict an out-of-plane vibration. Similarly, mode 3 and mode 5 of the three-dimensional model are vibrations in the x-y plane while mode 4 and 6 are the corresponding vibration modes in the x-z plane. This example illustrates that for most structures a three-dimensional model is needed if all the modes of vibration are to the predicted. For this example, a three-dimensional beam element (beam element transformed to three-dimensional coordinates with 6 DOFs per node) would also be able to predict the modes in the x-z plane. ■

EXAMPLE 8.13 *Tuning fork vibration*

Compute the natural frequencies and mode shapes of the tuning fork shown in the drawing in figure 8.22 with dimensions in millimeters. The material is chrome stainless steel whose elastic modulus $E = 200$ GPa, Poisson's ratio $= 0.28$, and density $= 7800$ kg/m^3.

SOLUTION   This tuning fork is known to produce a lower A sound which corresponds to a frequency of 440 Hz. The natural frequencies of a tuning fork are computed here using a three-dimensional model that uses quadratic tetrahedral elements as shown in figure 8.23.

When using this tuning fork, it is held at the handle or knob, but modeling the handle/knob as rigidly fixed would be incorrect because if it is hand-held, then a hand is too soft to prevent free vibration at the handle. In fact, if we assume that the fork is rigidly clamped, then its first natural frequency will not be the expected value. Therefore, we perform modal analysis both ways, first with no displacement boundary conditions on the tuning fork and then with the knob on the handle rigidly clamped.

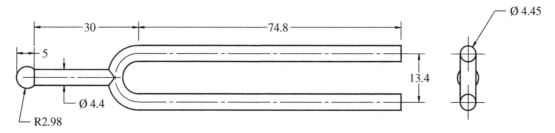

Figure 8.22 Tuning fork

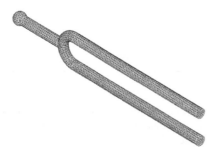

Figure 8.23 Finite element mesh for the tuning fork

Table 8.4   Natural frequencies (Hz) of the tuning fork

Free	$f_7 = 440$	$f_8 = 675$	$f_9 = 1627$	$f_{10} = 1750$	$f_{11} = 2774$	$f_{12} = 3612$
Clamped	$f_1 = 215$	$f_2 = 216$	$f_3 = 420$	$f_4 = 440$	$f_5 = 1568$	$f_6 = 1608$

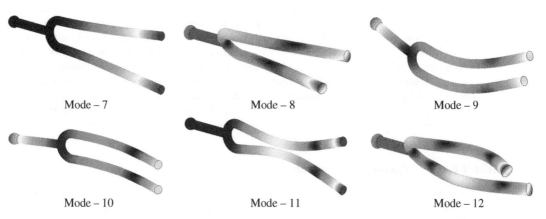

Figure 8.24 Mode shapes of the tuning fork with no boundary conditions

When no displacement boundary conditions are applied, the first six natural modes correspond to rigid body motions and have zero frequency. In table 8.4 the first row lists the next six frequencies in Hz for this case where the fork is essentially assumed to be free floating. The second row in the table shows the first six frequencies of the tuning fork when the knob is rigidly clamped. Figure 8.24 shows the first six non-rigid modes of vibration of the tuning fork when it is not subjected to any boundary conditions. ▬▬

EXAMPLE 8.14   *Forced vibration of fully clamped beam*

A beam clamped at both ends is subjected to a time-varying distributed load. The beam, shown in figure 8.25, has a length of 1 m and its cross section is a square of 0.025 m × 0.025 m. The material parameters are: Young's modulus $E = 120$ GPa, Poisson's ratio $\nu = 0.31$, and density $\rho = 7100$ kg/m^3. The distributed pressure applied on the beam is a harmonic function of time. Compute the response of the beam using the mode superposition method for the following distributed harmonic pressure load P (in Pa):

(a) $P = 1000 \sin(200\pi t)$
(b) $P = 1000 \sin(200\pi t) + 2000 \sin(1000\pi t) + 1500 \sin(300\pi t)$

Plot the displacements at the points A ($x = 0.25$ m) and B ($x = 0.5$ m) on the beam as a function of time.

SOLUTION   The beam in this example can be modeled using beam elements or using plane stress elements (for simplicity we will assume that only planar motion is permitted). To compute the dynamic response, either the time integration approach or the mode superposition approach could be used, but the latter would be computationally more efficient. Mode superposition also can give a more accurate solution if we include all the modes that are likely to be activated. The load case (a) has a single frequency harmonic load. Time-varying loads are defined as "load curves" in software, and if they are harmonic loads, then the load curves are defined by specifying the amplitudes and frequencies of all the harmonics involved. We will assume that there is no damping in this example and solve it using the mode superposition approach described in section 8.5. This method of dynamic analysis is available in most finite element analysis software and is usually linked to modal analysis because as a first step we need to compute the necessary mode shapes. Table 8.5 shows the lowest five natural frequencies for this beam.

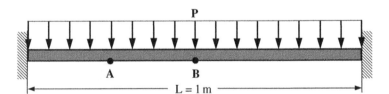

Figure 8.25 Beam with a harmonic distributed load and clamped at both ends

Table 8.5   Natural frequencies of the fully clamped beam

Mode	Natural Frequency (Hz)
Mode-1	105
Mode-2	289
Mode-3	562
Mode-4	920
Mode-5	1360

▬▬

The frequency for mode 5 is higher than the highest frequency component in the applied loads, and so we can reasonably assume that only these first five modes will be activated significantly by these two load cases. Figure 8.26 shows the deflection of the beam at A ($x = 0.25$ m) and B ($x = 0.5$ m) due to the single frequency load case (a). Note that the forcing frequency (100 Hz) is very close to the first natural frequency (105 Hz) and as a result, the amplitude of vibration increases quickly due to near-resonant vibration. The shape of the response is the same at A and B, and they differ only in amplitude. Since the excitation frequency is close to mode 1, the shape of the response will be similar to that of the first mode.

The load defined in part (b) has multiple frequencies. For linear problems, such as this example, the dynamic response due to a load that contains multiple frequencies can be computed as a sum of the response from each frequency component acting separately.

Figure 8.27 shows the dynamic response of the beam to the load case (b). The dominant mode for this response is still the first mode due to the proximity of the first resonant frequency and the harmonic

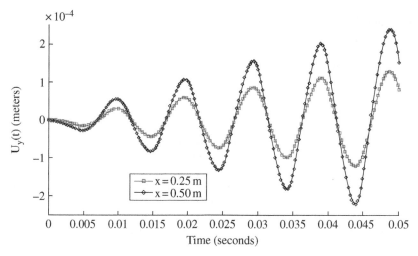

Figure 8.26 Beam with deflection at A and B due to load (a)

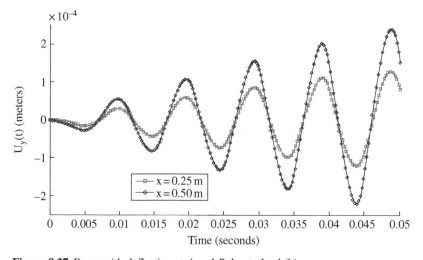

Figure 8.27 Beam with deflection at A and B due to load (b)

components of the load. As a result, the deflection due to this load looks somewhat similar to case (a), but the response to the load components with other frequencies are superimposed. The shape of the response is no longer identical for the two points A and B.

EXAMPLE 8.15   *Super convergence of bar impact*

An elastic rod is impacting onto a rigid wall with initial velocity of $v_0 = 1.0$ as shown in figure 8.28 (units are ignored in this problem). The rod is made of a linear elastic material. The wall is perfectly rigid and frictionless. The dimension and material properties of the rod are: length = 10.0, the cross-sectional area $A = 1.0$, density $\rho = 0.01$, and Young's modulus $E = 100.0$. Use 10 bar finite elements and Newmark's time integration method. (a) Study the effect of time-step size on the stability and accuracy of the explicit method. Especially discuss the superconvergent condition. (b) Study the effect of time-step size on the accuracy of the implicit method. Plot the stress history at the left-end, middle, and right-end element, and compare the numerical solution with the analytical solution.

SOLUTION   When an elastic rod impacts with a rigid wall, the stress wave propagates with a wave velocity $c = \sqrt{E/\rho}$. The governing equation of motion constitutes a wave equation, which is a conventional hyperbolic partial differential equation. We will study the effects of domain discretization and temporal discretization by solving a wave equation. The domain of the rod is discretized by 10 bar elements, and Newmark family method is used for time integration. Figure 8.28 shows the initial status of the rod impact problem. For simplicity, the right-end node is fixed to simulate impact phenomenon.

The initial and boundary conditions are

$$\begin{cases} u(x,0) = 0 \\ v(x,0) = v_0 \end{cases} \quad x \in [0,L],$$

$$\begin{cases} u(L,t) = 0 \\ \dfrac{du}{dx}(0,t) = 0 \end{cases} \quad t \in [0,1].$$

Analytical solution: Since this problem is simple, an analytical solution can be obtained by solving the wave equation:

$$\frac{\partial^2 u}{\partial t^2} = \frac{E}{\rho}\frac{\partial^2 u}{\partial x^2},$$

with the initial and boundary conditions. By using the conventional method of separation of variables, a solution is assumed to have the following form:

$$u(x,t) = X(x)T(t).$$

After substituting the assumed solution into the wave equation, the solution can be obtained in the form of infinite series as

$$u(x,t) = \sum_{n=1}^{\infty} b_n \sin\frac{(n-\frac{1}{2})\pi at}{L}\cos\frac{(n-\frac{1}{2})\pi x}{L}.$$

where the coefficients can be determined by

$$b_n = \frac{2}{(n-\frac{1}{2})\pi a}\int_0^L v_0 \cos\frac{(n-\frac{1}{2})\pi x}{L}dx = \frac{2(-1)^{n+1}v_0 L}{(n-\frac{1}{2})^2\pi^2 a}.$$

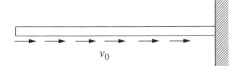

v_0

Figure 8.28 Elastic rod impact problem

In addition, stress can be obtained easily since it is a one-dimensional problem:

$$\sigma(x,t) = E\frac{\partial u(x,t)}{\partial x} = -\sum_{n=1}^{\infty}\frac{2(-1)^{n+1}v_0}{(n-\frac{1}{2})\pi a}\sin\frac{(n-\frac{1}{2})\pi at}{L}\sin\frac{(n-\frac{1}{2})\pi x}{L}.$$

Figure 8.29 shows analytical displacement and stress plots with 60 terms. Since the left end ($x = 0$) is the stress-free region, we can see that stress $\sigma(0,t) = 0$ and displacement at the right end ($x = 10.0$) is also zero. At the right end, the stress is supposed to be a step function, but we can see oscillation with a small amplitude due to the usage of finite terms.

Explicit time integration: When the time integration parameters are $\gamma = 1/2$ and $\beta = 0$, the Newmark method becomes the central difference method. If the lumped mass is used with the central difference, then explicit time integration is possible. However, this method is known as conditionally stable, and the critical time step size can be computed by

(a)

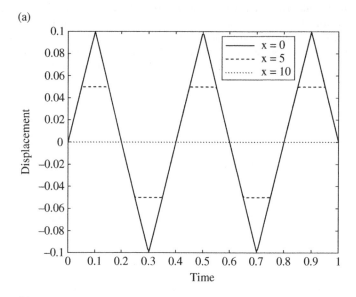

(b)

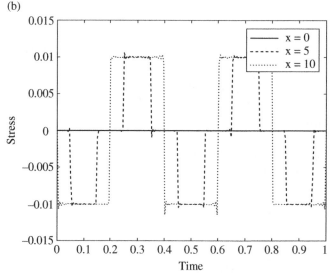

Figure 8.29 Analytical solutions of elastic rod impact problem; (a) displacements and (b) stresses

$$\Delta t_{cr} = \frac{\Delta x}{\sqrt{E/\rho}} = 0.01.$$

The following MATLAB script is used to calculate the displacements and stresses using Newmark time integration.

```
%Bar impact problem (Newmark method)
Area=1;E=100;rho=0.01;L=1;tmax=1;gamma=0.5; beta=0;
Dt=sqrt(rho/E); Nmax=round(tmax/Dt)+1;
t = linspace(0, tmax, Nmax);
%Global matrices
K=2*eye(10);M=2*eye(10);K(1,1)=1;M(1,1)=1;
for n=1:9, K(n,n+1)=-1; K(n+1,n)=-1; end
K=(Area*E/L)*K;
M=(rho*Area*L/2)*M;
Minv=inv(M + (beta*Dt^2)*K);
Q=zeros(10,Nmax);V=zeros(10,Nmax);A=zeros(10,Nmax);sigma=zeros(3,Nmax);
%Initial conditions
V(:,1)=ones(10,1); A(:,1)=-inv(M)*K*Q(:,1);sigma(:,1)=0;
%Time integration
for n=1:Nmax-1
 %Predictor
 Vpr = V(:,n)+(1-gamma)*Dt*A(:,n);
 Qpr = Q(:,n)+Dt*V(:,n)+(0.5-beta)*Dt^2*A(:,n);
 A(:,n+1)=-Minv*K*Qpr;
 %Corrector
 V(:,n+1)=Vpr + gamma*Dt*A(:,n+1);
 Q(:,n+1)=Qpr + (beta*Dt^2)*A(:,n+1);
 sigma(1,n+1)=E*(Q(2,n+1)-Q(1,n+1))/L;
 sigma(2,n+1)=E*(Q(6,n+1)-Q(5,n+1))/L;
 sigma(3,n+1)=E*(-Q(10,n+1))/L;
end
figure; plot(t,Q(1,:),t,Q(6,:)); axis([0 1 -0.1 0.1]);
figure; plot(t,sigma(1,:),t,sigma(2,:),t,sigma(3,:));
```

First, we try the explicit method with $\Delta t = \Delta t_{cr}$. At this particular combination of Newmark parameters with the critical time step, the displacement results are identical to the analytical solution in figure 8.29(a). When $\Delta t = \Delta t_{cr}$, since the wave front exactly corresponds to nodal position and all the masses are lumped to the nodes, we can see the results are exactly matched with the analytical solution, which is called superconvergence. Figure 8.30 also shows the stress results at three elements: left end, middle, and right end. The stress results are slightly different from those in figure 8.29(b). First, the stress is not zero for the left-end element (element 1). This is because the 2-node bar element approximates the constant stress within the element. In addition, due to the finite size of the element, the right-end element does not show step functions but trapezoidal shape. If a larger number of elements are used, the stress at the right-end element will converge to the analytical stress. Nonetheless, the superconvergent solution in figure 8.30 does not show any oscillation in displacement and stress results.

When $\Delta t > \Delta t_{cr}$, the results diverge suddenly for the explicit method. As the size of time step decreases, we can see the results are oscillating around the exact solution as shown in figure 8.31.

Implicit time integration: For the implicit time integration method, the displacement and velocity update are divided by two steps. The first one is called predictor, and displacement and velocity are estimated from the results of time t_n, and the second is called corrector, and displacement and velocity are corrected by the result of acceleration

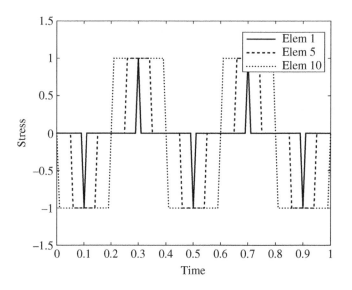

Figure 8.30 Stress history of elastic rod impact problem with explicit time integration (superconvergent solution)

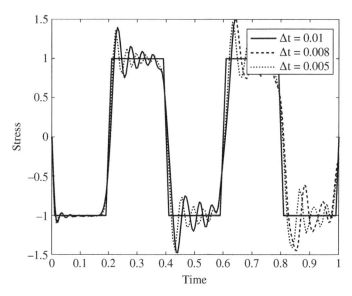

Figure 8.31 Stress history of element 10 of elastic rod impact problem with different time-step sizes

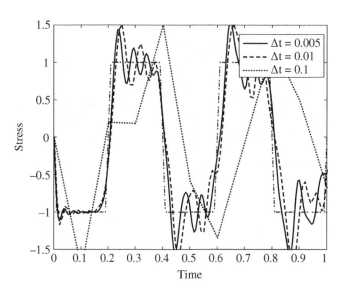

Figure 8.32 Stress history of elastic rod impact problem with implicit time integration

at time t_{n+1}. There are several methods corresponding to different γ and β. Here we use the average acceleration method ($\gamma = 1/2$ and $\beta = 1/4$), which is unconditionally stable. Figure 8.32 shows the stress results of the implicit method at the last element with various time-step sizes. Since this method is unconditionally stable, this method converged with a relatively large time-step size, such as 0.1, but the results are not accurate enough for a large time-step size. Unlike the explicit method, we cannot see the superconvergent solution for this case because a consistent mass is used. ▮▮

8.8 EXERCISES

1. Answer the following descriptive questions.

   (a) In the interpolation of displacement of a finite element, $u(x) = \sum N_I(x)u_I$, which one depends on time and which one depends on domain?

   (b) Explain the difference between the consistent mass and lumped mass matrix.

   (c) Will the natural frequency using lumped mass matrix be larger or smaller than the actual natural frequency? Explain why.

   (d) If the mass of the system increases, will the natural frequency increase or decrease?

   (e) If the stiffness of the system increases, will the natural frequency increase or decrease?

   (f) Explain the difference between the eigenvalue problem and the generalized eigenvalue problem.

   (g) Explain why the magnitude of mode shape is not important.

   (h) If a 1D bar is modeled using 10 bar elements, when both ends of the bar are free, what would be the lowest natural frequency of the bar?

   (i) If a numerical integration algorithm is conditionally stable, how should we choose the time-step size?

   (j) Explain what the method of mode superposition is.

2. Calculate the consistent and lumped mass matrices of a plane truss element in chapter 1. The cross-sectional area of the element is A, length L, and density ρ.

3. Calculate the consistent and lumped mass matrices of a plane beam element in chapter 2. The cross-sectional area of the element is A, length L, and density ρ.

4. Calculate the consistent and lumped mass matrices of a plane frame element in chapter 2. The cross-sectional area of the element is A, length L, and density ρ.

5. Calculate the consistent and lumped mass matrices of a triangular element with three nodes. The area of the element is A, thickness t, and density ρ.

6. Show that the Newmark method with $\gamma = 1/2$ and $\beta = 0$ becomes the central difference method in section 8.4.2.

7. Derive the displacement form of Newmark method. That is, define the effective stiffness and force vector such that the displacement is solved by $\hat{\mathbf{K}}\mathbf{Q}_{n+1} = \hat{\mathbf{F}}_{n+1}$ instead of acceleration in eq. (8.51).

8. Calculate the natural frequencies of a uniaxial bar clamped at both ends. Use: (a) two, (b) three, and (c) four elements of equal length. Perform the analysis using both consistent mass matrix and lumped mass matrix. Plot the frequencies as a function of the number of elements. Comment on your observations. Note that the exact frequencies for a clamped-clamped bar are: $\omega_n = (n\pi/L)\sqrt{E/\rho}$.

   The properties of the bar are: length $= 0.6$ m; area of cross section $= 10^{-3}$ m^2; Young's modulus $= 75$ GPa; density $= 3,000$ kg/m^3.

9. A gear of mass 0.6 kg is attached to the bar in problem 8 at the center. Calculate the natural frequencies for axial vibration using (a) two and (b) four elements of equal length. Use both consistent and lumped mass matrix approach. Comment on your results.

10. Use two elements of equal length to determine the natural frequencies and mode shapes in flexure of a shaft modeled as a simply supported beam. The length of the shaft is 1 m, $E = 100\,\text{GPa}$, $A = 10^{-3}\,\text{m}^2$, $I = 10^{-7}\,\text{m}^4$; $\rho = 4000\,\text{kg/m}^3$.

11. A gear of mass 1 kg is mounted at the center of the shaft in problem 10. Use two elements of equal length to determine how the natural frequencies and mode shapes are affected by the gear.

12. A shaft is modeled as clamped at the left end and on an elastic foundation on the right end. The elastic foundation is represented by a spring with stiffness k.

    (a) Calculate the natural frequencies and mode shapes of the shaft using two beam elements of equal length.

    (b) Compare the results from (a) with those for a clamped-hinged shaft.

    (c) Compare the results from (a) with those for a clamped-clamped shaft.

    In cases (b) and (c) there is no elastic foundation supporting the shaft. Use two beam elements for (b) and (c) also. Assume: $L = 1$ m, $E = 100\,\text{GPa}$, $A = 10^{-3}\,\text{m}^2$, $I = 10^{-7}\,\text{m}^4$; $\rho = 4000\,\text{kg/m}^3$, and $k = 2 \times 10^6\,\text{N/m}^2$.

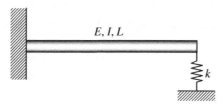

13. The mass matrix of a 2-node beam element undergoing only bending is:

$$[\mathbf{M_e}] = \frac{\rho A L}{420} \begin{bmatrix} 156 & 22L & 54 & -13L \\ 22L & 4L^2 & 13L & -3L^2 \\ 54 & 13L & 156 & -22L \\ -13L & -3L^2 & -22L & 4L^2 \end{bmatrix}$$

$$\rho = 4.2 \times 10^4, \ E = 2 \times 10^{10}, \ I = 10^{-8}, \ A = 10^{-4}, \ L = 10$$

    (a) Assume that the beam is cantilevered. State the generalized eigenvalue problem for computing the natural frequencies and mode shapes of vibration of this beam.

    (b) Write the characteristic equation for the eigenvalue problem, and solve for the natural frequencies.

    (c) Determine the mode shapes of vibration.

    (d) If the beam is subjected to a transverse load at the tip that is varying harmonically as: $F = \sin t$, use the modal superposition approach to write the displacement as a weighted sum of the first two modes of vibration, and obtain a decoupled set of equations of each mode.

    (e) Determine the forced response of this structure due to the applied forcing function assuming that at time $t = 0$ the beam was at rest with no deflection.

    (f) Redo the modal superposition using only the first mode, and compare the solutions from part (e).

14. Consider a two-DOF spring-mass system shown in the figure. When time-independent loads are applied, calculate the transient response (displacements of two masses) as a function of time. Use the central difference method with $t \in [0,10]$sec. Use $m_1 = 2$, $m_2 = 1$, $k_1 = 4$, $k_2 = 2$, and $k_3 = 2$.

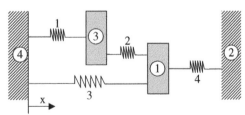

15. A single two-node beam element is used to model a cantilever beam. The beam is subjected to a transverse load at the tip that is varying harmonically as $F = \sin t$ and the equation of motion for this system is:

$$\begin{bmatrix} 15.6 & -22 \\ -22 & 40 \end{bmatrix} \begin{Bmatrix} \ddot{w} \\ \ddot{\theta} \end{Bmatrix} + \begin{bmatrix} 2.4 & -12 \\ -12 & 80 \end{bmatrix} \begin{Bmatrix} w \\ \theta \end{Bmatrix} = \begin{Bmatrix} \sin t \\ 0 \end{Bmatrix}$$

(a) Integrate this equation using Newmark method, and plot w and θ for two cycles of the lowest frequency oscillation. Plot w and θ as a function of time. Assume that the initial conditions are: $\{w \ \theta\}|_{t=0} = 0$ and $\{\dot{w} \ \dot{\theta}\}|_{t=0} = 0$. Use a Δt such that it is one tenth of the period associated with the highest frequency.

(b) Using the modal superposition approach, we can decouple the equations of motion to get the following two equations:

$$\begin{Bmatrix} \ddot{u}_1 \\ \ddot{u}_2 \end{Bmatrix} + \begin{bmatrix} 0.0594 & 0 \\ 0 & 5.769 \end{bmatrix} \begin{Bmatrix} u_1 \\ u_2 \end{Bmatrix} = \begin{Bmatrix} -0.31 \\ 0.43 \end{Bmatrix} \sin t$$

Integrate these two equations using Newmark method, assuming $u_i|_{t=0} = 0$ and $\dot{u}_i|_{t=0} = 0$ and compare with the analytically solution.

16. A cantilever beam has transverse force $f = 10 \sin(20\pi t)$ acting at the tip. It is modeled using a one-beam element. The properties of the beam are: $\rho = 0.1$, $E = 10^7$, $I = 0.08$, $A = 1.0$, $L = 10$.

(a) Write the equation of motion for the beam element $[M]\{\ddot{Q}\} + [K]\{Q\} = \{F\}$ and then apply boundary conditions to reduce it to a two–degree-of-freedom equation.

(b) Compute the natural frequency.

(c) Compute the mode shapes of vibration and plot them.

(d) Determine the forced response of this structure due to the applied forcing function. Assume that there is no damping in the system and phase angle is zero.

(e) Determine the complementary solution assuming that at time $t = 0$ the beam was at rest with no deflection.

17. Four rigid bodies, 1, 2, 3, and 4, are connected by four springs as shown in the figure. Bodies 2 and 4 are fixed. Assume the bodies can undergo only translation in the horizontal direction. The spring constants (N/mm) are: $k_1 = 400$, $k_2 = 500$, $k_3 = 500$, and $k_4 = 300$. Assume the mass of bodies 1 and 3 as $m_1 = 10 \text{ kg}$, $m_3 = 30 \text{ kg}$. Calculate the natural frequencies and corresponding mode shapes.

18. Consider the spring-mass system in example 8.4. It is subjected to a time-varying force (in newtons) given by $F_2 = 100 \sin \omega t$, where $\omega = 0.5\omega_{\min}$ and $\omega_{\min}$ is the smallest natural frequency of the system. Assume the system is initially at rest with no initial displacements.

(a) Calculate the displacement history $u_2(t)$ in closed form using the mode superposition method.

(b) Solve the problem using the central difference method for $0 < t < 2\,T$, where $T = 2\pi/\omega$. Use $\Delta t \approx 0.5 \Delta t_{cr}$.

(c) Solve the above problem (b) using Houbolt method with $\Delta t \approx 2 \Delta t_{cr}$.

(d) Use Newmark method to solve the above problem (b) with $\Delta t = 2 \Delta t_{cr}$.

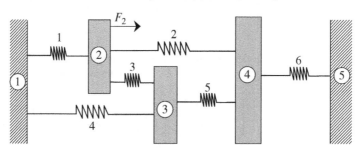

Chapter 9

Finite Element Procedure and Modeling

9.1 INTRODUCTION

The finite element method (FEM) is one of the numerical methods of solving differential equations that describe many engineering problems. The FEM, originated in the area of structural mechanics, has been extended to other areas of solid mechanics and later to other fields such as heat transfer, fluid dynamics, and electromagnetism. In fact, FEM has been recognized as a powerful tool for solving partial differential equations and integro-differential equations, and it is the numerical method of choice in many engineering and applied science areas. One of the reasons for its popularity is that the method results in computer programs versatile in nature that can solve many practical problems with the least amount of training. Obviously, there is a danger in using computer programs without properly understanding the assumptions and limitations of the method, so the objective of this textbook is to provide the readers a sound background on the underlying theory.

In the previous chapters, we developed a variety of finite elements and studied their application in solving problems in solid and structural mechanics. Different elements were used for different types of problems. When a structural problem is given, it is important to understand the following steps: (1) creation of the most appropriate FE model of the given problem; (2) applying the right boundary conditions and the loads; (3) solution techniques available for solving the finite element matrix equations; and (4) interpretation and verification of the FE results. In this chapter, we will learn some of the formal procedures of solving structural problems using finite elements and various modeling techniques. We will limit our interest to solid mechanics problems involving calculation of deflections and stresses. However, these procedures can be extended to problems in other engineering disciplines such as heat transfer.

9.2 FINITE ELEMENT ANALYSIS PROCEDURES

In general, the finite element analysis procedures can be divided into four stages: preliminary analysis, preprocessing, solution of equations, and postprocessing. In the following subsections, we will discuss these four stages of finite element procedures. In many cases, one may not be able to obtain a satisfactory solution from a single analysis. The model may have errors, or the accuracy of the solution may not be satisfactory. When the model has errors, it needs to be corrected, and the procedures should be repeated. If the accuracy of the solution is not satisfactory, the model needs to be refined and the procedure repeated until the solution converges. Figure 9.1 illustrates the sequence of finite element procedures.

Introduction to Finite Element Analysis and Design, Second Edition. Nam H. Kim, Bhavani V. Sankar, and Ashok V. Kumar.
© 2018 John Wiley & Sons Ltd. Published 2018 by John Wiley & Sons Ltd.
Companion website: www.wiley.com/go/kim/finite_element_analysis_design

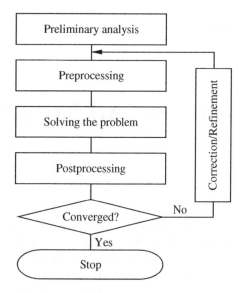

Figure 9.1 Finite element analysis procedures

9.2.1 Preliminary Analysis

Preliminary analysis should be an important part of the finite element analysis but is often ignored. Preliminary analysis will provide an insight into the problem at hand and predict the proper behavior of the model. At this stage, the given problem is idealized and analytical methods are used to obtain an approximate solution. The analytical procedures include, for example, drawing free-body diagrams of different components and analyzing force equilibrium and applying simplified strain-displacement relations and stress-strain relations. Obtaining analytical solutions of practical problems is often not feasible, but in many cases, the problem can be simplified. For example, 3D or 2D structures can be approximated as bars or beams. The goal is not solving the problem with precision but predicting the level of displacements and stresses as well as the locations of their maximum values. Before performing any numerical analysis, the engineer should at least know the range of the solution and the expected location of critical points. If the finite element solutions are far away from the results of the preliminary analysis, there must be a good explanation for the discrepancy. In many cases, small mistakes such as the use of inconsistent units may cause large errors.

EXAMPLE 9.1 *Preliminary analysis of a plane frame*

Consider a plane frame shown in figure 9.2. Using the analytical method, calculate the stress at point C and the maximum normal stress that the uniformly distributed load produces in the member.

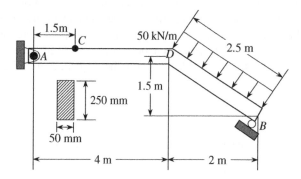

Figure 9.2 Frame structure under a uniformly distributed load

SOLUTION   The state of stress at point C can be determined by using the principle of superposition. The stress distribution due to each loading (axial, shear, and bending) is first determined, and then every contribution is superposed to obtain the final stress distribution. The principle of superposition can be used because of a linear relationship between the stress and loads.

Let us calculate the reaction forces of the member using the free-body diagram as shown below.

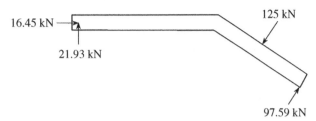

To find forces and moment at point C, let us examine the horizontal segment of the member. From the three equilibrium conditions we have

$$\sum F_X = 0 \Rightarrow \quad 16.45 - N = 0, \qquad N = 16.45\,\text{kN},$$

$$\sum F_Y = 0 \Rightarrow \quad 21.93 - V = 0, \qquad V = 21.93\,\text{kN},$$

$$\sum M = 0 \Rightarrow \quad -21.93x + M = 0, \qquad M = 21.93x\,\text{kN·m}.$$

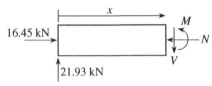

Normal Force: The normal stress caused by N at C is a compressive uniform stress.

$$\sigma_{\text{normal\_force}} = -\frac{16.45\ \text{kN}}{0.05 \times 0.25 m^2} = -1.32\ \text{MPa}.$$

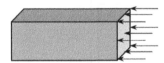

Shear Force: Since point C is located at the top of the beam, no shear stress exists at point C.
Bending Moment: Since point C is located 1.5 m from point A and 125 mm from the neutral axis, the normal stress caused by bending moment is compressive

$$\sigma_{\text{bending\_moment}} = -\frac{32.89\,\text{kN·m} \times 0.125\text{m}}{0.05 \times 0.25^3/12 m^4} = -63.15\,\text{MPa}.$$

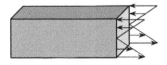

Superposition:
Thus, the stress at point C is obtained using superposition as

$$\sigma_C = -1.32\,\text{MPa} - 63.15\,\text{MPa} = -64.5\,\text{MPa}.$$

As the bending stress will increase along the horizontal segment of the member, the maximum normal stress for the horizontal segment will occur at point D, where $x = 4$ m.

$$\sigma_{normal\_force} = -\frac{16.45 \text{ kN}}{0.05 \times 0.25 m^2} = -1.32 \text{ MPa,}$$

$$\sigma_{bending\_moment} = -\frac{87.72 \text{ kN·m} \times 0.125 m}{0.05 \times 0.25^3/12 m^4} = -168.4 \text{ MPa,}$$

$$\sigma_D = -1.32 \text{ MPa} - 168.4 \text{ MPa} = -169.7 \text{ MPa.}$$

However, it is unclear whether the maximum stress occurs at point D. It needs to be verified using finite element analysis. ▪

EXAMPLE 9.2   *Preliminary analysis of a plate with a hole*

Figure 9.3 shows a plate with a hole under a uniaxial tension load. Because of the hole, stress concentration occurs at the hole edge. Using a stress concentration table and the analytical method, calculate the maximum stress.

SOLUTION   The first step of the analysis is to estimate the stress concentration factor using the analytical method. Far from the applied load location, stress is uniformly distributed throughout the cross section, which is called the nominal stress. From the assumption that the stress is constant in the cross section, the nominal stress can be calculated, as

$$\sigma_{nominal} = \frac{P}{A} = \frac{300}{(2-.75) \times .25} = 960 \text{ psi.}$$

In order to calculate the stress concentration factor at the hole, we need to calculate the geometric factor, which is the ratio between the diameter of the hole and the width of the plate. From figure 9.3, the geometric factor becomes $d/D = .75/2 = 0.375$. The stress concentration factor corresponding to the geometric factor of 0.375 can be obtained from the graph in figure 9.4, which is $K = 2.17$. Then, the maximum stress occurs at the top and bottom parts of the hole, and its value becomes

$$\sigma_{MAX} = K\sigma_{nominal} = 2.17 \times 960 = 2,083.2 \text{ psi.}$$

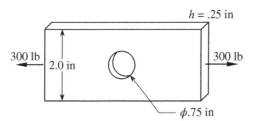

$h = .25$ in

300 lb

2.0 in

300 lb

$\phi.75$ in

Figure 9.3 Plate with a hole under tension

▪

Preliminary analysis is important in the sense that it helps to understand the physical problem better. Based on preliminary analysis results, engineers can plan a modeling strategy in preprocessing.

9.2.2   Preprocessing

Preprocessing is the stage of preparing a model for finite element analysis. It includes discretizing the structure into elements, as well as specifying displacement boundary conditions and applied loads. At this stage, the engineer considers the following modeling-related issues:

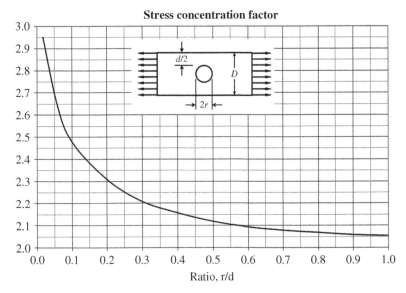

Figure 9.4 Stress concentration factor of plate with a hole

(a) Modeling a physical problem using finite elements

(b) The types and number of elements that should be used

(c) Applying displacement boundary conditions

(d) Applying external loads

We will discuss each issue in the following subsections.

A. Modeling a physical problem: It is important to understand the difference between the physical system and the finite element model. The finite element model is not a replication of the physical system but a mathematical representation of the physical system. The purpose of finite element analysis is to analyze mathematically the behavior of a physical system. In other words, the analysis must be an accurate mathematical model of a physical system. Thus, in order to perform a proper finite element modeling, the user must understand the physics of the problem.

One of the common mistakes in finite element modeling is that the engineers want to make the finite element model exactly the same as the physical system. However, the finite element model is a goal-oriented model. Depending on the purpose of analysis, the finite element model can be a simplified version of the physical system. For example, a complex space rocket system can be modeled using one- or two-beam elements if the interest of analysis is to calculate the maximum bending moment in the frame. The appropriate finite element model depends on the answers we seek, and it is important to only include those aspects of the physical system that affect the answer.

The other important aspect of a finite element model is to understand the difference between the behavior of the physical system and that of the finite element analysis. Many errors can be caused by a lack of understanding of the behavior of finite elements. For example, consider a plane truss with two elements as shown in figure 9.5. When the force is applied vertically, the physical system will support this force by deforming the two elements and producing axial forces in the members. However, a linear finite element model will produce an error because the two elements can only support force in the axial direction, and there is no stiffness in the vertical direction. In fact, the global stiffness matrix will be singular for this case.

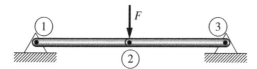

Figure 9.5 Singularity in finite element model

This aspect is clear from the following assembled matrix equation:

$$\frac{EA}{L}\begin{bmatrix} 1 & 0 & -1 & 0 & 0 & 0 \\ 0 & 0 & 0 & 0 & 0 & 0 \\ -1 & 0 & 2 & 0 & -1 & 0 \\ 0 & 0 & 0 & 0 & 0 & 0 \\ 0 & 0 & -1 & 0 & 1 & 0 \\ 0 & 0 & 0 & 0 & 0 & 0 \end{bmatrix}\begin{Bmatrix} u_1 \\ v_1 \\ u_2 \\ v_2 \\ u_3 \\ v_3 \end{Bmatrix} = \begin{Bmatrix} R_{1x} \\ R_{1y} \\ 0 \\ -F \\ R_{3x} \\ R_{3y} \end{Bmatrix}. \tag{9.1}$$

After applying displacement boundary conditions (striking rows and columns 1, 2, 5, and 6), the global matrix equation becomes

$$\frac{EA}{L}\begin{bmatrix} 2 & 0 \\ 0 & 0 \end{bmatrix}\begin{Bmatrix} u_2 \\ v_2 \end{Bmatrix} = \begin{Bmatrix} 0 \\ -F \end{Bmatrix}. \tag{9.2}$$

It is clear that the matrix is singular, and we cannot solve for unknown nodal displacements. More precisely, we cannot solve the problem using linear finite element analysis because in linear analysis the initial geometry is assumed to remain unmodified and is used to calculate the response of the structure. In nonlinear finite element analysis[1], the response of the structure can be calculated by accounting for the change in shape due to the deformation of the structure. The change in shape of the structure will modify its stiffness. Then, since the element stiffness is a function of deformation, the problem becomes nonlinear and can only be solved iteratively.

Units: In general, there are no embedded units in the finite element model. It is the engineer's responsibility to use consistent units throughout the entire analysis procedure. Most FEA software programs available commercially assume that the user has entered numerical values of properties and dimensions all in the same consistent system of units. The results computed would then also be in the same system of units. Some of the recently developed FEA software programs that are fully integrated with CAD software have the ability to convert units for the user. This means that the user must select the units while entering a numerical value of a property or applied loads. Care must be taken to make sure that the values entered are in fact in the units selected and that the results are also displayed in the desired units. Due to numerical nature of finite element analysis, non-standard units are often employed. For example, the length unit meter in the SI system is too big for finite element analysis if the magnitude of displacement is in the order of microns. In such cases, it is common to use millimeters rather than meters as the unit of length. On the other hand, the pressure unit pascal in the SI system is too small for engineering applications. For example, Young's modulus of steel is about 2×10^{11} pascal. The mega-pascal unit is often used for describing pressure. It is recommended that the user select the proper units for the various quantities before beginning the modeling.

Automatic mesh generation: In the second step, the physical system is approximated using the finite element model, which is composed of nodes and elements. For a simple model, it is relatively straightforward to create individual nodes by specifying its location and to define elements by connecting nodes. For a complex model, however, it would be laborious to define thousands of nodes and elements

---

[1] N. H. Kim. 2014. Introduction to Nonlinear Finite Element Method, Springer, NY

manually. Fortunately, many commercial preprocessing programs have mesh-generation capability so that nodes and elements are automatically generated. In such programs, the user first defines a solid model that has the same geometry as the physical system. Then, nodes and elements can be automatically created on the edges, surfaces, and within the volume of the solid model. Bar and beam elements can be generated when the model consists only of lines or curves. Plane solid or plate elements are used when the model consists of only surfaces. Solid elements can be generated only when a fully defined solid model is available. Sometimes, automatic mesh generation using solid models may not have nodes where needed. In such a case, manual creation of nodes and elements are often combined with automatic mesh generation. Often automatic mesh generation fails for very complex geometries, and some programs are not able to generate the mesh automatically for some type of elements. Then the user may have to manually sub-divide the geometry into simpler shapes to assist the mesh generator.

Using a solid model: The solid modeling capabilities are available in all modern computer-aided design (CAD) programs and also in many commercial preprocessing programs. Many CAD programs also have preprocessing capabilities such as mesh generation and application loads and boundary conditions. A solid model is a geometric representation of the physical system or component and can be used for various purposes. In general, a solid model is defined using points, lines/curves, and surfaces that represent the boundaries of a volume. Figure 9.6 shows the solid model of the plate with a hole in example 9.2. The preprocessing programs can either create the solid model or import it from CAD programs. To create a 2D model, the user typically specifies a face of the solid model on which the mesh is to be generated.

Mesh control: In order to create nodes and elements automatically on the solid model, the user must provide mesh parameters that define the size and type of elements and other attributes. The user can also control the global element size, local element size, and curvature-based element size. These parameters are chosen based on the engineering knowledge. For example, the user may want to have a fine mesh in the vicinity of a hole where there is stress concentration. In that case, the local element size near the hole is set smaller than the global element size. Then, the program will create small-sized elements near the hole and gradually increase to the global element size away from that point. The curvature-based element size control is good to represent curves. Figure 9.7 shows quadrilateral elements generated on the solid model in figure 9.6 using the global element size of $0.1''$ and $0.2''$. Note that the element size is a general guideline; not all elements have the same size. In addition, the circular hole is discretized using piecewise linear segments because the element edge is linear.

Mesh quality: Mesh quality is an important factor that determines the accuracy of the solution. A good quality mesh is necessary to obtain a good approximation of the solution in finite element analysis. As we learned in chapter 7, the accuracy of the numerical integration depends on the shape of elements.

    (a) The first criterion for checking the quality of the mesh is the shape of elements. For quadrilateral elements, the element performs best when the shape is a rectangle because the Jacobian matrix is diagonal and constant (see section 7.3). When an element is distorted too much such as shown in figure 9.8(a), numerical integration becomes inaccurate and the Jacobian is close to zero.

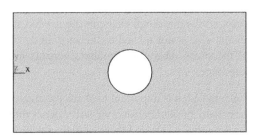

Figure 9.6 Solid model of plate with a hole

(a)

(b)

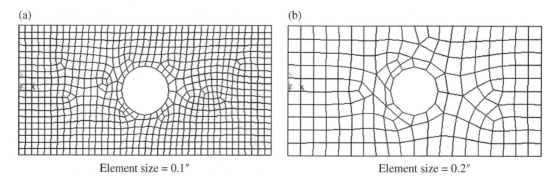

Element size = 0.1″

Element size = 0.2″

Figure 9.7 Automatically generated elements in a plate with hole

(a)

(b)

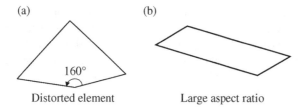

Distorted element

Large aspect ratio

Figure 9.8 Bad quality elements

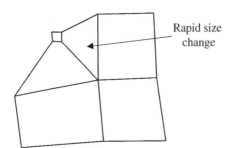

Rapid size change

Figure 9.9 Quick transition of element size

Although we cannot make all elements rectangular, the angle between adjacent edges should be close to 90 degrees.

(b) The second criterion for quality is the aspect ratio of the elements. For a rectangular element, the aspect ratio is the ratio of its length to height. A square element has an aspect ratio of one, which is the ideal aspect ratio. Elements with large aspect ratios should be avoided.

(c) The third criterion is related to the element size. In general, the element size is related to the dimensions of the model. However, quick transition from small elements to large elements should be avoided as shown in figure 9.9. This often happens when the ratio between the global and local element sizes is too large.

(d) One misconception in controlling element size is that small elements are needed at high-stress regions. However, the appropriate element size is not related to the magnitude of the stress. Instead, it depends on the gradient of the stress. Therefore, smaller elements must be used in regions where the stress gradients are large.

Checking the mesh: After nodes and elements are created, it is important to check the created model for any possible errors or defects. Although there are many possible errors, we will discuss three common mesh errors. The first one is duplicated nodes. All interior edges and faces of elements must be shared

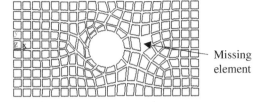

— Missing
element

Figure 9.10 Shrink plot of elements to find missing elements

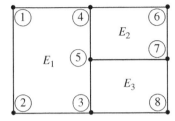

Figure 9.11 Error in element connection

with other elements. This can automatically be achieved when the nodes on the common edge/surface are shared with adjacent elements. However, during the mesh generation, it is possible that two nodes are created at the same location and each of them belongs to different elements. Even though graphically, two elements may seem to be connected, during the assembly process, these two elements are considered to be separated. This can cause an artificial crack in the model. Many preprocessing programs have the capability to check "free surface," which plots the edges/surfaces that are not shared with adjacent elements. If a free surface occurs at an unintended location, the user needs to check for duplicated nodes and merge them.

The second type of mesh error is missing elements. This error can occur due to an error in the mesh-generating software or when elements are created manually. In the regular graphical display, it is not easy to find one or two missing elements, especially in a 3D mesh. Many preprocessing programs can plot elements with shrunk size. As can be seen in figure 9.10, the missing element can be more easily identified in a shrunk plot at least in a 2D mesh.

The third type of mesh error is due to mismatched boundaries. Consider the three elements shown in figure 9.11. Element E_1 is defined by connecting nodes 1-2-3-4, E_2 by nodes 4-5-7-6, and E_3 by nodes 5-3-8-7. Graphically these three elements look connected to each other, but there is a crack between nodes 3 and 4. In order to explain this, let us consider that all nodes have zero displacements except for node 5, which has positive x-translation. Since the displacements of each element are interpolated by nodal displacements, E_1 has zero displacements along edge 3-4 since nodes 3 and 4 have zero displacements. In element E_2, however, since node 5 has nonzero displacement, the edge along nodes 4-5 has nonzero displacements. Thus, a gap is developed between E_1 and E_2. The same is true for E_1 and E_3. In order to connect two elements, all nodes on the same edge must be shared. Even if element E_1 where a 5-node element, there will be a gap or crack along edges 4-5 and 3-5 because the edge 3-5-4 would be deformed into a parabolic curve in a 5-node element while the edge 4-5 and 3-5 in the adjacent elements will remain straight lines.

Material properties: Although we only considered isotropic, linear elastic materials in this text, a variety of material models can be used in finite element analysis, including anisotropic materials, composite materials, nonlinear elastic materials, elastoplastic materials, viscoelastic materials, and so forth. The interested users are referred to advanced textbooks for a detailed discussion of these materials. For isotropic, linear elastic materials, the following three material properties are often used: Young's modulus, shear modulus, and Poisson's ratio. However, it is necessary to provide only two of the three material constants because the third can be calculated from the other two.[2] Sometimes, it is required to provide the

---

[2] For example, see eq. (5.56) in chapter 5.

yield strength or failure stress of the material in order to check the safety status of the finite element model under given loading conditions. It is important to note that the units of material constants should be consistent with the units used in the finite element model.

B. Choosing element types and size: There is no unique way of modeling the given problem using finite elements. Different models and hence different element types can be used for solving the same problem. However, that does not mean that any element and model can be used. An important issue is that the user should understand the capability of the elements and models so that a proper element is used.

Selection of element type is one of the most important steps. The same part can be modeled using different types of elements. For example, the rectangular structure in figure 9.12 can be modeled using beam, plane stress, shell, or solid elements. Although all these different elements can be used to solve the same problem, each element has different characteristics.

Solid element: In general, the best way of modeling the geometry of a structure is using solid elements, for this can represent geometric details, such as the sectional geometry of components, filet and rounded corners, detailed joint geometry, and so forth. However, the number of elements required for modeling a structure can be exorbitant in some cases so that it may not be computationally feasible. For example, let us consider a sheet metal component that is being modeled using solid elements. As we discussed in chapter 6, several elements will be required in the thickness direction in order to capture local bending effects accurately, and the other dimensions of the elements would have to be kept small so that the aspect ratios of the elements are acceptable. Thus, the size of the elements would have to be very small, and as a result, it is not feasible to model many thin-wall structures with solid elements.

Shell/plate element: This element is not studied in this text, but can be found in advanced finite element textbooks. Shell/plate elements were originally developed to efficiently represent thin sheets or plates, both flat and curved. They model the structure as a surface with the thickness of the structure provided as a geometric property or constant. They include out-of-plane bending effects in their fundamental formulation, as well as transverse shear, tension, and compression in the plane. Conceptually, it is a 2D analog of the beam element. For the example in figure 9.12, the required number of shell elements is larger than that of beam elements but smaller than that of solid elements. This element performs particularly well for thin-walled structures where bending and in-plane forces are important.

Beam/frame element: The beam/frame elements model the structure as a one-dimensional line with appropriate cross-sectional geometric properties provided as geometric constants. Beam/frame elements are even simpler and more efficient for the example in figure 9.12. As shown in chapter 3, the beam element has translational and rotational DOFs at each node. The beam element is good for predicting the overall deflection, stiffness, and bending moments of slender bar-like members with constant cross-sectional properties. The stresses in such structures can also be more accurately determined by beam elements. Structural steel tubing and rolled sections are often best modeled using beam elements. There are occasions when shells or solids are better at modeling some aspects of frame- or beam-like

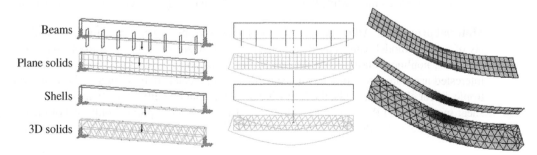

Figure 9.12 Finite element modeling using different element types

Table 9.1   Different types of finite elements

Element	Name
	1D element
	2D triangular element
	2D quadrilateral element
	3D tetrahedral element
	3D hexahedral element

structures, in order to examine some details such as stress concentration or interface/contact stresses where they are connected to other parts. Beam elements are not capable of accurately predicting stresses at joints and junctions due to the simplification of geometry by ignoring joint details such as fillets and welds.

Table 9.1 shows different types of finite elements. Beam and bar elements are one-dimensional elements, while plane solid and shell/plate elements can be 2D triangular or quadrilateral elements. Tetrahedral and hexahedral elements can be used to model three-dimensional solids.

Element order: In previous chapters, we introduced many types of elements including linear and higher-order elements. These elements can be categorized as follows.

(a) Linear elements: 2-node bar, 3-node triangular, 4-node quadrilateral, 4-node tetrahedral, 8-node hexahedral elements

(b) Quadratic elements: 3-node bar, 6-node triangular, 8-node quadrilateral, 10-node tetrahedral, 20-node hexahedral elements

(c) Cubic elements: 4-node bar or 2-node element with 2 DOFs per node, 9-node triangular, 12-node quadrilateral, 16-node tetrahedral, 32-node hexahedral elements

Linear elements have two nodes along each edge, quadratic elements have three, and cubic elements have four nodes along an edge. In general, a higher-order element is more accurate than a lower-order element because the former has more nodes than the latter and therefore can more accurately approximate the real solution.

Element size: Choosing a proper element size is extremely important in obtaining good results, and yet, there is no systematic method available that will help in determining the proper element size a priori. If the mesh is too coarse, finite element analysis results can contain large errors. If the mesh is too fine, solving the problem becomes computationally expensive due to an excessive number of DOFs. The proper mesh size depends on the problem at hand and the user's experience. The results from the preliminary analytical calculations can be a guide to estimate the accuracy of the obtained finite element analysis results. However, this is only possible when analytical solutions are available. Refining the mesh everywhere is also not a practical solution since it would needlessly increase the cost of computation. Critical regions where the stress gradients have a large magnitude require a fine mesh, whereas other regions can have a relatively coarse mesh.

Two common methods to determine whether the current mesh is appropriate are (1) error analysis and (2) convergence study. The error analysis is based on the error between original stress and averaged stress. If the error is larger than a threshold, it indicates that the current mesh is not acceptable and needs refinements. We will discuss error estimation in section 9.4.

A convergence study is a powerful tool when there is no analytical solution available. Since we do not know the exact solution in most cases, it is impossible to know how accurate the current analysis result is even if we used an extremely fine mesh. Only after a convergence study, we can have confidence in the analysis results. The convergence study consists of two sets of meshes: one is the original mesh and the other is the mesh with twice as many elements in critical regions. If the two meshes yield nearly the same results, then the mesh is probably adequate. If the two meshes yield substantially different results, then further mesh refinement might be required. Figure 9.13 shows a typical output from the convergence study. The vertical axis represents the quantity of interest, such as displacement, stress, temperature, and so forth. The horizontal axis shows the number of elements used for each of the trial solutions. The number of DOFs can also be used. In general, the number of elements is doubled with successive trial solutions.

The convergence study in figure 9.13 can go a step further to calculate the convergence rate. In this case, we have to calculate the function of interest at three different meshes. Let h_1, h_2, and h_3 be the sizes of elements, ordered such that $h_1 > h_2 > h_3$; that is, h_3 represents the finest mesh. It is recommended that $h_1 = 2h_2 = 4h_3$. In order to calculate the convergence rate, the function of interest, u, is calculated at each mesh. Then, the ratio in difference in the functions of interest is defined as

$$\frac{\|u_{h_3} - u_{h_2}\|}{\|u_{h_2} - u_{h_1}\|} \approx \left(\frac{h_2}{h_1}\right)^\alpha, \tag{9.3}$$

where α is the *convergence rate*. It indicates how fast the solution will converge to the exact one. A more detailed discussion on the convergence analysis will be presented in section 9.4.

One important question in the convergence study is the degree of accuracy the designer should seek. This is a very practical question and yet difficult to answer because in many cases it depends on the application. In some applications, we only need a rough estimate, and we can tolerate large errors. For design problems, we want to make sure that the errors are within the range where the safety factor used is adequate.

C. Applying displacement boundary conditions: Once the finite element mesh is generated, displacement boundary conditions need to be applied. It is important to note that the finite element model should be properly restrained from rigid body motions, both translation and rotation. Otherwise, a unique solution does not exist. Special techniques are available to analyze a finite element model under rigid body motion, but in general, all deformations are supposed to produce strains and thus stresses. Otherwise, the global stiffness matrix will be singular and will cause an error when solving the matrix equations. On the other hand, if the model is restrained at too many places, the analysis results will be different from the actual response of the physical model. For example,

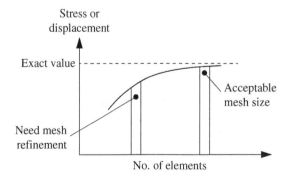

Figure 9.13 Convergence of finite element analysis results

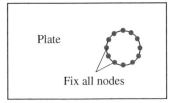

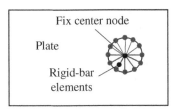

Figure 9.14 Applying displacement boundary conditions at a hole in a plate

consider a plate with a circular hole that is connected to the other parts of the structure through a pin as shown in figure 9.14. Depending on what actually happens, two different ways of modeling are possible. If the pin is tight so that the plate is not allowed to rotate, all the nodes along the circumference of the hole can be fixed. The second scenario is as follows. The pin is not tight so that the plate can rotate, but it is not allowed to move in the x- and y-direction. In this case, we can create a node at the center of the hole and connect all the nodes in the circumference of the hole with the center node using rigid-bar elements. Then, the x- and y-translations of the center node can be fixed.

In the previous section, we mentioned that the errors in the finite element solutions would decrease, as the mesh is refined. However, the errors in the boundary condition will not reduce no matter how much one refines the model. Any unexplained high stress may be due to a wrong boundary condition. Thus, it is important to check if the displacement boundary conditions are properly implemented.

Errors in the boundary conditions are sometimes subtle; therefore, they are not easy to identify. For example, consider two plane trusses as shown in figure 9.15. Since the problem is two-dimensional, overall three DOFs need to be fixed in order to eliminate all rigid-body motions. In figure 9.15(a), node 1 is fixed in both directions and node 3 is fixed in the y-direction. Thus, all three DOFs are fixed. However, the model is not free from rigid body motions because it is possible to move node 3 in the x-direction without deforming any element, although the movement is infinitesimal. Thus, in case (a), the stiffness matrix will be singular, and the matrix equation cannot be inverted to solve for displacements. On the other hand, the truss in figure 9.15(b) is properly restrained, and the global stiffness matrix will be positive definite.

When the mesh is generated from a solid model, many preprocessing software programs allow the boundary conditions to be applied directly to the solid model. In this case, the program automatically converts the boundary conditions and applied traction into equivalent nodal forces and displacements. The advantage of applying boundary conditions on the faces/edges of a solid model is that when the problem is solved multiple times with different mesh densities, the preprocessor automatically re-computes the nodal forces and displacement and applies them to the appropriate nodes. Therefore, it is not necessary to reapply boundary conditions for each mesh.

(a)

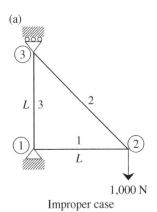

1,000 N

Improper case

(b)

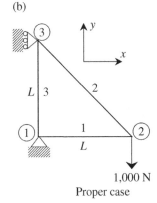

1,000 N

Proper case

Figure 9.15 Applying displacement boundary conditions on truss

D. Applying external forces: In practice, forces may be applied to a machine or a structure through complex mechanisms. Forces from one part are transferred to the other part through the contact pressure in the interface. However, if the region of interest is far from the location at which the force is applied, the complex mechanisms that transmit the force could be approximated using simpler ones. In most cases, the mechanisms can be completely eliminated, and the forces can be applied directly to the structure. In such cases, it is important to understand that the results in the vicinity of the force will not be accurate. In general, three different types of forces are applied to the finite element model: (1) concentrated forces at nodes, (2) distributed forces on the surface or edge, and (3) body forces. The body forces and surface forces usually vary over the volume or area in which they are acting.

Applying a concentrated force: In theory, if a concentrated force is applied at a point, the stress at that point becomes infinitely large because the area is zero. In reality, there is no way to apply a force at a point. All forces are distributed in a region. The concentrated force in finite elements is an idealization of distributed forces in a small region. When the region is relatively small compared to the size of elements, it can be idealized as a concentrated force at a point, as shown in figure 9.16(a). On the other hand, when the region is larger than the size of elements, it can be treated as distributed forces, as shown in figure 9.16(b). Note that the distributed forces are converted to the equivalent nodal forces. In fact, all applied forces must be converted to the equivalent nodal forces because the RHS of finite element matrix equations is the vector of nodal forces.

The effect of this idealization is limited to the immediate vicinity of the points of application of the force. If the interest region is relatively far from the force location, the stress distribution may be assumed independent of the actual mode of application of the force (St. Venant's principle). Figure 9.17 shows the distribution of stress due to a concentrated force at the top. In the section near the top, the stress is concentrated at the center. However, when the distance from the force application point is the same as the member cross-sectional dimension, the stress is almost uniformly distributed.

In a finite element model, the concentrated force can only be applied to a node. Thus, it is important to make sure that a node exists at the location of the force. If a concentrated force needs to be applied at the location where no node exists, it can be approximated by calculating equivalent nodal forces at the surrounding nodes using the shape functions (for example, see problem 27 in chapter 3). Even if the concentrated force is applied at a node, the stress at that point will not be singular but very large because the finite element distributes the effect of the force throughout the element.

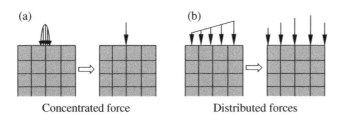

Concentrated force   Distributed forces

Figure 9.16 Concentrated and distributed forces in a finite element model

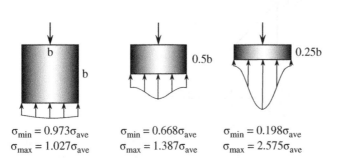

$\sigma_{min} = 0.973\sigma_{ave}$ $\sigma_{min} = 0.668\sigma_{ave}$ $\sigma_{min} = 0.198\sigma_{ave}$

$\sigma_{max} = 1.027\sigma_{ave}$ $\sigma_{max} = 1.387\sigma_{ave}$ $\sigma_{max} = 2.575\sigma_{ave}$

Figure 9.17 Stress distribution due to concentrated force

(a)

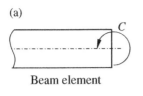

Beam element

(b)

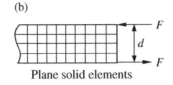

Plane solid elements

Figure 9.18 Applying a couple to different element types

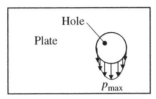

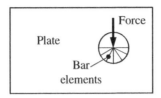

Figure 9.19 Modeling a shaft force using assumed pressure and bar elements

Applying a couple to a plane solid: The correct method for applying a load depends on the type of element used to model the structure. For example, consider applying a couple to a structure (figure 9.18). The structure can be modeled either using beam elements or plane solid elements. In the case of beam elements, shown in figure 9.18(a), it is straightforward to apply a couple because it is a force corresponding to the rotational DOF. At each node of a beam element, a transverse force and couple can be applied. However, in the case of plane solid elements, shown in figure 9.18(b), we can only apply x- and y-directional forces. Since there is no rotational DOF, a couple cannot be applied directly as a nodal force. In this case, a pair of equal and opposite forces has to be applied, and the forces will be separated by a distance d such that the equivalent couple $C = F \cdot d$. Of course, the effect will be different in the vicinity of the forces, but according to St. Venant's principle, the local effect of force application method will disappear at a short distance.

Force transmitted by a shaft: In a mechanical system, a force in a part is transferred to the other part through connections, such as shafts and joints. For example, one can model a plate and a shaft and apply contact conditions between them. Then the applied force on the shaft is transferred to the plate through the contact pressure at the interface. However, the problem is nonlinear when contact is specified between components and requiring iterative methods to solve. If the only stress distribution within the part due to the contact force is of interest, it is unnecessary to model the shaft. Instead, the contact force expected from the shaft can directly be applied on the surface of the hole. At this point, it is necessary to approximate the distribution of this force on the surface. Figure 9.19 explains two commonly used approximations. Note that only half the surface of the shaft is in contact with the plate, and the maximum contact pressure occurs at the center. The first method assumes that contact pressure distribution is elliptic, that is, given by the equation of an ellipse. The magnitude of the maximum distribution is calculated such that the total force supported by the contact pressure is the same as the force transmitted by the shaft. The second method approximates the effect of the shaft using bar elements that connect the center of the shaft to one side of the hole in the surface. Since the shaft is often stiffer than the plate, a large value of the axial stiffness is used for the bar elements, or sometimes, the bar elements are assumed to be rigid.

EXAMPLE 9.3   *Displacement boundary conditions for a plate model*

For the plate with a hole in example 9.2, apply displacement boundary conditions and external load such that it is under uniaxial tension.

SOLUTION   The displacement boundary conditions are given in figure 9.20. All nodes on the left edge are fixed in the x-direction. However, a rigid body motion in the y-direction is still possible. Therefore, the node at the center of

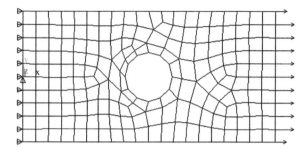

Figure 9.20 Displacement and forces of the plate model

the left edge is fixed in both x- and y-direction. Since the structure is modeled using a plane solid, its three rigid body motions are completely fixed by these boundary conditions.

A uniform pressure is applied to the right edge. The magnitude of the pressure is 600 psi, which is equivalent to the 300 lb. in example 9.2. Figure 9.20 shows the equivalent nodal forces corresponding to the uniform pressure. Note that the nodal forces are equal except for the two nodes at the top and bottom. As discussed in chapter 6, the uniform pressure load is equally divided between the two end nodes of the quadrilateral elements. Therefore, after assembly, the two end nodes will have half of the nodal force compared to the inside nodes. ■

9.2.3   Solution Techniques

After the model has been created with appropriate mesh, loads, and boundary conditions, the individual element stiffness matrices and the vector of nodal forces are assembled and solved for unknown DOFs. After solving the global matrix equation, two different types of results are obtained:

(a) Nodal DOF solutions – these are the primary unknowns.

(b) Derived solutions – such as stresses and strains in individual elements.

These solutions can be graphically displayed in the postprocessing stage. The primary solutions are available at each node, while the secondary or derived solutions are usually calculated at the integration points of individual elements.

Although we only discussed static analysis and steady-state heat conduction analysis, many different types of analyses are available, such as static, buckling or stability analysis, heat transfer, potential flow, dynamics, and nonlinear analysis. The engineer should understand the purpose of these analyses and expected outputs.

In most finite element analysis programs, the solution stage is opaque to the user, because the solutions steps such as constructing element stiffness matrices, assembling the global stiffness matrix, constructing the vector of applied forces, deleting rows and columns, and solving the matrix equations are performed automatically using the information that the user provided in the preprocessing stage. However, most failures in finite element analysis procedures occur during the solution stage. Although the failures are usually caused by errors or wrong information from the preprocessing stage, the effects of these errors can only be detected while solving the matrix equation. Thus, it is important to understand what types of errors or wrong information provided in the preprocessing stage can cause problems in the solution stage.

Singularities: One of the most common problems detected in the solution stage is that the global stiffness matrix is found to be singular. As we learned in the previous chapters, the element stiffness matrices and the structural stiffness matrix are singular. However, after applying displacement boundary conditions (striking rows and columns), the global stiffness matrix becomes positive definite, and it can be inverted. If singularity happens, the matrix solver cannot calculate nodal DOFs, and the solution stage stops with an error message. A singularity may also indicate the existence of an indeterminate or non-unique solution. Mathematically, as shown in the appendix, when the coefficient matrix is positive definite, there always exists an inverse of the matrix, and the equation has a unique solution. When the matrix is singular, the determinant of the matrix is zero, and the inverse of the matrix cannot be obtained. The following conditions may cause singularities in the global stiffness matrix:

(a) Insufficient/wrong displacement boundary conditions. This is the most common mistake/error that causes a singularity in the matrix. The displacement boundary conditions should be such that all rigid body modes are eliminated.

(b) Zero or negative value of Young's modulus. Zero value of the Poisson's ratio will not cause a singularity.

(c) Unconstrained DOFs. Lack of knowledge about the element being used may lead to errors in applying boundary conditions and cause singularities. For example, most commercial programs do not have one-dimensional or two-dimensional truss. A three-dimensional truss element is supposed to cover these elements. However, when the three-dimensional truss element is used to solve for one-dimensional bar problems, the displacements in the y- and z-axis for all nodes must be fixed even if no forces are applied in those directions. Some commercial programs automatically fix these DOFs.

(d) Coincident nodes causing cracks in the model. As explained earlier, coincident nodes that are not merged will lead to separated elements that can have a rigid body motion.

(e) A large difference in stiffness. When two different materials are used in the model, if the difference in stiffness is too large, the stiffness of the weaker material is practically zero. This happens because all data are stored with a limited number of significant digits. When the ratio between Young's modulus is greater than 10^6, the stiffness matrix is almost singular (or highly ill-conditioned), and that can lead to singularity errors.

(f) Irregular node numbering could also lead to singular stiffness matrices. For example, when using triangular elements it is important to number the nodes in the counterclockwise direction. Otherwise, the area of the triangle will be negative, and the stiffness matrix may not be positive definite.

Multiple load conditions: In this text, we only discussed the case when the RHS of the matrix equation is a single column vector, and the solution is also a single column vector. This corresponds to a situation when the structure is under a single loading condition. However, the engineer may want to evaluate the safety of the structure under different loading conditions. For example, bicycle design in section 3.8 of chapter 3 must satisfy two loading conditions, vertical bending and horizontal impact loads, but not simultaneously. Similarly, the safety of a building structure involves the wind load and the seismic load. Of course, it is possible to solve the finite element matrix equation multiple times, but it is possible to solve multiple loading conditions more efficiently. Let us consider the case that we want to calculate the responses of the system under N different load conditions. In this case, the global matrix equation can be written as

$$[\mathbf{K}]\left[\widetilde{\mathbf{Q}}\right] = \left[\widetilde{\mathbf{F}}\right], \tag{9.4}$$

where

$$\left[\widetilde{\mathbf{Q}}\right] = [\mathbf{Q}_1 \ \mathbf{Q}_2 \ \cdots \ \mathbf{Q}_N],$$

$$\left[\widetilde{\mathbf{F}}\right] = [\mathbf{F}_1 \ \mathbf{F}_2 \ \cdots \ \mathbf{F}_N].$$

Vector $\{\mathbf{Q}_1\}$ is the nodal responses corresponding to the nodal forces $\{\mathbf{F}_1\}$ and $\{\mathbf{Q}_2\}$, to $\{\mathbf{F}_2\}$, and so forth. This is different from solving the matrix equation N times. In fact, it is much more efficient than N finite element analyses. In general, solving a matrix equation can be decomposed into two stages. In the first stage, which is the most time-consuming part, the stiffness matrix is factorized using the LU decomposition. The stiffness matrix is expressed as the product of lower and upper triangular matrices. However, factorizing the stiffness matrix is independent of the force vectors. Once the stiffness matrix is factorized, solving for $\{\mathbf{Q}_I\}$ for a given $\{\mathbf{F}_I\}$ is computationally inexpensive. Thus, it is very efficient to decompose the stiffness matrix once and solve for multiple load cases.

Restarting solution stage: Occasionally, the user may need to restart an analysis after the initial result has been obtained. For example, the user may want to add more load conditions to the analysis. In this case, restarting the solution can be an effective tool. Let us assume that the matrix equation with initial load condition has been solved. Then, the LU-factorized stiffness matrix has already been computed and can be saved on the hard disk. When the user wants to restart the analysis with different load conditions, it is now easier to construct the new vector of global forces and solve for unknown DOFs using the already factorized stiffness matrix. This procedure seems attractive, but it has its own drawbacks. It requires storing the factorized stiffness matrix in the disk space. Thus, the users trade-off between saving computational time and disk space.

9.2.4   Postprocessing

After building the model and obtaining the solution, the user will want to review the analysis results and evaluate the performance of the structure. Postprocessing programs, which may be a separate program or integrated with the analysis program, provide tools to display and interpret the results. This is probably the most important step in the analysis because the purpose of the analysis is to understand the effect of applied loads on the structure and to evaluate the structural response. It is very important to correctly interpret the results displayed graphically by the postprocessing software, and this requires knowledge and experience in mechanics. The user must have the intuition and understanding of the physics to detect possible errors in the solution due to modeling mistakes or numerical errors.

Deformed shape display: After solving the finite element matrix equation, the values of all nodal DOFs are available. For structural (or thermal) problems, the nodal DOFs are displacement (or temperature). For a simple model, it may be easy to interpret the deformation of the structure by examining the nodal displacement values. For a more complex model with a large number of elements, however, it is not straightforward to interpret the deformation using numerical data. Most postprocessing programs can graphically display the deformed geometry of the model. The deformed shape plot helps the user visualize the structural behavior. Modeling errors can often be detected by examining the deformed shape plot. For example, the user can verify if the displacement boundary conditions and external forces are correctly applied.

Note that the magnitude of deformation for a linear elastic material is usually in the range of microns or millimeters. Such a small deformation is almost not visible when plotted to scale. For this reason, the deformation is often magnified such that it is visible. Many postprocessing programs automatically scale the deformation such that the maximum deformation is about 5 ~ 10% of the model size. Figure 9.21 shows magnified deformation of the plate model.

Contour display: The displacements are vectors, and hence a deformed shape of the mesh is the best way to depict the deformations. On the other hand, contour plots are suitable for scalar quantities. Although stress at a point is a matrix quantity, a single component of stress, say σ_{xx}, is a scalar, and the stress variation throughout the model can be displayed using a contour plot. Finite element programs output stress results for all elements in the mesh. The contour plot of the stress enables the user to visualize and understand the distribution of the stress in the structure and identify the critical locations where the stress is high and the structure is likely to fail. Ductile materials yield when the von Mises

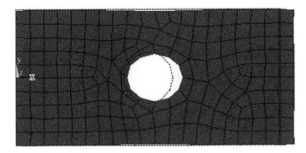

Figure 9.21 Deformed shape of the plate model

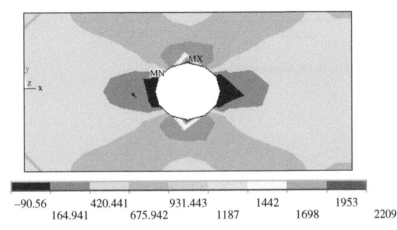

-90.56   420.441   931.443   1442   1953
     164.941   675.942   1187   1698   2209

Figure 9.22 Contour plot of σ_{xx} of the plate model (element size = 0.2 in)

stress exceeds the yield stress, and hence it makes sense to have a contour plot of von Mises stress in order to determine whether the structure would yield.

The plate model in figure 9.22 shows that the maximum stress occurs at the top and bottom of the hole, which was expected from the preliminary analysis. From the preliminary analysis in example 9.2, the stress concentration at these locations was estimated to be 2,083 psi, but the finite element analysis yields 2,209 psi, which is about 6% higher than the preliminary analysis results. Note that the stress results in figure 9.22 are obtained with the mesh of size 0.2 in.

There are several reasons for the mismatch in stress between preliminary analysis and finite element analysis. First, the stress concentration factor in the preliminary analysis is based on an approximate reading of the stress concentration factor graph. In addition, the stress in Figure 9.22 may not be accurate because of the coarse mesh. In finite element analysis, the stress value is most accurate at the Gauss quadrature points. The stresses at the nodes are obtained by extrapolation of stresses at the integration points to obtain values at the nodes and then averaging the stresses from adjacent elements. In order to see the effect of element size, the initial mesh size of 0.2 in. is refined to 0.1 in. Figure 9.23 shows stress contour plot with the refined mesh size. Note that the maximum stress is reduced to 2,198 psi. The maximum stress only changes by 0.5% from the coarse mesh stress. Thus, the stress results from finite element analysis can be considered to be reasonably accurate.

Stress averaging: One of the challenging tasks in postprocessing is to graphically display analysis results accurately. Since most contour-plotting algorithms are based on nodal values, it is necessary to calculate stress values at nodes. In the case of displacements, they are continuous between elements, and hence the displacements at the nodes are unique. In the case of stresses, however, they are continuous within the element but discontinuous across element boundaries. The stresses computed along a

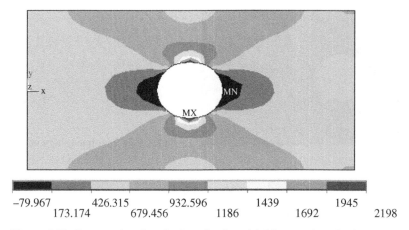

Figure 9.23 Contour plot of σ_{xx} in the refined model (element size = 0.1 in)

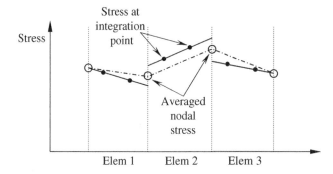

Figure 9.24 Averaging stresses at nodes

boundary or at a node will have different values for each adjacent element. In addition, the stress results are available at the Gauss quadrature points, and the values at nodes are obtained for each element through extrapolation. The nodal values of the stresses are computed by averaging the extrapolated values computed from all the adjacent elements at a node. Thus, extrapolation and averaging can affect the accuracy of stress results in the contour plot. The computed stresses are most accurate at the Gauss quadrature points. Figure 9.24 shows stress averaging for one-dimensional elements. Stress values are calculated at Gauss quadrature points (in this case, two per element). Then, these stress values are extrapolated to the nodes. The extrapolated stress values at the node for the two neighboring elements are in general different. Then, the averaged nodal stress is calculated at that node. If the difference in stress at the connecting node is small, the error due to averaging is small. The difference between the extrapolated and averaged stress values is often used as a criterion of the accuracy of the analysis results.

9.3 FINITE ELEMENT MODELING ISSUES

In this section, we discuss many issues to consider while constructing a finite element model, such as mesh-generation techniques, ideas for model simplification and selection of the most appropriate model for a given problem. Some of the techniques used for mesh generation and model creation vary between software programs. In the following subsections, we summarize a small list of modeling techniques.

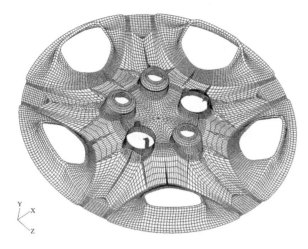

Figure 9.25 Detail model of a wheel cover

9.3.1   Model Abstraction

One of the common misunderstandings in finite element modeling is that the finite element model is more accurate if all the geometric details of the real physical system are included. However, small details that are unimportant to the analysis should not be included in the model, since they will only make the model more complicated than necessary. A good model should only include those aspects of the real physical system or component that are relevant to the results that we are interested in computing. Often it is better to gain insight from several simple models than to spend time making a single detailed model. Figure 9.25 shows a wheel cover model. If the purpose of the analysis is to estimate bending/torsional stiffness, the model shown in the figure is too detailed.

Many smaller features in the geometric model, such as fillets, chamfers, and holes can be suppressed in the CAD if these features do not significantly influence the stiffness or the maximum stress. This is often referred to as "de-featuring" in the CAD/preprocessing software programs. For some structures, however, small features such as fillets or holes can be at locations of maximum stress and therefore, be important to include in the model depending on the purpose of analysis. The user must have an adequate understanding of the structure's expected behavior in order to make proper decisions concerning how much detail should be included in the model. For example, if the purpose of the analysis is to estimate the maximum deflection of a bridge structure, it might be good enough to model the bridge using beam elements. However, if the purpose is to predict the stress concentration at the joints or connections, then it is necessary to model holes and pins. It is important to perform a trade-off between details in the model and the computational costs. For example, when the size of a hole is less than 1% of the entire dimension of the model, it would be better to ignore it in modeling. Instead, nominal stress values in conjunction with stress concentration factors can be used to predict the effects of the hole on the maximum stress in the structure.

9.3.2   Free Meshing vs. Mapped Meshing

In general, there are two ways of creating a mesh on the solid model: free meshing and mapped meshing. In free meshing, the user provides a general guideline for meshing, and the preprocessing programs will make the mesh accordingly. Thus, the mesh generation is automatic and requires very little input from the user. On the other hand, in mapped meshing, the user provides detailed instructions on how the mesh should be created. In two dimensions, all surfaces are divided into four-sided quadrilaterals. The user then specifies how many elements will be generated on each side of the quadrilateral. Consider the 2D model shown in figure 9.26. Its boundary consists of six lines/curves. In order to make a mapped

(a)

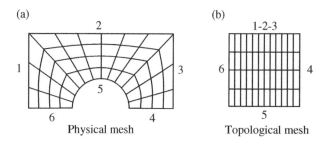

Physical mesh

(b)

Topological mesh

Figure 9.26 Mesh generation using mapping

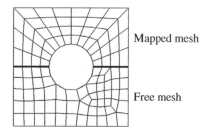

Mapped mesh

Free mesh

Figure 9.27 Mapped and free meshes

mesh, these six curves are divided into four groups. For example, we can group curves 1, 2, and 3 into one composite curve so that the boundary now consists of three single curves (4, 5, and 6) and one composite curve (1-2-3). Now, let us consider that four elements are assigned to curves 4 and 6, and twelve elements to curve 5 and the composite curve. Note that the opposite side should have the same number of elements. Then, 4 × 12 rectangular elements can be created in the topological square in figure 9.26(b). In the physical mesh, the locations of nodes can be found by assuming the topological square is morphed into the physical shape. Then, quadrilateral elements are created as shown in figure 9.26(a). A similar approach can be applied for mapped meshing in a three-dimensional volume. In this case, all volumes are grouped into six surfaces and twelve edges. Then, the number of elements in each edge is specified. The opposite sides should have the same number of elements.

There are advantages and disadvantages in both of these approaches for meshing. More user action is required for the mapped meshing approach, while more complex computer algorithms are needed for free meshing. In general, the mapped mesh looks better because the grid is more regular. However, the quality of elements should not be judged by the regularity of the overall mesh. For example, two meshes are compared in figure 9.27. Even though the mapped mesh looks more regular, the actual quality should be measured by the level of distortion in individual elements. The ideal element is a square, and therefore the angle between the edges should be 90 degrees. The quality of the element should, therefore, be measured by how much the angle between edges differs from 90 degrees.

9.3.3 Using Symmetry

Exploiting symmetry in finite element analysis was considered important to reduce the cost of computation. As computer hardware has developed rapidly and computational costs have reduced, the use of symmetry to reduce modeling effort is less important. However, it is still advantageous to use symmetry when creating models to simplify the application of boundary conditions. Consider a plate with a hole, as shown in figure 9.28. The plate is under uniform traction at both ends. As shown in the figure, there is no displacement boundary condition. Either traction is applied on boundaries or they are free boundaries. If this model is used as such, the stiffness matrix will be singular because the rigid body motion is not removed in the model. Although it is possible to remove the rigid body motion by fixing suitably selected

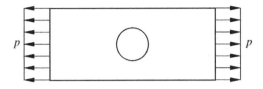

Figure 9.28 Full-sized model of a plate with a hole

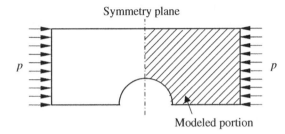

Figure 9.29 An example of a symmetric model with a symmetric load

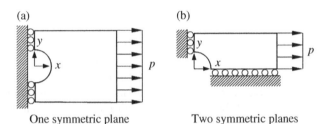

Figure 9.30 Symmetric models of a plate with a hole

points, it is better to make use of the symmetry. Note that the geometry, as well as the applied load, is symmetric with respect to both the x- and y-axis. Thus, the results of the finite element analysis must be symmetric.

In general, a problem can be considered to be symmetric only if the geometry, the applied loadings, and the boundary conditions are symmetric. If non-symmetric loads are applied on a symmetric geometry, the problem is not symmetric. The symmetric boundary condition is applied by setting the translation normal to the plane of symmetry to zero along the line of symmetry. For the structure shown in figure 9.29, the geometry and applied load are symmetric with respect to the vertical plane indicated by dotted lines. In such a case, only half of the structure needs to be modeled with appropriate symmetry boundary conditions on the plane of symmetry. If more than one symmetry plane exist, the model can further be reduced.

The symmetric model in figure 9.30(a) removes the x-translational and z-rotational rigid body motions, but the model can still undergo y-translational rigid body motion (remember a plane solid has three rigid body motions). Thus, the symmetric model in figure 9.30(a) cannot be solved as it is. We can further apply additional symmetry with respect to the x-axis, as shown in figure 9.30(b). Now all rigid body motions are fixed, and only a quarter of the elements are needed compared to using the full model.

When a concentrated force is applied along the symmetric line, the force must be halved in the symmetric model.

In some cases, a structure could be considered symmetric but for a few minor details that disrupt the symmetry. In such cases, those conditions that prevent symmetry could be removed as long as the user understands the effect of such deliberate changes. The user must weigh the gain in model simplification against the cost in reduced accuracy when deciding whether or not to deliberately ignore unsymmetrical features of an otherwise symmetric structure.

9.3.4 Connecting Beam with Plane Solids

For a complex system, it is often necessary to use different types of elements in different regions of the structure in order to model the physical problem. These different types of elements are connected to each other by sharing common nodes. For some types of elements, this connection is trivial, but for others, it is not. The issue is that not all elements have the same DOFs at the node. For example, quadrilateral elements can be connected to triangular elements without any modification because both types of elements have the same DOFs at the node. However, this is not the case when we want to connect plane solid elements with beam/frame elements. The plane solid element has x- and y-translation DOFs at the node, but the plane frame element has x- and y-translation as well as z-rotational DOFs at the node. When these two types of elements are connected, the rotational DOF of the frame is not constrained, and thus, the frame element will experience rigid body motion; that is, there is a singularity in the model (see figure 9.31).

When a plane solid is connected to a frame, as shown in figure 9.31, it means these the frame is welded to the solid. Thus, the rotation of the frame should be related to the rotation of the plane solid at the weld. However, the plane solid element does not have a rotational DOF. Therefore, special techniques will be required to ensure that the rotation of the frame element at the joint matches the rotation of the solid and that a moment is transmitted to the plane solid elements, as shown in figure 9.32. There are two possible modeling techniques to achieve this connection between the elements. In figure 9.32(a), the frame element is extended into the plane solid. Thus, the rotation of the beam can cause deformation of the solid. This technique is approximate but is useful when the mesh in the solid is parallel to the axial direction of the frame, such as a mapped mesh. The second method is illustrated in figure 9.32(b), in which the following constraint equation is imposed between the nodes on the solid and the rotational DOF of the frame so that the rotation of the frame can cause deformation in the solid:

$$u_1 + h\theta_2 - u_3 = 0. \tag{9.5}$$

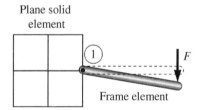

Figure 9.31 Singularity in connecting a plane solid with a frame

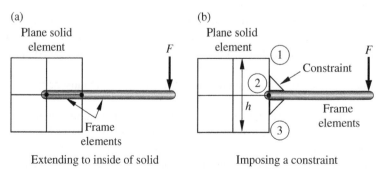

Figure 9.32 Connecting a plane solid with frame

The constraint is equivalent to the welded joint at the interface. This may not be a perfect solution either because this approach will result in high local stresses. A similar approach could be used to connect a three-dimensional solid to shell elements.

9.3.5   Modeling Bolted Joints

Many mechanical parts are connected using bolted joints. A frequent question is how detailed should a bolted joint model be for finite element analysis. Perhaps the most complex way of modeling such a joint is to use three-dimensional representations of the bolts and parts and connecting them through contact constraints. However, this approach is not practical unless the purpose of the analysis is to calculate the local stresses in the bolt. In addition, this approach will lead to difficulties because: (1) the model size will be huge due to the detailed representation of small features; (2) the problem becomes nonlinear due to the contact constraints; and (3) there will be rigid body motion, if an initial gap exists between the bolt and parts. Therefore, we need a more practical and simpler modeling technique for bolted joints.

Nodal coupling: The structural parts that are connected through the bolted joints are often thin plates. In addition, the size of the bolt hole is much smaller than the dimension of the parts. Hence, the bolt hole in the part can be ignored in modeling. The first simplified approach is to represent the bolted (or riveted) connection of overlapping shell structures by locating a node on each surface at the location of the bolt. The two nodes have to be located at the same $x-$, $y-$, and z-location in space, as shown in figure 9.33. This is an approximation because physically it is impossible for the two nodes on the two plates to be in the same location due to the finite thickness of the plates. Then, a constraint is imposed so that these two nodes are tied such that they have the same displacement. Most finite element programs support nodal-coupling capability. It will be desirable to tie two of the three rotations as well. The only rotation that is free is that about an axis perpendicular to the planes of elements (about the axis of the bolt).

When bolt holes are ignored in modeling, the user needs to do additional work to examine the stresses in the vicinity of the bolt holes, using the stress concentration factor to find the allowable net stress, bearing force, and total force in that zone.

Using rigid link: In order to remove the approximation of locating two nodes at the same location, the two plates are positioned properly in space and then a rigid element is used to link the pairs of nodes where bolt joints are located. The rotational DOF about the axis of the bolt must be free at the ends of the rigid element. Instead of the rigid element, it is possible to use a very stiff bar element to connect the two points.

Bolt preload: When bolts are used to tighten two plates, there exists a preload in the bolts. The previous two methods will not able to include bolt preload. When a preload needs to be modeled, a frame element with "initial strain" can be used. The intended preload must exist *before* the structure is loaded. In addition, the preload must be squeezing two surfaces together, which means surface contact elements must be in use between the separated plate element surfaces, and this makes the model nonlinear. If the bolts are

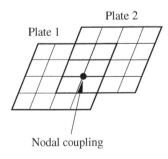

Nodal coupling

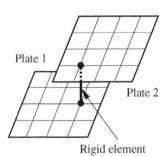

Rigid element

Figure 9.33 Modeling bolted joints

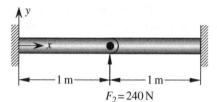

$F_2 = 240\,\text{N}$

Figure 9.34 Finite element models of stepped cantilevered beam

not overloaded when the structure is loaded, the bolt preload will be nearly unchanged when the structure is loaded. The use of a friction coefficient at the contact surfaces needs to be considered carefully, for it can reduce or prevent completely the shear loading on the bolts. It would be conservative to assume frictionless contact because the bolts will carry the entire load.

EXAMPLE 9.4 *Bolted Joint*

Two identical beams of length 1 m are clamped at one end and connected by a bolt, as shown in figure 9.34. An upward transverse force of 240 N is applied to the bolted joint. Assume $EI = 1000\,\text{Nm}^2$. Use two beam elements to determine the deflection at the bolted joint. Compare the deflection with that of example 3.6 of chapter 3.

SOLUTION   In order to make a bolted joint, two beam elements must be separated. Thus, the finite element model has four nodes as shown in the figure below.

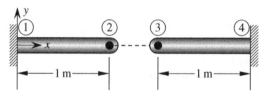

The element stiffness matrices of the two elements will be the same although the row and column addresses will be different. Using the formula in eq. (3.52) we can write the element stiffness matrices as

$$
\left[\mathbf{k}^{(1)}\right] = 1000
\begin{array}{cccc}
v_1 & \theta_1 & v_2 & \theta_2 \\
\left[\begin{array}{cccc}
12 & 6 & -12 & 6 \\
6 & 4 & -6 & 2 \\
-12 & -6 & 12 & -6 \\
6 & 2 & -6 & 4
\end{array}\right]
\begin{array}{c}
v_1 \\
\theta_1 \\
v_2 \\
\theta_2
\end{array}
\end{array},
$$

$$
\left[\mathbf{k}^{(2)}\right] = 1000
\begin{array}{cccc}
v_3 & \theta_3 & v_4 & \theta_4 \\
\left[\begin{array}{cccc}
12 & 6 & -12 & 6 \\
6 & 4 & -6 & 2 \\
-12 & -6 & 12 & -6 \\
6 & 2 & -6 & 4
\end{array}\right]
\begin{array}{c}
v_3 \\
\theta_3 \\
v_4 \\
\theta_4
\end{array}
\end{array}.
$$

Assembling the element stiffness matrices, we can obtain the 8×8 structural stiffness matrix [Ks]. However, to apply the boundary conditions, the rows and columns corresponding to zero deflections and rotations can be deleted from the above element stiffness matrices. Since nodes 1 and 4 are fixed, we can only keep those DOFs corresponding to nodes 2 and 3. In addition, since node 2 is connected with node 3 by a bolt, the two nodes have the same vertical displacement; i.e., $v_2 = v_3$. Therefore, it is also possible to combine v_2 and v_3 DOFs. The global matrix equation is then obtained as

$$1000 \begin{bmatrix} 12+12 & -6 & 6 \\ -6 & 4 & 0 \\ 6 & 0 & 4 \end{bmatrix} \begin{Bmatrix} v_2 \\ \theta_2 \\ \theta_3 \end{Bmatrix} = \begin{Bmatrix} 240 \\ 0 \\ 0 \end{Bmatrix} \tag{9.6}$$

Since we applied displacement boundary conditions, the above global matrix equation is positive definite, and we can solve for unknown nodal DOFs. The above matrix equation is solved for unknown nodal DOFs, to yield

$$v_2 = v_3 = 0.04, \theta_2 = 0.06, \theta_3 = -0.06 \tag{9.7}$$

Due to the constraint, it is expected that the vertical deflections at both nodes are identical. Due to the symmetry of the geometry, the rotations are equal and opposite. Note that the vertical deflection at node 2 is four times larger than that of the welded joint in example 3.6 of chapter 3.

Using an advanced optimization theory, it is also possible to formulate the effect of the bolt condition, $v_2 = v_3$, as a constraint. In this case, we keep both DOFs and add one more constraint relation to the global matrix equation.

$$1000 \begin{bmatrix} 12 & -6 & 0 & 0 & 1 \\ -6 & 4 & 0 & 0 & 0 \\ 0 & 0 & 12 & 6 & -1 \\ 0 & 0 & 6 & 4 & 0 \\ 1 & 0 & -1 & 0 & 0 \end{bmatrix} \begin{Bmatrix} v_2 \\ \theta_2 \\ v_3 \\ \theta_3 \\ \lambda \end{Bmatrix} = \begin{Bmatrix} 240 \\ 0 \\ 0 \\ 0 \\ 0 \end{Bmatrix} \tag{9.8}$$

where λ is the Lagrange multiplier to impose the bolt constraint. In eq. (9.8), we assume that the vertical load is applied at node 2. It could have been applied at node 3 instead, but the result will be identical. Solving eq. (9.8) yields the identical solutions in eq. (9.7). The Lagrange multiplier $1000\lambda = 120$ represents the force of constraint. It is the force transmitted by element 1 to element 2. ■

9.3.6 Method of Superposition

Throughout this book, we have assumed small displacements of the solid/structure such that the strains $\varepsilon_{ij} \ll 1$ and rotations $\omega_{ij} \ll 1$; that is, the deformation is infinitesimal. This assumption was relaxed in the study of buckling of beams and frames. Furthermore, we have been using linear stress-strain relations (Hooke's law). Due to the aforementioned assumptions, the behavior of the solid is governed by linear differential equations. As the model is linear, the solution scales linearly with load, and the solution due to multiple loads acting together can be obtained by adding or superposing the solutions due to the loads acting individually. This principle is called the *superposition* principle, and it was first illustrated for trusses in chapter 1. Here we extend the principle to the general case of three-dimensional solids. Before we discuss the method of superposition, it would be beneficial to further discuss the characteristics of linear problems first.

A system is said to be linear when the relationship between input and output is linear. Specifically, in structural mechanics, the relationship between applied loads (input) and displacements (output) is linear. When an applied load is doubled, the displacement will also be doubled. Thus, it is unnecessary to solve the linear system again when a different magnitude of load is applied. This property makes it possible to use the method of superposition. Mathematically, linearity can be explained using a linear operator. A general operator, A, is called linear when it satisfies $A(\alpha x_1 + \beta x_2) = \alpha A(x_1) + \beta A(x_2)$ for any scalars α and β, where x_1 and x_2 are inputs, such as applied loads.

Figure 9.35 shows a linear relationship between input x and output y. In structural mechanics, input x represents applied loads or applied heat, while output y symbolizes displacements, stresses, or temperatures. For example, let x_1 and x_2 be transverse loads applied at two different locations of a beam, and let y be the reaction moment at the wall. Let the reaction moment at the wall be y_1 when only x_1 is applied, and y_2 is the reaction when only x_2 is applied. Then, when a combined load $2x_1 + 3x_2$ is applied to the beam, there is no need to solve the system again. Because of linearity, the reaction moment under the combined load becomes $2y_1 + 3y_2$, which is the principle of superposition. This is very useful, especially when the magnitude of the load varies frequently.

When a structural system is linear, the above-mentioned superposition can be a very useful tool for design. In order to understand linear structural systems further, consider the diagram in figure 9.36, which illustrates the flow of physical quantities in structural systems. First, when loads are applied to the system, it generates local stresses in order to equilibrate against the globally applied loads. In an elastic system, stresses are generated by deforming its shape, which generates strains. Strains at every point are accumulated (or integrated) to yield displacements in the global level. In such a case, the structural system is called linear when all relationships among loads, stresses, strains, and displacements are linear. If any of them is not linear, then the structural system becomes nonlinear.

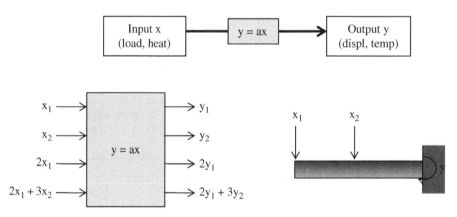

Figure 9.35 Illustration of linear systems

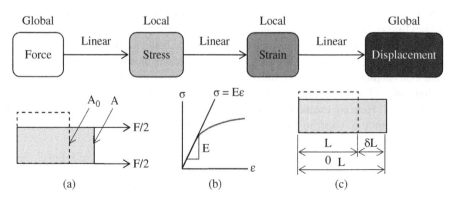

Figure 9.36 Structural linear systems

Then, let us consider a simple example of the uniaxial tension of a bar in order to understand the linear versus nonlinear relationship among the above-mentioned physical quantities. First, when a load, F, is applied as shown in figure 9.36(a), the bar elongates. In addition, because of Poisson's effect, the original cross-sectional area, A_0, of the bar shrinks to A. Then, the stress generated by the load, F, can be calculated by dividing the load by the cross-sectional area, A; that is, $\sigma = F/A$. However, the cross-sectional area depends on the load; as the load increases, the area decreases. Therefore, the relationship, $\sigma = F/A(F)$, is nonlinear between σ and F. However, if the load is small enough so that the difference between A_0 and A is ignorable, then it is possible to approximate the stress as $\sigma = F/A_0$. Based on this approximation, now the relationship between load and stress becomes linear. This approximation becomes invalid if the elongation increases significantly such that the change in cross-section cannot be ignored.

Next, consider the relationship between stress and strain. For general metallic materials, such as aluminum or steel, figure 9.36(b) illustrates the stress-strain curve that can normally be obtained from uniaxial tension tests. Initially, the stress is increased linearly proportional to the strain. In this region, the stress and strain relationship is linear and reversible; that is, if the stress varies, the strain also varies along the straight line. The slope of this straight line corresponds to Young's modulus. Therefore the relationship between stress and strain is linear; that is, $\sigma = E\varepsilon$. When the stress reaches a threshold, called the yield strength, the relationship becomes nonlinear, and its behavior is irreversible. Therefore, in order to be a linear relationship between stress and strain, the stress must be less than the yield strength.

Lastly, the relationship between strain and displacement must be linear. Consider the elongation of the bar, again, in figure 9.36(c). The original length, L_0, of the bar is increased by δL and ends up as the final length of L. In this case, δL is called the displacement or deformation. Then, the strain is defined as the ratio of the change in length and the length of the bar; that is, $\varepsilon = \delta L/L$. However, since the deformed length, L, already includes the displacement, δL, the relationship becomes nonlinear. As in the case of a force-stress relation, if the displacement is small, then the definition of strain can be approximated by $\varepsilon = \delta L/L_0$ so that the relationship between displacement and strain can be linear. This approximation is only valid when the displacement is small compared to the length of the bar.

As discussed above, many phenomena in physics show nonlinear behavior, and linear systems are approximations of nonlinear systems under limited conditions. For example, the relation between the deflection of a beam and applied load at its tip is linear when the deflection is small. This includes small strain, small displacement, and small rotation in solid mechanics. In this sense, a linear system is an approximation of a nonlinear one. Many engineering applications can be solved by considering them as linear. For example, large deflections are not expected in bridges or buildings. In such cases, linear analysis works well for estimating deflections and stresses.

As described before, solutions from a linear system under different loading conditions can be superimposed onto each other to produce a solution to the same system under combined loading conditions. Consider a solid subjected to displacement boundary conditions on the surface S_g and traction boundary condition on the surface S_T. When traction $\mathbf{t}^{(1)}$ is applied, the resulting displacement field is calculated as $\mathbf{u}^{(1)}(\mathbf{x})$, strain field $\varepsilon^{(1)}(\mathbf{x})$, and stress field $\sigma^{(1)}(\mathbf{x})$. Similarly, when traction $\mathbf{t}^{(2)}$ acts on the same boundary, the corresponding results are: $\mathbf{u}^{(2)}(\mathbf{x})$, $\varepsilon^{(2)}(\mathbf{x})$, and $\sigma^{(2)}(\mathbf{x})$. If tractions $\mathbf{t}^{(1)}$ and $\mathbf{t}^{(2)}$ act together as a combination of $\mathbf{t} = \alpha\mathbf{t}^{(1)} + \beta\mathbf{t}^{(2)}$ with constants α and β, then the response of the structure can be derived as the sum of the responses due to the two tractions:

$$\mathbf{u}(\mathbf{x}) = \alpha\mathbf{u}^{(1)}(\mathbf{x}) + \beta\mathbf{u}^{(2)}(\mathbf{x}),$$
$$\varepsilon(\mathbf{x}) = \alpha\varepsilon^{(1)}(\mathbf{x}) + \beta\varepsilon^{(2)}(\mathbf{x}), \tag{9.9}$$
$$\sigma(\mathbf{x}) = \alpha\sigma^{(1)}(\mathbf{x}) + \beta\sigma^{(2)}(\mathbf{x}).$$

Therefore, in linear structural systems, it is unnecessary to repeat finite element analysis when the magnitude of the load changes. In fact, it is enough to apply a unit load to calculate the resulting displacements and stress. For example, the resulting stress under a unit load is called *stress influence*

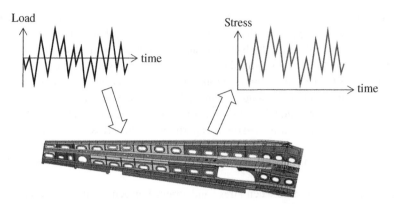

Figure 9.37 Fatigue analysis of airplane wing structure

coefficient (SIF). When the actual magnitude of the load is given, the resulting stress can be calculated by multiplying the load with SIF.

The stress influence coefficient can significantly reduce computational time especially when the load varies many times. A good example is fatigue analysis where the time history of the load is given. Then, the fatigue analysis requires calculating the time history of stress. Even if this is a dynamic problem, it is often considered as a quasi-static problem when the motion of the structure is small and the effect of inertia can be ignored. With a quasi-static assumption, a unit load is applied to the structure to calculate SIF first. Let the time history of the dynamic load given at discrete time steps be $\{\mathbf{t}^{(1)}, \mathbf{t}^{(2)}, \ldots, \mathbf{t}^{(N)}\}$. Then the stress history can be obtained as

$$\left\{\sigma^{(1)}, \sigma^{(2)}, \ldots, \sigma^{(N)}\right\} = SIF \times \left\{\mathbf{t}^{(1)}, \mathbf{t}^{(2)}, \ldots, \mathbf{t}^{(N)}\right\}. \tag{9.10}$$

Therefore, once SIF is available, it is very efficient to calculate the time history of stress using the method of superposition (for example, see figure 9.37).

9.3.7 Patch Test

As we have discussed in the previous chapters, finite element solutions are often an approximation of the exact one. We also discussed that more accurate solution can be obtained by reducing element size. In general, we expect that finite element solutions converge to the exact solution as the mesh is refined. Although this is true for most cases, it does not always apply. Some specially designed elements do not converge to the exact solution even if the mesh is gradually refined. In order to systematically address the convergent behavior of elements, we will discuss the requirements and methods of testing it.

For one-dimensional bars and two-dimensional plane solids, the strain energy involves the integral of first-order derivatives, from which we can deduce the following two requirements:

1. **Compatibility**: Displacements must be continuous across element boundaries—no gaps in materials. The derivatives of displacements, strains, must be continuous within the element but not necessarily across element boundaries.

2. **Completeness**: The element should be able to represent rigid body motions and constant strain conditions. For example, when an element is under rigid body translation, there should be no strains developed by that motion.

If an element satisfies the above two requirements, it is called a *conforming* or *compatible* element. Elements that violate these conditions are called nonconforming or incompatible elements. When an element is conforming, the solution converges monotonically as the mesh is refined.

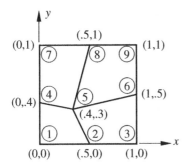

Figure 9.38 A patch of quadrilateral elements

A compatible element may become incompatible if a lower-order Gauss quadrature rule is used than the necessary one for numerical integration of the stiffness matrix. The extra zero-energy modes in section 6.7 are examples of incompatibility. Thus, not only the element formulation but also numerical integration can cause an element to be incompatible.

In order to guarantee the convergence of the solution, the element must pass a test, called *patch test*, which is based on the requirements that the element should be able to represent rigid body motions as well as constant strain conditions. The patch test is a simple test to determine if an element satisfies the basic convergence requirements. The patch test can find not only problems in the basic formulation but also problems in numerical integration or the computer implementation. The test procedure is as follows.

Consider a set of finite elements, called a patch, with at least one interior node, such as shown in figure 9.38. It is important to make sure that the shape of the elements is not regular because some errors in the formulation may not appear in the elements with a regular shape. Now we will discuss three different patch tests using these elements.

1. **Rigid body motion test**: The first test checks whether the elements can represent rigid body motions. The displacements of all nodes on the boundary are prescribed corresponding to a rigid body motion. With these prescribed displacements as boundary conditions, the displacements at the interior nodes are calculated. Since there are no applied loads, the calculated displacements at the interior nodes must be consistent with the rigid body motion. If this, in fact, is the case, the element is said to pass the rigid body patch test. For example, for the patch in figure 9.38, consider an x-translation of the boundary nodes; that is, $u_1 = u_2 = u_3 = u_4 = u_6 = u_7 = u_8 = u_9 = 1$ and the y-displacement of these nodes are all zero. Using these displacements, the finite element matrix equations are solved for unknown displacements at node 5. In order to pass the rigid body patch test, the computed horizontal displacement at node 5 should be $u_5 = 1$ and $v_5 = 0$. Note that all three rigid body motion cases (two translations and one rotation) must be verified before concluding the element passes the rigid body patch test.

2. **Constant strain patch test**: The second test checks whether the elements can represent constant strain conditions. Since strains are derivatives of displacements, constant strain conditions can be produced by linear displacements. First, linearly varying displacements are applied to the nodes on the boundary, and the displacements at the interior nodes are calculated. If the calculated displacements satisfy the constant strain conditions (i.e., linearly varying displacements), the element passes the constant strain patch test. For example, for the patch in figure 9.38, consider the condition of $\varepsilon_{xx} = 1$. This strain condition can be achieved by prescribing displacements of nodes at the boundary given by $u(x) = x$ and $v(x) = 0$; that is, the nodal x-displacement is the same as its x-coordinate. In order to pass the test, the computed x-displacement at node 5 must be equal to its x-coordinates; that is, $u_5 = 0.4$ and $v_5 = 0.0$. Again, several different linear displacements must be tried before concluding that the element passes the patch test.

3. **Generalized patch test**: In the previous two patch tests, displacements are specified only along the boundary. Thus, this form of the test cannot verify errors associated with the applied loads at the boundary. A more general form of patch test is needed to check the element formulation and implementation more thoroughly. In the generalized form, the patch of the element is supported by a minimum number of boundary conditions to prevent rigid body motion. For the rest of the boundary nodes, a set of loads consistent with constant stress state in the patch is applied. The computed stress state in the elements should obviously match with the assumed stress state. The procedure is illustrated for the patch in figure 9.39. The objective is to test whether the element can represent uniform stress of $\sigma_{xx} = 1$ Pa. For the minimum displacement boundary conditions, we apply $u_1 = v_1 = u_4 = u_7 = 0$. A distributed load is then applied on the right edge to produce a constant σ_{xx}. This distributed load must be converted to the equivalent nodal forces. The patch can now be analyzed using the finite element code. If the assumed stresses and displacements are recovered, the element passes the patch test. The generalized patch test requires a little more work but tests the implementation more thoroughly. Furthermore, the test does not need an interior node and therefore can be performed even on a single element.

Note that the patch test numerically tries to test whether linear and constant terms are present in the finite element approximation. If the shape functions are derived from an assumed polynomial and the analytical integration is performed, there is no need to go through the patch test. However, if shape functions are written by intuition, it may not be obvious whether the necessary terms have been included. The numerical integration scheme and the introduction of additional shape functions make the behavior of elements more complicated. The patch test has, therefore, become a standard tool for the development of new element formulations to test whether the element has the necessary convergence properties. Most commercial programs will only have elements that have already undergone extensive patch tests and satisfied convergence requirements. Therefore, the user does not need to test them.

Patch test for two-dimensional problems involves consideration of several cases. For example, table 9.2 summarizes various cases of patch tests.

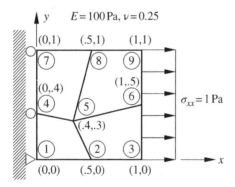

Figure 9.39 Generalized patch test for constant σ_{xx}

Table 9.2   Patch tests for plane solids ($E = 1$ GPa, $\nu = 0.3$)

Patch test type	u	v
x-translational rigid-body motion	1	0
y-translational rigid-body motion	0	1
xy-diagonal rigid-body motion	1	1
Constant $\sigma_{xx} = 200$ MPa	$0.2x$	$-0.06y$
Constant $\sigma_{yy} = 200$ MPa	$-0.06x$	$0.2y$
Constant $\tau_{xy} = 100$ MPa	$0.26y$	0

EXAMPLE 9.5   *Patch test for bar elements*

In one-dimensional bar elements, there is only one rigid body motion test (x-translation) and one constant strain test (ε_{xx} = constant). Consider a patch of two bar elements shown in figure 9.40 with $E = 10\,\text{GPa}$, $A = 10^{-6}\,\text{m}^2$. Show that the element passes the two patch tests using the following shape functions for element (i):

$$N_1(x) = \frac{x_{i+1} - x}{L^{(i)}}, \quad N_2(x) = \frac{x - x_i}{L^{(i)}}.$$

SOLUTION   Element stiffness matrices

Element 1:
$$\left[\mathbf{k}^{(1)}\right] = 10^4 \begin{bmatrix} 2 & -2 \\ 2 & -2 \end{bmatrix}.$$

Element 2:
$$\left[\mathbf{k}^{(2)}\right] = 10^4 \begin{bmatrix} 1 & -1 \\ 1 & -1 \end{bmatrix}.$$

Using the standard assembly procedure, the equations for the patch can be written as follows.

$$10^4 \begin{bmatrix} 2 & -2 & 0 \\ -2 & 3 & -1 \\ 0 & -1 & 1 \end{bmatrix} \begin{Bmatrix} u_1 \\ u_2 \\ u_3 \end{Bmatrix} = \begin{Bmatrix} F_1 \\ F_2 \\ F_3 \end{Bmatrix}.$$

(a) Rigid body motion test: In the rigid body motion test, the displacements at the boundary nodes are prescribed. In this case, we can specify $u_1 = u_3 = 1$ and need to check whether $u_2 = 1$. By applying displacement boundary conditions, we have

$$10^4 \begin{bmatrix} 2 & -2 & 0 \\ -2 & 3 & -1 \\ 0 & -1 & 1 \end{bmatrix} \begin{Bmatrix} 1 \\ u_2 \\ 1 \end{Bmatrix} = \begin{Bmatrix} R_1 \\ 0 \\ R_3 \end{Bmatrix},$$

where R_1 and R_3 are unknown reaction forces in order to impose the displacement boundary conditions. From the second equation, we have

$$10^4(-2 \cdot 1 + 3 \cdot u_2 - 1 \cdot 1) = 0 \quad \Rightarrow \quad u_2 = 1.$$

Thus, the element passes the patch test for rigid body motion.

(b) Constant strain test: In order to produce the constant strain, the linear displacements are assumed. We prescribe $u(x) = x$ for the nodes at the boundaries; that is, $u_1 = 0$ and $u_3 = 1.5$. We expect $u_2 = 0.5$ in order to pass the patch test. Then, the matrix equation becomes

$$10^4 \begin{bmatrix} 2 & -2 & 0 \\ -2 & 3 & -1 \\ 0 & -1 & 1 \end{bmatrix} \begin{Bmatrix} 0 \\ u_2 \\ 1.5 \end{Bmatrix} = \begin{Bmatrix} R_1 \\ 0 \\ R_3 \end{Bmatrix}.$$

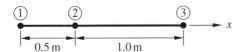

0.5 m   1.0 m   **Figure 9.40** Patch test for bar elements

From the second equation, we have

$$10^4(-2\cdot 0 + 3\cdot u_2 - 1\cdot 1.5) = 0 \quad \Rightarrow \quad u_2 = 0.5.$$

Thus, the element passes the patch test for the constant strain. ▪

9.4   ERROR ANALYSIS AND CONVERGENCE

Even though the finite element method is versatile and powerful, as with all other numerical methods, its results inevitably include errors. For example, the interpolation capability of a finite element may not be capable of approximating the true behavior of the structure, the idealized boundary conditions may be different from real boundary conditions, the material behavior may be different from the constitutive model used, and the numerical method may have round-off errors, integration errors, or equation solving errors. Therefore, engineers must be aware of the sources of error and their magnitudes in order to draw meaningful conclusions out of the computer simulation.

In general, the processes of identifying the sources of error in a computer simulation can be categorized into two stages. The first stage is whether the computer simulation represents the intended mathematical model accurately. Since most mathematical models are expressed in the form of differential equations, the equivalent statement is whether the computer simulation solves the differential equations accurately. This is called model verification, which is defined as the process of determining whether the FE model accurately solves the mathematical model of the structure or physical system.

The second stage is to check if the computer simulation represents the physical phenomenon accurately. Model validation is defined as the process of determining the degree to which a model is an accurate representation of the real phenomenon, from the perspective of the model's intended uses. Even if the computer simulation can accurately solve for the mathematical model, it may not predict the physical phenomenon accurately, due to approximation and simplification of the model. Model validation can be executed at the completion of model verification. It should be noted that model validation not only assesses the accuracy of a computational model but is also a process of improving the model based on the validation results. In general, model validation requires performing experiments to compare with the model prediction.

Since experiments are out of the scope of this textbook, we will not discuss model validation here. We will only focus on model verification. That is, we will focus on how numerical simulation accurately calculates the solution of the mathematical model. When the accurate solution to the mathematical model is available, this can be a trivial task. Most commercial finite element analysis programs have a verification manual, where the numerical solutions are verified against the mathematical or analytical solutions. However, these comparisons are limited to relatively simple problems where the mathematical solutions are available. Since most complex engineering problems do not have mathematical solutions, it is challenging to estimate the amount of error without having the exact solution. In this section, we introduce two ways of estimating errors: mesh accuracy and convergence analysis. Both of them are useful when the exact solution is not available.

9.4.1   Mesh Error

One of the most important sources of error in finite element analysis is related to the size of elements. The error caused by the element size is often called discretization error. The fundamental premise of finite element analysis is that the approximation error is gradually decreased as the element size is decreased. For a given size of elements, therefore, engineers may want to estimate the error. However, this may not be possible with a single set of elements. It may require multiple sets of elements, and we will discuss this topic in the next section.

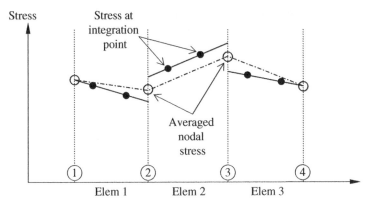

Figure 9.41 Stresses at integration points versus node-averages stresses

Error estimation is important in order to check the accuracy of the current analysis results. Since the exact solution is not available, the absolute magnitude of the error cannot be determined. Instead, an error estimate can be used as a criterion of mesh refinement. There are many error estimation methods, but we will discuss error estimation based on stress results. We will explain the error estimation using a uniaxial stress case. However, the same idea can be generalized to three-dimensional stress state.

Figure 9.41 illustrates stresses of three elements in one dimension. As we discussed in chapter 7, isoparametric elements use Gauss quadrature to integrate over the area or volume of the element to determine the stiffness matrix and distributed loads. Therefore, the computed stresses are most accurate at the integration points. The three elements in figure 9.41 have two integration points each, and thus, the stress within the element is approximated by a linear function (solid lines) by connecting the stresses at the two integration points. Since stress is extrapolated within each element in this fashion, it is discontinuous at the element boundary, that is, at a node. For example, at node 2, the stress from element 1 is different from the stress from element 2. In order to have a single value of stress at the node, we often average stresses from all elements connected to the node, which is called the averaged nodal stress. During post-processing, these node-averaged stresses are interpolated in order to make a smooth contour plot of stresses (dashed line). As shown in figure 9.41, the stresses at the integration points are different from interpolated node-averaged stresses. This difference depends on the size of the elements. The smaller the element size is, the smaller the difference is. Therefore, this difference can be used as a criterion to check if the current mesh is appropriate or not.

Let σ be the element stress that is interpolated/extrapolated using stresses at the integration points and σ^* be the stress that is interpolated using node-averaged stresses. If analysis results are accurate, there should no significant difference between σ and σ^*. Thus, the error estimation is based on the difference between element stress and node-averaged stress.

$$\sigma_E = \sigma - \sigma^*. \tag{9.11}$$

Since we are interested in the error in the entire model, we use the strain energy. The objective is to make the strain energy from σ_E small compared to the strain energy from the original stress. We define the two strain energies, as

$$U = \sum_{e=1}^{NE} \int_{V^{(e)}} \frac{\sigma^2}{2E} dV, \tag{9.12}$$

$$U_E = \sum_{e=1}^{NE} \int_{V^{(e)}} \frac{\sigma_E^2}{2E} dV. \tag{9.13}$$

Note that the strain energy is written in terms of stress, not strain. Equation (9.12) is regular strain energy of a structure under deformation, while eq. (9.13) is the contribution from the error in stresses. Then, the accuracy is measured by the mesh error, which is the ratio between U_E and U. In particular, since strain energy contains square of stresses, we define

$$\eta = \sqrt{\frac{U_E}{U + U_E}}. \tag{9.14}$$

It is suggested that the current mesh size is considered to be appropriate, if $\eta \approx 0.05$. That is, we are willing to accept the 5% root-mean-squared error of stress. The error estimate is often used for the adaptive mesh refinement. The mesh is gradually refined until η in eq. (9.14) is less than 0.05.

EXAMPLE 9.6   *Mesh error*

When the stresses at integration points in figure 9.41 are given in the following table, calculate the mesh error in eq. (9.14) and determine if mesh refinement is necessary or not based on the criterion of 5% root-mean-squared error.

Element	Integration point	Stress (MPa)
1	1	350
	2	300
2	1	400
	2	420
3	1	360
	2	340

SOLUTION   Since the values of stress, σ, at integration points are available, the real strain energy in eq. (9.12) can be calculated using two-point Gauss quadrature. In order to calculate the node-averaged stress, on the other hand, it is necessary to fit a linear function using the stresses at the integration points. Let s be the parametric coordinate in the reference element. The parametric coordinates of the two integration points are $-1/\sqrt{3}$ and $1/\sqrt{3}$, where stresses at these points are σ_a and σ_b, respectively. Then, by assuming a linear function for stress, $\sigma(s) = c_1 + c_2 s$ and by using the two stresses at the integration points, we can obtain the following function for the element stress:

$$\sigma(s) = \frac{1}{2}(\sigma_a + \sigma_b) + \frac{\sqrt{3}}{2}(\sigma_b - \sigma_a)s.$$

Then, by substituting $s = -1$ and $s = 1$ in the above equation, we can obtain the stresses at the two nodes. The following table shows stresses at the integration points (second column) and stresses at the two nodes (third column). The table also shows the node-averaged stresses (fourth column).

Element	Element stress at int. pts.	Element stress at nodes	Node-averaged stress at nodes	Node-averaged stress at int. pts.
1	350	368.3	368.3	361.7
	300	281.7	337.2	343.8
2	400	392.7		349.9
	420	427.3	397.3	384.6
3	360	367.3		383.6
	340	332.7	332.7	346.4

Note that nodes 1 and 4 are not averaged because they only belong to a single element. Using a similar interpolation equation, the node-averaged stresses at the integration points can also be calculated.

$$\sigma^*(s) = \frac{1}{2}\left(\sigma_i^* + \sigma_j^*\right) + \frac{1}{2}\left(\sigma_j^* - \sigma_i^*\right)s,$$

where σ_i^* and σ_j^* are node-averaged stress at the first and second node of the element, respectively. The fifth column of the table shows node-averaged stresses at the integration points. By comparing the third and fifth columns, we can see significant differences between the element stress and node-averaged stress.

The strain energy in eq. (9.12) can be calculated using Gauss quadrature, as

$$U = \sum_{e=1}^{NE} \sum_{I=1}^{2} \frac{\sigma(s_I)^2}{2E} w_I = \frac{1}{E}(1.245 + 1.352 + 1.336) \times 10^5 = \frac{3.933 \times 10^5}{E}.$$

In a similar way, the strain energy from $\sigma_E = \sigma - \sigma^*$ can be calculated as

$$U_E = \sum_{e=1}^{NE} \sum_{I=1}^{2} \frac{\sigma_E(s_I)^2}{2E} w_I = \frac{1}{E}(1.950 + 1.828 + 0.226) \times 10^3 = \frac{4.004 \times 10^3}{E}.$$

Note that the strain energy error in element 3 is much smaller than other elements. This is partly because the element is the last element of the mesh. The mesh error in eq. (9.14) can be calculated as

$$\eta = \sqrt{\frac{U_E}{U + U_E}} = 0.1004 = 10.04\%.$$

Since the mesh error is larger than 5%, the mesh needs to be refined with a smaller element size. It is also possible to calculate the mesh errors for individual elements. For the three elements: $\eta = 12.42\%$, 11.55%, and 4.11%, respectively. Based on the element mesh error, it is possible to change the size of the individual element. However, this method cannot provide the relationship between the element size and mesh error. ▬▬

As mentioned before, this method cannot give us the absolute magnitude of error, but it tells us if the current mesh is appropriate or not. In the following section, we will discuss how to estimate the actual magnitude of the error.

9.4.2   Convergence Analysis Using Richardson Extrapolation

The convergence analysis is based on the premise that the approximation error in finite element analysis results gradually decreases as the element size decreases, and eventually, the exact solution can be obtained if the element size approaches zero. Of course, this is only true when several conditions are satisfied, such as the appropriate boundary conditions must be used, the numerical error is small enough to ignore its effect on the solution accuracy, the used elements must the satisfy convergence requirement, and so forth. In this section, we assume that the effects of all other sources of error are small compared to the mesh error.

The basic idea of convergence analysis is to change element sizes and check how the analysis results change as a function of the element size, or equivalently, the number of elements. Figure 9.42 shows a common behavior of finite element solutions with respect to the number of elements. The horizontal solid line represents the exact solution, such as stress or displacement, which is unknown. The solid curve shows the finite element solution as a function of the number of elements. In general, since the finite element solutions are stiffer than the exact solution, the finite element solution curve is normally below the exact solution. As the number of elements increases, the finite element solutions converge to the exact solution. Therefore, without knowing the exact solution, it is possible to estimate the exact solution by increasing the number of elements. The question is how many elements are required to produce a solution that is close enough to the exact one.

Without too much theoretical study, it is possible to check if the analysis results change significantly when the number of elements changes. In figure 9.42, for example, when the number of elements is close to Na, two different element sizes produce a large change in finite element results. That means the current

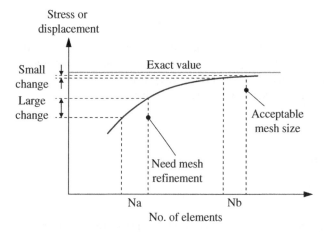

Figure 9.42 Converging to the exact solution with mesh refinement

number of elements is not enough, and the exact solution is far away from the finite element solution. On the other hand, when the number of elements is near *Nb*, two different element sizes produce a much smaller change. That is, even if the mesh is refined further, the finite element results will not change much, which means that the solution is almost converged.

However, this trend-based convergence analysis has a problem in practice. If the current mesh is not good enough, it requires refining the mesh continuously until the change in solution is less than the threshold. It can tell whether the current mesh is converged or not, but it cannot provide an estimate of the error based on the several sets of element sizes that are not converged. The idea of estimating the exact solution based on finite element solutions with two or three sets of element sizes is called extrapolation. In this section, we will introduce the Richardson extrapolation to estimate the exact solution using finite element solutions from two or three sets of unconverged meshes.

The original purpose of Richardson extrapolation is to accelerate the rate of convergence of a sequence in numerical analysis, named after Lewis F. Richardson[3]. It is also used to find the convergence rate of a computer simulation. In the following explanation, we use h as the size of elements. In general, finite element solutions are a function of h. For the purpose of explanation, let us consider displacement as a function of the element size. In general, the displacement function can be written as

$$u(h) = u_h = u_0 + gh^\alpha + O\left(h^{\alpha+1}\right), \tag{9.15}$$

where u_0 is the exact solution, g is a constant, and α is called the convergence rate. These terms are unknown. The last term, $O(h^{\alpha+1})$, is a higher-order error estimate. The convergence rate indicates how fast the solution will converge as a function of element size h. Most finite element solutions show the convergence rate between one and two.

The idea of Richardson extrapolation is to take three different values of h, and identify all unknown terms in eq. (9.15). Let h_1, h_2, and h_3 be the sizes of elements, ordered by $h_1 > h_2 > h_3$. Although there are different ways of selecting the different sizes of elements, it is customary to select them in such a way that they preserve the same ratio:

$$h_3 = ph_2 = p^2 h_1, \tag{9.16}$$

---

[3] Richardson, L. F. 1911. "The approximate arithmetical solution by finite differences of physical problems including differential equations, with an application to the stresses in a masonry dam." *Philosophical Transactions of the Royal Society* A. 210 (459–470): 307–357.

where $p = h_3/h_2 = h_2/h_1 < 1$.

By ignoring the higher-order error estimate, the displacements with three different element sizes can be written as

$$\begin{cases} u_{h1} = u_0 + gh_1^\alpha \\ u_{h2} = u_0 + gh_2^\alpha = u_0 + gp^\alpha h_1^\alpha \\ u_{h3} = u_0 + gh_3^\alpha = u_0 + gp^{2\alpha} h_1^\alpha. \end{cases} \tag{9.17}$$

The differences between the terms can be written as

$$\begin{cases} u_{h3} - u_{h2} = gp^\alpha (p^\alpha - 1) h_1^\alpha \\ u_{h2} - u_{h1} = g(p^\alpha - 1) h_1^\alpha. \end{cases} \tag{9.18}$$

The ratio between these two differences can be used to determine the convergence rate as

$$\frac{\| u_{h_3} - u_{h_2} \|}{\| u_{h_2} - u_{h_1} \|} \approx p^\alpha = \left(\frac{h_2}{h_1} \right)^\alpha. \tag{9.19}$$

Once the convergence rate is determined, the second term in eq. (9.18) can be used to determine the unknown constant g, and then, the first term in eq. (9.17) can be used to determine the estimated exact solution u_0.

In general, when the convergence rate is unknown, it is necessary to have at least three mesh sizes to determine the convergence rate and the estimate of the exact solution. However, when the convergence rate is known, only two mesh sizes are enough to determine the estimate of the exact solution. In order to show this, the first term in eq. (9.17) is multiplied by h_2^α and the second term by h_1^α, and the two terms are subtracted to obtain

$$u_0 = \frac{u(h_1) \, h_2^\alpha - u(h_2) \, h_1^\alpha}{h_2^\alpha - h_1^\alpha}. \tag{9.20}$$

Or, equivalently,

$$u_0 = \frac{u(h_1) p^\alpha - u(h_2)}{p^\alpha - 1}. \tag{9.21}$$

EXAMPLE 9.7   *Richardson's extrapolation*

Calculating the maximum stress in a finite element model, we used a 3×2, 6×4, 12×8, and 24×16 elements. The results are given in the table. Estimate the exact maximum stress. For convenience, you can assume that the largest element size h_1 is 1.

Mesh size	Max. stress (MPa)
3×2	71.53
6×4	75.63
12×8	77.35
24×16	77.92

SOLUTION   Since four data are available, we can use different combinations of data for the Richardson extrapolation. In this example, we will use the first three data, while the last three data will be given as an exercise problem.

As the mesh size is doubled at every data, you can say that $h_1 = 1, h_2 = 1/2, h_3 = 1/4$, and $p = 1/2$. From eq. (9.19), the convergence rate can be calculated as

$$\frac{\|77.35 - 75.63\|}{\|75.63 - 71.54\|} \approx \left(\frac{1}{2}\right)^\alpha \quad \Rightarrow \quad \alpha = 1.2532.$$

Now, using eq. (9.21), the estimated exact maximum stress can be

$$\sigma_{\max} = \frac{71.53 \times \left(\frac{1}{2}\right)^\alpha - 75.63}{\left(\frac{1}{2}\right)^\alpha - 1} = 78.59\,\text{MPa}.$$

9.5   PROJECT

Figure 9.43 shows a rough design of a bracket that carries a load at the circular hole whose diameter is 40 mm and is attached to the wall at the other end. Optimize its design to minimize weight subject to constraints that it should not yield. The final design must fit within the 400×200 mm^2 box shown in figure 9.43. The part is cast and then machined. Therefore, it is preferred that it is not less than 10 mm wide anywhere. The bracket is made of aluminum 6061, which has Young's modulus $E = 69$ GPa, Poisson's ratio $\nu = 0.3$, and yield strength $\sigma_Y = 378$ MPa. Assume that the bracket has a uniform thickness of 10 mm. Use the safety factor of 2.0. The bracket has to support a resultant maximum load $F = 15,000$ N. Write a project report describing the objective, preliminary analysis, finite element analysis results, convergence study, and conclusion. Carry out the analysis using both triangular and quadrilateral elements and turn in plots (mesh showing boundary conditions, deformed shape, stress plots) to justify the validity of your design.

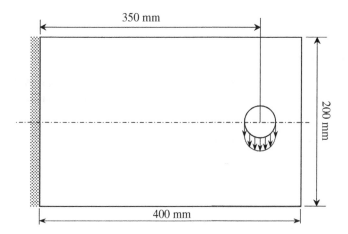

Figure 9.43 Design domain and boundary and loading condition for the bracket

9.6 EXERCISES

1. Solve the frame in example 9.1 using a commercial finite element program. Draw axial force, bending moment, and shear force diagrams. Compare the maximum stress value and its location from the preliminary analysis.

2. Consider a cantilevered beam shown in the figure. Solve the problem using: (a) five beam elements, (b) 4×20 plane stress solid elements, and (c) $4 \times 4 \times 20$ hexahedral elements. Compare the maximum stress and tip deflection. Assume $E = 72$ GPa and $\nu = 0.3$.

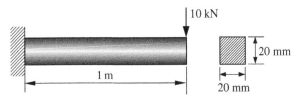

3. Consider a simply supported beam shown in the figure. Solve the problem using plane stress solids. Perform a convergence study for displacement at the center and maximum tensile stress. Increase the mesh size by a factor of two starting from an initial mesh size of 2×10. Compare the results with the exact solution. Assume $E = 72$ GPa and $\nu = 0.3$.

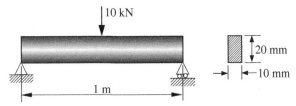

4. Consider a cantilevered beam shown in the figure. Solve the problem using plane stress solids. Perform a convergence study for displacement at the tip and maximum tensile stress. Increase the mesh size by a factor of two starting from an initial mesh size of 2×10. Compare the results with the exact solution. Assume $E = 72$ GPa and $\nu = 0.3$.

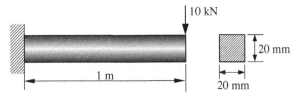

5. A cantilevered beam is modeled using: (a) two rectangular elements and (b) two quadrilateral elements as shown in the figure. Compare the accuracy of analysis results with exact stress and tip deflection. Assume $E = 72$ GPa and $\nu = 0.3$.

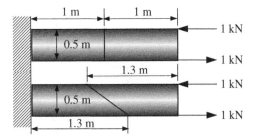

6. Using symmetric properties, draw an equivalent, simplified geometry of the structure shown in the figure with appropriate boundary conditions. Can the original geometry be solved using FEM? Can the simplified geometry be solved using FEM? Explain your answers.

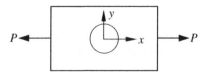

7. Using symmetric properties, draw an equivalent, simplified geometry of the beam shown in the figure with appropriate boundary conditions and applied loads.

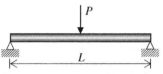

8. Using symmetry, find deflection, bending moment, and shear force in a continuous beam shown in the figure when $F = 8$ kN is applied at the center. Assume $E = 200$ GPa and $I = 10^5$ mm^4.

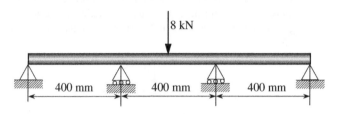

9. Using symmetry, find deflection, bending moment, and shear force in a continuous beam shown in the figure. Assume $E = 200$ GPa and $I = 10^5$ mm^4.

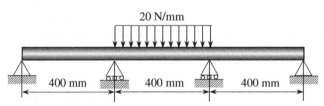

10. Consider a plane structure shown in the figure. Using symmetric modeling, draw the smallest geometry, and corresponding forces and boundary conditions that can provide the same analysis results as the original problem. Consider two cases:

    (a) when P_x and P_y are different;

    (b) when $P_x = P_y = P$.

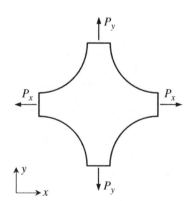

11. In one-dimensional bar elements, there is only one rigid body motion test (x-translation) and one constant strain test ($\varepsilon_{xx} =$ constant). Consider a patch of two bar elements shown in the figure with $E = 10\,\text{GPa}$, $A = 10^{-6}\,\text{m}^2$. Check if the element passes the two patch tests using the following shape functions for element (i):

$$N_1(x) = \frac{x_{i+1}^2 - x^2}{x_{i+1}^2 - x_i^2}, \quad N_2(x) = \frac{x^2 - x_i^2}{x_{i+1}^2 - x_i^2}.$$

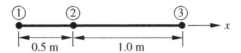

12. A patch test is often used to ensure that the solutions from the finite element method converge to the exact solution as the finite element mesh is refined. In the generalized form of the patch test, the patch of elements is supported by a minimum number of boundary conditions to prevent rigid body motion. For the remainder of the boundary nodes, a set of load consistent with constant stress state in the element is applied. The computed stress state should obviously match with the assumed stress state. Let a structure be approximated by four plane stress finite elements with thickness $= 0.1$, as shown in the figure with $E = 1000$ and $\nu = 0.3$.

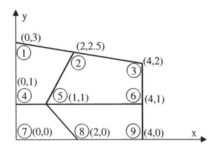

(a) For all boundary nodes (1, 2, 3, 4, 6, 7, 8, 9) apply displacement boundary conditions of $u = 0.2x$ and $v = -0.06y$. Using a commercial FE software, check the analysis results with the analytical solution of a constant stress field: $\sigma_{xx} = 200$, $\sigma_{yy} = 0$, and $\tau_{xy} = 0$. You need to make sure that incompatible shape functions are removed in the analysis. Provide a discussion of patch test results.

(b) Instead of applying a displacement boundary condition, apply distributed traction forces that represent a constant stress field $\sigma_{xx} = 200$. Convert the distributed traction forces into work–equivalent nodal forces. In the figure shown below, draw your nodal force vectors showing their directions and magnitudes. Carry out FE analysis with these nodal forces and verify whether all elements have the constant stress field.

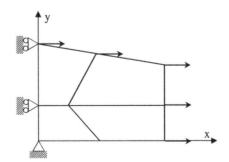

13. A rectangular solid under plane stress conditions is modeled using four finite elements as shown in the figure. It has been proved that the finite elements pass the patch test. (a) When a constant displacement $u = 0.01$ m is applied to nodes 3, 6, and 9, calculate the displacement $\{u, v\}$ at node 5 whose initial coordinates are (0.4, 0.4). (b) Explain how passing the patch test is important for the element's performance.

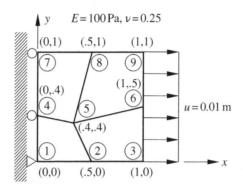

14. A cantilever beam can be analyzed using the three different finite element models shown in the figure.

 (a) Rank these three models based on which one you expect to give results closest to the analytic solution for the displacement at the tip of the beam (point of application of load). Give a brief explanation for the ranking you assign.

 (b) Plot the normal stress distribution you expect to compute at a section near the center of the beam for each model. Explain why you expect these plots.

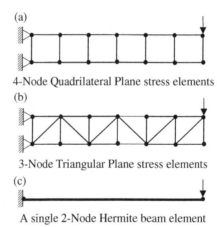

15. A thick cylinder is under the internal pressure $p = 1,000$ psi as illustrated in the figure. Young's modulus is 10^6 psi and Poisson's ratio is 0.2. Assuming plane strain conditions, calculate nodal displacements, element stress components, principal stresses and von Mises stress. Solve 1/4 model using symmetry boundary conditions. Nodal coordinates and element connectivity are given in the following table.

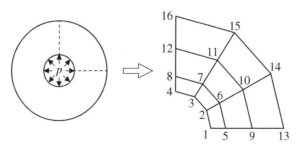

Node	x	y
1	2.0000	0.0
2	1.7052	1.0349
3	1.0349	1.7015
4	0.0000	2.0000
5	3.1111	0.0000
6	2.8105	1.7734
7	1.7734	2.8105
8	0.0000	3.1111
9	5.1111	0.0000
10	4.5120	2.8083
11	2.8083	4.5120
12	0.0000	5.1111
13	8.0000	0.0000
14	6.8061	4.1394
15	4.1394	6.8061
16	0.0000	8.0000

Element	Node 1	Node 2	Node 3	Node 4
1	1	5	6	2
2	2	6	7	3
3	3	7	8	4
4	5	9	10	6
5	6	10	11	7
6	7	11	12	8
7	9	13	14	10
8	10	14	15	11
9	11	15	16	12

16. Estimate the exact maximum stress in example 9.7 using the last three data. Compare the convergence rate and estimated exact stress with those of example 9.7.

Chapter 10

Structural Design Using Finite Elements

10.1 INTRODUCTION

Finite element analysis is concerned with determining the response (displacements and stresses) of a given structure for a given set of loads and boundary conditions. It is an analysis procedure in which the structural configuration—its geometry, material properties, boundary conditions and loads—is well defined, and the goal is to determine its response. On the other hand, engineering design is a process of synthesis in which parts are put together to build a structure that will perform a required set of functions satisfactorily. Analysis is very systematic and can be taught easily. Design is an intuitive and iterative process. Clearly, analysis is one of the several steps in the design process because we use analysis to evaluate the adequacy of the design. In this chapter, we will briefly discuss the basic steps in structural design and the use of finite element analysis (FEA) in the process of designing a structure.

There are two general approaches to design: creative design and adaptive design. The former is concerned with creating a new structure or machine that does not exist, whereas the latter is concerned with modifying an existing design to perform better. Although analysis techniques such as FEA play a crucial role, the designer's experience and creative ideas are important for the former. Adaptive design is an evolutionary process and is encountered much more frequently in practice. For example, how many times does an automotive company design a new car from scratch? The majority of engineers' work concentrates on improving the existing vehicle so that the new car will be more comfortable, more durable, safer, and more fuel efficient. In this chapter, we will discuss the role of FEA in the adaptive design process.

Structural design is a procedure to improve or enhance the performance of a structure by changing its parameters. *Performances* can be quite general in engineering fields and can include the weight, stiffness, or compliance; the fatigue life; noise and vibration levels; safety, and so forth. However, the performance does not include such aesthetic measures as attractiveness. The performances are measurable quantities. Especially, we are interested in the performance measures that can be computed by finite element analysis.

In structural design, two different types of performances are often considered. The first type is related to the criteria that the system must satisfy. As long as the performance satisfies the criteria, its level is not important. In engineering design, this type of performance is called a *constraint*. For example, the allowable strength is often used as a constraint in structural design so that the stresses are less than the allowable strength. The second type, called *goal*, is the performance that the engineer wants to improve as much as possible. The total weight of the structure or cost of manufacturing is an example of a goal. Since a goal is a function of design, it is often referred to as the *objective function*.

Introduction to Finite Element Analysis and Design, Second Edition. Nam H. Kim, Bhavani V. Sankar, and Ashok V. Kumar.
© 2018 John Wiley & Sons Ltd. Published 2018 by John Wiley & Sons Ltd.
Companion website: www.wiley.com/go/kim/finite_element_analysis_design

System parameters are variables that the engineer can change during the design process. For example, the thickness of a vehicle body panel can be changed to improve the stiffness of a vehicle. The cross section of a beam can be changed in designing a bridge structure. System parameters that can be changed during the design process are called *design variables*. The design variables can include the plate thickness, cross-sectional dimensions, location and size of cutouts, shape of the structure, and so forth. In this chapter, we will learn how to change the design variables to improve the goal, while satisfying system constraints.

A set of design variables that satisfies the constraints is called a *feasible design*, while a set that does not satisfy constraints is called an *infeasible design*. It is difficult to determine whether a current design is feasible unless the structural problem is analyzed. For complicated structural problems, it may not be easy to choose appropriate design constraints so that the feasible region is not empty.

In the process of structural design, finite element analysis is used to calculate the objective function and constraints. For example, if a design problem has a stress constraint, it is necessary to use finite element analysis to calculate the stress constraint. That is, a new finite element model is defined for a given set of design variables, from which the objective function and constraints are evaluated. In addition to the function values, the design process often requires the gradients (i.e., derivatives) of these functions with respect to design variables, which is called *sensitivity*. It is possible to calculate this sensitivity information using a variation of finite element analysis. A brief introduction of sensitivity analysis will also be provided in this chapter.

Since this is only an introduction to structural design, we will first present conventional design approaches using safety margin and intuitive design in the first two sections. Design parameterization in section 10.4 deals with the definition of design variables. The topics of parametric study and sensitivity analysis in section 10.5 investigate the effect of a change in design variables on the performance. Section 10.6 introduces structural optimization, which is a mathematical tool to find the best design.

10.2   CONSERVATISM IN STRUCTURAL DESIGN

Structural design anticipates various sources of uncertainty. For example, material properties show variability even if they are manufactured under the same process. In addition, it is difficult to accurately estimate the maximum load applied to an airplane wing under a gust wind environment. As we discussed in chapter 9, finite element analysis results may have an error in stress calculation. Therefore, it is important to make sure that the designed structure is safe under such uncertain material properties, environment, and calculation errors[1]. In this section, various ways of making the structure safe under uncertainty are discussed. In structural design, the major sources of uncertainty come from material properties, applied loads, and calculation errors.

10.2.1   Allowable Strength

The common approach to make the designed structure to be safe under uncertainty is to design based on conservative values. For example, when the failure strength of a material shows a statistical distribution, engineers are allowed to design a structure with a smaller strength than the nominal strength, which is called an allowable strength or stress allowables. In the design community, the allowable strength is sometimes included in the calculation of the factor of safety, but in this text, the allowable strength is not included in the calculation of the factor of safety.

---

[1] Although the calculation error is not random, since its value is unknown, it is considered as uncertain. This type of uncertainty is called epistemic uncertainty, which represents the lack of knowledge.

Before defining the allowable strength, it is necessary to define the failure strength first, as the former is based on the latter. For ductile materials (e.g., most metals), the failure strength means either the *yield* or *ultimate strength*, depending on the definition of structural failure. The yield strength is when the material starts to plastically deform, while the ultimate strength is when the material fracture occurs. Therefore, if the structure is considered to be failed when a permanent deformation occurs, then the yield strength is used for the failure strength. On the other hand, if the structure is not considered failed until fracture, then the ultimate strength can be used for the failure strength. On brittle materials, however, these two values are often so close as to be indistinguishable. Therefore, it is usually acceptable to use the ultimate strength as the failure strength.

In the aircraft structural design, for example, the Federal Aviation Administration (FAA) requires aircraft structure to be designed using either A-basis or B-basis allowable strength (MIL-HDBK-17 Vol. 1, FAR 25.613). In order to maintain safety, these allowable strengths are lower than the average failure strength of the material. The A-basis allowable strength is used when the failure of a structural component can lead to a catastrophic failure. On the other hand, the B-basis allowable strength is used when structural redundancy exists such that when a structural component fails, another component(s) can redistribute the load. In this chapter, we will use $\sigma_{\text{allowable}}$ to represent the allowable strength of the material.

In general, the allowable strength is calculated using *coupon* tests. A coupon represents a specimen for testing material properties, such as a tensile test specimen of metallic materials. A set of coupons is tested to estimate the statistical distribution of the material's failure strengths. Let us assume that N number of coupons are used, from which the average failure strength is σ_{average} and the *standard deviation* is s. Then, the allowable strength can be calculated based on the statistical distribution of failure strengths, whose histogram is shown in figure 10.1. The A-basis is the lowest first percentile of the distribution, while the B-basis is the lowest tenth percentile. That is, if 100 coupons are used, then at the A-basis allowable stress less than 1% of the coupons fail, while at the B-basis allowable stress a tenth of the coupons fail. From the assumption that the failure strength shows a *normal distribution*, these allowable strengths can be calculated by

$$\begin{aligned}
\left(\sigma_{\text{allowable}}\right)_{\text{A-basis}} &= \sigma_{\text{average}} - 2.33s, \\
\left(\sigma_{\text{allowable}}\right)_{\text{B-basis}} &= \sigma_{\text{average}} - 1.28s.
\end{aligned} \tag{10.1}$$

The allowable strength will be low when the coupon tests show a wide distribution, that is, a large standard deviation. That means, even if the average strength of a material is high, the allowable strength can be low when the coupon test results are scattered.

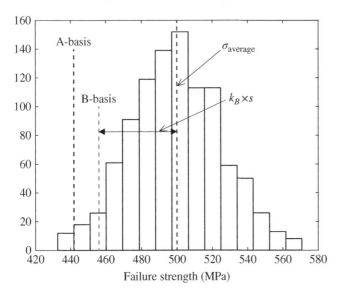

Figure 10.1 Histogram of failure strengths and allowable strengths

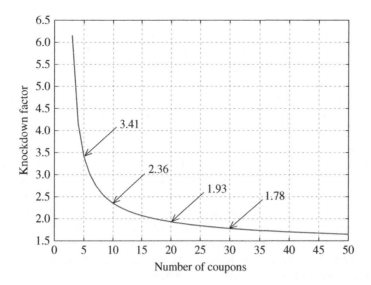

Figure 10.2 Knockdown factor for the B-basis allowable strength

If infinitely many coupons are used, then the average and the standard deviation are the true average and the true standard deviation of the material's strength. However, since N number of coupons are used, the calculated average and standard deviation are different from the true ones. Therefore, an additional *knockdown factor* k_B is used to reduce the allowable strength further, which can compensate for inaccuracy caused by the limited number of coupons. Therefore, using the coupon test results, the allowable strength can be defined as

$$\sigma_{\text{allowable}} = \sigma_{\text{average}} - k_B s. \tag{10.2}$$

where k_B is also called a *tolerance limit factor*.

It is obvious that the knockdown factor in eq. (10.2) depends on the number of coupons. When a small number of coupons are used, the errors in the calculated mean and standard deviation can be large, and therefore, a large value of the knockdown factor should be used. The knockdown factor will converge to the exact factor in eq. (10.1) as the number of coupons increases. Figure 10.2 shows the trend of the knockdown factor for the B-basis allowable strength as a function of the number of coupons. In fact, the knockdown factors in this figure are obtained from the 95% confidence of the tenth percentile of a normal distribution.

EXAMPLE 10.1 *Allowable strength*

Calculate the B-basis allowable strength of aluminum alloy 6061-T6 when the measured failure strengths of ten coupons are given as

$\sigma = \{351.5, 246.7, 318.1, 310.4, 350.2, 294.4, 292.7, 333.5, 309.3, 327.6\}$ MPa.

SOLUTION For the given set of coupons, the mean and the standard deviation can be calculated by

$$\sigma_{\text{average}} = \frac{1}{N} \sum_{i=1}^{N} \sigma_i = 313.4 \, \text{MPa},$$

$$s = \sqrt{\frac{1}{N-1} \sum_{i=1}^{N} (\sigma_i - \sigma_{\text{average}})^2} = 31.1 \, \text{MPa}.$$

From figure 10.2, the knockdown factor when $N = 10$ is 2.36. Therefore, the allowable strength for B-basis can be obtained as

$$\sigma_{\text{allowable}} = \sigma_{\text{average}} - k_B s = 313.4 - 2.36 \times 31.1 = 240 \text{MPa}.$$

Note that the original 10 coupon test results were obtained from a normal distribution with mean = 310 MPa and the standard deviation = 25 MPa. Therefore, the true tenth percentile of the distribution is 278 MPa. Therefore, the estimated allowable strength of 240 MPa is much lower than the true tenth percentile; that is, the estimated allowable strength is conservative. It may be a good practice to check if the estimated allowable strength converges to the true tenth percentile by increasing the number of coupons. ▄▄

Unlike the allowable strength, other types of performance measures do not have the concept of allowable values and do not have a specific method to establish it. In the case of displacement, for example, the allowable displacement is determined based on geometry or interference with other components. Although the design engineers can determine the allowable displacement with a certain margin, there is no randomness or uncertainty in the allowable displacement. Therefore, there is no need to apply additional conservatism to it.

10.2.2 Factor of Safety

In chapter 5, we defined the concept of a factor of safety and its application in determining whether a structure is safe or not. For a given loading condition, the structural analysis determines the stresses, and the ratio between the failure strength and the actual stress is defined as the factor of safety. In this section, the concept of the factor of safety is redefined in the perspective of structural design; that is, we want to design a structure that has a given level of the safety factor.

As we mentioned in the previous section, the randomness in material strength is taken into account by calculating a conservative allowable strength. In addition to the variability in material strength, however, many different factors need to be considered in order to make the structure safe, such as an error in stress calculation, unexpected loads, misuse, or degradation as time goes by. Many structural systems are purposefully built much stronger than needed for normal usage to take into account these factors. In structural design, the effect of all these factors is translated into a load carrying capacity of the structural system.

Factors of safety, also known as a *safety factor*, is a term describing the load carrying capacity of a system beyond the expected or actual loads. Essentially, the factor of safety is how much stronger the system is than it usually needs to be for an intended load. In order to consider the factor of safety in the design process, it is required to determine the maximum applied load that the system can experience during its normal operation. The *maximum applied load* can be measured from an existing similar system or can be estimated from experience. If the system is operated under a constant loading condition, such as gas turbine engines, the maximum load is the nominal operating load. Under the variable loading condition, on the other hand, it is the maximum load that the system can experience during its lifecycle. For example, the maximum load on the airplane wing is calculated when the airplane meets an extreme gust wind. Of course, the extreme condition can be subjective. The maximum load is often determined by government regulations.

Once the maximum applied load is determined, the *design load* can be determined using the factor of safety, S_F, as

$$S_F = \frac{\text{Design load}}{\text{Maximum load}}. \tag{10.3}$$

It is beneficial to emphasize two aspects that are different here from the definition of the factor of safety in chapter 5. First, the factor of safety is defined here in terms of loads, not stresses. Therefore, it is a global quantity, not local. This is because the factor of safety is used to measure the load carrying capacity of the entire structural system. Second, eq. (10.3) is not for calculating the factor of safety but for calculating the design load. Once the factor of safety is defined based on regulation or experience, the design load can be calculated by multiplying it with the maximum load. Therefore, the factor of safety is related to how much the structure can carry the load beyond the maximum load. In aircraft structural design, for example, the FAA requires designing the structure with $S_F = 1.5$, which means that the aircraft structure can support a load that is 1.5 times larger than the maximum load. In fact, during the certification of the airplane, the design load is applied on the structure, and no major failure should happen for it to pass the certification test.

Once the design load is determined for a given safety factor, the stress constraint can be defined as

$$\sigma_{\text{design}} = S_F \sigma_{\text{max}} \leq \sigma_{\text{allowable}}, \tag{10.4}$$

where σ_{design} is the calculated stress under the design load, and σ_{max} is the calculated stress under the maximum load. If the design load is used to calculate stress, there is no need to multiply the safety factor because the design load already includes it. On the other hand, if the maximum applied load is used, then the stress should be multiplied by the safety factor before comparing with the allowable strength.

The structure is considered to be failed when any part of the structure reaches its allowable strength. Due to the safety factor and the allowable strength, however, it is unlikely that the structure actually fails. Conservatism is the main difference between analysis and design. This level of conservatism is required to compensate for various factors in the design process. For example, the stress calculated from finite element analysis may differ from the actual stress due to various errors and also various assumptions made in the model, and the failure strength might be beyond the linear region of the material behavior.

Although we explained the factor of safety using the strength constraint, the above safety constraint can be applied to other types of performance measures. The above safety constraint can be restated as: the *response* (R_i) of a member under the maximum operating condition should be less than the *capacity* (C_i) of the member material. In the case of strength constraint, the calculated stress under the maximum load is the response, while the allowable strength is the capacity. Then, we can extend eq. (10.4) to general performance measures as

$$R_i(\mathbf{x}) \leq \frac{C_i}{S_F}. \tag{10.5}$$

When multiple loads are applied simultaneously, the response R_i should be the combination of the effects of all loads. Typically,

$$R_i = \sum_{j=1}^{N_L} R_i(F_j), \tag{10.6}$$

where F_j is the j-th load and N_L is the total number of applied loads.

Note that the safety constraint in eq. (10.5) is applied to a point $\mathbf{x}$ of the i-th member, and it must be satisfied everywhere. In practice, the location at which the response is critical is considered.

An alternative and useful measure of safety is the *safety margin*, which measures the excess capacity compared with the response; thus

$$Z_i = C_i - R_i. \tag{10.7}$$

In the case of the strength constraint, the member can afford additional Z_i stress before it fails.

From a different viewpoint, we can define a *sufficiency factor S_i* as a ratio of the allowable capacity to the response. Then, the safety constraint can be restated as

$$\frac{C_i}{S_F R_i} = S_i \geq 1. \tag{10.8}$$

For example, a sufficiency factor 0.8 means that R_i has to be multiplied by 0.8, or C_i is divided by 0.8 so that the sufficiency factor increases to one. In other words, this means that R_i has to be decreased by 20% $(1 - 0.8)$, or C_i has to be increased by 25% $((1/0.8) - 1)$ in order to achieve the required safety. The sufficiency factor is automatically normalized with respect to the capacity. The sufficiency factor is useful in estimating the resources needed to achieve the required safety factor. For example, if the current design has the sufficiency factor of 0.8, then this indicates that maximum stresses must be lowered by 20% to meet the safety. This permits the engineers to readily estimate the load to reduce stresses to a given level.

10.2.3 Load Factor

Instead of reducing the capacity by the factor of safety, it is possible to increase the applied loads, which is basic idea of the load factor. The *load factor* λ is the minimum factor by which a set of loads acting on the structure must be multiplied to cause the structure to fail. Commonly, the loads are taken as those acting on the structure during service conditions. For ductile materials, the strength of the structure is determined from the yield stress in the idealized elastic-plastic material model. For brittle materials, the tensile strength is used as the allowable strength.

Let us assume that a set of working loads F_j are applied to the structure, and the structure is safe. For a given failure mode (i.e., for a given ultimate strength), the structure is considered to have failed or collapsed when the capacity C_i are related to the factored load λF_j by

$$R_i(\lambda \mathbf{F}) = C_i, \tag{10.9}$$

where $\mathbf{F}$ is the vector of all applied loads. The load factor is the scale of the applied loads such that the response becomes equal to the capacity. If proportional loading is assumed, R_i is a linear function, and the load factor can be taken out of parentheses. Then, eq. (10.9) can be written in the following form:

$$\frac{C_i}{R_i(\mathbf{F})} = \lambda. \tag{10.10}$$

In the structural analysis viewpoint, the load factor is the ratio between capacity and response. From the design viewpoint, the structure should be designed in order to satisfy a given level of load factor.

Clearly, there is much similarity in formulation between the factor of safety and the load factor as measures of structural safety. The difference is the reference level at which the two measures operate: the first at the level of working loads, while the second, at the level of collapse loads.

EXAMPLE 10.2 *Factor of safety*

Consider a cantilevered beam shown in figure 10.3 with $E = 2.9 \times 10^4$ ksi. The width w is taken as 2.25 inches, while the height h is considered a design variable. The design should satisfy the factor of safety 1.5. (a) When the allowable tip displacement is $D_{\text{allowable}} = 2.5$ in., determine the height of the beam. (b) When the allowable strength of the material is 40 ksi, determine the height of the beam.

SOLUTION

    (a) We use one element to model the beam. From chapter 3 the matrix equation can be written as

$$\frac{EI}{L^3} \begin{bmatrix} 12 & 6L & -12 & 6L \\ 6L & 4L^2 & -6L & 2L^2 \\ -12 & -6L & 12 & -6L \\ 6L & 2L^2 & -6L & 4L^2 \end{bmatrix} \begin{Bmatrix} v_1 = 0 \\ \theta_1 = 0 \\ v_2 \\ \theta_2 \end{Bmatrix} = \begin{Bmatrix} R_1 \\ C_1 \\ F \\ 0 \end{Bmatrix}, \tag{10.11}$$

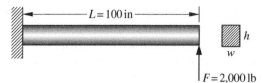

$F = 2,000\,\text{lb}$ **Figure 10.3** Cantilevered beam design

where R_1 and C_1 are the supporting force and couple at the wall. After deleting the first and second columns and rows, we obtain

$$\frac{EI}{L^3}\begin{bmatrix} 12 & -6L \\ -6L & 4L^2 \end{bmatrix}\begin{Bmatrix} v_2 \\ \theta_2 \end{Bmatrix} = \begin{Bmatrix} F \\ 0 \end{Bmatrix}. \tag{10.12}$$

The solutions of the above equation become

$$v_2 = \frac{4FL^3}{Ewh^3}, \quad \theta_2 = \frac{6FL^2}{Ewh^3}.$$

Since the allowable deflection is given, we do not need the safety factor. Thus, we have

$$v_2 = \frac{4FL^3}{Ewh^3} = D_{\text{allowable}} \quad \Rightarrow \quad h = \sqrt[3]{\frac{4FL^3}{EwD_{\text{allowable}}}} = 3.66 \text{ in.}$$

(b) The bending moment can be calculated from the second row of eq. (10.11). The supporting moment at the wall is

$$C_1 = \frac{EI}{L^3}\left[6Lv_1 + 4L^2\theta_1 - 6Lv_2 + 2L^2\theta_2\right] = -FL.$$

Thus, the bending moment at the wall is $M = FL$, which is consistent with the result from elementary mechanics. Then, the maximum stress at the wall becomes

$$\sigma_{\text{max}} = \frac{M\frac{h}{2}}{I} = \frac{6FL}{wh^2}.$$

Since the safety factor is given as 1.5, the failure stress is divided by the safety factor and compared to the maximum stress to calculate height, as

$$\frac{6FL}{wh^2} = \frac{\sigma_{\text{allowable}}}{S_F} \quad \Rightarrow \quad h = \sqrt{\frac{6FLS_F}{w\sigma_{\text{allowable}}}} = 4.47 \text{ in.}$$

Thus, the strength requirement is more severe than the displacement requirement. Note that in this case, we can use a load factor $\lambda = 1.5$ instead of the factor of safety because part (b) has only one load case and one performance. ▄

10.3 INTUITIVE DESIGN: FULLY STRESSED DESIGN

In example 10.2, we were able to determine the height of the beam that satisfies the stress or deflection constraint because the constraint was explicitly written in terms of the design variable. It is unlikely that we will have such an explicit relationship for more complex structures. Instead, we can evaluate the performance measures at a given set of values for the design variables. In such a case, an iterative process can be used to find the best design. Starting from the initial values of the design variables, we can update the design variables according to the calculated performances. In the case of the beam design problem, for example, if the stress is too high, we increase the height gradually. We can repeat this process until the stress becomes just right. The above-mentioned process is called *intuitive design*.

When structures are subject only to stress and minimum gauge constraints, the *fully stressed design* (FSD) is the best design. The basic concept can be used to design the structure as follows:

> For the best design, each member of the structure that is not at its minimum gauge is fully stressed under at least one of the design load conditions.

This implies that we should remove material from members that are not fully stressed unless prevented by minimum gauge constraints. This appears reasonable but it is based on an implicit assumption that the primary effect of adding or removing material from a structural member is to change the stresses in that member. If this assumption is not true, that is, if adding material to one part of the structure can have large effects on the stresses in other parts of the structure, we may want to have members that are not fully stressed because they help to relieve stresses in other members.

For *statically determinate* structures, the assumption that adding material to a member influences primarily the stresses in that member is correct. In fact, without an inertia or thermal load, there is no effect at all on stresses in other members. Therefore, we can expect that the FSD criterion will hold at the minimum weight design of such structures. However, for *statically indeterminate* structures, the minimum weight design may not be fully stressed. In most structures made of a single material, there is a fully stressed design near the optimum design, and so the method has been extensively used for metal structures. The FSD method may not do as well when several materials are used.

The FSD technique is usually complemented by a resizing algorithm based on the assumption that the load distribution in the structure is independent of member sizes. That is, the stress in each member is calculated, and then the member is resized to bring the stresses to their allowable values assuming that the loads carried by members remained constant (this is logical since the FSD criterion is based on a similar assumption). For example, for truss structures, where the design variables are often cross-sectional areas, the force in any member is $\sigma{\cdot}A$, where σ is the axial stress and A is the cross-sectional area. Assuming that $\sigma{\cdot}A$ is constant leads to the following stress ratio resizing technique:

$$A_{\text{new}} = A_{\text{old}} \left| \frac{\sigma}{\sigma_{\text{allowable}}} \right|, \tag{10.13}$$

which gives the resized area A_{new} in terms of the current area A_{old}, the current stress σ, and the allowable stress $\sigma_{\text{allowable}}$. For a statically determinate truss, the assumption that member forces are constant is valid, and therefore eq. (10.13) will bring the stress in each member to its allowable value. If the structure is not statically determinate, eq. (10.13) has to be applied repeatedly until convergence to any desired tolerance is achieved. Also, if A_{new} obtained by eq. (10.13) is smaller than the minimum gauge, the minimum gauge is selected rather than the value given by eq. (10.13).

Equation (10.13) works well for truss structures and under the assumption that the member force does not change significantly as the cross sections change. In the case of beam element, we can obtain a similar stress ratio resizing technique. In that case, we use the *section modulus*. The maximum stress in the beam occurs at the top or bottom of the cross section. Thus,

$$\sigma = \frac{M\frac{h}{2}}{I} = \frac{M}{S}, \tag{10.14}$$

where S is the section modulus. Assuming that the bending moment remains constant during the design process, we can obtain the following stress ratio resizing technique:

$$S_{\text{new}} = S_{\text{old}} \left| \frac{\sigma}{\sigma_{\text{allowable}}} \right|. \tag{10.15}$$

Note that the difference compared to the truss structure is the use of section modulus rather than cross-sectional area. Again, if the bending moment remains constant (i.e., statically determinate), then eq. (10.15) will yield the fully stress design in one iteration.

EXAMPLE 10.3   *Fully stressed design of a cantilevered beam*

Consider the cantilevered beam in figure 10.3. Let the initial height of the cross section be 3.5 in. Using fully stressed design, calculate the new height so that the maximum stress is equal to $\sigma_{allowable}$.

SOLUTION   First, we calculate the section modulus and the maximum stress at the initial design:

$$S_{old} = \frac{2I}{h} = \frac{wh^2}{6} = \frac{2.25 \times 3.5^2}{6} = 4.594 \text{ in}^3,$$

$$\sigma_{max} = \frac{M}{S_{old}} = 43.537 \text{ ksi}.$$

Thus, the new section modulus can be obtained using the stress ratio resizing technique, as

$$S_{new} = S_{old}\frac{\sigma_{max}}{\sigma_{allowable}} = 4.594 \times \frac{43.537}{26.667} = 7.5 \text{ in}^3.$$

Thus, the new height can be obtained from the definition of the section modulus, as

$$S_{new} = \frac{wh^2}{6} \quad \Rightarrow \quad h = \sqrt{\frac{6S_{new}}{w}} = 4.47 \text{ in}.$$

Note that the solution is identical with that of example 10.2. Thus, the fully stressed design converges to the optimum design in one iteration for the statically determinate system. ▄▄

EXAMPLE 10.4   *Fully stressed design of three-bar truss*

For the three-bar truss shown in figure 10.4, the goal is to find the cross-sectional areas of bars so that the weight of the truss is minimal while the stresses in all members are less than the yield strength with the factor of safety = 2.0. Perform one iteration using fully stressed design. The current cross-sectional areas are $b_1 = b_2 = b_3 = 10 \text{ mm}^2$. Use length $L = 1$ m, Young's modulus = 80 GPa, and yield stress $\sigma_Y = 250$ MPa.

SOLUTION   The element table is shown below.

Element	First Node i	Second Node j	AE/L	l	m
1	1	2	Eb_1	1	0
2	2	3	$Eb_2/\sqrt{2}$	$-1/\sqrt{2}$	$1/\sqrt{2}$
3	1	3	Eb_3	0	1

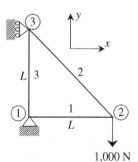

1,000 N   **Figure 10.4** Three-bar truss for fully stressed design

Using the above element table, the global element equations can be obtained after applying displacement boundary conditions as:

$$10^5 \begin{bmatrix} 10.83 & -2.83 & 2.83 \\ -2.83 & 2.83 & -2.83 \\ 2.83 & -2.83 & 10.83 \end{bmatrix} \begin{Bmatrix} u_2 \\ v_2 \\ v_3 \end{Bmatrix} = \begin{Bmatrix} 0 \\ -1000 \\ 0 \end{Bmatrix}. \tag{10.16}$$

By solving the above equation, we obtain the following nodal DOFs:

$$u_2 = -1.25 \text{mm}, \quad v_2 = -6.04 \text{mm}, \quad v_3 = -1.25 \text{mm}. \tag{10.17}$$

The stress in each element can be found from

Element 1: $\sigma^{(1)} = \dfrac{E}{L^{(1)}} \left[l^{(1)} (u_2 - u_1) - m^{(1)} (v_2 - v_1) \right] = -100 \text{MPa}.$

Element 2: $\sigma^{(2)} = \dfrac{E}{L^{(2)}} \left[l^{(2)} (u_3 - u_2) - m^{(2)} (v_3 - v_2) \right] = 141.4 \text{MPa}.$

Element 3: $\sigma^{(3)} = \dfrac{E}{L^{(3)}} \left[l^{(3)} (u_3 - u_1) - m^{(3)} (v_3 - v_1) \right] = -100 \text{MPa}.$

Now, using the stress ratio test, the new area of each element can be obtained, as

Element 1: $b_{\text{new}}^{(1)} = b_{\text{new}}^{(1)} \left| \dfrac{\sigma^{(1)}}{\sigma_{\text{allowable}}} \right| = 10 \left| \dfrac{-100}{250/2} \right| = 8 \text{mm}^2.$

Element 2: $b_{\text{new}}^{(2)} = b_{\text{new}}^{(2)} \left| \dfrac{\sigma^{(2)}}{\sigma_{\text{allowable}}} \right| = 10 \left| \dfrac{141.4}{250/2} \right| = 11.31 \text{mm}^2.$

Element 3: $b_{\text{new}}^{(3)} = b_{\text{new}}^{(3)} \left| \dfrac{\sigma^{(3)}}{\sigma_{\text{allowable}}} \right| = 10 \left| \dfrac{-100}{250/2} \right| = 8 \text{mm}^2.$

Using the new element areas, the new global element matrix equation can be obtained after applying displacement boundary conditions, as

$$10^5 \begin{bmatrix} 9.6 & -3.2 & 3.2 \\ -3.2 & 3.2 & -3.2 \\ 3.2 & -3.2 & 9.6 \end{bmatrix} \begin{Bmatrix} u_2 \\ v_2 \\ v_3 \end{Bmatrix} = \begin{Bmatrix} 0 \\ -1000 \\ 0 \end{Bmatrix}.$$

By solving the above equation, we have the following unknown nodal DOFs:

$$u_2 = -1.6 \text{mm}, \quad v_2 = -6.2 \text{mm}, \quad v_3 = -1.6 \text{mm}.$$

The stress in each element can be found from

Element 1: $\sigma^{(1)} = \dfrac{E}{L^{(1)}} \left[l^{(1)} (u_2 - u_1) - m^{(1)} (v_2 - v_1) \right] = -125 \text{MPa}.$

Element 2: $\sigma^{(2)} = \dfrac{E}{L^{(2)}} \left[l^{(2)} (u_3 - u_2) - m^{(2)} (v_3 - v_2) \right] = 125 \text{MPa}.$

Element 3: $\sigma^{(3)} = \dfrac{E}{L^{(3)}} \left[l^{(3)} (u_3 - u_1) - m^{(3)} (v_3 - v_1) \right] = -125 \text{MPa}.$

Now, all three members are in the allowable stress. Thus, the fully stressed design converged to the optimum design in one iteration. ∎

10.4 DESIGN PARAMETERIZATION

In the previous sections, we discussed simple design processes that can be applied to design variables that are related to cross-sectional geometry. However, when more complicated structures, such as plane solids or three-dimensional solids, are considered, it may not be trivial to define design variables. In this section, we discuss various types of design variables and procedures to define them.

Selecting design variables is called design parameterization. The design variables are assumed to vary during the design process. In some cases, it is relatively simple to choose them from analysis parameters and to vary their values. In other cases, however, it may not be easy because changing the design variables involves modifying finite element mesh. Based on their role in finite element analysis, we will discuss three different types of design variables.

Material property design variables: In structural analysis, material properties are used as a parameter. Young's modulus and Poisson's ratio, for example, are required in the analysis of an isotropic material. If these material properties are subject to change, then they are called *material property design variables*. These kinds of design variables do not appear in regular design problems since in most cases material properties are presumed to be constant. Analysis using constant material properties is called a deterministic approach. On the other hand, a probabilistic approach assumes that material properties are not constant but randomly distributed within certain ranges. This approach is more practical because if multiple specimens from the same material batch are tested, they will show different material properties. This happens because of the randomness in the microstructures during the manufacturing process. In this case, material properties are no longer considered to be constant and can, therefore, be used as design variables.

Sizing design variables: Sizing design variables are related to geometric parameters of a structure, and they are often called parametric design variables. For example, most automotive and airplane parts are made from plate/shell components. It is natural that engineers want to vary the thicknesses (or gauge) of the plate/shell in order to reduce the weight of the vehicle. In that case, the plate thicknesses are sizing design variables. During structural analysis, the thicknesses are considered as variable parameters. The sizing design variables are similar to the material property design variables in the sense that both types of variables change analysis parameters, not the overall shape and layout of the structure.

Another important type of sizing design variable is the cross-sectional geometry of bars and beams. Figure 10.5 provides some examples of the shapes and parameters that define these cross sections. In the structural analysis of bars, for example, the cross-sectional area is required to determine the axial rigidity. If a rectangular cross section is used, then the area would be defined as $A = b \times h$. Thus, the two parameters, b and h, can be considered design variables. Note that these variables contribute to the cross-sectional area for bars and the moment of inertia for beams.

Shape design variables: While material properties and sizing design variables are related to the parameters of structural analysis, shape design variables are related to the structure's geometry. Difficulty in this approach is that the shape of the structure does not explicitly appear as a parameter. Although the design variables in figure 10.5 determine the cross-sectional shape, they are not shape design variables,

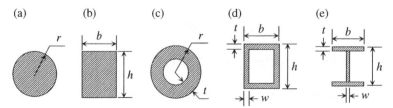

Figure 10.5 Sizing design variables for cross sections of bars and beams; (a) Solid circular cross section; (b) Rectangular cross section; (c) Circular tube; (d) Rectangular tube; (e) I–section

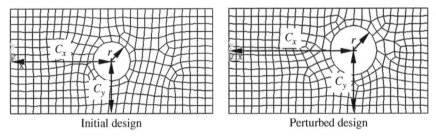

Figure 10.6 Shape design variables in a plate with a hole

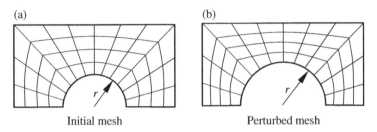

Figure 10.7 Design perturbation using isoparametric mapping method

since these cross-sectional shapes are considered parameters in structural analysis. However, the length of bars or beams should be treated as a shape design variable. Usually, the shape design variable defines the domain of integration in structural analysis. Thus, it is not convenient to extract shape design variables from a structural model and to use them as sizing design variables.

Consider a rectangular block with a hole, as shown in figure 10.6. The location and size of the hole are determined by the geometric values of C_x, C_y, and r, which are shape design variables. Different values of shape design variables yield different structural shapes. However, these shape design variables do not explicitly appear in structural analysis. If finite element analysis is used to perform structural analysis, then the shape design variables change the mesh, as shown in figure 10.6. Note that the mesh remains constant in the case of material properties and sizing design variables. Thus, the shape design problem is more difficult to solve than the sizing design problem.

It is important to note that inappropriate parameterization can lead to unacceptable shapes. This includes not only design variables but also the range of designs. For example, the ranges in C_x, C_y, and r should be limited such that the hole remains inside of the rectangle in figure 10.6.

Shape design parameterization describes the boundary shape of a structure as a function of design variables. There are many different methods, but we will only discuss two methods: isoparametric mapping method and solid model-based parameterization. Both methods have advantages and disadvantages. The first method works well with the mapped mesh in which the topological mesh remains constant throughout the design process. Only the physical mesh changes according to shape designs. For example, let us consider changing the radius of the structure shown in figure 10.7. The initial mesh is generated using the mapped mesh. Then when the radius is changed, the topology of the mesh remains unchanged, but due to change in physical dimension, the geometries of elements change accordingly. This type of parameterization is convenient because the number of elements and nodes will not change during the design process. For example, if the user specifies the performance as displacement at a node or stress at an element, it is easy to track the location and value of the performance. However, mesh distortion will be the bottleneck of this method when the changes in design variables are large. Initially, well-shaped elements will eventually be distorted as design variables change from their initial values.

The second method parameterizes the shape dimensions on the solid model. This method assumes that the preprocessing program has solid modeling and automatic mesh generation capabilities. When a solid model is generated, the user provides dimensions to geometric features, such as fillets, holes, cutouts, and so forth. It is also assumed that the preprocessing has a capability of automatically updating the solid model when the values of dimensions are changed. Most CAD software programs have these capabilities. As shown in figure 10.6, dimensions of solid models are usually selected as shape designs. It is unnecessary to select all dimensions. Only those dimensions that are supposed to change are considered as design variables. For a given design, the dimensions of the solid model are fixed, and the mesh is automatically generated on the solid model (refer to figure 10.7(a)). When design variables are changed, the solid model is updated to reflect the new designs, and the mesh is regenerated on the new model (refer to figure 10.7(b)). This type of parameterization can reduce mesh distortion problems because the preprocessing program will generate a free mesh for the given designs. However, a small change in design may end up in a completely different mesh. Thus, it is difficult to track the performance value at a particular location. Especially, numerical/discretization errors may dominate in finding the trend of performance.

10.5   PARAMETRIC STUDY – SENSITIVITY ANALYSIS

10.5.1   Parameter Study

Once the design variables are determined, structural analysis (i.e., finite element analysis) can be carried out to calculate performances. When the design variables are changed, we expect different values of the performance measures. Often we can estimate if a performance will increase or decrease based on our knowledge of mechanics. However, it would be nontrivial to estimate how much the performance will change. The *parameter study* investigates the effect of design variables on the performance. Usually, one design variable is changed at a time and the performance changes are plotted in a graph.

For example, the cantilevered beam in example 10.2 has two design variables: w (width) and h (height) of the cross section. Let each design variable (or parameter) have three levels. Then, table 10.1 shows nine cases for the parametric study. Since there are only two parameters, it is possible to plot the results as a three-dimensional surface as shown in figure 10.8. Note that the maximum stress decreases as both the width and height of the beam increase. In addition, the maximum stress is higher than the allowable stress when the height and width are small. Thus, a valid design can be chosen from the acceptable region.

10.5.2   Sensitivity Analysis

The parametric study is a useful tool to provide quantitative behavior of the performance as the design changes. However, when the number of design variables becomes large, the parameter study can be expensive. In the case of two design variables with three levels, nine analyses were required. Sometimes, multiple parameters are varied simultaneously in order to reduce the number of analyses, but still, this

Table 10.1   Parametric study of a cantilevered beam

w (in)	h (in)	σ_{max} (ksi)
2.0	4.0	37.5
2.0	4.5	29.6
2.0	5.0	24.0
2.5	4.0	30.0
2.5	4.5	23.7
2.5	5.0	19.2
3.0	4.0	25.0
3.0	4.5	19.8
3.0	5.0	16.0

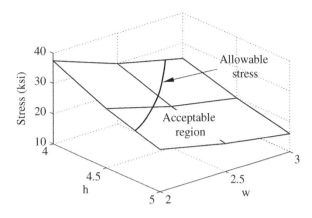

Figure 10.8 Parametric study plot for the cantilevered beam

can be expensive. In addition, when the performance changes rapidly, the parametric study cannot capture the local change unless the interval is small enough.

In many cases, we cannot afford to perform the parametric study with all variables. In addition, we do not need to find the performance changes throughout the entire range. It is often enough to find the effect of design variables in the vicinity of the current design point. In such a case, design sensitivity analysis can be used to effectively find it. *Design sensitivity analysis computes the rate of performance change with respect to design variables.* In conjunction with structural analysis, design sensitivity analysis generates a critical information, the gradient, for design optimization. Obviously, the performance is presumed to be a continuous function of the design, at least in the neighborhood of the current design point.

Explicit dependency on design: In general, a structural performance depends on the design. For example, a change in the cross-sectional area of a beam would affect the structural weight. This type of dependence is simple if the expression of weight in terms of the design variables is known. For example, the weight of a beam with a circular cross section can be expressed as

$$W(r) = \pi r^2 l, \tag{10.18}$$

where r is the radius, and l is the length of the beam. If the radius is a design variable, then the design sensitivity of W with respect to r would be

$$\frac{dW}{dr} = 2\pi r l. \tag{10.19}$$

This type of function is *explicitly dependent* on the design variable since the function can be explicitly written in terms of that design. Consequently, only algebraic manipulation is involved, and no finite element analysis is required to obtain the design sensitivity of an explicitly dependent performance.

Implicit dependence on design: However, in most cases, a structural performance does not explicitly depend on the design. For example, when the stress in a beam is considered as a performance, there is no simple way to express the design sensitivity of stress explicitly in terms of the design variable r. In the linear elastic problem, the stress of the structure is determined from the displacement, which is a solution to the finite element analysis. Thus, the sensitivity of stress $\sigma(\mathbf{q})$ can be written as

$$\frac{d\sigma}{dr} = \frac{d\sigma}{d\mathbf{q}} \cdot \frac{d\mathbf{q}}{dr}, \tag{10.20}$$

where $\mathbf{q}$ is the vector of nodal DOFs of the beam element. Since the expression of stress as a function of displacement is known, $d\sigma/d\mathbf{q}$ can easily be obtained. The only difficulty is the computation of $d\mathbf{q}/dr$, which is the displacement sensitivity with respect to the design variable r.

When a design engineer wants to compute the design sensitivity of performance such as stress $\sigma(\mathbf{q})$ in eq. (10.20), structural analysis (finite element analysis, for example) has presumably already been carried out. We will use the symbol b for generic design variable, and the nodal DOFs $\mathbf{q}$ is a part of the global DOF vector $\{\mathbf{Q}\}$. Assume that the structural problem is governed by the following linear algebraic equation

$$[\mathbf{K}(b)]\{\mathbf{Q}\} = \{\mathbf{F}(b)\}. \tag{10.21}$$

Equation (10.21) is a matrix equation of finite elements. Suppose the explicit expressions of $[\mathbf{K}(b)]$ and $\{\mathbf{F}(b)\}$ are known and differentiable with respect to design variable b. Since the stiffness matrix $[\mathbf{K}(b)]$ and load vector $\{\mathbf{F}(b)\}$ depend on the design b, solution $\{\mathbf{Q}\}$ also depends on the design b. However, it is important to note that this dependency is implicit, which is why we need to develop a design sensitivity analysis methodology. As shown in eq. (10.20), $d\mathbf{q}/db$ must be computed using the governing equation of eq. (10.21). This can be achieved by differentiating eq. (10.21) with respect to b as

$$[\mathbf{K}]\left\{\frac{d\mathbf{Q}}{db}\right\} = \left\{\frac{d\mathbf{F}}{db}\right\} - \left[\frac{d\mathbf{K}}{db}\right]\{\mathbf{Q}\}. \tag{10.22}$$

Assuming that the explicit expressions of $[\mathbf{K}(b)]$ and $\{\mathbf{F}(b)\}$ are known, $[d\mathbf{K}/db]$ and $\{d\mathbf{F}/db\}$ can be evaluated. Thus, if solution $\{\mathbf{Q}\}$ in eq. (10.21) is known, then $\{d\mathbf{Q}/db\}$ can be computed from eq. (10.22), which can then be substituted into eq. (10.20) to compute $d\sigma/db$. Note that the stress is *implicitly dependent* on the design through nodal DOFs $\mathbf{q}$.

When more than one design variable is defined, the above sensitivity equation must be solved for each design variable. Thus, the sensitivity analysis can be expensive when the problem has a large number of design variables. However, the sensitivity equation (10.22) uses the same stiffness matrix as the original finite element analysis. The difference is on the RHS. As we discussed in chapter 9, eq. (10.22) is similar to the finite element analysis with multiple load cases. The RHS of eq. (10.22) can be considered as a pseudo-force vector. The best way of solving eq. (10.22) might be to construct the RHS of eq. (10.22) for different design variables and to use the restart procedure. Then, the sensitivity equation can be solved with the stiffness matrix that is already factorized during the finite element analysis. Thus, the computational cost in solving eq. (10.22) is usually less than 5% of that of finite element analysis.

In general, it can be assumed that a general performance measure H depends on the design explicitly and implicitly. That is, the performance measure H is presumed to be a function of design b, and nodal DOFs $\mathbf{q}(b)$ as

$$H = H(\mathbf{q}(b), b). \tag{10.23}$$

The sensitivity of H can thus be expressed as

$$\frac{dH(\mathbf{q}(b), b)}{db} = \frac{\partial H}{\partial b}\bigg|_{\mathbf{q}=const} + \frac{\partial H}{\partial \mathbf{q}}\bigg|_{b=const} \cdot \frac{d\mathbf{q}}{db}. \tag{10.24}$$

The only unknown term in eq. (10.24) is $d\mathbf{q}/db$, which can be obtained from eq. (10.22). When $[d\mathbf{K}/db]$ and $\{d\mathbf{F}/db\}$ are not available, we can calculate the sensitivity using a finite difference method as follows.

Finite difference method: The easiest way to compute sensitivity information of the performance is by using the finite difference method. Different designs yield different analysis results and, thus, different performance values. The finite difference method actually computes design sensitivity of performance by evaluating performance at different stages in the design process. If b is the current design, then the analysis results provide the value of performance measure $H(b)$. In addition, if the design is perturbed to $b + \Delta b$, where Δb represents a small change in the design, then the sensitivity of $H(b)$ can be approximated as

$$\frac{dH}{db} \approx \frac{H(b + \Delta b) - H(b)}{\Delta b}. \tag{10.25}$$

Equation (10.25) is called the *forward difference method* since the design is perturbed in the direction of $+ \Delta b$. If $-\Delta b$ is substituted in eq. (10.25) for Δb, then the equation is defined as the *backward*

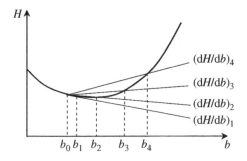

Figure 10.9 Influence of step size in the forward finite difference method

difference method. Additionally, if the design is perturbed in both directions, such that the design sensitivity is approximated by

$$\frac{dH}{db} \approx \frac{H(b+\Delta b) - H(b-\Delta b)}{2\Delta b},$$

(10.26)

then the equation is defined as the *central difference method.*

The advantage of the finite difference method is obvious. If structural analysis can be performed and the performance measure can be computed as a result of the structural analysis, then the expressions in eqs. (10.25) and (10.26) are virtually independent of the problem types considered. Consequently, this method is still popular in engineering design.

However, sensitivity computation costs become the dominant concern in the design process if the finite difference method is used. If n represents the number of designs, then $n+1$ analyses have to be carried out for the forward and backward difference methods, and $2n+1$ analyses are required for the central difference method. Unlike the sensitivity analysis in eq. (10.22), we cannot use the restart procedure in the finite difference method because the stiffness matrix at the perturbed design is different from that of the original design. For practical engineering applications using the finite element method, the cost of structural analysis is rather expensive. Hence, this method is not feasible for large-scale problems containing many design variables.

Another major disadvantage of the finite difference method is the accuracy of its sensitivity results. In eq. (10.25), accurate results can be expected when Δb approaches zero. Figure 10.9 shows some sensitivity results using the finite difference method. The tangential slope of the curve at b_0 is the exact sensitivity value. Depending on perturbation size, we can see that sensitivity results are quite different. For a mildly nonlinear performance measure, relatively large perturbation provides a reasonable estimation of sensitivity results. However, for highly nonlinear performance measures, a large perturbation yields completely inaccurate results. Thus, the determination of perturbation size greatly affects the sensitivity result. And even though it may be necessary to choose a very small perturbation, numerical noise becomes dominant for a too-small perturbation size. That is, with a too-small perturbation, no reliable difference can be found in the analysis results. For example, if up to five digits of significant numbers are valid in a structural analysis, then any design perturbation in the finite difference that is smaller than the first five significant digits cannot provide meaningful results. As a consequence, it is very difficult to determine design perturbation sizes that work for all problems

EXAMPLE 10.5 *Sensitivity analysis of a cantilevered beam*

Consider the cantilevered beam in example 10.2. At the optimum design with the strength constraint, we have $w = 2.25$ in. and $h = 4.47$ in. Calculate the sensitivity of the tip displacement with respect to the height of the beam. Compare the sensitivity results with the exact sensitivity

$$\left.\frac{dv_2}{dh}\right|_{\text{exact}} = -\frac{12FL^3}{Ewh^4} = -\frac{12 \times 2{,}000 \times 100^3}{2.9 \times 10^7 \times 2.25 \times 4.47^3} = -4.118.$$

(10.27)

SOLUTION   The design sensitivity of nodal DOFs can be obtained by differentiating the finite element matrix equation in eq. (10.11) with respect to the design. The height of the cross section is the design variable in this problem (i.e., $b = h$). The design sensitivity equation is given in eq. (10.22). In order to solve the design sensitivity equation, we need to calculate the RHS of eq. (10.22). Since the applied load $\{\mathbf{F}\}$ is independent of the design, the first term $\{d\mathbf{F}/db\} = \{\mathbf{0}\}$. The stiffness matrix in eq. (10.11) depends on design through the moment of inertia $I = wh^3/12$. Thus, it can be differentiated with respect to design. After multiplying with $\{\mathbf{Q}\}$, we have

$$\left[\frac{d\mathbf{K}}{db}\right]\{\mathbf{Q}\} = \frac{F}{4Lh}\begin{bmatrix} 12 & 6L & -12 & 6L \\ 6L & 4L^2 & -6L & 2L^2 \\ -12 & -6L & 12 & -6L \\ 6L & 2L^2 & -6L & 4L^2 \end{bmatrix}\begin{Bmatrix} 0 \\ 0 \\ 4L \\ 6 \end{Bmatrix} = \frac{F}{4h}\begin{Bmatrix} -12 \\ -12L \\ 12 \\ 0 \end{Bmatrix}. \tag{10.28}$$

Thus, the RHS of eq. (10.22) can be computed as

$$\left\{\frac{d\mathbf{F}}{db}\right\} - \left[\frac{d\mathbf{K}}{db}\right]\{\mathbf{Q}\} = \frac{F}{4h}\begin{Bmatrix} 12 \\ 12L \\ -12 \\ 0 \end{Bmatrix}. \tag{10.29}$$

Then, the design sensitivity equation can be obtained as

$$\frac{EI}{L^3}\begin{bmatrix} 12 & 6L & -12 & 6L \\ 6L & 4L^2 & -6L & 2L^2 \\ -12 & -6L & 12 & -6L \\ 6L & 2L^2 & -6L & 4L^2 \end{bmatrix}\begin{Bmatrix} dv_1/db = 0 \\ d\theta_1/db = 0 \\ dv_2/db \\ d\theta_2/db \end{Bmatrix} = \frac{F}{4h}\begin{Bmatrix} 12 \\ 12L \\ -12 \\ 0 \end{Bmatrix}. \tag{10.30}$$

When a displacement is fixed, the sensitivity is also zero, and therefore the first two rows and columns are deleted as we do in the finite element method for zero displacement boundary conditions. After removing these rows and columns, we have

$$\frac{EI}{L^3}\begin{bmatrix} 12 & -6L \\ -6L & 4L^2 \end{bmatrix}\begin{Bmatrix} dv_2/db \\ d\theta_2/db \end{Bmatrix} = \frac{F}{4h}\begin{Bmatrix} -12 \\ 0 \end{Bmatrix}. \tag{10.31}$$

The above equation can be solved for the unknown nodal DOFs. Now we have

$$\frac{dv_2}{db} = -\frac{12FL^3}{Ewh^4}, \quad \frac{d\theta_2}{db} = -\frac{18FL^2}{Ewh^3}. \tag{10.32}$$

Note that dv_2/db is the same with eq. (10.27). Thus, the sensitivity we calculated is exact. Note that in differentiating the stiffness matrix in eq. (10.28), only the moment of inertia I was differentiated. The basic form of the matrix remains unchanged. This type of design variable is called a sizing design variable. ■

EXAMPLE 10.6   *Finite difference sensitivity of a three-bar truss*

Calculate the sensitivity of the vertical displacement (v_2) at node 2 in the three-bar truss problem in example 10.4 with respect to b_2. Compare the accuracy of calculated sensitivity with the finite difference sensitivity. Use 1.0% perturbation size for the forward finite difference.

SOLUTION   Consider the three-bar truss example shown in figure 10.4. The finite element matrix equation, after applying displacement boundary conditions, is given in eq. (10.16), along with the nodal solutions in eq. (10.17). In order to build the sensitivity equation, we need to calculate the RHS of eq. (10.22). Since the applied load $\{\mathbf{F}\}$ is independent of the design, the first term $\{d\mathbf{F}/db\} = \{\mathbf{0}\}$. Out of three element stiffness matrices, only $[\mathbf{k}^{(2)}]$ depends on design b_2. Thus,

$$\left[\frac{d\mathbf{k}^{(1)}}{db_2}\right] = \left[\frac{d\mathbf{k}^{(3)}}{db_2}\right] = [\mathbf{0}], \quad \left[\frac{d\mathbf{k}^{(2)}}{db_2}\right] = \frac{E}{2L^{(2)}}\begin{bmatrix} 1 & -1 & -1 & 1 \\ -1 & 1 & 1 & -1 \\ -1 & 1 & 1 & -1 \\ 1 & -1 & -1 & 1 \end{bmatrix}\begin{matrix} u_2 \\ v_2 \\ u_3 \\ v_3 \end{matrix}.$$

These three matrices are assembled in the same way with the stiffness matrix and then, multiplied by the vector of nodal displacements to obtain

$$\left\{\frac{d\mathbf{F}}{db_2}\right\} - \left[\frac{d\mathbf{K}}{db_2}\right]\{\mathbf{Q}\} = -2.828 \times 10^{10} \begin{bmatrix} 0 & 0 & 0 & 0 & 0 & 0 \\ 0 & 0 & 0 & 0 & 0 & 0 \\ 0 & 0 & 1 & -1 & -1 & 1 \\ 0 & 0 & -1 & 1 & 1 & -1 \\ 0 & 0 & -1 & 1 & 1 & -1 \\ 0 & 0 & 1 & -1 & -1 & 1 \end{bmatrix} \left\{ \begin{array}{c} 0 \\ 0 \\ -.0013 \\ -.0060 \\ 0 \\ -.0013 \end{array} \right\} = \left\{ \begin{array}{c} 0 \\ 0 \\ -10^8 \\ 10^8 \\ 10^8 \\ -10^8 \end{array} \right\}.$$

Then, the design sensitivity equation, after applying displacement boundary conditions, becomes

$$10^5 \begin{bmatrix} 10.83 & -2.83 & 2.83 \\ -2.83 & 2.83 & -2.83 \\ 2.83 & -2.83 & 10.83 \end{bmatrix} \left\{ \begin{array}{c} du_2/db_2 \\ dv_2/db_2 \\ dv_3/db_2 \end{array} \right\} = \left\{ \begin{array}{c} -10^8 \\ 10^8 \\ -10^8 \end{array} \right\}. \tag{10.33}$$

The solution of the above sensitivity equation yields

$$\frac{du_2}{db_2} = 0, \quad \frac{dv_2}{db_2} = 353.55, \quad \frac{dv_3}{db_2} = 0. \tag{10.34}$$

Thus, the change in the cross-sectional area of member 2 will only change the vertical displacement of node 2.

Let us compute the design sensitivity of v_2 by using the finite difference method. The original displacements in eq. (10.17) are saved as $H(b_2) = -6.036$ mm. Then, design b_2 is perturbed by 1.0%, that is, $b_2 = 10.1$ mm. A new global matrix equation is produced with new design, as

$$10^5 \begin{bmatrix} 10.86 & -2.86 & 2.86 \\ -2.86 & 2.86 & -2.86 \\ 2.86 & -2.86 & 10.86 \end{bmatrix} \left\{ \begin{array}{c} u_2 \\ v_2 \\ v_3 \end{array} \right\} = \left\{ \begin{array}{c} 0 \\ -1000 \\ 0 \end{array} \right\}. \tag{10.35}$$

Note that the matrix is slightly different from the one in eq. (10.16). By solving the above equation, we have the following unknown nodal DOFs:

$$u_2 = -1.25\text{mm}, \quad v_2 = -6.00\text{mm}, \quad v_3 = -1.25\text{mm}. \tag{10.36}$$

Note that u_2 and v_3 did not change, which is consistent with the zero sensitivity in eq. (10.34). With the vertical displacement at node 2, we have the performance at the perturbed design, $H(b_2 + \Delta b_2) = -6.001$ mm. From the finite difference sensitivity formula in eq. (10.25), we have

$$\frac{dH}{db_2} \approx \frac{H(b_2 + \Delta b_2) - H(b_2)}{\Delta b_2} = \frac{-6.001 \times 10^{-3} + 6.306 \times 10^{-3}}{0.1 \times 10^{-5}} = 350.05. \tag{10.37}$$

Note that the finite difference sensitivity in eq. (10.37) is slightly different from the one in eq. (10.34). This is because of the influence of finite perturbation size. When 0.1% perturbation is used, the finite difference sensitivity becomes 353.2, which is much closer to the one in eq. (10.34). However, as we can see in eqs. (10.17) and (10.35), the difference in stiffness matrix is small. Thus, it is required to maintain high accuracy in matrix solution in order to have a small perturbation size. ▪

10.6   STRUCTURAL OPTIMIZATION

The purpose of many structural design problems is to find the best design among many possible candidates. As will be discussed in this section, at least one possible candidate should exist within a feasible design region that satisfies problem constraints. Every design in the feasible region is an acceptable design, even if it is not the best one. The best design is usually the one that minimizes (or maximizes) the objective function (goal) of the design problem. Thus, the goal of the design optimization problem is to find the design that minimizes the objective function among all feasible designs. Unfortunately, there is no mathematical theory that can find the global optimum design for general nonlinear functions. In this section, simple optimization methods are briefly introduced. However, this brief discussion is by no means the complete treatment of optimization methods. For a more detailed treatment, refer to Haftka and Gurdal[2] or Arora[3].

Most gradient-based optimization algorithms are based on mathematical programming methods, which require performance and sensitivity information for given values of the design variables. For a given design that defines the structural model, the structural analysis provides the values of the objective and constraint functions to the algorithm. Gradients (design sensitivities) of the objective and constraint functions must also be supplied to the optimization algorithm. Then, the optimization algorithm calculates the best possible design of the problem. In this section, we will introduce how the optimization problem can be formulated and how it can be solved using graphical and mathematical programming methods.

10.6.1   Optimization Problem Formulation

An important step in optimization is to transcribe the verbal statement of the optimization problem into a well-defined mathematical statement. In the verbal statement, the goal of the optimization problem is to find the best design that minimizes (or maximizes) the objective function (or cost function) under the given constraints by changing design parameters. Examples of the objective function in structural analysis are the weight, stiffness or compliance; the fatigue life; the noise level; amplitude and frequency of vibration; safety, and so forth. Constraints are similar to objective functions, but they are given as limits so that a design must satisfy them in order to be a candidate. Therefore, as long as a constraint satisfies the limit, the magnitude of the constraint is not important. However, the design that can yield a smaller value of the objective function is better than other designs with larger values of the objective function. Design variables are parameters or geometry of the structure that can be varied during design. Figure 10.10 shows the flow chart of the design optimization process.

Three-step problem formulation: In order to formulate the optimization problem, it is important to properly define three components—design variables, objective, and constraint functions.

1. Identification and precise definition of design variables. As we discussed in section 10.4, different types of design variables are available. Some designs only change some parameters in the finite element analysis, but some designs change the finite element model itself. It is important to make sure that the finite element model, including properties and parameters, must completely be defined with a given set of design variables. For example, the cross section in figure 10.11(a) requires four parameters to define it: w, h, t_1, and t_2. Of course, it is unnecessary to choose all four as design variables. We can fix t_1 and t_2 and define w and h as design variables. The important thing is that once the values of design variables are given, the model should be defined uniquely. In addition, it is important to provide the lower and upper bounds of design variables and to make sure that the finite element analysis can be carried out successfully over the entire range of design variables. When finite element analysis cannot be carried out for a given design, the objective function or constraint is often assigned with very large values so that the optimum design occurs somewhere else.

   Another important aspect of selecting design variables is independence. All design variables must be independent of each other. If design variables are dependent, it may not be possible to

---

[2] Haftka, R. T., and Gurdal, Z. 1992. *Elements of Structural Optimization*. Kluwer Academic Publishers.

[3] Arora, J. S. 2004. *Introduction to Optimum Design*. Elsevier Academic Press.

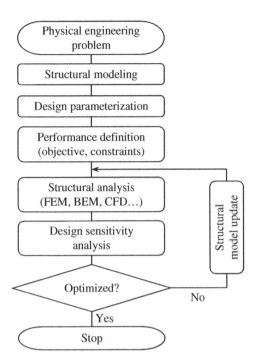

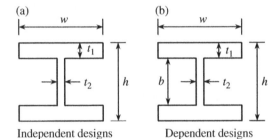

Figure 10.10 Structural design optimization procedure

Independent designs   Dependent designs   **Figure 10.11** Design parameters for beam cross section

generate the model, and the optimization algorithm may fail. Figure 10.11(b) shows an example of improperly defined design variables. If w, h, b, t_1, and t_2 are defined as design variables, then the following relation exists: $h = b + 2\,t_1$. It is possible to define six design variables and provide this relation as a constraint, but it will make the optimization problem unnecessarily complicated. In addition, selection of design variables is not unique. In figure 10.11(b), for example, we can choose w, h, t_1, and t_2 as independent design variables, or w, b, t_1, and t_2.

2. Defining an objective function (cost function): Once design variables are defined, the next step is to define the objective function of the problem. The objective function defines the ranking of different designs. The goal of optimization is to find the design that has the best ranking in terms of the objective function. It is obvious that the objective function must depend on design variables. The absolute magnitude of the objective function is not important. A constant can be added or multiplied to the objective function without changing the optimization result.

When more than one objective functions are involved, it is possible to combine them using weights as

$$F(\mathbf{b}) = \sum_{I=1}^{N_{OBJ}} w_I f_I(\mathbf{b}), \qquad (10.38)$$

where $f_l(\mathbf{b})$ is an individual objective function, and w_l is the corresponding weight. The weights must be chosen based on the importance of each objective function. Different optimum designs will be expected for a different set of weights. When one is unclear about the weights, the multi-objective optimization problem can be solved, which will provide all possible sets of optimum designs[4].

3. Identification and definition of constraints. Unlike the objective functions, constraints do not rank the designs, but they validate whether the design is feasible or not. Once the design is determined to be usable, it is not important how much safety margin the design has. In general, there are two types of constraints: equality and inequality. Equality constraints provide relations between design variables or impose conditions that a usable design must satisfy. When the relation is linear, it is possible to remove one design variable for each equality constraint. Because of that, they are considered as strong constraints. In general, however, it is not easy to remove a design variable when the equality constraint is nonlinear. Inequality constraints are more popular in structural problems, and they provide limits on performances. For example, the maximum stress of usable design should be less than the allowable strength of the material. As long as the maximum stress is lower than the allowable strength, it is not important how low the stress is.

Some constraints limit the lower or upper bounds of design variables. These constraints are called *side constraints*. Since it is relatively easy to impose the lower or upper bounds of design variables, these constraints are often treated separately.

Usually, an optimization problem has one objective function with many constraints. In general, there is no limit on the number of inequality constraints. For example, one can choose stresses of all elements and make them less than the allowable stress. In such a case, many elements will have stress much less than the allowable stress, and those elements will be ignored during optimization until their values become close to the allowable stress. However, the number of equality constraints should be less than that of design variables. This is obvious from the fact that we can theoretically remove one design variable for each equality constraint.

Standard form: The above description of three components in optimization needs to be written in mathematical form so that there will be no ambiguity in the problem definition. The standard form of design optimization problem can be written as

$$
\begin{aligned}
&\text{minimize} \quad f(\mathbf{b}) \\
&\text{subject to} \quad g_i(\mathbf{b}) \leq 0, \quad i = 1, \cdots, N \\
&\qquad\qquad\ h_j(\mathbf{b}) = 0, \quad j = 1, \cdots, M \\
&\qquad\qquad\ b_l^L \leq b_l \leq b_l^U, \quad l = 1, \cdots, K,
\end{aligned}
\tag{10.39}
$$

where $\mathbf{b} = \{b_1, b_2, \ldots, b_K\}^{\mathrm{T}}$ is the vector of design variables, $f(\mathbf{b})$ is the objective function, $g_i(\mathbf{b})$ $(i = 1, \ldots, N)$ are inequality constraints, $h_j(\mathbf{b})$ $(j = 1, \ldots, M)$ are equality constraints, and b_l^L and b_l^U are, respectively, the lower and upper bounds of design variables. Note that the objective function is minimized, and the inequality constraints are written in "less than or equal to" form.

The standard form in eq. (10.39) is to find the minimum value of the objective function within the region that satisfies constraints. Thus, it is convenient to define the following feasible set S:

$$
S = \left\{ \mathbf{b} \,|\, g_i(\mathbf{b}) \leq 0, \ i = 1, \cdots N, \ h_j(\mathbf{b}) = 0, \ j = 1, \cdots M \right\}.
\tag{10.40}
$$

Then, the optimization problem is to find the minimum $f(\mathbf{b})$ in the feasible set S. Once an optimization problem is written in the standard form, solving the optimization problem is independent of applications. Whether it is a structural or financial problem, the same optimization technique can be used to solve it.

---

[4] Arora, J. S. 2004. *Introduction to Optimum Design.* Elsevier Academic Press.

EXAMPLE 10.7   *Beer can design formulation*[5]

Formulate a standard optimization problem for a beer can design problem. We want to design the beer can that can hold at least specific amount of beer and meet other design requirements. The goal is to minimize the manufacturing cost as they are produced in billions. Since the cost can be related directly to the surface area of the sheet metal used, it is reasonable to minimize the sheet metal required to fabricate the can. Fabrication, handling, aesthetic, and shipping considerations impose the following restrictions on the size of the can:

1. The can is required to hold at least 400 ml of fluid.

2. The diameter of the can should be no more than 8 cm. In addition, it should not be less than 3.5 cm.

3. The height of the can should be no more than 18 cm and no less than 8 cm.

Following the three-step procedure, write a standard form of the optimization problem.

SOLUTION   By following three-step procedure, we first choose design variables. In order to design the volume and surface area of the can, it is necessary to choose the height of the can as well as either the radius or diameter of the can. Thus, we choose $\mathbf{b} = \{D, H\}^T$ from figure 10.12 as design variables.

The objective function will be the surface area of the can, which is the amount of sheet metal used. Since the surface area is composed of top, bottom, and side, we have

$$f(D,H) = \pi DH + \frac{\pi}{2}D^2.$$

The three constraints also can be written in terms of design variables. Thus, the standard form of optimization problem becomes

$$\text{minimize}\quad f(\mathbf{b}) = \pi DH + \frac{\pi}{2}D^2 \quad \text{cm}^2$$

$$\text{subject to}\quad 400 - \frac{\pi}{4}D^2 H \le 0 \tag{10.41}$$

$$3.5 \le D \le 8 \qquad \text{cm}$$

$$8.0 \le H \le 18 \qquad \text{cm}.$$

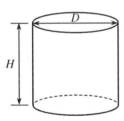

Figure 10.12 Design of a beer can

It is important to distinguish two different types of minima in order to understand the nature of optimum design that we can obtain using practical methods.

Global minimum: In an optimization problem, the best design is the one that yields the smallest value of the objective function within the feasible set, which is called a global minimum. Formally, a point $\mathbf{b}^*$ is called a global minimum for $f(\mathbf{b})$ if

$$f(\mathbf{b}^*) \le f(\mathbf{b}), \tag{10.42}$$

---

[5] Arora, J. S. 2004. *Introduction to Optimum Design*. Elsevier Academic Press.

for all $\mathbf{b} \in S$. The global minimum is the eventual goal of optimization. It is easy to define the global minimum, but unfortunately, it is not trivial to find one. If $f(\mathbf{b})$ is continuous, and the set S is closed and bounded, then there is a global minimum. However, there is no mathematical method to find it. In addition, it is possible that the problem may have multiple global minima. Figure 10.13 shows an objective function f as a function of design b. The point b_3 corresponds to the global minimum because $f(b_3)$ is smallest within the feasible set. In this case, a unique global minimum exists.

Local minimum: Unlike the global minimum, a local minimum is easier to find. There are many mathematical theories to find it. Formally, a point $\mathbf{b}^*$ is called a local minimum for $f(\mathbf{b})$ if

$$f(\mathbf{b}^*) \leq f(\mathbf{b}), \tag{10.43}$$

for all $\mathbf{b} \in S$ in a small neighborhood of $\mathbf{b}^*$. In the case of a single design variable, a small neighborhood is $(b^* - \Delta) \leq b \leq (b^* + \Delta)$ for an arbitrarily small Δ. If we can find any Δ such that f satisfies eq. (10.43), then it is a local minimum. In figure 10.13, b_1, b_3, and b_5 are local minima. Note that b_3 is a local minimum as well as a global minimum.

10.6.2 Graphical Optimization

After the optimization problem is written in the standard form, it can be solved using numerical methods. Before we discuss the various numerical methods to solve the optimization problem, let us consider simple cases in which we can solve the optimization problem graphically. When the optimization problem has only one or two design variables, graphical methods can be used to solve the problem. Graphical methods are often more expensive than other numerical methods because they require a large number of evaluations of the objective functions and constraints. However, they help engineers to visualize the design space and to understand the nature of design problem.

The first step in graphical optimization is to set up the graphical domain using the side constraints. The range of design variables defines all possible combinations of the two design variables. Since any design outside the side constraints is not feasible, it is unnecessary to plot functions in the infeasible region. The feasible region is typically referred to as the *design space*. The next step is to plot the objective function and constraints on the graph. In the case of constraints, it is possible to draw a boundary curve for each of the constraints. An inequality constraint will define the feasible region as one side (i.e., constraint is negative) of the curve, while an equality constraint will define the feasible region along the curve of the constraint. The difference between the equality and inequality constraints is in the interpretation of the design space.

Once the feasible set is clearly defined using constraints, the next step is to plot the objective function. Unlike the constraints, the objective function does not have a fixed value. One way of representing the objective function is to plot contour lines. They are similar to equal-temperature or equal-pressure curves in weather maps. They can be created by setting the objective function equal to a constant. This

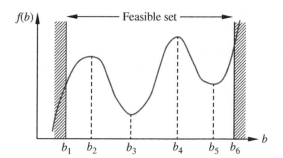

Figure 10.13 Local and global minima of a function

will provide an equation for one of the curves. Along this curve, the objective function has the same value. By gradually increasing or decreasing the constant, we can plot other contour curves. Sometimes, different colors are used for different constant values so that the values of the objective function can be easily found.

The final step of graphical optimization is to inspect the optimal solution. The optimal solution is the point in the feasible set that has the smallest value of the objective function. The optimal design may be located on the constraint curves or inside the feasible set. The latter is rarely encountered in structural optimization because most problems have important constraints that limit the objective function. If the lowest contour line intersects the boundary of one or more inequality constraints, then some or all of these constraints dictate the location of the optimal solution and are called active constraints.

EXAMPLE 10.8 *Graphical optimization of beer can problem*

Find the optimum design of the beer can problem in example 10.7 using a graphical method.

SOLUTION   Figure 10.14 depicts the optimization problem. Since two constraints are side constraints, we plot the graph between lower and upper bounds of design variables. Then the only constraint that we need to take care of is the volume constraint, which is plotted as a solid curve. Then, we can identify the feasible set as shown in figure 10.14. Now, we gradually reduce the objective function and plot the contour lines in dashed curves. When the value of the objective function becomes 300, there is no feasible design. In fact, the optimum design occurs when $H = 8$ cm and the volume constraint is just satisfied, which means the inequality becomes equality. Thus, we can calculate the diameter at the optimum design, as

$$400 - \frac{\pi}{4}D^2 H = 0.$$

From the above relation, we can solve for the diameter; $D = 7.98$ cm. Thus, we have

$$\mathbf{b}^* = \{D^*, H^*\}^T = \{7.98,\ 8\}^T.$$

At the optimum design, the objective function becomes $f(\mathbf{b}^*) = 300.53 \text{ cm}^2$.

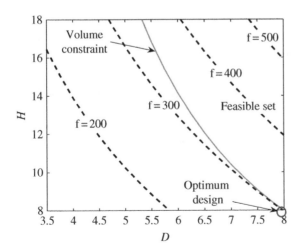

Figure 10.14 Graphical optimization of the beer can problem

10.6.3 Numerical Methods

As we discussed before, there is no mathematical method to find a global minimum, except for very simple cases. Almost all numerical methods provide a local minimum. They are basically, starting from an initial design, finding a new design that reduces the objective function while satisfying all constraints. The method repeats until there is no design in the vicinity that can reduce the objective function further.

Basic algorithm: Most of gradient-based optimization algorithms take the following steps:

1. Start with $\mathbf{b}^{(0)}$ and $K = 0$.
2. Evaluate function values and their gradients.
3. Using information from step 2, determine $\Delta \mathbf{b}^{(K)}$.
4. Check for termination.
5. Update design

$$\mathbf{b}^{(K+1)} = \mathbf{b}^{(K)} + \Delta \mathbf{b}^{(K)}. \tag{10.44}$$

6. Increase $K = K + 1$ and go to step 2.

Change in design: In the above algorithm, the change in design is further decomposed into two steps. First, the direction of design change is found, and then the amount of design change in that direction is determined. Thus, the change in design can be written as

$$\Delta \mathbf{b}^{(K)} = \alpha_K \mathbf{d}^{(K)}, \tag{10.45}$$

where α_K is called the *step size*, and $\mathbf{d}^{(K)}$ is called the *search direction* vector. Since the search direction also reduces the objective function, it is also called the *descent direction*. Various algorithms have different methods to calculate the search direction so that the optimization problem can converge fast. In this introductory chapter, we will not present any algorithms in detail. Interested users are referred to advanced optimization textbooks by Haftka and Gurdal[6].

10.6.4 Optimization Using Excel™ Solver

There are many commercially available optimization programs. Some of them are based on mathematical programming, while others are based on heuristic approaches. However, in this section, we demonstrate the optimization process using Microsoft® Excel spreadsheet. We explain the process using a simple structural example.

Problem definition: Consider the minimum weight design of the four-bar truss shown in figure 10.15. For the sake of simplicity, we assume that members 1 through 3 have the same area A_1 and member 4 has an area A_2. The constraints are limits on the stresses in the members and on the vertical displacement at the right end of the truss. Under the specified loading, the member forces and the vertical displacement δ at the end are found to be

$$P^{(1)} = 5F, \quad P^{(2)} = -F, \quad P^{(3)} = 4F, \quad P^{(4)} = -2\sqrt{3}F$$

$$\delta = v_4 = \frac{6FL}{E}\left(\frac{3}{A_1} + \frac{\sqrt{3}}{A_2}\right).$$

---

[6] Haftka, R. T., and Gurdal, Z. 1992. *Elements of Structural Optimization*. Kluwer Academic Publishers.

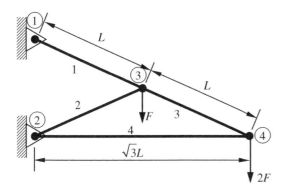

Figure 10.15 Minimum weight design of four-bar truss

We assume the allowable stresses in tension and compression to be $8.74 \times 10^{-4}E$ and $4.83 \times 10^{-4}E$, respectively, and limit the displacement to be no greater than $3 \times 10^{-3}L$. First, convert the design variables into non-dimensional ones as

$$b_1 = 10^{-3}\frac{A_1 E}{F}, \quad b_2 = 10^{-3}\frac{A_2 E}{F}.$$

Then, the minimum weight design subject to stress and displacement constraints can be formulated as

$$\text{Minimize} \quad f(\mathbf{b}) = 3b_1 + \sqrt{3}b_2$$

$$\text{subject to} \quad g_1 = \frac{18}{b_1} + \frac{6\sqrt{3}}{b_2} - 3 \le 0$$

$$g_2 = 5.73 - b_1 \le 0$$

$$g_3 = 7.17 - b_2 \le 0.$$

Loading the Solver add-in: Before using *Solver*, the user must first load the *Solver* add-in into the memory. When installing Excel, the user was given the option of installing the add-ins that ship with Excel. If the add-ins were installed, *Solver* can be loaded into the memory using *Add-ins* menu in *File →
Options*. If add-ins is not installed, it must be installed first in order to use *Solver*. To load *Solver* into the memory, follow these steps:

1. Select *File → Options* on the main menu. Figure 10.16 shows *Excel Options* window.
2. If *Solver Add-in* is shown in Excel Options window, then *Solver* is already loaded into the memory. Skip next steps.
3. If *Solver Add-in* is not shown, click *Go...* button at the bottom to open *Add-In* window. Figure 10.17 shows the currently installed *Add-ins*.
4. From the list of installed add-ins, check *Solver Add-In*.
5. Choose *OK* or press Enter. Now, *Solver Add-in* should appear in *Excel Options* window as shown in figure 10.16. The Solver command is located in the main menu *Data → Solver*.

Setting up the example problem: To use *Solver* in Excel worksheets, the optimization problem must be defined first. The user must specify the cells corresponding to design variables, objective function, and constraints. With *Solver*, the objective and constraint cells should be based on formulas in terms of design variable cells. Thus, if the values of design variables cells are changed, the objective and constraints are changed accordingly. Therefore, to set up the problem, determine which of the cells will be used as the objective and constraints and make sure that they contain formulas. The worksheet shown in figure 10.18 illustrates the example problem that the Solver will use.

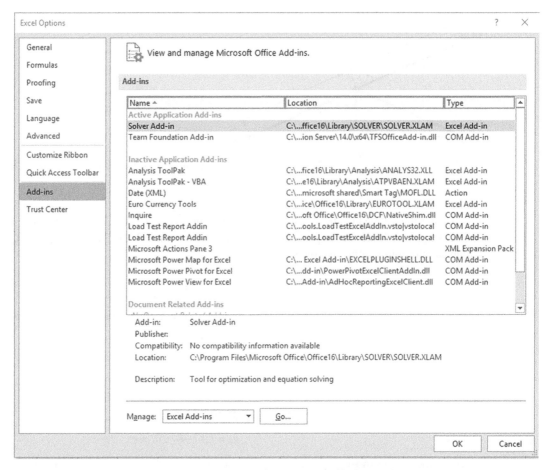

Figure 10.16 List of the submenu in Tools menu (Solver appears in Tools menu)

In figure 10.18, cells C4 and C5 are selected for design variables b_1 and b_2, respectively (*Changing Cell* in *Solver Parameters* dialog box). The objective function is formulated in cell C7 (*Target Cell* in *Solver Parameters* dialog box). The constraint equation g_1, g_2, and g_3 are formulated in cell C10, C11, and C12, respectively.

1. Type text in cells A1, A3, A7, A9, B4, B5, B10, B11, and B12, as shown in figure 10.18. These are unnecessary for *Solver* but can help the user understand the meaning of each cell.
2. Select cell C4 and type 10. This is an arbitrary initial value of design variable b_1.
3. Select cell C5 and type 10. This is an arbitrary initial value of design variable b_2.
4. Click C7 and key in objective function formulas as follows
   ```
   =3*C4+sqrt(3)*C5
   ```
5. Click C10 and key in constraint equation formulas g_1 as follows
   ```
   =18/C4+6*sqrt(3)/C5-3
   ```
6. Click C11 and key in constraint equation formulas g_2 as follows
   ```
   =5.73-C4
   ```
7. Click C12 and type constraint equation formulas g_3 as follows
   ```
   =7.17-C5
   ```

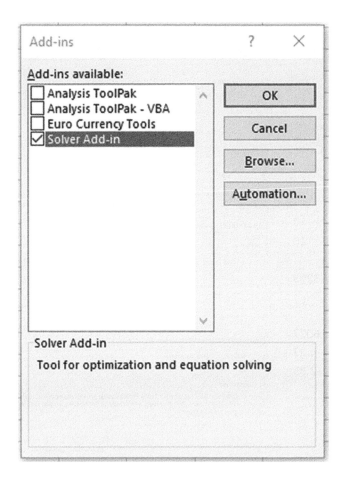

Figure 10.17 Add-in dialog box with installed Solver add-in

Running Solver: After the locations and formulas of design variables, objective, and constraints are set up, follow these steps to run *Solver*:

1. Select *Solver* in *Data* menu to start the solver add-in. The *Solver Parameter* dialog box will be displayed as shown in figure 10.18.

2. Indicate the cell that contains the objective function formula in the *Set Objective* text box. The objective cell can directly type in as C7 or can be selected using the cell selection method in Excel.

3. In the *To* section of the dialog box, select *Min* button as the objective function will be minimized.

4. In the *By Changing Variable Cells* text box, indicate the cell or range of cells that will be used as design variables. In this case, choose C4 and C5.

5. To specify constraints, select the *Add* button to add each constraint to the problem. Figure 10.19 shows the *Add Constraint* dialog box.

6. To create a constraint, specify the cell containing the formula on which the constraint is based on the *Cell Reference* text box (for example, C10 for the first constraint). Click the drop-down arrow to display the list of constraint operators, and select the appropriate operator (choose < = symbol). In the final text box, enter the value the constraint must meet (type in 0). Choose the *Add* button to add the current constraint to the problem and create another, or choose *OK* to add the constraint and return to the *Solver Parameters* dialog box. The three constraints should appear in the *Subject to the Constraints* list box in the *Solver Parameters* dialog box (See figure 10.18).

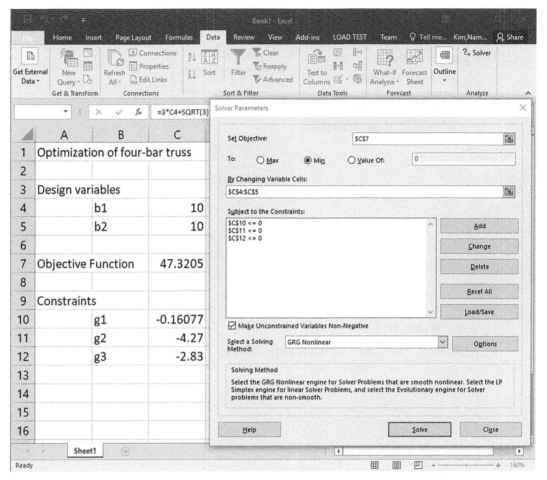

Figure 10.18 Excel worksheet for minimum weight design of the four-bar truss and Solver Parameters dialog box

Figure 10.19 Add Constraint dialog box

7. Choose *Options* button in the *Solver Parameters* dialog box to control maximum computing time, iteration, convergence, etc. The *Options* dialog box will be displayed as shown in figure 10.20. Choose appropriate settings in the *Options* dialog box. Check *Show Iteration Results* buttons and click *OK* button. The *Solver Parameters* dialog box in figure 10.18 will be redisplayed.

8. Click *Solve* button to start the Solver. The *Solver* begins calculating the optimal solutions. Intermediate solution values are displayed on the worksheet and *Show Trial Solution* dialog box as shown in figure 10.21 will appear.

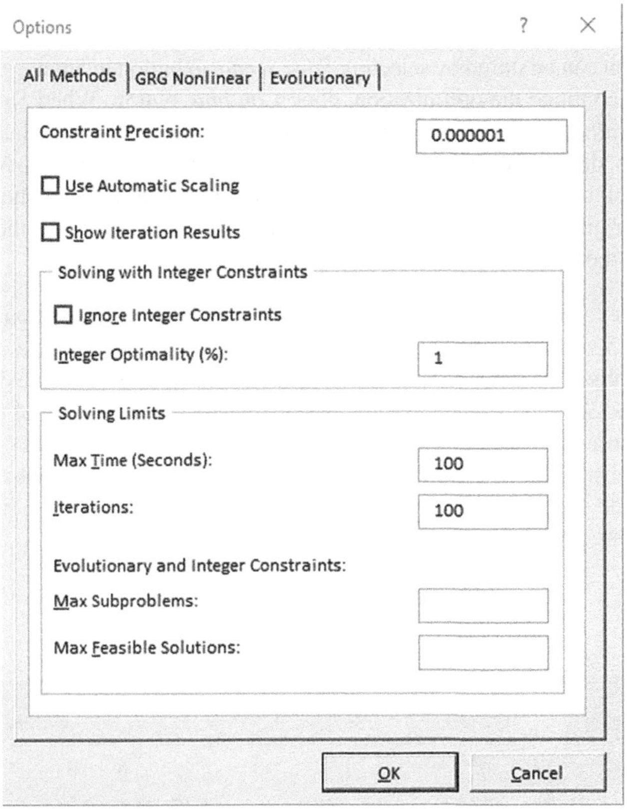

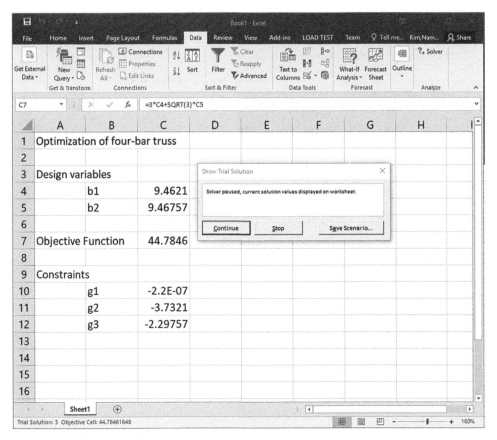

Figure 10.20 Solver Options dialog box

Figure 10.21 Show Trial Solution dialog box

9. The intermediate solution can be stored by selecting *Save Scenario* button in *Show Trial Solution* dialog box. In order to continue the optimization, click *Continue* button. When *Solver* finds a solution, the *Solver Results* dialog box appears, as shown in figure 10.22. Note that the values of the objective function, design variables, and constraints are changed. Select *Keep Solver Solution* to use the offered solutions. If the *Restore Original Values* button is selected, then the worksheet will return to the original values. Figure 10.22 also shows the worksheet after the *Solver* has found the solutions for the problem.

Creating Solver Reports: Solver can generate reports summarizing the results of its solutions. There are three types of reports: *Answer*, *Sensitivity*, and *Limit* report. The *Answer* report shows the original and final values for the target cell (objective function) and the adjustable cells (design variables), as well as the status of each constraint. The *Sensitivity* report shows the sensitivity of each element of the solution to changes in input cells or constraints. The *Limits* report shows the upper and lower values of the design variables within the specified constraints. To create a report, select the reports from the list that appears in the *Solver Results* dialog box, and choose *OK*. Excel creates the reports in a separate sheet. Figure 10.23 shows *Answer Report* of the four bar truss example.

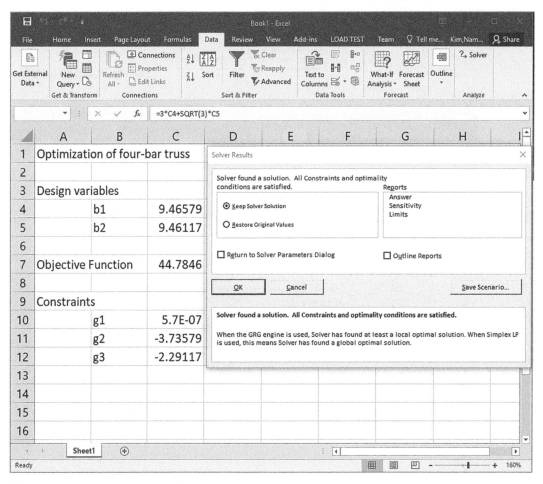

Figure 10.22 Solver Results dialog box

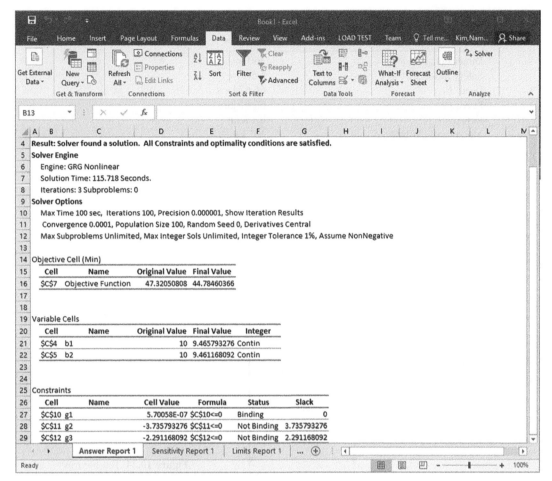

Figure 10.23 Answer Report worksheet

10.7   PROJECTS

Project 10.1 – Fully stressed design of ten-bar truss
The fully stressed design is often used for truss structures. The idea is that we should remove material from members that are not fully stressed unless prevented by minimum cross-sectional area constraint. A 10-bar truss shown in figure 10.24 is under two maximum loads, P_1 and P_2. The design goal is to minimize the weight, W, by varying the cross-sectional areas, A_i, of the truss members. The stress of the member should be less than the allowable strength with the safety factor. For manufacturing reasons, the cross-sectional areas should be greater than the minimum value. Input data are summarized in table 10.2. Find optimum design using fully stressed design.

Project 10.2 – Design Optimization of a Bracket
A bracket shown in figure 10.25 has the following properties: Young's modulus $E = 2.068 \times 10^{11} \, \text{N/m}^2$, Poisson's ratio $\nu = 0.29$, density $\rho = 7.82 \times 10^3 \, \text{kg/m}^3$, thickness = 3 mm. A horizontal force $F_x = 15,000$

Table 10.2   Input data for ten-bar truss

Parameters	Values
Dimension, b	360 inches
Safety factor, S_F	1.5
Load, P_1	66.67 kips
Load, P_2	66.67 kips
Density, ρ	0.1 lb/in^3
Modulus of elasticity, E	10^4 ksi
Allowable stress, $\sigma_{\text{allowable}}$	25 ksi*
Initial area A_i	1.0 in^2
Minimum cross-sectional area	0.1 in^2

*Member 9 has the allowable strength of 75 ksi

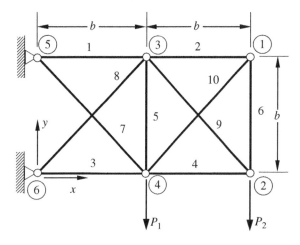

Figure 10.24 Ten-bar truss

N is applied at the center of the upper hole, and two bottom holes are fixed to the ground. The design goal is to minimize the mass of the bracket, while the maximum stress is less than 800 MPa.[7]

(a) In the report, clearly state the units for mass, length, force, and stress.

(b) Using a CAD tool, create a solid model with appropriate dimensions as shown in figure 10.25. It is important to define dimensions and relations so that the solid model can update properly when the values in dimensions are changed.

(c) Provide a plot of the finite element model that includes boundary conditions and applied force. Use proper modeling techniques in chapter 9 to approximate the load application method and the displacement boundary conditions.

(d) Carry out a parametric study by changing design b_1 (55, 60, 65, 70, 75). Provide mass and maximum stress plots as functions of design b_1.

(e) Carry out design optimization with design boundaries given in table 10.3 to minimize the mass, while the maximum stress is less than 800 MPa. Provide an optimum geometry plot and optimum stress plot. Provide a history of design goal, stress constraints, and design parameters.

---

[7] Bennett and Botkin. 1985. *AIAA Journal* 23:458–464.

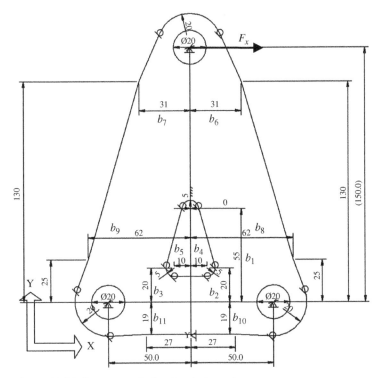

Figure 10.25 Geometry of a bracket (unit mm)

Table 10.3   Lower and upper bounds of design parameters (unit mm)

Design	Name	Lower Bound	Initial Value	Upper Bound
b_1	Slot height	54	55	120
b_2	Slot Vr 1	9	20	21
b_3	Slot Vr 2	9	20	21
b_4	Slot Hr 1	9	10	30
b_5	Slot Hr 2	9	10	30
b_6	Out Hr 1	17	31	32
b_7	Out Hr 2	17	31	32
b_8	Out Hr 3	43	62	63
b_9	Out Hr 4	43	62	63
b_{10}	Bottom 1	1	19	21
b_{11}	Bottom 2	1	19	21

10.8   EXERCISES

1. Answer the following descriptive questions.

   (a) Explain the relationship between performance, objective, and constraints in design optimization.

   (b) Explain the usage of a knockdown factor in calculating allowable strength.

   (c) Based on 10 coupon tests, the strength of material A is normally distributed by $N(500, 40^2)$, while material B is by $N(480, 20^2)$. Which material has a higher B-basis allowable strength?

(d) In structural design, why do we need to consider both the allowable strength and the safety factor simultaneously?

(e) When a truss is statically determinate, how many iterations would be necessary to obtain the fully stressed design? Explain why.

(f) If a beam structure is modeled using 3D solid elements and if the cross-sectional geometry is design, is it sizing design or shape design?

(g) What are the three steps of optimization formulation?

(h) Write the standard form of an optimization problem.

(i) At an optimum design, the gradient of the objective may not be zero. Explain in what situation that will happen.

2. Calculate the B-basis allowable strengths with 5, 10, 20, and 30 coupons, and show the calculated allowable strengths converge to the true allowable strength in example 10.1. The coupon test results can be simulated by generating random numbers from a normal distribution with mean = 310 MPa and standard deviation = 25 MPa. The true tenth percentile of the distribution is 278 MPa. Plot a graph of the allowable strength versus the number of coupons, and show that the curve converges to 278 MPa.

3. Determine the height of the beam in example 10.2 when the load factor of $\lambda = 2.0$ is used with the failure stress of 40 ksi.

4. Determine the height of the beam in example 10.2 so that the safety margin is 10 ksi with the failure stress of 40 ksi.

5. A two-dimensional truss shown in the figure is made of aluminum with Young's modulus $E = 80$ GPa and failure stress $\sigma_Y = 150$ MPa. Determine the minimum cross-sectional area of each member so that the truss is safe with safety factor 1.5.

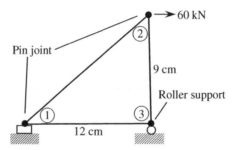

6. Consider a stepped beam structure modeled using two beam elements. The cross-sections are circular. Use Young's modulus $E = 80$ GPa, yield stress $\sigma_Y = 250$ MPa, and $L = 1$ m. When $F_2 = F_3 = 1,000$ N, calculate the minimum diameters of two sections so that the beam does not fail with a safety margin of 100 MPa.

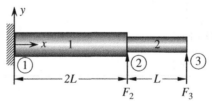

7. A cantilever beam of length 1 m is subjected to a uniformly distributed load $p(x) = p_0 = 12,000$ N/m and a clockwise couple 5,000 N·m at the tip. The load factors for the distributed load and couple are, respectively, 1.5 and 2.0. When the cross section is circular, calculate the minimum diameter. Use Young's modulus 80 GPa and yield strength 250 MPa.

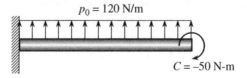

8. The frame shown in the figure is clamped at the left end and supported on a hinged roller at the right end. An axial force P and a couple C act at the right end. The load factor for the axial force is 1.5 and that of the couple is 2.0. Determine the radius of the circular cross section. Assume the following numerical values: $L = 1$ m, $E = 80$ GPa, $P = 15,000$ N, $C = 1,000$ Nm.

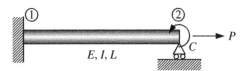

9. All members of the truss shown in the figure initially have a circular cross section with a diameter of $2''$. Using a commercial finite element analysis program, calculate the minimum diameter of each member using fully stressed design. Assume that Young's modulus is 10^4 psi.

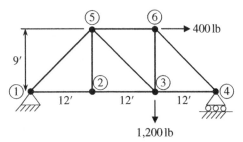

10. Consider a two-bar structure in the figure with Young's modulus $E = 100$ GPa, yield stress $\sigma_Y = 250$ MPa, and $F = 10,000$ N. Design variables are $b_1 =$ area of section AB and $b_2 =$ area of section BC. Starting from the initial design of $b_1 = 1 \times 10^{-4}$ m^2 and $b_2 = 2 \times 10^{-4}$ m^2, perform fully stressed design to obtain minimum cross-sectional areas.

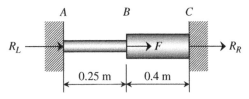

11. Repeat problem 10 with the initial design of $b_1 = 2 \times 10^{-4}$ m^2 and $b_2 = 1 \times 10^{-4}$ m^2. Discuss the results with that of problem 10.

12. For the clamped beam shown in the figure, two design variables are defined, as $I_1 = b_1$ and $I_2 = b_2$. Using the finite element method and sensitivity analysis, calculate the sensitivity of the vertical displacement v_2 with respect to b_1 and b_2. Use the following values for the current design: $E = 30 \times 10^6$ psi, $b_1 = 0.1$ in^4, and $b_2 = 0.05$ in^4. Compare the results with the exact sensitivity.

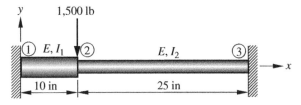

13. Calculate the sensitivity of the vertical displacement v_2 in problem 12 using forward finite difference method with perturbation size 1%. Compare the results with the exact sensitivity.

14. Consider a simply supported beam of length $L = 1$ m subjected to a uniformly distributed transverse load $p_0 = 100$ N/m. The cross section is rectangular with width $w = 0.01$ m and height $h = 0.02$ m. Calculate the sensitivity of the vertical displacement at the center with respect to h. Use one finite element with Young's modulus $= 80$ GPa. Compare the sensitivity against the finite difference method with perturbation size 1%.

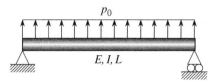

15. Repeat problem 14 to calculate the sensitivity of maximum tensile stress with respect to h.

16. A cantilevered beam shown in the figure is under a couple of 500 lb.·in. at the end. The optimization problem is to find a design that minimizes the cross-sectional area, while the maximum stress is less than 2,000 psi. The thicknesses of the flange and the web of the cross section are fixed with $t = 0.1$ in. The design variables are the width w and the height h of the cross section. Determine graphically the optimal design. The width and the height are constrained to remain in the range $0.1 \le w \le 10$ in and $0.2 \le h \le 10$ in.

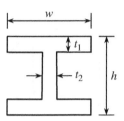

Appendix

Mathematical Preliminaries

Since vector calculus and linear algebra are used extensively in finite element analysis, it is worth reviewing some fundamental concepts and recalling some important results that will be used in this book. A brief summary of concepts and results pertinent to the development of the subject are provided for the convenience of students. For a thorough understanding of the mathematical concepts, readers are advised to refer to any standard textbook, such as Kreyszig[1] and Strang.[2]

A.1 VECTORS AND MATRICES

A.1.1 Vector

A *vector* is a collection of scalars and is defined using a bold typeface[3] inside a pair of braces, such as

$$\{\mathbf{a}\} = \begin{Bmatrix} a_1 \\ a_2 \\ \vdots \\ a_N \end{Bmatrix}. \tag{A.1}$$

In eq. (A.1), $\{\mathbf{a}\}$ is an N-dimensional column vector. When the context is clear, we will remove the braces and simply use the letter "$\mathbf{a}$" to denote the vector. The transpose of $\mathbf{a}$ above will be a row vector and will be denoted by $\mathbf{a}^{\mathrm{T}}$.

$$\{\mathbf{a}\}^{\mathrm{T}} = \{ a_1 \quad a_2 \quad \cdots \quad a_N \}. \tag{A.2}$$

By default, in this text all vectors are considered as column vectors unless specified. For simplicity of notation, a geometric vector in two- or three-dimensional space is denoted by a bold typeface without braces:

$$\mathbf{a} = \begin{Bmatrix} a_1 \\ a_2 \\ a_3 \end{Bmatrix}, \text{ or } \mathbf{a} = \begin{Bmatrix} a_1 \\ a_2 \end{Bmatrix}, \tag{A.3}$$

---

[1] Kreyszig, E. 1983. *Advanced Engineering Mathematics*, 5th Ed. John Wiley & Sons, New York.
[2] Strang, G. 1980. *Linear Algebra and its Applications*, 2nd Ed. Academic Press, New York.
[3] In the classroom one can use an underscore ($\underline{a}$) to denote vectors on the blackboard.

Introduction to Finite Element Analysis and Design, Second Edition. Nam H. Kim, Bhavani V. Sankar, and Ashok V. Kumar.
© 2018 John Wiley & Sons Ltd. Published 2018 by John Wiley & Sons Ltd.
Companion website: www.wiley.com/go/kim/finite_element_analysis_design

where a_1, a_2, and a_3 are components of the vector $\mathbf{a}$ in the x-, y-, and z-direction, respectively, as shown in figure A.1. To save space, the above column vector $\mathbf{a}$ can be written as $\mathbf{a} = \{a_1, a_2, a_3\}^\mathrm{T}$, in which $\{\,\bullet\,\}^\mathrm{T}$ denotes the *transpose*. The above three-dimensional geometric vector can also be denoted using a unit vector in each coordinate direction. Let $\mathbf{i} = \{1, 0, 0\}^\mathrm{T}$, $\mathbf{j} = \{0, 1, 0\}^\mathrm{T}$, and $\mathbf{k} = \{0, 0, 1\}^\mathrm{T}$ be the unit vectors in the x-, y-, and z-direction, respectively. Then,

$$\mathbf{a} = a_1\mathbf{i} + a_2\mathbf{j} + a_3\mathbf{k}. \tag{A.4}$$

The magnitude of the vector $\mathbf{a}$, $\|\mathbf{a}\|$, is given by

$$\|\mathbf{a}\| = \sqrt{a_1^2 + a_2^2 + a_3^2}. \tag{A.5}$$

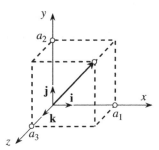

Figure A.1 Three-dimensional geometric vector

A.1.2 Matrix

A *matrix* is a collection of vectors and is defined using a bold typeface within square brackets. For example, let the matrix $[\mathbf{M}]$ be a collection of K number of column vectors $\{\mathbf{m}^i\}$, $i = 1, 2, \ldots, K$. Then, the matrix $[\mathbf{M}]$ is denoted by

$$[\mathbf{M}] = \left[\{\mathbf{m}^1\}\ \{\mathbf{m}^2\}\ \cdots\ \{\mathbf{m}^K\}\right], \tag{A.6}$$

where

$$\{\mathbf{m}^i\} = \begin{Bmatrix} m_1^i \\ m_2^i \\ \vdots \\ m_N^i \end{Bmatrix}, \qquad i = 1, \cdots, K. \tag{A.7}$$

By expanding each component of $\{\mathbf{m}^i\}$, the matrix $[\mathbf{M}]$ can be denoted using the $N \times K$ number of components as

$$[\mathbf{M}] = \begin{bmatrix} M_{11} & M_{12} & \cdots & M_{1K} \\ M_{21} & M_{22} & \cdots & M_{2K} \\ \vdots & \vdots & \ddots & \vdots \\ M_{N1} & M_{N2} & \cdots & M_{NK} \end{bmatrix}, \tag{A.8}$$

where $M_{ij} = m_i^j$ is a component of the matrix. The notation for the subscripts M_{ij} is such that the first index denotes the position in the row, while the second index denotes the position in the column. The components that have the same indices are called diagonal components, e.g., M_{11}, M_{22}, etc. In eq. (A.8), the dimensions of the matrix $[\mathbf{M}]$ are $N \times K$. When $N = K$, the matrix is called a *square* matrix.

A column vector can be considered as a matrix containing only one column. The column vector $\{\mathbf{m}^i\}$ in eq. (A.7) is an $N \times 1$ matrix.

A.1.3   Transpose of a Matrix

The *transpose* of a matrix can be obtained by switching the rows and columns of the matrix. For example, the transpose of the matrix $[\mathbf{M}]$ in eq. (A.8) can be written as

$$[\mathbf{M}]^{\mathrm{T}} = \begin{bmatrix} M_{11} & M_{21} & \cdots & M_{N1} \\ M_{12} & M_{22} & \cdots & M_{N2} \\ \vdots & \vdots & \ddots & \vdots \\ M_{1K} & M_{2K} & \cdots & M_{NK} \end{bmatrix}, \tag{A.9}$$

which is now a matrix of size $K \times N$.

A.1.4   Symmetric Matrix

A matrix is called *symmetric* when the matrix and its transpose are identical. It is clear from the definition that only a square matrix can be a symmetric matrix. For example, if $[\mathbf{S}]$ is symmetric, then

$$[\mathbf{S}] = [\mathbf{S}]^{\mathrm{T}} = \begin{bmatrix} S_{11} & S_{12} & \cdots & S_{1N} \\ S_{12} & S_{22} & \cdots & S_{2N} \\ \vdots & \vdots & \ddots & \vdots \\ S_{1N} & S_{2N} & \cdots & S_{NN} \end{bmatrix}. \tag{A.10}$$

Matrix $[\mathbf{A}]$ is called *skew-symmetric* when $[\mathbf{A}]^{\mathrm{T}} = -[\mathbf{A}]$. It is clear that the diagonal components of skew-symmetric matrix are zero. A typical skew-symmetric matrix can be defined as

$$[\mathbf{A}] = \begin{bmatrix} 0 & A_{12} & \cdots & A_{1N} \\ -A_{12} & 0 & \cdots & A_{2N} \\ \vdots & \vdots & \ddots & \vdots \\ -A_{1N} & -A_{2N} & \cdots & 0 \end{bmatrix}. \tag{A.11}$$

A.1.5   Diagonal and Identity Matrix

A *diagonal matrix* is a special case of a symmetric matrix in which all off-diagonal components are zero. An identity matrix is a diagonal matrix in which all diagonal components are equal to unity. For example, the (3×3) *identity matrix* is given by

$$[\mathbf{I}_3] = \begin{bmatrix} 1 & 0 & 0 \\ 0 & 1 & 0 \\ 0 & 0 & 1 \end{bmatrix}. \tag{A.12}$$

A.2 VECTOR-MATRIX CALCULUS

A.2.1 Vector and Matrix Operations

Addition and subtraction of vectors and matrices are possible when their dimensions are the same. Let $\{\mathbf{a}\}$ and $\{\mathbf{b}\}$ be two N-dimensional vectors. Then, the addition and subtraction of these two vectors are defined as

$$
\begin{aligned}
\{\mathbf{c}\} &= \{\mathbf{a}\} + \{\mathbf{b}\}, &\Rightarrow&\quad c_i = a_i + b_i, \quad i = 1, \cdots, N, \\
\{\mathbf{d}\} &= \{\mathbf{a}\} - \{\mathbf{b}\}, &\Rightarrow&\quad d_i = a_i - b_i, \quad i = 1, \cdots, N.
\end{aligned}
\tag{A.13}
$$

Note that the dimensions of the resulting vectors $\{\mathbf{c}\}$ and $\{\mathbf{d}\}$ are the same as those of $\{\mathbf{a}\}$ and $\{\mathbf{b}\}$.

A scalar multiple of a vector is obtained by multiplying all of its components by a constant. For example, k times a vector $\{\mathbf{a}\}$ is obtained by multiplying each component of the vector by the constant k:

$$
k\{\mathbf{a}\} = \{\, ka_1 \quad ka_2 \quad \cdots \quad ka_N \,\}^{\mathrm{T}}.
\tag{A.14}
$$

Similar operations can be defined for matrices. Let $[\mathbf{A}]$ and $[\mathbf{B}]$ be $N \times K$ matrices. Then, the addition and subtraction of these two matrices are defined as

$$
\begin{aligned}
[\mathbf{C}] &= [\mathbf{A}] + [\mathbf{B}], &\Rightarrow&\quad C_{ij} = A_{ij} + B_{ij}, \quad i = 1, \cdots, N, \quad j = 1, \cdots, K, \\
[\mathbf{D}] &= [\mathbf{A}] - [\mathbf{B}], &\Rightarrow&\quad D_{ij} = A_{ij} - B_{ij}, \quad i = 1, \cdots, N, \quad j = 1, \cdots, K.
\end{aligned}
\tag{A.15}
$$

Note that the dimensions of matrices $[\mathbf{C}]$ and $[\mathbf{D}]$ are the same as those of matrices $[\mathbf{A}]$ and $[\mathbf{B}]$. Similarly, one also can define the scalar multiple of a matrix.

Although the above matrix addition and subtraction are very similar to those of scalars, the multiplication and division of vectors and matrices are quite different from those of scalars.

A.2.2 Scalar Product

Since scalar products between two vectors will frequently appear in this text, it is necessary to clearly understand their definitions and notations used. Let $\mathbf{a}$ and $\mathbf{b}$ be two three-dimensional geometric vectors defined by

$$
\mathbf{a} = \{\, a_1 \quad a_2 \quad a_3 \,\}^{\mathrm{T}}, \quad \text{and} \quad \mathbf{b} = \{\, b_1 \quad b_2 \quad b_3 \,\}^{\mathrm{T}}.
\tag{A.16}
$$

The scalar product of $\mathbf{a}$ and $\mathbf{b}$ is defined as

$$
\mathbf{a} \cdot \mathbf{b} = a_1 b_1 + a_2 b_2 + a_3 b_3,
\tag{A.17}
$$

which is the summation of component-by-component products. Often notations in matrix product can be used such that $\mathbf{a} \cdot \mathbf{b} = \mathbf{a}^{\mathrm{T}} \mathbf{b} = \mathbf{b}^{\mathrm{T}} \mathbf{a}$. If $\mathbf{a}$ and $\mathbf{b}$ are two geometric vectors, then the scalar product can be written as

$$
\mathbf{a} \cdot \mathbf{b} = \|\mathbf{a}\|\|\mathbf{b}\|\cos\theta.
\tag{A.18}
$$

where θ is the angle between two vectors. Note that the scalar product of two vectors is a scalar, and hence the name *scalar product*. A scalar product can also be defined as the matrix product of one of the vectors and transpose of the other. In order for the scalar product to exist, the dimensions of the two vectors must be the same.

A.2.3 Norm

The *norm* or the magnitude of a vector [see eq. (A.5)] can also be defined using the scalar product. For example, the norm of a three-dimensional vector $\mathbf{a}$ can be defined as

$$
\|\mathbf{a}\| = \sqrt{\mathbf{a} \cdot \mathbf{a}}.
\tag{A.19}
$$

Note that the norm is always a non-negative scalar and is the length of the geometric vector. When $\|\mathbf{a}\| = 1$, the vector $\mathbf{a}$ is called the *unit vector*.

A.2.4   Determinant of a Matrix

Determinant is an important concept, and it is useful in solving a linear system of equations. If the determinant of a matrix is zero, then it is not invertible and it is called a singular matrix. The determinant is defined only for square matrices. The formula for calculating the *determinant* of any square matrix can be easily understood by considering a 2×2 or 3×3 matrix. The determinant of a 2×2 matrix is defined as

$$|\mathbf{A}| = \begin{vmatrix} a_{11} & a_{12} \\ a_{21} & a_{22} \end{vmatrix} = a_{11}a_{22} - a_{12}a_{21}. \tag{A.20}$$

The determinant of a 3×3 matrix is defined as

$$\begin{aligned} |\mathbf{A}| &= \begin{vmatrix} a_{11} & a_{12} & a_{13} \\ a_{21} & a_{22} & a_{23} \\ a_{31} & a_{32} & a_{33} \end{vmatrix} \\ &= a_{11}\begin{vmatrix} a_{22} & a_{23} \\ a_{32} & a_{33} \end{vmatrix} - a_{12}\begin{vmatrix} a_{21} & a_{23} \\ a_{31} & a_{33} \end{vmatrix} + a_{13}\begin{vmatrix} a_{21} & a_{22} \\ a_{31} & a_{32} \end{vmatrix} \\ &= a_{11}(a_{22}a_{33} - a_{23}a_{32}) - a_{12}(a_{21}a_{33} - a_{23}a_{31}) + a_{13}(a_{21}a_{32} - a_{22}a_{31}). \end{aligned} \tag{A.21}$$

A matrix is called *singular* when its determinant is zero.

A.2.5   Vector Product

Different from the scalar product, the result of the *vector product* is another vector. In the three-dimensional space, the vector product of two vectors $\mathbf{a}$ and $\mathbf{b}$ can be defined by the determinant as

$$\begin{aligned} \mathbf{a} \times \mathbf{b} &= \begin{vmatrix} \mathbf{i} & \mathbf{j} & \mathbf{k} \\ a_1 & a_2 & a_3 \\ b_1 & b_2 & b_3 \end{vmatrix} \\ &= (a_2 b_3 - a_3 b_2)\mathbf{i} + (a_3 b_1 - a_1 b_3)\mathbf{j} + (a_1 b_2 - a_2 b_1)\mathbf{k} \\ &= \begin{Bmatrix} a_2 b_3 - a_3 b_2 \\ a_3 b_1 - a_1 b_3 \\ a_1 b_2 - a_2 b_1 \end{Bmatrix}. \end{aligned} \tag{A.22}$$

In eq. (A.22), we consider unit vectors $\mathbf{i}, \mathbf{j}$, and $\mathbf{k}$ as components of a matrix. As with the scalar product, the vector product can be defined only when the dimensions of two vectors are the same.

In the conventional notation, the vector product of two geometric vectors is defined by

$$\mathbf{a} \times \mathbf{b} = \|\mathbf{a}\| \|\mathbf{b}\| \sin\theta\, \mathbf{n}, \tag{A.23}$$

where θ is the angle between two vectors and $\mathbf{n}$ is the unit vector that is perpendicular to the plane that contains both vectors $\mathbf{a}$ and $\mathbf{b}$. The right-hand rule is used to determine the positive direction of vector $\mathbf{n}$ as shown in figure A.2. It is clear from its definitions in eqs. (A.22) and (A.23), $\mathbf{a} \times \mathbf{a} = \mathbf{0}$, and $\mathbf{b} \times \mathbf{a} = -\mathbf{a} \times \mathbf{b}$.

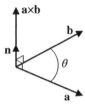

Figure A.2 Illustration of vector product

A.2.6 Matrix-vector Multiplication

The matrix-vector multiplication often appears in the finite element analysis. Let $[\mathbf{M}]$ be a 3×3 matrix defined by

$$[\mathbf{M}] = \begin{bmatrix} m_{11} & m_{12} & m_{13} \\ m_{21} & m_{22} & m_{23} \\ m_{31} & m_{32} & m_{33} \end{bmatrix}.$$

The multiplication between a matrix $[\mathbf{M}]$ and a vector $\mathbf{a}$ is defined by

$$\mathbf{c} = [\mathbf{M}] \cdot \mathbf{a} = \begin{bmatrix} m_{11} & m_{12} & m_{13} \\ m_{21} & m_{22} & m_{23} \\ m_{31} & m_{32} & m_{33} \end{bmatrix} \cdot \begin{Bmatrix} a_1 \\ a_2 \\ a_3 \end{Bmatrix} = \begin{Bmatrix} m_{11}a_1 + m_{12}a_2 + m_{13}a_3 \\ m_{21}a_1 + m_{22}a_2 + m_{23}a_3 \\ m_{31}a_1 + m_{32}a_2 + m_{33}a_3 \end{Bmatrix}, \tag{A.24}$$

where $\mathbf{c}$ is a 3×1 column vector. Using a conventional summation notation, eq. (A.24) can be written as

$$c_i = \sum_{j=1}^{3} m_{ij} a_j, \quad i = 1, 2, 3. \tag{A.25}$$

Since the result of eq. (A.24) is a vector, it is possible to obtain the scalar product of $\mathbf{c}$ with a vector $\mathbf{b}$, yielding

$$\begin{aligned} \mathbf{b} \cdot [\mathbf{M}] \cdot \mathbf{a} = &\, b_1 (m_{11}a_1 + m_{12}a_2 + m_{13}a_3) \\ &+ b_2 (m_{21}a_1 + m_{22}a_2 + m_{23}a_3) \\ &+ b_3 (m_{31}a_1 + m_{32}a_2 + m_{33}a_3), \end{aligned} \tag{A.26}$$

which is a scalar.

The above matrix-vector multiplication can be generalized to arbitrary dimensions. For example, let $[\mathbf{M}]$ be an $N \times K$ matrix and $\{\mathbf{a}\}$ be an $L \times 1$ vector. The multiplication of $[\mathbf{M}]$ and $\{\mathbf{a}\}$ can be defined if and only if $K = L$. In addition, the result $\{\mathbf{c}\}$ will be a vector of $N \times 1$ dimension.

$$\{\mathbf{c}\}_{N \times 1} = [\mathbf{M}]_{N \times K} \{\mathbf{a}\}_{K \times 1}$$

$$c_i = \sum_{j=1}^{K} m_{ij} a_j, \quad i = 1, \ldots N. \tag{A.27}$$

A.2.7 Matrix-matrix Multiplication

The matrix-matrix multiplication is a more general case of eq. (A.24). For 3×3 matrices, the matrix-matrix multiplication can be defined as

$$[\mathbf{C}] = [\mathbf{A}][\mathbf{B}], \tag{A.28}$$

where $[\mathbf{C}]$ is also a 3×3 matrix. Using the component notation, eq. (A.28) is equivalent to

$$C_{ij} = \sum_{k=1}^{3} A_{ik} B_{kj}, \quad i = 1,2,3, \quad j = 1,2,3. \tag{A.29}$$

The above matrix-matrix multiplication can be generalized to arbitrary dimensions. For example, let the dimensions of matrices $[\mathbf{A}]$ and $[\mathbf{B}]$ be $N \times K$ and $L \times M$, respectively. The multiplication of $[\mathbf{A}]$ and $[\mathbf{B}]$ can be defined if and only if $K = L$, that is, the number of columns in the first matrix must be equal to the number of rows in the second matrix. In addition, the dimension of the resulting matrix $[\mathbf{C}]$ will be $N \times M$.

$$C_{ij} = \sum_{k=1}^{K} A_{ik} B_{kj}, \quad i = 1, \ldots, N, \quad j = 1, \ldots, M. \tag{A.30}$$

EXAMPLE A.1   *Determinant*

The reader is encouraged to derive the following results using eq. (A.20).

$$\begin{vmatrix} a & b \\ 0 & 0 \end{vmatrix} = 0$$

$$\begin{vmatrix} ka & kb \\ c & d \end{vmatrix} = k \begin{vmatrix} a & b \\ c & d \end{vmatrix}$$

$$\begin{vmatrix} a & b \\ c & d \end{vmatrix} = - \begin{vmatrix} c & d \\ a & b \end{vmatrix} = - \begin{vmatrix} b & a \\ d & c \end{vmatrix}$$

$$\begin{vmatrix} a & b \\ ka & kb \end{vmatrix} = 0, \quad \begin{vmatrix} a & ka \\ b & kb \end{vmatrix} = 0$$

$$\begin{vmatrix} a+e & b+f \\ c & d \end{vmatrix} = \begin{vmatrix} a & b \\ c & d \end{vmatrix} + \begin{vmatrix} e & f \\ c & d \end{vmatrix} = (ad - bc) + (ed - cf)$$

A.2.8 Inverse of a Matrix

If a square matrix [**A**] is invertible, then one can find another square matrix [**B**] such that [**A**][**B**] = [**B**][**A**] = [**I**], and then [**B**] is called the inverse of [**A**] and vice versa. A simple expression can be obtained for the *inverse* of a matrix when the dimension is 2×2, as

$$[\mathbf{A}]^{-1} = \begin{bmatrix} a_{11} & a_{12} \\ a_{21} & a_{22} \end{bmatrix}^{-1} = \frac{1}{|\mathbf{A}|} \begin{bmatrix} a_{22} & -a_{12} \\ -a_{21} & a_{11} \end{bmatrix}. \tag{A.31}$$

For procedures of inverting a general $N \times N$ matrix, the reader should refer to textbooks such as Kreyszig[4] or Strang.[5] If a matrix is *singular* ($|\mathbf{A}| = 0$), then the inverse does not exist.

A.2.9 Rules of Matrix Multiplication

The following rules of matrix multiplication will be useful in manipulating matrices and their functions. We present some results without proof.

$$\text{Associative rule}: \quad (\mathbf{AB})\mathbf{C} = \mathbf{A}(\mathbf{BC}). \tag{A.32}$$

$$\text{Distributive rule}: \quad \mathbf{A}(\mathbf{B} + \mathbf{C}) = \mathbf{AB} + \mathbf{AC}. \tag{A.33}$$

$$\text{Non} - \text{commutative}: \quad \mathbf{AB} \neq \mathbf{BA}. \tag{A.34}$$

$$\text{Transpose of product}: \quad (\mathbf{AB})^{\mathrm{T}} = \mathbf{B}^{\mathrm{T}}\mathbf{A}^{\mathrm{T}}, \quad (\mathbf{ABC})^{\mathrm{T}} = \mathbf{C}^{\mathrm{T}}\mathbf{B}^{\mathrm{T}}\mathbf{A}^{\mathrm{T}}. \tag{A.35}$$

$$\text{Inverse of product}: \quad (\mathbf{AB})^{-1} = \mathbf{B}^{-1}\mathbf{A}^{-1}, \quad (\mathbf{ABC})^{-1} = \mathbf{C}^{-1}\mathbf{B}^{-1}\mathbf{A}^{-1}. \tag{A.36}$$

A.3 MATRIX EQUATIONS AND SOLUTION

Consider the following simultaneous linear equations:

$$\begin{aligned}
a_{11}x_1 + a_{12}x_2 + \cdots + a_{1N}x_N &= b_1 \\
a_{21}x_1 + a_{22}x_2 + \cdots + a_{2N}x_N &= b_2 \\
&\vdots \\
a_{N1}x_1 + a_{N2}x_2 + \cdots + a_{NN}x_N &= b_N.
\end{aligned} \tag{A.37}$$

Equation (A.37) has N number of unknowns $(x_1, x_2, \ldots, x_N)$, and there are N number of equations. If all equations are independent, then eq. (A.37) has a unique solution. Equation (A.37) can be equivalently denoted using the matrix notation, as

$$[\mathbf{A}] \cdot \{\mathbf{x}\} = \{\mathbf{b}\}, \tag{A.38}$$

---

[4] Kreyszig, E. 1983. *Advanced Engineering Mathematics*, 5th Ed. John Wiley & Sons, New York.
[5] Strang, G. 1980. *Linear Algebra and its Applications*, 2nd Ed. Academic Press, New York.

where

$$[\mathbf{A}] = \begin{bmatrix} a_{11} & a_{12} & \cdots & a_{1N} \\ a_{21} & a_{22} & \cdots & a_{2N} \\ \vdots & \vdots & \ddots & \vdots \\ a_{N1} & a_{N2} & \cdots & a_{NN} \end{bmatrix}, \quad \{\mathbf{x}\} = \begin{Bmatrix} x_1 \\ x_2 \\ \vdots \\ x_N \end{Bmatrix}, \quad \{\mathbf{b}\} = \begin{Bmatrix} b_1 \\ b_2 \\ \vdots \\ b_N \end{Bmatrix}.$$

When the matrix $[\mathbf{A}]$ and the vector $\{\mathbf{b}\}$ are known, the solution $\{\mathbf{x}\}$ can be obtained by multiplying both sides of the equation by $[\mathbf{A}]^{-1}$ to obtain

$$[\mathbf{A}]^{-1}[\mathbf{A}]\cdot\{\mathbf{x}\} = [\mathbf{A}]^{-1}\cdot\{\mathbf{b}\},$$

$$[\mathbf{I}]\cdot\{\mathbf{x}\} = [\mathbf{A}]^{-1}\cdot\{\mathbf{b}\}, \tag{A.39}$$

$$\{\mathbf{x}\} = [\mathbf{A}]^{-1}\cdot\{\mathbf{b}\}.$$

Note that $[\mathbf{I}]\,\{\mathbf{x}\} = \{\mathbf{x}\}$. Thus, a unique solution can be obtained if $[\mathbf{A}]^{-1}$ exists or, equivalently, if the matrix $[\mathbf{A}]$ is not singular.

Although the multiplication of the inverse of a matrix can solve for the linear system of equations, calculating the inverse of the stiffness matrix is often computationally very expensive, especially when the size of matrix is large. In practical application, the number of DOFs can easily be in the order of $10^5 \sim 10^6$. In such a big matrix, it is impractical to calculate the inverse of the matrix. Therefore, a different method is required calculating the solution $\{\mathbf{x}\}$ without calculating the inverse of matrix $[\mathbf{A}]$. In the following, we will explain how to calculate the solution in the context of finite element matrix equation.

The finite element method reduced the partial differential equations describing the physics into a set of linear simultaneous equations for the form as

$$[\mathbf{K}]\{\mathbf{Q}\} = \{\mathbf{F}\}. \tag{A.40}$$

Such equations can be solved using Gauss elimination approach. For implementation in a software a modified version of this algorithm, called LU decomposition method, is widely used. The following steps are used.

  (i) Forward reduction of $[\mathbf{K}]$: Decompose $[\mathbf{K}]$ into a lower triangular matrix and an upper triangular matrix as: $[\mathbf{K}] = [\mathbf{L}][\mathbf{U}]$, where $[\mathbf{L}]$ is a lower triangular matrix, and $[\mathbf{U}]$ is an upper triangular matrix.

 (ii) Forward reduction of $\{\mathbf{F}\}$: Solve the equation $[\mathbf{L}]\{\bar{\mathbf{F}}\} = \{\mathbf{F}\}$ to calculate the forward reduced right hand side vector $\{\bar{\mathbf{F}}\}$.

(iii) Backward substitution: Solve the equation $[\mathbf{U}]\{\mathbf{Q}\} = \{\bar{\mathbf{F}}\}$ to obtain $\{\mathbf{Q}\}$.

The key idea is that equations involving lower and upper triangular matrices are very easy to solve. This algorithm is very general but is applied below for the case of a 3×3 matrix to explain the method.

A.3.1 Forward Reduction of a Matrix Equation

Consider a 3×3 $[\mathbf{K}]$ matrix:

$$[\mathbf{K}] = [\mathbf{K}_0] = \begin{bmatrix} k_{11} & k_{12} & k_{13} \\ k_{21} & k_{21} & k_{23} \\ k_{31} & k_{32} & k_{33} \end{bmatrix}. \tag{A.41}$$

To simplify the stiffness matrix, we want to make all terms below the diagonal component zero. In order to do that for the first column, for example, the following form of operator $[\mathbf{L}_1]^{-1}$ is defined:

$$[\mathbf{L_1}]_{3x3}^{-1} = \begin{bmatrix} 1 & 0 & 0 \\ -m_{21} & 1 & 0 \\ -m_{31} & 0 & 1 \end{bmatrix}, \tag{A.42}$$

where m_{21} and m_{31} will be determined to make all terms below the diagonal component zero. The reason for defining the inverse of the matrix, not the original matrix, will be clear at the end of derivations. By multiplying $[\mathbf{L}_1]^{-1}$ with $[\mathbf{K}_0]$, we have

$$[\mathbf{L}_1]^{-1}[\mathbf{K}_0] = \begin{bmatrix} k_{11} & k_{12} & k_{13} \\ -m_{21}k_{11}+k_{21} & -m_{21}k_{12}+k_{22} & -m_{21}k_{13}+k_{23} \\ -m_{31}k_{11}+k_{31} & -m_{31}k_{12}+k_{32} & -m_{31}k_{13}+k_{33} \end{bmatrix}.$$

We can determine m_{21} and m_{31} from the condition that all terms of the first column below the diagonal should be zero, which yields

$$-m_{21}k_{11}+k_{21}=0, \Rightarrow m_{21}=\frac{k_{21}}{k_{11}},$$

$$-m_{31}k_{11}+k_{31}=0, \Rightarrow m_{31}=\frac{k_{31}}{k_{11}}.$$

In general, we want,

$$m_{ji} = \frac{k_{ji}}{k_{ii}}. \tag{A.43}$$

For an arbitrary sized matrix, $[\mathbf{K}_0]$, the operator for forward reduction of the ith column can be defined as:

$$[\mathbf{L}_i]_{n \times n}^{-1} = \begin{bmatrix} 1 & 0 & . & . & {}^{i^{th}} & . & 0 \\ 0 & 1 & 0 & . & . & . & . \\ . & 0 & . & . & . & . & . \\ . & . & . & . & 0 & . & . \\ . & . & . & . & 1 & . & . \\ . & . & . & . & -m_{i+1,i} & . & 0 \\ 0 & . & . & . & -m_{i+2,i} & . & 1 \end{bmatrix}. \tag{A.44}$$

By applying the forward reduction operations for all the columns of the matrix we can reduce matrix $[\mathbf{K}_0]$ to an upper triangular matrix $[\mathbf{U}]$, as

$$[\mathbf{L}_{N-1}]^{-1}[\mathbf{L}_{N-2}]^{-1}\cdots[\mathbf{L}_1]^{-1}[\mathbf{K}_0] = [\mathbf{U}]. \tag{A.45}$$

In order to obtain the expression of matrix $[\mathbf{K}_0]$ from the above equation, the inverses of the matrices are combined together to define the lower triangular matrix as

$$[\mathbf{L}] = [\mathbf{L}_1]\cdots[\mathbf{L}_{N-2}][\mathbf{L}_{N-1}]. \tag{A.46}$$

Then, by multiplying $[\mathbf{L}]$ both sides of the equation, we obtain

$$[\mathbf{K}_0] = [\mathbf{L}][\mathbf{U}]. \tag{A.47}$$

The original equation $[\mathbf{K}_0]\{\mathbf{Q}\} = \{\mathbf{F}\}$ can now be written as

$$[\mathbf{L}][\mathbf{U}]\{\mathbf{Q}\} = \{\mathbf{F}\}. \tag{A.48}$$

Or,

$$[\mathbf{U}]\{\mathbf{Q}\} = \{\bar{\mathbf{F}}\}, \tag{A.49}$$

where

$$\{\bar{\mathbf{F}}\} = [\mathbf{L}]^{-1}\{\mathbf{F}\}. \tag{A.50}$$

It is interesting to note that even if the stiffness matrix is decomposed into the lower- and upper-triangular matrices, it is actually unnecessary to calculate the lower triangular matrix because its inverse in eq. (A.44) is used to reduce the force vector in eq. (A.50).

A.3.2   Properties of Forward Reduction Operators

The matrix multiplications in eq. (A.46) are obviously very expensive for a large system of equations. Similarly the inverses of large matrices are also not very easy to compute. However, the following properties of lower triangular matrices make these two operations rather easy.

Property–I: The inverse of $[\mathbf{L}_i]$ can easily be calculated because

$$\text{If } [\mathbf{L}_i]^{-1} = \begin{bmatrix} 1 & 0 & . & . & .^{ith} & . & 0 \\ 0 & 1 & 0 & . & . & & . & . \\ . & 0 & . & . & . & & . & . \\ . & . & . & . & 0 & & . & . \\ . & . & . & . & 1 & & . & . \\ . & . & . & . & -m_{i+1,i} & . & 0 \\ 0 & . & . & . & -m_{i+2,i} & . & 1 \end{bmatrix} \text{ then } [\mathbf{L}_i] = \begin{bmatrix} 1 & 0 & . & . & .^{ith} & . & 0 \\ 0 & 1 & 0 & . & . & & . & . \\ . & 0 & . & . & . & & . & . \\ . & . & . & . & 0 & & . & . \\ . & . & . & . & 1 & & . & . \\ . & . & . & . & m_{i+1,i} & . & 0 \\ 0 & . & . & . & m_{i+2,i} & . & 1 \end{bmatrix}.$$

In other words, the inverse of $[\mathbf{L}_i]^{-1}$ can be obtained simply by changing the signs of the numbers below the diagonal of the matrix. For example, let us consider the following 3×3 matrix:

$$[\mathbf{L}_1]^{-1} = \begin{bmatrix} 1 & 0 & 0 \\ -m_{21} & 1 & 0 \\ -m_{31} & 0 & 1 \end{bmatrix}.$$

It is straightforward to show that the following multiplication yield an identity matrix:

$$[\mathbf{L}_1]^{-1}[\mathbf{L}_1] = \begin{bmatrix} 1 & 0 & 0 \\ -m_{21} & 1 & 0 \\ -m_{31} & 0 & 1 \end{bmatrix} \begin{bmatrix} 1 & 0 & 0 \\ m_{21} & 1 & 0 \\ m_{31} & 0 & 1 \end{bmatrix} = [\mathbf{I}],$$

which proves that the inverse of the lower triangular matrix can be calculated by simply changing the sign of terms below the diagonal component.

Property–II: The product of the lower triangular matrices $[\mathbf{L}] = [\mathbf{L}_1] \cdots [\mathbf{L}_{N-2}][\mathbf{L}_{N-1}]$ for forward reducing the columns of an $N \times N$ matrix can be obtained by adding all terms below the diagonals as:

$$[\mathbf{L}] = [\mathbf{L}_1] \cdots [\mathbf{L}_{N-2}][\mathbf{L}_{N-1}] = \begin{bmatrix} 1 & 0 & . & . & . & 0 \\ m_{21} & 1 & 0 & . & . & . \\ m_{31} & m_{32} & 1 & . & . & . \\ . & . & . & . & . & . \\ . & . & . & . & . & 0 \\ m_{N1} & . & . & . & m_{N,N-1} & 1 \end{bmatrix}, \tag{A.51}$$

where $m_{ji} = k_{ji}^{(i-1)} / k_{ii}^{(i-1)}$. In eq. (A.51), $k_{ji}^{(i-1)}$ is obtained by applying the previous $(i-1)$ forward reductions operation on k_{ji}. To illustrate the process, let us consider the 3×3 matrix example again. In this case, the two lower triangular matices are

$$[\mathbf{L}_1] = \begin{bmatrix} 1 & 0 & 0 \\ m_{21} & 1 & 0 \\ m_{31} & 0 & 1 \end{bmatrix} \quad [\mathbf{L}_2] = \begin{bmatrix} 1 & 0 & 0 \\ 0 & 1 & 0 \\ 0 & m_{32} & 1 \end{bmatrix}.$$

Therefore, the combined lower triangular matrix becomes

$$[\mathbf{L}] = [\mathbf{L}_1][\mathbf{L}_2] = \begin{bmatrix} 1 & 0 & 0 \\ m_{21} & 1 & 0 \\ m_{31} & m_{32} & 1 \end{bmatrix},$$

where m_{ji} terms are calculated by

$$m_{21} = \frac{k_{21}}{k_{11}}, \quad m_{31} = \frac{k_{31}}{k_{11}}, \quad m_{32} = \frac{k_{32}^{(1)}}{k_{22}^{(1)}}.$$

In the above equation, the last term m_{32} is calculated using the following updated stiffness matrix after applying for the first forward update as

$$k_{22}^{(1)} = k_{22} - \frac{k_{21}}{k_{11}} k_{12}, \quad k_{32}^{(1)} = k_{32} - \frac{k_{31}}{k_{11}} k_{12}.$$

A.3.3 Forward Reduction of a Vector

The forward reduction of the right-hand side vector $\{\bar{\mathbf{F}}\}$ is computed by solving

$$[\mathbf{L}]\{\bar{\mathbf{F}}\} = \{\mathbf{F}\}. \tag{A.52}$$

At first glance this looks as difficult as solving the original problem. But in fact this equation is easy to solve because $[\mathbf{L}]$ is a lower triangular matrix. For a 3×3 example this equation looks as follows:

$$\begin{bmatrix} 1 & 0 & 0 \\ m_{21} & 1 & 0 \\ m_{31} & m_{32} & 1 \end{bmatrix} \begin{Bmatrix} \bar{F}_1 \\ \bar{F}_2 \\ \bar{F}_3 \end{Bmatrix} = \begin{Bmatrix} F_1 \\ F_2 \\ F_3 \end{Bmatrix}. \tag{A.53}$$

The first equation involves only one unknown; so it can be solved easily. The second equation involves only the first and second unknown and so on.

A.3.4 Backward Substitution

The last step involves solving for the original unknowns by solving the following equations. These equations are again easy to solve, because $[U]$ is an upper triangular matrix.

$$[U]\{Q\} = \{\bar{F}\}, \tag{A.54}$$

$$\begin{bmatrix} u_{11} & u_{12} & . & . & k_{1N} \\ 0 & u_{22} & . & . & . \\ 0 & 0 & u_{33} & . & . \\ . & . & . & . & . \\ 0 & . & . & 0 & \bar{k}_{NN} \end{bmatrix} \begin{Bmatrix} q_1 \\ q_2 \\ . \\ . \\ q_N \end{Bmatrix} = \begin{Bmatrix} \bar{F}_1 \\ \bar{F}_2 \\ . \\ . \\ \bar{F}_N \end{Bmatrix}. \tag{A.55}$$

In these equations, the last equation has only one unknown, the second to last equation has two, and so on. Therefore, the last row is used to solve for q_N first, and then, the second to the last equation is used to solve for q_{N-1} with known q_N. Therefore, solving for eq. (A.54) is computationally inexpensive.

EXAMPLE A.2   *Forward reduction and backward substitution*

Solve $[K]\{Q\} = \{F\}$, where,

$$[K] = \begin{bmatrix} 2 & -1 & 0 \\ -1 & 2 & -1 \\ 0 & -1 & 1 \end{bmatrix}, \quad \{F\} = \begin{Bmatrix} 0 \\ 0 \\ 1 \end{Bmatrix}.$$

(i) Forward reduction of $[K]$:

From matrix $[K]$, we calculate $m_{21} = k_{21}/k_{11} = -1/2$ and $m_{31} = k_{31}/k_{11} = 0$. Therefore, we have

$$[K_1] = [L_1]^{-1}[K] = \begin{bmatrix} 1 & 0 & 0 \\ 1/2 & 1 & 0 \\ 0 & 0 & 1 \end{bmatrix} \begin{bmatrix} 2 & -1 & 0 \\ -1 & 2 & -1 \\ 0 & -1 & 1 \end{bmatrix} = \begin{bmatrix} 2 & -1 & 0 \\ 0 & 3/2 & -1 \\ 0 & -1 & 1 \end{bmatrix}.$$

For the second column, we can calculate $m_{32} = k_{32}^{(1)}/k_{22}^{(1)} = -2/3$. Therefore, we have the following upper triangular matrix:

$$[U] = [L_2]^{-1}[K_1] = \begin{bmatrix} 1 & 0 & 0 \\ 0 & 1 & 0 \\ 0 & 2/3 & 1 \end{bmatrix} \begin{bmatrix} 2 & -1 & 0 \\ 0 & 3/2 & -1 \\ 0 & -1 & 1 \end{bmatrix} = \begin{bmatrix} 2 & -1 & 0 \\ 0 & 3/2 & -1 \\ 0 & 0 & 1/3 \end{bmatrix}.$$

Also, the lower triangular matrix can be obtained as

$$[L] = [L_1][L_2] = \begin{bmatrix} 1 & 0 & 0 \\ -1/2 & 1 & 0 \\ 0 & -2/3 & 1 \end{bmatrix}.$$

(ii) Forward reduction of $\{F\}$:

In order to reduce the vector $\{F\}$, the lower triangular matrix can be used as in eq. (A.52) as

$$[L]\{\bar{F}\} = \{F\},$$

$$\begin{bmatrix} 1 & 0 & 0 \\ -1/2 & 1 & 0 \\ 0 & -2/3 & 1 \end{bmatrix} \begin{Bmatrix} \bar{F}_1 \\ \bar{F}_2 \\ \bar{F}_3 \end{Bmatrix} = \begin{Bmatrix} 0 \\ 0 \\ 1 \end{Bmatrix} \Rightarrow \begin{cases} \bar{F}_1 = 0 \\ -\frac{1}{2}\bar{F}_1 + \bar{F}_2 = 0 \rightarrow \bar{F}_2 = 0 \\ -\frac{2}{3}\bar{F}_2 + \bar{F}_3 = 1 \rightarrow \bar{F}_3 = 1. \end{cases}$$

(iii) Backward substitution
Using $\{\bar{\mathbf{F}}\}$, the solution can be found as

$$[\mathbf{U}]\{\mathbf{Q}\} = \{\bar{\mathbf{F}}\},$$

$$\begin{bmatrix} 2 & -1 & 0 \\ 0 & 3/2 & -1 \\ 0 & 0 & 1/3 \end{bmatrix} \begin{Bmatrix} u_1 \\ u_2 \\ u_3 \end{Bmatrix} = \begin{Bmatrix} 0 \\ 0 \\ 1 \end{Bmatrix} \Rightarrow \begin{cases} u_3 = 3 \\ \frac{3}{2}u_2 - u_3 = 0 \rightarrow u_2 = 2 \\ 2u_1 - u_2 = 0 \rightarrow u_1 = 1. \end{cases}$$

A.4 EIGENVALUES AND EIGENVECTORS

Eigenvalue problems occur when the principal values of stress or strain need to be calculated. Also, the free vibration analysis of a structural system requires eigenvalue analysis. Consider the equation shown below for a square matrix $[\mathbf{A}]$

$$[\mathbf{A}]\cdot\{\mathbf{x}\} = \lambda\{\mathbf{x}\}, \tag{A.56}$$

where λ is a scalar. The above equation can be thought of as a matrix equation similar to that in eq. (A.38) except that $\{\mathbf{b}\}$ is replaced by a scalar multiple of $\{\mathbf{x}\}$ itself. Such equations, which arise in many engineering applications, are interesting and have physical significance. There are only certain special values for λ that will satisfy eq. (A.56), and they are called the *eigenvalues*[6] of the square matrix $[\mathbf{A}]$. For each eigenvalue there will be a corresponding vector $\{\mathbf{x}\}$ called the *eigenvector*. Of course the null vector $\{\mathbf{x}\} = \{\mathbf{0}\}$ is a solution of eq. (A.56), and we do not consider it, as it is trivial. It can be shown that the maximum number of eigenvalues will be equal to the number of rows (or columns) of $[\mathbf{A}]$.

The procedure for computing the eigenvalues and eigenvectors is as follows. Equation (A.56) can be written as

$$[\mathbf{A}]\cdot\{\mathbf{x}\} - \lambda\{\mathbf{x}\} = \{\mathbf{0}\}, \tag{A.57}$$

or

$$[\mathbf{A} - \lambda\mathbf{I}]\cdot\{\mathbf{x}\} = \{\mathbf{0}\}, \tag{A.58}$$

where $[\mathbf{I}]$ is the identity matrix of same dimensions as $[\mathbf{A}]$. There are two possibilities for eq. (A.58). Obviously $\{\mathbf{x}\} = \{\mathbf{0}\}$ is a solution of eq. (A.58), but we have already declared it as trivial. If we want a nontrivial solution for $\{\mathbf{x}\}$, then the matrix $[\mathbf{A} - \lambda\mathbf{I}]$ must be singular. Otherwise one can invert that matrix and multiply with $\{\mathbf{0}\}$ on the RHS[7] to obtain the trivial solution $\{\mathbf{x}\} = \{\mathbf{0}\}$. Letting $[\mathbf{A} - \lambda\mathbf{I}]$ singular will open up new possibilities for $\{\mathbf{x}\}$. In order for the coefficient matrix to be singular, its determinant must be equal to zero, that is,

$$|\mathbf{A} - \lambda\mathbf{I}| = 0, \tag{A.59}$$

or

$$\begin{vmatrix} a_{11}-\lambda & a_{12} & \dots & a_{1n} \\ a_{21} & a_{22}-\lambda & \cdots & a_{2n} \\ \vdots & \vdots & \ddots & \vdots \\ a_{n1} & \cdots & \cdots & a_{nn}-\lambda \end{vmatrix} = 0. \tag{A.60}$$

---

[6] *Eigen* means "own," and Eigenschaft means "characteristic" in German.
[7] In this book, RHS and LHS mean the right-hand side and left-hand side of an equation, respectively.

The determinant in eq. (A.60) can be expanded to obtain a polynomial equation in λ as

$$\lambda^n + C_1\lambda^{n-1} + \ldots\ldots C_{n-1}\lambda + C_n = 0. \tag{A.61}$$

The n-th degree polynomial on LHS of eq. (A.61) is called the characteristic polynomial of matrix [**A**]. The n roots of the polynomial equation are the n eigenvalues of [**A**].

Each one of the n eigenvalues can be substituted back in eq. (A.58) to obtain a set of simultaneous equations for the unknown {**x**}. The solutions are called eigenvectors. The eigenvector corresponding the i-th eigenvalue is denoted by {$\mathbf{x}_i$}. One cannot obtain a unique solution for {$\mathbf{x}_i$}, as the set of equations are not linearly independent (remember that the determinant of the coefficient matrix was set to zero in order to solve for the eigenvalues). We will discuss this further as we find applications for the concepts of eigenvalues and eigenvectors.

The numerical method for solving the eigenvalue problem can be found in the literature[8]. An analytical method is available to solve eq. (A.58) when $n = 2$ or 3. Here we introduce an analytical method when $n = 3$ and the coefficient matrix is symmetric. In such a case, the characteristic polynomial in eq. (A.61) is cubic and can be written as

$$\lambda^3 + C_1\lambda^2 + C_2\lambda + C_3 = 0, \tag{A.62}$$

where

$$
\begin{aligned}
C_1 &= -(a_{11} + a_{22} + a_{33}), \\
C_2 &= a_{11}a_{22} + a_{22}a_{33} + a_{33}a_{11} - a_{12}^2 - a_{23}^2 - a_{13}^2, \\
C_3 &= -(a_{11}a_{22}a_{33} + 2a_{12}a_{23}a_{13} - a_{11}a_{23}^2 - a_{22}a_{13}^2 - a_{33}a_{12}^2).
\end{aligned}
\tag{A.63}
$$

A general analytical solution for the above cubic equation can be written as

$$
\begin{aligned}
\lambda_1 &= g\cos\frac{\phi}{3} - \frac{C_1}{3}, \\
\lambda_2 &= g\cos\left(\frac{\phi+2\pi}{3}\right) - \frac{C_1}{3}, \\
\lambda_3 &= g\cos\left(\frac{\phi+4\pi}{3}\right) - \frac{C_1}{3}.
\end{aligned}
\tag{A.64}
$$

where

$$
\left\{
\begin{aligned}
\phi &= \cos^{-1}\left[-\frac{b}{2\sqrt{-a^3/27}}\right], \\
g &= 2\sqrt{-a/3}, \\
a &= \frac{1}{3}(3C_2 - C_1^2), \\
b &= \frac{1}{27}(2C_1^3 - 9C_1C_2 + 27C_3).
\end{aligned}
\right.
\tag{A.65}
$$

It can be shown that eigenvalues of a real symmetric matrix are always real, and hence one can always compute three (not necessarily different) eigenvalues.

---

[8] Press, W. H., Flannery, B. P., Teukolsky, S. A., and Vetterling, W. T. 1986. *Numerical Recipes*. Cambridge University Press, Cambridge.

There is no analytical way of calculating the eigenvalues and eigenvectors for the matrix whose dimension is larger than 3×3. There are many numerical methods to calculate eigenvalues and eigenvectors of a square matrix. A MATLAB command $[Q, D] = eig(A)$ can be used to calculate eigenvalues and eigenvectors of matrix $[A]$. The diagonal matrix $[D]$ contains eigenvalues, whose corresponding eigenvectors are the columns of matrix $[Q]$.

Equation (A.57) can be considered as the standard form of an eigenvalue problem. When two matrices are involved, it is called a generalized eigenvalue problem, whose definition is given as

$$[A] \cdot \{x\} = \lambda [B] \cdot \{x\}. \tag{A.66}$$

In the viewpoint of the generalized eigenvalue problem, the standard form is when $[B] = [I]$. Similar to the zero determinant condition in eq. (A.59), the eigenvalues can be found by satisfying the following condition:

$$|A - \lambda B| = 0. \tag{A.67}$$

Once eigenvalues are found, they can be used to find eigenvectors by substituting each eigenvalue into eq. (A.66). Because the determinant is zero, eq. (A.66) does not have a unique solution for the eigenvectors. Among infinitely many possible eigenvectors, the one that satisfies the condition of $\{x\}^T [B] \{x\} = 1$ is chosen, which is called normalization. A MATLAB command $[Q,D] = eig(A,B)$ can be used to calculate eigenvalues and eigenvectors of the generalized eigenvalue problem.

EXAMPLE A.3 *Eigenvalues and eigenvectors*

Find the eigenvalues and eigenvectors of the 3×3 matrix A given below.

$$A = \begin{bmatrix} 1 & 0 & 2 \\ 0 & 1 & 0 \\ 2 & 0 & 4 \end{bmatrix}.$$

SOLUTION The first step is to derive the characteristic equation for the matrix A similar to that shown in eq. (A.62). It can be derived as

$$\lambda^3 - 6\lambda^2 + 5\lambda = 0.$$

The solution of the above cubic equation is the eigenvalues, and they are $\lambda = 0$, 1, and 5. The eigenvectors for each of the above eigenvalues are calculated using eq. (A.58).

For $\lambda = 0$, we obtain

$$\begin{bmatrix} (1-0) & 0 & 2 \\ 0 & (1-0) & 0 \\ 2 & 0 & (4-0) \end{bmatrix} \begin{Bmatrix} x_1 \\ x_2 \\ x_3 \end{Bmatrix} = \begin{Bmatrix} 0 \\ 0 \\ 0 \end{Bmatrix}.$$

The above equation yields three simultaneous equations for x_1, x_2, and x_3, as follows:

$$x_1 + 2x_3 = 0$$
$$x_2 = 0$$
$$2x_1 + 4x_3 = 0.$$

As we mentioned earlier, there is no unique solution to the above set of equations. The solution can be written as $x_2 = 0$ and $x_1 = -2x_3$. One possible solution is $x_1 = -2$, $x_2 = 0$, and $x_3 = 1$. Thus, the eigenvector corresponding to the eigenvalue 0 can be written as $(-2, 0, 1)$. Usually the eigenvector is normalized, such that its norm is equal to unity. Then the above eigenvector takes the form $x^{(1)} = (-2,0,1)/\sqrt{5}$, where the superscript denotes that this is the first eigenvector.

For $\lambda = 1$, we obtain

$$\begin{bmatrix} 0 & 0 & 2 \\ 0 & 0 & 0 \\ 2 & 0 & 3 \end{bmatrix} \begin{Bmatrix} x_1 \\ x_2 \\ x_3 \end{Bmatrix} = \begin{Bmatrix} 0 \\ 0 \\ 0 \end{Bmatrix}.$$

First we note that the second equation is not useful as it will be satisfied by any set of **x**. The first equation clearly yields $x_1 = 0$. Substituting for x_1 in the third equation, we obtain $x_3 = 0$. We note that x_2 is arbitrary, and hence the eigenvector can be taken as $\mathbf{x}^{(2)} = (0, 1, 0)$.

Next, consider $\lambda = 5$. Following the same procedure as for the other eigenvalues, we obtain

$$\begin{bmatrix} -4 & 0 & 2 \\ 0 & -4 & 0 \\ 2 & 0 & -1 \end{bmatrix} \begin{Bmatrix} x_1 \\ x_2 \\ x_3 \end{Bmatrix} = \begin{Bmatrix} 0 \\ 0 \\ 0 \end{Bmatrix}.$$

The solution can be derived as $x_2 = 0$ and $x_3 = 2x_1$. After normalizing, the eigenvector takes the form $\mathbf{x}^{(3)} = (1, 0, 2)/\sqrt{5}$.

It may be noted that the scalar product of any two eigenvectors of a symmetric matrix is equal to zero. That is,

$$\mathbf{x}^{(1)} \cdot \mathbf{x}^{(2)} = \mathbf{x}^{(2)} \cdot \mathbf{x}^{(3)} = \mathbf{x}^{(3)} \cdot \mathbf{x}^{(1)} = 0$$

Physically it means that the eigenvectors are orthogonal to each other. In geometric terms, the three directions represented by the eigenvectors are perpendicular to each other. ■

EXAMPLE A.4 *Repeated eigenvalues*

Find the eigenvalues and eigenvectors of the 3×3 matrix **A** given below.

$$\mathbf{A} = \begin{bmatrix} 9 & 4 & 0 \\ 4 & 3 & 0 \\ 0 & 0 & 1 \end{bmatrix}.$$

SOLUTION   The characteristic equation is

$$\lambda^3 - 13\lambda^2 + 23\lambda - 11 = 0.$$

The roots of the above cubic equation are $\lambda = 1, 1$, and 11. Thus, we note that the equation has repeated roots or the matrix has repeated eigenvalues. Let us first determine the eigenvector for $\lambda = 11$. The set of simultaneous equations are

$$\begin{bmatrix} -2 & 4 & 0 \\ 4 & -8 & 0 \\ 0 & 0 & -10 \end{bmatrix} \begin{Bmatrix} x_1 \\ x_2 \\ x_3 \end{Bmatrix} = \begin{Bmatrix} 0 \\ 0 \\ 0 \end{Bmatrix}.$$

The solution for x_3 is uniquely obtained as $x_3 = 0$, and we also obtain $x_1 = 2x_2$. Thus we can write the eigenvector as $\mathbf{x}^{(1)} = (2, 1, 0)/\sqrt{5}$. Next we consider the repeating eigenvalue $\lambda_2 = \lambda_3 = 1$.

The set of simultaneous equations are

$$\begin{bmatrix} 8 & 4 & 0 \\ 4 & 2 & 0 \\ 0 & 0 & 0 \end{bmatrix} \begin{Bmatrix} x_1 \\ x_2 \\ x_3 \end{Bmatrix} = \begin{Bmatrix} 0 \\ 0 \\ 0 \end{Bmatrix}.$$

We note that the third equation is not useful, and the first two equations are essentially the same, that is, they are not linearly independent. The first (or the second) equation yields $x_2 = -2x_1$. We do not have any information to determine x_3, and hence it can be considered arbitrary. Thus, the eigenvector can be written as $\mathbf{x}^{(2)} = \mathbf{x}^{(3)} = (1, -2, \alpha)$, where α is an arbitrary number. Thus, there is an infinite number of eigenvectors. Such a situation arises whenever there are repeating eigenvalues. It may be noted that $\mathbf{x}^{(1)} \cdot \mathbf{x}^{(2)} = 0$ is satisfied for any value of α. Thus, we can state that any direction perpendicular to $\mathbf{x}^{(1)}$ is also an eigenvector. ▄▄

A.5 QUADRATIC FORMS

The sum of products of variables x_i of the form

$$F \equiv a_{11}x_1^2 + a_{22}x_2^2 + \cdots + a_{nn}x_n^2$$
$$+ a_{12}x_1x_2 + a_{13}x_1x_3 + \cdots + a_{n,n-1}x_nx_{n-1} \tag{A.68}$$

is called the *quadratic form* in x_1, x_2, ..., x_n, where a_{ij} are real constants. The quadratic form can be written in matrix form as

$$F = \{\mathbf{x}\}^{\mathrm{T}}[\mathbf{A}]\{\mathbf{x}\}$$

$$= \{x_1, x_2, \ldots, x_n\} \begin{bmatrix} a_{11} & a_{12} & \cdots & a_{1n} \\ a_{21} & a_{22} & \cdots & a_{2n} \\ \vdots & \vdots & \ddots & \vdots \\ a_{n1} & a_{n2} & \cdots & a_{nn} \end{bmatrix} \begin{Bmatrix} x_1 \\ x_2 \\ \vdots \\ x_n \end{Bmatrix} \tag{A.69}$$

$$= \sum_{i=1}^{n}\sum_{j=1}^{n} a_{ij}x_ix_j,$$

where $\{\mathbf{x}\} = \{x_1, x_2, \ldots, x_n\}^{\mathrm{T}}$ and $[\mathbf{A}] = [a_{ij}]$ is the coefficient matrix of the quadratic form. The quadratic form appears often in engineering applications, such as strain energy of a solid or structure.

For a general $n \times n$ matrix $[\mathbf{B}]$ that is not necessarily symmetric, consider the quadratic form

$$F = \{\mathbf{x}\}^{\mathrm{T}}[\mathbf{B}]\{\mathbf{x}\}. \tag{A.70}$$

The matrix $[\mathbf{B}]$ can be decomposed into symmetric ($\mathbf{B}_S$) and skew-symmetric ($\mathbf{B}_A$) parts as

$$[\mathbf{B}] = \frac{1}{2}\left[\mathbf{B} + \mathbf{B}^{\mathrm{T}}\right] + \frac{1}{2}\left[\mathbf{B} - \mathbf{B}^{\mathrm{T}}\right]$$
$$= [\mathbf{B}_S] + [\mathbf{B}_A]. \tag{A.71}$$

The quadratic form using the symmetric part can be obtained as

$$F_S = \{\mathbf{x}\}^{\mathrm{T}}[\mathbf{B}_S]\{\mathbf{x}\}. \tag{A.72}$$

It can easily be shown that F and F_S are identical; that is,

$$F = \{\mathbf{x}\}^{\mathrm{T}}[\mathbf{B}]\{\mathbf{x}\} = \{\mathbf{x}\}^{\mathrm{T}}[\mathbf{B}_S]\{\mathbf{x}\} = F_S. \tag{A.73}$$

The reader can show that the quadratic form of the skew-symmetric part is identically equal to zero. Thus, a non-symmetric matrix $[\mathbf{B}]$ in a quadratic form can always be replaced by the symmetric part of the matrix without affecting the value of the quadratic form.

A.5.1 Positive Definite Quadratic Form

If $\{\mathbf{x}\}^T[\mathbf{A}]\{\mathbf{x}\} \geq 0$ for all real vectors $\{\mathbf{x}\}$ and if $\{\mathbf{x}\}^T[\mathbf{A}]\{\mathbf{x}\} = 0$ only if $\{\mathbf{x}\} = \{\mathbf{0}\}$, then the quadratic form, hence the symmetric matrix $[\mathbf{A}]$, is said to be *positive definite*; that is,

$$\begin{aligned} \{\mathbf{x}\}^T[\mathbf{A}]\{\mathbf{x}\} &\geq 0, \quad \text{for } all \; \{\mathbf{x}\} \; in \; R^n, \\ \{\mathbf{x}\}^T[\mathbf{A}]\{\mathbf{x}\} &= 0, \quad \text{only } if \; \{\mathbf{x}\} = \{\mathbf{0}\}. \end{aligned} \tag{A.74}$$

A quadratic form with the symmetric matrix $[\mathbf{A}]$ is said to be *positive semidefinite* when it takes on only nonnegative values for all values of the variables x but vanishes for some nonzero value of the variables; that is,

$$\begin{aligned} \{\mathbf{x}\}^T[\mathbf{A}]\{\mathbf{x}\} &\geq 0, \quad \text{for } all \; \{\mathbf{x}\} \; in \; R^n, \\ \{\mathbf{x}\}^T[\mathbf{A}]\{\mathbf{x}\} &= 0, \quad \text{for } some \; \{\mathbf{x}\} \neq \{\mathbf{0}\}. \end{aligned} \tag{A.75}$$

Positive definiteness is an important property in structural analysis. When a matrix is positive definite, each column of the matrix is linearly independent, and as discussed in section A.3, the matrix can be inverted.

EXAMPLE A.5 *Quadratic form*

Consider the following quadratic form:

$$F(x,y) = \{x \; y\}[\mathbf{A}]\begin{Bmatrix} x \\ y \end{Bmatrix} = \{x \; y\}\begin{bmatrix} 1 & -1 \\ -1 & 2 \end{bmatrix}\begin{Bmatrix} x \\ y \end{Bmatrix}.$$

F can be expanded as $F(x,y) = x^2 - 2xy + 2y^2 = (x-y)^2 + y^2$. Since F is the sum of two squared quantities, it is always positive, and $F = 0$ only if $x = y = 0$. Thus, $[\mathbf{A}]$ is positive definite.

Now consider different matrix $[\mathbf{A}]$ defined as

$$[\mathbf{A}] = \begin{bmatrix} 1 & -1 \\ -1 & 1 \end{bmatrix}.$$

Then, $F(x,y) = x^2 - 2xy + y^2 = (x-y)^2$. One can note that F is always positive except when $x = y$. Hence $[\mathbf{A}]$ is positive semidefinite. ∎

A.6 MAXIMA AND MINIMA OF FUNCTIONS

Consider the function $F(\mathbf{x})$ given by $F(\mathbf{x}) = \frac{1}{2}\{\mathbf{x}\}^T[\mathbf{A}]\{\mathbf{x}\} - \{\mathbf{x}\}^T\{\mathbf{b}\}$, where $\{\mathbf{x}\} = \{x_1, x_2, \ldots, x_n\}^T$ and $[\mathbf{A}]$ is a symmetric positive definite matrix of size $n \times n$. Then, we can show that the function $F(\mathbf{x}) = F(x_1, x_2, \ldots, x_n)$ has its minimum value at the point where $[\mathbf{A}]\{\mathbf{x}\} = \{\mathbf{b}\}$.

Proof[9]

Let $\{\mathbf{y}\}$ be an arbitrary vector of size n. Consider the following expression:

$$F(\mathbf{y}) - F(\mathbf{x}) = \frac{1}{2}\{\mathbf{y}\}^T[\mathbf{A}]\{\mathbf{y}\} - \{\mathbf{y}\}^T\{\mathbf{b}\} - \frac{1}{2}\{\mathbf{x}\}^T[\mathbf{A}]\{\mathbf{x}\} + \{\mathbf{x}\}^T\{\mathbf{b}\}. \tag{A.76}$$

---

[9] Strang, G. 1976. *Linear Algebra and its Applications*. Academic Press, New York.

Substituting $\{b\} = [A]\{x\}$ in the above equation, we obtain

$$F(\mathbf{y}) - F(\mathbf{x}) = \frac{1}{2}\{\mathbf{y}\}^T[\mathbf{A}]\{\mathbf{y}\} - \{\mathbf{y}\}^T[\mathbf{A}]\{\mathbf{x}\} + \frac{1}{2}\{\mathbf{x}\}^T[\mathbf{A}]\{\mathbf{x}\}$$

$$= \frac{1}{2}\{\mathbf{y} - \mathbf{x}\}^T[\mathbf{A}]\{\mathbf{y} - \mathbf{x}\}.$$

(A.77)

In deriving the above, we have used the relation $\{\mathbf{x}\}^T[\mathbf{A}]\{\mathbf{y}\} = \{\mathbf{y}\}^T[\mathbf{A}]\{\mathbf{x}\}$, which is due to the fact that $[\mathbf{A}]$ is symmetric, that is, $[\mathbf{A}]^T = [\mathbf{A}]$. Since $[\mathbf{A}]$ is positive definite, the quantity on the RHS of the above equation is always positive except when $\{\mathbf{x}\} = \{\mathbf{y}\}$. That is, $F(\mathbf{y})$ is always greater than $F(\mathbf{x})$ except at $\{\mathbf{y}\} = \{\mathbf{x}\}$. Hence, the minimum value of F occurs at $\{\mathbf{x}\}$.

A.7 EXERCISES

1. Consider the following 3×3 matrix $[\mathbf{T}]$:

$$[\mathbf{T}] = \begin{bmatrix} 2 & 7 & 2 \\ 3 & 4 & 5 \\ 6 & 3 & 7 \end{bmatrix}.$$

(a) Write the transpose $\mathbf{T}^T$.
(b) Show that the matrix $[\mathbf{S}] = [\mathbf{T}] + [\mathbf{T}]^T$ is a symmetric matrix.
(c) Show that the matrix $[\mathbf{A}] = [\mathbf{T}] - [\mathbf{T}]^T$ is a skew-symmetric matrix. What are the diagonal components of the matrix $[\mathbf{A}]$?

2. Consider the following two 3×3 matrices $[\mathbf{A}]$ and $[\mathbf{B}]$:

$$[\mathbf{A}] = \begin{bmatrix} 1 & 3 & 7 \\ 3 & 4 & 3 \\ 6 & 2 & 7 \end{bmatrix}, \quad [\mathbf{B}] = \begin{bmatrix} 3 & 7 & 2 \\ 4 & 2 & 8 \\ 7 & 4 & 5 \end{bmatrix}.$$

(a) Calculate $[\mathbf{C}] = [\mathbf{A}] + [\mathbf{B}]$.
(b) Calculate $[\mathbf{D}] = [\mathbf{A}] - [\mathbf{B}]$.
(c) Calculate the scalar multiple $[\mathbf{D}] = 3[\mathbf{A}]$.

3. Consider the following two three-dimensional vectors $\mathbf{a}$ and $\mathbf{b}$:

$$\mathbf{a} = \{2, \ 4, \ 5\}^T \ \text{and} \ \mathbf{b} = \{2, \ 7, \ 3\}^T.$$

(a) Calculate the scalar product $c = \mathbf{a} \cdot \mathbf{b}$.
(b) Calculate the norm of vector $\mathbf{a}$.
(c) Calculate the vector product of $\mathbf{a}$ and $\mathbf{b}$.

4. For the matrix $[\mathbf{T}]$ in Problem 1 and the two vectors $\mathbf{a}$ and $\mathbf{b}$ in Problem 3, answer the following questions.
(a) Calculate the product of the matrix-vector multiplication $[\mathbf{T}] \cdot \mathbf{a}$.
(b) Calculate $\mathbf{b} \cdot [\mathbf{T}] \cdot \mathbf{a}$.

5. For the two matrices $[\mathbf{A}]$ and $[\mathbf{B}]$ in problem 2, answer the following questions.
(a) Evaluate the matrix-matrix multiplication $[\mathbf{C}] = [\mathbf{A}][\mathbf{B}]$.
(b) Evaluate the matrix-matrix multiplication $[\mathbf{D}] = [\mathbf{B}][\mathbf{A}]$.

6. Calculate the determinant of the following matrices:

$$[\mathbf{A}] = \begin{bmatrix} 2 & 3 \\ 3 & 7 \end{bmatrix}, \ [\mathbf{B}] = \begin{bmatrix} 1 & 3 & 2 \\ 3 & 6 & 5 \\ 1 & 2 & 7 \end{bmatrix}.$$

7. Calculate the inverse of the matrix [**A**] in problem 6.

8. Matrices [**A**] and [**B**] are defined below. If $\mathbf{B} = \mathbf{A}^{-1}$, determine the values of p, q, r, and s.

$$[\mathbf{A}] = \begin{bmatrix} 2 & 6 & p \\ 1 & 2 & 3 \\ q & 5 & 6 \end{bmatrix}, \ [\mathbf{B}] = \begin{bmatrix} 1.5 & 8 & -5 \\ 0 & -2 & 1 \\ -0.5 & r & s \end{bmatrix}.$$

9. Solve the following simultaneous system of equations using the matrix method:

$$4x_1 + 2x_2 = 2$$
$$2x_1 + 3x_2 = 4.$$

10. Consider the following row vectors and matrices

$$\mathbf{a} = \begin{bmatrix} 5 & 2 & 3 \end{bmatrix} \ \mathbf{b} = \begin{bmatrix} 1 & 3 & 4 \end{bmatrix}$$
$$\mathbf{A} = \begin{bmatrix} 4 & 1 & 3 \\ 2 & 6 & 5 \end{bmatrix} \ \mathbf{B} = \begin{bmatrix} 2 & 3 & 4 \\ 1 & 2 & 0 \end{bmatrix}.$$

Using MATLAB, calculate $\mathbf{A}^\mathrm{T}$, $\mathbf{a}^\mathrm{T}$, $\mathbf{A}+\mathbf{B}$, $\mathbf{A}-\mathbf{B}$, $\mathbf{a}\mathbf{b}^\mathrm{T}$, $\mathbf{a}^\mathrm{T}\mathbf{b}$, $\mathbf{A}^\mathrm{T}\mathbf{B}$, $\mathbf{C} = \mathbf{B}\mathbf{A}^\mathrm{T}$, $\mathbf{A}\mathbf{B}$, $\mathbf{C}^{-1}$, $\det(\mathbf{C})$. Test the commands $\mathbf{a}.*\mathbf{b}$, $\mathbf{A}.*\mathbf{B}$ and explain the difference between them and $\mathbf{a}*\mathbf{b}$, $\mathbf{A}*\mathbf{B}$, respectively.

11. Find the eigenvalues and eigenvectors of the following matrices:

(a) $[\mathbf{A}] = \begin{bmatrix} 4 & 2 & 0 \\ 2 & 1 & 0 \\ 0 & 0 & -1 \end{bmatrix}.$

(b) $[\mathbf{B}] = \begin{bmatrix} 1 & 0 & 2 \\ 0 & -1 & 0 \\ 2 & 0 & 4 \end{bmatrix}.$

12. Construct the quadratic form for the matrix [**A**] in problem 2, that is, $\{\mathbf{x}\}^\mathrm{T}[\mathbf{A}]\{\mathbf{x}\}$, and compare with the quadratic form that is calculated using symmetric part $[\mathbf{A}_S]$.

13. Consider the matrix equation $[\mathbf{A}]\{\mathbf{x}\} = \{\mathbf{b}\}$ given by

$$\begin{bmatrix} 2 & -1 & 0 \\ -1 & 4 & -1 \\ 0 & -1 & 2 \end{bmatrix} \begin{Bmatrix} x_1 \\ x_2 \\ x_3 \end{Bmatrix} = \begin{Bmatrix} 4 \\ 0 \\ 4 \end{Bmatrix}.$$

(a) Construct the quadratic form $F(\mathbf{x}) = \{\mathbf{x}\}^\mathrm{T}[\mathbf{A}]\{\mathbf{x}\} - 2\{\mathbf{x}\}^\mathrm{T}\{\mathbf{b}\}$.

(b) Find $\{\mathbf{x}\} = \{\mathbf{x}^*\}$ by minimizing $F(\mathbf{x})$.

(c) Verify that the vector $\{\mathbf{x}^*\}$ satisfies $[\mathbf{A}]\{\mathbf{x}\} = \{\mathbf{b}\}$.

14. A function $f(x_1, x_2)$ of two variables x_1 and x_2 is given by

$$f(x_1, x_2) = \frac{1}{2}\{x_1 \ \ x_2\} \begin{bmatrix} 1 & -1 \\ -1 & 2 \end{bmatrix} \begin{Bmatrix} x_1 \\ x_2 \end{Bmatrix} - \{x_1 \ \ x_2\} \begin{Bmatrix} 0 \\ 2 \end{Bmatrix} - 1.$$

(a) Multiply the matrices and express f as a polynomial in x_1 and x_2.

(b) Determine the extreme (maximum or minimum) value of the function and corresponding x_1 and x_2.

(c) Is this a maxima or minima?

15. A function $f(x, y, z)$ of x, y, and z is defined as

$$f(x,y,z) = \frac{1}{2}\{x \ \ y \ \ z\}[\mathbf{K}]\begin{Bmatrix} x \\ y \\ z \end{Bmatrix} - \{x \ \ y \ \ z\}\{\mathbf{R}\} + 10,$$

where

$$[\mathbf{K}] = \begin{bmatrix} 1 & 2 & 3 \\ 2 & 4 & 5 \\ 3 & 5 & 6 \end{bmatrix} \quad \text{and} \quad \{\mathbf{R}\} = \begin{Bmatrix} 1 \\ 2 \\ 1 \end{Bmatrix}.$$

(a) Multiply the matrices and express f as a polynomial in x, y, and z.

(b) Write down the three equations necessary to find the extreme value of the function in the form

$$\underset{3\times 3}{[\mathbf{A}]}\underset{3\times 1}{\begin{Bmatrix} x \\ y \\ z \end{Bmatrix}} = \underset{3\times 1}{\{\mathbf{b}\}}.$$

(c) Solve the equations in (b) to determine x, y, and z corresponding to the extreme value of f.

(d) Compute the extreme value of f.

(e) Is this a maxima or minima?

(f) Compute the determinant of $[\mathbf{K}]$.

16. Consider a function $f(x,y)$ given in the following expression:

$$f(x,y) = \frac{1}{2}\{x \ \ y\}\begin{bmatrix} 6 & 4 \\ 4 & 0 \end{bmatrix}\begin{Bmatrix} x \\ y \end{Bmatrix} - \{x \ \ y\}\begin{Bmatrix} 4 \\ 0 \end{Bmatrix}.$$

The eigenvalues of $[\mathbf{A}]$ are -2 and 8. It is found that the function is extremum at $\{x, y\} = \{0, 1\}$. Determine if it is (a) maximum, (b) minimum, or (c) neither.

17. Solve the matrix equation $[\mathbf{K}]\{\mathbf{Q}\} = \{\mathbf{F}\}$ using forward reduction and backward substitution, where,

$$[\mathbf{K}] = \begin{bmatrix} 5 & -1 & 0 \\ -1 & 3 & -1 \\ 0 & -1 & 1 \end{bmatrix}, \quad \{\mathbf{F}\} = \begin{Bmatrix} 2 \\ 0 \\ 1 \end{Bmatrix}.$$

Index

Introduction to Finite Element Analysis and Design, Second Edition. Nam H. Kim, Bhavani V. Sankar, and Ashok V. Kumar.
© 2018 John Wiley & Sons Ltd. Published 2018 by John Wiley & Sons Ltd.
Companion website: www.wiley.com/go/kim/finite_element_analysis_design

Printed and bound by CPI Group (UK) Ltd, Croydon, CR0 4YY

16/04/2025

14658398-0005